T0227362

The Designer's Guide
to VHDL
Third Edition

The Morgan Kaufmann Series in Systems on Silicon

Series Editor: Wayne Wolf, Georgia Institute of Technology

The Designer's Guide to VHDL, Second Edition
Peter J. Ashenden

The System Designer's Guide to VHDL-AMS
Peter J. Ashenden, Gregory D. Peterson, and Darrell A. Teegarden

Modeling Embedded Systems and SoCs
Axel Jantsch

ASIC and FPGA Verification: A Guide to Component Modeling
Richard Munden

Multiprocessor Systems-on-Chips
Edited by Ahmed Amine Jerraya and Wayne Wolf

Functional Verification
Bruce Wile, John Goss, and Wolfgang Roesner

Customizable and Configurable Embedded Processors
Edited by Paolo Ienne and Rainer Leupers

Networks-on-Chips: Technology and Tools
Edited by Giovanni De Micheli and Luca Benini

VLSI Test Principles & Architectures
Edited by Laung-Terng Wang, Cheng-Wen Wu, and Xiaoqing Wen

Designing SoCs with Configured Processors
Steve Leibson

ESL Design and Verification
Grant Martin, Andrew Piziali, and Brian Bailey

Aspect-Oriented Programming with e
David Robinson

Reconfigurable Computing: The Theory and Practice of FPGA-Based Computation
Edited by Scott Hauck and André DeHon

System-on-Chip Test Architectures
Edited by Laung-Terng Wang, Charles E. Stroud, and Nur A. Touba

Verification Techniques for System-Level Design
Masahiro Fujita, Indradeep Ghosh, and Mukul Prasad

VHDL-2008: Just the New Stuff
Peter J. Ashenden and Jim Lewis

On-Chip Communication Architectures: System on Chip Interconnect
Sudeep Pasricha and Nikil Dutt

Embedded DSP Processor Design: Application Specific Instruction Set Processors
Dake Liu

Processor Description Languages
Prabhat Mishra

The Designer's Guide to VHDL

Third Edition

Peter J. Ashenden

EDA CONSULTANT, ASHENDEN DESIGNS PTY. LTD.

ADJUNCT ASSOCIATE PROFESSOR, ADELAIDE UNIVERSITY

AMSTERDAM • BOSTON • HEIDELBERG • LONDON
NEW YORK • OXFORD • PARIS • SAN DIEGO
SAN FRANCISCO • SINGAPORE • SYDNEY • TOKYO

Morgan Kaufmann Publishers is an imprint of Elsevier

MORGAN KAUFMANN PUBLISHERS

Morgan Kaufmann Publishers is an imprint of Elsevier.
30 Corporate Drive, Suite 400, Burlington, MA 01803, USA

♾ This book is printed on acid-free paper.

© 2008 by Elsevier Inc. All rights reserved.

Designations used by companies to distinguish their products are often claimed as trademarks or registered trademarks. In all instances in which Morgan Kaufmann Publishers is aware of a claim, the product names appear in initial capital or all capital letters. Readers, however, should contact the appropriate companies for more complete information regarding trademarks and registration.

No part of this publication may be reproduced, stored in a retrieval system, or transmitted in any form or by any means—electronic, mechanical, photocopying, scanning, or otherwise—without prior written permission of the publisher.

Permissions may be sought directly from Elsevier's Science & Technology Rights Department in Oxford, UK: phone: (+44) 1865 843830, fax: (+44) 1865 853333, E-mail: permissions@elsevier.com. You may also complete your request online via the Elsevier homepage (http://elsevier.com), by selecting "Support & Contact" then "Copyright and Permission" and then "Obtaining Permissions."

Library of Congress Cataloging-in-Publication Data
Ashenden, Peter J.
    The designer's guide to VHDL / Peter J. Ashenden. -- 3rd ed.
        p. cm. -- (The Morgan Kaufmann series in systems on silicon)
    Includes bibliographical references and index.
    ISBN 978-0-12-088785-9 (hardcover : alk. paper) 1. VHDL (Computer hardware description language)
2. Electronic digital computers--Computer simulation. I. Title.

    TK7888.3.A863 2008
    621.39'2--dc22

                                        2008011059

ISBN: 978-0-12-088785-9

For information on all Morgan Kaufmann publications,
visit our Web site at *www.mkp.com* or *www.books.elsevier.com*

Printed and bound by CPI Group (UK) Ltd, Croydon, CR0 4YY

Transferred to Digital Print 2011

Working together to grow
libraries in developing countries

www.elsevier.com | www.bookaid.org | www.sabre.org

ELSEVIER   BOOK AID
            International   Sabre Foundation

To my wife Katrina

Contents

Preface

VHDL is a language for describing digital electronic systems. It arose out of the United States government's Very High Speed Integrated Circuits (VHSIC) program. In the course of this program, it became clear that there was a need for a standard language for describing the structure and function of integrated circuits (ICs). Hence the VHSIC Hardware Description Language (VHDL) was developed. It was subsequently developed further under the auspices of the Institute of Electrical and Electronic Engineers (IEEE) and adopted in the form of the IEEE Standard 1076, Standard VHDL Language Reference Manual, in 1987. This first standard version of the language is often referred to as VHDL-87.

Like all IEEE standards, the VHDL standard is subject to review from time to time. Comments and suggestions from users of the 1987 standard were analyzed by the IEEE working group responsible for VHDL, and in 1992 a revised version of the standard was proposed. This was eventually adopted in 1993, giving us VHDL-93. A second round of revision of the standard was started in 1998. That process was completed in 2001, giving us VHDL-2002. After that, further development took place in the IEEE working group and in a technical committee of an organization, Accellera, whose charter is to promote standards for electronics design. These efforts led to the current version of the language, VHDL-2008, described in this book.

VHDL is designed to fill a number of needs in the design process. First, it allows description of the structure of a system, that is, how it is decomposed into subsystems and how those subsystems are interconnected. Second, it allows the specification of the function of a system using familiar programming language forms. Third, as a result, it allows the design of a system to be simulated before being manufactured, so that designers can quickly compare alternatives and test for correctness without the delay and expense of hardware prototyping. Fourth, it allows the detailed structure of a design to be synthesized from a more abstract specification, allowing designers to concentrate on more strategic design decisions and reducing time to market.

This book presents a structured guide to the modeling facilities offered by the VHDL language, showing how they can be used for the design of digital systems. The book does not purport to teach digital design, since that topic is large enough by itself to warrant several textbooks covering its various aspects. Instead, the book assumes that the reader has at least a basic grasp of digital design concepts, such as might be gained from a first course in digital design in an engineering degree program. Some exposure to computer programming and to concepts of computer organization will also be beneficial. This book is suitable for use in a course in digital or computer design and will also serve practicing engineers who need to acquire VHDL fluency as part of their changing job requirements.

One pervasive theme running through the presentation in this book is that modeling a system using a hardware description language is essentially a software design exercise. This implies that good software engineering practice should be applied. Hence the treatment in this book draws directly from experience in software engineering. There are numerous hints and techniques from small-scale and large-scale software engineering presented throughout the book, with the sincere intention that they might be of use to readers.

I am particularly pleased to be able to include this book in the Morgan Kaufmann Series in Systems on Silicon. Modeling for simulation and synthesis is a vital part of a design methodology for large-scale systems. VHDL allows models to be expressed at a range of levels of abstraction, from gate-level up to algorithmic and architectural levels. It will continue to play an important role in the design of silicon-based systems for some time to come.

Structure of the Book

The Designer's Guide to VHDL is organized so that it can be read linearly from front to back. This path offers a graduated development, with each chapter building on ideas introduced in the preceding chapters. Each chapter introduces a number of related concepts or language facilities and illustrates each one with examples. Scattered throughout the book are three case studies, which bring together preceding material in the form of extended worked examples.

Chapter 1 introduces the idea of a hardware description language and outlines the reasons for its use and the benefits that ensue. It then proceeds to introduce the basic concepts underlying VHDL, so that they can serve as a basis for examples in subsequent chapters. The next three chapters cover the aspects of VHDL that are most like conventional programming languages. These may be used to describe the behavior of a system in algorithmic terms. Chapter 2 explains the basic type system of the language and introduces the scalar data types. Chapter 3 describes the sequential control structures, and Chapter 4 covers composite data structures used to represent collections of data elements. In Chapter 5, the main facilities of VHDL used for modeling hardware are covered in detail. These include facilities for modeling the basic behavioral elements in a design, the signals that interconnect them and the hierarchical structure of the design.

The next group of chapters extends this basic set of facilities with language features that make modeling of large systems more tractable. Chapter 6 introduces procedures and functions, which can be used to encapsulate behavioral aspects of a design. Chapter 7 introduces the package as a means of collecting together related parts of a design or of creating modules that can be reused in a number of designs. Chapter 8 deals with the important topic of resolved signals, and Chapter 9 describes a number of predefined and standard packages for use in VHDL designs. The combination of facilities described in these early chapters is sufficient for many modeling tasks, so Chapter 10 brings them together in the first case study, in which a multiplier/accumulator circuit is designed.

The third group of chapters covers advanced modeling features in VHDL. Chapter 11 covers aliases as a way of managing the large number of names that arise in a large model. Chapter 12 describes generics as a means of parameterizing the behavior and structure of a design and enhancing the resusability of designs. This leads to a discussion of abstract data types as a means of managing the complexity associated with large designs. Chapter 13 deals with the topics of component instantiation and configuration. These features are important in large real-world models, but they can be difficult to understand. Hence this book introduces structural modeling through the mechanism of direct instantiation in earlier chapters and leaves the more general case of component instantiation and configuration until this later chapter. In Chapter 14, generated regular structures are covered.

The fourth group of chapters covers language facilities generally used for system-level modeling. Chapter 15 introduces the notion of access types (or pointers) and uses them to develop linked data structures. The topic of abstract data types is revisited in the context of container data types. Chapter 16 covers the language facilities for input and output using files, including binary files and text files. Chapter 17 is a case study in which a package for designing memories is developed. The package draws upon features described in the third and fourth groups of chapters.

In the fifth group of chapters, we introduce language features for advanced design and verification. Chapter 18 deals with features for test bench support and verification. It describes how specifications written in the IEEE standard Property Specification Language (PSL) can be embedded in VHDL models. Chapter 19 covers protected types and their use as a means of concurrency control. Chapter 20 describes how we can annotate items in a design with attributes to specify information to be used by design automation tools. This leads into Chapter 21, which covers guidelines for writing synthesizable models. This group of chapters is drawn together in a further case study, Chapter 22, showing development of a synthesizable processor core and its use in a small embedded system, a digital alarm clock.

The final chapter, Chapter 23, is a miscellany of advanced topics not covered in the previous chapters. It includes a discussion of blocks and guarded signals, which are not as widely used in modern designs as previously. Nonetheless, we describe them here for completeness. The chapter also covers use of features for encrypting the source text of models as a means of protecting intellectual property (IP), and use of features of the VHDL Procedureall Interface (VHPI) for incorporating models and applications written in non-VHDL programming languages.

Each chapter in the book is followed by a set of exercises designed to help the reader develop understanding of the material. Where an exercise relates to a particular topic described in the chapter, the section number is included in square brackets. An approximate "difficulty" rating is also provided, expressed using the following symbols:

❶   quiz-style exercise, testing basic understanding

❷   basic modeling exercise—10 minutes to half an hour effort

❸   advanced modeling exercise—one half to two hours effort

❹   modeling project—half a day or more effort

Answers for the first category of exercises are provided in Appendix C. The remaining categories involve developing VHDL models. Readers are encouraged to test correctness of their models by running them on a VHDL simulator. This is a much more effective learning exercise than comparing paper models with paper solutions.

Changes in the Second and Third Editions

The first edition of this book was published in 1995, just as VHDL-93 was gaining acceptance. The second edition was updated to reflect the changes in VHDL-2002. Many of the changes in the language standard corrected ambiguities in the previous standard that caused incompatibility between VHDL tools from different vendors. There were also changes that enhanced the usability of the language. The text and examples in the second edition were revised where necessary to reflect the changes in the language. Furthermore, following publication of the first edition, a number of VHDL-related standards were published and gained widespread acceptance. The second edition added descriptions of the IEEE 1076.3 synthesis and IEEE 1076.2 math packages, and was revised to cover the IEEE 1076.6 Synthesis Interoperability Standard.

The latest revision of the language, VHDL-2008, adds a number of significant new language features, making this edition of *The Designer's Guide to VHDL* significantly bigger than its predecessors. VHDL-2008 also specifies numerous minor new features and changes to existing features to enhance the usability of the language. This edition integrates descriptions of all of the new and revised features into the text. The differences between the various versions are highlighted in call-outs within the text, headed with "VHDL-2002," "VHDL-93," or "VHDL-87," as appropriate. In addition, some of the material has been removed or rearranged. The case study on a package for arithmetic on bit-vector operands has been deleted because the standard numeric packages have now become widespread. The first case study in this book is a revised version of the MAC case study in previous editions, and shows how the standard packages can be used. The chapter on blocks and guarded signals has been contracted and moved to a section in the last chapter, since the features are now little used in practice. There is a greater emphasis on synthesis in this edition. What was an appendix on the topic in previous editions has been substantially revised and promoted to full chapter status. The large case study showing development of a 32-bit processor model has been revised to show a smaller synthesizable model of an 8-bit microcontroller core and its use in an embedded system. This is much more relevant, both for educational purposes and professional practice. Finally, this edition includes a listing of all of the VHDL standard packages as an appendix for reference.

Resources for Help and Information

Although this book attempts to be comprehensive in its coverage of VHDL, there will no doubt be questions that it does not answer. For these, the reader will need to seek other resources. A valuable source of experience and advice, often overlooked, is one's colleagues, either at the workplace or in user groups. User groups generally hold regular meetings that either formally or informally include a time for questions and answers. Many also run e-mail lists and on-line discussion groups for problem solving.

Accellera is one of a number of organizations that sponsors the EDA Industry Working Groups Web server (www.eda.org). The server has links to Web pages and repositories of several VHDL standards groups and user groups.

Readers who have access to the Usenet electronic news network will find the news group comp.lang.vhdl a valuable resource. This discussion group is a source of announcements, sample models, questions and answers and useful software. Participants include VHDL users and people actively involved in the language standard working group and in VHDL tool development. The "frequently asked questions" (FAQ) file for this group is a mine of useful pointers to books, products and other information. It is archived at www.eda.org.

One resource that must be mentioned is IEEE Standard 1076, IEEE Standard VHDL Language Reference Manual, sometimes referred to as the "VHDL Bible." It is the authoritative source of information about VHDL. However, since it is a definitional document, not a tutorial, it is written in a complex legalistic style. This makes it very difficult to use to answer the usual questions that arise when writing VHDL models. It should only be used once you are somewhat familiar with VHDL. It can be ordered from the IEEE at standards.ieee.org.

This book contains numerous examples of VHDL models that may also serve as a resource for resolving questions. The VHDL source code for these examples and the case studies, as well as other related information, is available on the companion website for the book at *books.elsevier.com/companions/9780120887859*.

Although I have been careful to avoid errors in the example code, there are no doubt some that I have missed. I would be pleased to hear about them, so that I can correct them in the on-line code and in future printings of this book. Errata and general comments can be e-mailed to me at *vhdl-book@ashenden.com.au*.

Acknowledgments

The seeds for this book go back to 1990 when I developed a brief set of notes, *The VHDL Cookbook*, for my computer architecture class at the University of Adelaide. At the time, there were few books on VHDL available, so I made my booklet available for on-line access. News of its availability spread quickly around the world, and within days, my e-mail in-box was bursting. At the time of writing this, nearly 20 years later, I still regularly receive messages about the *Cookbook*. Many of the respondents urged me to write a full textbook version. With that encouragement, I embarked upon the exercise that led to the first edition of *The Designer's Guide to VHDL*. Two years after publication of *The Designer's Guide*, the need for a book specifically for students became evident. That led to publication of the first edition of *The Student's Guide to VHDL*. I am grateful to the many engineers, students and teachers around the world who gave me the impetus to write these books and who made them such a success. I hope this new edition will continue to meet the need for a comprehensive guide to VHDL.

In the previous editions of *The Designer's Guide* and *The Student's Guide*, I had the opportunity to extend thanks to the many people who assisted in development of the books. They included my colleagues at the University of Adelaide; my research collaborators, Phil Wilsey at the University of Cincinnati and Perry Alexander at the University of Kansas; the staff at Morgan Kaufmann Publishers, including, in particular, Denise Penrose; the reviewers of the manuscript for the first edition, namely, Poras Balsara of the University of Texas, Paul Menchini of Menchini & Associates, David Pitts of GTE Labs and the University of Lowell and Philip Wilsey of the University of Cincinnati; David Bishop for his contribution to the material on synthesis in the first edition of *The Designer's Guide*; and Mentor Graphics Corporation, for use of their ModelSim simulator to check the example models. I remain grateful to all of these people and organizations for their valuable contributions to the earlier editions and to this edition.

For the current edition, I would also like to thank Jim Lewis, who collaborated on a recent book, *VHDL-2008: Just the New Stuff*. Much of the material from that book has found its way into this book in one form or another. Thanks also to Mentor Graphics Corporation for continued use of the ModelSim simulator to check the example code. I continue to enjoy an excellent working relationship with the staff at Morgan Kaufmann Publishers and their parent company, Elsevier. Thanks to Chuck Glaser, Senior Acquisitions Editor, for his support in the continued development of these VHDL books; to Dawn-marie Simpson, Senior Project Manager in the Production Department, for her meticulous attention to detail; and to Denise Penrose, Publisher, for her longstanding support of my writing endeavors.

The previous editions of *The Designer's Guide to VHDL* were dedicated to my wife Katrina. As I said in the first edition preface, I used to think that authors dedicating their books to their partners was somewhat contrived, but that Katrina's understanding, encouragement and support taught me otherwise. I remain deeply grateful for her continued support and am honored to also dedicate this third edition to her.

Chapter 1

Fundamental Concepts

In this introductory chapter, we describe what we mean by digital system modeling and see why modeling and simulation are an important part of the design process. We see how the hardware description language VHDL can be used to model digital systems and introduce some of the basic concepts underlying the language. We complete this chapter with a description of the basic lexical and syntactic elements of the language, to form a basis for the detailed descriptions of language features that follow in later chapters.

1.1 Modeling Digital Systems

If we are to discuss the topic of modeling digital systems, we first need to agree on what a digital system is. Different engineers would come up with different definitions, depending on their background and the field in which they were working. Some may consider a single VLSI circuit to be a self-contained digital system. Others might take a larger view and think of a complete computer, packaged in a cabinet with peripheral controllers and other interfaces.

For the purposes of this book, we include any digital circuit that processes or stores information as a digital system. We thus consider both the system as a whole and the various parts from which it is constructed. Therefore, our discussions cover a range of systems from the low-level gates that make up the components to the top-level functional units.

If we are to encompass this range of views of digital systems, we must recognize the complexity with which we are dealing. It is not humanly possible to comprehend such complex systems in their entirety. We need to find methods of dealing with the complexity, so that we can, with some degree of confidence, design components and systems that meet their requirements.

The most important way of meeting this challenge is to adopt a systematic methodology of design. If we start with a requirements document for the system, we can design an abstract structure that meets the requirements. We can then decompose this structure into a collection of components that interact to perform the same function. Each of these components can in turn be decomposed until we get to a level where we have some ready-made, primitive components that perform a required function. The result of this process is a hierarchically composed system, built from the primitive elements.

The advantage of this methodology is that each subsystem can be designed independently of others. When we use a subsystem, we can think of it as an abstraction rather than having to consider its detailed composition. So at any particular stage in the design process, we only need to pay attention to the small amount of information relevant to the current focus of design. We are saved from being overwhelmed by masses of detail.

We use the term *model* to mean our understanding of a system. The model represents that information which is relevant and abstracts away from irrelevant detail. The implication of this is that there may be several models of the same system, since different information is relevant in different contexts. One kind of model might concentrate on representing the function of the system, whereas another kind might represent the way in which the system is composed of subsystems. We will come back to this idea in more detail in the next section.

There are a number of important motivations for formalizing this idea of a model. First, when a digital system is needed, the requirements of the system must be specified. The job of the engineers is to design a system that meets these requirements. To do that, they must be given an understanding of the requirements, hopefully in a way that leaves them free to explore alternative implementations and to choose the best according to some criteria. One of the problems that often arises is that requirements are incompletely and ambiguously spelled out, and the customer and the design engineers disagree on what is meant by the requirements document. This problem can be avoided by using a formal model to communicate requirements.

A second reason for using formal models is to communicate understanding of the function of a system to a user. The designer cannot always predict every possible way in which a system may be used, and so is not able to enumerate all possible behaviors. If the designer provides a model, the user can check it against any given set of inputs and determine how the system behaves in that context. Thus a formal model is an invaluable tool for documenting a system.

A third motivation for modeling is to allow testing and verification of a design using simulation. If we start with a requirements model that defines the behavior of a system, we can simulate the behavior using test inputs and note the resultant outputs of the system. According to our design methodology, we can then design a circuit from subsystems, each with its own model of behavior. We can simulate this composite system with the same test inputs and compare the outputs with those of the previous simulation. If they are the same, we know that the composite system meets the requirements for the cases tested. Otherwise we know that some revision of the design is needed. We can continue this process until we reach the bottom level in our design hierarchy, where the components are real devices whose behavior we know. Subsequently, when the design is manufactured, the test inputs and outputs from simulation can be used to verify that the physical circuit functions correctly. This approach to testing and verification of course assumes that the test inputs cover all of the circumstances in which the final circuit will be used. The issue of test coverage is a complex problem in itself and is an active area of research.

A fourth motivation for modeling is to allow formal verification of the correctness of a design. Formal verification requires a mathematical statement of the required function of a system. This statement may be expressed in the notation of a formal logic system, such as temporal logic. Formal verification also requires a mathematical definition of the meaning of the modeling language or notation used to describe a design. The process of verification involves application of the rules of inference of the logic system to prove that the design implies the required function. While formal verification is not yet in everyday use, it is steadily becoming a more important part of the design process. There have already been significant demonstrations of formal verification techniques in real design projects, and the promise for the future is bright.

One final, but equally important, motivation for modeling is to allow automatic synthesis of circuits. If we can formally specify the function required of a system, it is in theory possible to translate that specification into a circuit that performs the function. The advantage of this approach is that the human cost of design is reduced, and engineers are free to explore alternatives rather than being bogged down in design detail. Also, there is less scope for errors being introduced into a design and not being detected. If we automate the translation from specification to implementation, we can be more confident that the resulting circuit is correct.

The unifying factor behind all of these arguments is that we want to achieve maximum reliability in the design process for minimum cost and design time. We need to ensure that requirements are clearly specified and understood, that subsystems are used correctly and that designs meet the requirements. A major contributor to excessive cost is having to revise a design after manufacture to correct errors. By avoiding errors, and by providing better tools for the design process, costs and delays can be contained.

1.2 Domains and Levels of Modeling

In the previous section, we mentioned that there may be different models of a system, each focusing on different aspects. We can classify these models into three domains: *function*, *structure* and *geometry*. The functional domain is concerned with the operations performed by the system. In a sense, this is the most abstract domain of description, since it does not indicate how the function is implemented. The structural domain deals with how the system is composed of interconnected subsystems. The geometric domain deals with how the system is laid out in physical space.

Each of these domains can also be divided into levels of abstraction. At the top level, we consider an overview of function, structure or geometry, and at lower levels we introduce successively finer detail. Figure 1.1 (devised by Gajski and Kuhn, see reference [8]) represents the domains for digital systems on three independent axes and represents the levels of abstraction by the concentric circles crossing each of the axes.

Let us look at this classification in more detail, showing how at each level we can create models in each domain. As an example, we consider a single-chip microcontroller system used as the controller for some measurement instrument, with data input connections and some form of display outputs.

1.2.1 Modeling Example

At the most abstract level, the function of the entire system may be described in terms of an algorithm, much like an algorithm for a computer program. This level of functional modeling is often called *behavioral modeling*, a term we shall adopt when presenting abstract descriptions of a system's function. A possible algorithm for our instrument control-

FIGURE 1.1

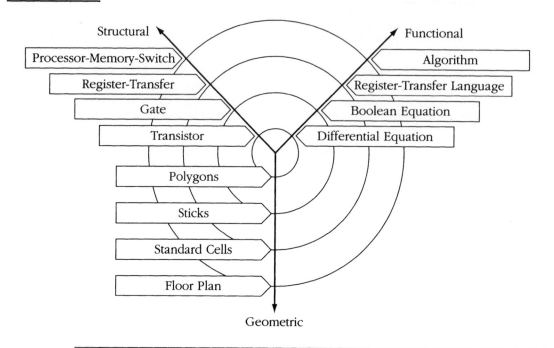

Domains and levels of abstraction. The radial axes show the three different domains of model-ing. The concentric rings show the levels of abstraction, with the more abstract levels on the out-side and more detailed levels toward the center.

ler is shown below. This model describes how the controller repeatedly scans each data input and writes a scaled display of the input value.

```
loop
   for each data input loop
      read the value on this input;
      scale the value using the current scale factor
         for this input;
      convert the scaled value to a decimal string;
      write the string to the display output corresponding
         to this input;
   end loop;
   wait for 10 ms;
end loop;
```

At this top level of abstraction, the structure of a system may be described as an in-terconnection of such components as processors, memories and input/output devices. This level is sometimes called the Processor Memory Switch (PMS) level, named after the notation used by Bell and Newell (see reference [3]). Figure 1.2 shows a structural model of the instrument controller drawn using this notation. It consists of a processor connected

FIGURE 1.2

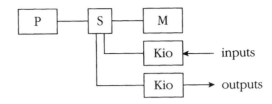

A PMS model of the controller structure. It is constructed from a processor (P), a memory (M), an interconnection switch (S) and two input/output controllers (Kio).

via a switch to a memory component and to controllers for the data inputs and display outputs.

In the geometric domain at this top level of abstraction, a system to be implemented as a VLSI circuit may be modeled using a floor plan. This shows how the components described in the structural model are arranged on the silicon die. Figure 1.3 shows a possible floor plan for the instrument controller chip. There are analogous geometric descriptions for systems integrated in other media. For example, a personal computer system might be modeled at the top level in the geometric domain by an assembly diagram showing the positions of the motherboard and plug-in expansion boards in the desktop cabinet.

The next level of abstraction in modeling, depicted by the second ring in Figure 1.1, describes the system in terms of units of data storage and transformation. In the structural domain, this is often called the register-transfer level (RTL), composed of a data path and a control section. The data path contains data storage registers, and data is transferred between them through transformation units. The control section sequences operation of the data path components. For example, a register-transfer-level structural model of the processor in our controller is shown in Figure 1.4.

FIGURE 1.3

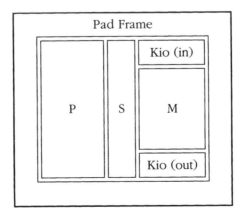

A floor plan model of the controller geometry.

FIGURE 1.4

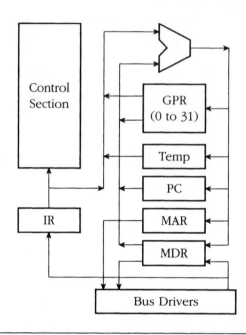

A register-transfer-level structural model of the controller processor. It consists of a general-purpose register (GPR) file; registers for the program counter (PC), memory address (MAR), memory data (MDR), temporary values (Temp) and fetched instructions (IR); an arithmetic unit; bus drivers and the control section.

In the functional domain, a register-transfer language is often used to specify the operation of a system at this level. Storage of data is represented using register variables, and transformations are represented by arithmetic and logical operators. For example, an RTL model for the processor in our example controller might include the following description:

```
MAR ← PC,  memory_read ← 1
PC ← PC + 1
wait until ready = 1
IR ← memory_data
memory_read ← 0
```

This section of the model describes the operations involved in fetching an instruction from memory. The contents of the PC register are transferred to the memory address register, and the **memory_read** signal is asserted. Then the value from the PC register is transformed (incremented in this case) and transferred back to the PC register. When the **ready** input from the memory is asserted, the value on the memory data input is transferred to the instruction register. Finally, the **memory_read** signal is negated.

In the geometric domain, the kind of model used depends on the physical medium. In our example, standard library cells might be used to implement the registers and data transformation units, and these must be placed in the areas allocated in the chip floor plan.

The third level of abstraction shown in Figure 1.1 is the conventional logic level. At this level, structure is modeled using interconnections of gates, and function is modeled by Boolean equations or truth tables. In the physical medium of a custom integrated circuit, geometry may be modeled using a virtual grid, or "sticks," notation.

At the most detailed level of abstraction, we can model structure using individual transistors, function using the differential equations that relate voltage and current in the circuit, and geometry using polygons for each mask layer of an integrated circuit. Most designers do not need to work at this detailed level, as design tools are available to automate translation from a higher level.

1.3 Modeling Languages

In the previous section, we saw that different kinds of models can be devised to represent the various levels of function, structure and physical arrangement of a system. There are also different ways of expressing these models, depending on the use made of the model.

As an example, consider the ways in which a structural model may be expressed. One common form is a circuit schematic. Graphical symbols are used to represent subsystems, and instances of these are connected using lines that represent wires. This graphical form is generally the one preferred by designers. However, the same structural information can be represented textually in the form of a net list.

When we move into the functional domain, we usually see textual notations used for modeling. Some of these are intended for use as specification languages, to meet the need for describing the operation of a system without indicating how it might be implemented. These notations are usually based on formal mathematical methods, such as temporal logic or abstract state machines. Other notations are intended for simulating the system for test and verification purposes and are typically based on conventional programming languages. Yet other notations are oriented toward hardware synthesis and usually have a more restricted set of modeling facilities, since some programming language constructs are difficult to translate into hardware.

The purpose of this book is to describe the modeling language VHDL. VHDL includes facilities for describing structure and function at a number of levels, from the most abstract down to the gate level. It also provides an attribute mechanism that can be used to annotate a model with information in the geometric domain. VHDL is intended, among other things, as a modeling language for specification and simulation. We can also use it for hardware synthesis if we restrict ourselves to a subset that can be automatically translated into hardware.

1.4 VHDL Modeling Concepts

In Section 1.2, we looked at the three domains of modeling: function, structure and geometry. In this section, we look at the basic modeling concepts in each of these domains and introduce the corresponding VHDL elements for describing them. This will provide a feel for VHDL and a basis from which to work in later chapters.

EXAMPLE 1.1   *A four-bit register design*

Figure 1.5 shows a schematic symbol for a four-bit register. Using VHDL terminology, we call the module **reg4** a design *entity*, and the inputs and outputs are *ports*.

FIGURE 1.5

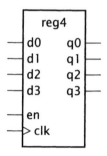

A four-bit register module. The register is named reg4 *and has six inputs,* d0, d1, d2, d3, en *and* clk, *and four outputs,* q0, q1, q2 *and* q3.

We write a VHDL description of the interface to this entity as follows:

```
entity reg4 is
  port ( d0, d1, d2, d3, en, clk : in bit;
         q0, q1, q2, q3 : out  bit );
end entity reg4;
```

This is an example of an *entity declaration*. It introduces a name for the entity and lists the input and output ports, specifying that they carry bit values ('0' or '1') into and out of the entity. From this we see that an entity declaration describes the external view of the entity.

1.4.1   Elements of Behavior

In VHDL, a description of the internal implementation of an entity is called an *architecture body* of the entity. There may be a number of different architecture bodies of the one interface to an entity, corresponding to alternative implementations that perform the same function. We can write a *behavioral* architecture body of an entity, which describes the function in an abstract way. Such an architecture body includes only *process statements*, which are collections of actions to be executed in sequence. These actions are called *sequential statements* and are much like the kinds of statements we see in a conventional programming language. The types of actions that can be performed include evaluating expressions, assigning values to variables, conditional execution, repeated execution and subprogram calls. In addition, there is a sequential statement that is unique to hardware modeling languages, the *signal assignment* statement. This is similar to variable assignment, except that it causes the value on a signal to be updated at some future time.

EXAMPLE 1.2   *Behavioral architecture for the four-bit register*

To illustrate these ideas, let us look at a behavioral architecture body for the **reg4** entity of Example 1.1:

```
architecture behav of reg4 is
begin

  storage : process is
    variable stored_d0, stored_d1, stored_d2, stored_d3 : bit;
  begin
    wait until clk;
    if en then
      stored_d0 := d0;
      stored_d1 := d1;
      stored_d2 := d2;
      stored_d3 := d3;
    end if;
    q0 <= stored_d0 after 5 ns;
    q1 <= stored_d1 after 5 ns;
    q2 <= stored_d2 after 5 ns;
    q3 <= stored_d3 after 5 ns;
  end process storage;

end architecture behav;
```

In this architecture body, the part after the first **begin** keyword includes one process statement, which describes how the register behaves. It starts with the process name, storage, and finishes with the keywords **end process**.

The process statement defines a sequence of actions that are to take place when the system is simulated. These actions control how the values on the entity's ports change over time; that is, they control the behavior of the entity. This process can modify the values of the entity's ports using signal assignment statements.

The way this process works is as follows. When the simulation is started, the signal values are set to '0', and the process is activated. The process's variables (listed after the keyword **variable**) are initialized to '0', then the statements are executed in order. The first statement is a *wait statement* that causes the process to *suspend*, that is, to become inactive. It stays suspended until one of the signals to which it is *sensitive* changes value. In this case, the process is sensitive only to the signal clk, since that is the only one named in the wait statement. When that signal changes value, the process is resumed and continues executing statements. The next statement is a condition that tests whether the value of the en signal is '1'. If it is, the statements between the keywords **then** and **end if** are executed, updating the process's variables using the values on the input signals. After the conditional if statement, there are four signal assignment statements that cause the output signals to be updated 5 ns later.

When all of these statements in the process have been executed, the process starts again from the keyword **begin**, and the cycle repeats. Notice that while the process is suspended, the values in the process's variables are not lost. This is how the process can represent the state of a system.

1.4.2   Elements of Structure

An alternative way of describing the implementation of an entity is to specify how it is composed of subsystems. We can give a structural description of the entity's implementation. An architecture body that is composed only of interconnected subsystems is called a structural architecture body.

EXAMPLE 1.3   *Structural architecture for the four-bit register*

Figure 1.6 shows how the **reg4** entity might be composed of flipflops and gates.

FIGURE 1.6

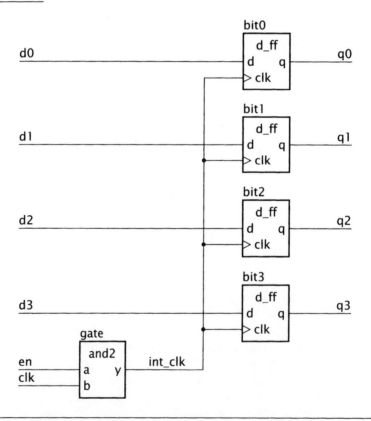

A structural composition of the reg4 entity.

If we are to describe this in VHDL, we will need entity declarations and architecture bodies for the subsystems. For the flipflops, the entity and architecture are

```vhdl
entity d_ff is
  port ( d, clk : in bit;  q : out bit );
end d_ff;

architecture basic of d_ff is
begin

  ff_behavior : process is
  begin
    wait until clk;
    q <= d after 2 ns;
  end process ff_behavior;

end architecture basic;
```

For the two-input and gate, the entity and architecture are

```vhdl
entity and2 is
  port ( a, b : in bit;  y : out bit );
end and2;

architecture basic of and2 is
begin

  and2_behavior : process is
  begin
    y <= a and b after 2 ns;
    wait on a, b;
  end process and2_behavior;

end architecture basic;
```

We can now proceed to a VHDL architecture body declaration that describes the reg4 structure shown in Figure 1.6:

```vhdl
architecture struct of reg4 is

  signal int_clk : bit;

begin
  bit0 : entity work.d_ff(basic)
    port map (d0, int_clk, q0);
  bit1 : entity work.d_ff(basic)
    port map (d1, int_clk, q1);
  bit2 : entity work.d_ff(basic)
    port map (d2, int_clk, q2);
  bit3 : entity work.d_ff(basic)
    port map (d3, int_clk, q3);
```

```
    gate : entity work.and2(basic)
      port map (en, clk, int_clk);

end architecture struct;
```

The *signal declaration*, before the keyword **begin**, defines the internal signals of the architecture. In this example, the signal int_clk is declared to carry a bit value ('0' or '1'). In general, VHDL signals can be declared to carry arbitrarily complex values. Within the architecture body the ports of the entity are also treated as signals.

In the second part of the architecture body, a number of *component instances* are created, representing the subsystems from which the **reg4** entity is composed. Each component instance is a copy of the entity representing the subsystem, using the corresponding **basic** architecture body. (The name **work** refers to the current working library, in which all of the entity and architecture body descriptions are assumed to be held.)

The *port map* specifies the connection of the ports of each component instance to signals within the enclosing architecture body. For example, bit0, an instance of the d_ff entity, has its port d connected to the signal d0, its port clk connected to the signal int_clk and its port q connected to the signal q0.

1.4.3   Mixed Structural and Behavioral Models

Models need not be purely structural or purely behavioral. Often it is useful to specify a model with some parts composed of interconnected component instances, and other parts described using processes. We use signals as the means of joining component instances and processes. A signal can be associated with a port of a component instance and can also be assigned to or read in a process.

We can write such a hybrid model by including both component instance and process statements in the body of an architecture. These statements are collectively called *concurrent statements*, since the corresponding processes all execute concurrently when the model is simulated.

EXAMPLE 1.4   *A mixed structural and behavioral model for a multiplier*

A sequential multiplier consists of a data path and a control section. An outline of a mixed structural and behavioral model for the multiplier is:

```
entity multiplier is
  port ( clk, reset : in bit;
         multiplicand, multiplier : in integer;
         product : out integer );
end entity multiplier;

--------------------------------------------------

architecture mixed of multiplier is
```

```
      signal partial_product, full_product : integer;
      signal arith_control, result_en, mult_bit, mult_load : bit;

begin -- mixed

  arith_unit : entity work.shift_adder(behavior)
    port map ( addend => multiplicand, augend => full_product,
               sum => partial_product,
               add_control => arith_control);

  result : entity work.reg(behavior)
    port map ( d => partial_product, q => full_product,
               en => result_en, reset => reset);

  multiplier_sr : entity work.shift_reg(behavior)
    port map ( d => multiplier, q => mult_bit,
               load => mult_load, clk => clk);

  product <= full_product;

  control_section : process is
    -- variable declarations for control_section
    -- ...
  begin -- control section
    -- sequential statements to assign values to control signals
    -- ...
    wait on clk, reset;
  end process control_section;

end architecture mixed;
```

The data path is described structurally, using a number of component instances. The control section is described behaviorally, using a process that assigns to the control signals for the data path.

1.4.4 Test Benches

In our introductory discussion, we mentioned testing through simulation as an important motivation for modeling. We often test a VHDL model using an enclosing model called a *test bench*. The name comes from the analogy with a real hardware test bench, on which a device under test is stimulated with signal generators and observed with signal probes. A VHDL test bench consists of an architecture body containing an instance of the component to be tested and processes that generate sequences of values on signals connected to the component instance. The architecture body may also contain processes that test that the component instance produces the expected values on its output signals. Alternatively, we may use the monitoring facilities of a simulator to observe the outputs.

EXAMPLE 1.5   *Test bench for the four-bit register*

A test bench model for the behavioral implementation of the **reg4** register is:

```
entity test_bench is
end entity test_bench;

architecture test_reg4 of test_bench is

    signal d0, d1, d2, d3, en, clk, q0, q1, q2, q3 : bit;

begin

    dut : entity work.reg4(behav)
      port map ( d0, d1, d2, d3, en, clk, q0, q1, q2, q3 );

    stimulus : process is
    begin
      d0 <= '1';   d1 <= '1';   d2 <= '1';   d3 <= '1';
      en <= '0';   clk <= '0';
      wait for 20 ns;
      en <= '1';   wait for 20 ns;
      clk <= '1';   wait for 20 ns;
      d0 <= '0';   d1 <= '0';   d2 <= '0';   d3 <= '0';   wait for 20 ns;
      en <= '0';   wait for 20 ns;
      ...
      wait;
    end process stimulus;

end architecture test_reg4;
```

The entity declaration has no port list, since the test bench is entirely self-contained. The architecture body contains signals that are connected to the input and output ports of the component instance **dut**, the device under test. The process labeled **stimulus** provides a sequence of test values on the input signals by performing signal assignment statements, interspersed with wait statements. Each wait statement specifies a 20 ns pause during which the register device determines its output values. We can use a simulator to observe the values on the signals **q0** to **q3** to verify that the register operates correctly. When all of the stimulus values have been applied, the stimulus process waits indefinitely, thus completing the simulation.

1.4.5   Analysis, Elaboration and Execution

One of the main reasons for writing a model of a system is to enable us to simulate it. This involves three stages: *analysis,* *elaboration* and *execution.* Analysis and elaboration are also required in preparation for other uses of the model, such as logic synthesis.

In the first stage, analysis, the VHDL description of a system is checked for various kinds of errors. Like most programming languages, VHDL has rigidly defined *syntax* and *semantics.* The syntax is the set of grammatical rules that govern how a model is written.

The rules of semantics govern the meaning of a program. For example, it makes sense to perform an addition operation on two numbers but not on two processes.

During the analysis phase, the VHDL description is examined, and syntactic and static semantic errors are located. The whole model of a system need not be analyzed at once. Instead, it is possible to analyze *design units*, such as entity and architecture body declarations, separately. If the analyzer finds no errors in a design unit, it creates an intermediate representation of the unit and stores it in a library. The exact mechanism varies between VHDL tools.

The second stage in simulating a model, elaboration, is the act of working through the design hierarchy and creating all of the objects defined in declarations. The ultimate product of design elaboration is a collection of signals and processes, with each process possibly containing variables. A model must be reducible to a collection of signals and processes in order to simulate it.

We can see how elaboration achieves this reduction by starting at the top level of a model, namely, an entity, and choosing an architecture of the entity to simulate. The architecture comprises signals, processes and component instances. Each component instance is a copy of an entity and an architecture that also comprises signals, processes and component instances. Instances of those signals and processes are created, corresponding to the component instance, and then the elaboration operation is repeated for the sub-component instances. Ultimately, a component instance is reached that is a copy of an entity with a purely behavioral architecture, containing only processes. This corresponds to a primitive component for the level of design being simulated. Figure 1.7 shows how elaboration proceeds for the structural architecture body of the **reg4** entity from Example 1.3. As each instance of a process is created, its variables are created and given initial values. We can think of each process instance as corresponding to one instance of a component.

The third stage of simulation is the execution of the model. The passage of time is simulated in discrete steps, depending on when events occur. Hence the term *discrete event simulation* is used. At some simulation time, a process may be stimulated by changing the value on a signal to which it is sensitive. The process is resumed and may schedule new values to be given to signals at some later simulated time. This is called *scheduling a transaction* on that signal. If the new value is different from the previous value on the signal, an *event* occurs, and other processes sensitive to the signal may be resumed.

The simulation starts with an *initialization phase*, followed by repetitive execution of a *simulation cycle*. During the initialization phase, each signal is given an initial value, depending on its type. The simulation time is set to zero, then each process instance is activated and its sequential statements executed. Usually, a process will include a signal assignment statement to schedule a transaction on a signal at some later simulation time. Execution of a process continues until it reaches a wait statement, which causes the process to be suspended.

During the simulation cycle, the simulation time is first advanced to the next time at which a transaction on a signal has been scheduled. Second, all the transactions scheduled for that time are performed. This may cause some events to occur on some signals. Third, all processes that are sensitive to those events are resumed and are allowed to continue until they reach a wait statement and suspend. Again, the processes usually execute signal assignments to schedule further transactions on signals. When all the processes have suspended again, the simulation cycle is repeated. When the simulation gets to the stage

FIGURE 1.7

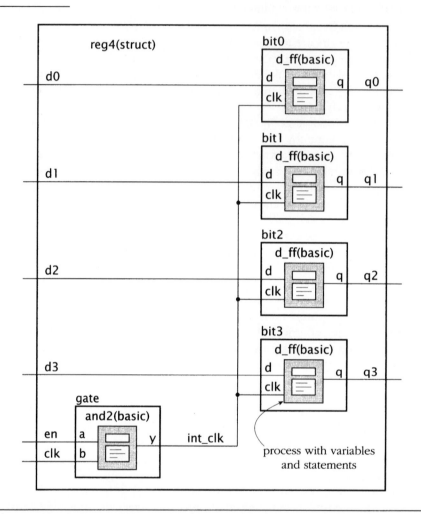

The elaboration of the **reg4** *entity using the structural architecture body. Each instance of the* **d_ff** *and* **and2** *entities is replaced with the contents of the corresponding* **basic** *architecture. These each consist of a process with its variables and statements.*

where there are no further transactions scheduled, it stops, since the simulation is then complete.

1.5 Learning a New Language: Lexical Elements and Syntax

When we learn a new natural language, such as Greek, Chinese or English, we start by learning the alphabet of symbols used in the language, then form these symbols into words. Next, we learn the way to put the words together to form sentences and learn the meaning of these combinations of words. We reach fluency in a language when we can easily express what we need to say using correctly formed sentences.

The same ideas apply when we need to learn a new special-purpose language, such as VHDL for describing digital systems. We can borrow a few terms from language theory to describe what we need to learn. First, we need to learn the alphabet with which the language is written. The VHDL alphabet consists of all of the characters in the ISO 8859 Latin-1 8-bit character set. This includes uppercase and lowercase letters (including letters with diacritical marks, such as "à", "ä" and so forth), digits 0 to 9, punctuation and other special characters. Second, we need to learn the *lexical elements* of the language. In VHDL, these are the identifiers, reserved words, special symbols and literals. Third, we need to learn the *syntax* of the language. This is the grammar that determines what combinations of lexical elements make up legal VHDL descriptions. Fourth, we need to learn the *semantics*, or meaning, of VHDL descriptions. It is the semantics that allow a collection of symbols to describe a digital design. Fifth, we need to learn how to develop our own VHDL descriptions to describe a design we are working with. This is the creative part of modeling, and fluency in this part will greatly enhance our design skills.

In the remainder of this chapter, we describe the lexical elements used in VHDL and introduce the notation we use to describe the syntax rules. Then in subsequent chapters, we introduce the different facilities available in the language. For each of these, we show the syntax rules, describe the corresponding semantics and give examples of how they are used to model particular parts of a digital system. We also include some exercises at the end of each chapter to provide practice in the fifth stage of learning described above.

VHDL-87

VHDL-87 uses the ASCII character set, rather than the full ISO character set. ASCII is a subset of the ISO character set, consisting of just the first 128 characters. This includes all of the unaccented letters, but excludes letters with diacritical marks.

1.5.1 Lexical Elements

In the following section, we discuss the lexical elements of VHDL: *comments*, *identifiers*, *reserved words*, *special symbols*, *numbers*, *characters*, *strings* and *bit strings*.

Comments

When we are writing a hardware model in VHDL, it is important to annotate the code with comments. The reason for doing this is to help readers understand the structure and logic behind the model. It is important to realize that although we only write a model once, it may subsequently be read and modified many times, both by its author and by other engineers. Any assistance we can give to understanding the model is worth the effort. In this book, we set comments in slanted text to make them visually distinct.

A VHDL model consists of a number of lines of text. One form of comment, called a single-line comment, can be added to a line by writing two hypens together, followed by the comment text. For example:

```
... a line of VHDL code ...      -- a descriptive comment
```

The comment extends from the two hyphens to the end of the line and may include any text we wish, since it is not formally part of the VHDL model. The code of a model can include blank lines and lines that only contain comments, starting with two hyphens. We can write long comments on successive lines, each line starting with two hyphens, for example:

```
-- The following code models
-- the control section of the system
... some VHDL code ...
```

Another form of comment, called a delimited comment, starts with the characters "/*" and extends to the closing characters "*/". The opening and closing characters can be on different lines, or can be on the same line. Moreover, there can be further VHDL code on the line after the closing characters. Some examples are:

```
/* This is a comment header that describes
   the purpose of the design unit. It contains
   all you ever wanted to know, plus more.
*/


entity thingumy is
   port ( clk    : in bit; -- keeps it going
          reset : in bit /* start over */
          /* other ports to be added later */ );
end entity thingumy;
```

Since the text in comments is ignored, it may contain comment delimiters. Mixing comment styles can be quite useful. For example, if we use delimited comments in a section of code, and we want to "comment out" the section, we can use single-line comments:

```
-- This section commented out because it doesn't work
-- /* Process to do a complicated computation
-- involving lots of digital signal processing.
-- */
-- dsp_stuff : process is
-- begin
--    ...
-- end process dsp_stuff;
```

However, we should be aware that comments do not nest. For example, the following is ill-formed:

```
-- Here is the start of the comment: /* A comment extending
                                         over two lines */
```

The opening "/*" characters occur in a single-line comment, and so are ignored. Similarly, we cannot reliably use delimited comments to comment out a section of code, since the section might already contain a delimited comment:

```
/* Comment out the following code:
signal count_en : bit; /* counter enable */
*/
```

In this case, the occurrence of the characters "*/" on the second line closes the comment started on the first line, making the orphaned delimiter "*/" on the third line illegal. Provided we avoid pitfalls such as these, single-line and delimited comments are useful language features.

VHDL-87, -93, and -2002

These versions of VHDL only allow single-line comments, not delimited comments.

Identifiers

Identifiers are used to name items in a VHDL model. It is good practice to use names that indicate the purpose of the item, so VHDL allows names to be arbitrarily long. However, there are some rules about how identifiers may be formed. A basic identifier

- may only contain alphabetic letters ('A' to 'Z' and 'a' to 'z'), decimal digits ('0' to '9') and the underline character ('_');

- must start with an alphabetic letter;

- may not end with an underline character; and

- may not include two successive underline characters.

Some examples of valid basic identifiers are

```
A  X0  counter  Next_Value  generate_read_cycle
```

Some examples of invalid basic identifiers are

```
last@value      -- contains an illegal character for an identifier
5bit_counter    -- starts with a non-alphabetic character
_A0             -- starts with an underline
A0_             -- ends with an underline
clock__pulse    -- two successive underlines
```

Note that the case of letters is not considered significant, so the identifiers cat and Cat are the same. Underline characters in identifiers are significant, so This_Name and This-Name are different identifiers.

In addition to the basic identifiers, VHDL allows *extended identifiers*, which can contain any sequence of characters. Extended identifiers are included to allow communication between computer-aided engineering tools for processing VHDL descriptions and other tools that use different rules for identifiers. An extended identifier is written by enclosing the characters of the identifier between '\' characters. For example:

```
\data bus\ \global.clock\ \923\ \d#1\ \start__\
```

If we need to include a '\' character in an extended identifier, we do so by doubling the character, for example:

```
\A:\\name\    -- contains a '\' between the ':' and the 'n'
```

Note that the case of letters is significant in extended identifiers and that all extended identifiers are distinct from all basic identifiers. So the following are all distinct identifiers:

```
name \name\ \Name\ \NAME\
```

VHDL-87

VHDL-87 only allows basic identifiers, not extended identifiers. The rules for forming basic identifiers are the same as those for VHDL-93 and VHDL-2002.

Reserved Words

Some identifiers, called reserved words or keywords, are reserved for special use in VHDL. They are used to denote specific constructs that form a model, so we cannot use them as identifiers for items we define. The full list of reserved words is shown in Table 1.1. Often, when a VHDL program is typeset, reserved words are printed in boldface. This convention is followed in this book.

VHDL-2002

The following identifiers are not used as reserved words in VHDL-2002. They may be used as identifiers for other purposes, although it is not advisable to do so, as this may cause difficulties in porting the models to VHDL-2008.

	fairness	restrict_guarantee
assume	force	sequence
assume_guarantee	parameter	strong
context	property	vmode
cover	release	vprop
default	restrict	vunit

TABLE 1.1   *VHDL reserved words*

abs	default	label	package	sla
access	disconnect	library	parameter	sll
after	downto	linkage	port	sra
alias		literal	postponed	srl
all	else	loop	procedure	strong
and	elsif		process	subtype
architecture	end	map	property	
array	entity	mod	protected	then
assert	exit		pure	to
assume		nand		transport
assume_guarantee	fairness	new	range	type
attribute	file	next	record	
	for	nor	register	unaffected
begin	force	not	reject	units
block	function	null	release	until
body			rem	use
buffer	generate	of	report	
bus	generic	on	restrict	variable
	group	open	restrict_guarantee	vmode
	guarded	or	return	vprop
case		others	rol	vunit
component		out	ror	
configuration	if			wait
constant	impure			when
context	in		select	while
cover	inertial		sequence	with
	inout		severity	
	is		shared	xnor
			signal	xor

VHDL-93

In addition to those listed for VHDL-2002, the identifier **protected** is not used as a reserved word in VHDL-93.

VHDL-87

In addition to those listed for VHDL-2002 and VHDL-93, the following identifiers are not used as reserved words in VHDL-87:

group	protected	ror	sra
impure	pure	shared	srl
inertial	reject	sla	unaffected
literal	rol	sll	xnor
postponed			

Special Symbols

VHDL uses a number of special symbols to denote operators, to delimit parts of language constructs and as punctuation. Some of these special symbols consist of just one character. They are

    " # & ' ( ) * + - , . / : ; < = > ? @ [ ] ` |

Other special symbols consist of pairs of characters. The two characters must be typed next to each other, with no intervening space. These symbols are

    => ** := /= >= <= <> ?? ?= ?/= ?> ?< ?>= ?<= << >>

Numbers

There are two forms of numbers that can be written in VHDL code: *integer literals* and *real literals*. An integer literal simply represents a whole number and consists of digits without a decimal point. Real literals, on the other hand, can represent fractional numbers. They always include a decimal point, which is preceded by at least one digit and followed by at least one digit. Real literals represent an approximation to real numbers.

Some examples of decimal integer literals are

    23 0 146

Note that –10, for example, is not an integer literal. It is actually a combination of a negation operator and the integer literal 10.

Some examples of real literals are

    23.1 0.0 3.14159

Both integer and real literals can also use exponential notation, in which the number is followed by the letter 'E' or 'e', and an exponent value. This indicates a power of 10 by which the number is multiplied. For integer literals, the exponent must not be negative, whereas for real literals, it may be either positive or negative. Some examples of integer literals using exponential notation are

    46E5 1E+12 19e00

Some examples of real literals using exponential notation are

    1.234E09 98.6E+21 34.0e-08

Integer and real literals may also be expressed in a base other than base 10. In fact, the base can be any integer between 2 and 16. To do this, we write the number surrounded by sharp characters ('#'), preceded by the base. For bases greater than 10, the letters 'A' through 'F' (or 'a' through 'f') are used to represent the digits 10 through 15. For example, several ways of writing the value 253 are as follows:

    2#11111101# 16#FD# 16#0fd# 8#0375#

Similarly, the value 0.5 can be represented as

    2#0.100# 8#0.4# 12#0.6#

Note that in all these cases, the base itself is expressed in *decimal.*

Based literals can also use exponential notation. In this case, the exponent, expressed in decimal, is appended to the based number after the closing sharp character. The exponent represents the power of the base by which the number is multiplied. For example, the number 1024 could be represented by the integer literals:

    2#1#E10 16#4#E2 10#1024#E+00

Finally, as an aid to readability of long numbers, we can include underline characters as separators between digits. The rules for including underline characters are similar to those for identifiers; that is, they may not appear at the beginning or end of a number, nor may two appear in succession. Some examples are

    123_456 3.141_592_6 2#1111_1100_0000_0000#

Characters

A character literal can be written in VHDL code by enclosing it in single quotation marks. Any of the printable characters in the standard character set (including a space character) can be written in this way. Some examples are

```
'A'    -- uppercase letter
'z'    -- lowercase letter
','    -- the punctuation character comma
'''    -- the punctuation character single quote
' '    -- the separator character space
```

Strings

A string literal represents a sequence of characters and is written by enclosing the characters in double quotation marks. The string may include any number of characters (including zero), but it must fit entirely on one line. Some examples are

```
"A string"
"A string can include any printing characters (e.g., &%@^*)."
"00001111ZZZZ"
""   -- empty string
```

If we need to include a double quotation mark character in a string, we write two double quotation mark characters together. The pair is interpreted as just one character in the string. For example:

```
"A string in a string: ""A string"". "
```

If we need to write a string that is longer than will fit on one line, we can use the concatenation operator ("&") to join two substrings together. (This operator is discussed in Chapter 4.) For example:

```
"If a string will not fit on one line, "
& "then we can break it into parts on separate lines."
```

Bit Strings

VHDL includes values that represent bits (binary digits), which can be either '0' or '1'. A bit-string literal represents a string of these bit values. It is represented by a string of digits, enclosed by double quotation marks and preceded by a character that specifies the base of the digits. The base specifier can be one of the following:

- B for binary,

- O for octal (base 8) and

- X for hexadecimal (base 16).

- D for decimal (base 10).

For example, some bit-string literals specified in binary are

```
B"0100011"  B"10"  b"1111_0010_0001"  B""
```

Notice that we can include underline characters in bit-string literals to separate adjacent digits. The underline characters do not affect the meaning of the literal; they simply make the literal more readable. The base specifier can be in uppercase or lowercase. The last of the examples above denotes an empty bit string.

If the base specifier is octal, the digits '0' through '7' can be used. Each digit represents exactly three bits in the bit string. Some examples are

```
O"372"   -- equivalent to B"011_111_010"
o"00"    -- equivalent to B"000_000"
```

If the base specifier is hexadecimal, the digits '0' through '9' and 'A' through 'F' or 'a' through 'f' (representing 10 through 15) can be used. In hexadecimal, each digit represents exactly four bits. Some examples are

```
X"FA"    -- equivalent to B"1111_1010"
x"0d"    -- equivalent to B"0000_1101"
```

Notice that O"372" is not the same as X"FA", since the former is a string of nine bits, whereas the latter is a string of eight bits.

If the base specifier is decimal, the digits '0' through '9' can be used. The digits in the literal are interpreted as a decimal number and are converted to the equivalent binary value. The number of bits in the string is the minimal number needed to represent the value. Some examples are

```
D"23"      -- equivalent to B"10111"
D"64"      -- equivalent to B"1000000"
D"0003"    -- equivalent to B"11"
```

In some cases, it is convenient to include characters other than digits in bit string literals. As we will see later, many VHDL models use characters such as 'Z', 'X', and '–' to represent high-impedance states, unknown values, and don't-care conditions. Models may use other characters for similar purposes. We can include such non-binary characters in bit-string literals. In an octal literal, any non-octal-digit character is expanded to three occurrences of that character in the bit string. Similarly, in a hexadecimal literal any non-hexadecimal-digit character is expanded to four occurrences of the character. In a binary literal, any non-bit character just represents itself in the vector. Some examples are:

```
O"3XZ4"    -- equivalent to B"011XXXZZZ100"
X"A3--"    -- equivalent to B"10100011--------"
X"0#?F"    -- equivalent to B"0000####????1111"
B"00UU"    -- equivalent to B"00UU"
```

While allowing this for binary literals might seem vacuous at first, the benefit will become clear shortly. Note that expansion of non-digit characters does not extend to embedded underscores, which we might add for readability. Thus, O"3_X" represents "011XXX", not "011___XXX". Also, non-digit characters are not allowed in decimal literals, since it would be unclear which bits of the resulting string correspond to the non-digit characters. Thus, the literal D"23Z9" is illegal.

In all of the preceding cases, the number of bits in the string is determined from the base specifier and the number of characters in the literal. We can, however, specify the exact length of bit string that we require from a literal. This allows us to specify strings whose length is not a multiple or three (for octal) or four (for hexadecimal). We do so by writing the length immediately before the base specifier, with no intervening space. Some examples are:

```
7X"3C"     -- equivalent to B"0111100"
80"5"      -- equivalent to B"00000101"
10B"X"     -- equivalent to B"000000000X"
```

If the final length of the string is longer than that implied by the digits, the string is padded on the left with '0' bits. If the final length is less than that implied by the digits, the left-most elements of the string are truncated, provided they are all '0'. An error occurs if any non-'0' bits are truncated, as they would be in the literal 8X"90F".

A further feature of bit-string literals is provision for specifying whether the literal represents an unsigned or signed number. We represent an unsigned number using one of the base specifiers UB, UO, or UX. These are the same as the ordinary base specifiers B, O, and X. When a sized unsigned literal is extended, it is padded with '0' bits, and when bits are truncated, they must be all '0'. Decimal literals are always interpreted as unsigned, so D is the only base specifier for decimal. We can extend a decimal literal by padding with '0' bits. However, we cannot truncate a decimal literal from its default size, since the default size always gives a '1' as the leftmost bit, which must not be truncated.

We represent a signed number using one of the base specifiers SB, SO, or SX. The rules for extension and truncation are based on those for sign extension and truncation of 2s-complement binary numbers. When a sized signed literal is extended, each bit of padding on the left is a replication of the leftmost bit prior to padding. For example:

```
10SX"71"    -- equivalent to B"0001110001"
10SX"88"    -- equivalent to B"1110001000"
10SX"W0"    -- equivalent to B"WWWWWW0000"
```

When a sized signed literal is truncated, all of the bits removed at the left must be the same as the leftmost remaining bit. For example:

```
6SX"16"    -- equivalent to B"010110"
6SX"E8"    -- equivalent to B"101000"
6SX"H3"    -- equivalent to B"HH0011"
```

However, 6SX"28" is invalid, since, prior to truncation, the bit string would be "00101000". The two leftmost bits removed are each '0', which differ from the leftmost remaining '1' bit. The literal would have to be written as 6SX"E8" for this reason. The rationale for this rule is that it prevents the signed numeric value represented by the literal being inadvertently changed by the truncation.

VHDL-87, -93, and -2002

These versions of VHDL only allow the base specifiers B, O, and X. They do not allow unsigned and signed specifiers UB, UO, UX, SB, SO, and SX; nor do they allow the decimal specifier D. They do not allow the size to be specified; thus, octal literals are always a multiple of three in length, and hexidecimal literals are always a multiple of four in length. Finally, non-digit characters, other than underlines for readability, are not allowed.

1.5.2   Syntax Descriptions

In the remainder of this book, we describe rules of syntax using a notation based on the Extended Backus-Naur Form (EBNF). These rules govern how we may combine lexical elements to form valid VHDL descriptions. It is useful to have a good working knowledge of the syntax rules, since VHDL analyzers expect valid VHDL descriptions as input. The error messages they otherwise produce may in some cases appear cryptic if we are unaware of the syntax rules.

The idea behind EBNF is to divide the language into *syntactic categories*. For each syntactic category we write a rule that describes how to build a VHDL clause of that category by combining lexical elements and clauses of other categories. These rules are analogous to the rules of English grammar. For example, there are rules that describe a sentence in terms of a subject and a predicate, and that describe a predicate in terms of a verb and an object phrase. In the rules for English grammar, "sentence", "subject", "predicate", and so on, are the syntactic categories.

In EBNF, we write a rule with the syntactic category we are defining on the left of a "⇐" sign (read as "is defined to be"), and a pattern on the right. The simplest kind of pattern is a collection of items in sequence, for example:

variable_assignment ⇐ target := expression ;

This rule indicates that a VHDL clause in the category "variable_assignment" is defined to be a clause in the category "target", followed by the symbol ":=", followed by a clause in the category "expression", followed by the symbol ";". To find out whether the VHDL clause

    d0 := 25 + 6;

is syntactically valid, we would have to check the rules for "target" and "expression". As it happens, "d0" and "25+6" are valid subclauses, so the whole clause conforms to the pattern in the rule and is thus a valid variable assignment. On the other hand, the clause

    25 fred := x if := .

cannot possibly be a valid variable assignment, since it doesn't match the pattern on the right side of the rule.

The next kind of rule to consider is one that allows for an optional component in a clause. We indicate the optional part by enclosing it between the symbols "⟦" and "⟧". For example:

function_call ⇐ name ⟦ (association_list) ⟧

This indicates that a function call consists of a name that may be followed by an association list in parentheses. Note the use of the outline symbols for writing the pattern in the rule, as opposed to the normal solid symbols that are lexical elements of VHDL.

In many rules, we need to specify that a clause is optional, but if present, it may be repeated as many times as needed. For example, in this simplified rule for a process statement:

process_statement ⇐
    process is
        ⦃ process_declarative_item ⦄
    begin
        ⦃ sequential_statement ⦄
    end process ;

the curly braces specify that a process may include zero or more process declarative items and zero or more sequential statements. A case that arises frequently in the rules of VHDL is a pattern consisting of some category followed by zero or more repetitions of that category. In this case, we use dots within the braces to represent the repeated category, rather than writing it out again in full. For example, the rule

case_statement ⇐
    case expression **is**

case_statement_alternative
⦃ ₒₒₒ ⦄
end case ;

indicates that a case statement must contain at least one case statement alternative, but may contain an arbitrary number of additional case statement alternatives as required. If there is a sequence of categories and symbols preceding the braces, the dots represent only the last element of the sequence. Thus, in the example above, the dots represent only the case statement alternative, not the sequence "**case** expression **is** case_statement_alternative".

We also use the dots notation where a list of one or more repetitions of a clause is required, but some delimiter symbol is needed between repetitions. For example, the rule

identifier_list ⇐ identifier ⦃ , ₒₒₒ ⦄

specifies that an identifier list consists of one or more identifiers, and that if there is more than one, they are separated by comma symbols. Note that the dots always represent a repetition of the category immediately preceding the left brace symbol. Thus, in the above rule, it is the identifier that is repeated with comma delimiters; it is not just the comma that is repeated.

Many syntax rules allow a category to be composed of one of a number of alternatives. One way to represent this is to have a number of separate rules for the category, one for each alternative. However, it is often more convenient to combine alternatives using the "❘" symbol. For example, the rule

mode ⇐ **in** ❘ **out** ❘ **inout**

specifies that the category "mode" can be formed from a clause consisting of one of the reserved words chosen from the alternatives listed.

The final notation we use in our syntax rules is parenthetic grouping, using the symbols "⦅" and "⦆". These simply serve to group part of a pattern, so that we can avoid any ambiguity that might otherwise arise. For example, the inclusion of parentheses in the rule

term ⇐ factor ⦃ ⦅ * ❘ / ❘ **mod** ❘ **rem** ⦆ factor ⦄

makes it clear that a factor may be followed by one of the operator symbols, and then another factor. Without the parentheses, the rule would be

term ⇐ factor ⦃ * ❘ / ❘ **mod** ❘ **rem** factor ⦄

indicating that a factor may be followed by one of the operators "*", "/" or **mod** alone, or by the operator **rem** and then another factor. This is certainly not what is intended. The reason for this incorrect interpretation is that there is a *precedence*, or order of priority, in the EBNF notation we are using. In the absence of parentheses, a sequence of pattern components following one after the other is considered as a group with higher precedence than components separated by "❘" symbols.

This EBNF notation is sufficient to describe the complete grammar of VHDL. However, there are often further constraints on a VHDL description that relate to the meaning of the lexical elements used. For example, a description specifying connection of a signal to a named object that identifies a component instead of a port is incorrect, even though it may conform to the syntax rules. To avoid such problems, many rules include additional information relating to the meaning of a language feature. For example, the rule shown above describing how a function call is formed is augmented thus:

function_call ⇐ function_name ⟦ (*parameter*_association_list) ⟧

The italicized prefix on a syntactic category in the pattern simply provides semantic information. This rule indicates that the name cannot be just any name, but must be the name of a function. Similarly, the association list must describe the parameters supplied to the function. (We will describe the meaning of functions and parameters in a later chapter.) The semantic information is for our benefit as designers reading the rule, to help us understand the intended semantics. So far as the syntax is concerned, the rule is equivalent to the original rule without the italicized parts.

In the following chapters, we will introduce each new feature of VHDL by describing its syntax using EBNF rules, and then we will describe the meaning and use of the feature through examples. In many cases, we will start with a simplified version of the syntax to make the description easier to learn and come back to the full details in a later chapter. For reference, Appendix B contains a complete listing of VHDL syntax in EBNF notation.

Exercises

1. [❶ 1.4] Briefly outline the purposes of the following VHDL modeling constructs: entity declaration, behavioral architecture body, structural architecture body, process statement, signal assignment statement and port map.

2. [❶ 1.5] Single-line comment symbols are often used to make lines of a model temporarily ineffective. The symbol is added at the front of the line, turning the line into a comment. The comment symbol can be simply removed to reactivate the statement. The following process statement includes a line to assign a value to a test signal, to help debug the model. Modify the process to make the assignment ineffective.

```
apply_transform : process is
begin
  d_out <= transform(d_in) after 200 ps;
  debug_test <= transform(d_in);
  wait on enable, d_in;
end process apply_transform;
```

3. [❶ 1.5] Which of the following are valid VHDL basic identifiers? Which are reserved words? Of the invalid identifiers, why are they invalid?

```
last_item    prev item    value-1    buffer

element#5    _control     93_999     entry_
```

4. [❶ 1.5] Rewrite the following decimal literals as hexadecimal literals.

```
1    34    256.0    0.5
```

5. [❶ 1.5] What decimal numbers are represented by the following literals?

    8#14#   2#1000_0100#   16#2C#

    2.5E5   2#1#E15   2#0.101#

6. [❶ 1.5] What is the difference between the literals 16#23DF# and X"23DF"?

7. [❶ 1.5] Express the following octal and hexadecimal bit strings as binary bit-string literals.

    O"747"   O"377"   O"1_345"

    X"F2"   X"0014"   X"0000_0001"

8. [❶ 1.5] Express the following octal and hexadecimal bit strings as binary bit-string literals, or, if they are illegal, say why.

    10UO"747"   10UO"377"   10UO"1_345"

    10SO"747"   10SO"377"   10SO"1_345"

    12UX"F2"   12SX"F2"   10UX"F2"   10SX"F2"

9. [❶ 1.5] Express the following decimal bit strings as binary bit-string literals, or, if they are illegal, say why.

    D"24"   12D"24"   4D"24"

10. [❷ 1.4] Write an entity declaration and a behavioral architecture body for a two-input multiplexer, with input ports **a**, **b** and **sel** and an output port **z**. If the **sel** input is '0', the value of a should be copied to z, otherwise the value of b should be copied to z. Write a test bench for the multiplexer model, and test it using a VHDL simulator.

11. [❷ 1.4] Write an entity declaration and a structural architecture body for a 4-bit-wide multiplexer, using instances of the 2-bit multiplexer from Exercise 10. The input ports are a0, a1, a2, a3, b0, b1, b2, b3 and sel, and the output ports are z0, z1, z2 and z3. When **sel** is '0', the inputs a0 to a3 are copied to the outputs, otherwise the inputs b0 to b3 are copied to the outputs. Write a test bench for the multiplexer model, and test it using a VHDL simulator.

Chapter 2

Scalar Data Types
and Operations

The concept of type is very important when describing data in a VHDL model. The type of a data object defines the set of values that the object can assume, as well as the set of operations that can be performed on those values. A scalar type consists of single, indivisible values. In this chapter we look at the basic scalar types provided by VHDL and see how they can be used to define data objects that model the internal state of a module.

2.1 Constants and Variables

An object is a named item in a VHDL model that has a value of a specified type. There are four classes of objects: constants, variables, signals and files. In this chapter, we look at constants and variables; signals are described fully in Chapter 5, and files in Chapter 16. Constants and variables are objects in which data can be stored for use in a model. The difference between them is that the value of a constant cannot be changed after it is created, whereas a variable's value can be changed as many times as necessary using variable assignment statements.

2.1.1 Constant and Variable Declarations

Both constants and variables need to be declared before they can be used in a model. A declaration simply introduces the name of the object, defines its type and may give it an initial value. The syntax rule for a constant declaration is

constant_declaration ⇐
    constant identifier ⟨ , ⟩ : subtype_indication ⟦ := expression ⟧ ;

The identifiers listed are the names of the constants being defined (one per name), and the subtype indication specifies the type of all of the constants. We look at ways of specifying the type in detail in subsequent sections of this chapter. The optional part shown in the syntax rule is an expression that specifies the value that each constant assumes. This part can only be omitted in certain cases that we discuss in Chapter 7. Until then, we always include it in examples. Here are some examples of constant declarations:

```
constant number_of_bytes : integer := 4;
constant number_of_bits : integer := 8 * number_of_bytes;
constant e : real := 2.718281828;
constant prop_delay : time := 3 ns;
constant size_limit, count_limit : integer := 255;
```

The reason for using a constant is to have a name and an explicitly defined type for a value, rather than just writing the value as a literal. This makes the model more intelligible to the reader, since the name and type convey much more information about the intended use of the object than the literal value alone. Furthermore, if we need to change the value as the model evolves, we only need to update the declaration. This is much easier and more reliable than trying to find and update all instances of a literal value throughout a model. It is good practice to use constants rather than writing literal values within a model.

The form of a variable declaration is similar to a constant declaration. The syntax rule is

variable_declaration ⇐
    variable identifier ⟨ , ₀₀₀ ⟩ : subtype_indication ⟦ := expression ⟧ ;

Here also the initialization expression is optional. If we omit it, the default initial value assumed by the variable when it is created depends on the type. For scalar types, the default initial value is the leftmost value of the type. For example, for integers it is the smallest representable integer. Some examples of variable declarations are

```
variable index : integer := 0;
variable sum, average, largest : real;
variable start, finish : time := 0 ns;
```

If we include more than one identifier in a variable declaration, it is the same as having separate declarations for each identifier. For example, the last declaration above is the same as the two declarations

```
variable start : time := 0 ns;
variable finish : time := 0 ns;
```

This is not normally significant unless the initialization expression is such that it potentially produces different values on two successive evaluations. The only time this may occur is if the initialization expression contains a call to a function with side effects (see Chapter 6).

Constant and variable declarations can appear in a number of places in a VHDL model, including in the declaration parts of processes. In this case, the declared object can be used only within the process. One restriction on where a variable declaration may occur is that it may not be placed so that the variable would be accessible to more than one process. This is to prevent the strange effects that might otherwise occur if the processes were to modify the variable in indeterminate order. The exception to this rule is if a variable is declared specially as a *shared* variable. We will leave discussion of shared variables until Chapter 19.

EXAMPLE 2.1   *Constants and variables in an architecture*

The following outline of an architecture body shows how constant and variable declarations may be included in a VHDL model. It includes declarations of a constant **pi** and a variable **counter**.

```
entity ent is

end entity ent;

architecture sample of ent is

  constant pi : real := 3.14159;

begin

  process is
    variable counter : integer;
  begin
    -- ...            -- statements using pi and counter
  end process;

end architecture sample;
```

2.1.2   Variable Assignment

Once a variable has been declared, its value can be modified by an assignment statement. The syntax of a variable assignment statement is given by the rule

variable_assignment_statement ⇐ ⟦ label : ⟧ name := expression ;

The optional label provides a means of identifying the assignment statement. We will discuss reasons for labeling statements in Chapter 20. Until then, we will simply omit the label in our examples. The name in a variable assignment statement identifies the variable to be changed, and the expression is evaluated to produce the new value. The type of this value must match the type of the variable. The full details of how an expression is formed are covered in the rest of this chapter. For now, just think of expressions as the usual combinations of identifiers and literals with operators. Here are some examples of assignment statements:

```
program_counter := 0;
index := index + 1;
```

The first assignment sets the value of the variable **program_counter** to zero, overwriting any previous value. The second example increments the value of **index** by one.

   It is important to note the difference between a variable assignment statement, shown here, and a signal assignment statement, introduced in Chapter 1. A variable assignment immediately overwrites the variable with a new value. A signal assignment, on the other hand, schedules a new value to be applied to a signal at some later time. We will return to signal assignments in Chapter 5. Because of the significant difference between the two kinds of assignment, VHDL uses distinct symbols: ":=" for variable assignment and "<=" for signal assignment.

VHDL-87

Variable assignment statements may not be labeled in VHDL-87.

2.2   Scalar Types

The notion of *type* is very important in VHDL. We say that VHDL is a *strongly typed* language, meaning that every object may only assume values of its nominated type. Furthermore, the definition of each operation includes the types of values to which the operation may be applied. The aim of strong typing is to allow detection of errors at an early stage of the design process, namely, when a model is analyzed.

In this section, we show how a new type is declared. We then show how to define different scalar types. A *scalar* type is one whose values are indivisible. In Chapter 4 we will show how to declare types whose values are composed of collections of element values.

2.2.1   Type Declarations

We introduce new types into a VHDL model by using type declarations. The declaration names a type and specifies which values may be stored in objects of the type. The syntax rule for a type declaration is

type_declaration ⇐ **type** identifier **is** type_definition ;

One important point to note is that if two types are declared separately with identical type definitions, they are nevertheless distinct and incompatible types. For example, if we have two type declarations:

type apples **is range** 0 **to** 100;
type oranges **is range** 0 **to** 100;

we may not assign a value of type **apples** to a variable of type **oranges**, since they are of different types.

An important use of types is to specify the allowed values for ports of an entity. In the examples in Chapter 1, we saw the type name bit used to specify that ports may take only the values '0' and '1'. If we define our own types for ports, the type names must be declared in a *package*, so that they are visible in the entity declaration. We will describe packages in more detail in Chapter 7; we introduce them here to enable us to write entity declarations using types of our own devising. For example, suppose we wish to define an adder entity that adds small integers in the range 0 to 255. We write a package containing the type declaration, as follows:

```
package int_types is

  type small_int is range 0 to 255;

end package int_types;
```

This defines a package named **int_types**, which provides the type named **small_int**. The package is a separate design unit and is analyzed before any entity declaration that needs to use the type it provides. We can use the type by preceding an entity declaration with a *use clause*, for example:

```
use work.int_types.all;

entity small_adder is
  port ( a, b : in small_int;  s : out small_int );
end entity small_adder;
```

When we discuss packages in Chapter 7, we will explain the precise meaning of use clauses such as this. For now, we treat it as "magic" needed to declare types for use in entity declarations.

2.2.2 Integer Types

In VHDL, integer types have values that are whole numbers. An example of an integer type is the predefined type **integer**, which includes all the whole numbers representable on a particular host computer. The language standard requires that the type **integer** include at least the numbers $-2{,}147{,}483{,}647$ to $+2{,}147{,}483{,}647$ ($-2^{31} + 1$ to $+2^{31} - 1$), but VHDL implementations may extend the range.

We can define a new integer type using a range-constraint type definition. The simplified syntax rule for an integer type definition is

integer_type_definition ⟸
    range simple_expression ⟦ **to** ⟧ **downto** ⟧ simple_expression

which defines the set of integers between (and including) the values given by the two expressions. The expressions must evaluate to integer values. If we use the keyword **to**, we are defining an *ascending range*, in which values are ordered from the smallest on the left to the largest on the right. On the other hand, using the keyword **downto** defines a *descending range*, in which values are ordered left to right from largest to smallest. The reasons for distinguishing between ascending and descending ranges will become clear later.

An an example, here are two integer type declarations:

```
type day_of_month is range 0 to 31;
type year is range 0 to 2100;
```

These two types are quite distinct, even though they include some values in common. Thus if we declare variables of these types:

```
variable today : day_of_month := 9;
variable start_year : year := 1987;
```

it would be illegal to make the assignment

```
start_year := today;
```

Even though the number 9 is a member of the type **year**, in context it is treated as being of type **day_of_month**, which is incompatible with type **year**. This type rule helps us to avoid inadvertently mixing numbers that represent different kinds of things.

If we wish to use an arithmetic expression to specify the bounds of the range, the values used in the expression must be *locally static*; that is, they must be known when the model is analyzed. For example, we can use constant values in an expression as part of a range definition:

```
constant number_of_bits : integer := 32;
type bit_index is range 0 to number_of_bits - 1;
```

The operations that can be performed on values of integer types include the familiar arithmetic operations:

+	addition, or identity
–	subtraction, or negation
*	multiplication
/	division
mod	modulo
rem	remainder
abs	absolute value
**	exponentiation

The result of an operation is an integer of the same type as the operand or operands. For the binary operators (those that take two operands), the operands must be of the same type. The right operand of the exponentiation operator must be a non-negative integer.

The identity and negation operators are unary, meaning that they only take a single, right operand. The result of the identity operator is its operand unchanged, while the negation operator produces zero minus the operand. So, for example, the following all produce the same result:

```
A + (-B)     A - (+B)     A - B
```

The division operator produces an integer that is the result of dividing, with any fractional part truncated toward zero. The remainder operator is defined such that the relation

```
A  =  (A / B) * B + (A rem B)
```

is satisfied. The result of **A rem** B is the remainder left over from division of A by B. It has the same sign as A and has absolute value less than the absolute value of B. For example:

```
5 rem 3    = 2      (-5) rem 3    = -2
5 rem (-3) = 2      (-5) rem (-3) = -2
```

Note that in these expressions, the parentheses are required by the grammar of VHDL. The two operators, rem and negation, may not be written side by side. The modulo operator conforms to the mathematical definition satisfying the relation

```
A  =  B * N  +  (A mod B)     -- for some integer N
```

The result of **A mod** B has the same sign as B and has absolute value less than the absolute value of B. For example:

```
5 mod 3    = 2      (-5) mod 3    = 1
5 mod (-3) = -1     (-5) mod (-3) = -2
```

In addition to these operations, VHDL defines operations to find the larger (**maximum**) and the smaller (**minimum**) of two integers. For example

```
maximum(3, 20) = 20    minimum(3, 20) = 3
```

While we could use an if statement for this purpose, such as the following:

```
if A > B then
  greater := A;
else
  greater := B;
end if;
```

using the **maximum** or **minimum** operation is much more convenient:

```
greater := maximum(A, B);
```

When a variable is declared to be of an integer type, the default initial value is the leftmost value in the range of the type. For ascending ranges, this will be the least value, and for descending ranges, it will be the greatest value. If we have these declarations:

```
type set_index_range is range 21 downto 11;
type mode_pos_range is range 5 to 7;
variable set_index : set_index_range;
variable mode_pos : mode_pos_range;
```

the initial value of **set_index** is 21, and that of **mode_pos** is 5. The initial value of a variable of type **integer** is –2,147,483,647 or less, since this type is predefined as an ascending range that must include –2,147,483,647.

VHDL-87, -93, and -2002

The maximum and minimum operations are not predefined in these versions of VHDL.

2.2.3 Floating-Point Types

Floating-point types in VHDL are used to represent real numbers. Mathematically speaking, there is an infinite number of real numbers within any interval, so it is not possible to represent real numbers exactly on a computer. Hence floating-point types are only an approximation to real numbers. The term "floating point" refers to the fact that they are represented using a mantissa part and an exponent part. This is similar to the way in which we represent numbers in scientific notation.

Floating-point types in VHDL conform to IEEE Standard 754 or 854 for floating-point computation and are represented using at least 64 bits. This gives approximately 15 decimal digits of precision, and a range of approximately −1.8E+308 to +1.8E+308. An implementation may choose to use a larger representation, providing correspondingly greater precision or range. There is a predefined floating-point type called **real**, which includes the greatest range allowed by the implementation's floating-point representation. In most implementations, this will be the range of the IEEE 64-bit double-precision representation.

We define a new floating-point type using a range-constraint type definition. The simplified syntax rule for a floating-point type definition is

floating_type_definition ⇐
    range simple_expression (**to** ⟦ **downto** ⟧) simple_expression

This is similar to the way in which an integer type is declared, except that the bounds must evaluate to floating-point numbers. Some examples of floating-point type declarations are

```
type input_level is range -10.0 to +10.0;
type probability is range 0.0 to 1.0;
```

The operations that can be performed on floating-point values include the arithmetic operations addition and identity ("+"), subtraction and negation ("−"), multiplication ("*"), division ("/"), absolute value (**abs**), exponentiation ("**"), maximum, and minimum. The result of an operation is of the same floating-point type as the operand or operands. For the binary operators (those that take two operands), the operands must be of the same type. The exception is that the right operand of the exponentiation operator must be an integer. The identity and negation operators are unary (meaning that they only take a single, right operand).

Variables that are declared to be of a floating-point type have a default initial value that is the leftmost value in the range of the type. So if we declare a variable to be of the type **input_level** shown above:

```
variable input_A : input_level;
```

its initial value is −10.0.

VHDL-87, -93, and -2002

The **maximum** and **minimum** operations are not predefined in these versions of VHDL.

VHDL-87 and VHDL-93

In VHDL-87 and VHDL-93, the precision of floating-point types is only guaranteed to be at least six decimal digits, and the range at least –1.0E+38 to +1.0E+38. This corresponds to IEEE 32-bit single-precision representation. Implementations are allowed to use larger representations. The predefined type **real** is only guaranteed to have at least six digits precision and a range of at least –1.0E+38 to +1.0E+38, regardless of the size of the representation chosen by the implementation.

2.2.4 Physical Types

The remaining numeric types in VHDL are physical types. They are used to represent real-world physical quantities, such as length, mass, time and current. The definition of a physical type includes the *primary unit* of measure and may also include some *secondary units*, which are integral multiples of the primary unit. The simplified syntax rule for a physical type definition is

physical_type_definition ⇐
    range simple_expression 〔 **to** ▯ **downto** 〕 simple_expression
      units
        identifier ;
        〔 identifier = physical_literal ; 〕
      end units 〔 identifier 〕

physical_literal ⇐ 〔 decimal_literal ▯ based_literal 〕 *unit*_name

A physical type definition is like an integer type definition, but with the units definition part added. The primary unit (the first identifier after the **units** keyword) is the smallest unit that is represented. We may then define a number of secondary units, as we shall see in a moment. The range specifies the multiples of the primary unit that are included in the type. If the identifier is included at the end of the units definition part, it must repeat the name of the type being defined.

To illustrate, here is a declaration of a physical type representing electrical resistance:

```
type resistance is range 0 to 1E9
  units
    ohm;
  end units resistance;
```

Literal values of this type are written as a numeric literal followed by the unit name, for example:

```
5 ohm     22 ohm     471_000 ohm
```

Notice that we must include a space before the unit name. Also, if the number is the literal 1, it can be omitted, leaving just the unit name. So the following two literals represent the same value:

```
ohm     1 ohm
```

Note that values such as –5 ohm and 1E16 ohm are not included in the type **resistance**, since the values –5 and 1E16 lie outside of the range of the type.

Now that we have seen how to write physical literals, we can look at how to specify secondary units in a physical type declaration. We do this by indicating how many primary units comprise a secondary unit. Our declaration for the resistance type can now be extended:

```
type resistance is range 0 to 1E9
  units
    ohm;
    kohm = 1000 ohm;
    Mohm = 1000 kohm;
  end units resistance;
```

Notice that once one secondary unit is defined, it can be used to specify further secondary units. Of course, the secondary units do not have to be powers of 10 times the primary unit; however, the multiplier must be an integer. For example, a physical type for length might be declared as

```
type length is range 0 to 1E9
  units
    um;                    -- primary unit: micron
    mm = 1000 um;          -- metric units
    m = 1000 mm;
    inch = 25400 um;       -- imperial units
    foot = 12 inch;
  end units length;
```

We can write physical literals of this type using the secondary units, for example:

```
23 mm     2 foot     9 inch
```

When we write physical literals, we can write non-integral multiples of primary or secondary units. If the value we write is not an exact multiple of the primary unit, it is rounded down to the nearest multiple. For example, we might write the following literals of type length, each of which represents the same value:

```
0.1 inch     2.54 mm     2.540528 mm
```

The last of these is rounded down to 2540 um, since the primary unit for length is um. If we write the physical literal 6.8 um, it is rounded down to the value 6 um.

Many of the arithmetic operators can be applied to physical types, but with some restrictions. The addition, subtraction, identity, negation, **abs**, **mod**, **rem**, maximum, and
minimum operations can be applied to values of physical types, in which case they yield
results that are of the same type as the operand or operands. In the case of **mod** and **rem**,
the operations are based on the number of primary units in each of the operand values,
for example:

```
20 mm rem   6 mm  =    2 mm
 1 m   rem 300 mm  =  100 mm
```

A value of a physical type can be multiplied by a number of type integer or real to
yield a value of the same physical type, for example:

```
5 mm * 6  =   30 mm
```

A value of a physical type can be divided by a number of type integer or real to yield
a value of the same physical type. Furthermore, two values of the same physical type can
be divided to yield an integer, for example:

```
18 kohm / 2.0     =  9 kohm
33 kohm / 22 ohm  =  1500
```

Also, the **abs** operator may be applied to a value of a physical type to yield a value
of the same type, for example:

```
abs 2 foot     =  2 foot
abs (-2 foot)  =  2 foot
```

The restrictions make sense when we consider that physical types represent actual
physical quantities, and arithmetic should be done so as to produce results of the correct
dimensions. It doesn't make sense to multiply two lengths together to yield a length; the
result should logically be an area. So VHDL does not allow direct multiplication of two
physical types. Instead, we must convert the values to abstract integers to do the calculation, then convert the result back to the final physical type. (See the discussion of the 'pos
and 'val attributes in Section 2.4.)

A variable that is declared to be of a physical type has a default initial value that is the
leftmost value in the range of the type. For example, the default initial values for the types
declared above are 0 ohm for **resistance** and 0 um for **length**.

VHDL-87, -93, and -2002

The **maximum** and **minimum** operations are not predefined in these versions of
VHDL. Moreover, the **mod** and **rem** operations are not applicable to values of
physical types in these versions.

VHDL-87

A physical type definition in VHDL-87 may not repeat the type name after the keywords **end units**.

Time

The predefined physical type **time** is very important in VHDL, as it is used extensively to specify delays. Its definition is

```
type time is range implementation defined
  units
    fs;
    ps = 1000 fs;
    ns = 1000 ps;
    us = 1000 ns;
    ms = 1000 us;
    sec = 1000 ms;
    min = 60 sec;
    hr = 60 min;
  end units;
```

EXAMPLE 2.2   *Waveform generation*

We can use the **mod** operator on values of type **time** to simplify generation of a periodic waveform. For example, the following process creates a triangle wave on the real signal **triangle_wave**. The constant **t_period_wave** defines the period of the output wave, **t_offset** defines the offset within the triangle wave, and **t_period_sample** defines how many points are in the waveform. The value **now** defines the current time as simulation progresses.

```
    signal triangle_wave : real;
    ...

    wave_proc : process is
      variable phase : time;
    begin
      phase := (now + t_offset) mod t_period_wave;
      if phase <= t_period_wave/2 then
        triangle_wave <= 4.0 * real(phase/t_period_wave) - 1.0;
      else
        triangle_wave <= 3.0 - 4.0 * real(phase/t_period_wave);
      end if;
      wait for tperiod_sample;
    end process wave_proc;
```

By default, the primary unit **fs** is the *resolution limit* used when a model is simulated. Time values smaller than the resolution limit are rounded down to zero units. A simulator may allow us to select a secondary unit of **time** as the resolution limit. In this case, the unit of all physical literals of type **time** in the model must not be less than the resolution limit. When the model is executed, the resolution limit is used to determine the precision with which time values are represented. The reason for allowing reduced precision in this way is to allow a greater range of time values to be represented. This may allow a model to be simulated for a longer period of simulation time.

2.2.5 Enumeration Types

Often when writing models of hardware at an abstract level, it is useful to use a set of names for the encoded values of some signals, rather than committing to a bit-level encoding straightaway. VHDL *enumeration types* allow us to do this. For example, suppose we are modeling a processor, and we want to define names for the function codes for the arithmetic unit. A suitable type declaration is

```
type alu_function is
  (disable, pass, add, subtract, multiply, divide);
```

Such a type is called an enumeration, because the literal values used are enumerated in a list. The syntax rule for enumeration type definitions in general is

enumeration_type_definition ⇐ (⟦ identifier ⟦ character_literal ⟧ ⟦ , ₒₒₒ ⟧)

There must be at least one value in the type, and each value may be either an identifier, as in the above example, or a character literal. An example of this latter case is

```
type octal_digit is ('0', '1', '2', '3', '4', '5', '6', '7');
```

Given the above two type declarations, we could declare variables:

```
variable alu_op : alu_function;
variable last_digit : octal_digit := '0';
```

and make assignments to them:

```
alu_op := subtract;
last_digit := '7';
```

Different enumeration types may include the same identifier as a literal (called *overloading*), so the context of use must make it clear which type is meant. To illustrate this, consider the following declarations:

```
type logic_level is (unknown, low, undriven, high);
variable control : logic_level;
type water_level is (dangerously_low, low, ok);
variable water_sensor : water_level;
```

Here, the literal **low** is overloaded, since it is a member of both types. However, the assignments

```
control := low;
water_sensor := low;
```

are both acceptable, since the types of the variables are sufficient to determine which **low** is being referred to.

When a variable of an enumeration type is declared, the default initial value is the leftmost element in the enumeration list. So **unknown** is the default initial value for type logic_level, and **dangerously_low** is that for type water_level.

There are three predefined enumeration types defined as

```
type severity_level is
  (note, warning, error, failure);
type file_open_status is
  (open_ok, status_error, name_error, mode_error);
type file_open_kind is
  (read_mode, write_mode, append_mode);
```

The type **severity_level** is used in assertion statements, which we will discuss in Chapter 3, and the types **file_open_status** and **file_open_kind** are used for file operations, which we will discuss in Chapter 16. For the remainder of this section, we look at the other predefined enumeration types and the operations applicable to them.

VHDL-87

The types **file_open_status** and **file_open_kind** are not predefined in VHDL-87.

Characters

In Chapter 1 we saw how to write literal character values. These values are members of the predefined enumeration type **character**, which includes all of the characters in the ISO 8859 Latin-1 8-bit character set. The type definition is shown below. Note that this type is an example of an enumeration type containing a mixture of identifiers and character literals as elements.

```
type character is (
    nul,   soh,   stx,   etx,   eot,   enq,   ack,   bel,
    bs,    ht,    lf,    vt,    ff,    cr,    so,    si,
    dle,   dc1,   dc2,   dc3,   dc4,   nak,   syn,   etb,
    can,   em,    sub,   esc,   fsp,   gsp,   rsp,   usp,
    ' ',   '!',   '"',   '#',   '$',   '%',   '&',   ''',
    '(',   ')',   '*',   '+',   ',',   '-',   '.',   '/',
    '0',   '1',   '2',   '3',   '4',   '5',   '6',   '7',
    '8',   '9',   ':',   ';',   '<',   '=',   '>',   '?',
    '@',   'A',   'B',   'C',   'D',   'E',   'F',   'G',
```

```
'H',   'I',   'J',   'K',   'L',   'M',   'N',   'O',
'P',   'Q',   'R',   'S',   'T',   'U',   'V',   'W',
'X',   'Y',   'Z',   '[',   '\',   ']',   '^',   '_',
'`',   'a',   'b',   'c',   'd',   'e',   'f',   'g',
'h',   'i',   'j',   'k',   'l',   'm',   'n',   'o',
'p',   'q',   'r',   's',   't',   'u',   'v',   'w',
'x',   'y',   'z',   '{',   '|',   '}',   '~',   del,
c128,  c129,  c130,  c131,  c132,  c133,  c134,  c135,
c136,  c137,  c138,  c139,  c140,  c141,  c142,  c143,
c144,  c145,  c146,  c147,  c148,  c149,  c150,  c151,
c152,  c153,  c154,  c155,  c156,  c157,  c158,  c159,
' ',   '¡',   '¢',   '£',   '¤',   '¥',   '¦',   '§',
'¨',   '©',   'ª',   '«',   '¬',   '',   '®',   '¯',
'°',   '±',   '²',   '³',   '´',   'µ',   '¶',   '·',
'¸',   '¹',   'º',   '»',   '¼',   '½',   '¾',   '¿',
'À',   'Á',   'Â',   'Ã',   'Ä',   'Å',   'Æ',   'Ç',
'È',   'É',   'Ê',   'Ë',   'Ì',   'Í',   'Î',   'Ï',
'Ð',   'Ñ',   'Ò',   'Ó',   'Ô',   'Õ',   'Ö',   '×',
'Ø',   'Ù',   'Ú',   'Û',   'Ü',   'Ý',   'Þ',   'ß',
'à',   'á',   'â',   'ã',   'ä',   'å',   'æ',   'ç',
'è',   'é',   'ê',   'ë',   'ì',   'í',   'î',   'ï',
'ð',   'ñ',   'ò',   'ó',   'ô',   'õ',   'ö',   '÷',
'ø',   'ù',   'ú',   'û',   'ü',   'ý',   'þ',   'ÿ');
```

The first 128 characters in this enumeration are the ASCII characters, which form a subset of the Latin-1 character set. The identifiers from **nul** to **usp** and **del** are the non-printable ASCII control characters. Characters c128 to c159 do not have any standard names, so VHDL just gives them nondescript names based on their position in the character set. The character at position 160 is a non-breaking space character, distinct from the ordinary space character, and the character at position 173 is a soft hyphen.

To illustrate the use of the **character** type, we declare variables as follows:

```
variable cmd_char, terminator : character;
```

and then make the assignments

```
cmd_char := 'P';
terminator := cr;
```

VHDL-87

Since VHDL-87 uses the ASCII character set, the predefined type **character** includes only the first 128 characters shown above.

Booleans

One of the most important predefined enumeration types in VHDL is the type **boolean**, defined as

    type boolean **is** (false, true);

This type is used to represent condition values, which can control execution of a behavioral model. There are a number of operators that we can apply to values of different types to yield Boolean values, namely, the relational and logical operators. The relational operators equality ("=") and inequality ("/=") can be applied to operands of any type (except files), including the composite types that we will see later in this chapter. The operands must both be of the same type, and the result is a Boolean value. For example, the expressions

    123 = 123   'A' = 'A'   7 ns = 7 ns

all yield the value **true**, and the expressions

    123 = 456   'A' = 'z'   7 ns = 2 us

yield the value **false**.

    The relational operators that test ordering are the less-than ("<"), less-than-or-equal-to ("<="), greater-than (">") and greater-than-or-equal-to (">=") operators. These can only be applied to values of types that are ordered, including all of the scalar types described in this chapter. As with the equality and inequality operators, the operands must be of the same type, and the result is a Boolean value. For example, the expressions

    123 < 456   789 ps <= 789 ps   '1' > '0'

all result in **true**, and the expressions

    96 >= 102   2 us < 4 ns   'X' < 'X'

all result in **false**.

    The logical operators **and**, **or**, **nand**, **nor**, **xor**, **xnor** and **not** take operands that are Boolean values and produce Boolean results. Table 2.1 shows the results produced by the binary logical operators. The result of the unary **not** operator is **true** if the operand is **false**, and **false** if the operand is **true**. The operators **and**, **or**, **nand** and **nor** are called "short-

TABLE 2.1   *The truth table for binary logical operators*

A	B	A **and** B	A **nand** B	A **or** B	A **nor** B	A **xor** B	A **xnor** B
false	false	false	true	false	true	false	true
false	true	false	true	true	false	true	false
true	false	false	true	true	false	true	false
true	true	true	false	true	false	false	true

circuit" operators, as they only evaluate the right operand if the left operand does not determine the result. For example, if the left operand of the **and** operator is false, we know that the result is false, so we do not need to consider the other operand. This is useful where the left operand is a test that guards against the right operand causing an error. Consider the expression

    (b /= 0) **and** (a/b > 1)

If **b** were zero and we evaluated the right-hand operand, we would cause an error due to dividing by zero. However, because and is a short-circuit operator, if **b** were zero, the left-hand operand would evaluate to false, so the right-hand operand would not be evaluated. For the **nand** operator, the right-hand operand is similarly not evaluated if the left-hand is false. For **or** and **nor**, the right-hand operand is not evaluated if the left-hand is true.

VHDL-87

The logical operator **xnor** is not provided in VHDL-87.

Bits

Since VHDL is used to model digital systems, it is useful to have a data type to represent bit values. The predefined enumeration type **bit** serves this purpose. It is defined as

    type bit **is** ('0', '1');

Notice that the characters '0' and '1' are overloaded, since they are members of both **bit** and **character**. Where '0' or '1' occurs in a model, the context is used to determine which type is being used.

The logical operators that we mentioned for Boolean values can also be applied to values of type **bit**, and they produce results of type **bit**. The value '0' corresponds to false, and '1' corresponds to true. So, for example:

    '0' **and** '1' = '0', '1' **xor** '1' = '0'

The operands must still be of the same type as each other. Thus it is not legal to write

    '0' **and** true

The difference between the types **boolean** and **bit** is that **boolean** values are used to model abstract conditions, whereas **bit** values are used to model hardware logic levels. Thus, '0' represents a low logic level and '1' represents a high logic level. The logical operators, when applied to **bit** values, are defined in terms of positive logic, with '0' representing the negated state and '1' representing the asserted state. If we need to deal with negative logic, we need to take care when writing logical expressions to get the correct logic sense. For example, if write_enable_n, select_reg_n and write_reg_n are negative logic bit variables, we perform the assignment

```
       write_reg_n := not ( not write_enable_n and not select_reg_n );
```

The variable **write_reg_n** is asserted ('0') only if **write_enable_n** is asserted and **select_reg_n** is asserted. Otherwise it is negated ('1').

Standard Logic

Since VHDL is designed for modeling digital hardware, it is necessary to include types to represent digitally encoded values. The predefined type bit shown above can be used for this in more abstract models, where we are not concerned about the details of electrical signals. However, as we refine our models to include more detail, we need to take account of the electrical properties when representing signals. There are many ways we can define data types to do this, but the IEEE has standardized one way in a package called **std_logic_1164**. The full details of the package are included in Appendix A. One of the types defined in this package is an enumeration type called **std_ulogic**, defined as

```
type std_ulogic is ( 'U',          -- Uninitialized
                     'X',          -- Forcing Unknown
                     '0',          -- Forcing zero
                     '1',          -- Forcing one
                     'Z',          -- High Impedance
                     'W',          -- Weak Unknown
                     'L',          -- Weak zero
                     'H',          -- Weak one
                     '-' );        -- Don't care
```

This type can be used to represent signals driven by active drivers (forcing strength), resistive drivers such as pull-ups and pull-downs (weak strength) or three-state drivers including a high-impedance state. Each kind of driver may drive a "zero," "one" or "unknown" value. An "unknown" value is driven by a model when it is unable to determine whether the signal should be "zero" or "one." For example, the output of an and gate is unknown when its inputs are driven by high-impedance drivers. In addition to these values, the leftmost value in the type represents an "uninitialized" value. If we declare signals of **std_ulogic** type, by default they take on 'U' as their initial value. If a model tries to operate on this value instead of a real logic value, we have detected a design error in that the system being modeled does not start up properly. The final value in **std_ulogic** is a "don't care" value. This is sometimes used by logic synthesis tools and may also be used when defining test vectors, to denote that the value of a signal to be compared with a test vector is not important.

Even though the type **std_ulogic** and the other types defined in the **std_logic_1164** package are not actually built into the VHDL language, we can write models as though they were, with a little bit of preparation. For now, we describe some "magic" to include at the beginning of a model that uses the package; we explain the details in Chapter 7. If we include the line

```
library ieee;  use ieee.std_logic_1164.all;
```

preceding each entity or architecture body that uses the package, we can write models as though the types were built into the language.

With this preparation in hand, we can now create constants, variables and signals of type **std_ulogic**. As well as assigning values of the type, we can also use the logical operators **and**, **or**, **not** and so on. Each of these operates on **std_ulogic** values and returns a **std_ulogic** result of 'U', 'X', '0' or '1'. The operators are "optimistic," in that if they can determine a '0' or '1' result despite inputs being unknown, they do so. Otherwise they return 'X' or 'U'. For example '0' **and** 'Z' returns '0', since one input to an and gate being '0' always causes the output to be '0', regardless of the other input.

One important point to note about comparing **std_ulogic** values using the "=" and "/=" operations is that it is not the logic levels that are compared, but the enumeration literals. Thus, the expression '0' = 'L' yields false, even though both values represent low logic levels. If we want to compare logic levels without taking account of drive strength, we should use the matching operators "?=" and "?/=". These operators perform logical equivalence and unequivalence comparisons, respectively. If both operands are '0', '1', 'L', or 'H', the operations yield a '0' or '1' result. For example:

```
'1' ?= 'H'  =  '1'      '1' ?/= 'H'  =  '0'
'0' ?= 'H'  =  '0'      '0' ?/= 'H'  =  '1'
```

The "?=" and "?/=" operators yield 'X' when either operand is 'X', 'Z', or 'W'. However, if either operand is 'U', the result is 'U'. The final exception is for don't care ('–') operands. For these, "?=" always yeilds '1' and "?/=" always yields '0'. Some examples are:

```
'Z' ?= 'H'  =  'X'      'W' ?/= 'H'  =  'X'
'0' ?= 'U'  =  'U'      '0' ?/= 'U'  =  'U'
'1' ?= '-'  =  '1'      '-' ?/= '-'  =  '0'
```

We note briefly here, for completeness, that VHDL also defines the matching operators "?<", "?<=", "?>" and "?=>" for comparing **std_ulogic** values. They treat a logic low level ('0' or 'L') as being less than a logic high level ('1' or 'H'), yield 'X' for comparison with a non-logic-level value other than 'U', and yield 'U' for comparison with 'U'. Comparison with '–' is not allowed. We also note that all of the matching operators are defined for operands of type **bit**, yielding '0' or '1' where an ordinary comparison would yield **false** or **true**, respectively. We will return to the way in which these operators are used in later chapters.

Condition Conversion

We mentioned above that Boolean values are used as condition values to control execution in VHDL models. We have seen this in if statements in previous examples, where Boolean conditions control whether groups of statements are executed or not. When we are modeling digital systems, we often use signals and variables of type **bit** or **std_ulogic** to represent logical conditions. It would seem reasonable to want to use such values in conditions controlling execution. VHDL allows us to do this, as it implicitly converts **bit** and **std_ulogic** values to **boolean** values when they occur as conditions. For example, if we have control signals declared as

```
signal cs1, ncs2, cs3 : std_ulogic;
```

then we can write an if statement as follows:

```
if cs1 and not ncs2 and cs3 then
  . . .
end if;
```

The logical **and** and **not** operators applied to the signals yield a result of type std_ulogic. However, VHDL implicitly converts this to **boolean**, treating '1' and 'H' as true, and all other values as false. Had we declared the signals to be of type **bit**, the implicit conversion would also occur, with '1' treated as true and '0' as false. Implicit conversion occurs in this way in any place where a Boolean condition is required. Another example of such a place that we have seen is a wait statement, for example:

```
wait until clk;
```

If **clk** is of type **std_ulogic**, the wait statement suspends until **clk** changes to '1' or 'H'.

The way in which VHDL does the conversion is by applying the predefined operator "??" to the result of the condition. This operator takes a **bit** or **std_ulogic** value and yields a **boolean** result. We could, if we wanted to, make the conversion explicit, for example:

```
if ?? (cs1 and not ncs2 and cs3) then
  . . .
end if;
```

Note that, if the condition is more involved than just a signal or variable name, we must enclose it in parentheses, as shown here. Normally, we would not need to write the conversion operation in a condition explicitly. However, we describe it here to show how the conversion mechanism works. If we were to define our own data type representing logic levels, we could also define a version of the condition operator that VHDL would use to perform implicit conversion. We describe the way in which we define multiple versions of an operation in Section 6.5.

One final point to note is that implicit conversions don't allow us to write a condition that mixes **bit** or **std_ulogic** operands with **boolean** operands for logical operators. For example, the following is illegal:

```
if cs1 and cs3 and alu_op = pass then ... -- illegal
```

The result of the "=" comparison is of type **boolean**, which cannot be mixed with the std_ulogic operand for the **and** operator. The conversion is only applied to the overall condition. We would have to write the condition as:

```
if cs1 = '1' and cs3 = '1' and alu_op = pass then ...
```

VHDL-87, -93, and -2002

These versions of VHDL do not perform implicit conversion of conditions to **boolean**. Conditions must yield **boolean** results without conversion. Hence, we must write conditions such as:

```
if cs1 = '1' and ncs2 = '0' and cs3 = '1' then
    ...
end if;
```

2.3   Type Classification

In the preceding sections we have looked at the scalar types provided in VHDL. Figure 2.1 illustrates the relationships between these types, the predefined scalar types and the types we look at in later chapters.

FIGURE 2.1

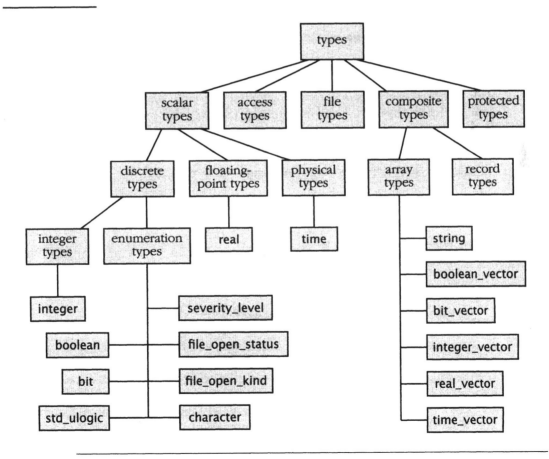

A classification of VHDL types.

The scalar types are all those composed of individual values that are ordered. Integer and floating-point types are ordered on the number line. Physical types are ordered by the number of primary units in each value. Enumeration types are ordered by their declaration. The discrete types are those that represent discrete sets of values and comprise the integer types and enumeration types. Floating-point and physical types are not discrete, as they approximate a continuum of values.

2.3.1 Subtypes

In Section 2.2 we saw how to declare a type, which defines a set of values. Often a model contains objects that should only take on a restricted range of the complete set of values. We can represent such objects by declaring a *subtype*, which defines a restricted set of values from a *base type*. The condition that determines which values are in the subtype is called a *constraint*. Using a subtype declaration makes clear our intention about which values are valid and makes it possible to check that invalid values are not used. The simplified syntax rules for a subtype declaration are

> subtype_declaration ⇐ **subtype** identifier **is** subtype_indication ;

> subtype_indication ⇐
>   type_mark 〖 **range** simple_expression (**to** ⫾ **downto**) simple_expression 〗

We will look at more advanced forms of subtype indications in later chapters. The subtype declaration defines the identifier as a subtype of the base type specified by the type mark, with the range constraint restricting the values for the subtype. The constraint is optional, which means that it is possible to have a subtype that includes all of the values of the base type.

Here is an example of a declaration that defines a subtype of **integer**:

```
subtype small_int is integer range -128 to 127;
```

Values of small_int are constrained to be within the range −128 to 127. If we declare some variables:

```
variable deviation : small_int;
variable adjustment : integer;
```

we can use them in calculations:

```
deviation := deviation + adjustment;
```

Note that in this case, we can mix the subtype and base type values in the addition to produce a value of type **integer**, but the result must be within the range −128 to 127 for the assignment to succeed. If it is not, an error will be signaled when the variable is assigned. All of the operations that are applicable to the base type can also be used on values of a subtype. The operations produce values of the base type rather than the subtype. However, the assignment operation will not assign a value to a variable of a subtype if the value does not meet the constraint.

Another point to note is that if a base type is a range of one direction (ascending or descending), and a subtype is specified with a range constraint of the opposite direction, it is the subtype specification that counts. For example, the predefined type integer is an ascending range. If we declare a subtype as

```
subtype bit_index is integer range 31 downto 0;
```

this subtype is a descending range.

The VHDL standard includes two predefined integer subtypes, defined as

```
subtype natural is integer range 0 to highest_integer;
subtype positive is integer range 1 to highest_integer;
```

Where the logic of a design indicates that a number should not be negative, it is good style to use one of these subtypes rather than the base type **integer**. In this way, we can detect any design errors that incorrectly cause negative numbers to be produced. There is also a predefined subtype of the physical type **time**, defined as

```
subtype delay_length is time range 0 fs to highest_time;
```

This subtype should be used wherever a non-negative time delay is required.

VHDL-87

The subtype **delay_length** is not predefined in VHDL-87.

2.3.2 Type Qualification

Sometimes it is not clear from the context what the type of a particular value is. In the case of overloaded enumeration literals, it may be necessary to specify explicitly which type is meant. We can do this using *type qualification*, which consists of writing the type name followed by a single quote character, then an expression enclosed in parentheses. For example, given the enumeration types

```
type logic_level is (unknown, low, undriven, high);
type system_state is (unknown, ready, busy);
```

we can distinguish between the common literal values by writing

```
logic_level'(unknown)    system_state'(unknown)
```

Type qualification can also be used to narrow a value down to a particular subtype of a base type. For example, if we define a subtype of **logic_level**

```
subtype valid_level is logic_level range low to high;
```

we can explicitly specify a value of either the type or the subtype

```
logic_level'(high)    valid_level'(high)
```

Of course, it is an error if the expression being qualified is not of the type specified.

VHDL-87, -93, and -2002

In these earlier versions of VHDL, if a type qualification specified a subtype, the value qualified had to be of that specific subtype. It was not sufficient for it just to be of the base type of the specified subtype. In VHDL-2008, the value is converted to be of the subtype. While there is no distinction in the case of scalar types, when we come to composite types, the distinction can be significant.

2.3.3 Type Conversion

When we introduced the arithmetic operators in previous sections, we stated that the operands must be of the same type. This precludes mixing integer and floating-point values in arithmetic expressions. Where we need to do mixed arithmetic, we can use *type conversions* to convert between integer and floating-point values. The form of a type conversion is the name of the type we want to convert to, followed by a value in parentheses. For example, to convert between the types **integer** and **real**, we could write

```
real(123)    integer(3.6)
```

Converting an integer to a floating-point value is simply a change in representation, although some loss of precision may occur. Converting from a floating-point value to an integer involves rounding to the nearest integer. Numeric type conversions are not the only conversion allowed. In general, we can convert between any closely related types. Other examples of closely related types are certain array types, discussed in Chapter 4.

One thing to watch out for is the distinction between type qualification and type conversion. The former simply states the type of a value, whereas the latter changes the value, possibly to a different type. One way to remember this distinction is to think of "*qu*ote for *qu*alification."

2.4 Attributes of Scalar Types

A type defines a set of values and a set of applicable operations. There is also a predefined set of *attributes* that are used to give information about the values included in the type. Attributes are written by following the type name with a quote mark (') and the attribute name. The value of an attribute can be used in calculations in a model. We now look at some of the attributes defined for the types we have discussed in this chapter.

First, there are a number of attributes that are applicable to all scalar types and provide information about the range of values in the type. If we let T stand for any scalar type or subtype, x stand for a value of that type and s stand for a string value, the attributes are

T'left	first (leftmost) value in T
T'right	last (rightmost) value in T
T'low	least value in T

T'high   greatest value in T

T'ascending   **true** if T is an ascending range, **false** otherwise

T'image(x)   a string representing the value of **x**

T'value(s)   the value in T that is represented by **s**

The string produced by the **'image** attribute is a correctly formed literal according to the rules shown in Chapter 1. The strings allowed in the **'value** attribute must follow those rules and may include leading or trailing spaces. These two attributes are useful for input and output in a model, as we will see when we come to that topic.

To illustrate the attributes listed above, recall the following declarations from previous examples:

```
type resistance is range 0 to 1E9
  units
    ohm;
    kohm = 1000 ohm;
    Mohm = 1000 kohm;
  end units resistance;

type set_index_range is range 21 downto 11;

type logic_level is (unknown, low, undriven, high);
```

For these types:

```
resistance'left = 0 ohm
resistance'right = 1E9 ohm
resistance'low = 0 ohm
resistance'high = 1E9 ohm
resistance'ascending = true
resistance'image(2 kohm) = "2000 ohm"
resistance'value("5 Mohm") = 5_000_000 ohm

set_index_range'left = 21
set_index_range'right = 11
set_index_range'low = 11
set_index_range'high = 21
set_index_range'ascending = false
set_index_range'image(14) = "14"
set_index_range'value("20") = 20

logic_level'left = unknown
logic_level'right = high
logic_level'low = unknown
logic_level'high = high
logic_level'ascending = true
logic_level'image(undriven) = "undriven"
logic_level'value("Low") = low
```

Next, there are attributes that are applicable to just discrete and physical types. For any such type T, a value x of that type and an integer n, the attributes are

T'pos(x)	position number of x in T
T'val(n)	value in T at position n
T'succ(x)	value in T at position one greater than that of x
T'pred(x)	value in T at position one less than that of x
T'leftof(x)	value in T at position one to the left of x
T'rightof(x)	value in T at position one to the right of x

For enumeration types, the position numbers start at zero for the first element listed and increase by one for each element to the right. So, for the type **logic_level** shown above, some attribute values are

```
logic_level'pos(unknown) = 0
logic_level'val(3) = high
logic_level'succ(unknown) = low
logic_level'pred(undriven) = low
```

For integer types, the position number is the same as the integer value, but the type of the position number is a special anonymous type called *universal integer*. This is the same type as that of integer literals and, where necessary, is implicitly converted to any other declared integer type. For physical types, the position number is the integer number of base units in the physical value. For example:

```
time'pos(4 ns) = 4_000_000
```

since the base unit is fs.

We can use the **'pos** and **'val** attributes in combination to perform mixed-dimensional arithmetic with physical types, producing a result of the correct dimensionality. Suppose we define physical types to represent length and area, as follows:

```
type length is range integer'low to integer'high
  units
    mm;
  end units length;

type area is range integer'low to integer'high
  units
    square_mm;
  end units area;
```

and variables of these types:

```
variable L1, L2 : length;
variable A : area;
```

The restrictions on multiplying values of physical types prevents us from writing something like

```
A := L1 * L2;    -- this is incorrect
```

To achieve the correct result, we can convert the length values to abstract integers using the 'pos attribute, then convert the result of the multiplication to an area value using 'val, as follows:

```
A := area'val( length'pos(L1) * length'pos(L2) );
```

Note that in this example, we do not need to include a scale factor in the multiplication, since the base unit of **area** is the square of the base unit of **length**.

For ascending ranges, T'succ(x) and T'rightof(x) produce the same value, and T'pred(x) and T'leftof(x) produce the same value. For descending ranges, T'pred(x) and T'rightof(x) produce the same value, and T'succ(x) and T'leftof(x) produce the same value. For all ranges, T'succ(T'high), T'pred(T'low), T'rightof(T'right) and T'leftof(T'left) cause an error to occur.

The last attribute we introduce here is T'base. For any subtype T, this attribute produces the base type of T. The only context in which this attribute may be used is as the prefix of another attribute. For example, if we have the declarations

```
type opcode is
    (nop, load, store, add, subtract, negate, branch, halt);
subtype arith_op is opcode range add to negate;
```

then

```
arith_op'base'left = nop
arith_op'base'succ(negate) = branch
```

VHDL-87

The attributes 'ascending, 'image and 'value are not provided in VHDL-87.

2.5 Expressions and Predefined Operations

In Section 2.1 we showed how the value resulting from evaluation of an expression can be assigned to a variable. In this section, we summarize the rules governing expressions. We can think of an expression as being a formula that specifies how to compute a value. As such, it consists of primary values combined with operators and other operations. The precise syntax rules for writing expressions are shown in Appendix B. The primary values that can be used in expressions include

- literal values,

- identifiers representing data objects (constants, variables and so on),

- attributes that yield values,
- qualified expressions,
- type-converted expressions,
- operation functions, and
- expressions in parentheses.

We have seen examples of these in this chapter and in Chapter 1. For reference, all of the predefined operators and the types they can be applied to are summarized in Table 2.2. We will discuss array operators in Chapter 4. The operators in this table are grouped by precedence, with "**", **abs**, and the unary logical reduction operators having highest precedence and the binary logical operators lowest. In addition, the condition operator "??" can only be applied to a primary value at the outermost level of an expression. The precedence rules mean that if an expression contains a combination of operators, those with highest precedence are applied first. Parentheses can be used to alter the order of evaluation, or for clarity.

TABLE 2.2   *VHDL operators in order of precedence, from most binding to least binding*

Operator	Operation	Left operand type	Right operand type	Result type
**	exponentiation	integer or floating-point	integer	same as left operand
abs	absolute value		numeric	same as operand
not	logical negation		boolean, bit, std_ulogic, 1-D array of **boolean** or bit, std_ulogic_vector	same as operand
and	logical and reduction		1-D array of **boolean** or bit, std_ulogic_vector	element type of operand
or	logical or reduction			
nand	negated logical and reduction			
nor	negated logical or reduction			
xor	exclusive or reduction			
xnor	negated exclusive or reduction			

Operator	Operation	Left operand type	Right operand type	Result type
*	multiplication	integer or floating-point	same as left operand	same as operands
		physical	**integer** or **real**	same as left operand
		integer or **real**	physical	same as right operand
/	division	integer or floating-point	same as left operand	same as operands
		physical	integer or real	same as left operand
		physical	same as left operand	universal integer
mod	modulo	integer or physical	same as left operand	same as operands
rem	remainder			
+	identity		numeric	same as operand
–	negation			
+	addition	numeric	same as left operand	same as operands
–	subtraction			
&	concatenation	1-D array	same as left operand	same as operands
		1-D array	element type of left operand	same as left operand
		element type of right operand	1-D array	same as right operand
		element type of result	element type of result	1-D array
sll	shift-left logical	1-D array of **boolean** or bit, std_ulogic_vector	integer	same as left operand
srl	shift-right logical			
rol	rotate left			
ror	rotate right			
sla	shift-left arithmetic	1-D array of **boolean** or bit	integer	same as left operand
sra	shift-right arithmetic			
=	equality	any except file or protected type	same as left operand	**boolean**
/=	inequality			
<	less than	scalar or 1-D array of any discrete type	same as left operand	**boolean**
<=	less than or equal to			
>	greater than			
>=	greater than or equal to			

Operator	Operation	Left operand type	Right operand type	Result type
?= ?/=	matching equality matching inequality	bit, std_ulogic or 1-D array of bit or std_ulogic	same as left operand	bit or std_ulogic
?< ?<= ?> ?>=	matching less than matching less than or equal to matching greater than matching greater than or equal to	bit or std_ulogic	same as left operand	bit or std_ulogic
and **or** **nand** **nor** **xor** **xnor**	logical and logical or negated logical and negated logical or exclusive or negated exclusive or	boolean, bit, std_ulogic, 1-D array of **boolean** or bit, std_ulogic_vector	same as left operand	same as operands
??	condition conversion		bit or std_ulogic	boolean

In Section 2.2, we described the **maximum** and **minimum** operations for numeric types. These are examples of function operations that can form primary values in expressions. In fact, these operations are defined for all scalar types, as are the relational operators "<", "<=", ">", and ">=". For numeric types, the **maximum** and **minimum** operations compare the numeric values to determine the result. For enumeration types, the operations compare the position numbers. Thus, enumeration values declared earlier in the list of an enumeration type declaration are considered to be less than those declared later in the list. For example, in the **character** type, 'A' < 'Z', and the minimum of 'c' and 'q' is 'c'.

Another function operation that can be applied to any scalar value is the **to_string** operation. It yields a character-string representation of the value that can be used in any place where we can use a string literal. For example:

```
to_string(123)      = "123"
to_string(456.78)   = "4.5678e+2"
to_string(warning)  = "warning"
```

For floating-point types, **to_string** represents the value using exponential notation, with the number of digits depending on the particular VHDL tool being used. There are, however, two alternate forms of **to_string** for floating-point values that give us more control over the formatting. First, we can specify that the value be represented with a given number of post-decimal digits rather than in exponential form. For example:

```
to_string(456.78, 4) = "456.7800"
to_string(456.78, 1) = "456.8"
```

Second, we can provide a format specification string of the same form as that used in the C **printf** function. For example:

```
to_string(456.78, "%10.3f")  = "   456.780"
to_string(456.78, "%-12.3E") = "4.568E+02    "
```

For physical types, **to_string** represents the value using the primary unit of the type. For example, given the declaration of type **resistance** on page 39:

```
to_string(2.2 kohm) = "2200 ohm"
```

The exception is the physical type **time**, for which **to_string** represents the value as a multiple of the resolution limit. So, for example, if the resolution limit for a simulation is set to **ns**:

```
to_string(29.5 us) = "29500 ns"
```

There is also an alternate form of **to_string** for **time** that allows us to control the unit used to represent the value. For example, even if the resolution limit is set to **ns**, we can represent a time value in microseconds as follows:

```
to_string(29500 ns, us) = "29.5 us"
```

The final predefined function operations on scalar types are the **rising_edge** and **falling_edge** operations, which can be applied to signals of type **boolean**, **bit**, or **std_ulogic**. These allow us to describe edge-triggered behavior in a natural manner. For example, we can write a process describing a D-flipflop as follows:

```
signal clk, d, q : bit;
...

dff : process is
begin
  if rising_edge(clk) then
    q <= d;
  end if;
  wait on clk;
end process dff;
```

For a **boolean** signal s, rising_edge(s) is **true** when s changes from **false** to **true** and **false** at other times (including when s changes from **true** to **false** and when s is not changing). Similarly, falling_edge(s) is **true** when s changes from **true** to **false** and **false** at other times. For a **bit** signal s, rising_edge(s) is **true** when s changes from '0' to '1' and **false** at other times; and falling_edge(s) is **true** when s changes from '1' to '0', and **false** at other times. Finally, for a **std_ulogic** signal s, rising_edge(s) is **true** when s changes from '0' or 'L' to '1' or 'H' and **false** at other times (including changes from or to 'U', 'X', 'W', 'Z', or '–'); and falling_edge(s) is **true** when s changes from '1' or 'H' to '0' or 'L' and **false** at other times.

VHDL-87, -93, and -2002

The unary logical operators (**and**, **or**, **nand**, **nor**, **xor** and **xnor**), the matching relational operators ("?=", "?/=", "?<","?<=", "?>", and "?>="), the condition conversion operator ("??"), and the function operations maximum, minimum, to_string, rising_edge, and falling_edge are not provided in earlier versions of VHDL. Also, the **mod** and **rem** operators are not applicable to physical types in earlier versions.

VHDL-87

The shift operators (**sll**, **srl**, **sla**, **sra**, **rol** and **ror**) and the binary **xnor** operator are not provided in VHDL-87.

Exercises

1. [**❶** 2.1] Write constant declarations for the number of bits in a 32-bit word and for the number π (3.14159).

2. [**❶** 2.1] Write variable declarations for a counter, initialized to 0; a status flag used to indicate whether a module is busy; and a standard-logic value used to store a temporary result.

3. [**❶** 2.1] Given the declarations in Exercise 2, write variable assignment statements to increment the counter, to set the status flag to indicate the module is busy and to indicate a weak unknown temporary result.

4. [**❶** 2.2] Write a package declaration containing type declarations for small non-negative integers representable in eight bits; fractional numbers between −1.0 and +1.0; electrical currents, with units of nA, μA, mA and A; and traffic light colors.

5. [**❶** 2.2] Given the following declarations:

   ```
   signal a, b, c : std_ulogic;
   type state_type is (idle, req, ack);
   signal state : state_type;
   ```

   indicate whether each of the following expressions is legal as a Boolean condition, and if not, correct it:

   a. a **and not** b **and** c

   b. a **and not** b **and** state = idle

   c. a = '0' **and** b **and** state = idle

   d. a = '1' **and** b = '0' **and** state = idle

6. [**❶** 2.4] Given the subtype declarations

```
subtype pulse_range is time range 1 ms to 100 ms;
subtype word_index is integer range 31 downto 0;
```

what are the values of 'left, 'right, 'low, 'high and 'ascending attributes of each of these subtypes?

7. [● 2.4] Given the type declaration

```
type state is (off, standby, active1, active2);
```

what are the values of

```
state'pos(standby)       state'val(2)
state'succ(active2)      state'pred(active1)
state'leftof(off)        state'rightof(off)
```

8. [● 2.5] For each of the following expressions, indicate whether they are syntactically correct, and if so, determine the resulting value.

```
2 * 3 + 6 / 4              3 + -4
"cat" & character'('0')    true and x and not y or z
B"101110" sll 3            (B"100010" sra 2) & X"2C"
```

9. [❷ 2.1] Write a counter model with a clock input clk of type bit, and an output q of type integer. The behavioral architecture body should contain a process that declares a count variable initialized to zero. The process should wait for changes on clk. When clk changes to '1', the process should increment the count and assign its value to the output port.

10. [❷ 2.2] Write a model that represents a simple ALU with integer inputs and output, and a function select input of type bit. If the function select is '0', the ALU output should be the sum of the inputs; otherwise the output should be the difference of the inputs.

11. [❷ 2.2] Write a model for a digital integrator that has a clock input of type bit and data input and output each of type real. The integrator maintains the sum of successive data input values. When the clock input changes from '0' to '1', the integrator should add the current data input to the sum and provide the new sum on the output.

12. [❷ 2.2] Following is a process that generates a regular clock signal.

```
clock_gen : process is
begin
  clk <= '1';  wait for 10 ns;
  clk <= '0';  wait for 10 ns;
end process clock_gen;
```

Use this as the basis for experiments to determine how your simulator behaves with different settings for the resolution limit. Try setting the resolution limit to 1 ns (the default for many simulators), 1 ps and 1 µs.

13. [❷ 2.2] Write a model for a tristate buffer using the standard-logic type for its data and enable inputs and its data output. If the enable input is '0' or 'L', the output should be 'Z'. If the enable input is '1' or 'H' and the data input is '0' or 'L', the output should be '0'. If the enable input is '1' or 'H' and the data input is '1' or 'H', the output should be '1'. In all other cases, the output should be 'X'.

Chapter 3

Sequential Statements

In the previous chapter we saw how to represent the internal state of models using VHDL data types. In this chapter we look at how that data may be manipulated within processes. This is done using *sequential statements*, so called because they are executed in sequence. We have already seen one of the basic sequential statements, the variable assignment statement, when we were looking at data types and objects. The statements we look at in this chapter deal with controlling actions within a model; hence they are often called *control structures*. They allow selection between alternative courses of action as well as repetition of actions.

3.1 If Statements

In many models, the behavior depends on a set of conditions that may or may not hold true during the course of simulation. We can use an *if statement* to express this behavior. The syntax rule for an if statement is

> if_statement ⇐
>   〚 *if*_label : 〛
>   **if** condition **then**
>   { sequential_statement }
>   { **elsif** condition **then**
>   { sequential_statement } }
>   〚 **else**
>   { sequential_statement } 〛
>   **end if** 〚 *if*_label 〛 ;

At first sight, this may appear somewhat complicated, so we start with some simple examples and work up to examples showing the general case. The label may be used to identify the if statement. We will discuss labeled statements in Chapter 20. A simple example of an if statement is

```
if en then
   stored_value := data_in;
end if;
```

The condition after the keyword **if** is used to control whether or not the statement after the keyword **then** is executed. If the condition is true, the statement is executed. In this example, if the value of the object en is '1', the assignment is made; otherwise it is skipped. We can also specify actions to be performed if the condition is false. For example:

```
if sel = 0 then
   result <= input_0;   -- executed if sel = 0
else
   result <= input_1;   -- executed if sel /= 0
end if;
```

Here, as the comments indicate, the first signal assignment statement is executed if the condition is true, and the second signal assignment statement is executed if the condition is false.

In many models, we may need to check a number of different conditions and execute a different sequence of statements for each case. We can construct a more elaborate form of if statement to do this, for example:

```
if mode = immediate then
   operand := immed_operand;
elsif opcode = load or opcode = add or opcode = subtract then
   operand := memory_operand;
else
   operand := address_operand;
end if;
```

In this example, the first condition is evaluated, and if true, the statement after the first **then** keyword is executed. If the first condition is false, the second condition is evaluated, and if it evaluates to true, the statement after the second **then** keyword is executed. If the second condition is false, the statement after the **else** keyword is executed.

In general, we can construct an if statement with any number of **elsif** clauses (including none), and we may include or omit the **else** clause. Execution of the if statement starts by evaluating the first condition. If it is false, successive conditions are evaluated, in order, until one is found to be true, in which case the corresponding statements are executed. If none of the conditions is true, and we have included an **else** clause, the statements after the **else** keyword are executed.

We are not restricted to just one statement in each part of the if statement. This is illustrated by the following if statement:

```
if opcode = halt_opcode then
   PC := effective_address;
   executing := false;
   halt_indicator <= true;
end if;
```

If the condition is true, all three statements are executed, one after another. On the other hand, if the condition is false, none of the statements are executed. Furthermore, each statement contained in an if statement can be any sequential statement. This means we can nest if statements, for example:

```
if phase = wash then
  if cycle_select = delicate_cycle then
    agitator_speed <= slow;
  else
    agitator_speed <= fast;
  end if;
  agitator_on <= true;
end if;
```

In this example, the condition **phase = wash** is first evaluated, and if true, the nested if statement and the following signal assignment statement are executed. Thus the assignment **agitator_speed <= slow** is executed only if both conditions evaluate to true, and the assignment **agitator_speed <= fast** is executed only if the first condition is true and the second condition is false.

EXAMPLE 3.1   *A heater thermostat*

Let us develop a behavioral model for a simple heater thermostat. The device can be modeled as an entity with two integer inputs, one that specifies the desired temperature and another that is connected to a thermometer, and one Boolean output that turns a heater on and off. The thermostat turns the heater on if the measured temperature falls below two degrees less than the desired temperature, and turns the heater off if the measured temperature rises above two degrees greater than the desired temperature. The entity and architecture body for the thermostat are:

```
entity thermostat is
  port ( desired_temp, actual_temp : in integer;
         heater_on : out boolean );
end entity thermostat;

-------------------------------------------------

architecture example of thermostat is
begin

  controller : process (desired_temp, actual_temp) is
  begin
    if actual_temp < desired_temp - 2 then
      heater_on <= true;
    elsif actual_temp > desired_temp + 2 then
      heater_on <= false;
    end if;
  end process controller;
```

```
end architecture example;
```

The entity declaration defines the input and output ports. Since it is a behavioral model, the architecture body contains only a process statement that implements the required behavior. The process statement includes a *sensitivity list* after the keyword **process**. This is a list of signals to which the process is sensitive. When any of these signals changes value, the process resumes and executes the sequential statements. After it has executed the last statement, the process suspends again. In this example, the process is sensitive to changes on either of the input ports. Thus, if we adjust the desired temperature, or if the measured temperature from the thermometer varies, the process is resumed. The body of the process contains an if statement that compares the actual temperature with the desired temperature. If the actual temperature is too low, the process executes the first signal assignment to turn the heater on. If the actual temperature is too high, the process executes the second signal assignment to turn the heater off. If the actual temperature is within the range, the state of the heater is not changed, since there is no **else** clause in the if statement.

VHDL-87, -93, and -2002

These versions of VHDL do not perform implicit conversion of conditions to **boolean**. Conditions in if statements must yield **boolean** results without conversion. Hence, we must write the if statement on page 66 as:

```
if en = '1' then
    ...
end if;
```

VHDL-87

If statements may not be labeled in VHDL-87.

3.1.1   Conditional Variable Assignments

In many models, we want to assign different values to a variable depending on one or more conditions. We could write this using an if statement, with each part containing a different assignment to the variable. However, VHDL provides a shorthand notation, called a *conditional variable assignment*, that we can use in these cases. The syntax rule is

conditional_variable_assignment ⇐
    〚 label : 〛
    name := expression **when** condition
            〖 **else** expression **when** condition 〗
            〚 **else** expression 〛 ;

For example, we could write:

```
result := a - b when mode = subtract else a + b;
```

instead of the longer equivalent if statement:

```
if mode = subtract then
  result := a - b;
else
  result := a + b;
end if;
```

Of course, if we need to perform more than one assignment for each condition, or if the actions required are not assignments, we would still use an if statement.

VHDL-87, -93, and -2002

These versions of VHDL do not provide the conditional variable assignment shorthand notation. We must write an if statement with a separate variable assignment for each condition.

3.2 Case Statements

If we have a model in which the behavior is to depend on the value of a single expression, we can use a *case statement*. The syntax rules are as follows:

```
case_statement ⇐
    〚 case_label : 〛
    case expression is
        ( when choices => { sequential_statement } )
        { ⸰⸰⸰ }
    end case 〚 case_label 〛 ;
```

choices ⇐ (simple_expression ▯ discrete_range ▯ **others**) { | ⸰⸰⸰ }

The label may be used to identify the case statement. We will discuss labeled statements in Chapter 20. We start with some simple examples of case statements and build up from them. First, suppose we are modeling an arithmetic/logic unit, with a control input, **func**, declared to be of the enumeration type:

```
type alu_func is (pass1, pass2, add, subtract);
```

We could describe the behavior using a case statement:

```
case func is
  when pass1 =>
    result := operand1;
  when pass2 =>
    result := operand2;
  when add =>
```

```
      result := operand1 + operand2;
   when subtract =>
      result := operand1 - operand2;
end case;
```

At the head of this case statement is the *selector expression*, between the keywords **case** and **is**. In this example it is a simple expression consisting of just a primary value. The value of this expression is used to select which statements to execute. The body of the case statement consists of a series of *alternatives*. Each alternative starts with the keyword **when** and is followed by one or more *choices* and a sequence of statements. The choices are values that are compared with the value of the selector expression. There must be exactly one choice for each possible value. The case statement finds the alternative whose choice value is equal to the value of the selector expression and executes the statements in that alternative. In this example, the choices are all simple expressions of type alu_func. If the value of func is pass1, the statement result := operand1 is executed; if the value is pass2, the statement result := operand2 is executed; and so on.

A case statement bears some similarity to an if statement in that they both select among alternative groups of sequential statements. The difference lies in how the statements to be executed are chosen. We saw in the previous section that an if statement evaluates successive Boolean expressions in turn until one is found to be true. The group of statements corresponding to that condition is then executed. A case statement, on the other hand, evaluates a single selector expression to derive a selector value. This value is then compared with the choice values in the case statement alternatives to determine which statement to execute. An if statement provides a more general mechanism for selecting between alternatives, since the conditions can be arbitrarily complex Boolean expressions. However, case statements are an important and useful modeling mechanism, as the examples in this section show.

The selector expression of a case statement must result in a value of a discrete type, or a one-dimensional array of character elements, such as a character string or bit string (see Chapter 4). Thus, we can have a case statement that selects an alternative based on an integer value. If we assume index_mode and instruction_register are declared as

```
subtype index_mode is integer range 0 to 3;

variable instruction_register : integer range 0 to 2**16 - 1;
```

then we can write a case statement that uses a value of this type:

```
case index_mode'((instruction_register / 2**12) rem 2**2) is
   when 0 =>
      index_value := 0;
   when 1 =>
      index_value := accumulator_A;
   when 2 =>
      index_value := accumulator_B;
   when 3 =>
      index_value := index_register;
end case;
```

Notice that in this example, we use a qualified expression in the selector expression. If we had omitted this, the result of the expression would have been **integer**, and we would have had to include alternatives to cover all possible integer values. The type qualification avoids this need by limiting the possible values of the expression.

Another rule to remember is that the type of each choice must be the same as the type resulting from the selector expression. Thus in the above example, it is illegal to include an alternative such as

```
when 'a' => ...    -- illegal!
```

since the choice listed cannot be an integer. Such a choice does not make sense, since it can never match a value of type **integer**.

We can include more than one choice in each alternative by writing the choices separated by the "|" symbol. For example, if the type **opcodes** is declared as

```
type opcodes is
   (nop, add, subtract, load, store, jump, jumpsub, branch, halt);
```

we could write an alternative including three of these values as choices:

```
when load | add | subtract =>
   operand := memory_operand;
```

If we have a number of alternatives in a case statement and we want to include an alternative to handle all possible values of the selector expression not mentioned in previous alternatives, we can use the special choice **others**. For example, if the variable opcode is a variable of type **opcodes**, declared above, we can write

```
case opcode is
   when load | add | subtract =>
      operand := memory_operand;
   when store | jump | jumpsub | branch =>
      operand := address_operand;
   when others =>
      operand := 0;
end case;
```

In this example, if the value of **opcode** is anything other than the choices listed in the first and second alternatives, the last alternative is selected. There may only be one alternative that uses the **others** choice, and if it is included, it must be the last alternative in the case statement. An alternative that includes the **others** choice may not include any other choices. Note that, if all of the possible values of the selector expression are covered by previous choices, we may still include the **others** choice, but it can never be matched.

The remaining form of choice that we have not yet mentioned is a *discrete range*, specified by these simplified syntax rules:

discrete_range ⇐
    *discrete*_subtype_indication
    ▯ simple_expression (**to** ▯ **downto**) simple_expression

```
subtype_indication ⇐
    type_mark
        〚 range simple_expression ( to ▯ downto ) simple_expression 〛
```

These forms allow us to specify a range of values in a case statement alternative. If the value of the selector expression matches any of the values in the range, the statements in the alternative are executed. The simplest way to specify a discrete range is just to write the left and right bounds of the range, separated by a direction keyword. For example, the case statement above could be rewritten as

```
case opcode is
  when add to load =>
    operand := memory_operand;
  when branch downto store =>
    operand := address_operand;
  when others =>
    operand := 0;
end case;
```

Another way of specifying a discrete range is to use the name of a discrete type, and possibly a range constraint to narrow down the values to a subset of the type. For example, if we declare a subtype of **opcodes** as

```
subtype control_transfer_opcodes is opcodes range jump to branch;
```

we can rewrite the second alternative as

```
when control_transfer_opcodes | store =>
  operand := address_operand;
```

Note that we may only use a discrete range as a choice if the selector expression is of a discrete type. We may not use a discrete range if the selector expression is of an array type, such as a bit-vector type. If we specify a range by writing the bounds and a direction, the direction has no significance except to identify the contents of the range.

An important point to note about the choices in a case statement is that they must all be written using *locally static* values. This means that the values of the choices must be determined during the analysis phase of design processing. All of the above examples satisfy this requirement. To give an example of a case statement that fails this requirement, suppose we have an integer variable N, declared as

```
variable N : integer := 1;
```

If we wrote the case statement

```
case expression is              -- example of an illegal case statement
  when N | N+1 => ...
  when N+2 to N+5 => ...
  when others => ...
end case;
```

the values of the choices depend on the value of the variable **N**. Since this might change during the course of execution, these choices are not locally static. Hence the case statement as written is illegal. On the other hand, if we had declared **C** to be a constant integer, for example with the declaration

```
constant C : integer := 1;
```

then we could legally write the case statement

```
case expression is
  when C | C+1 => ...
  when C+2 to C+5 => ...
  when others => ...
end case;
```

This is legal, since we can determine, by analyzing the model, that the first alternative includes choices 1 and 2, the second includes numbers between 3 and 6 and the third covers all other possible values of the expression.

The previous examples all show only one statement in each alternative. As with the if statement, we can write an arbitrary number of sequential statements of any kind in each alternative. This includes writing nested case statements, if statements or any other form of sequential statements in the alternatives.

Although the preceding rules governing case statements may seem complex, in practice there are just a few things to remember, namely:

- all possible values of the selector expression must be covered by one and only one choice,

- the values in the choices must be locally static and

- if the **others** choice is used it must be in the last alternative and must be the only choice in that alternative.

EXAMPLE 3.2   *A four-input multiplexer*

We can write a behavioral model of a multiplexer with a select input **sel**; four data inputs **d0**, **d1**, **d2** and **d3**; and a data output **z**. The data inputs and outputs are of the IEEE standard-logic type, and the select input is of type **sel_range**, which we assume to be declared elsewhere as

```
type sel_range is range 0 to 3;
```

We described in Section 2.2 how we define a type in a package for use in an entity declaration. The entity declaration defining the ports and a behavioral architecture body are:

```
library ieee;  use ieee.std_logic_1164.all;

entity mux4 is
  port ( sel : in sel_range;
```

```
            d0, d1, d2, d3 : in std_ulogic;
            z : out std_ulogic );
end entity mux4;

-------------------------------------------------

architecture demo of mux4 is
begin

  out_select : process (sel, d0, d1, d2, d3) is
  begin
    case sel is
      when 0 =>
        z <= d0;
      when 1 =>
        z <= d1;
      when 2 =>
        z <= d2;
      when 3 =>
        z <= d3;
    end case;
  end process out_select;

end architecture demo;
```

The architecture body contains just a process declaration. Since the output of the multiplexer must change if any of the data or select inputs change, the process must be sensitive to all of the inputs. It makes use of a case statement to select which of the data inputs is to be assigned to the data output.

VHDL-87

Case statements may not be labeled in VHDL-87.

3.2.1 Selected Variable Assignments

Just as there is a shorthand notation for an if statement containing variable assignments, there is also a shorthand for a case statement containing variable assignments. It is called a *selected variable assignment*, and the syntax rule is

selected_variable_assignment ⇐
    ⟦ label : ⟧
    with expression **select**
        name := ⟦ expression **when** choices , ⟧
                expression **when** choices ;

The first expression is the selector expression, and its value is compared with the choices to determine which expression value to assign to the named variable.

As an example, we could rewrite the case statement on page 69 as:

```
with func select
   result := operand1           when pass1,
            operand2            when pass2,
            operand1 + operand2 when add,
            operand1 - operand2 when subtract;
```

As with the conditional shorthand, if we need to perform more than one assignment for each alternative, or if the actions required are not assignments, we would still use a case statement.

VHDL-87, -93, and -2002

These versions of VHDL do not provide the selected variable assignment shorthand notation. We must write a case statement with a separate variable assignment for each alternative.

3.3 Null Statements

Sometimes when writing models we need to state that when some condition arises, no action is to be performed. This need often arises when we use case statements, since we must include an alternative for every possible value of the selector expression. Rather than just leaving the statement part of an alternative blank, we can use a *null statement* to state explicitly that nothing is to be done. The syntax rule for the null statement is simply

null_statement ⇐ ⟦ label : ⟧ **null** ;

The optional label serves to identify the statement. We discuss labeled statements in Chapter 20. A simple, unlabeled null statement is

```
null;
```

An example of its use in a case statement is

```
case opcode is
  when add =>
    Acc := Acc + operand;
  when subtract =>
    Acc := Acc - operand;
  when nop =>
    null;
end case;
```

We can use a null statement in any place where a sequential statement is required, not just in a case statement alternative. A null statement may be used during the development phase of model writing. If we know, for example, that we will need an entity as part of a system, but we are not yet in a position to write a detailed model for it, we can write a behavioral model that does nothing. Such a model just includes a process with a null statement in its body:

```
control_section : process ( sensitivity_list ) is
begin
  null;
end process control_section;
```

Note that the process must include the sensitivity list, for reasons that are explained in Chapter 5.

VHDL-87

Null statements may not be labeled in VHDL-87.

3.4 Loop Statements

Often we need to write a sequence of statements that is to be repeatedly executed. We use a *loop statement* to express this behavior. There are several different forms of loop statements in VHDL; the simplest is a loop that repeats a sequence of statements indefinitely, often called an *infinite loop*. The syntax rule for this kind of loop is

loop_statement ⇐
    ⟦ *loop*_label : ⟧
    loop
        ⟪ sequential_statement ⟫
    end loop ⟦ *loop*_label ⟧ ;

In most computer programming languages, an infinite loop is not desirable, since it means that the program never terminates. However, when we are modeling digital systems, an infinite loop can be useful, since many hardware devices repeatedly perform the same function until we turn off the power. Typically a model for such a system includes a loop statement in a process body; the loop, in turn, contains a wait statement.

EXAMPLE 3.3   *A modulo-16 counter*

The following is a model for a counter that starts from zero and increments on each clock transition from '0' to '1'. When the counter reaches 15, it wraps back to zero on the next clock transition. The architecture body for the counter contains a process that first initializes the count output to zero, then repeatedly waits for a clock transition before incrementing the count value.

```
entity counter is
  port ( clk : in bit;  count : out natural );
end entity counter;
```

```
------------------------------------------------
architecture behavior of counter is
begin

  incrementer : process is
    variable count_value : natural := 0;
  begin
    count <= count_value;
    loop
      wait until clk;
      count_value := (count_value + 1) mod 16;
      count <= count_value;
    end loop;
  end process incrementer;

end architecture behavior;
```

The wait statement in this example causes the process to suspend in the middle of the loop. When the **clk** signal changes from '0' to '1', the process resumes and updates the count value and the **count** output. The loop is then repeated starting with the wait statement, so the process suspends again.

Another point to note in passing is that the process statement does not include a sensitivity list. This is because it includes a wait statement. A process may contain either a sensitivity list or wait statements, but not both. We will return to this in detail in Chapter 5.

3.4.1 Exit Statements

In the previous example, the loop repeatedly executes the enclosed statements, with no way of stopping. Usually we need to exit the loop when some condition arises. We can use an *exit statement* to exit a loop. The syntax rule is

exit_statement ⟸
    ⟦ label : ⟧ **exit** ⟦ *loop*_label ⟧ ⟦ **when** condition ⟧ ;

The optional label at the start of the exit statement serves to identify the statement. We discuss labeled statements in Chapter 20. The simplest form of exit statement is just

```
exit;
```

When this statement is executed, any remaining statements in the loop are skipped, and control is transferred to the statement after the **end loop** keywords. So in a loop we can write

```
if condition then
  exit;
end if;
```

Since this is perhaps the most common use of the exit statement, VHDL provides a short-hand way of writing it, using the when clause. We use an exit statement with the when clause in a loop of the form

```
loop
  ...
  exit when condition;
  ...
end loop;
...              -- control transferred to here
                 -- when condition becomes true within the loop
```

EXAMPLE 3.4   *A modulo-16 counter with reset*

We now revise the counter model from Example 3.3 to include a **reset** input that, when '1', causes the **count** output to be reset to zero. The output stays at zero as long as the **reset** input is '1' and resumes counting on the next clock transition after **reset** changes to '0'. The revised entity declaration includes the new input port.

```
entity counter is
  port ( clk, reset : in bit;  count : out natural );
end entity counter;

--------------------------------------------------

architecture behavior of counter is
begin

  incrementer : process is
    variable count_value : natural := 0;
  begin
    count <= count_value;
    loop
      loop
        wait until clk or reset;
        exit when reset;
        count_value := (count_value + 1) mod 16;
        count <= count_value;
      end loop;
      -- at this point, reset = '1'
      count_value := 0;
      count <= count_value;
      wait until not reset;
    end loop;
  end process incrementer;

end architecture behavior;
```

The architecture body is revised by nesting the loop inside another loop statement and adding the **reset** signal to the original wait statement. The inner loop performs the same function as before, except that when **reset** changes to '1', the process is resumed, and the exit statement causes the inner loop to be terminated. Control is transferred to the statement just after the end of the inner loop. As the comment indicates, we know that this point can only be reached when **reset** is '1'. The count value and count outputs are reset, and the process then waits for **reset** to return to '0'. While it is suspended at this point, any changes on the clock input are ignored. When **reset** changes to '0', the process resumes, and the outer loop repeats.

This example also illustrates another important point. When we have nested loop statements, with an exit statement inside the inner loop, the exit statement causes control to be transferred out of the inner loop only, not the outer loop. By default, an exit statement transfers control out of the immediately enclosing loop.

In some cases, we may wish to transfer control out of an inner loop and also a containing loop. We can do this by labeling the outer loop and using the label in the exit statement. We can write

```
loop_name : loop
  ...
  exit loop_name;
  ...
end loop loop_name;
```

This labels the loop with the name **loop_name**, so that we can indicate which loop to exit in the exit statement. The loop label can be any valid identifier. The exit statement referring to this label can be located within nested loop statements.

To illustrate how loops can be nested, labeled and exited, let us consider the following statements:

```
outer : loop
  ...
  inner : loop
    ...
    exit outer when condition_1;   -- exit 1
    ...
    exit when condition_2;         -- exit 2
    ...
  end loop inner;
  ...                              -- target A
  exit outer when condition_3;     -- exit 3
  ...
end loop outer;
  ...                              -- target B
```

This example contains two loop statements, one labeled **inner** nested inside another labeled **outer**. The first exit statement, tagged with the comment **exit 1**, transfers control to the statement tagged **target B** if its condition is true. The second exit statement, tagged **exit 2**, transfers control to **target A**. Since it does not refer to a label, it only exits the immediately enclosing loop statement, namely, loop **inner**. Finally, the exit statement tagged **exit 3** transfers control to **target B**.

VHDL-87, -93, and -2002

Since these versions of VHDL do not perform implicit conversion of conditions to **boolean**, conditions in exit statements must yield **boolean** results without conversion. Hence, we must write the exit statement in Example 3.4 as:

```
exit when reset = '1';
```

VHDL-87

Exit statements may not be labeled in VHDL-87.

3.4.2   Next Statements

Another kind of statement that we can use to control the execution of loops is the *next statement*. When this statement is executed, the current iteration of the loop is completed without executing any further statements, and the next iteration is begun. The syntax rule is

next_statement ⇐
    〚 label : 〛 **next** 〚 *loop*_label 〛 〚 **when** condition 〛 ;

The optional label at the start of the next statement serves to identify the statement. We discuss labeled statements in Chapter 20. A next statement is very similar in form to an exit statement, the difference being the keyword **next** instead of **exit**. The simplest form of next statement is

```
next;
```

which starts the next iteration of the immediately enclosing loop. We can also include a condition to test before completing the iteration:

```
next when condition;
```

and we can include a loop label to indicate for which loop to complete the iteration:

```
next loop-label;
```

or

```
next loop-label when condition;
```

A next statement that exits the immediately enclosing loop can be easily rewritten as an equivalent loop with an if statement replacing the next statement. For example, the following two loops are equivalent:

```
loop                              loop
   statement-1;                      statement-1;
   next when condition;              if not condition then
   statement-2;                         statement-2;
end loop;                            end if;
                                  end loop;
```

However, nested labeled loops that contain next statements referring to outer loops cannot be so easily rewritten. As a matter of style, if we find ourselves about to write such a collection of loops and next statements, it's probably time to think more carefully about what we are trying to express. If we check the logic of the model, we may be able to find a simpler formulation of loop statements. Complicated loop/next structures can be confusing, making the model hard to read and understand.

VHDL-87, -93, and -2002

Since these versions of VHDL do not perform implicit conversion of conditions to **boolean**, conditions in next statements must yield **boolean** results without conversion.

VHDL-87

Next statements may not be labeled in VHDL-87.

3.4.3 While Loops

We can augment the basic loop statement introduced previously to form a while loop, which tests a condition before each iteration. If the condition is true, iteration proceeds. If it is false, the loop is terminated. The syntax rule for a while loop is

```
loop_statement ⇐
    〚 loop_label : 〛
    while condition loop
        { sequential_statement }
    end loop 〚 loop_label 〛 ;
```

The only difference between this form and the basic loop statement is that we have added the keyword **while** and the condition before the **loop** keyword. All of the things we said about the basic loop statement also apply to a while loop. We can write any sequential statements in the body of the loop, including exit and next statements, and we can label the loop by writing the label before the **while** keyword.

There are three important points to note about while loops. The first point is that the condition is tested before each iteration of the loop, including the first iteration. This means that if the condition is false before we start the loop, it is terminated immediately, with no iterations being executed. For example, given the while loop

```
while index > 0 loop
    ...             -- statement A: do something with index
end loop;
    ...             -- statement B
```

if we can demonstrate that **index** is not greater than zero before the loop is started, then we know that the statements inside the loop will not be executed, and control will be transferred straight to **statement B**.

The second point is that in the absence of exit statements within a while loop, the loop terminates only when the condition becomes false. Thus, we know that the negation of the condition must hold when control reaches the statement after the loop. Similarly, in the absence of next statements within a while loop, the loop performs an iteration only when the condition is true. Thus, we know that the condition holds when we start the statements in the loop body. In the above example, we know that **index** must be greater then zero when we execute the statement tagged **statement A**, and also that index must be less than or equal to zero when we reach **statement B**. This knowledge can help us reason about the correctness of the model we are writing.

The third point is that when we write the statements inside the body of a while loop, we must make sure that the condition will eventually become false, or that an exit statement will eventually exit the loop. Otherwise the while loop will never terminate. Presumably, if we had intended to write an infinite loop, we would have used a simple loop statement.

EXAMPLE 3.5   *A cosine module*

We can develop a model for an entity **cos** that might be used as part of a specialized signal processing system. The entity has one input, **theta**, which is a real number representing an angle in radians, and one output, **result**, representing the cosine function of the value of **theta**. We can use the relation

$$\cos\theta = 1 - \frac{\theta^2}{2!} + \frac{\theta^4}{4!} - \frac{\theta^6}{6!} + \cdots$$

by adding successive terms of the series until the terms become smaller than one millionth of the result. The entity and architecture body declarations are:

```
entity cos is
  port ( theta : in real;  result : out real );
end entity cos;
```

```vhdl
architecture series of cos is
begin

  summation : process (theta) is
    variable sum, term : real;
    variable n : natural;
  begin
    sum := 1.0;
    term := 1.0;
    n := 0;
    while abs term > abs (sum / 1.0E6) loop
      n := n + 2;
      term := (-term) * theta**2 / real(((n-1) * n));
      sum := sum + term;
    end loop;
    result <= sum;
  end process summation;

end architecture series;
```

The architecture body consists of a process that is sensitive to changes in the input signal **theta**. Initially, the variables **sum** and **term** are set to 1.0, representing the first term in the series. The variable **n** starts at 0 for the first term. The cosine function is computed using a while loop that increments **n** by two and uses it to calculate the next term based on the previous term. Iteration proceeds as long as the last term computed is larger in magnitude than one millionth of the sum. When the last term falls below this threshold, the while loop is terminated. We can determine that the loop will terminate, since the values of successive terms in the series get progressively smaller. This is because the factorial function grows at a greater rate than the exponential function.

VHDL-87, -93, and -2002

Since these versions of VHDL do not perform implicit conversion of conditions to **boolean**, conditions in while loops must yield **boolean** results without conversion.

3.4.4 For Loops

Another way we can augment the basic loop statement is the for loop. A for loop includes a specification of how many times the body of the loop is to be executed. The syntax rule for a for loop is

loop_statement ⇐
    [[*loop*_label :]]
    for identifier **in** discrete_range **loop**

```
  { sequential_statement }
end loop [ loop_label ] ;
```

We saw on page 71 that a discrete range can be of the form

simple_expression (**to** ⎮ **downto**) simple_expression

representing all the values between the left and right bounds, inclusive. The identifier is called the *loop parameter*, and for each iteration of the loop, it takes on successive values of the discrete range, starting from the left element. For example, in this for loop:

```
for count_value in 0 to 127 loop
  count_out <= count_value;
  wait for 5 ns;
end loop;
```

the identifier **count_value** takes on the values 0, 1, 2 and so on, and for each value, the assignment and wait statements are executed. Thus the signal **count_out** will be assigned values 0, 1, 2 and so on, up to 127, at 5 ns intervals.

We also saw that a discrete range can be specified using a discrete type or subtype name, possibly further constrained to a subset of values by a range constraint. For example, if we have the enumeration type

```
type controller_state is (initial, idle, active, error);
```

we can write a for loop that iterates over each of the values in the type:

```
for state in controller_state loop
  ...
end loop;
```

Within the sequence of statements in the for loop body, the loop parameter is a constant whose type is the base type of the discrete range. This means we can use its value by including it in an expression, but we cannot make assignments to it. Unlike other constants, we do not need to declare it. Instead, the loop parameter is implicitly declared over the for loop. It only exists when the loop is executing, and not before or after it. For example, the following process statement shows how not to use the loop parameter:

```
erroneous : process is
  variable i, j : integer;
begin
  i := loop_param;                    -- error!
  for loop_param in 1 to 10 loop
    loop_param := 5;                  -- error!
  end loop;
  j := loop_param;                    -- error!
end process erroneous;
```

The assignments to i and j are illegal since the loop parameter is defined neither before nor after the loop. The assignment within the loop body is illegal because **loop_param** is a constant and thus may not be modified.

A consequence of the way the loop parameter is defined is that it *hides* any object of the same name defined outside the loop. For example, in this process:

```
hiding_example : process is
  variable a, b : integer;
begin
  a := 10;
  for a in 0 to 7 loop
    b := a;
  end loop;
  -- a = 10, and b = 7
  ...
end process hiding_example;
```

the variable **a** is initially assigned the value 10, and then the for loop is executed, creating a loop parameter also called **a**. Within the loop, the assignment to **b** uses the loop parameter, so the final value of **b** after the last iteration is 7. After the loop, the loop parameter no longer exists, so if we use the name **a**, we are referring to the variable object, whose value is still 10.

As we mentioned above, the for loop iterates with the loop parameter assuming successive values from the discrete range starting from the leftmost value. An important point to note is that if we specify a null range, the for loop body does not execute at all. A null range can arise if we specify an ascending range with the left bound greater than the right bound, or a descending range with the left bound less than the right bound. For example, the for loop

```
for i in 10 to 1 loop
  ...
end loop;
```

completes immediately, without executing the enclosed statements. If we really want the loop to iterate with i taking values 10, 9, 8 and so on, we should write

```
for i in 10 downto 1 loop
  ...
end loop;
```

One final thing to note about for loops is that, like basic loop statements, they can enclose arbitrary sequential statements, including next and exit statements, and we can label a for loop by writing the label before the **for** keyword.

EXAMPLE 3.6 *A revised cosine module*

We now rewrite the cosine model in Example 3.5 to calculate the result by summing the first 10 terms of the series. The entity declaration is unchanged. The revised architecture body, shown below, consists of a process that uses a for loop instead of a while loop. As before, the variables **sum** and **term** are set to 1.0, representing the first term in the series. The variable **n** is replaced by the for loop parameter. The loop iterates nine times, calculating the remaining nine terms of the series.

```
architecture fixed_length_series of cos is
begin

  summation : process (theta) is
    variable sum, term : real;
  begin
    sum := 1.0;
    term := 1.0;
    for n in 1 to 9 loop
      term := (-term) * theta**2 / real(((2*n-1) * 2*n));
      sum := sum + term;
    end loop;
    result <= sum;
  end process summation;

end architecture fixed_length_series;
```

3.4.5 Summary of Loop Statements

The preceding sections describe the various forms of loop statements in detail. It is worth summarizing this information in one place, to show the few basic points to remember. First, the syntax rule for all loop statements is

```
loop_statement ⇐
    〚 loop_label : 〛
    〚 while condition ▯ for identifier in discrete_range 〛 loop
        ⦃ sequential_statement ⦄
    end loop 〚 loop_label 〛 ;
```

Second, in the absence of exit and next statements, the while loop iterates as long as the condition is true, and the for loop iterates with the loop parameter assuming successive values from the discrete range. If the condition in a while loop is initially false, or if the discrete range in a for loop is a null range, then no iterations occur.

Third, the loop parameter in a for loop cannot be explicitly declared, and it is a constant within the loop body. It also shadows any other object of the same name declared outside the loop.

Finally, an exit statement can be used to terminate any loop, and a next statement can be used to complete the current iteration and commence the next iteration. These statements can refer to loop labels to terminate or complete iteration for an outer level of a nested set of loops.

3.5　Assertion and Report Statements

One of the reasons for writing models of computer systems is to verify that a design functions correctly. We can partially test a model by applying sample inputs and checking that the outputs meet our expectations. If they do not, we are then faced with the task of determining what went wrong inside the design. This task can be made easier using *assertion statements* that check that expected conditions are met within the model. An assertion statement is a sequential statement, so it can be included anywhere in a process body. The full syntax rule for an assertion statement is

> assertion_statement ⇐
> 　〖 label : 〗 **assert** condition
> 　　　　〖 **report** expression 〗〖 **severity** expression 〗 ;

The optional label allows us to identify the assertion statement. We will discuss labeled statements in Chapter 20. The simplest form of assertion statement just includes the keyword **assert** followed by a condition that we expect to be true when the assertion statement is executed. If the condition is not met, we say that an *assertion violation* has occurred. If an assertion violation arises during simulation of a model, the simulator reports the fact. During synthesis, the condition in an assertion statement may be interpreted as a condition that the synthesizer may assume to be true. During formal verification, the condition may be interpreted as a condition to be proven by the verifier. For example, if we write

```
assert initial_value <= max_value;
```

and initial_value is larger than **max_value** when the statement is executed during simulation, the simulator will let us know. During synthesis, the synthesizer may assume that initial_value <= max_value and optimize the circuit based on that information. During formal verification, the verifier may attempt to prove initial_value <= max_value for all possible input stimuli and execution paths leading to the assertion statement.

If we have a number of assertion statements throughout a model, it is useful to know which assertion is violated. We can get the simulator to provide extra information by including a **report** clause in an assertion statement, for example:

```
assert initial_value <= max_value
  report "initial value too large";
```

The string that we provide is used to form part of the assertion violation message. We can write any expression in the **report** clause provided it yields a string value, for example:

```
assert current_character >= '0' and current_character <= '9'
  report "Input number " & input_string & " contains a non-digit";
```

Here the message is derived by concatenating three string values together. We can use the to_string operation to get a string representation of a value to include in a message, for example:

```vhdl
assert initial_value <= max_value
  report "initial value " & to_string(initial_value)
        & " too large";
```

In Section 2.2, we mentioned a predefined enumeration type **severity_level**, defined as

```vhdl
type severity_level is (note, warning, error, failure);
```

We can include a value of this type in a **severity** clause of an assertion statement. This value indicates the degree to which the violation of the assertion affects operation of the model. The value **note** can be used to pass informative messages out from a simulation, for example:

```vhdl
assert free_memory >= low_water_limit
  report "low on memory, about to start garbage collect"
  severity note;
```

The severity level **warning** can be used if an unusual situation arises in which the model can continue to execute, but may produce unusual results, for example:

```vhdl
assert packet_length /= 0
  report "empty network packet received"
  severity warning;
```

We can use the severity level **error** to indicate that something has definitely gone wrong and that corrective action should be taken, for example:

```vhdl
assert clock_pulse_width >= min_clock_width
  severity error;
```

Finally, the value **failure** can be used if we detect an inconsistency that should never arise, for example:

```vhdl
assert (last_position - first_position + 1) = number_of_entries
  report "inconsistency in buffer model"
  severity failure;
```

We have seen that we can write an assertion statement with either a **report** clause or a **severity** clause, or both. If both are present, the syntax rule shows us that the **report** clause must come first. If we omit the **report** clause, the default string in the error message is "Assertion violation." If we omit the **severity** clause, the default value is error. The severity value is usually used by a simulator to determine whether or not to continue execution after an assertion violation. Most simulators allow the user to specify a severity threshold, beyond which execution is stopped. The VHDL standard recommends that, in the absence of such a specification, simulation continue for assertion violations with severity error or less.

Usually, failure of an assertion means either that the entity is being used incorrectly as part of a larger design or that the model for the entity has been incorrectly written. We illustrate both cases.

EXAMPLE 3.7 *A set/reset flipflop including a check for correct usage*

A set/reset (SR) flipflop has two inputs, S and R, and an output Q. When S is '1', the output is set to '1', and when R is '1', the output is reset to '0'. However, S and R may not both be '1' at the same time. If they are, the output value is not specified. A behavioral model for an SR flipflop that includes a check for this illegal condition is:

```
entity SR_flipflop is
  port ( S, R : in bit;  Q : out bit );
end entity SR_flipflop;


-------------------------------------------------


architecture checking of SR_flipflop is
begin

    set_reset : process (S, R) is
    begin
      assert S nand R;
      if S then
        Q <= '1';
      end if;
      if R then
        Q <= '0';
      end if;
    end process set_reset;

end architecture checking;
```

The architecture body contains a process sensitive to the S and R inputs. Within the process body we write an assertion statement that requires that S and R not both be '1'. If both are '1', the assertion is violated, so the simulator writes an "Assertion violation" message with severity error. If execution continues after the violated assertion, the value '1' will first be assigned to Q, followed by the value '0'. The resulting value is '0'. This is allowed, since the state of Q was not specified for this illegal condition, so we are at liberty to choose any value. If the assertion is not violated, then at most one of the following if statements is executed, correctly modeling the behavior of the SR flipflop.

EXAMPLE 3.8 *Sanity check on calculation of the maximum value*

To illustrate the use of an assertion statement as a "sanity check," let us look at a model for an entity that has three integer inputs, a, b and c, and produces an integer output z that is the largest of its inputs.

```vhdl
entity max3 is
  port ( a, b, c : in integer;  z : out integer );
end entity max3;

-----------------------------------------------------

architecture check_error of max3 is
begin
    maximizer : process (a, b, c)
      variable result : integer;
    begin
      if a > b then
        if a > c then
          result := a;
        else
          result := a;   -- Oops!  Should be: result := c;
        end if;
      elsif  b > c then
        result := b;
      else
        result := c;
      end if;
      assert result >= a and result >= b and result >= c
        report "inconsistent result for maximum: "
               & to_string(result)
        severity failure;
      z <= result;
    end process maximizer;

end architecture check_error;
```

The architecture body is written using a process containing nested if statements. For this example we have introduced an "accidental" error into the model. If we simulate this model and put the values **a** = 7, **b** = 3 and **c** = 9 on the ports of this entity, we expect that the value of **result**, and hence the output port, is 9. The assertion states that the value of **result** must be greater than or equal to all of the inputs. However, our coding error causes the value 7 to be assigned to **result**, and so the assertion is violated. This violation causes us to examine our model more closely, and correct the error.

Another important use for assertion statements is in checking timing constraints that apply to a model. For example, most clocked devices require that the clock pulse be longer than some minimum duration. We can use the predefined primary **now** in an expression to calculate durations. We return to **now** in a later chapter. Suffice it to say that it yields the current simulation time when it is evaluated.

EXAMPLE 3.9   *An edge-triggered register with timing check*

An edge-triggered register has a data input and a data output of type **real** and a clock input of type **bit**. When the clock changes from '0' to '1', the data input is sampled, stored and transmitted through to the output. Let us suppose that the clock input must remain at '1' for at least 5 ns. The following is a model for this register, including a check for legal clock pulse width.

```vhdl
entity edge_triggered_register is
  port ( clock : in bit;
         d_in : in real;  d_out : out real );
end entity edge_triggered_register;

--------------------------------------------------

architecture check_timing of edge_triggered_register is
begin

  store_and_check : process (clock) is
    variable stored_value : real;
    variable pulse_start : time;
  begin
    if rising_edge(clock) then
      pulse_start := now;
      stored_value := d_in;
      d_out <= stored_value;
    else
      assert now = 0 ns or (now - pulse_start) >= 5 ns
        report "clock pulse too short: "
                & to_string(now - pulse_start);
    end if;
  end process store_and_check;

end architecture check_timing;
```

The architecture body contains a process that is sensitive to changes on the clock input. When the clock changes from '0' to '1', the input is stored, and the current simulation time is recorded in the variable **pulse_start**. Otherwise, when the clock changes from '1' to '0', the difference between **pulse_start** and the current simulation time is checked by the assertion statement.

VHDL-87

Assertion statements may not be labeled in VHDL-87.

VHDL also provides us with a *report statement*, which is similar to an assertion statement. The syntax rule for the report statement shows this similarity:

report_statement ⇐
    ⟦ label : ⟧ **report** expression ⟦ **severity** expression ⟧ ;

The differences are that there is no condition, and if the severity level is not specified, the default is **note**. Indeed, the report statement can be thought of as an assertion statement in which the condition is the value **false** and the severity is **note**, hence it always produces the message. One way in which the report statement is useful is as a means of including "trace writes" in a model as an aid to debugging.

EXAMPLE 3.10   *Trace messages using a report statement*

Suppose we are writing a complex model and we are not sure that we have got the logic quite right. We can use report statements to get the processes in the model to write out messages, so that we can see when they are activated and what they are doing. An example process is

```
transmit_element : process (transmit_data) is
   ...          -- variable declarations
begin
   report "transmit_element: data = "
          & to_string(transmit_data);
   ...
end process transmit_element;
```

Both assertion statements and report statements allow inclusion of a message string to provide useful information. In some cases, the information we wish to provide may be extensive and not fit entirely on a single line when displayed. We can include the line-feed character in the message string to break the message over multiple lines. A line-feed is represented by the identifier LF of type **character**. For example:

```
assert data = expected_data
   report "%%%ERROR data value miscompare." & LF &
          " Actual value = " & to_string(data) & LF &
          " Expected value = " & to_string(expdata) & LF &
          " at time: " & to_string(now);
```

The message produced when this assertion is violated consists of four lines of text. A VHDL tool interprets the line feed characters using the appropriate convention for the host operating system. For example, if a Unix-based system were to write the message to a file, it would just include the line-feed characters. A Windows-based system, on the other hand, would write a carriage-return/line-feed pair for every line-feed in the message.

VHDL-87, -93, and -2002

These versions of VHDL do not make any recommendation about continuing simulation based on the severity of an assertion violation from an assertion statement or report statement. Different simulators take different approaches, which can make it difficult to write portable models with consistent simulation behavior on different tools.

These versions also do not necessarily interpret line-feed characters in message strings as denoting line breaks. Interpretation of line feeds is implementation defined.

VHDL-87

Report statements are not provided in VHDL-87. We achieve the same effect by writing an assertion statement with the condition **false** and a severity level of **note**. For example, the VHDL-93 or VHDL-2002 report statement

```
report "Initialization complete";
```

can be written in VHDL-87 as

```
assert false
    report "Initialization complete" severity note;
```

Exercises

1. [❶ 3.1] Write an if statement that sets a variable **odd** to '1' if an integer **n** is odd, or to '0' if it is even. Rewrite your if statement as a conditional variable assignment.

2. [❶ 3.1] Write an if statement that, given the year of today's date in the variable **year**, sets the variable **days_in_February** to the number of days in February. A year is a leap year if it is divisible by four, except for years that are divisible by 100. However, years that are divisible by 400 are leap years. February has 29 days in a leap year and 28 days otherwise. Rewrite your if statement as a conditional variable assignment.

3. [❶ 3.2] Write a case statement that strips the strength information from a standard-logic variable **x**. If **x** is '0' or 'L', set it to '0'. If **x** is '1' or 'H', set it to '1'. If **x** is 'X', 'W', 'Z', 'U' or '–', set it to 'X'. (This is the conversion performed by the standard-logic function **to_X01**.) Rewrite your case statement as a selected variable assignment.

4. [❶ 3.2] Write a case statement that sets an integer variable **character_class** to 1 if the character variable **ch** contains a letter, to 2 if it contains a digit, to 3 if it contains some other printable character or to 4 if it contains a non-printable character. Note that the VHDL character set contains accented letters, as shown in Section 2.2.5 on page 44. Rewrite your case statement as a selected variable assignment.

5. [❶ 3.4] Write a loop statement that samples a bit input **d** when a clock input **clk** changes to '1'. So long as **d** is '0', the loop continues executing. When **d** is '1', the loop exits.

6. [❶ 3.4] Write a while loop that calculates the exponential function of **x** to an accuracy of one part in 10^4 by summing terms of the following series:

$$e^x = 1 + \frac{x}{1} + \frac{x^2}{2!} + \frac{x^3}{3!} + \frac{x^4}{4!} + \cdots$$

7. [❶ 3.4] Write a for loop that calculates the exponential function of **x** by summing the first eight terms of the series in Exercise 6.

8. [❶ 3.5] Write an assertion statement that expresses the requirement that a flipflop's two outputs, **q** and **q_n**, of type **std_ulogic**, are complementary.

9. [❶ 3.5] We can use report statements in VHDL to achieve the same effect as using "trace writes" in software programming languages, to report a message when part of the model is executed. Insert a report statement in the model of Example 3.4 to cause a trace message when the counter is reset.

10. [❷ 3.1] Develop a behavioral model for a limiter with three integer inputs, **data_in**, **lower** and **upper**; an integer output, **data_out**; and a bit output, **out_of_limits**. The **data_out** output follows **data_in** so long as it is between **lower** and **upper**. If **data_in** is less than **lower**, **data_out** is limited to **lower**. If **data_in** is greater than **upper**, **data_out** is limited to **upper**. The **out_of_limit** output indicates when **data_out** is limited.

11. [❷ 3.2] Develop a model for a floating-point arithmetic unit with data inputs **x** and **y**, data output **z** and function code inputs **f1** and **f0** of type **bit**. Function codes **f1** = '0' and **f0** = '0' produce addition; **f1** = '0' and **f0** = '1' produce subtraction of **y** from **x**; **f1** = '1' and **f0** = '0' produce multiplication; and **f1** = '1' and **f0** = '1' produce division of **x** by **y**.

12. [❷ 3.4] Write a model for a counter with an output port of type **natural**, initially set to 15. When the **clk** input changes to '1', the counter decrements by one. After counting down to zero, the counter wraps back to 15 on the next clock edge.

13. [❷ 3.4] Modify the counter of Exercise 12 to include an asynchronous load input and a data input. When the load input is '1', the counter is preset to the data input value. When the load input changes back to '0', the counter continues counting down from the preset value.

14. [❷ 3.4] Develop a model of an averaging module that calculates the average of batches of 16 real numbers. The module has clock and data inputs and a data output. The module accepts the next input number when the clock changes to '1'. After 16 numbers have been accepted, the module places their average on the output port, then repeats the process for the next batch.

15. [❷ 3.5] Write a model that causes assertion violations with different severity levels. Experiment with your simulator to determine its behavior when an assertion violation occurs. See if you can specify a severity threshold above which it stops execution.

Chapter 4

Composite Data Types
and Operations

Now that we have seen the basic data types and sequential operations from which the behavioral part of a VHDL model is formed, it is time to look at composite data types. We first mentioned them in the classification of data types in Chapter 2. Composite data objects consist of related collections of data elements in the form of either an *array* or a *record*. We can treat an object of a composite type as a single object or manipulate its constituent elements individually. In this chapter, we see how to define composite types and how to manipulate them using operators and sequential statements.

4.1 Arrays

An *array* consists of a collection of values, all of which are of the same type as each other. The position of each element in an array is given by a scalar value called its *index*. To create an array object in a model, we first define an array type in a type declaration. The syntax rule for an array type definition is

array_type_definition ⇐
    array (discrete_range ⦃ , ⦄) **of** *element*_subtype_indication

This defines an array type by specifying one or more index ranges (the list of discrete ranges) and the element type or subtype.

Recall from previous chapters that a discrete range is a subset of values from a discrete type (an integer or enumeration type), and that it can be specified as shown by the simplified syntax rule

discrete_range ⇐
    *discrete*_subtype_indication
    ⎸ simple_expression ⦅ **to** ⎸ **downto** ⦆ simple_expression

Recall also that a subtype indication can be just the name of a previously declared type (a type mark) and can include a range constraint to limit the set of values from that type, as shown by the simplified rule

subtype_indication ⇐
    type_mark 〖 **range** simple_expression 《 **to** ▯ **downto** 》 simple_expression 〗

We illustrate these rules for defining arrays with a series of examples. We start with single-dimensional arrays, in which there is just one index range. Here is a simple example to start off with, showing the declaration of an array type to represent words of data:

```
type word is array (0 to 31) of bit;
```

Each element is a bit, and the elements are indexed from 0 up to 31. An alternative declaration of a word type, more appropriate for "little-endian" systems, is

```
type word is array (31 downto 0) of bit;
```

The difference here is that index values start at 31 for the leftmost element in values of this type and continue down to 0 for the rightmost.

The index values of an array do not have to be numeric. For example, given this declaration of an enumeration type:

```
type controller_state is (initial, idle, active, error);
```

we could then declare an array as follows:

```
type state_counts is array (idle to error) of natural;
```

This kind of array type declaration relies on the type of the index range being clear from the context. If there were more than one enumeration type with values **idle** and **error**, it would not be clear which one to use for the index type. To make it clear, we can use the alternative form for specifying the index range, in which we name the index type and include a range constraint. The previous example could be rewritten as

```
type state_counts is
  array (controller_state range idle to error) of natural;
```

If we need an array element for every value in an index type, we need only name the index type in the array declaration without specifying the range. For example:

```
subtype coeff_ram_address is integer range 0 to 63;
type coeff_array is array (coeff_ram_address) of real;
```

Once we have declared an array type, we can define objects of that type, including constants, variables and signals. For example, using the types declared above, we can declare variables as follows:

```
variable buffer_register, data_register : word;
variable counters : state_counts;
variable coeff : coeff_array;
```

Each of these objects consists of the collection of elements described by the corresponding type declaration. An individual element can be used in an expression or as the target of an assignment by referring to the array object and supplying an index value, for example:

```
coeff(0) := 0.0;
```

If active is a variable of type controller_state, we can write

```
counters(active) := counters(active) + 1;
```

An array object can also be used as a single composite object. For example, the assignment

```
data_register := buffer_register;
```

copies all of the elements of the array buffer_register into the corresponding elements of the array data_register.

EXAMPLE 4.1 *A memory module for real-number coefficients*

The following is a model for a memory that stores 64 real-number coefficients, initialized to 0.0. We assume the type coeff_ram_address is previously declared as above. The architecture body contains a process with an array variable representing the coefficient storage. When the process starts, it initializes the array using a for loop. It then repetitively waits for any of the input ports to change. When rd is '1', the array is indexed using the address value to read a coefficient. When wr is '1', the address value is used to select which coefficient to change.

```
entity coeff_ram is
  port ( rd, wr : in bit;  addr : in coeff_ram_address;
         d_in : in real;  d_out : out real );
end entity coeff_ram;

--------------------------------------------------

architecture abstract of coeff_ram is
begin
  memory : process is
    type coeff_array is array (coeff_ram_address) of real;
    variable coeff : coeff_array;
  begin
    for index in coeff_ram_address loop
      coeff(index) := 0.0;
    end loop;
    loop
      wait on rd, wr, addr, d_in;
      if rd then
        d_out <= coeff(addr);
      end if;
      if wr then
```

```
            coeff(addr) := d_in;
          end if;
        end loop;
      end process memory;

    end architecture abstract;
```

4.1.1 Multidimensional Arrays

VHDL also allows us to create multidimensional arrays, for example, to represent matrices or tables indexed by more than one value. A multidimensional array type is declared by specifying a list of index ranges, as shown by the syntax rule on page 95. For example, we might include the following type declarations in a model for a finite-state machine:

```
type symbol is ('a', 't', 'd', 'h', digit, cr, error);
type state is range 0 to 6;

type transition_matrix is array (state, symbol) of state;
```

Each index range can be specified as shown above for single-dimensional arrays. The index ranges for each dimension need not all be from the same type, nor have the same direction. An object of a multidimensional array type is indexed by writing a list of index values to select an element. For example, if we have a variable declared as

```
variable transition_table : transition_matrix;
```

we can index it as follows:

```
transition_table(5, 'd');
```

EXAMPLE 4.2 *Transformation matrices*

In three-dimensional graphics, a point in space may be represented using a three-element vector [x, y, z] of coordinates. Transformations, such as scaling, rotation and reflection, may be done by multiplying a vector by a 3×3 transformation matrix to get a new vector representing the transformed point. We can write VHDL type declarations for points and transformation matrices:

```
type point is array (1 to 3) of real;
type matrix is array (1 to 3, 1 to 3) of real;
```

We can use these types to declare point variables p and q and a matrix variable transform:

```
variable p, q : point;
variable transform : matrix;
```

The transformation can be applied to the point p to produce a result in q with the following statements:

```
      for i in 1 to 3 loop
        q(i) := 0.0;
        for j in 1 to 3 loop
          q(i) := q(i) + transform(i, j) * p(j);
        end loop;
      end loop;
```

4.1.2 Array Aggregates

We have seen how we can write literal values of scalar types. Often we also need to write literal array values, for example, to initialize a variable or constant of an array type. We can do this using a VHDL construct called an array aggregate, according to the syntax rule

aggregate ⇐ ((⟦ choices => ⟧ expression) { , ... })

Let us look first at the form of aggregate without the choices part. It simply consists of a list of the elements enclosed in parentheses, for example:

```
type point is array (1 to 3) of real;
constant origin : point := (0.0, 0.0, 0.0);
variable view_point : point := (10.0, 20.0, 0.0);
```

This form of array aggregate uses *positional association* to determine which value in the list corresponds to which element of the array. The first value is the element with the left-most index, the second is the next index to the right, and so on, up to the last value, which is the element with the rightmost index. There must be a one-to-one correspondence between values in the aggregate and elements in the array.

An alternative form of aggregate uses *named association*, in which the index value for each element is written explicitly using the choices part shown in the syntax rule. The choices may be specified in exactly the same way as those in alternatives of a case statement, discussed in Chapter 3. As a reminder, here is the syntax rule for choices:

choices ⇐ (simple_expression ❘ discrete_range ❘ **others**) { | ... }

For example, the variable declaration and initialization could be rewritten as

```
variable view_point : point := (1 => 10.0, 2 => 20.0, 3 => 0.0);
```

The main advantage of named association is that it gives us more flexibility in writing aggregates for larger arrays. To illustrate this, let us return to the coefficient memory described above. The type declaration was

```
type coeff_array is array (coeff_ram_address) of real;
```

Suppose we want to declare the coefficient variable, initialize the first few locations to some non-zero value and initialize the remainder to zero. Following are a number of ways of writing aggregates that all have the same effect:

```
variable coeff : coeff_array
        := (0 => 1.6, 1 => 2.3, 2 => 1.6, 3 to 63 => 0.0);
```

Here we are using a range specification to initialize the bulk of the array value to zero.

```
variable coeff : coeff_array
        := (0 => 1.6, 1 => 2.3, 2 => 1.6, others => 0.0);
```

The keyword **others** stands for any index value that has not been previously mentioned in the aggregate. If the keyword **others** is used, it must be the last choice in the aggregate.

```
variable coeff : coeff_array
        := (0 | 2 => 1.6, 1 => 2.3, others => 0.0);
```

The "|" symbol can be used to separate a list of index values, for which all elements have the same value.

Note that we may not mix positional and named association in an array aggregate, except for the use of an **others** choice in the final postion. Thus, the following aggregate is illegal:

```
variable coeff : coeff_array
        := (1.6, 2.3, 2 => 1.6, others => 0.0);   -- illegal
```

We can also use aggregates to write multidimensional array values. In this case, we treat the array as though it were an array of arrays, writing an array aggregate for each of the leftmost index values first.

EXAMPLE 4.3   *Transition matrix for a modem finite-state machine*

We can use a two-dimensional array to represent the transition matrix of a finite-state machine (FSM) that interprets simple modem commands. A command must consist of the string "atd" followed by a string of digits and a cr character, or the string "ath" followed by cr. The state transition diagram is shown in Figure 4.1. The symbol "other" represents a character other than 'a', 't', 'd', 'h', a digit or cr.

An outline of a process that implements the FSM is:

```
modem_controller : process is

    type symbol is ('a', 't', 'd', 'h', digit, cr, other);
    type symbol_string is array (1 to 20) of symbol;
    type state is range 0 to 6;
    type transition_matrix is array (state, symbol) of state;

    constant next_state : transition_matrix :=
      ( 0 => ('a' => 1, others => 6),
        1 => ('t' => 2, others => 6),
        2 => ('d' => 3, 'h' => 5, others => 6),
        3 => (digit => 4, others => 6),
        4 => (digit => 4, cr => 0, others => 6),
```

FIGURE 4.1

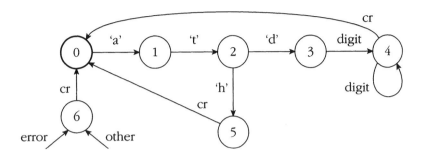

The state transition diagram for a modem command finite-state machine. State 0 is the initial state. The machine returns to this state after recognizing a correct command. State 6 is the error state, to which the machine goes if it detects an illegal or unexpected character.

```
    5 => (cr => 0, others => 6),
    6 => (cr => 0, others => 6) );

variable command : symbol_string;
variable current_state : state := 0;

begin
  ...
  for index in 1 to 20 loop
    current_state := next_state( current_state, command(index) );
    case current_state is

      ...
    end case;
  end loop;
  ...
end process modem_controller;
```

The type declarations for **symbol** and **state** represent the command symbols and the states for the FSM. The transition matrix, **next_state**, is a two-dimensional array constant indexed by the state and symbol type. An element at position (i, j) in this matrix indicates the next state the FSM should move to when it is in state i and the next input symbol is j. The matrix is initialized according to the transition diagram. The process uses the **current_state** variable and successive input symbols as indices into the transition matrix to determine the next state. For each transition, it performs some action based on the new state. The actions are implemented within the case statement.

In the array aggregates we have seen so far, each value in the aggregate corresponds to a single array element. VHDL also allows an alternate form of aggregate in which we write a combination of individual element values and sub-array values. The element values and sub-array values are joined together to form a complete array value. For example, given the declarations

```
type byte is array (7 downto 0) of bit;
variable d_reg : byte;
variable a, b : bit;
```

we could assign an aggregate value as follows:

```
d_reg := (a, "1001", b, "00");
```

The aggregate represents a value consisting of an element taken from the bit variable **a**, followed by a sub-array of the four bits, an element taken from the bit variable **b**, and a further sub-array of two bits. We could also write this using named association:

```
d_reg := (7 => a, 6 downto 3 => "1001",
          2 => b, 1 downto 0 => "00");
```

Note that when we write an aggregate containing a sub-array using named association, the choice for the sub-array must take the form of a discrete range, the number of choice values in the range must be the same as the number of elements in the sub-array, and the direction of the range must match the context in which the aggregate appears. Thus, in the preceding example, we used descending ranges for the choices, since the aggregate is assigned to a variable with a descending range.

Another place in which we may use an aggregate is the target of a variable assignment or a signal assignment. The full syntax rule for a variable assignment statement is

variable_assignment_statement ⇐
    〖 label : 〗〘 name 〙 aggregate 〗 := expression ;

Aggregate target names can also be used in the conditional and selected forms of variable assignments. If the target is an aggregate, it must contain a variable name at each position. Furthermore, the expression on the right-hand side of the assignment must produce a composite value of the same type as the target aggregate. Each element of the right-hand side is assigned to the corresponding variable in the target aggregate. The variable names in the target aggregate can represent a combination of array elements and sub-arrays. For a sub-array variable, the corresponding elements of the right-hand side are assigned to the variable elements.

The full syntax rule for a signal assignment also allows the target to be in the form of an aggregate, with a signal name at each position in the aggregate. We can use assignments of this form to split a composite value among a number of scalar signals. For example, if we have a variable **flag_reg**, which is a four-element bit vector, we can perform the following signal assignment to four signals of type **bit**:

```
( z_flag, n_flag, v_flag, c_flag ) <= flag_reg;
```

Since the right-hand side is a bit vector, the target is taken as a bit-vector aggregate. The leftmost element of **flag_reg** is assigned to **z_flag**, the second element of **flag_reg** is assigned to **n_flag**, and so on. This form of multiple assignment is much more compact to write than four separate assignment statements.

As another example, suppose we have signals declared as follows:

```
signal status_reg : bit_vector(7 downto 0);
signal int_priority, cpu_priority : bit_vector(2 downto 0);
signal int_enable, cpu_mode : bit;
```

where the type **bit_vector** is declared to be an array of bit elements (see Section 4.2). We can then write the assignment:

```
(2 downto 0 => int_priority,
 6 downto 4 => cpu_priority,
 3 => int_en, 7 => cpu_mode) <= status_reg;
```

This specifies that the bits of the **status_reg** value are assigned in left-to-right order to **cpu_mode**, **cpu_priority**, **int_enable**, and **int_priority**, respectively.

VHDL-87, -93, and -2002

These earlier versions to not allow aggregates with sub-arrays. Instead, each value in an aggregate must be an individual element value. Similarly, where an aggregate of names is used as the target of an assignment, each name must be an object of the same type as the elements of the right-hand side expression.

4.1.3 Array Attributes

In Chapter 2 we saw that attributes could be used to refer to information about scalar types. There are also attributes applicable to array types; they refer to information about the index ranges. Array attributes can also be applied to array objects, such as constants, variables and signals, to refer to information about the types of the objects. Given some array type or object A, and an integer N between 1 and the number of dimensions of A, VHDL defines the following attributes:

A'left(N)	Left bound of index range of dimension N of A
A'right(N)	Right bound of index range of dimension N of A
A'low(N)	Lower bound of index range of dimension N of A
A'high(N)	Upper bound of index range of dimension N of A
A'range(N)	Index range of dimension N of A
A'reverse_range(N)	Reverse of index range of dimension N of A
A'length(N)	Length of index range of dimension N of A
A'ascending(N)	true if index range of dimension N of A is an ascending range, false otherwise

A'element   The element subtype of A

For example, given the array declaration

type A **is array** (1 **to** 4, 31 **downto** 0) **of** boolean;

some attribute values are

A'left(1) = 1   A'low(1) = 1
A'right(2) = 0   A'high(2) = 31

A'range(1) *is* 1 **to** 4   A'reverse_range(2) *is* 0 **to** 31

A'length(1) = 4   A'length(2) = 32

A'ascending(1) = true   A'ascending(2) = false

A'element *is* boolean

For all of these attributes (except **'element**), to refer to the first dimension (or if there is only one dimension), we can omit the dimension number in parentheses, for example:

A'low = 1   A'length = 4

In the next section, we see how these array attributes may be used to deal with array ports. We will also see, in Chapter 6, how they may be used with subprogram parameters that are arrays. Another major use is in writing for loops to iterate over elements of an array. For example, given an array variable **free_map** that is an array of bits, we can write a for loop to count the number of '1' bits without knowing the actual size of the array:

```
count := 0;
for index in free_map'range loop
  if free_map(index) then
    count := count + 1;
  end if;
end loop;
```

The **'range** and **'reverse_range** attributes can be used in any place in a VHDL model where a range specification is required, as an alternative to specifying the left and right bounds and the range direction. Thus, we may use the attributes in type and subtype definitions, in subtype constraints, in for loop parameter specifications, in case statement choices and so on. The advantage of taking this approach is that we can specify the size of the array in one place in the model and in all other places use array attributes. If we need to change the array size later for some reason, we need only change the model in one place.

The **'element** attribute allows us to declare objects of the same type as the elements of an array. We will see examples where that is useful in the next section.

VHDL-87, -93, and -2002

The array attribute **'element** is not provided in these earlier versions of VHDL.

VHDL-87

The array attribute **'ascending** is not provided in VHDL-87.

4.2 Unconstrained Array Types

The array types we have seen so far in this chapter are called *constrained* arrays, since the type definition constrains index values to be within a specific range. VHDL also allows us to define *unconstrained* array types, in which we just indicate the type of the index values, without specifying bounds. An unconstrained array type definition is described by the alternate syntax rule

array_type_definition ⇐
    array (⟨ type_mark **range** <> ⟩ ⟨ , ₀₀₀ ⟩) **of** *element*_subtype_indication

The symbol "<>", often called "box," can be thought of as a placeholder for the index range, to be filled in later when the type is used. An example of an unconstrained array type declaration is

```
type sample is array (natural range <>) of integer;
```

An important point to understand about unconstrained array types is that when we declare an object of such a type, we need to provide a constraint that specifies the index bounds. We can do this in several ways. One way is to provide the constraint when an object is created, for example:

```
variable short_sample_buf : sample(0 to 63);
```

This indicates that index values for the variable **short_sample_buf** are natural numbers in the ascending range 0 to 63. Another way to specify the constraint is to declare a subtype of the unconstrained array type. Objects can then be created using this subtype, for example:

```
subtype long_sample is sample(0 to 255);
variable new_sample_buf, old_sample_buf : long_sample;
```

These are both examples of a new form of subtype indication that we have not yet seen. The syntax rule is

subtype_indication ⇐ type_mark ⟦ (discrete_range ⟨ , ₀₀₀ ⟩) ⟧

The type mark is the name of the unconstrained array type, and the discrete range specifications constrain the index type to a subset of values used to index array elements. Each discrete range must be of the same type as the corresponding index type.

When we declare a constant of an unconstrained array type, there is a third way in which we can provide a constraint. We can infer it from the expression used to initialize the constant. If the initialization expression is an array aggregate written using named association, the index values in the aggregate imply the index range of the constant. For example, in the constant declaration

```
constant lookup_table : sample
          := ( 1 => 23, 3 => -16, 2 => 100, 4 => 11);
```

the index range is 1 to 4.

If the expression is an aggregate using positional association, the index value of the first element is assumed to be the leftmost value in the array subtype. For example, in the constant declaration

```
constant beep_sample : sample
          := ( 127, 63, 0, -63, -127, -63, 0, 63 );
```

the index range is 0 to 7, since the index subtype is **natural**. The index direction is ascending, since **natural** is defined to be an ascending range.

4.2.1   Predefined Array Types

VHDL predefines a number of unconstrained array types. In many models, these types are sufficient to represent our data. We list the predefined array types in this section.

Strings

VHDL provides a predefined unconstrained array type called **string**, declared as

```
type string is array (positive range <>) of character;
```

In principle the index range for a constrained string may be either an ascending or descending range, with any positive integers for the index bounds. However, most applications simply use an ascending range starting from 1. For example:

```
constant LCD_display_len : positive := 20;
subtype LCD_display_string is string(1 to LCD_display_len);
variable LCD_display : LCD_display_string := (others => ' ');
```

Boolean Vectors, Integer Vectors, Real Vectors, and Time Vectors

VHDL also provides predefined unconstrained types for arrays of **boolean**, **integer**, **real**, and **time** elements, respectively. They are declared as:

```
type boolean_vector is array (natural range <>) of boolean;
type integer_vector is array (natural range <>) of integer;
type real_vector is array (natural range <>) of real;
type time_vector is array (natural range <>) of time;
```

These types can be used to represent collections of data of the respective element types. For example, a subtype representing a collection of comparator thresholds can be declared as:

```
subtype thresholds is integer_vector(15 downto 0);
```

Alternatively, we can supply the constraint for a vector when an object is declared, for example:

```
variable max_temperatures : real_vector(1 to 10);
```

The time_vector type, in particular, is useful for specifying collection of timing parameters. For example, we can declare a constant representing individual propagation delays for each of eight output bits as follows:

```
constant Tpd_result : time_vector
  := (0 to 3 => 100 ps, 4 to 7 => 150 ps);
```

VHDL-87, -93, and -2002

The types **boolean_vector**, **integer_vector**, **real_vector**, and **time_vector** are not predefined in these earlier versions of VHDL. Instead, they (or similar types) must be explicitly defined.

Bit Vectors

VHDL provides a further predefined unconstrained array type called **bit_vector**, declared as

```
type bit_vector is array (natural range <>) of bit;
```

This type can be used to represent words of data at the architectural level of modeling. For example, subtypes for representing bytes of data in a little-endian processor might be declared as

```
subtype byte is bit_vector(7 downto 0);
```

Alternatively, we can supply the constraint when an object is declared, for example:

```
variable channel_busy_register : bit_vector(1 to 4);
```

Standard-Logic Arrays

The standard-logic package **std_logic_1164** provides an unconstrained array type for vectors of standard-logic values. It is declared as

type std_ulogic_vector **is array** (natural **range** <>) **of** std_ulogic;

This type can be used in a way similar to bit vectors, but provides more detail in representing the electrical levels used in a design. We can define subtypes of the standard-logic vector type, for example:

subtype std_ulogic_word **is** std_ulogic_vector(0 **to** 31);

Or we can directly create an object of the standard-logic vector type:

signal csr_offset : std_ulogic_vector(2 **downto** 1);

String and Bit-String Literals

In Chapter 1, we saw that a string literal may be used to write a value representing a sequence of characters. We can use a string literal in place of an array aggregate for a value of type string. For example, we can initialize a string constant as follows:

constant ready_message : string := "Ready   ";

We can also use string literals for any other one-dimensional array type whose elements are of an enumeration type that includes characters. The IEEE standard-logic array type **std_ulogic_vector** is an example. Thus we could declare and initialize a variable as follows:

variable current_test : std_ulogic_vector(0 **to** 13)
          := "ZZZZZZZZZZ----";

In Chapter 1 we also saw bit-string literals as a way of writing a sequence of bit values. Bit strings can be used in place of array aggregates to write values of bit-vector types. For example, the variable **channel_busy_register** defined above may be initialized with an assignment:

channel_busy_register := b"0000";

We can also use bit-string literals for other one-dimensional array types whose elements are of an enumeration type that includes the characters '0' and '1'. Each character in the bit-string literal represents one, three or four successive elements of the array value, depending on whether the base specified in the literal is binary, octal or hexadecimal. Again, using **std_ulogic_vector** as an example type, we can write a constant declaration using a bit-string literal:

constant all_ones : std_ulogic_vector(15 **downto** 0) := X"FFFF";

VHDL-87

In VHDL-87, bit-string literals may only be used as literals for array types in which the elements are of type **bit**. The predefined type **bit_vector** is such a type. However, the standard-logic type **std_ulogic_vector** is not. We may use string literals for array types such as **std_ulogic_vector**.

4.2.2 Unconstrained Array Element Types

In the preceding examples of unconstrained array types, the elements were all of scalar subtypes. In general, arrays can have elements of almost any type, including other array types. The array element types can themselves be constrained or unconstrained. Strictly, we just use the term unconstrained to refer to an array type in which the top-level type has its index range unspecified and the element type, if an array type, is also unconstrained. For example, the type **sample** that we declared as

type sample **is array** (natural **range** <>) **of** integer;

is unconstrained, since its index range is unspecified and the element type is scalar rather than an array type. If we declare a type for a collection of samples:

type sample_set **is array** (positive **range** <>) **of** sample;

this type also is unconstrained. The top-level array type, **sample_set**, has an unspecified index range, and the element type is unconstrained. There is no constraint information specified at any level of the type's hierarchy.

We use the term *fully constrained* for a type in which all index ranges are constrained at all levels of the type's hierarchy. For example, the type that we declared for points,

type point **is array** (1 **to** 3) **of** real;

is fully constrained, since there is only one index range, and it is constrained to be 1 to 3. Similarly, if we declare a type for a line segment determined by two points:

type line_segment **is array** (1 **to** 2) **of** point;

that type is also fully constrained. It constrains the top-level index range to 1 to 2 and the element index range to 1 to 3.

In between unconstrained and fully constrained types, we have what are called *partially constrained* types. Such types have one or more index ranges unspecified and others that are constrained. For example, we can declare a type for fixed-sized collections of samples:

type dozen_samples **is array** (1 **to** 12) **of** sample;

The top-level index range is constrained to 1 to 12, but the element index range is unspecified. Similarly, we can declare a type for a path consisting of an unspecified number of points:

```
type path is array (positive range <>) of point;
```

The top-level index range is unspecified, but the element index range is 1 to 3.

We mentioned that when we declare an object of an array type, we must provide constraints for index ranges. If the type used in the declaration is fully constrained, the constraints from the type provide sufficient information. On the other hand, if the type is unconstrained or partially constrained, we need to provide the missing constraints. We saw how to do that for the simple case of an unconstrained array of scalar elements. We do something similar for an unconstrained array with array elements. For example, to declare a variable of type **sample_set** to store 100 samples each of 20 values, we can write

```
variable main_sample_set : sample_set(1 to 100)(1 to 20);
```

The first index range, 1 to 100, is used for the top level of the type's hierarchy, and the second index range, 1 to 20, is used for the second level. We can use the same type for a second object with a different size:

```
variable alt_sample_set : sample_set(1 to 50)(1 to 10);
```

This represents a collection of 50 samples, each with 10 values. Both variables are of the same type, **sample_set**, but they are of different sizes and have different element array sizes.

If we are to use a partially constrained type to declare an object, we specify constraints only for the index ranges not specified by the type. We can use the preceding notation for the case of the top-level index range being the one we have to specify. For example, we can declare a variable for a path connecting five points:

```
variable short_path : path(5 downto 1);
```

The index range 5 down to 1 is used for the top level of the type's hierarchy, and the constraint 1 to 3 for the second level comes from the type itelf.

For the case in which we need to specify only a second-level index range, we need to use the reserved word **open** in place of the top-level index range. An example is

```
variable bakers_samples : dozen_samples(open)(0 to 9);
```

In this example, we write **open** for the top-level index range, since the type itself specifies 1 to 12 for that index range. Using the word **open** allows us to advance to the second-level index range, where we specify 0 to 9 as the constraint.

We can use this notation in subtype declarations as well as in declarations of objects. For example, if we want to declare several objects with the same constraints, we could write the subtype declaration

```
subtype dozen_short_samples is dozen_samples(open)(0 to 9);
```

Since this subtype is fully constrained, we can use it to declare objects without adding any further constraint information.

When we declare a subtype, we do not need to constrain all of the index ranges to make a fully constrained subtype. We can add index constraints for some of the index ranges to declare partially constrained subtypes. For example:

subtype short_sample_set **is** sample_set(**open**)(0 **to** 9);

In this case, we use the word **open** to skip over the unspecified index range at the top level and add a constraint at the second level. The top-level index range remains unspecified, and so the subtype is partially constrained. When using this subtype to declare an object, we would have to specify the top-level index range for the object, for example:

variable unprocessed_samples : short_sample_set(1 **to** 20);

We mentioned earlier that, when declaring a constant of an unconstrained array type, we can infer the index range of the constant from the value of the expression used to initialize the constant. That applies not only to unconstrained arrays of scalar elements, but also to unconstrained arrays of array elements and to partially constrained arrays. For example, if we write:

```
constant default_sample_pair : sample_set
        := ( 1 => (39, 25, 6, -4, 5),
             2 => (9, 52, 100, 0, -1),
             3 => (0, 0, 0, 0, 0) );
```

the top-level index range is inferred to be 1 to 3 (from the choices in the outer level of the aggregate), and the second-level index range is inferred to be 0 to 4 (from the fact that the index subtype of **sample** is natural). Similarly, if we write:

```
constant default_samples : short_sample_set
        := ( 1 to 20 => (0, 0, 0, 0, 0, 0, 0, 0, 0, 0) );
```

the top-level index range is inferred to be 1 to 20 (from the choice in the outer level of the aggregate), but the second-level index range is 0 to 9, specified in the subtype.

4.2.3 Unconstrained Array Ports

An important use of an unconstrained array type is to specify the type of an array port. This use allows us to write an entity interface in a general way, so that it can connect to array signals of any size or with any range of index values. When we instantiate the entity, the index bounds of the array signal connected to the port are used as the bounds of the port. If the port type is an unconstrained array of unconstrained array elements, the index ranges at all levels of the type's hierarchy come from the connected signal.

EXAMPLE 4.4 *Multiple-input and gates*

Suppose we wish to model a family of and gates, each with a different number of inputs. We declare the entity interface as shown below. The input port is of the unconstrained type **bit_vector**. The architecture body includes a process that is sensitive to changes on the input port. When any element changes, the process performs a logical and operation across the input array. It uses the 'range attribute to determine the index range of the array, since the index range is not known until the entity is instantiated.

```
entity and_multiple is
  port ( i : in bit_vector;  y : out bit );
end entity and_multiple;

--------------------------------------------------

architecture behavioral of and_multiple is
begin

  and_reducer : process ( i ) is
    variable result : bit;
  begin
    result := '1';
    for index in i'range loop
      result := result and i(index);
    end loop;
    y <= result;
  end process and_reducer;

end architecture behavioral;
```

To illustrate the use of the multiple-input gate entity, suppose we have the following signals:

```
signal count_value : bit_vector(7 downto 0);
signal terminal_count : bit;
```

We instantiate the entity, connecting its input port to the bit-vector signal:

```
tc_gate : entity work.and_multiple(behavioral)
  port map ( i => count_value, y => terminal_count);
```

For this instance, the input port is constrained by the index range of the signal. The instance acts as an eight-input and gate.

EXAMPLE 4.5   *Multiple-input and-or-invert gates*

We can model a family of and-or-invert gates with varying numbers of groups of input bits. Each group of input bits is and-ed, the results are or-ed, and the final result negated to derive the output. The inputs are represented using an array of arrays:

```
type bv_array is array (natural range <>) of bit_vector;
```

Each array element of type **bit_vector** is a group of bits that are to be and-ed together. The entity and architecture are:

```vhdl
entity and_or_invert is
  port ( a : in bv_array; z : out bit );
end entity and_or_invert;

architecture behavioral of and_or_invert is

  begin
  reducer : process ( a ) is
    variable and_result, result : bit;
  begin
    result := '0';
    for i in a'range loop
      and_result := '1';
      for j in a'element'range loop
        and_result := and_result and a(i)(j);
      end loop;
      result := result or and_result;
    end loop;
    z <= not result;
  end process reducer;

end architecture behavioral;
```

The **reducer** process contains an outer loop that iterates over the array of groups. It uses the input array's top-level index range to govern the loop parameter. The inner loop iterates over the bits in a group represented by a particular element of **a**. This loop uses the index range of the elements of **a**, given by the 'range attribute of the 'element attribute of **a**. Note that we could have used the reducing **and** operator, described in Section 4.3.1, in place of the inner loop, as follows:

```vhdl
and_result := and a(i);
```

For this example, however, we have included the loop to illustrate use of the 'element attribute in cases where the index ranges of array elements are determined by the signal connected to the port of an instance.

We can also use partially constrained types for ports of entities. In such cases, the index ranges of the signal connected to the port are used to determine the those bounds of the port that are not specified in the port's subtype. For example, if we declare an entity using the subtype **short_sample_set** subtype from Section 4.2.2:

```vhdl
entity sample_processor is
  port ( samples : in short_sample_set; ... );
end entity sample_processor;
```

and instantiate it as follows:

```vhdl
signal sensor_samples : short_sample_set(1 to 3);
...
```

```
preprocessor : entity "work." sample_processor
  port map ( samples => sensor_samples, ... );
```

then, for the **preprocessor** instance, the top-level index range of the samples port would be 1 to 3, obtained from the **samples** signal. The second-level index range for the **samples** port would be 0 to 9, obtained from the subtype **short_sample_set**, the declared subtype of the port.

4.3   Array Operations and Referencing

Although an array is a collection of values, much of the time we operate on arrays one element at a time, using the operators described in Chapter 2. However, if we are working with one-dimensional arrays of scalar values, we can use some of the operators to operate on whole arrays. In this section, we describe the operators that can be applied to arrays and introduce a number of other ways in which arrays can be referenced.

4.3.1   Logical Operators

The logical operators (**and**, **or**, **nand**, **nor**, **xor** and **xnor**) can be applied to two one-dimensional arrays of bit or Boolean elements. That includes the predefined types **bit_vector** and **boolean** vector, as well as any array types with **bit** or **boolean** elements that we might declare. The operators can also be applied to two arrays of type **std_ulogic_vector** defined by the **std_logic_1164** package. In each case, the operands must be of the same length and type, and the result is computed by applying the operator to matching elements from each array to produce an array of the same length. Elements are matched starting from the leftmost position in each array. An element at a given position from the left in one array is matched with the element at the same position from the left in the other array. The operator **not** can also be applied to a single array of any of these types, with the result being an array of the same length and type as the operand. The following declarations and statements illustrate this use of logical operators when applied to bit vectors:

```
subtype pixel_row is bit_vector (0 to 15);
variable current_row, mask : pixel_row;

current_row := current_row and not mask;
current_row := current_row xor X"FFFF";
```

The logical operators **and**, **or**, **nand**, **nor**, **xor** and **xnor** can also be applied to a pair of operands, one of which is a one-dimensional arrays of the types mentioned above, and the other of which is a scalar of the same type as the array elements. The result is an array of the same length and type as the array operand. Each element of the result is computed by applying the operator to the scalar operand and the corresponding element of the array operand. Some examples are

```
B"0011" and '1' = B"0011"      B"0011" and '0' = B"0000"
B"0011" xor '1' = B"1100"      B"0011" xor '0' = B"0011"
```

EXAMPLE 4.6 *Select signal for a data bus*

Suppose the outputs of three registers are provided on std_ulogic_vector signals a, b, and c. We can use three select signals, a_sel, b_sel and c_sel, to select among the register outputs for assignment to a data bus signal. The declarations and statements are

```
signal a, b, c, data_bus : std_ulogic_vector(31 downto 0);
signal a_sel, b_sel, c_sel : std_ulogic;
...

data_bus <= (a and a_sel) or (b and b_sel) or (c and c_sel);
```

Further, these operators can be applied in unary form to a single one-dimensional array of the types mentioned above, reducing the array to a single scalar result of the same type as the array elements. The reduction **and**, **or**, and **xor** operators form the logical and, or, and exclusive or, respectively of the array elements. Thus:

```
and "0110" = '0' and '1' and '1' and '0' = '0'

 or "0110" = '0'  or '1'  or '1'  or '0' = '1'

xor "0110" = '0' xor '1' xor '1' xor '0' = '0'
```

In each case, if the array has only one element, the result is the value of that element. If the array is a null array (that is, it has no elements), the result of the **and** operator is '1', and the result of the **or** and xor operators is '0'.

The reduction **nand**, **nor**, and **xnor** operators are the negation of the reduction **and**, **or**, and **xor** operators, respectively. Thus:

```
nand "0110" = not ('0' and '1' and '1' and '0') = not '0' = '1'

 nor "0110" = not ('0'  or '1'  or '1'  or '0') = not '1' = '0'

xnor "0110" = not ('0' xor '1' xor '1' xor '0') = not '0' = '1'
```

In each case, application to a single-element array produces the negation of the element value. Application of **nand** to a null array produces '0' and application of **nor** or **xnor** to a null array produces '1'.

The logical reduction operators have the same precedence as the unary **not** and **abs** operators. In the absence of parentheses, they are evaluated before binary operators. So the expression:

and A **or** B

involves applying the reduction **and** operator to A, then applying the binary **or** operator to the result and B. In some cases, we need to include parentheses to make an expression legal. For example, the expression:

and not X

is not legal without parentheses, since we cannot chain unary operators. Instead, we must write the expression as:

```
and (not X)
```

EXAMPLE 4.7 *Parity reduction*

Given a signal **data** of type bit_vector, we can calculate parity using the reduction **xor** operator:

```
parity <= xor data;
```

Since reduction operators have higher precedence than binary logical operators, the following two statements produce the same value:

```
parity_error1 <= (xor data) and received_parity;
parity_error2 <= xor data and received_parity;
```

However, the parentheses make the first form more readable.

VHDL-87

The logical operator **xnor** is not provided in VHDL-87.

4.3.2   Shift Operators

The shift operators introduced in Chapter 2 (**sll**, **srl**, **sla**, **sra**, **rol** and **ror**) can be used with a one-dimensional array of bit or Boolean values as the left operand and an integer value as the right operand. The **sll**, **srl**, **rol** and **ror** operators can be used with a left operand of type **std_ulogic_vector**, defined in the **std_logic_1164** package. A shift-left logical operation shifts the elements in the array n places to the left (n being the right operand), filling in the vacated positions with '0' or **false** and discarding the leftmost n elements. If n is negative, the elements are instead shifted to the right. Some examples are

```
B"10001010" sll 3 = B"01010000"     B"10001010" sll -2 = B"00100010"
```

The shift-right logical operation similarly shifts elements n positions to the right for positive n, or to the left for negative n, for example:

```
B"10010111" srl 2 = B"00100101"     B"10010111" srl -6 = B"11000000"
```

The next two shift operations, shift-left arithmetic and shift-right arithmetic, operate similarly, but instead of filling vacated positions with '0' or **false**, they fill them with a copy of the element at the end being vacated, for example:

```
B"01001011" sra 3 = B"00001001"     B"10010111" sra 3 = B"11110010"
B"00001100" sla 2 = B"00110000"     B"00010001" sla 2 = B"01000111"
```

As with the logical shifts, if *n* is negative, the shifts work in the opposite direction, for example:

B"00010001" **sra** -2 = B"01000111"   B"00110000" **sla** -2 = B"00001100"

A rotate-left operation moves the elements of the array *n* places to the left, transferring the *n* elements from the left end of the array around to the vacated positions at the right end. A rotate-right operation does the same, but in the opposite direction. As with the shift operations, a negative right argument reverses the direction of rotation. Some examples are

B"10010011" **rol** 1 = B"00100111"   B"10010011" **ror** 1 = B"11001001"

VHDL-87

The shift operators **sll**, **srl**, **sla**, **sra**, **rol** and **ror** are not provided in VHDL-87.

4.3.3 Relational Operators

Relational operators can also be applied to one-dimensional arrays. For the ordinary relation operators ("=", "/=", "<", "<=", ">" and "=>"), the array elements can be of any discrete type. The two operands need not be of the same length, so long as they have the same element type. The "=" and "/=" operators are quite straightforward. If two arrays have the same length and the corresponding elements are pairwise equal, then the arrays are equal. The "/=" operator is simply the negation of the "=" operator. The way the remaining operators work can be most easily seen when they are applied to strings of characters, in which case they are compared according to case-sensitive dictionary ordering.

To see how dictionary comparison can be generalized to one-dimensional arrays of other element types, let us consider the "<" operator applied to two arrays **a** and **b**. If both **a** and **b** have length 0, **a** < **b** is false. If **a** has length 0, and **b** has non-zero length, then **a** < **b**. Alternatively, if both **a** and **b** have non-zero length, then **a** < **b** if a(1) < b(1), or if a(1) = b(1) and the rest of **a** < the rest of **b**. In the remaining case, where **a** has non-zero length and **b** has length 0, **a** < **b** is false. Comparison using the other relational operators is performed analogously.

An important point that follows from this definition of "<" ordering is that, if we apply it to bit vectors representing binary-coded numbers, it does not correspond to numeric ordering. For example, consider the comparison B"001000" < B"10". Since the first element of the left operand is '0' and the first element of the right operand is '1', and since '0' < '1', the result of the comparison is true. However, if we interpret the vectors as unsigned binary numbers, the left operand represents 8 and the right operand represents 2. We would then expect the comparison to yield false. Clearly, the predefined relational operators are inappropriate for this interpretation of bit vectors. We will see in Chapter 9 how we can perform the right kind of comparisons on binary-coded numbers.

In Section 2.2.5, we introduced the matching equality ("?=") and inequality ("?/=") operators for comparing **bit** or **std_ulogic** operands. These operators can also be applied to operands that are one-dimensional arrays of **bit** or **std_ulogic** elements to yield a std_ulogic result. The operands must be of the same type and length. For the "?=" operator, corresponding elements are compared using the scalar version of the operator, and the results reduced using the logical **and** reduction operator. The "?/=" operator is computed similarly, with the **not** operator applied to the reduced **and** result.

EXAMPLE 4.8   *Chip-select and address decoding*

We can write a Boolean equation for a **std_ulogic** select signal that includes chip-select control signals and an address signal. We can use the "?=" operator, which returns a **std_ulogic** result. We can combine that result with the **std_ulogic** control signals to produce a **std_ulogic** form of the Boolean equation:

```
dev_sel1 <= cs1 and not ncs2 and addr ?= X"A5";
```

We can also use this form of expression in a condition, since the value is implictly converted to **boolean** (see Section 2.2.5):

```
if cs1 and not ncs2 and addr ?= X"A5" then
   . . .
```

or similarly:

```
if cs1 and ncs2 ?= '0' and addr ?= X"A5" then
   . . .
```

VHDL-87, -93, and -2002

The matching relational operators are not provided in these versions of VHDL. The assignment in the above example would have to be written as:

```
dev_sel1 <= '1' when cs1 = '1' and
                     ncs2 = '0' and addr = X"A5" else '0';
```

Similarly, since condition values are not implicitly converted to **boolean**, the conditions in the if statements would be written as:

```
if cs1 = '1' and ncs2 = '0' and addr = X"A5" then
   . . .
```

or

```
if (cs1 and not ncs2) = '1' and addr = X"A5" then
   . . .
```

Maximum and Minimum Operations

In Chapter 2, we introduced the **maximum** and **minimum** operations for scalar types. These operations can also be applied to arrays of discrete-type elements. The "<" operator is defined for such arrays, and the **maximum** and **minimum** operations are defined in terms of the "<" operator. Thus, for example:

```
minimum(string'(B"0001"), string'(B"0110")) = B"0001"
maximum(string'(B"001000"), string'(B"10")) = B"10"
```

Note that the same argument that we made above about comparing binary-coded numbers using the "<" operator applies to use of the maximum and minimum operations. Again, we will see in Chapter 9 how to perform the operation correctly for binary-coded numbers.

The **maximum** and **minimum** operations are further defined as reduction operations on one-dimensional arrays of any scalar type. The **maximum** function of this form returns the largest element in the array, and the **minimum** function returns the smallest element in the array. Again, the comparisons are performed using the predefined "<" operator for the element type. Thus,

```
maximum(string'("WXYZ")) = 'Z'

minimum(string'("WXYZ")) = 'W'

maximum(time_vector'(10 ns, 50 ns, 20 ns)) = 50 ns

minimum(time_vector'(10 ns, 50 ns, 20 ns)) = 10 ns
```

For a null array (one with no elements), the **maximum** function returns the smallest value of the element type, and the **minimum** function returns the largest value of the element type.

VHDL-87, -93, and -2002

The **maximum** and **minimum** operations are not provided in earlier versions of VHDL.

4.3.4 The Concatenation Operator

The one remaining operator that can be applied to one-dimensional arrays is the concatenation operator ("&"), which joins two array values end to end. For example, when applied to bit vectors, it produces a new bit vector with length equal to the sum of the lengths of the two operands. Thus, b"0000" & b"1111" produces b"0000_1111".

The concatenation operator can be applied to two operands, one of which is an array and the other of which is a single scalar element. It can also be applied to two scalar values to produce an array of length 2. Some examples are

```
"abc" & 'd'   =  "abcd"
'w'   & "xyz"  =  "wxyz"
'a'   & 'b'    =  "ab"
```

4.3.5 **To_String** Operations

In Section 2.5, we described the predefined **to_string** operation on scalar types. **To_string** can also be applied to values of one-dimensional array types that contain only character-literal elements. Examples of such types include the predefined types **bit_vector** and **string** and the type **std_ulogic_vector** defined in std_logic_1164. Applying **to_string** to a **string** value is not of much use, since it just yields the operand unchanged. For other array types, the operation yields a **string** value with the same characters as the operand. This can be useful for including such array values in message strings. For example:

```
signal x : bit_vector(7 downto 0);
...

report "Trace: x = " & to_string(x);
```

The value of **x** is an array of **bit** elements, whereas the report statement expects a **string** value, which is an array of **character** elements. The **to_string** operation deals with the conversion. Thus, if **x** has the **bit_vector** value "00110101", the result of the **to_string** operation would be the character string "00110101".

VHDL also provides operations, **to_ostring** and **to_hstring**, for converting **bit_vector** operands to strings in octal and hexadecimal form, respectively. **To_ostring** takes each group of three bits from the operand, starting from the right, and includes the corresponding octal-digit character in the result. If the operand is not a multiple of three in length, additional '0' bits are assumed on the left of the operand. Some examples are:

```
to_ostring(B"101_011_000") = "530"
to_ostring( B"11_000_111") = "307"
```

The **to_hstring** operation similarly takes each group of four bits and includes the corresponding hexadecimal-digit character in the result. The digits A to F appear in uppercase. Some examples are:

```
to_hstring(B"0110_1100") = "6C"
to_hstring(   B"11_0101") = "35"
```

Note that we don't necessarily need a separate **to_bstring** operation, since the **to_string** operation would serve that purpose. However, in the interest of completeness and consistency, VHDL does provide a **to_bstring** operation with exactly the same behavior. (In fact, **to_bstring** is an alias for **to_string**. We discuss aliases in detail in Chapter 11.) In addition, VHDL provides alternate names for all of these operations: **to_binary_string**, **to_octal_string** and **to_hex_string**. Some designers may consider these to be more readable than the shorter names. Their use is a matter of taste or organizational coding style.

4.3.6 **Array Slices**

Often we want to refer to a contiguous subset of elements of an array, but not the whole array. We can do this using *slice* notation, in which we specify the left and right index values of part of an array object. For example, given arrays **a1** and **a2** declared as follows:

```
type array1 is array (1 to 100) of integer;
type array2 is array (100 downto 1) of integer;

variable a1 : array1;
variable a2 : array2;
```

we can refer to the array slice a1(11 **to** 20), which is an array of 10 elements having the indices 11 to 20. Similarly, the slice a2(50 **downto** 41) is an array of 10 elements but with a descending index range. Note that the slices a1(10 **to** 1) and a2(1 **downto** 10) are null slices, since the index ranges specified are null. Furthermore, the ranges specified in the slice must have the same direction as the original array. Thus we may not legally write a1(10 **downto** 1) or a2(1 **to** 10).

EXAMPLE 4.9   *A byte-swapper module*

We can write a behavioral model for a byte-swapper that has one input port and one output port, each of which is a bit vector of subtype **halfword**, declared as follows:

```
subtype halfword is bit_vector(0 to 15);
```

The entity and architecture are:

```
entity byte_swap is
  port (input : in halfword;  output : out halfword);
end entity byte_swap;

-------------------------------------------------

architecture behavior of byte_swap is

begin

  swap : process (input)
  begin
    output(8 to 15) <= input(0 to 7);
    output(0 to 7) <= input(8 to 15);
  end process swap;

end architecture behavior;
```

The process in the architecture body swaps the two bytes of input with each other. It shows how the slice notation can be used for signal array objects in signal assignment statements.

VHDL-87

In VHDL-87, the range specified in a slice may have the opposite direction to that of the index range of the array. In this case, the slice is a null slice.

4.3.7 Array Type Conversions

In Chapter 2 we introduced the idea of type conversion of a numeric value of one type to a value of a so-called *closely related* type. A value of an array type can also be converted to a value of another array type, provided the array types are closely related. Two array types are closely related if they have the same number of dimensions and their element types can be converted (that is, the element types are closely related). The type conversion simply produces a new array value of the specified type containing the converted elements of the original array in the same order.

To illustrate the idea of type-converting array values, suppose we have the following declarations in a model:

```
subtype name is string(1 to 20);
type display_string is array (integer range 0 to 19) of character;

variable item_name : name;
variable display : display_string;
```

We cannot directly assign the value of **item_name** to **display**, since the types are different. However, we can write the assignment using a type conversion:

```
display := display_string(item_name);
```

This produces a new array, with the left element having index 0 and the right element having index 19, which is compatible with the assignment target.

The rule that the element types of the converted expression and the target type must be convertible allows us to perform the following conversions:

```
subtype integer_vector_10 is integer_vector(1 to 10);
subtype real_vector_10 is real_vector(1 to 10);

variable i_vec : integer_vector_10;
variable r_vec : real_vector_10;
...

i_vec := integer_vector_10(r_vec);
r_vec := real_vector_10(i_vec);
```

Since we can convert between values of type **integer** and **real**, we can also convert between arrays with **integer** and **real** elements, respectively. Each element of the converted expression is converted to the element subtype of the target type.

The index ranges need not all be numeric for a type conversion to be performed. For example, suppose we have array types and signals declared as follows:

```
type exception_type is (int, ovf, div0, undef, trap);
type exception_vector is array (exception_type) of bit;
signal d_in, d_out: bit_vector(31 downto 0);
signal exception_reg : exception_vector;
```

Then the type conversion:

```
exception_vector( d_in(4 downto 0) )
```

yields a vector of bits indexed from **int** to **trap**, with each element being the matching element of the slice of **d_in**, from left to right. Since the element types for the expression and the target type are both **bit**, conversion of the elements is trivial.

The above examples illustrate the case of a type conversion in which the target type is fully constrained, specifying the index range for the result. In general, we can convert to a target type that is fully constrained, partially constrained, or unconstrained. (We described array index constraints in Section 4.2.2.) If the target type of the conversion specifies an index range at any level of the type's hierarchy, that index range is used, as in the examples. In this case, for each dimension at that level of the hierarchy, the array being converted must a matching element for each target index value. On the other hand, if the target type leaves some index range or ranges unspecified (that is, the target type is unconstrained or partially constrained), the correponding index range or ranges for the result depend on the index subtypes. If the index subtypes of the original array and the target type are both numeric, the index range of the result has the same numeric bounds and direction as the index range of the original array. For example, in the type conversion

```
integer_vector(r_vec)
```

the index range of the result is 1 to 10, taken from the index range of **r_vec**. If, however, one or both index ranges are enumeration types, the index range of the result is determined from the index subtype of the target type. The direction is the same as that of the index subtype, the left bound is the leftmost value in the index subtype, and the right bound depends on the number of elements. For example, in the type conversion

```
bit_vector( exception_reg )
```

the index range of the result comes from the index subtype defined for **bit_vector**, namely, **natural**. The subtype **natural** is declared to be an ascending range with a left bound of 0. This direction and left bound are used as the direction and left bound of the type-conversion result. The right bound comes from the number of elements. Thus, the result is a **bit_vector** value indexed from 0 to 4.

A common case in which we do not need a type conversion is the assignment of an array value of one subtype to an array signal or variable of a different subtype of the same base type. This occurs where the index ranges of the target and the operand have different bounds or directions. VHDL automatically includes an implicit subtype conversion in the assignment. For example, given the subtypes and variables declared thus:

```
subtype big_endian_upper_halfword is bit_vector(0 to 15);
subtype little_endian_upper_halfword is bit_vector(31 downto 16);

variable big : big_endian_upper_halfword;
variable little : little_endian_upper_halfword;
```

we could make the following assignments without including explicit type conversions:

```
big := little;
little := big;
```

A final point to make about conversions relates to type qualification, introduced in Section 2.3.2. There, we mentioned that the operand of a type qualification has to be of the specified type, but not necessarily of the specific subtype. In the case of type qualification for arrays, we can qualify an array operand as being of some different subtype to that of the operand. The operand is then converted to that subtype, as described above.

VHDL-87, -93, and -2002

The rules for array type conversions in these earlier versions of VHDL were more strict. Array types were only closely related if they had the same element type, the same number of dimensions and index types that could be type converted. The restriction on index subtypes effectively meant that both had to be numeric or both had to be the same enumeration type. In addition, for type qualification, no subtype conversion was performed. If the qualified type was a constrained array subtype, the operand had to have exactly the index range or ranges specified in the subtype.

4.3.8　Arrays in Case Statements

In Section 3.2, we introduced case statements, and described a number of requirements on the selector expression and choices. In our examples, we just used scalar types for the selector expression. However, VHDL also allows us to write a selector expression of a one-dimensional character array type, that is, an array type whose element type includes character literals. Examples of such types are **bit_vector**, **std_ulogic_vector**, and similar types. The choices are typically string literals, bit-string literals, or, less commonly, aggregates, and all must have the same length. When the case statement is executed, the value of the selector expression is checked to ensure it has the same length as the choices, and then compared with the choices to determine which alternative to execute. When the case statement is executed, the value of the expression must have the same length as the choices. For example, in the following:

```
variable s : bit_vector(3 downto 0);
variable c : bit;
. . .
case c & s is
  when "00000" => . . .
  when "10000" => . . .
  when others  => . . .
end case;
```

all of the choices (except the **others** choice) are of length five, so that determines the required length for the result of the concatenation.

We saw in Section 4.1.2 that we can include the word others in an array aggregate to refer to all elements not identified in the preceding part of the aggregate. If we write an aggregate containing **others** as a choice in a case statement, the index range of the case expression must be locally static, that is, determined during the analysis phase of design processing. For example, we can write a case statement as follows:

```
variable s : bit_vector(3 downto 0);
...

case s is
  when ('0', others => '1') => ...
  when ('1', others => '0') => ...
  ...
end case;
```

In this example, the index range of the expression **s** can be determined at analysis time as being 3 **downto** 0. That means the analyzer can use the index range for the choice values. If the analyzer cannot work out the index range for the case expression, it cannot determine the index values represented by **others** in the aggregates.

VHDL-87, -93, and -2002

In these versions of VHDL, if the selector expression in a case statement was of an array type, the index range had to be locally static, regardless of whether the choices used the word **others** or not. Thus, the expression **c & s** in the example above would be illegal. Instead, we would have to write the example as:

```
variable s : bit_vector(3 downto 0);
variable c : bit;
subtype bv5 is bit_vector(0 to 4);
...

case bv5'(c & s) is
  ...
end case;
```

4.3.9   Matching Case Statements

In the previous subsection, we saw how we can write a case statement with a selector expression of an array type. The choices are compared for exact equality with the expression value to select an alternative to execute. If the type of the case expression and choices is a vector of **std_ulogic** values, exact comparison is not always what we want. In particular, we would like to be able to include don't care elements ('–') in the choices to indicate that we don't care about some elements of the selector expression when selecting an alternative. We can do this with a new form of case statement, called a *matching case statement*, that uses the predefined "?=" operator described in Section 4.3.3 to compare choice values with the expression value. We include a question mark symbol after the keyword **case**, as follows:

```
case? expression is
  . . .
end case?;
```

The most common use of a matching case statement is with an expression of a vector type whose elements are **std_ulogic** values, such as the standard type **std_ulogic_vector** defined in the **std_logic_1164** package. It also includes vector types that we might define. With a case expression of such a type, we can write choice values that include '–' elements to specify don't care matching.

EXAMPLE 4.10   *A priority arbiter*

Suppose we have vectors of request and grant values, declared as follows:

```
variable request, grant : std_ulogic_vector(0 to 3);
```

We can use a matching case statement in a priority arbiter, with request 0 having highest priority:

```
case? request is
  when "1---" => grant := "1000";
  when "01--" => grant := "0100";
  when "001-" => grant := "0010";
  when "0001" => grant := "0001";
  when others => grant := "0000";
end case?;
```

Each choice is compared with the case expression using the predefined "?=" operator. Thus, the first choice matches values "1000", "1001", "100X", "H000", and so on, and similarly for the remaining choices. This is a much more succinct way of describing the arbiter than using an ordinary case statement. Moreover, unlike a sequence of tests in an if statement, it does not imply chained decision logic.

When we use a matching case statement with a vector-type expression, the value of the expression must not include any '–' elements. (This is different from the choice values, which can include '–' elements.) The reason is that an expression value with a '–' element would match multiple choice values, making selection of an alternative ambiguous. Normally, this rule is not a problem, since we don't usually assign '–' values to signals or variables. They usually just occur in literal values for comparison and in test bench assertions.

In an ordinary case statement, we need to include choices for all possible values of the case expression. A related rule applies in a matching case statement. Each possible value of the case expression, except those that include any '–' elements, must be represented by exactly one choice. By "represented," we mean that comparison of the choice and the expression value using the "?=" operator yields '1'. Hence, our choice values would generally just include '0', '1', and '–' elements, matching with '0', 'L', '1', 'H' elements in the case expression value. We could also include 'L' and 'H' elements in a choice. However, we would not include 'U', 'X', 'W', or 'Z' choice elements, since they only ever produce 'U' or 'X' results, and so never match. As with an ordinary case statement, we can include an **others** choice to represent expression values not otherwise represented. Unlike an ordinary case statement, a choice can represent multiple expression values if it contains a '–' element.

We mentioned that a vector type including **std_ulogic** values is the most common type for a matching case statement. Less commonly, we can write a selector expression of type **std_ulogic**, **bit**, or a vector of **bit** elements (such as **bit_vector**). These are the other types for which the "?=" operator is predefined. For **std_ulogic** expressions, the choice values would typically be either '0' (matching an expression value of '0' or 'L') or '1' (matching an expression value of '1' or 'H'). We would not write a choice of '–', since that would match all expression values, preventing us from selecting distinct alternatives. For case expressions of type **bit** or a vector of **bit** elements, a matching case statement has exactly the same behavior as an ordinary case statement. VHDL allows matching case statements of this form to allow synthesizable models to be written uniformly regardless of whether **bit** or **std_ulogic** data types are used.

VHDL-87, -93, and -2002

These versions of VHDL do not provide matching case statements.

Matching Selected Variable Assignments

Selected variable assignments, introduced in Section 3.2.1, are a shorthand notation for variable assignments within case statements. There is an analogous shorthand for variable assignments within matching case statements. We simply include a "?" symbol after the **select** keyword to indicate that the implied case statement is a matching case statement instead of an ordinary case statement. The rules covering the type of the selector expression and the way in which choices are matched then apply to the selected assignment.

EXAMPLE 4.11 *A revised model for the priority arbiter*

We can rewrite the priority arbiter from Example 4.10 using a matching selected assignment as follows:

```
with request select?
  grant := "1000" when "1---",
           "0100" when "01--",
           "0010" when "001-",
           "0001" when "0001",
           "0000" when others;
```

VHDL-87, -93, and -2002

These versions of VHDL do not provide the matching selected variable assignment notation.

4.4   Records

In this section, we discuss the second class of composite types, *records*. We start with record types, and return to record natures subsequently, since there are some significant differences between them. A record is a composite value comprising elements that may be of different types from one another. Each element is identified by a name, which is unique within the record. This name is used to select the element from the record value. The syntax rule for a record type definition is

record_type_definition ⇐
   record
     (identifier ⟨ , ... ⟩ : subtype_indication ;)
     ⟨ ... ⟩
   end record ⟦ identifier ⟧

Each of the names in the identifier lists declares an element of the indicated type or subtype. Recall that the curly brackets in the syntax rule indicate that the enclosed part may be repeated indefinitely. Thus, we can include several elements of different types within the record. The identifier at the end of the record type definition, if included, must repeat the name of the record type.

VHDL-87

The record type name may not be included at the end of a record type definition in VHDL-87.

The following is an example record type declaration and variable declarations using the record type:

```
type time_stamp is record
    seconds : integer range 0 to 59;
    minutes : integer range 0 to 59;
    hours : integer range 0 to 23;
  end record time_stamp;

variable sample_time, current_time : time_stamp;
```

Whole record values can be assigned using assignment statements, for example:

```
sample_time := current_time;
```

We can also refer to an element in a record using a *selected name*, for example:

```
sample_hour := sample_time.hours;
```

In the expression on the right of the assignment symbol, the prefix before the dot names the record value, and the suffix after the dot selects the element from the record. A selected name can also be used on the left side of an assignment to identify a record element to be modified, for example:

```
current_time.seconds := clock mod 60;
```

EXAMPLE 4.12   *Representing CPU instructions and data using records*

In the early stages of designing a new instruction set for a CPU, we don't want to commit to an encoding of opcodes and operands within an instruction word. Instead we use a record type to represent the components of an instruction. We illustrate this in an outline of a system-level behavioral model of a CPU and memory that uses record types to represent instructions and data:

```
architecture system_level of computer is

  type opcodes is
          (add, sub, addu, subu, jmp, breq, brne, ld, st, ...);
  type reg_number is range 0 to 31;
  constant r0 : reg_number := 0;   constant r1 : reg_number := 1;
  ...

  type instruction is record
      opcode : opcodes;
      source_reg1, source_reg2, dest_reg : reg_number;
      displacement : integer;
    end record instruction;

  type word is record
      instr : instruction;
      data : bit_vector(31 downto 0);
    end record word;

  signal address : natural;
  signal read_word, write_word : word;
  signal mem_read, mem_write : bit := '0';
  signal mem_ready : bit := '0';

begin

  cpu : process is
    variable instr_reg : instruction;
    variable PC : natural;
    ...       -- other declarations for register file, etc.
  begin
```

```
      address <= PC;
      mem_read <= '1';
      wait until mem_ready;
      instr_reg := read_word.instr;
      mem_read <= '0';
      PC := PC + 4;
      case instr_reg.opcode is   -- execute the instruction
        ...
      end case;
    end process cpu;

    memory : process is
      subtype address_range is natural range 0 to 2**14 - 1;
      type memory_array is array (address_range) of word;
      variable store : memory_array :=
        ( 0  =>     ( ( ld, r0, r0, r2, 40 ), X"00000000" ),
          1  =>     ( ( breq, r2, r0, r0, 5 ), X"00000000" ),
          ...
          40  =>    ( ( nop, r0, r0, r0, 0 ),  X"FFFFFFFE"),
         others => ( ( nop, r0, r0, r0, 0 ),  X"00000000") );
    begin
      ...
    end process memory;

  end architecture system_level;
```

The record type **instruction** represents the information to be included in each instruction of a program and includes the opcode, source and destination register numbers and a displacement. The record type **word** represents a word stored in memory. Since a word might represent an instruction or data, elements are included in the record for both possibilities. Unlike many conventional programming languages, VHDL does not provide variant parts in record values. The record type **word** illustrates how composite data values can include elements that are themselves composite values. The signals in the model are used for the address, data and control connections between the CPU and the memory.

Within the CPU process the variable **instr_reg** represents the instruction register containing the current instruction to be executed. The process fetches a word from memory and copies the instruction element from the record into the instruction register. It then uses the opcode field of the value to determine how to execute the instruction.

The memory process contains a variable that is an array of word records representing the memory storage. The array is initialized with a program and data. Words representing instructions are initialized with a record aggregate containing an instruction record aggregate and a bit vector, which is ignored. Similarly, words representing data are initialized with an aggregate containing an instruction aggregate, which is ignored, and the bit vector of data.

4.4.1 Record Aggregates

We can use a record aggregate to write a literal value of a record type, for example, to initialize a record variable or constant. Using a record aggregate is analogous to using an array aggregate for writing a literal value of an array type (see Section 4.1.2). A record aggregate is formed by writing a list of the elements enclosed in parentheses. An aggregate using positional association lists the elements in the same order as they appear in the record type declaration. For example, given the record type time_stamp shown above, we can initialize a constant as follows:

```
constant midday : time_stamp := (0, 0, 12);
```

We can also use named association, in which we identify each element in the aggregate by its name. The order of elements identified using named association does not affect the aggregate value. The example above could be rewritten as

```
constant midday : time_stamp
            := (hours => 12, minutes => 0, seconds => 0);
```

Unlike array aggregates, we can mix positional and named association in record aggregates, provided all of the named elements follow any positional elements. We can also use the symbols "|" and **others** when writing choices. Here are some more examples, using the types instruction and time_stamp declared above:

```
constant nop_instr : instruction :=
    ( opcode => addu,
      source_reg1 | source_reg2 | dest_reg => 0,
      displacement => 0 );

variable latest_event : time_stamp
            := (others => 0); -- initially midnight
```

Note that unlike array aggregates, we can't use a range of values to identify elements in a record aggregate, since the elements are identified by names, not indexed by a discrete range.

4.4.2 Unconstrained Record Element Types

In Section 4.2.2, we showed how the element type of an array type can be unconstrained, leading to unconstrained and partially constrained array types. In a similar way, an element of a record type can be of an unconstrained or partially constrained composite type. We use the terms unconstrained, partially constrained, and fully constrained to describe a record type in an analogous way to the use of the terms for array types. Specifically, an unconstrained record type is one that has one or more elements of composite types, all of which are unconstrained. For example, the following type

```
type test_vector is record
    id : natural;
    stimulus : bit_vector;
```

```
      response : bit_vector;
    end record test_vector;
```

is unconstrained, since the elements stimulus and response are both of an unconstrained array type.

A fully constrained record type is one in which all composite elements (if any) are fully constrained. For example, the types **instruction** and **word** in Example 4.12 are both fully constrained. The type **instruction** has no composite elements, so there is no place in the type's hierarchy where a constraint is needed. The type **word** has one element, **data**, of a composite subtype, which is fully constrained.

As with array types, a partially constrained record type is one that is neither unconstrained nor fully constrained. It may have a mix of unconstrained, partially constrained, and fully constrained elements, or it may just have one or more partially constrained elements. For example, if we declare a fully constrained record type

```
    type test_times is record
        stimulus_time : time;
        response_delay : delay_length;
      end record test_times;
```

we can use this and the **test_vector** type as element types for a larger record type:

```
    type test_application is record
        test_to_apply : test_vector;
        application_times : test_times;
      end record;
```

Since the **test_to_apply** element is unconstrained and the **application_times** element is fully constrained, the **test_application** type is partially constrained.

As another example, recall the types **sample** and **dozen_samples** defined in Section 4.2.2 as

```
    type sample is array (natural range <>) of integer;
    type dozen_samples is array (1 to 12) of sample;
```

If we declare a type as follows:

```
    type analyzed_samples is record
        samples : dozen_samples;
        result : real;
      end record analyzed_samples;
```

this type is partially constrained, since it has just one composite element, **samples**, that is itself partially constrained.

We have mentioned that, when we declare an object of an array type, we must provide constraints for any index ranges that remain unspecified by the type. This rule applies to composite types in general, including record types. If any of the record elements are arrays, their index ranges must be specified. For a fully constrained record type, the type itself specifies the index ranges, if any. Thus, in Example 4.12, we were able to declare signals and variables of the fully constrained types **instruction** and **word** without supplying any further information beyond the type names.

For an unconstrained or partially constrained record type, we need to fill in any unspecified index ranges, and we need to specify which element of the type each index range constrains. The way in which we do so is illustrated by the following declaration using the unconstrained type **test_vector**:

```
variable next_test_vector : test_vector(stimulus(0 to 7),
                                        response(0 to 9));
```

For those elements of the type that we need to constrain, we write the element name followed by the index constraint. If the index range to be constrained is nested more deeply within the type's hierarchy, we can nest the constraint notation. For example, to declare a variable of type **test_application**, we could write:

```
variable scheduled_test :
        test_application(test_to_apply(stimulus(0 to 7),
                                       response(0 to 9)));
```

Since the **application_times** element of the **test_application** type is fully constrained, we do not mention it in the record constraint. As a second example, the declaration

```
variable analysis : analyzed_samples(samples(open)(1 to 100));
```

constrains the samples element of the record type. The constraint uses the word **open** to skip over the top-level index range of the element, since that index range is specified by the element type **dozen_samples**. The nested index range is constrained to be 1 to 100.

Just as we did for array types, we can use this notation to declare subtypes of record types. Some examples are:

```
subtype byte_test_vector is test_vector(stimulus(7 downto 0),
                                        response(7 downto 0));
```

```
subtype analyzed_short_samples is
        analyzed_samples(samples(open)(1 to 100));
```

We do not need to constrain every element. For example, if we write:

```
subtype test_application_word is
        test_application(test_to_apply(stimulus(0 to 31)));
```

only the **stimulus** element of the nested record element is constrained, and so the subtype **test_application_word** remains partially constrained. This is equivalent to writing

```
subtype test_application_word is
        test_application(test_to_apply(stimulus(0 to 31),
                                       response(open)));
```

We could then further constrain it as follows:

```
subtype test_application_word_byte is
        test_application_word(test_to_apply(response(0 to 7)));
```

or equivalently:

```
subtype test_application_word_byte is
        test_application_word(test_to_apply(stimulus(open),
                                            response(0 to 7)));
```

Finally, the rule that index ranges for a constant can be inferred from the initial value also applies to a constant of an unconstrained or partially constrained record type. For example, if we declare a constant as follows:

```
constant first_test_vector : test_vector
        := (id => 0,
            stimulus => B"100010",
            response => B"00000001");
```

the index ranges for the **stimulus** and **response** elements are inferred to be 0 to 5 and 0 to 7, respectively, since they are both of type **bit_vector**, which has natural as its index subtype.

Exercises

1. [● 4.1] Write an array type declaration for an array of 30 integers, and a variable declaration for a variable of the type. Write a for loop to calculate the average of the array elements.

2. [● 4.1] Write an array type declaration for an array of bit values, indexed by standard-logic values. Then write a declaration for a constant, **std_ulogic_to_bit**, of this type that maps standard-logic values to the corresponding bit value. (Assume unknown values map to '0'.) Given a standard-logic vector **v1** and a bit-vector variable **v2**, both indexed from 0 to 15, write a for loop that uses the constant **std_ulogic_to_bit** to map the standard-logic vector to the bit vector.

3. [● 4.1] The data on a diskette is arranged in 18 sectors per track, 80 tracks per side and two sides per diskette. A computer system maintains a map of free sectors. Write a three-dimensional array type declaration to represent such a map, with a '1' element representing a free sector and a '0' element representing an occupied sector. Write a set of nested for loops to scan a variable of this type to find the location of the first free sector.

4. [● 4.2] Write a declaration for a subtype of **std_ulogic_vector**, representing a byte. Declare a constant of this subtype, with each element having the value 'Z'.

5. [● 4.2] Write a type declaration for an unconstrained array of **time_vector** elements, indexed by **positive** values. Then write a subtype declaration representing an array with 4 unconstrained elements. Last, write a variable declaration using the subtype with each element having 10 subelements.

6. [❶ 4.2] Write a for loop to count the number of '1' elements in a bit-vector variable **v**.

7. [❶ 4.3] An 8-bit vector **v1** representing a two's-complement binary integer can be sign-extended into a 32-bit vector **v2** by copying it to the leftmost eight positions of **v2**, then performing an arithmetic right shift to move the eight bits to the rightmost eight positions. Write variable assignment statements that use slicing and shift operations to express this procedure.

8. [❶ 4.4] Write a record type declaration for a test stimulus record containing a stimulus bit vector of three bits, a delay value and an expected response bit vector of eight bits.

9. [❷ 4.1] Develop a model for a register file that stores 16 words of 32 bits each. The register file has data input and output ports, each of which is a 32-bit word; read-address and write-address ports, each of which is an integer in the range 0 to 15; and a write-enable port of type **bit**. The data output port reflects the content of the location whose address is given by the read-address port. When the write-enable port is '1', the input data is written to the register file at the location whose address is given by the write-address port.

10. [❷ 4.1] Develop a model for a priority encoder with a 16-element bit-vector input port, an output port of type **natural** that encodes the index of the leftmost '1' value in the input and an output of type **bit** that indicates whether any input elements are '1'.

11. [❷ 4.2] Write a package that declares an unconstrained array type whose elements are integers. Use the type in an entity declaration for a module that finds the maximum of a set of numbers. The entity has an input port of the unconstrained array type and an integer output. Develop a behavioral architecture body for the entity. How should the module behave if the actual array associated with the input port is empty (i.e., of zero length)?

12. [❷ 4.2/4.3] Develop a model for a general and-or-invert gate, with two standard-logic vector input ports **a** and **b** and a standard-logic output port **y**. The output of the gate is

$$\overline{a_0 \cdot b_0 + a_1 \cdot b_1 + \cdots + a_{n-1} \cdot b_{n-1}}$$

13. [❷ 4.4] Develop a model of a 3-to-8 decoder and a test bench to exercise the decoder. In the test bench, declare the record type that you wrote for Exercise 8 and a constant array of test record values. Initialize the array to a set of test vectors for the decoder, and use the vectors to perform the test.

Chapter 5

Basic Modeling Constructs

The description of a module in a digital system can be divided into two facets: the external view and the internal view. The external view describes the interface to the module, including the number and types of inputs and outputs. The internal view describes how the module implements its function. In VHDL, we can separate the description of a module into an *entity declaration*, which describes the external interface, and one or more *architecture bodies*, which describe alternative internal implementations. These were introduced in Chapter 1 and are discussed in detail in this chapter. We also look at how a design is processed in preparation for simulation or synthesis.

5.1 Entity Declarations and Architecture Bodies

Let us first examine the syntax rules for an entity declaration and then show some examples. We start with a simplified description of entity declarations and move on to a full description later in this chapter. The syntax rules for this simplified form of entity declaration are

entity_declaration ⟸
    entity identifier **is**
        ⟦ **port** (*port*_interface_list) ; ⟧
        ❴ entity_declarative_item ❵
    end ⟦ **entity** ⟧ ⟦ identifier ⟧ ;

interface_list ⟸
    ⟮ identifier ❴ , ₒₒₒ ❵ : ⟦ mode ⟧ subtype_indication ⟦ := expression ⟧ ⟯ ❴ ; ₒₒₒ ❵

mode ⟸ **in** ∣ **out** ∣ **buffer** ∣ **inout**

The identifier in an entity declaration names the module so that it can be referred to later. If the identifier is included at the end of the declaration, it must repeat the name of the entity. The port clause names each of the *ports*, which together form the interface to the entity. We can think of ports as being analogous to the pins of a circuit; they are the means by which information is fed into and out of the circuit. In VHDL, each port of an entity has a *type*, which specifies the kind of information that can be communicated, and a *mode*, which specifies how information flows into or out from the entity through the port. These aspects of type and direction are in keeping with the strong typing philosophy of VHDL, which helps us avoid erroneous circuit descriptions. A simple example of an entity declaration is

```
entity adder is
  port ( a : in word;
          b : in word;
          sum : out word );
end entity adder;
```

This example describes an entity named **adder**, with two input ports and one output port, all of type **word**, which we assume is defined elsewhere. We can list the ports in any order; we do not have to put inputs before outputs. Also, we can include a list of ports of the same mode and type instead of writing them out individually. Thus the above declaration could equally well be written as follows:

```
entity adder is
  port ( a, b : in word;
          sum : out word );
end entity adder;
```

In this example we have seen input and output ports. An input port allows us to model a device that senses data provided externally on a pin. An output port allows us to model a device that drives a pin to provide data to external connections. VHDL also allows us to specify the mode **buffer** for a port that not only provides data to external connections, but also provides that same data for use internally. We could achieve the same effect with an **out**-mode port, since VHDL does allow us to read the driven value of an **out**-mode port internally. However, we prefer to use an **out**-mode port when the value is read internally just for verification purposes, for example, using assertion statements. The real device we are modeling in this case does not have an internal connection to implement its functionality. For a device that does have such an internal connection, we use a **buffer** port to make our design intent clear.

As an example, suppose we want to model an SR-latch and allow for an implementation using cross-coupled gates. Since each output is also fed back internally to an input of a gate, we use buffer-mode ports for the outputs. The entity declaration is:

```
entity SR_latch is
  port ( S, R : in bit; Q, Q_n : buffer bit );
end entity SR_latch;
```

The syntax rules for entities show that we can also have bidirectional ports, with mode **inout**. These can be used to model devices that alternately sense and drive data through a pin. Such models must deal with the possibility of more than one connected device driving a given signal at the same time. VHDL provides a mechanism for this, *signal resolution*, which we will return to in Chapter 8.

The similarity between the description of a port in an entity declaration and the declaration of a variable may be apparent. This similarity is not coincidental, and we can extend the analogy by specifying a default value on a port description; for example:

```
entity and_or_inv is
  port ( a1, a2, b1, b2 : in bit := '1';
         y : out bit );
end entity and_or_inv;
```

The default value, in this case the '1' on the input ports, indicates the value each port should assume if it is left unconnected in an enclosing model. We can think of it as describing the value that the port "floats to." On the other hand, if the port is used, the default value is ignored. We say more about use of default values when we look at the execution of a model.

Another point to note about entity declarations is that the port clause is optional. So we can write an entity declaration such as

```
entity top_level is
end entity top_level;
```

which describes a completely self-contained module. As the name in this example implies, this kind of module usually represents the top level of a design hierarchy.

Finally, if we return to the first syntax rule on page 137, we see that we can include declarations of items within an entity declaration. These include declarations of constants, types, signals and other kinds of items that we will see later in this chapter. The items can be used in all architecture bodies corresponding to the entity. Thus, it makes sense to include declarations that are relevant to the entity and all possible implementations. Anything that is part of only one particular implementation should instead be declared within the corresponding architecture body.

EXAMPLE 5.1 *A ROM entity including the ROM contents*

Suppose we are designing an embedded controller using a microprocessor with a program stored in a read-only memory (ROM). The program to be stored in the ROM is fixed, but we still need to model the ROM at different levels of detail. We can include declarations that describe the program in the entity declaration for the ROM:

```
entity program_ROM is
  port ( address : in std_ulogic_vector(14 downto 0);
         data : out std_ulogic_vector(7 downto 0);
         enable : in std_ulogic );

  subtype instruction_byte is bit_vector(7 downto 0);
  type program_array is
          array (0 to 2**14 - 1) of instruction_byte;
  constant program : program_array
    := ( X"32", X"3F", X"03",    -- LDA  $3F03
         X"71", X"23",           -- BLT  $23
```

```
        . . .
    );

end entity program_ROM;
```

The declarations within the ROM entity are not directly accessible to a user of the entity, but serve to document the contents of the ROM. Each architecture body corresponding to the entity can use the constant program to initialize whatever structure it uses internally to implement the ROM.

The internal operation of a module is described by an architecture body. An architecture body generally applies some operations to values on input ports, generating values to be assigned to output ports. The operations can be described either by processes, which contain sequential statements operating on values, or by a collection of components representing sub-circuits. Where the operation requires generation of intermediate values, these can be described using *signals*, analogous to the internal wires of a module. The syntax rule for architecture bodies shows the general outline:

architecture_body ⇐
    architecture identifier **of** *entity*_name **is**
        ⟦ block_declarative_item ⟧
    begin
        ⟦ concurrent_statement ⟧
    end ⟦ **architecture** ⟧ ⟦ identifier ⟧ ;

The identifier names this particular architecture body, and the entity name specifies which module has its operation described by this architecture body. If the identifier is included at the end of the architecture body, it must repeat the name of the architecture body. There may be several different architecture bodies corresponding to a single entity, each describing an alternative way of implementing the module's operation. The block declarative items in an architecture body are declarations needed to implement the operations. The items may include type and constant declarations, signal declarations and other kinds of declarations that we will look at in later chapters.

VHDL-87, -93, and -2002

These versions of VHDL do not allow reading of an **out**-mode port within an entity or a corresponding architecture body. Instead, the model must declare an internal signal for the driven value and read that internally. The model can then assign the value of the signal to the **out**-mode port using a concurrent signal assignment statement (see Section 5.2.7).

    In VHDL-2002, another alternative would be to use a **buffer**-mode port instead of an **out**-mode port. However, if the port is only read for verification purposes, using a **buffer**-mode port would be counter to the design intent.

VHDL-87

In VHDL-87, the keyword **entity** may not be included at the end of an entity declaration, and the keyword **architecture** may not be included at the end of an architecture body.

5.1.1 Concurrent Statements

The *concurrent statements* in an architecture body describe the module's operation. One form of concurrent statement, which we have already seen, is a process statement. Putting this together with the rule for writing architecture bodies, we can look at a simple example of an architecture body corresponding to the **adder** entity on page 138:

```
architecture abstract of adder is
begin

  add_a_b : process (a, b) is
  begin
    sum <= a + b;
  end process add_a_b;

end architecture abstract;
```

The architecture body is named **abstract**, and it contains a process **add_a_b**, which describes the operation of the entity. The process assumes that the operator "+" is defined for the type **word**, the type of **a** and **b**. We will see in Chapter 6 how such a definition may be written. We could also envisage additional architecture bodies describing the adder in different ways, provided they all conform to the external interface laid down by the entity declaration.

We have looked at processes first because they are the most fundamental form of concurrent statement. All other forms can ultimately be reduced to one or more processes. Concurrent statements are so called because conceptually they can be activated and perform their actions together, that is, concurrently. Contrast this with the sequential statements inside a process, which are executed one after another. Concurrency is useful for modeling the way real circuits behave. If we have two gates whose inputs change, each evaluates its new output independently of the other. There is no inherent sequencing governing the order in which they are evaluated. We look at process statements in more detail in Section 5.2. Then, in Section 5.3, we look at another form of concurrent statement, the component instantiation statement, used to describe how a module is composed of interconnected sub-modules.

5.1.2 Signal Declarations

When we need to provide internal signals in an architecture body, we must define them using *signal declarations*. The syntax for a signal declaration is very similar to that for a variable declaration:

signal_declaration ⇐
    signal identifier ⟦ , … ⟧ : subtype_indication ⟦ := expression ⟧ ;

This declaration simply names each signal, specifies its type and optionally includes an initial value for all signals declared by the declaration.

EXAMPLE 5.2   *An and-or-invert architecture with internal signals*

The following is an example of an architecture body for the entity **and_or_inv**, defined on page 139. The architecture body includes declarations of some signals that are internal to the architecture body. They can be used by processes within the architecture body but are not accessible outside, since a user of the module need not be concerned with the internal details of its implementation. Values are assigned to signals using signal assignment statements within processes. Signals can be sensed by processes to read their values.

```
architecture primitive of and_or_inv is

  signal and_a, and_b : bit;
  signal or_a_b : bit;

begin

  and_gate_a : process (a1, a2) is
  begin
    and_a <= a1 and a2;
  end process and_gate_a;

  and_gate_b : process (b1, b2) is
  begin
    and_b <= b1 and b2;
  end process and_gate_b;

  or_gate : process (and_a, and_b) is
  begin
    or_a_b <= and_a or and_b;
  end process or_gate;

  inv : process (or_a_b) is
  begin
    y <= not or_a_b;
  end process inv;

end architecture primitive;
```

An important point illustrated by this example is that the ports of the entity are also visible to processes inside the architecture body and are used in the same way as signals. This corresponds to our view of ports as external pins of a circuit: from the internal point of view, a pin is just a wire with an external connection. So it makes sense for VHDL to treat ports like signals inside an architecture of the entity.

5.2 Behavioral Descriptions

At the most fundamental level, the behavior of a module is described by signal assignment statements within processes. We can think of a process as the basic unit of behavioral description. A process is executed in response to changes of values of signals and uses the present values of signals it reads to determine new values for other signals. A signal assignment is a sequential statement and thus can only appear within a process. In this section, we look in detail at the interaction between signals and processes.

5.2.1 Signal Assignment

In all of the examples we have looked at so far, we have used a simple form of signal assignment statement. Each assignment just provides a new value for a signal. The value is determined by evaluating an expression, the result of which must match the type of the signal. What we have not yet addressed is the issue of timing: when does the signal take on the new value? This is fundamental to modeling hardware, in which events occur over time. First, let us look at the syntax for a basic signal assignment statement in a process:

signal_assignment_statement ⇐
    ⟦ label : ⟧ name <= ⟦ delay_mechanism ⟧ waveform ;

waveform ⇐ ⟦ *value*_expression ⟦ **after** *time*_expression ⟧ ⟧ ⟦ , ₀₀₀ ⟧
    ⟦ **unaffected**

The optional label allows us to identify the statement. We will discuss labeled statements in Chapter 20. The syntax rules tell us that we can specify a delay mechanism, which we come to soon, and one or more waveform elements, each consisting of a new value and an optional delay time. We will return to the use of the reserved word **unaffected** as a waveform shortly. It is these delay times in a signal assignment that allow us to specify when the new value should be applied. For example, consider the following assignment:

    y <= **not** or_a_b **after** 5 ns;

This specifies that the signal y is to take on the new value at a time 5 ns later than that at which the statement executes. The delay can be read in one of two ways, depending on whether the model is being used purely for its descriptive value or for simulation. In the first case, the delay can be considered in an abstract sense as a specification of the module's propagation delay: whenever the input changes, the output is updated 5 ns later. In the second case, it can be considered in an operational sense, with reference to a host machine simulating operation of the module by executing the model. Thus if the above assignment is executed at time 250 ns, and or_a_b has the value '1' at that time, then the signal y will take on the value '0' at time 255 ns. Note that the statement itself executes in zero modeled time.

The time dimension referred to when the model is executed is *simulation time*, that is, the time in which the circuit being modeled is deemed to operate. This is distinct from real execution time on the host machine running a simulation. We measure simulation time starting from zero at the start of execution and increasing in discrete steps as events occur in the model. Not surprisingly, this technique is called *discrete event simulation*. A discrete event simulator must have a simulation time clock, and when a signal assignment statement is executed, the delay specified is added to the current simulation time to determine when the new value is to be applied to the signal. We say that the signal assignment schedules a *transaction* for the signal, where the transaction consists of the new value and the simulation time at which it is to be applied. When simulation time advances to the time at which a transaction is scheduled, the signal is updated with the new value. We say that the signal is *active* during that simulation cycle. If the new value is not equal to the old value it replaces on a signal, we say an *event* occurs on the signal. The importance of this distinction is that processes respond to events on signals, not to transactions.

The syntax rules for signal assignments show that we can schedule a number of transactions for a signal, to be applied after different delays. For example, a clock driver process might execute the following assignment to generate the next two edges of a clock signal (assuming T_pw is a constant that represents the clock pulse width):

```
clk <= '1' after T_pw, '0' after 2*T_pw;
```

If this statement is executed at simulation time 50 ns and T_pw has the value 10 ns, one transaction is scheduled for time 60 ns to set clk to '1', and a second transaction is scheduled for time 70 ns to set clk to '0'. If we assume that clk has the value '0' when the assignment is executed, both transactions produce events on clk.

This signal assignment statement shows that when more than one transaction is included, the delays are all measured from the current time, not the time in the previous element. Furthermore, the transactions in the list must have strictly increasing delays, so that the list reads in the order that the values will be applied to the signal.

EXAMPLE 5.3   *A clock generator process*

We can write a process declaration for a clock generator using the above signal assignment statement to generate a symmetrical clock signal with pulse width T_pw. The difficulty is to get the process to execute regularly every clock cycle. One way to do this is by making it resume whenever the clock changes and scheduling the next two transitions when it changes to '0'. A process using this approach is

```
clock_gen : process (clk) is
begin
  if not clk then
    clk <= '1' after T_pw, '0' after 2*T_pw;
  end if;
end process clock_gen;
```

Since a process is the basic unit of a behavioral description, it makes intuitive sense to be allowed to include more than one signal assignment statement for a given signal within a single process. We can think of this as describing the different ways in which a signal's value can be generated by the process at different times.

EXAMPLE 5.4   *A process for a two-input multiplexer*

We can write a process that models a two-input multiplexer as shown below. The value of the **sel** port is used to select which signal assignment to execute to determine the output value.

```
mux : process (a, b, sel) is
begin
  case sel is
    when '0' =>
      z <= a after prop_delay;
    when '1' =>
      z <= b after prop_delay;
  end case;
end process mux;
```

We say that a process defines a *driver* for a signal if and only if it contains at least one signal assignment statement for the signal. So this example defines a driver for the signal z. If a process contains signal assignment statements for several signals, it defines drivers for each of those signals. A driver is a *source* for a signal in that it provides values to be applied to the signal. An important rule to remember is that for normal signals, there may only be one source. This means that we cannot write two different processes, each containing signal assignment statements for the one signal. If we want to model such things as buses or wired-or signals, we must use a special kind of signal called a *resolved signal*, which we will discuss in Chapter 8.

We now return to the use of the reserved word **unaffected** as a waveform in a signal assignment statement, as shown in the syntax rule on page 143. This simply represents no change to the value of the assigned signal. The assignment is equivalent to a null statement, except that it allows us to explicitly document the intention of not changing the target signal. For example, in the following:

```
if device_busy then
  collision_count := collision_count + 1;
  device_req <= unaffected;
else
  accepted_count := accepted_count + 1;
  device_req <= '1';
end if;
```

the assignment with **unaffected** explicitly document that we are not changing device_req
in the alternative where device_busy is true. Had we omitted the assignment, on later ex-
amination of the model, the omission might appear inadvertent.

VHDL-87, -93, and -2002

These versions of VHDL do not allow the reserved word **unaffected** to appear as a
waveform in a sequential signal assignment. It can only be used in concurrent signal
assignments (see Section 5.2.7). If a target signal is to remain unchanged as a result
of executing sequential statements, a model in the earlier versions of VHDL must either omit an assignment or use a null statement.

VHDL-87

Signal assignment statements may not be labeled in VHDL-87.

Conditional Signal Assignments

In Section 3.1.1, we introduced conditional variable assignments, which are a shorthand
notation for variable assignments within if statements. VHDL similarly provides conditional
signal assignments as a shorthand for signal assignment statements within if statements.
The syntax rule is similar:

conditional_signal_assignment ⇐
 〚 label : 〛
 name <= 〚 delay_mechanism 〛
   waveform **when** condition
   〚 **else** waveform **when** condition 〛
   〚 **else** waveform 〛 ;

We will return to the delay mechanism part in Section 5.2.5. As an example, we can
model a register with synchronous reset using a conditional assignment within a process
as follows:

```
reg : process (clk) is
begin
  if rising_edge(clk) then
    q <= (others => '0') when reset else d;
  end if;
end process reg;
```

The conditional assignment in the process is equivalent to the if statement:

```
if reset then
  q <= (others => '0');
else
```

```
    q <= d;
end if;
```

This form of signal assignment is particularly useful for describing the next-state logic of a finite-state machine, as is shown by the following process outline:

```
next_state_logic : process (all) is
begin
  case current_state is
    when idle =>
      next_state <= pending1 when request and busy     else
                    active1  when request and not busy else
                    idle;
    when pending1 =>
      . . .
    . . .
  end case;
end process next_state_logic;
```

The waveforms in a conditional signal assignment are the same as those described earlier, consisting of one or more values to be assigned after successive delays, or the reserved word **unaffected** if we want to leave the signal unchanged. For example, in a stimulus-generator process, we could write the assignment:

```
req <= '1', '0' after T_fixed when fixed_delay_mode else
       '1', '0' after next_random_delay
```

VHDL-87, -93, and -2002

These versions of VHDL do not provide the conditional signal assignment shorthand notation. We must write an if statement with a separate signal assignment for each condition.

Selected Signal Assignments

VHDL also provides a selected signal assignment statement that is short for a case statement containing simple assignments to the target signal. This mirrors the selected variable assignment that we introduced in Section 3.2.1. The syntax rule is:

selected_signal_assignment ⇐
    [[label :]]
    with expression **select**
        name <= [[delay_mechanism]]
            {{ waveform **when** choices , }}
            waveform **when** choices ;

We will return to the delay mechanism part in Section 5.2.5. As an example, we can write a process containing a selected signal assignment for a multiplexer as follows:

```
with d_sel select
    q <= source0 when "00",
         source1 when "01",
         source2 when "10",
         source3 when "11";
```

The assignment is equivalent to the following case statement and nested assignments:

```
case d_sel is
  when "00" =>
    q <= source0;
  when "01" =>
    q <= source1;
  when "10" =>
    q <= source2;
  when "11" =>
    q <= source3;
end case;
```

We can also include the reserved word **unaffected** as a waveform in an alternative of a selected signal assignment, for example:

```
with dut_state select
    dut_req <= '1' when ready,
               '0' when ack,
               unaffected when others;
```

As a further variant of the selected signal assignment, we can write the "?" symbol after the reserved word **select** to specify a matching case statement in the equivalent form. For example, if we write

```
with request select?
    grant <= "1000" when "1---",
             "0100" when "01--",
             "0010" when "001-",
             "0001" when "0001",
             "0000" when others;
```

the equivalent statement is

```
case? request is
  when "1---" =>
    grant <= "1000";
  when "01--" =>
    grant <=  "0100";
  when "001-" =>
    grant <=  "0010";
```

```
  when "0001" =>
    grant <= "0001";
  when others =>
    grant <= "0000";
end case?;
```

The rules for case statements covering the type of the selector expression and the way in which choices are matched apply to the selected signal assignment.

VHDL-87, -93, and -2002

These versions of VHDL do not provide the selected signal assignment shorthand notation. We must write a case statement with a separate signal assignment for each alternative.

5.2.2 Signal Attributes

In Chapter 2 we introduced the idea of attributes of types, which give information about allowed values for the types. Then, in Chapter 4, we saw how we could use attributes of array objects to get information about their index ranges. We can also refer to attributes of signals to find information about their history of transactions and events. Given a signal S, and a value T of type **time**, VHDL defines the following attributes:

S'delayed(T)   A signal that takes on the same values as S but is delayed by time T.

S'stable(T)   A Boolean signal that is true if there has been no event on S in the time interval T up to the current time, otherwise false.

S'quiet(T)   A Boolean signal that is true if there has been no transaction on S in the time interval T up to the current time, otherwise false.

S'transaction   A signal of type bit that changes value from '0' to '1' or vice versa each time there is a transaction on S.

S'event   True if there is an event on S in the current simulation cycle, false otherwise.

S'active   True if there is a transaction on S in the current simulation cycle, false otherwise.

S'last_event   The time interval since the last event on S.

S'last_active   The time interval since the last transaction on S.

S'last_value   The value of S just before the last event on S.

The first three attributes take an optional time parameter. If we omit the parameter, the value 0 fs is assumed. These attributes are often used in checking the timing behavior within a model. For example, we can verify that a signal **d** meets a minimum setup time requirement of **Tsu** before a rising edge on a clock **clk** of type **std_ulogic** as follows:

```
if clk'event and (clk = '1' or clk = 'H')
           and (clk'last_value = '0'
             or clk'last_value = 'L') then
    assert d'last_event >= Tsu
      report "Timing error: d changed within setup time of clk";
  end if;
```

The condition in the if statement performs the same test as the **rising_edge** operation on std_ulogic signals. Similarly, we might check that the pulse width of a clock signal input to a module doesn't exceed a maximum frequency by testing its pulse width:

```
assert (not clk'event) or clk'delayed'last_event >= Tpw_clk
  report "Clock frequency too high";
```

Note that we test the time since the last event on a delayed version of the clock signal. When there is currently an event on a signal, the **'last_event** attribute returns the value 0 fs. In this case, we determine the time since the previous event by applying the **'last_event** attribute to the signal delayed by 0 fs. We can think of this as being an infinitesimal delay. We will return to this idea later in this chapter, in our discussion of delta delays.

EXAMPLE 5.5   *An edge-triggered flipflop using the* **'event** *attribute*

We can use a similar test for the rising edge of a clock signal to model an edge-triggered module, such as a flipflop. The flipflop should load the value of its **D** input on a rising edge of **clk**, but asynchronously clear the outputs whenever **clr** is '1'. The entity declaration and a behavioral architecture body are:

```
entity edge_triggered_Dff is
  port ( D : in bit;  clk : in bit;  clr : in bit;
         Q : out bit );
end entity edge_triggered_Dff;

--------------------------------------------------

architecture behavioral of edge_triggered_Dff is
begin
  state_change : process (clk, clr) is
  begin
    if clr then
      Q <= '0' after 2 ns;
    elsif clk'event and clk = '1' then
      Q <= D after 2 ns;
    end if;
  end process state_change;

end architecture behavioral;
```

The condition in this if statement performs the same test as the predefined rising_edge operation on bit signals. If the flipflop did not have the asynchronous clear input, the model could have used a simple wait statement such as

wait until clk;

to trigger on a rising edge. However, with the clear input present, the process must be sensitive to changes on both clk and clr at any time. Hence it uses the 'event attribute to distinguish between clk changing to '1' and clr going back to '0' while clk is stable at '1'. Note that we cannot write the condition as

clk'event **and** clk

since clk'event yields a **boolean** value, and clk is of type **bit**. The **and** operator is not defined for this mixture of operand types. Instead, we compare the clk value with '1' using the "=" operator, as shown in the if statement.

VHDL-87

In VHDL-87, the 'last_value attribute for a composite signal returns the aggregate of last values for each of the scalar elements of the signal. For example, suppose a bit-vector signal s initially has the value B"00" and changes to B"01" and then B"11" in successive events. After the last event, the result of s'last_value is B"00" in VHDL-87. In VHDL-93 and VHDL-2002 it is B"01", since that is the last value of the entire composite signal.

5.2.3 Wait Statements

Now that we have seen how to change the values of signals over time, the next step in behavioral modeling is to specify when processes respond to changes in signal values. This is done using *wait statements*. A wait statement is a sequential statement with the following syntax rule:

wait_statement ⇐
    ⟦ label : ⟧ **wait** ⟦ **on** *signal*_name ⟨ , ⁀⁀ ⟩ ⟧
                    ⟦ **until** condition ⟧
                    ⟦ **for** *time*_expression ⟧ ;

The optional label allows us to identify the statement. We will discuss labeled statements in Chapter 20. The purpose of the wait statement is to cause the process that executes the statement to suspend execution. The *sensitivity* clause, *condition* clause and *timeout* clause specify when the process is subsequently to resume execution. We can include any combination of these clauses, or we may omit all three. Let us go through each clause and describe what it specifies.

The sensitivity clause, starting with the word **on**, allows us to specify a list of signals to which the process responds. If we just include a sensitivity clause in a wait statement, the process will resume whenever any one of the listed signals changes value, that is, whenever an event occurs on any of the signals. This style of wait statement is useful in a process that models a block of combinatorial logic, since any change on the inputs may result in new output values; for example:

```
half_add : process is
begin
  sum <= a xor b after T_pd;
  carry <= a and b after T_pd;
  wait on a, b;
end process half_add;
```

The process starts execution by generating values for **sum** and **carry** based on the initial values of **a** and **b**, then suspends on the wait statement until either **a** or **b** (or both) change values. When that happens, the process resumes and starts execution from the top.

This form of process is so common in modeling digital systems that VHDL provides the shorthand notation that we have seen in many examples in preceding chapters. A process with a sensitivity list in its heading is exactly equivalent to a process with a wait statement at the end, containing a sensitivity clause naming the signals in the sensitivity list. So the **half_add** process above could be rewritten as

```
half_add : process (a, b) is
begin
  sum <= a xor b after T_pd;
  carry <= a and b after T_pd;
end process half_add;
```

EXAMPLE 5.6   *A revised model of a two-input multiplexer*

Let us return to the model of a two-input multiplexer in Example 5.4. The process in that model is sensitive to all three input signals. This means that it will resume on changes on either data input, even though only one of them is selected at any time. If we are concerned about this slight lack of efficiency in simulation, we can write the process differently, using wait statements to be more selective about the signals to which the process is sensitive each time it suspends. The revised model is shown below. In this model, when input **a** is selected, the process only waits for changes on the select input and on **a**. Any changes on **b** are ignored. Similarly, if **b** is selected, the process waits for changes on **sel** and on **b**, ignoring changes on **a**.

```
entity mux2 is
  port ( a, b, sel : in bit;
         z : out bit );
end entity mux2;

-------------------------------------------------------

architecture behavioral of mux2 is

  constant prop_delay : time := 2 ns;
```

```
begin
  slick_mux : process is
  begin
    case sel is
      when '0' =>
        z <= a after prop_delay;
        wait on sel, a;
      when '1' =>
        z <= b after prop_delay;
        wait on sel, b;
    end case;
  end process slick_mux;

end architecture behavioral;
```

The condition clause in a wait statement, starting with the word **until**, allows us to specify a condition that must be true for the process to resume. For example, the wait statement

wait until clk;

causes the executing process to suspend until the value of the signal **clk** changes to '1'. The condition expression is tested while the process is suspended to determine whether to resume the process. A consequence of this is that even if the condition is true when the wait statement is executed, the process will still suspend until the appropriate signals change and cause the condition to be true again. If the wait statement doesn't include a sensitivity clause, the condition is tested whenever an event occurs on any of the signals mentioned in the condition.

EXAMPLE 5.7   *A revised clock generator process*

The clock generator process in Example 5.3 can be rewritten using a wait statement with a condition clause, as shown below. Each time the process executes the wait statement, **clk** has the value '0'. However, the process still suspends, and the condition is tested each time there is an event on **clk**. When **clk** changes to '1', nothing happens, but when it changes to '0' again, the process resumes and schedules transactions for the next cycle.

```
clock_gen : process is
begin
  clk <= '1' after T_pw, '0' after 2*T_pw;
  wait until not clk;
end process clock_gen;
```

If a wait statement includes a sensitivity clause as well as a condition clause, the condition is only tested when an event occurs on any of the signals in the sensitivity clause. For example, if a process suspends on the following wait statement:

wait on clk **until not** reset;

the condition is tested on each change in the value of clk, regardless of any changes on reset.

The timeout clause in a wait statement, starting with the word **for**, allows us to specify a maximum interval of simulation time for which the process should be suspended. If we also include a sensitivity or condition clause, these may cause the process to be resumed earlier. For example, the wait statement

wait until trigger **for** 1 ms;

causes the executing process to suspend until **trigger** changes to '1', or until 1 ms of simulation time has elapsed, whichever comes first. If we just include a timeout clause by itself in a wait statement, the process will suspend for the time given.

EXAMPLE 5.8   *The clock generator process further revised*

We can rewrite the clock generator process in Example 5.3 yet again, this time using a wait statement with a timeout clause, as shown in below. In this case we specify the clock period as the timeout, after which the process is to be resumed.

```
clock_gen : process is
begin
  clk <= '1' after T_pw, '0' after 2*T_pw;
  wait for 2*T_pw;
end process clock_gen;
```

If we refer back to the syntax rule for a wait statement shown on page 151, we note that it is legal to write

wait;

This form causes the executing process to suspend for the remainder of the simulation. Although this may at first seem a strange thing to want to do, in practice it is quite useful. One place it is used is in a process whose purpose is to generate stimuli for a simulation. Such a process should generate a sequence of transactions on signals connected to other parts of a model and then stop. For example, the process

```
test_gen : process is
begin
  test0 <= '0' after 10 ns, '1' after 20 ns,
           '0' after 30 ns, '1' after 40 ns;
  test1 <= '0' after 10 ns, '1' after 30 ns;
```

```
    wait;
end process test_gen;
```

generates all four possible combinations of values on the signals **test0** and **test1**. If the final wait statement were omitted, the process would cycle forever, repeating the signal assignment statements without suspending, and the simulation would make no progress.

VHDL-87, -93, and -2002

These versions of VHDL do not perform implicit conversion of conditions to **boolean**. Conditions in the **until** clauses of wait statements must yield **boolean** results without conversion.

VHDL-87

Wait statements may not be labeled in VHDL-87.

5.2.4 Delta Delays

Let us now return to the topic of delays in signal assignments. In many of the example signal assignments in previous chapters, we omitted the delay part of waveform elements. This is equivalent to specifying a delay of 0 fs. The value is to be applied to the signal at the current simulation time. However, it is important to note that the signal value does not change as soon as the signal assignment statement is executed. Rather, the assignment schedules a transaction for the signal, which is applied after the process suspends. Thus the process does not see the effect of the assignment until the next time it resumes, even if this is at the same simulation time. For this reason, a delay of 0 fs in a signal assignment is called a *delta delay*.

To understand why delta delays work in this way, it is necessary to review the simulation cycle, introduced in Chapter 1 on page 15. Recall that the simulation cycle consists of two phases: a signal update phase followed by a process execution phase. In the signal update phase, simulation time is advanced to the time of the earliest scheduled transaction, and the values in all transactions scheduled for this time are applied to their corresponding signals. This may cause events to occur on some signals. In the process execution phase, all processes that respond to these events are resumed and execute until they suspend again on wait statements. The simulator then repeats the simulation cycle.

Let us now consider what happens when a process executes a signal assignment statement with delta delay, for example:

```
data <= X"00";
```

Suppose this is executed at simulation time *t* during the process execution phase of the current simulation cycle. The effect of the assignment is to schedule a transaction to put the value X"00" on **data** at time *t*. The transaction is not applied immediately, since the simulator is in the process execution phase. Hence the process continues executing, with

data unchanged. When all processes have suspended, the simulator starts the next simulation cycle and updates the simulation time. Since the earliest transaction is now at time
t, simulation time remains unchanged. The simulator now applies the value X"00" in the
scheduled transaction to **data**, then resumes any processes that respond to the new value.

Writing a model with delta delays is useful when we are working at a high level of
abstraction and are not yet concerned with detailed timing. If all we are interested in is
describing the order in which operations take place, delta delays provide a means of ignoring the complications of timing. We have seen this in many of the examples in previous
chapters. However, we should note a common pitfall encountered by most beginner
VHDL designers when using delta delays: they forget that the process does not see the
effect of the assignment immediately. For example, we might write a process that includes
the following statements:

```
s <= '1';
...
if s then ...
```

and expect the process to execute the if statement assuming **s** has the value '1'. We would
then spend fruitless hours debugging our model until we remembered that **s** still has its
old value until the next simulation cycle, after the process has suspended.

EXAMPLE 5.9   *Using delta delays to synchronize communication*

The following is an outline of an abstract model of a computer system. The CPU and
memory are connected with address and data signals. They synchronize their operation with the **mem_read** and **mem_write** control signals and the **mem_ready** status signal. No delays are specified in the signal assignment statements, so synchronization
occurs over a number of delta delay cycles, as shown in Figure 5.1.

```
architecture abstract of computer_system is

  subtype word is bit_vector(31 downto 0);

  signal address : natural;
  signal read_data, write_data : word;
  signal mem_read, mem_write : bit := '0';
  signal mem_ready : bit := '0';

begin

  cpu : process is
    variable instr_reg : word;
    variable PC : natural;
    ...       -- other declarations
  begin
    loop
      address <= PC;
      mem_read <= '1';
      wait until mem_ready;
```

FIGURE 5.1

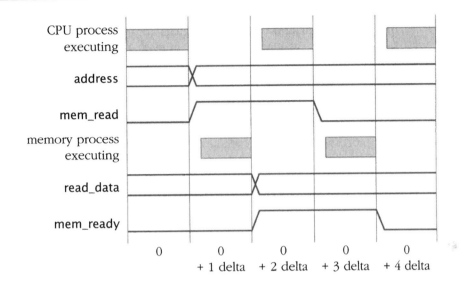

Synchronization over successive delta cycles in a simulation of a read operation between the CPU and memory.

```
      instr_reg := read_data;
      mem_read <= '0';
      wait until not mem_ready;
      PC := PC + 4;
      ...      -- execute the instruction
    end loop;
  end process cpu;

  memory : process is
    type memory_array is array (0 to 2**14 - 1) of word;
    variable store : memory_array := ( ... );
  begin
    wait until mem_read or mem_write;
    if mem_read then
      read_data <= store( address / 4 );
      mem_ready <= '1';
      wait until not mem_read;
      mem_ready <= '0';
    else
      ...      -- perform write access
    end if;
  end process memory;

end architecture abstract;
```

When the simulation starts, the CPU process begins executing its statements and the memory suspends. The CPU schedules transactions to assign the next instruction address to the **address** signal and the value '1' to the **mem_read** signal, then suspends. In the next simulation cycle, these signals are updated and the memory process resumes, since it is waiting for an event on **mem_read**. The memory process schedules the data on the **read_data** signal and the value '1' on **mem_ready**, then suspends. In the third cycle, these signals are updated and the CPU process resumes. It schedules the value '0' on **mem_read** and suspends. Then, in the fourth cycle, **mem_read** is updated and the memory process is resumed, scheduling the value '0' on **mem_ready** to complete the handshake. Finally, on the fifth cycle, **mem_ready** is updated and the CPU process resumes and executes the fetched instruction.

5.2.5   Transport and Inertial Delay Mechanisms

So far in our discussion of signal assignments, we have implicitly assumed that there were no pending transactions scheduled for a signal when a signal assignment statement was executed. In many models, particularly at higher levels of abstraction, this will be the case. If, on the other hand, there are pending transactions, the new transactions are merged with them in a way that depends on the *delay mechanism* used in the signal assignment statement. This is an optional part of the signal assignment syntax shown on page 143. The syntax rule for the delay mechanism is

delay_mechanism ⇐ **transport** ⫴ ⟦ **reject** *time*_expression ⟧ **inertial**

A signal assignment with the delay mechanism part omitted is equivalent to specifying **inertial**. We look at the *transport* delay mechanism first, since it is simpler, and then return to the *inertial* delay mechanism.

We use the transport delay mechanism when we are modeling an ideal device with infinite frequency response, in which any input pulse, no matter how short, produces an output pulse. An example of such a device is an ideal transmission line, which transmits all input changes delayed by some amount. A process to model a transmission line with delay 500 ps is

```
transmission_line : process (line_in) is
begin
  line_out <= transport line_in after 500 ps;
end process transmission_line;
```

In this model the output follows any changes in the input, but delayed by 500 ps. If the input changes twice or more within a period shorter than 500 ps, the scheduled transactions are simply queued by the driver until the simulation time at which they are to be applied, as shown in Figure 5.2.

In this example, each new transaction that is generated by a signal assignment statement is scheduled for a simulation time that is later than the pending transactions queued by the driver. The situation gets a little more complex when variable delays are used, since we can schedule a transaction for an earlier time than a pending transaction. The semantics of the transport delay mechanism specify that if there are pending transactions on a

FIGURE 5.2

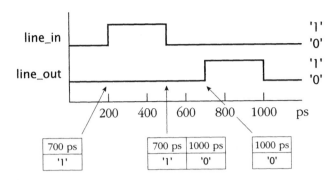

Transactions queued by a driver using transport delay. At time 200 ps the input changes, and a transaction is scheduled for 700 ps. At time 500 ps, the input changes again, and another transaction is scheduled for 1000 ps. This is queued by the driver behind the earlier transaction. When simulation time reaches 700 ps, the first transaction is applied, and the second transaction remains queued. Finally, simulation time reaches 1000 ps, and the final transaction is applied, leaving the driver queue empty.

driver that are scheduled for a time later than or equal to a new transaction, those later transactions are deleted.

EXAMPLE 5.10   *An asymmetric delay element using transport delay*

The following is a process that describes the behavior of an asymmetric delay element, with different delay times for rising and falling transitions. The delay for rising transitions is 800 ps and for falling transitions, 500 ps.

```
asym_delay : process (a) is
  constant Tpd_01 : time := 800 ps;
  constant Tpd_10 : time := 500 ps;
begin
  if a then
    z <= transport a after Tpd_01;
  else   -- not a
    z <= transport a after Tpd_10;
  end if;
end process asym_delay;
```

If we apply an input pulse of only 200 ps duration, we would expect the output not to change, since the delayed falling transition should "overtake" the delayed rising transition. If we were simply to add each transition to the driver queue when a signal assignment statement is executed, we would not get this behavior. However, the semantics of the transport delay mechanism produce the desired behavior, as Figure 5.3 shows.

FIGURE 5.3

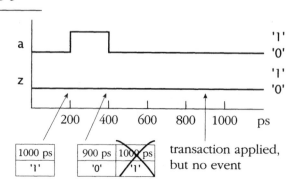

Transactions in a driver using asymmetric transport delay. At time 200 ps the input changes, and a transaction is scheduled for 1000 ps. At time 400 ps, the input changes again, and another transaction is scheduled for 900 ps. Since this is earlier than the pending transaction at 1000 ps, the pending transaction is deleted. When simulation time reaches 900 ps, the remaining transaction is applied, but since the value is '0', no event occurs on the signal.

Most real electronic circuits don't have infinite frequency response, so it is not appropriate to model them using transport delay. In real devices, changing the values of internal nodes and outputs involves moving electronic charge around in the presence of capacitance, inductance and resistance. This gives the device some inertia; it tends to stay in the same state unless we force it by applying inputs for a sufficiently long duration. This is why VHDL includes the inertial delay mechanism, to allow us to model devices that reject input pulses too short to overcome their inertia. Inertial delay is the mechanism used by default in a signal assignment, or we can specify it explicitly by including the word **inertial**.

To explain how inertial delay works, let us first consider a model in which all the signal assignments for a given signal use the same delay value, say, 3 ns, as in the following inverter model:

```
inv : process (a) is
begin
  y <= inertial not a after 3 ns;
end process inv;
```

So long as input events occur more than 3 ns apart, this model does not present any problems. Each time a signal assignment is executed, there are no pending transactions, so a new transaction is scheduled, and the output changes value 3 ns later. However, if an input changes less than 3 ns after the previous change, this represents a pulse less than the propagation delay of the device, so it should be rejected. This behavior is shown at the top of Figure 5.4. In a simple model such as this, we can interpret inertial delay by saying if a signal assignment produces an output pulse shorter than the propagation delay, then the output pulse does not happen.

FIGURE 5.4

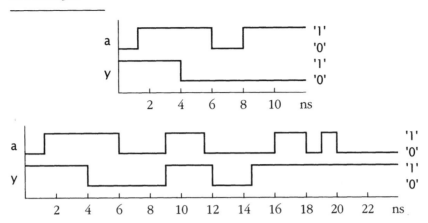

Results of signal assignments using the inertial delay mechanism. In the top waveform, an inertial delay of 3 ns is specified. The input change at time 1 ns is reflected in the output at time 4 ns. The pulse from 6 to 8 ns is less than the propagation delay, so it doesn't affect the output. In the bottom waveform, an inertial delay of 3 ns and a pulse rejection limit of 2 ns are specified. The input changes at 1, 6, 9 and 11.5 ns are all reflected in the output, since they occur greater than 2 ns apart. However, the subsequent input pulses are less than or equal to the pulse rejection limit in length, and so do not affect the output.

Next, let us extend this model by specifying a pulse rejection limit, after the word **reject** in the signal assignment:

```
inv : process (a) is
begin
  y <= reject 2 ns inertial not a after 3 ns;
end process inv;
```

We can interpret this by saying if a signal assignment produces an output pulse shorter than (or equal to) the pulse rejection limit, the output pulse does not happen. In this simple model, so long as input changes occur more than 2 ns apart, they produce output changes 3 ns later, as shown at the bottom of Figure 5.4. Note that the pulse rejection limit specified must be between 0 fs and the delay specified in the signal assignment. Omitting a pulse rejection limit is the same as specifying a limit equal to the delay, and specifying a limit of 0 fs is the same as specifying transport delay.

Now let us look at the full story of inertial delay, allowing for varying the delay time and pulse rejection limit in different signal assignments applied to the same signal. As with transport delay, the situation becomes more complex, and it is best to describe it in terms of deleting transactions from the driver. Those who are unlikely to be writing models that deal with timing at this level of detail may wish to move on to the next section.

An inertially delayed signal assignment involves examining the pending transactions on a driver when adding a new transaction. Suppose a signal assignment schedules a new transaction for time t_{new}, with a pulse rejection limit of t_r. First, any pending transactions scheduled for a time later than or equal to t_{new} are deleted, just as they are when trans-

FIGURE 5.5

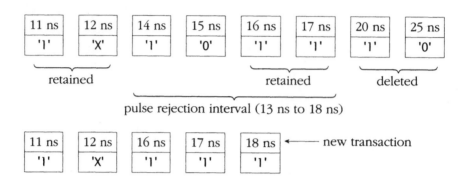

Transactions before (top) and after (bottom) an inertial delay signal assignment. The transactions at 20 and 25 ns are deleted because they are scheduled for later than the new transaction. Those at 11 and 12 ns are retained because they fall before the pulse rejection interval. The transactions at 16 and 17 ns fall within the rejection interval, but they form a run leading up to the new transaction, with the same value as the new transaction; hence they are also retained. The other transactions in the rejection interval are deleted.

port delay is used. Then the new transaction is added to the driver. Second, any pending transactions scheduled in the interval $t_{new} - t_r$ to t_{new} are examined. If there is a run of consecutive transactions immediately preceding the new transaction with the same value as the new transaction, they are kept in the driver. All other transactions in the interval are deleted.

An example will make this clearer. Suppose a driver for signal s contains pending transactions as shown at the top of Figure 5.5, and the process containing the driver executes the following signal assignment statement at time 10 ns:

```
s <= reject 5 ns inertial '1' after 8 ns;
```

The pending transactions after this assignment are shown at the bottom of Figure 5.5.

A further point to note about specifying the delay mechanism in a signal assignment statement is that if a number of waveform elements are included, the specified mechanism only applies to the first element. All the subsequent elements schedule transactions using transport delay. Since the delays for multiple waveform elements must be in ascending order, this means that all of the transactions after the first are just added to the driver transaction queue in the order written.

EXAMPLE 5.11   *An and gate with detailed timing*

A detailed model of a two-input and gate is shown below. The process **gate** implements the logical function of the entity, and the process **delay** implements its detailed timing characteristics using inertially delayed signal assignments. A delay of 1.5 ns is used for rising transitions, and 1.2 ns for falling transitions. When a change on either of the input signals results in a change scheduled for the output, the **delay** process determines the propagation delay to be used. On a rising output transition, spikes of less than 400 ps are rejected, and on a falling or unknown transition, spikes of less than 300 ps are rejected.

```vhdl
library ieee;  use ieee.std_logic_1164.all;

entity and2 is
  port ( a, b : in std_ulogic;  y : out std_ulogic );
end entity and2;

--------------------------------------------------

architecture detailed_delay of and2 is

  signal result : std_ulogic;

begin

  gate : process (a, b) is
  begin
    result <= a and b;
  end process gate;

  delay : process (result) is
  begin
    if result then
      y <= reject 400 ps inertial '1' after 1.5 ns;
    elsif not result then
      y <= reject 300 ps inertial '0' after 1.2 ns;
    else
      y <= reject 300 ps inertial 'X' after 500 ps;
    end if;
  end process delay;

end architecture detailed_delay;
```

The final point to make about delay mechanisms is that they can be specified in the conditional and selected forms of signal assignment, as shown by the syntax rules on pages 146 and 147. For example, we can write a sequential conditional assignment using transport delay as:

```vhdl
wire_out <= transport
  wire_in after T_wire_delay when delay_mode = fixed else
  wire_in after delay_lookup("wire_out");
```

Likewise, we can write a sequential conditional assignment using inertial delay as:

```vhdl
with speed_grade select
  z <= reject Tpr inertial
    result after Tpd_std when std_grade,
```

```
        result after Tpd_fast when fast_grade,
        result after Tpd_redhot when redhot_grade;
```

If we include a delay mechanism in a conditional or selected assignment containing multiple waveform elements in the alternatives, the delay mechanism applies to the first waveform element in each alternative, with any subsequent waveform elements using transport delay.

VHDL-87

VHDL-87 does not allow specification of the pulse rejection limit in a delay mechanism. The syntax rule in VHDL-87 is

delay_mechanism ⇐ **transport**

If the delay mechanism is omitted, inertial delay is used, with a pulse rejection limit equal to the delay specified in the waveform element.

5.2.6   Process Statements

We have been using processes quite extensively in examples in this and previous chapters, so we have seen most of the details of how they are written and used. To summarize, let us now look at the formal syntax for a process statement and review the operation of processes. The syntax rule is

```
process_statement ⇐
    〚 process_label : 〛
    process 〚 ( signal_name 〚 , … 〛 〚 all ) 〛 〚 is 〛
        〚 process_declarative_item 〛
    begin
        〚 sequential_statement 〛
    end process 〚 process_label 〛 ;
```

Recall that a process statement is a concurrent statement that can be included in an architecture body to implement all or part of the behavior of a module. The process label identifies the process. While it is optional, it is a good idea to include a label on each process. A label makes it easier to debug a simulation of a system, since most simulators provide a way of identifying a process by its label. Most simulators also generate a default name for a process if we omit the label in the process statement. Having identified a process, we can examine the contents of its variables or set breakpoints at statements within the process.

The declarative items in a process statement may include constant, type and variable declarations, as well as other declarations that we will come to later. Note that ordinary variables may only be declared within process statements, not outside of them. The variables are used to represent the state of the process, as we have seen in the examples. The sequential statements that form the process body may include any of those that we introduced in Chapter 3, plus signal assignment and wait statements. When a process is activated during simulation, it starts executing from the first sequential statement and continues until it reaches the last. It then starts again from the first. This would be an infinite loop, with no progress being made in the simulation, if it were not for the inclusion of wait statements, which suspend process execution until some relevant event occurs. Wait statements are the only statements that take more than zero simulation time to execute. It is only through the execution of wait statements that simulation time advances.

A process may include a sensitivity list in parentheses after the keyword **process**. The sensitivity list identifies a set of signals that the process monitors for events. If the sensitivity list is omitted, the process should include one or more wait statements. On the other hand, if the sensitivity list is included, then the process body cannot include any wait statements. Instead, there is an implicit wait statement, just before the **end process** keywords, that includes the signals listed in the sensitivity list as signals in an **on** clause.

The sensitivity list can be just the reserved word **all** instead of a list of signals, in which case the process is sensitive to all of the signals that it reads as inputs. This form of sensitivity list allows us to write models for combinational logic much more simply. If we change the model to include more inputs, we don't need to remember to revise the sensitivity list of the process.

EXAMPLE 5.12 *A combinational process for a finite-state machine*

One place where assembling the sensitivity list for a combinational process often causes problems is the next-state and output logic for a finite-state machine. The logic has, as inputs, the current state signal and signals whose values determine the next state and the output values. An example is:

```
next_state_logic : process (all) is
begin
  out1 <= '0'; out2 <= '0'; ...
  case current_state is
    when idle =>
      out1 <= '1';
      if in1 and not in2 then
        out2 <= '1';
        next_state <= busy1;
      elsif in1 and in2 then
        next_state <= busy2;
      else
        next_state_logic <= idle;
      end if;
    ...
  end case;
end process next_state;
```

As we revise the finite-state machine, we might include more signals as inputs. Using the reserved word **all** instead of explicitly listing the input signals makes the process easier to write and maintain.

VHDL-87, -93, and -2002

These versions of VHDL do not allow the reserved word **all** in a sensitivity list. Instead, the list of signals to which a process is sensitive must be written out explicitly.

VHDL-87

The keyword **is** may not be included in the header of a process statement in VHDL-87.

5.2.7 Concurrent Signal Assignment Statements

The form of process statement that we have been using is the basis for all behavioral modeling in VHDL, but for simple cases, it can be a little cumbersome and verbose. For this reason, VHDL provides us with some useful shorthand notations for *functional modeling*, that is, behavioral modeling in which the operation to be described is a simple combinatorial transformation of inputs to an output. *Concurrent signal assignment statements* are equivalent to sequential signal assignments contained in process statements. Unlike ordinary signal assignments, concurrent signal assignment statements can be included in the statement part of an architecture body. There are simple, conditional, and selected forms of concurrent signal assignments. The syntax rules are essentially the same as those for the corresponding sequential signal assignments.

Concurrent Simple Signal Assignments

A very common case in function modeling is to write a simple unconditional signal assignment, as in the following example:

```
PC_incr : next_PC <= PC + 4 after 5 ns;
```

At first sight this appears to be an ordinary sequential signal assignment statement, which by rights ought to be inside a process body. However, if it appears as a concurrent statement in an architecture body, it is equivalent to the following process statement:

```
PC_incr : process is
begin
  next_PC <= PC + 4 after 5 ns;
  wait on PC;
end process PC_incr;
```

Looking at the equivalent process shows us something important about the concurrent signal assignment statement, namely, that it is sensitive to the PC signal. In fact, a concurrent signal assignment is sensitive to all of the signals mentioned in the waveform. So whenever any of those signals change value, the assignment is reevaluated and a new transaction is scheduled on the driver for the target signal.

EXAMPLE 5.13   *A flipflop with complementary outputs*

We can model a flipflop with a process that assigns to a data output port. If the flipflop also has a complementary data output port, we can use a conditional signal assignment with no condition to drive that output. The entity and architecture body are

```
entity Dff is
  port ( clk, d : in bit; q, q_n : out bit );
end entity Dff;

-------------------------------------------------

architecture rtl of Dff is
begin
  ff : process (clk) is
  begin
    if clk then
      q <= d;
    end if;
  end process ff;
  q_n <= not q;
end architecture rtl;
```

Concurrent Conditional Signal Assignment

The concurrent conditional signal assignment statement is a shorthand for a collection of simple signal assignments contained in an if statement, which is in turn contained in a process statement. Let us look at some examples and show how each conditional signal assignment can be transformed into an equivalent process statement.

EXAMPLE 5.14   *A multiplexer using a concurrent conditional assignment*

First, here is a functional description of a multiplexer, with four data inputs (d0, d1, d2 and d3), two select inputs (sel0 and sel1) and a data output (z). All of these signals are of type bit.

```
zmux : z <= d0 when not sel1 and not sel0 else
            d1 when not sel1 and     sel0 else
            d2 when     sel1 and not sel0 else
            d3 when     sel1 and     sel0;
```

This statement has exactly the same meaning as the following process statement:

```
zmux : process is
begin
  if not sel1 and not sel0 then
    z <= d0;
```

```
        elsif not sel1 and sel0 then
          z <= d1;
        elsif sel1 and not sel0 then
          z <= d2;
        elsif sel1 and sel0 then
          z <= d3;
        end if;
        wait on d0, d1, d2, d3, sel0, sel1;
      end process zmux;
```

If we look at the equivalent process, we see that the concurrent conditional signal assignment statement is sensitive to all of the signals mentioned in the waveforms and in the conditions. So whenever any of these change value, the assignment is reevaluated and a new transaction is scheduled on the driver for the target signal.

On closer inspection, we note that the last condition is redundant, since the signals **sel0** and **sel1** are of type **bit**. If none of the previous conditions are true, the signal should always be assigned the last waveform. So we can rewrite the example as follows:

```
  zmux : z <= d0 when not sel1 and not sel0 else
              d1 when not sel1 and     sel0 else
              d2 when         sel1 and not sel0 else
              d3;
```

The equivalent process is

```
  zmux : process is
  begin
    if not sel1 and not sel0 then
      z <= d0;
    elsif not sel1 and sel0 then
      z <= d1;
    elsif sel1 and not sel0 then
      z <= d2;
    else
      z <= d3;
    end if;
    wait on d0, d1, d2, d3, sel0, sel1;
  end process zmux;
```

Another case that sometimes arises when writing functional models is the need for a process that schedules an initial set of transactions and then does nothing more for the remainder of the simulation.

EXAMPLE 5.15   *Reset signal generation*

An example of "one-shot" activity is the generation of a reset signal. One way of doing this is as follows:

```
reset_gen : reset <= '1', '0' after 200 ns when extended_reset else
                     '1', '0' after 50 ns;
```

The thing to note here is that there are no signals named in any of the waveforms or the conditions (assuming that **extended_reset** is a constant). This means that the statement is executed once when simulation starts, schedules two transactions on reset and remains quiescent thereafter. The equivalent process is

```
reset_gen : process is
begin
  if extended_reset then
    reset <= '1', '0' after 200 ns;
  else
    reset <= '1', '0' after 50 ns;
  end if;
  wait;
end process reset_gen;
```

Since there are no signals involved, the wait statement has no sensitivity clause. Thus, after the if statement has executed, the process suspends forever.

If we include a delay mechanism specification in a conditional signal assignment statement, it is used whichever waveform is chosen. So we might rewrite the model for the asymmetric delay element shown in Example 5.10 as

```
asym_delay : z <= transport a after Tpd_01 when a else
                            a after Tpd_10;
```

In Section 5.2.1 we saw the use of the reserved word **unaffected** as a waveform to specify that the target signal not be changed by an assignment. We can use **unaffected** in a concurrent assignement also.

EXAMPLE 5.16   *A scheduler for a server*

A scheduler selects among requests for a server, but when the server is busy, no request is scheduled. We can use **unaffected** to indicate that there should be no change on the signal representing the selected request, as follows:

```
scheduler :
  request <=
    first_priority_request after scheduling_delay
      when priority_waiting and server_status = ready else
```

```
       first_normal_request after scheduling_delay
         when not priority_waiting and server_status = ready else
       unaffected
         when server_status = busy else
       reset_request after scheduling_delay;
```

The equivalent process is

```
   scheduler : process is
   begin
     if priority_waiting and server_status = ready then
       request <= first_priority_request after scheduling_delay;
     elsif not priority_waiting and server_status = ready then
       request <= first_normal_request after scheduling_delay;
     elsif server_status = busy then
       null;
     else
       request <= reset_request after scheduling_delay;
     end if;
     wait on first_priority_request, priority_waiting,
             server_status, first_normal_request, reset_request;
   end process scheduler;
```

The effect of the **unaffected** waveform is to include a null statement in the equivalent process, causing it to bypass scheduling a transaction when the corresponding condition is true. (Recall that the effect of the null sequential statement is to do nothing.)

VHDL-87

In VHDL-87 the syntax rule for a conditional signal assignment statement is

conditional_signal_assignment ⇐
    name <= ⟦ **transport** ⟧
        ⦃ waveform **when** condition **else** ⦄
        waveform ;

The delay mechanism is restricted to the keyword **transport**, as discussed on page 164. The final waveform may not be conditional. Furthermore, we may not use the keyword **unaffected**. If the required behavior cannot be expressed with these restrictions, we must write a full process statement instead of a conditional signal assignment statement.

Concurrent Selected Signal Assignments

The concurrent selected signal assignment statement is shorthand for a number of simple signal assignments embedded in a case statement, which is in turn contained in a process. Let us look at some examples.

EXAMPLE 5.17   *An ALU using a concurrent selected assignment*

The following concurrent selected signal assignment models an ALU that performs addition, subtraction, logical and, logical or, and pass operations:

```
alu : with alu_function select
        result <= a + b after Tpd    when alu_add |
                                          alu_add_unsigned,
                  a - b after Tpd    when alu_sub |
                                          alu_sub_unsigned,
                  a and b after Tpd when alu_and,
                  a or b after Tpd  when alu_or,
                  a after Tpd        when alu_pass_a;
```

This has the same meaning as the following process statement containing a case statement:

```
alu : process is
begin
  case alu_function is
    when alu_add |
         alu_add_unsigned =>  result <= a + b after Tpd;
    when alu_sub |
         alu_sub_unsigned =>  result <= a - b after Tpd;
    when alu_and             =>  result <= a and b after Tpd;
    when alu_or              =>  result <= a or b after Tpd;
    when alu_pass_a          =>  result <= a after Tpd;
  end case;
  wait on alu_function, a, b;
end process alu;
```

A concurrent selected signal assignment statement is sensitive to all of the signals in the selector expression and in the waveforms. This means that the selected signal assignment for the ALU is always sensitive to **b** and will resume if **b** changes value, even if the value of **alu_function** is **alu_pass_a** and the value of **b** is not used.

Apart from the difference in the equivalent process, the concurrent selected signal assignment is similar to the concurrent conditional assignment. Thus the special waveform **unaffected** can be used to specify that no assignment take place for some values of the selector expression. Also, if a delay mechanism is specified in the statement, that mechanism is used on each sequential signal assignment within the equivalent process.

EXAMPLE 5.18   *A full adder in truth-table form*

We can use a selected signal assignment to express a combinatorial logic function in truth-table form. In the following entity declaration and an architecture body for a full adder, the selected signal assignment statement has, as its selector expression, a bit vector formed by aggregating the input signals. The choices list all possible values of inputs, and for each, the values for the **c_out** and **s** outputs are given.

```
entity full_adder is
  port ( a, b, c_in : bit;  s, c_out : out bit );
end entity full_adder;

-------------------------------------------------

architecture truth_table of full_adder is
begin

  with bit_vector'(a, b, c_in) select
    (c_out, s) <= bit_vector'("00") when "000",
                  bit_vector'("01") when "001",
                  bit_vector'("01") when "010",
                  bit_vector'("10") when "011",
                  bit_vector'("01") when "100",
                  bit_vector'("10") when "101",
                  bit_vector'("10") when "110",
                  bit_vector'("11") when "111";

end architecture truth_table;
```

This example illustrates the most common use of aggregate targets in signal assignments. Note that the type qualification is required in the selector expression to specify the type of the aggregate. The type qualification is needed in the output values to distinguish the bit-vector string literals from character string literals.

We can include the symbol "?" after the reserved word select in a concurrent selected signal assignment to specify matching case statement in the equivalent form, just as we did in the sequential form of the statement. Thus, we could write the assignment

```
with request select?
  grant <= "1000" when "1---",
           "0100" when "01--",
           "0010" when "001-",
           "0001" when "0001",
           "0000" when others;
```

as a concurrent statement in an architecture body.

VHDL-87, -93, and -2002

These versions of VHDL do not provide the matching concurrent selected signal assignment notation.

VHDL-87

In VHDL-87, the delay mechanism is restricted to the keyword **transport**, as discussed on page 164. Furthermore, the keyword **unaffected** may not be used. If the required behavior cannot be expressed without using the keyword **unaffected**, we must write a full process statement instead of a selected signal assignment statement.

5.2.8 Concurrent Assertion Statements

VHDL provides another shorthand process notation, the *concurrent assertion statement*, which can be used in behavioral modeling. As its name implies, a concurrent assertion statement represents a process whose body contains an ordinary sequential assertion statement. The syntax rule is

concurrent_assertion_statement ⇐
    ⟦ label : ⟧
    assert condition
        ⟦ **report** expression ⟧ ⟦ **severity** expression ⟧ ;

This syntax appears to be exactly the same as that for a sequential assertion statement, but the difference is that it may appear as a concurrent statement. The optional label on the statement serves the same purpose as that on a process statement: to provide a way of referring to the statement during simulation or synthesis. The process equivalent to a concurrent assertion contains a sequential assertion with the same condition, report clause and severity clause. The sequential assertion is then followed by a wait statement whose sensitivity list includes the signals mentioned in the condition expression. Thus the effect of the concurrent assertion statement is to check that the condition holds true each time any of the signals mentioned in the condition change value. Concurrent assertions provide a very compact and useful way of including timing and correctness checks in a model.

EXAMPLE 5.19 *A set/reset flipflop with usage check*

We can use concurrent assertion statements to check for correct use of a set/reset flipflop, with two inputs s and r and two outputs q and q_n, all of type **bit**. The requirement for use is that s and r are not both '1' at the same time. The entity and architecture body are

```
entity S_R_flipflop is
  port ( s, r : in bit;  q, q_n : out bit );
end entity S_R_flipflop;
```

```
--------------------------------------------------
architecture functional of S_R_flipflop is

begin

  q <= '1' when s else
       '0' when r;

  q_n <= '0' when s else
         '1' when r;

  check : assert not (s and r)
                report "Incorrect use of S_R_flip_flop: " &
                       "s and r both '1'";

end architecture functional;
```

The first and second concurrent statements implement the functionality of the model. The third checks for correct use and is resumed when either **s** or **r** changes value, since these are the signals mentioned in the Boolean condition. If both of the signals are '1', an assertion violation is reported. The equivalent process for the concurrent assertion is

```
check : process is
begin
  assert not (s  and r)
    report "Incorrect use of S_R_flip_flop: " &
           "s and r both '1'";
  wait on s, r;
end process check;
```

5.2.9 Entities and Passive Processes

We complete this section on behavioral modeling by returning to declarations of entities. We can include certain kinds of concurrent statements in an entity declaration, to monitor use and operation of the entity. The extended syntax rule for an entity declaration that shows this is

entity_declaration ⇐
    entity identifier **is**
        ⟦ **port** (*port*_interface_list) ; ⟧
        ❴ entity_declarative_item ❵
    ⟦ **begin**
        ❴ concurrent_assertion_statement
        ❘ *passive*_concurrent_procedure_call_statement
        ❘ *passive*_process_statement ❵ ⟧
    end ⟦ **entity** ⟧ ⟦ identifier ⟧ ;

The concurrent statements included in an entity declaration must be *passive*; that is, they may not affect the operation of the entity in any way. A concurrent assertion statement meets this requirement, since it simply tests a condition whenever events occur on signals to which it is sensitive. A process statement is passive if it contains no signal assignment statements or calls to procedures containing signal assignment statements. Such a process can be used to trace events that occur on the entity's inputs. We will describe the remaining alternative, concurrent procedure call statements, when we discuss procedures in Chapter 6. A concurrent procedure call is passive if the procedure called contains no signal assignment statements or calls to procedures containing signal assignment statements.

EXAMPLE 5.20 *A revised set/reset flipflop with usage check*

We can rewrite the entity declaration for the set/reset flipflop of Example 5.19 as shown below, using a concurrent assertion statement for the usage check. If we do this, the check is included for every possible implementation of the flipflop and does not need to be included in the corresponding architecture bodies.

```
entity S_R_flipflop is
  port ( s, r : in bit;  q, q_n : out bit );

begin

  check : assert not (s and r)
            report "Incorrect use of S_R_flip_flop: " &
                "s and r both '1'";

end entity S_R_flipflop;
```

EXAMPLE 5.21 *A ROM that traces read operations*

The following entity declaration for a read-only memory (ROM) includes a passive process, **trace_reads**, that is sensitive to changes on the **enable** port. When the value of the port changes to '1', the process reports a message tracing the time and address of the read operation. The process does not affect the course of the simulation in any way, since it does not include any signal assignments.

```
entity ROM is
  port ( address : in natural;
          data : out bit_vector(0 to 7);
          enable : in bit );

begin

  trace_reads : process (enable) is
  begin
    if enable then
      report "ROM read at time " & to_string(now)
```

```
                        & " from address " & to_string(address);
            end if;
        end process trace_reads;

    end entity ROM;
```

5.3   Structural Descriptions

A structural description of a system is expressed in terms of subsystems interconnected by signals. Each subsystem may in turn be composed of an interconnection of sub-subsystems, and so on, until we finally reach a level consisting of primitive components, described purely in terms of their behavior. Thus the top-level system can be thought of as having a hierarchical structure. In this section, we look at how to write structural architecture bodies to express this hierarchical organization.

We have seen earlier in this chapter that the concurrent statements in an architecture body describe an implementation of an entity interface. In order to write a structural implementation, we must use a concurrent statement called a *component instantiation statement*, the simplest form of which is governed by the syntax rule

> component_instantiation_statement ⇐
>   *instantiation*_label :
>   **entity** *entity*_name ⟦ (*architecture*_identifier) ⟧
>   ⟦ **port map** (port_association_list) ⟧ ;

This form of component instantiation statement performs *direct instantiation* of an entity. We can think of component instantiation as creating a copy of the named entity, with the corresponding architecture body substituted for the component instance. The port map specifies which ports of the entity are connected to which signals in the enclosing architecture body. The simplified syntax rule for a port association list is

> port_association_list ⇐
>   ⟨ ⟦ *port*_name => ⟧ ⟨ *signal*_name ▯ expression ▯ **open** ⟩ ⟩ ⟨ , ₒₒₒ ⟩

Each element in the association list associates one port of the entity either with one signal of the enclosing architecture body or with the value of an expression, or leaves the port unassociated, as indicated by the keyword **open**.

Let us look at some examples to illustrate component instantiation statements and the association of ports with signals. Suppose we have an entity declared as

```
entity DRAM_controller is
  port ( rd, wr, mem : in bit;
            ras, cas, we, ready : out bit  );
end entity DRAM_controller;
```

and a corresponding architecture called **fpld**. We might create an instance of this entity as follows:

```
main_mem_controller : entity work.DRAM_controller(fpld)
  port map ( cpu_rd, cpu_wr, cpu_mem,
             mem_ras, mem_cas, mem_we, cpu_rdy );
```

In this example, the name **work** refers to the current working library in which entities and architecture bodies are stored. We return to the topic of libraries in the next section. The port map of this example lists the signals in the enclosing architecture body to which the ports of the copy of the entity are connected. *Positional association* is used: each signal listed in the port map is connected to the port at the same position in the entity declaration. So the signal **cpu_rd** is connected to the port **rd**, the signal **cpu_wr** is connected to the port **wr** and so on.

One of the problems with positional association is that it is not immediately clear which signals are being connected to which ports. Someone reading the description must refer to the entity declaration to check the order of the ports in the entity interface. A better way of writing a component instantiation statement is to use *named association*, as shown in the following example:

```
main_mem_controller : entity work.DRAM_controller(fpld)
  port map ( rd => cpu_rd, wr => cpu_wr,
             mem => cpu_mem, ready => cpu_rdy,
             ras => mem_ras, cas => mem_cas, we => mem_we );
```

Here, each port is explicitly named along with the signal to which it is connected. The order in which the connections are listed is immaterial. The advantage of this approach is that it is immediately obvious to the reader how the entity is connected into the structure of the enclosing architecture body.

In the preceding example we have explicitly named the architecture body to be used corresponding to the entity instantiated. However, the syntax rule for component instantiation statements shows this to be optional. If we wish, we can omit the specification of the architecture body, in which case the one to be used may be chosen when the overall model is processed for simulation, synthesis or some other purpose. At that time, if no other choice is specified, the most recently analyzed architecture body is selected. We return to the topic of analyzing models in the next section.

EXAMPLE 5.22   *A structural model for a two-digit counter*

In Example 5.5 we looked at a behavioral model of an edge-triggered flipflop. We can use the flipflop as the basis of a 4-bit edge-triggered register, described by the following entity declaration and structural architecture body.

```
entity reg4 is
  port ( clk, clr, d0, d1, d2, d3 : in bit;
         q0, q1, q2, q3 : out bit );
end entity reg4;
```

```vhdl
architecture struct of reg4 is
begin

  bit0 : entity work.edge_triggered_Dff(behavioral)
    port map (d0, clk, clr, q0);
  bit1 : entity work.edge_triggered_Dff(behavioral)
    port map (d1, clk, clr, q1);
  bit2 : entity work.edge_triggered_Dff(behavioral)
    port map (d2, clk, clr, q2);
  bit3 : entity work.edge_triggered_Dff(behavioral)
    port map (d3, clk, clr, q3);

end architecture struct;
```

We can use the register entity, along with other entities, as part of a structural architecture for the two-digit decimal counter represented by the schematic of Figure 5.6.

FIGURE 5.6

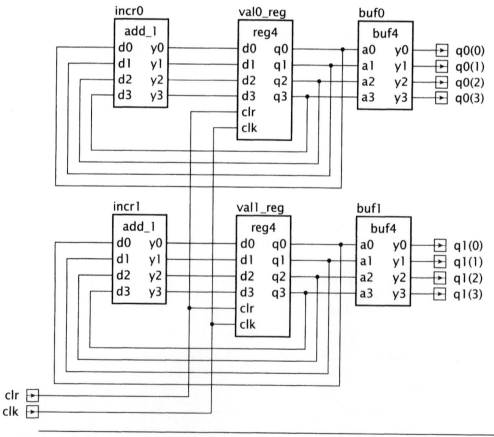

A schematic for a two-digit counter using the reg4 entity.

Suppose a digit is represented as a bit vector of length four, described by the subtype declaration

subtype digit **is** bit_vector(3 **downto** 0);

An entity declaration for the counter, along with an outline of the structural architecture body, are:

```
entity counter is
  port ( clk, clr : in bit;
         q0, q1 : out digit );
end entity counter;

-------------------------------------------------

architecture registered of counter is

  signal current_val0, current_val1, next_val0, next_val1 : digit;

begin

  val0_reg : entity work.reg4(struct)
    port map ( d0 => next_val0(0), d1 => next_val0(1),
               d2 => next_val0(2), d3 => next_val0(3),
               q0 => current_val0(0), q1 => current_val0(1),
               q2 => current_val0(2), q3 => current_val0(3),
               clk => clk, clr => clr );

  val1_reg : entity work.reg4(struct)
    port map ( d0 => next_val1(0), d1 => next_val1(1),
               d2 => next_val1(2), d3 => next_val1(3),
               q0 => current_val1(0), q1 => current_val1(1),
               q2 => current_val1(2), q3 => current_val1(3),
               clk => clk, clr => clr );

  incr0 : entity work.add_1(boolean_eqn) ...;

  incr1 : entity work.add_1(boolean_eqn) ...;

  buf0 : entity work.buf4(basic) ...;

  buf1 : entity work.buf4(basic) ...;

end architecture registered;
```

This example illustrates a number of important points about component instances and port maps. First, the two component instances **val0_reg** and **val1_reg** are both instances of the same entity/architecture pair. This means that two distinct copies of the architecture **struct** of **reg4** are created, one for each of the component instances. We return to this point when we discuss the topic of elaboration in the next section. Second, in each of the port maps, ports of the entity being instantiated are associated with separate elements of array signals. This is allowed, since a signal that is of a composite type, such as an array, can be treated as a collection of signals, one per element.

Third, some of the signals connected to the component instances are signals declared within the enclosing architecture body, **registered**, whereas the **clk** signal is a port of the entity **counter**. This again illustrates the point that within an architecture body, the ports of the corresponding entity are treated as signals.

We saw in the above example that we can associate separate ports of an instance with individual elements of an actual signal of a composite type, such as an array or record type. If an instance has a composite port, we can write associations the other way around; that is, we can associate separate actual signals with individual elements of the port. This is sometimes called *subelement association*. For example, if the instance **DMA_buffer** has a port **status** of type **FIFO_status**, declared as

```
type FIFO_status is record
    nearly_full, nearly_empty, full, empty : bit;
  end record FIFO_status;
```

we could associate a signal with each element of the port as follows:

```
DMA_buffer : entity work.FIFO
  port map ( ...,
              status.nearly_full => start_flush,
              status.nearly_empty => end_flush,
              status.full => DMA_buffer_full,
              status.empty => DMA_buffer_empty, ... );
```

This illustrates two important points about subelement association. First, all elements of the composite port must be associated with an actual signal. We cannot associate some elements and leave the rest unassociated. Second, all of the associations for a particular port must be grouped together in the association list, without any associations for other ports among them.

We can use subelement association for ports of an array type by writing an indexed element name on the left side of an association. Furthermore, we can associate a slice of the port with an actual signal that is a one-dimensional array, as the following example shows.

EXAMPLE 5.23   *Using subelement association for bus connections*

Suppose we have a register entity, declared as follows. The ports **d** and **q** are arrays of bits.

```
entity reg is
  port ( d : in bit_vector(7 downto 0);
          q : out bit_vector(7 downto 0);
          clk : in bit );
  end entity reg;
```

The architecture body for a microprocessor, outlined below, instantiates this entity as the program status register (PSR). Individual bits within the register represent condition and interrupt flags, and the field from bit 6 down to bit 4 represents the current interrupt priority level.

```
architecture RTL of microprocessor is

  signal interrupt_req : bit;
  signal interrupt_level : bit_vector(2 downto 0);
  signal carry_flag, negative_flag,
         overflow_flag, zero_flag : bit;
  signal program_status : bit_vector(7 downto 0);
  signal clk_PSR : bit;
  ...

begin

  PSR : entity work.reg
    port map ( d(7) => interrupt_req,
               d(6 downto 4) => interrupt_level,
               d(3) => carry_flag,      d(2) => negative_flag,
               d(1) => overflow_flag,   d(0) => zero_flag,
               q => program_status,
               clk => clk_PSR );

  ...

end architecture RTL;
```

In the port map of the instance, subelement association is used for the input port d to connect individual elements of the port with separate actual signals of the architecture. A slice of the port is connected to the **interrupt_level** signal. The output port q, on the other hand, is associated wholly with the bit-vector signal **program_status**.

We may also use subelement association for a port that is of an unconstrained or partially constrained array type whose top-level index bounds are not defined. The index bounds of the port are determined by the least and greatest index values used in the association list, and the index range direction is determined by the port type. For example, suppose we declare an and gate entity:

```
entity and_gate is
  port ( i : in bit_vector;  y : out bit );
end entity and_gate;
```

and a number of signals:

```
signal serial_select, write_en, bus_clk, serial_wr : bit;
```

We can instantiate the entity as a three-input and gate:

```
serial_write_gate : entity work.and_gate
  port map ( i(1) => serial_select,
             i(2) => write_en,
             i(3) => bus_clk,
             y => serial_wr );
```

Since the input port i is unconstrained, the index values in the subelement associations determine the index bounds for this instance. The least value is one and the greatest value is three. The port type is bit_vector, which has an ascending index range. Thus, the index range for the port in the instance is an ascending range from one to three.

This method of determining index ranges also applies to subelements with undefined index bounds. For example, given a type

```
type bv_pair is array (1 to 2) of bit_vector;
```

and a port declared in an entity as

```
entity ent3 is
  port ( p : in bv_pair );
end entity ent3;
```

we can write an instance of the entity:

```
signal s1, s2 : bit;
signal sv1, sv2 : bit_vector(4 to 7);
...

inst3 : entity work.ent3
  port map ( p(1)(0) => s1, p(1)(1 to 4) => sv1,
             p(2)(0) => s2, p(2)(1 to 4) => sv2 );
```

Here, the index range for the top level of p is 1 to 2, determined from the subtype bv_pair. However, for the elements, the subtype index range is not defined, so the index range for p comes from the formal element and slice names. Combining these effects, the index ranges for p are 1 to 2 for the top level, and 0 to 4 for the elements. Note that the index range determined for the two elements p(1) and p(2) must be the same. It would be illegal to write the instance as:

```
signal s1, s2 : bit;
signal sv1, sv2 : bit_vector(4 to 7);
...

inst3 : entity work.ent3
  port map ( p(1)(0) => s1, p(1)(1 to 4) => sv1,
             p(2)(15) => s2, p(2)(11 to 14) => sv2 ); -- illegal
```

since that would imply two different index ranges: 0 to 4 for p(1) and 11 to 15 for p(2). An array must have the same index ranges for all elements.

The syntax rule for a port association list shows that a port of a component instance may be associated with an expression instead of a signal. There are two possibilities for this case. First, if the expression is *globally static*, the value of the expression is used as a constant value for the port throughout the simulation. For an expression to be globally static, we must be able to determine the value from constants defined when the model is elaborated. So, for example, the expression must not include references to any signals. If real hardware is synthesized from the model, the port of the component instance would be tied to a fixed value determined by the expression. Association with an expression of this form is useful when we have an entity provided as part of a library, but we do not need to use all of the functionality provided by the entity.

EXAMPLE 5.24 *Using a four-input multiplexer as a two-input multiplexer*

Given a four-input multiplexer described by the entity declaration

```
entity mux4 is
  port ( i0, i1, i2, i3, sel0, sel1 : in bit;
         z : out bit );
end entity mux4;
```

we can use it as a two-input multiplexer by instantiating it as follows:

```
a_mux : entity work.mux4
  port map ( sel0 => select_line, i0 => line0, i1 => line1,
             z => result_line,
             sel1 => '0', i2 => '1', i3 => '1' );
```

For this component instance, the high-order select bit is fixed at '0', ensuring that only one of line0 or line1 is passed to the output. We have also followed the practice, recommended for many logic families, of tying unused inputs to a fixed value, in this case '1'.

In the second case of association with an expression, the expression is not static, but instead involves the values of signals. This allows us to include a small amount of functional logic in a port map, and avoids the need to express the logic with a separate assignment statement and an intermediate signal. If the expression is not static, the port association is defined to be equivalent to association with an anonymous signal that is the target of a signal assignment with the expression on the right-hand side.

EXAMPLE 5.25 *Including select logic in a port map*

Suppose an I/O controller connected to a CPU bus is to be enabled when bus control signals indicate a read from I/O address space and the bus address matches the controller's address. We can include the select logic in the port map for the controller instance:

```
io_ctrl_1 : entity work.io_controller(rtl)
  port map ( en => rd_en and io_sel and addr ?= io_base,
             ... );
```

This is a much more succinct way of expressing the model than the equivalent:

```
signal en_tmp : std_ulogic;
...

en_tmp <= rd_en and io_sel and addr ?= io_base;

io_ctrl_1 : entity work.io_controller(rtl)
  port map ( en => en_tmp,
             ... );
```

Some entities may be designed to allow inputs to be left open by specifying a default value for a port. When the entity is instantiated, we can specify that a port is to be left open by using the keyword **open** in the port association list, as shown in the syntax rule on page 176.

EXAMPLE 5.26   *And-or-invert with an unconnected input*

The and_or_inv entity declaration on page 139 includes a default value of '1' for each of its input ports, as again shown here:

```
entity and_or_inv is
  port ( a1, a2, b1, b2 : in bit := '1';
         y : out bit );
end entity and_or_inv;
```

We can write a component instantiation to perform the function **not** ((A **and** B) **or** C) using this entity as follows:

```
f_cell : entity work.and_or_inv
  port map ( a1 => A, a2 => B, b1 => C, b2 => open, y => F );
```

The port b2 is left open, so it assumes the default value '1' specified in the entity declaration.

There is some similarity between specifying a default value for an input port and associating an input port with a globally static expression. In both cases we must be able to determine the expression's value when the model is elaborated. The difference is that a default value is only used if the port is left open when the entity is instantiated, whereas association with a globally static expression specifies that the expression value is to be used to drive the port for the entire simulation or life of the component instance. If a port is declared with a default value and then associated with an expression, the expression value is used, overriding the default value.

Output and bidirectional ports may also be left unassociated using the **open** keyword, provided they are not of an unconstrained or partially constrained composite type. If a port of mode **out** or **buffer** is left open, any value driven by the entity is ignored. If a port of mode **inout** is left open, the value used internally by the entity (the *effective value*) is the value that it drives on to the port.

A final point to make about unassociated ports is that we can simply omit a port from a port association list to specify that it remain open. So, given an entity declared as follows:

```
entity and3 is
  port ( a, b, c : in bit := '1';
         z, not_z : out bit);
end entity and3;
```

the component instantiation

```
g1 : entity work.and3 port map ( a => s1, b => s2, not_z => ctrl1 );
```

has the same meaning as

```
g1 : entity work.and3 port map ( a => s1, b => s2, not_z => ctrl1,
                                 c => open, z => open );
```

The difference is that the second version makes it clear that the unused ports are deliberately left open, rather than being accidentally overlooked in the design process. This is useful information for someone reading the model.

VHDL-87, -93, and -2002

These earlier versions of VHDL do not allow a non-static expression in a port map. Instead, we must write the equivalent intermediate signal declaration and assignment explicitly.

VHDL-87 and VHDL-93

VHDL-87 and VHDL-93 impose a number of restrictions on how **buffer**-mode ports may be interconnected with other ports in structural designs. First, if the actual object associated with a buffer port of a component instance is a port of the enclosing entity, it must also be a **buffer** port. Second, if we associate a **buffer** as an actual object with some formal port of a component instance, the formal port must be of mode **in** or **buffer**. It may not be a port of mode **out**. There are also restrictions on the number of sources contributing to a **buffer** port or to a signal that is associated with a **buffer** port. These restrictions severely limit the uses of **buffer** ports, so they are not commonly used in VHDL-87 or VHDL-93.

VHDL-87

VHDL-87 does not allow direct instantiation. Instead, we must declare a *component* with a similar interface to the entity, instantiate the component and *bind* each component instance to the entity and an associated architecture body. Component declarations and binding are described in Chapter 13.

VHDL-87 does not allow association of an expression with a port in a port map. However, we can achieve a similar effect by declaring a signal, initializing it to the value of the expression and associating the signal with the port. For example, if we declare two signals

```
signal tied_0 : bit := '0';
signal tied_1 : bit := '1';
```

we can rewrite the port map shown in Example 5.24 as

```
port map ( sel0 => select_line, i0 => line0, i1 => line1,
           z => result_line,
           sel1 => tied_0, i2 => tied_1, i3 => tied_1 );
```

5.4   Design Processing

Now that we have seen how a design may be described in terms of entities, architectures, component instantiations, signals and processes, it is time to take a practical view. A VHDL description of a design is usually used to simulate the design and perhaps to synthesize the hardware. This involves processing the description using computer-based tools to create a simulation program to run or a hardware net-list to build. Both simulation and synthesis require two preparatory steps: analysis and elaboration. Simulation then involves executing the elaborated model, whereas synthesis involves creating a net-list of primitive circuit elements that perform the same function as the elaborated model. In this section, we look at the analysis, elaboration and execution operations introduced in Chapter 1. We will leave a discussion of synthesis to Chapter 21.

5.4.1   Analysis

The first step in processing a design is to analyze the VHDL descriptions. A correct description must conform to the rules of syntax and semantics that we have discussed at length. An *analyzer* is a tool that verifies this. If a description fails to meet a rule, the analyzer provides a message indicating the location of the problem and which rule was broken. We can then correct the error and retry the analysis. Another task performed by the analyzer in most VHDL systems is to translate the description into an internal form more easily processed by the remaining tools. Whether such a translation is done or not, the analyzer places each successfully analyzed description into a *design library*.

A complete VHDL description usually consists of a number of entity declarations and their corresponding architecture bodies. Each of these is called a *design unit*. Organizing a design as a hierarchy of modules, rather than as one large flat design, is good engineering practice. It makes the description much easier to understand and manage.

The analyzer analyzes each design unit separately and places the internal form into the library as a *library unit*. If a unit being analyzed uses another unit, the analyzer extracts information about the other unit from the library to check that the unit is used correctly. For example, if an architecture body instantiates an entity, the analyzer needs to check the number, type and mode of ports of the entity to make sure it is instantiated correctly. To do this, it requires that the entity be previously analyzed and stored in the library. Thus, we see that there are dependency relations between library units in a complete description that enforce an order of analysis of the original design units.

To clarify this point, we divide design units into *primary units*, which include entity declarations, and *secondary units*, which include architecture bodies. There are other kinds of design units in each class, which we come to in later chapters. A primary unit defines the external view or interface to a module, whereas a secondary unit describes an implementation of the module. Thus the secondary unit depends on the corresponding primary unit and must be analyzed after the primary unit has been analyzed. In addition, a library unit may draw upon the facilities defined in some other primary unit, as in the case of an architecture body instantiating some other entity. In this case, there is a further dependency between the secondary unit and the referenced primary unit. Thus we may build up a network of dependencies of units upon primary units. Analysis must be done in such an order that a unit is analyzed before any of its dependents. Furthermore, whenever we change and reanalyze a primary unit, all of the dependent units must also be reanalyzed. Note, however, that there is no way in which any unit can be dependent upon a secondary unit; that is what makes a secondary unit "secondary." This may seem rather complicated, and indeed, in a large design, the dependency relations can form a complex network. For this reason, most VHDL systems include tools to manage the dependencies, automatically reanalyzing units where necessary to ensure that an outdated unit is never used.

EXAMPLE 5.27 *Dependencies in the counter model*

The structural architecture of the **counter** module, described in Example 5.22, leads to the network of dependencies shown in Figure 5.7. One possible order of compilation for this set of design units is

    entity edge_triggered_Dff
    architecture behav of edge_triggered_Dff

    entity reg4
    architecture struct of reg4

    entity add_1
    architecture boolean_eqn of add_1

    entity buf4
    architecture basic of buf4

FIGURE 5.7

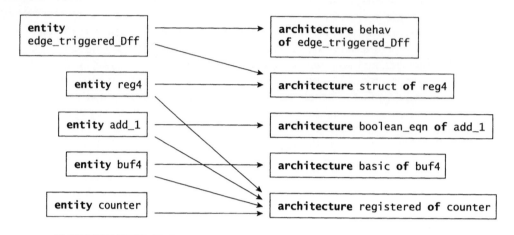

The dependency network for the counter module. The arrows point from a primary unit to a dependent secondary unit.

    entity **counter**
    architecture **registered** of **counter**

In this order, each primary unit is analyzed immediately before its corresponding secondary unit, and each primary unit is analyzed before any secondary unit that instantiates it. This is not the only possible order. Another alternative is to analyze all of the entity declarations first, then analyze the architecture bodies in arbitrary order.

5.4.2 Design Libraries and Contexts

So far, we have not actually said what a design library is, other than that it is where library units are stored. Indeed, this is all that is defined by the VHDL language specification, since to go further is to enter into the domain of the host operating system under which the VHDL tools are run. Some systems may use a database to store analyzed units, whereas others may simply use a directory in the host file system as the design library. The documentation for each VHDL tool suite indicates what we need to know about how the suite deals with design libraries.

A VHDL tool suite must also provide some means of using a number of separate design libraries. When a design is analyzed, we nominate one of the libraries as the *working library*, and the analyzed design is stored in this library. We use the special library name **work** in our VHDL models to refer to the current working library. We have seen examples of this in this chapter's component instantiation statements, in which a previously analyzed entity is instantiated in an architecture body.

If we need to access library units stored in other libraries, we refer to the libraries as *resource libraries*. We do this by including a *library clause* immediately preceding a design unit that accesses the resource libraries. The syntax rule for a library clause is library_clause ⇐ **library** identifier ⟨ , ₀₀₀ ⟩ ;

The identifiers are used by the analyzer and the host operating system to locate the design libraries, so that the units contained in them can be used in the description being analyzed. The exact way that the identifiers are used varies between different tool suites and is not defined by the VHDL language specification. Note that we do not need to include the library name **work** in a library clause; the current working library is automatically available.

EXAMPLE 5.28   *Using library cells*

Suppose we are working on part of a large design project code-named Wasp, and we are using standard cell parts supplied by Widget Designs, Inc. Our system administrator has loaded the design library for the Widget cells in a directory called /local/widget/cells in our workstation file system, and our project leader has set up another design library in /projects/wasp/lib for some in-house cells we need to use. We consult the manual for our VHDL analyzer and use operating system commands to set up the appropriate mapping from the identifiers **widget_cells** and **wasp_lib** to these library directories. We can then instantiate entities from these libraries, along with entities we have previously analyzed, into our own working library, as follows:

```
library widget_cells, wasp_lib;

architecture cell_based of filter is

  -- declaration of signals, etc
  ...

begin
  clk_pad : entity wasp_lib.in_pad
    port map ( i => clk, z => filter_clk );

  accum : entity widget_cells.reg32
    port map ( en => accum_en, clk => filter_clk, d => sum,
               q => result );

  alu : entity work.adder
    port map ( a => alu_op1, b => alu_op2, y => sum, c => carry );

  -- other component instantiations
  ...

end architecture cell_based;
```

If we need to make frequent reference to library units from a design library, we can include a use clause in our model to avoid having to write the library name each time. The simplified syntax rules are

use_clause ⇐ **use** selected_name ⟨ , ₀₀₀ ⟩ ;

selected_name ⇐ name . ⟦ identifier ▯ **all** ⟧

If we include a use clause with a library name as the prefix of the selected name (preceding the dot), and a library unit name from the library as the suffix (after the dot), the library unit is made *directly visible*. This means that subsequent references in the model to the library unit need not prefix the library unit name with the library name. For example, we might precede the architecture body in the previous example with the following library and use clauses:

```
library widget_cells, wasp_lib;

use widget_cells.reg32;
```

This makes **reg32** directly visible within the architecture body, so we can omit the library name when referring to it in component instantiations; for example:

```
accum : entity reg32
   port map ( en => accum_en, clk => filter_clk, d => sum,
              q => result );
```

If we include the keyword **all** in a use clause, all of the library units within the named library are made directly visible. For example, if we wanted to make all of the Wasp project library units directly visible, we might precede a library unit with the use clause

```
use wasp_lib.all;
```

Care should be taken when using this form of use clause with several libraries at once. If two libraries contain library units with the same name, VHDL avoids ambiguity by making neither of them directly visible. The solution is either to use the full selected name to refer to the particular library unit required, or to include in use clauses only those library units really needed in a model.

Use clauses can also be included to make names from packages directly visible. We will return to this idea when we discuss packages in detail in Chapter 7.

Context Declarations

Complex designs often call upon design units from several libraries and make use of several packages. As a consequence, we would need to precede each design unit with a long list of library and use clauses, many of which are common to all of the design units. VHDL provides a further form of design unit, a *context declaration*, in which we can gather a collection of library and use clauses. We can refer to a context declaration before a design unit, rather than having to repeat the collection of library and use clauses. The syntax rule for a context declaration is

context_declaration ⇐
    context identifier **is**
        ⟨ library_clause ▯ use_clause ▯ context_reference ⟩
    end ⟦ **context** ⟧ ⟦ identifier ⟧ ;

Within a context declaration, we write library and use clauses in the same form as we have seen earlier. We refer to a declared context with a *context reference*. The syntax is similar to that of a use clause:

context_reference ⇐
    context selected_name ⟦ , ₀₀₀ ⟧ ;

We can write a context reference preceding a design unit, or nested within another context declaration. In each case, the context reference is equivalent to replacement by the list of library clauses and use clauses contained within the named context declaration.

EXAMPLE 5.29   *Using contexts for library management*

Suppose the methodology support team in Widgets, Inc., has assembled a library of reusable entities in a library with logical name **widget_lib**. The entities refer to types defined in the standard **std_logic_1164** package in library **ieee**, so designs that instantiate the entities will also need to refer to those types. The methodology team can provide a context declaration for use by projects in the organization:

```
context widget_context is
  library ieee;
  use ieee.std_logic_1164.all;
  use widget_lib.all;
end context widget_context;
```

This context declaration is analyzed into the **widget_lib** library. Given that a design needs to include a library clause for **widget_lib** in order to refer to the context declaration, there is no need to include that library clause in the context declaration itself. A design unit could reference the context declaration as follows:

```
library widget_lib;
context widget_lib.widget_context;
entity sample is
  ...
end entity sample;
```

Now suppose the Dongle project within Widgets, Inc., uses additional entities provided by a third party, Gizmos Corp., in library **gizmo_IP_lib**. The project also maintains a library **dongle_lib** for verified design units to be used in the project design flow. The project's EDA support person can provide a context declaration for these libraries, as well as referring to the organization's context declaration:

```
context dongle_context is
  library widget_lib;
  context widget_lib.widget_context;
  library gizmo_IP_lib;
  use gizmo_IP_lib.all;
```

```
    use dongle_lib.all;
  end context dongle_context;
```

The EDA support person analyzes this context declaration into the **dongle_lib** library. A designer can then refer to the context in a design unit as follows:

```
library dongle_lib;
context dongle_lib.dongle_context;
entity frobber is
  ...
end entity frobber;
```

The reference to **dongle_context** expands to include the reference to the organization's context and the library and use clauses for the third-party IP and the project repository. The reference to the organization's context in turn expands to include the library and use clauses for the standard packages and the organization's packages. Thus, the context clause written is equivalent to the following expanded context clause:

```
library dongle_lib;
library widget_lib;
library ieee;
use ieee.std_logic_1164.all;
use widget_lib.all;
library gizmo_IP_lib;
use gizmo_IP_lib.all;
use dongle_lib.all;
entity frobber is
  ...
end entity frobber;
```

As we have seen, VHDL uses library names to refer to physical design libraries. The mapping from a library name to a physical library is implementation defined, and may vary between analysis of different design units. In order to avoid confusion when using context declarations, VHDL requires that a library name map to the same physical library during analysis of a context declaration and analysis of a reference to that context declaration. For example, if the library name **gizmo_IP_lib** in the preceding example refers to **/home/dongle/gizmo/gizmo_IP_lib** when **dongle_context** is analyzed, the library name must refer to the same physical library when entity **frobber** is analyzed.

As further reinforcement of this principle, we can't include library clauses, use clauses, or context references before a context declaration, as we can for other design units. Thus, the following would be illegal:

```
library fizz_lib;  -- Illegal: precedes context declaration
context frazzle_ctx is
  use fizz_lib.fizz_pkg.all;
end context frazzle_ctx;
```

Instead, we should write the library clause inside the context declaration, so that it is included for any design unit that references the context declaration. Another related rule is that we cannot include a library clause referring to the working library, **work**, within a context declaration. Nor can we refer to the library name **work** in a use clause. The reason is that **work** is not defined for a context declaration, since context declarations don't have preceding context clauses.

VHDL-87, -93, and -2002

These versions of VHDL do not provide context declarations or context references. Instead, each design unit must be preceded with all of the library and use clauses required.

5.4.3   Elaboration

Once all of the units in a design hierarchy have been analyzed, the design hierarchy can be *elaborated*. The effect of elaboration is to "flesh out" the hierarchy, producing a collection of processes interconnected by *nets*. This is done by substituting the contents of an architecture body for every instantiation of its corresponding entity. Each net in the elaborated design consists of a signal and the ports of the substituted architecture bodies to which the signal is connected. (Recall that a port of an entity is treated as a signal within a corresponding architecture body.) Let us outline how elaboration proceeds, illustrating it step by step with an example.

Elaboration is a recursive operation, started at the topmost entity in a design hierarchy. We use the **counter** design from Example 5.22 as our topmost entity. The first step is to create the ports of the entity. Next, an architecture body corresponding to the entity is chosen. If we do not explicitly specify which architecture body to choose, the most recently analyzed architecture body is used. For this illustration, we use the architecture **registered**. This architecture body is then elaborated, first by creating any signals that it declares, then by elaborating each of the concurrent statements in its body. Figure 5.8 shows the **counter** design with the signals created.

The concurrent statements in this architecture are all component instantiation statements. Each of them is elaborated by creating new instances of the ports specified by the instantiated entity and joining them into the nets represented by the signals with which they are associated. Then the internal structure of the specified architecture body of the instantiated entity is copied in place of the component instance, as shown in Figure 5.9. The architectures substituted for the instances of the **add_1** and **buf4** entities are both behavioral, consisting of processes that read the input ports and make assignments to the output ports. Hence elaboration is complete for these architectures. However, the architecture **struct**, substituted for each of the instances of **reg4**, contains further signals and component instances. Hence they are elaborated in turn, producing the structure shown in Figure 5.10 for each instance. We have now reached a stage where we have a collection of nets comprising signals and ports, and processes that sense and drive the nets.

Each process statement in the design is elaborated by creating new instances of the variables it declares and by creating a driver for each of the signals for which it has signal

FIGURE 5.8

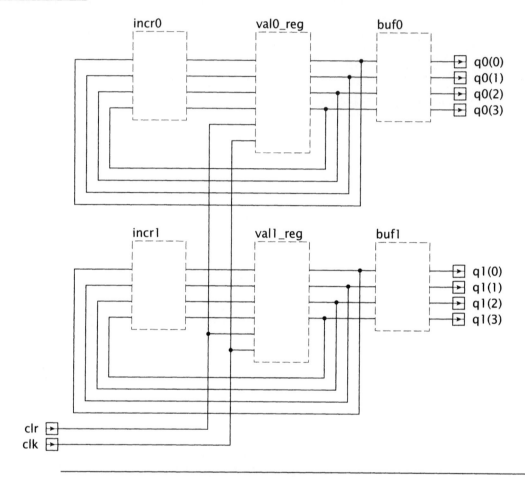

The first stage of elaboration of the **counter** *entity. The ports have been created, the architecture* **registered** *selected and the signals of the architecture created.*

assignment statements. The drivers are joined to the nets containing the signals they drive. For example, the **storage** process within **bit0** of **val0_reg** has a driver for the port **q**, which is part of the net based on the signal **current_val0(0)**.

Once all of the component instances and all of the resulting processes have been elaborated, elaboration of the design hierarchy is complete. We now have a fully fleshed-out version of the design, consisting of a number of process instances and a number of nets connecting them. Note that there are several distinct instances of some of the processes, one for each use of an entity containing the process, and each process instance has its own distinct version of the process variables. Each net in the elaborated design consists of a signal, a collection of ports associated with it and a driver within a process instance.

FIGURE 5.9

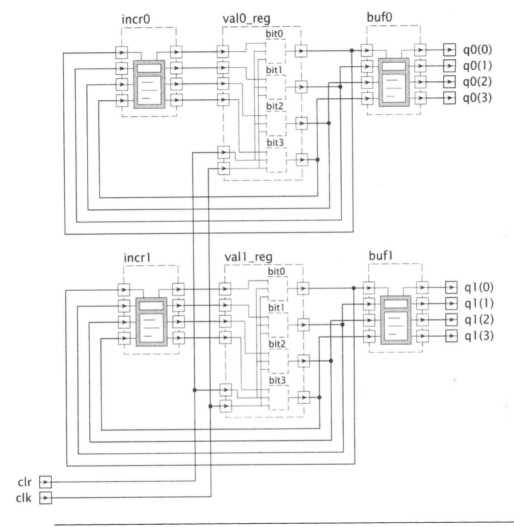

The counter design further elaborated. Behavioral architectures, consisting of just processes, have been substituted for instances of the **add_1** *and* **buf4** *entities. A structural architecture has been substituted for each instance of the* **reg4** *entity.*

5.4.4 Execution

Now that we have an elaborated design hierarchy, we can execute it to simulate operation of the system it describes. Much of our previous discussion of VHDL statements was in terms of what happens when they are executed, so we do not go over statement execution again here. Instead, we concentrate on the simulation algorithm introduced in Chapter 1.

FIGURE 5.10

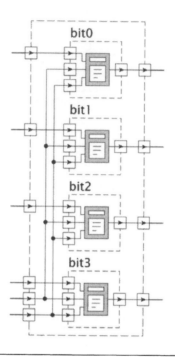

A register within the counter structure elaborated down to architectures that consist only of processes and signals.

Recall that the simulation algorithm consists of an initialization phase followed by a repeated simulation cycle. The simulator keeps a clock to measure out the passage of simulation time. In the initialization phase, the simulation time is set to zero. Each driver is initialized to drive its signal with the initial value declared for the signal or the default value for the signal if no initial value was declared. Next, each of the process instances in the design is started and executes the sequential statements in its body. We usually write a model so that at least some of these initial statements schedule some transactions to get the simulation under way, then suspend by executing a wait statement. When all of the process instances have suspended, initialization is complete and the simulator can start the first simulation cycle.

At the beginning of a simulation cycle, there may be a number of drivers with transactions scheduled on them and a number of process instances that have scheduled timeouts. The first step in the simulation cycle is to advance the simulation time clock to the earliest time at which a transaction or process timeout has been scheduled. Second, all of the transactions scheduled for this time are performed, updating the corresponding signals and possibly causing events on those signals. Third, all process instances that are sensitive to any of these events are resumed. In addition, process instances whose timeout expires at the current simulation time are resumed during this step. All of these processes execute their sequential statements, possibly scheduling more transactions or timeouts, and eventually suspend again by executing wait statements. When they have all suspended, the simulation cycle is done and the next cycle can start. If there are no more transactions or timeouts scheduled, or if simulation time reaches **time'high** (the largest representable time value), the simulation is complete.

Describing the operation of a simulator in this way is a little like setting a play in a theater without any seats—nobody is there to watch it, so what's the point! In reality, a simulator is part of a suite of VHDL tools and provides us with various means to control and monitor the progress of the simulation. Typical simulators allow us to step through the model one line at a time or to set breakpoints, causing the simulation to stop when a line of the model is executed or a signal is assigned a particular value. They usually provide commands to display the value of signals or variables. Many simulators also provide a graphical waveform display of the history of signal values similar to a logic analyzer display, and allow storage and subsequent redisplay of the history for later analysis. It is these facilities that make the simulation useful. Unfortunately, since there is a great deal of variation between the facilities provided by different simulators, it is not practical to go into any detail in this book. Simulator vendors usually provide training documentation and lab courses that explain how to use the facilities provided by their products.

Exercises

1. [❶ 5.1] Write an entity declaration for a lookup table ROM modeled at an abstract level. The ROM has an address input of type **lookup_index**, which is an integer range from 0 to 31, and a data output of type **real**. Include declarations within the declarative part of the entity to define the ROM contents, initialized to numbers of your choice.

2. [❶ 5.2] Write an architecture body for the ROM described in Exercise 1. Include a signal assignment that uses the address input to index the ROM content and assigns the resulting value to the data output with a delay of 200 ps.

3. [❶ 5.2] Trace the transactions applied to the signal s in the following process. At what times is the signal active, and at what times does an event occur on it?

```
process is
begin
  s <= 'Z', '0' after 10 ns, '1' after 30 ns;
  wait for 50 ns;
  s <= '1' after 5 ns; 'H' after 15 ns;
  wait for 50 ns;
  s <= 'Z';
  wait;
end process;
```

4. [❶ 5.2] Given the assignments to the signal s made by the process in Exercise 3, trace the values of the signals s'delayed(5 ns), s'stable(5 ns), s'quiet(5 ns) and s'transaction. What are the values of s'last_event, s'last_active and s'last_value at time 60 ns?

5. [❶ 5.2] Write a wait statement that suspends a process until a signal s changes from '1' to '0' while an enable signal en is '1'.

6. [❶ 5.2] Write a wait statement that suspends a process until a signal **ready** changes to '1' or until a maximum of 5 ms has elapsed.

7. [❶ 5.2] Suppose the signal **s** currently has the value '0'. What is the value of the Boolean variables **v1** and **v2** after execution of the following statements within a process?

    ```
    s <= '1';
    v1 := s = '1';
    wait on s;
    v2 := s = '1';
    ```

8. [❶ 5.2] Trace the transactions scheduled on the driver for **z** by the following statements, and show the values taken on by **z** during simulation.

    ```
    z <= transport '1' after 6 ns;
    wait for 3 ns;
    z <= transport '0' after 4 ns;
    wait for 5 ns;
    z <= transport '1' after 6 ns;
    wait for 1 ns;
    z <= transport '0' after 4 ns;
    ```

9. [❶ 5.2] Trace the transactions scheduled on the driver for **x** by the following statements, and show the values taken on by **x** during simulation. Assume **x** initially has the value zero.

    ```
    x <= reject 5 ns inertial   1 after 7 ns,
                               23 after 9 ns,
                                5 after 10 ns,
                               23 after 12 ns,
                               -5 after 15 ns;
    wait for 6 ns;
    x <= reject 5 ns inertial 23 after 7 ns;
    ```

10. [❶ 5.2] Identify the signals to which the following process is sensitive:

    ```
    logic_block : process (all) is
    begin
      out1 <= '0'; out2 <= '0';
      case current_state is
        when s0 => if in1 then
                       next_state <= s1;
                       out1 <= '1';
                   else
                       next_state <= idle;
                   end if;
        when s1 => next_state <= s2;
                   out2 <= '1';
        when s2 => next_state <= idle;
    ```

```
    end case;
  end process logic_block;
```

11. [● 5.2] Write the equivalent process for the conditional signal assignment statement

```
mux_logic :
  z <= a and not b after 5 ns
          when enable and not sel else
       x or y after 6 ns
          when enable and sel else
       '0' after 4 ns;
```

12. [● 5.2] Write the equivalent process for the selected signal assignment statement

```
with bit_vector'(s, r) select
  q <= unaffected when "00",
       '0' when "01",
       '1' when "10" | "11";
```

13. [● 5.2] Write a concurrent assertion statement that verifies that the time between changes of a clock signal, clk, is at least T_pw_clk.

14. [● 5.3] Write component instantiation statements to model the structure shown by the schematic diagram in Figure 5.11. Assume that the entity **ttl_74x74** and the corresponding architecture **basic** have been analyzed into the library **work**.

FIGURE 5.11

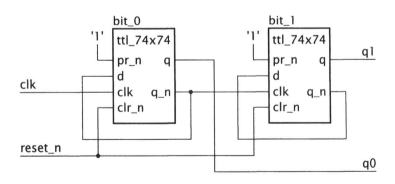

A schematic diagram of a 2-bit counter.

15. [● 5.3] Sketch a schematic diagram of the structure modeled by the following component instantiation statements.

```
decode_1 : entity work.ttl_74x138(basic)
  port map ( c => a(2), b => a(1), a => a(0),
             g1 => a(3), g2a_n => sel_n, g2b_n => '0',
             y7_n => en_n(15), y6_n => en_n(14),
```

```
            y5_n => en_n(13),  y4_n => en_n(12),
            y3_n => en_n(11),  y2_n => en_n(10),
            y1_n => en_n(9),   y0_n => en_n(8)  );
decode_0 : entity work.ttl_74x138(basic)
   port map ( c => a(2), b => a(1), a => a(0),
              g1 => '1', g2a_n => sel_n, g2b_n => a(3),
              y7_n => en_n(7),  y6_n => en_n(6),
              y5_n => en_n(5),  y4_n => en_n(4),
              y3_n => en_n(3),  y2_n => en_n(2),
              y1_n => en_n(1),  y0_n => en_n(0)  );
```

16. [❶ 5.4] Example 5.27 shows one possible order of analysis of the design units in the counter of Figure 5.7. Show two other possible orders of analysis.

17. [❶ 5.4] Write a context clause that makes the resource libraries company_lib and project_lib accessible and that makes directly visible the entities in_pad and out_pad from company_lib and all entities from project_lib.

18. [❶ 5.4] Write a context declaration that includes a library clause for libraries ieee, IP_worx, and phantom_lib, and use clauses for the package std_logic_1164 from ieee and for all library units from the other two libraries.

19. [❷ 5.2] Develop a behavioral model for a four-input multiplexer, with ports of type bit and a propagation delay from data or select input to data output of 4.5 ns. You should declare a constant for the propagation delay, rather than writing it as a literal in signal assignments in the model.

20. [❷ 5.2] Develop a behavioral model for a negative-edge-triggered 4-bit counter with asynchronous parallel load inputs. The entity declaration is

```
entity counter is
  port ( clk_n, load_en : in std_ulogic;
           d : in std_ulogic_vector(3 downto 0);
           q : out std_ulogic_vector(3 downto 0) );
end entity counter;
```

21. [❷ 5.2] Develop a behavioral model for a D-latch with a clock-to-output propagation delay of 3 ns and a data-to-output propagation delay of 4 ns.

22. [❷ 5.2] Develop a behavioral model for an edge-triggered flipflop that includes tests to verify the following timing constraints: data setup time of 3 ns, data hold time of 2 ns and minimum clock pulse width of 5 ns.

23. [❷ 5.2] Develop a model of an adder whose interface is specified by the following entity declaration:

```
entity adder is
  port ( a, b : in integer;  s : out integer );
end entity adder;
```

For each pair of integers that arrive on the inputs, the adder produces their sum on the output. Note that successive integers on each input may have the same value, so the adder must respond to transactions rather than to events. While integers in a pair may arrive in the inputs at different times, you may assume that neither value of the following pair will arrive until both values of the first pair have arrived. The adder should produce the sum only when both input values of a pair have arrived.

24. [❷ 5.2] Develop a behavioral model for a two-input Muller-C element, with two input ports and one output, all of type **bit**. The inputs and outputs are initially '0'. When both inputs are '1', the output changes to '1'. It stays '1' until both inputs are '0', at which time it changes back to '0'. Your model should have a propagation delay for rising output transitions of 3.5 ns, and for falling output transitions of 2.5 ns.

25. [❷ 5.2] The following process statement models a producer of data:

```
producer : process is
  variable next_data : natural := 0;
begin
  data <= next_data;  next_data := next_data + 1;
  data_ready <= '1';
  wait until data_ack;
  data_ready <= '0';
  wait until not data_ack;
end process producer;
```

The process uses a four-phase handshaking protocol to synchronize data transfer with a consumer process. Develop a process statement to model the consumer. It, too, should use delta delays in the handshaking protocol. Include the process statements in a test-bench architecture body, and experiment with your simulator to see how it deals with models that use delta delays.

26. [❷ 5.2] Develop a behavioral model for a multitap delay line, with the following interface:

```
entity delay_line is
  port ( input : in std_ulogic;
         output : out std_ulogic_vector );
end entity delay_line;
```

Each element of the output port is a delayed version of the input. The delay to the leftmost output element is 5 ns, to the next element is 10 ns and so on. The delay to the rightmost element is 5 ns times the length of the output port. Assume the delay line acts as an ideal transmission line.

27. [❷ 5.2] Develop a functional model using conditional signal assignment statements of an address decoder for a microcomputer system. The decoder has an address input port of type natural and a number of active-low select outputs, each activated when the address is within a given range. The outputs and their corresponding ranges are

ROM_sel_n   16#0000# to 16#3FFF#

RAM_sel_n	16#4000# to 16#5FFF#
PIO_sel_n	16#8000# to 16#8FFF#
SIO_sel_n	16#9000# to 16#9FFF#
INT_sel_n	16#F000# to 16#FFFF#

28. [❷ 5.2] Develop a functional model of a BCD-to-seven-segment decoder for a light-emitting diode (LED) display. The decoder has a 4-bit input that encodes a numeric digit between 0 and 9. There are seven outputs indexed from 'a' to 'g', corresponding to the seven segments of the LED display as shown in the margin. An output bit being '1' causes the corresponding segment to illuminate. For each input digit, the decoder activates the appropriate combination of segment outputs to form the displayed representation of the digit. For example, for the input "0010", which encodes the digit 2, the output is "1101101". Your model should use a selected signal assignment statement to describe the decoder function in truth-table form.

29. [❷ 5.2] Write an entity declaration for a 4-bit counter with an asynchronous reset input. Include a process in the entity declaration that measures the duration of each reset pulse and reports the duration at the end of each pulse.

30. [❷ 5.3] Develop a structural model of an 8-bit odd-parity checker using instances of an exclusive-or gate entity. The parity checker has eight inputs, i0 to i7, and an output, p, all of type **std_ulogic**. The logic equation describing the parity checker is

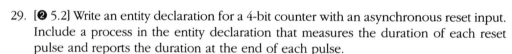

$$P = ((I_0 \oplus I_1) \oplus (I_2 \oplus I_3)) \oplus ((I_4 \oplus I_5) \oplus (I_6 \oplus I_7))$$

31. [❸ 5.3] Develop a structural model of a 14-bit counter with parallel load inputs, using instances of the 4-bit counter described in Exercise 20. Ensure that any unused inputs are properly connected to a constant driving value.

32. [❸ 5.2] Develop a behavioral model for a D-latch with tristate output. The entity declaration is

```
entity d_latch is
   port ( latch_en, out_en, d : in std_ulogic;
          q : out std_ulogic );
end entity d_latch;
```

When **latch_en** is asserted, data from the **d** input enters the latch. When **latch_en** is negated, the latch maintains the stored value. When **out_en** is asserted, data passes through to the output. When **out_en** is negated, the output has the value 'Z' (high-impedance). The propagation delay from **latch_en** to **q** is 3 ns and from **d** to **q** is 4 ns. The delay from **out_en** asserted to **q** active is 2 ns and from **out_en** negated to **q** high-impedance is 5 ns.

33. [❸ 5.2] Develop a functional model of a 4-bit carry-look-ahead adder. The adder has two 4-bit data inputs, a(3 **downto** 0) and b(3 **downto** 0); a 4-bit data output, s(3 **downto** 0); a carry input, c_in; a carry output, c_out; a carry generate output, g;

and a carry propagate output, **p**. The adder is described by the logic equations and associated propagation delays:

$$S_i = A_i \oplus B_i \oplus C_{i-1} \quad \text{(delay is 5 ns)}$$

$$G_i = A_i B_i \quad \text{(delay is 2 ns)}$$

$$P_i = A_i + B_i \quad \text{(delay is 3 ns)}$$

$$C_i = G_i + P_i C_{i-1}$$

$$= G_i + P_i G_{i-1} + P_i P_{i-1} G_{i-2} + \cdots + P_i P_{i-1} \ldots P_0 C_{-1} \quad \text{(delay is 5 ns)}$$

$$G = G_3 + P_3 G_2 + P_3 P_2 G_1 + P_3 P_2 P_1 G_0 \quad \text{(delay is 5 ns)}$$

$$P = P_3 P_2 P_1 P_0 \quad \text{(delay is 3 ns)}$$

where the G_i are the intermediate carry generate signals, the P_i are the intermediate carry propagate signals and the C_i are the intermediate carry signals. C_{-1} is **c_in** and C_3 is **c_out**. Your model should use the expanded equation to calculate the intermediate carries, which are then used to calculate the sums.

34. [❸ 5.2] Develop a behavioral model for a four-input arbiter with the following entity interface:

    ```
    entity arbiter is
      port ( request : in bit_vector(0 to 3);
             acknowledge : out bit_vector(0 to 3) );
    end entity arbiter;
    ```

    The arbiter should use a round-robin discipline for responding to requests. Include a concurrent assertion statement that verifies that no more than one acknowledgment is issued at once and that an acknowledgment is only issued to a requesting client.

35. [❸ 5.2] Write an entity declaration for a 7474 positive edge-triggered JK-flipflop with asynchronous active-low preset and clear inputs, and Q and $\overline{Q}$ outputs. Include concurrent assertion statements and passive processes as necessary in the entity declaration to verify that

    • the preset and clear inputs are not activated simultaneously,

    • the setup time of 6 ns from the J and K inputs to the rising clock edge is observed,

    • the hold time of 2 ns for the J and K inputs after the rising clock edge is observed and

    • the minimum pulse width of 5 ns on each of the clock, preset and clear inputs is observed.

    Write a behavioral architecture body for the flipflop and a test bench that exercises the statements in the entity declaration.

36. [❸ 5.3] Define entity interfaces for a microprocessor, a ROM, a RAM, a parallel I/O controller, a serial I/O controller, an interrupt controller and a clock generator. Use

instances of these entities and an instance of the address decoder described in Exercise 27 to develop a structural model of a microcomputer system.

37. **[❸ 5.3]** Develop a structural model of a 16-bit carry-look-ahead adder, using instances of the 4-bit adder described in Exercise 33. You will need to develop a carry-look-ahead generator with the following interface:

```
entity carry_look_ahead_generator is
  port ( p0, p1, p2, p3, g0, g1, g2, g3 : in bit;
         c_in : in bit;  c1, c2, c3 : out bit );
end entity carry_look_ahead_generator;
```

The carry-look-ahead generator is connected to the 4-bit adders as shown in Figure 5.12. It calculates the carry output signals using the generate, propagate and carry inputs in the same way that the 4-bit counters calculate their internal carry signals.

FIGURE 5.12

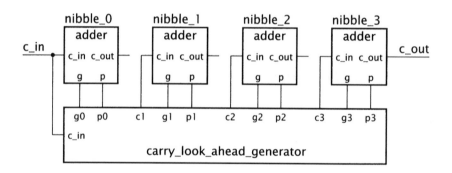

Connections between a carry-look-ahead generator and adders.

38. **[❹ 5.2]** Develop a behavioral model for a household burglar alarm. The alarm has inputs for eight sensors, each of which is normally '0'. When an intruder is detected, one of the sensors changes to '1'. There is an additional input from a key-switch and an output to a siren. When the key-switch input is '0', the alarm is disabled and the siren output is '0'. When the key-switch input changes to '1', there is a 30 s delay before the alarm is enabled. Once enabled, detection of an intruder starts another 30 s delay, after which time the siren output is set to '1'. If the key-switch input changes back to '0', the alarm is immediately disabled.

39. **[❹ 5.2]** In his book *Structured Computer Organization*, Tanenbaum describes the use of a Hamming code for error detection and correction of 16-bit data ([16], pages 44–48). Develop behavioral models for a Hamming code generator and for an error detector and corrector. Devise a test bench that allows you to introduce single-bit errors into the encoded data, to verify that the error corrector works properly.

40. **[❹ 5.2]** Develop a behavioral model of a 4K × 8-bit serial-input/output RAM. The device has a chip-enable input ce, a serial clock clk, a data input d_in and a data output

d_out. When **ce** is '1', the data input is sampled on 23 successive rising clock edges to form the 23 bits of a command string. A string of the form

$$1 \, A_{11} \, A_{10} \, ... \, A_0 \, 0 \, 1 \, D_7 \, D_6 \, ... \, D_0$$

is a write command, in which the bits A_i are the address and the bits D_j are the data to be written. A string of the form

$$1 \, A_{11} \, A_{10} \, ... \, A_0 \, 1 \, 1 \, X \, X \, X \, X \, X \, X \, X \, X$$

is a read command, in which the bits denoted by X are ignored. The RAM produces the successive bits of read data synchronously with the last eight rising clock edges of the command.

41. [❹ 5.2/5.3] Develop a model of a device to count the number of cars in a parking lot. The lot has a gate through which only one car at a time may enter or leave. There are two pairs, labeled A and B, each comprising an LED and a photodetector, mounted on the gate as shown in Figure 5.13. Each detector produces a '1' output when a car obscures the corresponding LED. When a car enters the yard, the front of the car obscures LED A, then LED B. When the car has advanced sufficiently, LED A becomes visible again, followed by LED B. The process is reversed for a car leaving the lot. Note that a car may partially enter or leave the lot and then reverse.

FIGURE 5.13

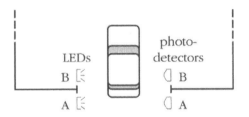

Arrangement of LEDs and photodetectors on a parking lot gate.

Your model should include a clocked finite-state machine (FSM) with two inputs, one from each detector, and increment and decrement outputs that pulse to '1' for one clock cycle when a car has totally entered or left the lot. The FSM outputs should drive a three-digit chain of BCD up/down counters, whose outputs are connected to seven-segment decoders.

Chapter 6

Subprograms

When we write complex behavioral models it is useful to divide the code into sections, each dealing with a relatively self-contained part of the behavior. VHDL provides a *subprogram* facility to let us do this. In this chapter, we look at the two kinds of subprograms: *procedures* and *functions*. The difference between the two is that a procedure encapsulates a collection of sequential statements that are executed for their effect, whereas a function encapsulates a collection of statements that compute a result. Thus a procedure is a generalization of a statement, whereas a function is a generalization of an expression.

6.1 Procedures

We start our discussion of subprograms with procedures. There are two aspects to using procedures in a model: first the procedure is declared, then elsewhere the procedure is *called*. The syntax rule for a procedure declaration is

```
subprogram_body ⇐
    procedure identifier [ ( parameter_interface_list ) ] is
        { subprogram_declarative_part }
    begin
        { sequential_statement }
    end [ procedure ] [ identifier ] ;
```

For now we will just look at procedures without the parameter list part; we will come back to parameters in the next section.

The *identifier* in a procedure declaration names the procedure. The name may be repeated at the end of the procedure declaration. The sequential statements in the body of a procedure implement the algorithm that the procedure is to perform and can include any of the sequential statements that we have seen in previous chapters. A procedure can declare items in its declarative part for use in the statements in the procedure body. The declarations can include types, subtypes, constants, variables and nested subprogram declarations. The items declared are not accessible outside of the procedure; we say they are *local* to the procedure.

EXAMPLE 6.1   *Averaging an array of data samples*

The following procedure calculates the average of a collection of data values stored
in an array called **samples** and assigns the result to a variable called **average**. This
procedure has a local variable **total** for accumulating the sum of array elements. Unlike variables in processes, procedure local variables are created anew and initialized
each time the procedure is called.

```vhdl
procedure average_samples is
  variable total : real := 0.0;
begin
  assert samples'length > 0 severity failure;
  for index in samples'range loop
    total := total + samples(index);
  end loop;
  average := total / real(samples'length);
end procedure average_samples;
```

The actions of a procedure are invoked by a *procedure call* statement, which is yet
another VHDL sequential statement. A procedure with no parameters is called simply by
writing its name, as shown by the syntax rule

procedure_call_statement ⇐ [label :] *procedure*_name ;

The optional label allows us to identify the procedure call statement. We will discuss labeled statements in Chapter 20. As an example, we might include the following statement
in a process:

```vhdl
average_samples;
```

The effect of this statement is to invoke the procedure **average_samples**. This involves
creating and initializing a new instance of the local variable **total**, then executing the statements in the body of the procedure. When the last statement in the procedure is
completed, we say the procedure *returns*; that is, the thread of control of statement execution returns to the process from which the procedure was called, and the next statement
in the process after the call is executed.

We can write a procedure declaration in the declarative part of an architecture body
or a process. We can also declare procedures within other procedures, but we will leave
that until a later section. If a procedure is included in an architecture body's declarative
part, it can be called from within any of the processes in the architecture body. On the
other hand, declaring a procedure within a process hides it away from use by other processes.

EXAMPLE 6.2 *A procedure to implement behavior within a process*

The outline below illustrates a procedure defined within a process. The procedure do_arith_op encapsulates an algorithm for arithmetic operations on two values, producing a result and a flag indicating whether the result is zero. It has a variable **result**, which it uses within the sequential statements that implement the algorithm. The statements also use the signals and other objects declared in the architecture body. The process **alu** invokes do_arith_op with a procedure call statement. The advantage of separating the statements for arithmetic operations into a procedure in this example is that it simplifies the body of the **alu** process.

```
architecture rtl of control_processor is

  type func_code is (add, subtract);

  signal op1, op2, dest : integer;
  signal Z_flag : boolean;
  signal func : func_code;
  ...

begin

  alu : process is

    procedure do_arith_op is
      variable result : integer;
    begin
      case func is
        when add =>
          result := op1 + op2;
        when subtract =>
          result := op1 - op2;
      end case;
      dest <= result after Tpd;
      Z_flag <= result = 0 after Tpd;
    end procedure do_arith_op;

  begin
    ...
    do_arith_op;
    ...
  end process alu;

  ...

end architecture rtl;
```

Another important use of procedures arises when some action needs to be performed several times at different places in a model. Instead of writing several copies of the statements to perform the action, the statements can be encapsulated in a procedure, which is then called from each place.

EXAMPLE 6.3   *A memory read procedure invoked from several places in a model*

The process outlined below is taken from a behavioral model of a CPU. The process fetches instructions from memory and interprets them. Since the actions required to fetch an instruction and to fetch a data word are identical, the process encapsulates them in a procedure, **read_memory**. The procedure copies the address from the memory address register to the address bus, sets the read signal to '1', then activates the request signal. When the memory responds, the procedure copies the data from the data bus signal to the memory data register and acknowledges to the memory by setting the request signal back to '0'. When the memory has completed its operation, the procedure returns.

```
instruction_interpreter : process is

  variable mem_address_reg, mem_data_reg,
           prog_counter, instr_reg, accumulator, index_reg : word;
  ...

  procedure read_memory is
  begin
    address_bus <= mem_address_reg;
    mem_read <= '1';
    mem_request <= '1';
    wait until mem_ready;
    mem_data_reg := data_bus_in;
    mem_request <= '0';
    wait until not mem_ready;
  end procedure read_memory;

begin
  ...  -- initialization
  loop
    -- fetch next instruction
    mem_address_reg := prog_counter;
    read_memory;        -- call procedure
    instr_reg := mem_data_reg;
    ...
    case opcode is
      ...
      when load_mem =>
        mem_address_reg := index_reg + displacement;
        read_memory;  -- call procedure
        accumulator := mem_data_reg;
      ...
    end case;
```

```
    end loop;
  end process instruction_interpreter;
```

The procedure is called in two places within the process. First, it is called to fetch an instruction. The process copies the program counter into the memory address register and calls the procedure. When the procedure returns, the process copies the data from the memory data register, placed there by the procedure, to the instruction register. The second call to the procedure takes place when a "load memory" instruction is executed. The process sets the memory address register using the values of the index register and some displacement, then calls the memory read procedure to perform the read operation. When it returns, the process copies the data to the accumulator.

Since a procedure call is a form of sequential statement and a procedure body implements an algorithm using sequential statements, there is no reason why one procedure cannot call another procedure. In this case, control is passed from the calling procedure to the called procedure to execute its statements. When the called procedure returns, the calling procedure carries on executing statements until it returns to its caller.

EXAMPLE 6.4 *Nested procedure calls in a control sequencer*

The process outlined below is a control sequencer for a register-transfer-level model of a CPU. It sequences the activation of control signals with a two-phase clock on signals **phase1** and **phase2**. The process contains two procedures, **control_write_back** and **control_arith_op**, that encapsulate parts of the control algorithm. The process calls **control_arith_op** when an arithmetic operation must be performed. This procedure sequences the control signals for the source and destination operand registers in the data path. It then calls **control_write_back**, which sequences the control signals for the register file in the data path, to write the value from the destination register. When this procedure is completed, it returns to the first procedure, which then returns to the process.

```
control_sequencer : process is

  procedure control_write_back is
  begin
    wait until phase1;
    reg_file_write_en <= '1';
    wait until not phase2;
    reg_file_write_en <= '0';
  end procedure control_write_back;

  procedure control_arith_op is
  begin
    wait until phase1;
    A_reg_out_en <= '1';
    B_reg_out_en <= '1';
```

```
            wait until not phase1;
            A_reg_out_en <= '0';
            B_reg_out_en <= '0';
            wait until phase2;
            C_reg_load_en <= '1';
            wait until not phase2;
            C_reg_load_en <= '0';
            control_write_back;   -- call procedure
        end procedure control_arith_op;

    ...

    begin

    ...

    control_arith_op;   -- call procedure

    ...

    end process control_sequencer;
```

VHDL-87

The keyword **procedure** may not be included at the end of a procedure declaration in VHDL-87. Procedure call statements may not be labeled in VHDL-87.

6.1.1 Return Statement in a Procedure

In all of the examples above, the procedures completed execution of the statements in their bodies before returning. Sometimes it is useful to be able to return from the middle of a procedure, for example, as a way of handling an exceptional condition. We can do this using a *return* statement, described by the simplified syntax rule

return_statement ⇐ ⟦ label : ⟧ return ;

The optional label allows us to identify the return statement. We will discuss labeled statements in Chapter 20. The effect of the return statement, when executed in a procedure, is that the procedure is immediately terminated and control is transferred back to the caller.

EXAMPLE 6.5 *A revised memory read procedure*

The following is a revised version of the instruction interpreter process from Example 6.3. The procedure to read from memory is revised to check for the reset signal becoming active during a read operation. If it does, the procedure returns immediately, aborting the operation in progress. The process then exits the fetch/execute loop and starts the process body again, reinitializing its state and output signals.

```
instruction_interpreter : process is

    ...
```

```
                   procedure read_memory is
                   begin
                     address_bus <= mem_address_reg;
                     mem_read <= '1';
                     mem_request <= '1';
                     wait until mem_ready or reset;
                     if reset then
                       return;
                     end if;
                     mem_data_reg := data_bus_in;
                     mem_request <= '0';
                     wait until not mem_ready;
                   end procedure read_memory;

                 begin
                   ...      -- initialization
                   loop
                     ...
                   read_memory;
                   exit when reset;
                     ...
                   end loop;
                 end process instruction_interpreter;
```

VHDL-87

Return statements may not be labeled in VHDL-87.

6.2 Procedure Parameters

Now that we have looked at the basics of procedures, we will discuss procedures that include parameters. A *parameterized procedure* is much more general in that it can perform its algorithm using different data objects or values each time it is called. The idea is that the caller passes parameters to the procedure as part of the procedure call, and the procedure then executes its statements using the parameters.

When we write a parameterized procedure, we include information in the *parameter interface list* (or *parameter list*, for short) about the parameters to be passed to the procedure. The syntax rule for a procedure declaration on page 207 shows where the parameter list fits in. Following is the syntax rule for a parameter list:

interface_list ⇐
    (⟦ **constant** ⟦ **variable** ⟦ **signal** ⟧
        identifier ⦃ , ₒₒₒ ⦄ : ⟦ mode ⟧ subtype_indication
                          ⟦ := *static*_expression ⟧) ⦃ ; ₒₒₒ ⦄

mode ⇐ **in** ⟦ **out** ⟦ **inout**

As we can see, it is similar to the port interface list used in declaring entities. This similarity is not coincidental, since they both specify information about objects upon which the user and the implementation must agree. In the case of a procedure, the user is the caller of the procedure, and the implementation is the body of statements within the procedure. The objects defined in the parameter list are called the *formal parameters* of the procedure. We can think of them as placeholders that stand for the *actual parameters*, which are to be supplied by the caller when it calls the procedure. Since the syntax rule for a parameter list is quite complex, let us start with some simple examples and work up from them.

EXAMPLE 6.6   *Using a parameter to select an arithmetic operation to perform*

Let's rewrite the procedure **do_arith_op** from Example 6.2 so that the function code is passed as a parameter. The new version is

```
procedure do_arith_op ( op : in func_code ) is
  variable result : integer;
begin
  case op is
    when add =>
      result := op1 + op2;
    when subtract =>
      result := op1 - op2;
  end case;
  dest  <=  result after Tpd;
  Z_flag  <=  result = 0 after Tpd;
end procedure do_arith_op;
```

In the parameter interface list we have identified one formal parameter named **op**. This name is used in the statements in the procedure to refer to the value that will be passed as an actual parameter when the procedure is called. The mode of the formal parameter is **in**, indicating that it is used to pass information into the procedure from the caller. This means that the statements in the procedure can use the value but cannot modify it. In the parameter list we have specified the type of the parameter as **func_code**. This indicates that the operations performed on the value in the statements must be appropriate for a value of this type, and that the caller may only pass a value of this type as an actual parameter.

Now that we have parameterized the procedure, we can call it from different places passing different function codes each time. For example, a call at one place might be

```
do_arith_op ( add );
```

The procedure call simply includes the actual parameter value in parentheses. In this case we pass the literal value **add** as the actual parameter. At another place in the model we might pass the value of the signal **func** shown in the model in Example 6.2:

```
do_arith_op ( func );
```

In this example, we have specified the mode of the formal parameter as **in**. Note that the syntax rule for a parameter list indicates that the mode is an optional part. If we leave it out, mode **in** is assumed, so we could have written the procedure as

procedure do_arith_op (op : func_code) **is** ...

While this is equally correct, it's not a bad idea to include the mode specification for **in** parameters, to make our intention explicitly clear.

The syntax rule for a parameter list also shows us that we can specify the *class* of a formal parameter, namely, whether it is a constant, a variable or a signal within the procedure. If the mode of the parameter is **in**, the class is assumed to be *constant*, since a constant is an object that cannot be updated by assignment. It is just a quirk of VHDL that we can specify both **constant** and **in**, even though to do so is redundant. Usually we simply leave out the keyword **constant**, relying on the mode to make our intentions clear. (The exceptions are parameters of access types, discussed in Chapter 15, and file types, discussed in Chapter 16.) For an **in**-mode constant-class parameter, we write an expression as the actual parameter. The value of this expression must be of the type specified in the parameter list. The value is passed to the procedure for use in the statements in its body.

Let us now turn to formal parameters of mode **out**. Such a parameter lets us transfer information out from the procedure back to the caller. Here is an example, before we delve into the details.

EXAMPLE 6.7   *A procedure for addition with overflow output*

The procedure below performs addition of two unsigned numbers represented as bit vectors of type **word32**, which we assume is defined elsewhere. The procedure has two **in**-mode parameters **a** and **b**, allowing the caller to pass two bit-vector values. The procedure uses these values to calculate the sum and overflow flag. Within the procedure, the two **out**-mode parameters, **sum** and **overflow**, appear as variables. The procedure performs variable assignments to update their values, thus transferring information back to the caller.

```
procedure addu ( a, b : in word32;
                 sum : out word32;
                 overflow : out bit ) is
  variable carry : bit := '0';
begin
  for index in sum'reverse_range loop
    sum(index) := a(index) xor b(index) xor carry;
    carry := ( a(index) and b(index) )
             or ( carry and ( a(index) xor b(index) ) );
  end loop;
  overflow := carry;
end procedure addu;
```

A call to this procedure may appear as follows:

```
variable PC, next_PC : word32;
variable overflow_flag : bit;
...

addu ( PC, X"0000_0004", next_PC, overflow_flag);
```

In this procedure call statement, the first two actual parameters are expressions whose values are passed in through the formal parameters **a** and **b**. The third and fourth actual parameters are the names of variables. When the procedure returns, the values assigned by the procedure to the formal parameters **sum** and **overflow** are used to update the variables **next_PC** and **overflow_flag**.

In the above example, the **out**-mode parameters are of the class *variable*. Since this class is assumed for **out** parameters, we usually leave out the class specification **variable**, although it may be included if we wish to state the class explicitly. We will come back to signal-class parameters in a moment. The mode **out** indicates that the procedure may update the formal parameters by variable assignment to transfer information back to the caller. The procedure may also read the values of the parameters, just as it can with **in**-mode parameters. The difference is that an **out**-mode parameter is not initialized with the value of the actual parameter. Instead, it is initalized in the same way as a locally declared variable, with the default initial value for the type of the parameter. When the procedure reads the parameter, it reads the parameter's current value, yielding the value most recently assigned within the procedure, or the initial value if no assignments have been made. For an **out** mode, variable-class parameter, the caller must supply a variable as an actual parameter. Both the actual parameter and the value returned must be of the type specified in the parameter list. When the procedure returns, the value of the formal parameter is copied back to the actual parameter variable.

The third mode we can specify for formal parameters is **inout**, which is a combination of **in** and **out** modes. It is used for objects that are to be both read and updated by a procedure. As with **out** parameters, they are assumed to be of class variable if the class is not explicitly stated. For **inout**-mode variable parameters, the caller supplies a variable as an actual parameter. The value of this variable is used to initialize the formal parameter, which may then be used in the statements of the procedure. The procedure may also perform variable assignments to update the formal parameter. When the procedure returns, the value of the formal parameter is copied back to the actual parameter variable, transferring information back to the caller.

EXAMPLE 6.8   *A procedure to negate a binary-coded number*

The following procedure negates a number represented as a bit vector, using the "complement and add one" method:

```
procedure negate ( a : inout word32 ) is
  variable carry_in : bit := '1';
  variable carry_out : bit;
```

```
begin
  a := not a;
  for index in a'reverse_range loop
    carry_out :=  a(index) and carry_in;
    a(index) := a(index) xor carry_in;
    carry_in := carry_out;
  end loop;
end procedure negate;
```

Since **a** is an **inout**-mode parameter, we can refer to its value in expressions in the procedure body. (This differs from the parameter **result** in the **addu** procedure of the previous example.) We might include the following call to this procedure in a model:

```
variable op1 : word32;
...

negate ( op1 );
```

This uses the value of **op1** to initialize the formal parameter **a**. The procedure body is then executed, updating **a**, and when it returns, the final value of **a** is copied back into **op1**.

VHDL-87, -93, and -2002

These versions of VHDL do not allow an **out**-mode parameter to be read. Instead, if the value must be read within the procedure, the procedure must declare and read a local variable. The final value of the local variable can then be assigned to the **out**-mode parameter immediately before the procedure returns.

6.2.1 Signal Parameters

The third class of object that we can specify for formal parameters is *signal*, which indicates that the algorithm performed by the procedure involves a signal passed by the caller. A signal parameter can be of any of the modes **in**, **out** or **inout**. The way that signal parameters work is somewhat different from constant and variable parameters, so it is worth spending a bit of time understanding them.

When a caller passes a signal as a parameter of mode **in**, instead of passing the value of the signal, it passes the signal object itself. Any reference to the formal parameter within the procedure is exactly like a reference to the actual signal itself. The statements within the procedure can read the signal value, include it in sensitivity lists in wait statements, and query its attributes. A consequence of passing a reference to the signal is that if the procedure executes a wait statement, the signal value may be different after the wait statement completes and the procedure resumes. This behavior differs from that of constant parameters of mode **in**, which have the same value for the whole of the procedure.

EXAMPLE 6.9   *A procedure to receive network packets*

Suppose we wish to model the receiver part of a network interface. It receives fixed-
length packets of data on the signal **rx_data**. The data is synchronized with changes,
from '0' to '1', of the clock signal **rx_clock**. An outline of part of the model is

```
architecture behavioral of receiver is

    ...      -- type declarations, etc

    signal recovered_data : bit;
    signal recovered_clock : bit;
    ...

    procedure receive_packet ( signal rx_data : in bit;
                               signal rx_clock : in bit;
                               data_buffer : out packet_array ) is
    begin
      for index in packet_index_range loop
        wait until rx_clock;
        data_buffer(index) := rx_data;
      end loop;
    end procedure receive_packet;

begin

    packet_assembler : process is
      variable packet : packet_array;
    begin
      ...
      receive_packet ( recovered_data, recovered_clock, packet );
      ...
    end process packet_assembler;

    ...

end architecture behavioral;
```

The **receive_packet** procedure has signal parameters of mode **in** for the network-
data and clock signals. During execution of the model, the process **packet_assembler**
calls the procedure **receive_packet**, passing the signals **recovered_data** and
recovered_clock as actual parameters. We can think of the procedure as executing "on
behalf of" the process. When it reaches the wait statement, it is really the calling process that suspends. The wait statement mentions **rx_clock**, and since this stands for
recovered_clock, the process is sensitive to changes on **recovered_clock** while it is
suspended. Each time it resumes, it reads the current value of **rx_data** (which represents the actual signal **recovered_data**) and stores it in an element of the array parameter data_buffer.

Now let's look at signal parameters of mode **out**. In this case, the caller must name a signal as the actual parameter, and the procedure is passed a reference to the driver for the signal. The procedure is not allowed to read the formal parameter. When the procedure performs a signal assignment statement on the formal parameter, the transactions are scheduled on the driver for the actual signal parameter. In Chapter 5, we said that a process that contains a signal assignment statement contains a driver for the target signal, and that an ordinary signal may only have one driver. When such a signal is passed as an actual **out**-mode parameter, there is still only the one driver. We can think of the signal assignments within the procedure as being performed on behalf of the process that calls the procedure.

EXAMPLE 6.10   *A procedure to generate pulses on a signal*

The following is an outline of an architecture body for a signal generator. The procedure generate_pulse_train has **in**-mode constant parameters that specify the characteristics of a pulse train and an **out**-mode signal parameter on which it generates the required pulse train. The process raw_signal_generator calls the procedure, supplying raw_signal as the actual signal parameter for s. A reference to the driver for raw_signal is passed to the procedure, and transactions are generated on it.

```
library ieee;  use ieee.std_logic_1164.all;

architecture top_level of signal_generator is

  signal raw_signal : std_ulogic;
  ...

  procedure generate_pulse_train
    ( width, separation : in delay_length;
      number : in natural;
      signal s : out std_ulogic ) is
  begin
    for count in 1 to number loop
      s <= '1', '0' after width;
      wait for width + separation;
    end loop;
  end procedure generate_pulse_train;

begin

  raw_signal_generator : process is
  begin
    ...
    generate_pulse_train ( width => period / 2,
                           separation => period - period / 2,
                           number => pulse_count,
                           s => raw_signal );
    ...
  end process raw_signal_generator;
```

```
   . . .
end architecture top_level;
```

An incidental point to note is the way we have specified the actual value for the **separation** parameter in the procedure call. This ensures that the sum of the **width** and **separation** values is exactly equal to **period**, even if **period** is not an even multiple of the time resolution limit. This illustrates an approach sometimes called "defensive programming," in which we try to ensure that the model works correctly in all possible circumstances.

As with variable-class parameters, we can also have a signal-class parameter of mode **inout**. When the procedure is called, both the signal and a reference to its driver are passed to the procedure. The statements within it can read the signal value, include it in sensitivity lists in wait statements, query its attributes, and schedule transactions using signal assignment statements.

An important point to note about procedures with signal parameters relates to procedure calls within processes with the reserved word **all** in their sensitivity lists. Such a process is sensitive to all signals read within the process. That includes signals used as actual **in**-mode and **inout**-mode parameters in procedure calls within the process. It also includes other signals that aren't parameters but that are read within the procedure body. (We will see in Section 6.6 how a procedure can reference such signals.) Since it could become difficult to determine which signals are read by such a process when procedure calls are involved, VHDL simplifies things somewhat by requiring that a procedure called by the process only read signals that are formal parameters or that are declared in the same design unit as the process. In most models this is not a problem.

A final point to note about signal parameters relates to procedures declared immediately within an architecture body. The target of any signal assignment statements within such a procedure must be a signal parameter, rather than a direct reference to a signal declared in the enclosing architecture body. The reason for this restriction is that the procedure may be called by more than one process within the architecture body. Each process that performs assignments on a signal has a driver for the signal. Without the restriction, we would not be able to tell easily by looking at the model where the drivers for the signal were located. The restriction makes the model more comprehensible and, hence, easier to maintain.

6.2.2 Default Values

The one remaining part of a procedure parameter list that we have yet to discuss is the optional default value expression, shown in the syntax rule on page 213. Note that we can only specify a default value for a formal parameter of mode **in**, and the parameter must be of the class constant or variable. If we include a default value in a parameter specification, we have the option of omitting an actual value when the procedure is called. We can either use the keyword **open** in place of an actual parameter value or, if the actual value would be at the end of the parameter list, simply leave it out. If we omit an actual value, the default value is used instead.

EXAMPLE 6.11   *A procedure to increment an integer*

The procedure below increments an unsigned integer represented as a bit vector. The amount to increment by is specified by the second parameter, which has a default value of the bit-vector representation of 1.

```
procedure increment ( a : inout word32;
                      by : in word32 := X"0000_0001" ) is
  variable sum : word32;
  variable carry : bit := '0';
begin
  for index in a'reverse_range loop
    sum(index) := a(index) xor by(index) xor carry;
    carry := ( a(index) and by(index) )
             or ( carry and ( a(index) xor by(index) ) );
  end loop;
  a := sum;
end procedure increment;
```

If we have a variable **count** declared to be of type **word32**, we can call the procedure to increment it by 4, as follows:

```
increment(count, X"0000_0004");
```

If we want to increment the variable by 1, we can make use of the default value for the second parameter and call the procedure without specifying an actual value to increment by, as follows:

```
increment(count);
```

This call is equivalent to

```
increment(count, by => open);
```

6.2.3 Unconstrained Array Parameters

In Chapter 4 we described unconstrained and partially constrained types, in which index ranges of arrays or array elements were left unspecified. For such types, we constrain the index bounds when we create an object, such as a variable or a signal, or when we associate an actual signal with a port. Another use of an unconstrained or partially constrained type is as the type of a formal parameter to a procedure. This use allows us to write a procedure in a general way, so that it can operate on composite values of any size or with any ranges of index values. When we call the procedure and provide a constrained array or record as the actual parameter, the index bounds of the actual parameter are used as the bounds of the formal parameter. The same rules apply as those we described in Section 4.2.3 for ports. Let us look at an example to show how unconstrained parameters work.

EXAMPLE 6.12   *A procedure to find the first set bit*

Following is a procedure that finds the index of the first bit set to '1' in a bit vector. The formal parameter v is of type **bit_vector**, which is an unconstrained array type. Note that in writing this procedure, we do not explicitly refer to the index bounds of the formal parameter v, since they are not known. Instead, we use the **'range** attribute.

```
procedure find_first_set ( v : in bit_vector;
                           found : out boolean;
                           first_set_index : out natural ) is
begin
  for index in v'range loop
    if v(index) then
      found := true;
      first_set_index := index;
      return;
    end if;
  end loop;
  found := false;
end procedure find_first_set;
```

When the procedure is executed, the formal parameters stand for the actual parameters provided by the caller. So if we call this procedure as follows:

```
variable int_req : bit_vector (7 downto 0);
variable top_priority : natural;
variable int_pending : boolean;
...

find_first_set ( int_req, int_pending, top_priority );
```

v'range returns the range 7 **downto** 0, which is used to ensure that the loop parameter **index** iterates over the correct index values for v. If we make a different call:

```
variable free_block_map : bit_vector(0 to block_count-1);
variable first_free_block : natural;
variable free_block_found : boolean;
...

find_first_set ( free_block_map,
                 free_block_found, first_free_block );
```

v'range returns the index range of the array **free_block_map**, since that is the actual parameter corresponding to v.

When we have formal parameters that are of array types, whether fully constrained, partially constrained, or unconstrained, we can use any of the array attributes mentioned in Chapter 4 to refer to the index bounds and range of the actual parameters. We can use the attribute values to define new local constants or variables whose index bounds and ranges depend on those of the parameters. The local objects are created anew each time the procedure is called.

EXAMPLE 6.13   *A procedure to compare binary-coded signed integers*

The following procedure has two bit-vector parameters, which it assumes represent signed integer values in two's-complement form. It performs an arithmetic comparison of the numbers.

```
procedure bv_lt ( bv1, bv2 : in bit_vector;
                     result : out boolean ) is
   variable tmp1 : bit_vector(bv1'range) := bv1;
   variable tmp2 : bit_vector(bv2'range) := bv2;
begin
   tmp1(tmp1'left) := not tmp1(tmp1'left);
   tmp2(tmp2'left) := not tmp2(tmp2'left);
   result :=  tmp1 < tmp2;
end procedure bv_lt;
```

The procedure operates by taking temporary copies of each of the bit-vector parameters, inverting the sign bits and performing a lexical comparison using the built-in "<" operator. This is equivalent to an arithmetic comparison of the original numbers. Note that the temporary variables are declared to be of the same size as the parameters by using the 'range attribute, and the sign bits (the leftmost bits) are indexed using the 'left attribute.

EXAMPLE 6.14   *A procedure to swap array values*

Given an unconstrained type representing arrays of bit vectors declared as follows:

```
type bv_vector is array (natural range <>) of bit_vector;
```

we can declare a procedure to swap the values of two variables of the type:

```
procedure swap ( a1, a2 : inout bv_array ) is
   variable temp : a1'subtype;
begin
   assert a1'length = a2'length and
          a1'element'length = a2'element'length;
   temp := a1; a1 := a2; a2 := temp;
end procedure swap;
```

Since the type bv_array is not fully constrained, we cannot use it as the type of the variable temp. Instead, we use the 'subtype attribute to get a fully constrained subtype with the same shape as a1. Once we've verified that a1 and a2 are the same shape, we can then swap their values in the usual way using temp as the intermediate variable. We use the 'length attribute to refer to the lengths of the top-level arrays, and the 'length attribute applied to the 'subtype attribute to refer to the lengths of the element arrays.

6.2.4   Summary of Procedure Parameters

Let us now summarize all that we have seen in specifying and using parameters for procedures. The syntax rule on page 213 shows that we can specify five aspects of each formal parameter. First, we may specify the class of object, which determines how the formal parameter appears within the procedure, namely, as a constant, a variable or a signal. Second, we give a name to the formal parameter so that it can be referred to in the procedure body. Third, we may specify the mode, **in**, **out** or **inout**, which determines the direction in which information is passed between the caller and the procedure and whether the procedure can assign to the formal parameter. Fourth, we must specify the type or subtype of the formal parameter, which restricts the type of actual parameters that can be provided by the caller. This is important as a means of preventing inadvertent misuse of the procedure. Fifth, we may include a default value, giving a value to be used if the caller does not provide an actual parameter. These five aspects clearly define the interface between the procedure and its callers, allowing us to partition a complex behavioral model into sections and concentrate on each section without being distracted by other details.

Once we have encapsulated some operations in a procedure, we can then call that procedure from different parts of a model, providing actual parameters to specialize the operation at each call. The syntax rule for a procedure call is

> procedure_call_statement ⇐
> 〚 label : 〛 *procedure*_name 〚 (*parameter*_association_list) 〛 ;

This is a sequential statement, so it may be used in a process or inside another subprogram body. If the procedure has formal parameters, the call can specify actual parameters to associate with the formal parameters. The actual associated with a constant-class formal is the value of an expression. The actual associated with a variable-class formal must be a variable, and the actual associated with a signal-class formal must be a signal. The simplified syntax rule for the parameter association list is

> parameter_association_list ⇐
> (〚 *parameter*_name => 〛
>   expression ∣ *signal*_name ∣ *variable*_name ∣ **open**) { , … }

This is in fact the same syntax rule that applies to port maps in component instantiations, seen in Chapter 5. Most of what we said there also applies to procedure parameter association lists. For example, we can use positional association in the procedure call by providing one actual parameter for each formal parameter in the order listed in the procedure declaration. Alternatively, we can use named association by identifying explicitly which formal corresponds to which actual parameter in the call. In this case, the parameters can be in any order. Also, we can use a mix of positional and named association, provided all of the positional parameters come first in the call.

EXAMPLE 6.15   *Positional and named association for parameters*

Suppose we have a procedure declared as

```
procedure p ( f1 : in t1;  f2 : in t2;
              f3 : out t3; f4 : in t4 := v4 ) is
begin
  ...
end procedure p;
```

We could call this procedure, providing actual parameters in a number of ways, including

```
p ( val1, val2, var3, val4 );
p ( f1 => val1, f2 => val2, f4 => val4, f3 => var3 );
p ( val1, val2, f4 => open, f3 => var3 );
p ( val1, val2, var3 );
```

6.3   Concurrent Procedure Call Statements

In Chapter 5 we saw that VHDL provides concurrent signal assignment statements and concurrent assertions as shorthand notations for commonly used kinds of processes. Now that we have looked at procedures and procedure call statements, we can introduce another shorthand notation, the *concurrent procedure call statement*. As its name implies, it is short for a process whose body contains a sequential procedure call statement. The syntax rule is

concurrent_procedure_call_statement ⇐
   ⟦ label : ⟧ *procedure*_name ⟦ (*parameter*_association_list) ⟧ ;

This looks identical to an ordinary sequential procedure call, but the difference is that it appears as a concurrent statement, rather than as a sequential statement. A concurrent procedure call is exactly equivalent to a process that contains a sequential procedure call to the same procedure with the same actual parameters. For example, a concurrent procedure call of the form

```
call_proc : p ( s1, s2, val1 );
```

where s1 and s2 are signals and val1 is a constant, is equivalent to the process

```
call_proc : process is
begin
  p ( s1, s2, val1 );
  wait on s1, s2;
end process call_proc;
```

This also shows that the equivalent process contains a wait statement, whose sensitivity clause includes the signals mentioned in the actual parameter list. This is useful, since it results in the procedure being called again whenever the signal values change. Note that only signals associated with **in**-mode or **inout**-mode parameters are included in the sensitivity list.

EXAMPLE 6.16   *A procedure to check setup time*

We can write a procedure that checks setup timing of a data signal with respect to a clock signal, as shown follows:

```
procedure check_setup ( signal data, clock : in bit;
                        constant Tsu : in time ) is
begin
  if rising_edge(clock) then
    assert data'last_event >= Tsu
      report "setup time violation" severity error;
  end if;
end procedure check_setup;
```

When the procedure is called, it tests to see if there is a rising edge on the clock signal, and if so, checks that the data signal has not changed within the setup time interval. We can invoke this procedure using a concurrent procedure call; for example:

```
check_ready_setup : check_setup ( data => ready,
                                  clock => phi2,
                                  Tsu => Tsu_rdy_clk );
```

The procedure is called whenever either of the signals in the actual parameter list, **ready** or phi2, changes value. When the procedure returns, the concurrent procedure call statement suspends until the next event on either signal. The advantage of using a concurrent procedure call like this is twofold. First, we can write a suite of commonly used checking procedures and reuse them whenever we need to include a check in a model. This is potentially a great improvement in productivity. Second, the statement that invokes the check is more compact and readily understandable than the equivalent process written in-line.

Another point to note about concurrent procedure calls is that if there are no signals associated with **in**-mode or **inout**-mode parameters, the wait statement in the equivalent process does not have a sensitivity clause. If the procedure ever returns, the process suspends indefinitely. This may be useful if we want the procedure to be called only once at startup time. On the other hand, we may write the procedure so that it never returns. If we include wait statements within a loop in the procedure, it behaves somewhat like a process itself. The advantage of this is that we can declare a procedure that performs some commonly needed behavior and then invoke one or more instances of it using concurrent procedure call statements.

EXAMPLE 6.17 *A procedure to generate a clock waveform*

The following procedure generates a periodic clock waveform on a signal passed as a parameter. The **in**-mode constant parameters specify the shape of a clock waveform. The procedure waits for the initial phase delay, then loops indefinitely, scheduling a new rising and falling transition on the clock signal parameter on each iteration. It never returns to its caller.

```
procedure generate_clock ( signal clk : out std_ulogic;
                           constant Tperiod,
                                    Tpulse,
                                    Tphase : in time ) is
begin
  wait for Tphase;
  loop
    clk <= '1', '0' after Tpulse;
    wait for Tperiod;
  end loop;
end procedure generate_clock;
```

We can use this procedure to generate a two-phase non-overlapping pair of clock signals, as follows:

```
signal phi1, phi2 : std_ulogic := '0';
...

gen_phi1 : generate_clock ( phi1, Tperiod => 50 ns,
                                   Tpulse => 20 ns,
                                   Tphase => 0 ns );

gen_phi2 : generate_clock ( phi2, Tperiod => 50 ns,
                                   Tpulse => 20 ns,
                                   Tphase => 25 ns );
```

Each of these calls represents a process that calls the procedure, which then executes the clock generation loop on behalf of its parent process. The advantage of this approach is that we only had to write the loop once in a general-purpose procedure. Also, we have made the model more compact and understandable.

6.4 Functions

Let us now turn our attention to the second kind of subprogram in VHDL: *functions*. We can think of a function as a generalization of expressions. The expressions that we described in Chapter 2 combined values with operators to produce new values. A function is a way of defining a new operation that can be used in expressions. We define how the new operation works by writing a collection of sequential statements that calculate the result. The syntax rule for a function declaration is very similar to that for a procedure declaration:

```
subprogram_body ⇐
    ⟦ pure ▯ impure ⟧
    function identifier ⟦ ( parameter_interface_list ) ⟧ return type_mark is
        { subprogram_declarative_item }
    begin
        { sequential_statement }
    end ⟦ function ⟧ ⟦ identifier ⟧ ;
```

The identifier in the declaration names the function. It may be repeated at the end of the declaration. Unlike a procedure subprogram, a function calculates and returns a result that can be used in an expression. The function declaration specifies the type of the result after the keyword **return**. The parameter list of a function takes the same form as that for a procedure, with two restrictions. First, the parameters of a function may not be of the class variable. If the class is not explicitly mentioned, it is assumed to be constant. Second, the mode of each parameter must be **in**. If the mode is not explicitly specified, it is assumed to be **in**. We come to the reasons for these restrictions in a moment. Like a procedure, a function can declare local items in its declarative part for use in the statements in the function body.

A function passes the result of its computation back to its caller using a return statement, given by the syntax rule

```
return_statement ⇐ ⟦ label : ⟧ return expression ;
```

The optional label allows us to identify the return statement. We will discuss labeled statements in Chapter 20. The form described by this syntax rule differs from the return statement in a procedure subprogram in that it includes an expression to provide the function result. Furthermore, a function must include at least one return statement of this form, and possibly more. The first to be executed causes the function to complete and return its result to the caller. A function cannot simply run into the end of the function body, since to do so would not provide a way of specifying a result to pass back to the caller.

A function call looks exactly like a procedure call. The syntax rule is

```
function_call ⇐ function_name ⟦ ( parameter_association_list ) ⟧
```

The difference is that a function call is part of an expression, rather than being a sequential statement on its own, like a procedure call. Since a function is called as part of evaluation of an expression, a function is not allowed to include a wait statement (nor call a procedure that includes a wait statement). Expressions must always be evaluated within a single simulation cycle.

EXAMPLE 6.18   *A function to limit a value to be within bounds*

The following function calculates whether a value is within given bounds and returns a result limited to those bounds.

```
function limit ( value, min, max : integer ) return integer is
begin
    if value > max then
```

```
      return max;
   elsif value < min then
      return min;
   else
      return value;
   end if;
end function limit;
```

A call to this function might be included in a variable assignment statement, as follows:

```
new_temperature := limit ( current_temperature
                           + increment, 10, 100 );
```

In this statement, the expression on the right-hand side of the assignment consists of just the function call, and the result returned is assigned to the variable new_temperature. However, we might also use the result of a function call in further computation, for example:

```
new_motor_speed := old_motor_speed
                   + scale_factor * limit ( error, -10, +10 );
```

EXAMPLE 6.19   *A bit-vector to numeric conversion function*

The function below determines the number represented in binary by a bit-vector value. The algorithm scans the bit vector from the most-significant end. For each bit, it multiplies the previously accumulated value by two and then adds in the integer value of the bit. The accumulated value is then used as the result of the function, passed back to the caller by the return statement.

```
function bv_to_natural ( bv : in bit_vector ) return natural is
   variable result : natural := 0;
begin
   for index in bv'range loop
     result := result * 2 + bit'pos(bv(index));
   end loop;
   return result;
end function bv_to_natural;
```

As an example of using this function, consider a model for a read-only memory, which represents the stored data as an array of bit vectors, as follows:

```
type rom_array is array (natural range 0 to rom_size-1)
                  of bit_vector(0 to word_size-1);
variable rom_data : rom_array;
```

If the model has an address port that is a bit vector, we can use the function to convert the address to a natural value to index the ROM data array, as follows:

```
    data <= rom_data ( bv_to_natural(address) ) after Taccess;
```

VHDL-87

The keyword **function** may not be included at the end of a function declaration in VHDL-87. Return statements may not be labeled in VHDL-87.

6.4.1   Functional Modeling

In Chapter 5 we looked at concurrent signal assignment statements for functional modeling of designs. We can use functions in VHDL to help us write functional models more expressively by defining a function that encapsulates the data transformation to be performed and then calling the function in a concurrent signal assignment statement. For example, given a declaration of a function to add two bit vectors:

```
function bv_add ( bv1, bv2 : in bit_vector ) return bit_vector is
begin
  ...
end function bv_add;
```

and signals declared in an architecture body:

```
signal source1, source2, sum : bit_vector(0 to 31);
```

we can write a concurrent signal assignment statement as follows:

```
adder : sum <= bv_add(source1, source2) after T_delay_adder;
```

6.4.2   Pure and Impure Functions

Let us now return to the reason for the restrictions on the class and mode of function formal parameters stated above. These restrictions are in keeping with our idea that a function is a generalized form of operator. If we pass the same values to an operator, such as the addition operator, in different expressions, we expect the operator to return the same result each time. By restricting the formal parameters of a function in the way described above, we go part of the way to ensuring the same property for function calls. One additional restriction we need to make is that the function may not refer to any variables or signals declared by its parents, that is, by any process, subprogram or architecture body in which the function declaration is nested. Otherwise the variables or signals might change values between calls to the function, thus influencing the result of the function. We call a function that makes no such reference a *pure* function. We can explicitly declare a function to be pure by including the keyword **pure** in its definition, as shown by the syntax rule on page 228. If we leave it out, the function is assumed to be pure. Both of the above examples of function declarations are pure functions.

On the other hand, we may deliberately relax the restriction about a function referencing its parents' variables or signals by including the keyword **impure** in the function declaration. This is a warning to any caller of the function that it might produce different results on different calls, even when passed the same actual parameter values.

EXAMPLE 6.20 *A function returning unique sequence numbers*

Many network protocols require a sequence number in the packet header so that they can handle packets getting out of order during transmission. We can use an impure function to generate sequence numbers when creating packets in a behavioral model of a network interface. The following is an outline of a process that represents the output side of the network interface.

```
network_driver : process is

  constant seq_modulo : natural := 2**5;
  subtype seq_number is natural range 0 to seq_modulo-1;
  variable next_seq_number : seq_number := 0;
  . . .

  impure function generate_seq_number return seq_number is
    variable number : seq_number;
  begin
    number := next_seq_number;
    next_seq_number := (next_seq_number + 1) mod seq_modulo;
    return number;
  end function generate_seq_number;

begin  -- network_driver
  . . .
  new_header := pkt_header'( dest => target_host_id,
                            src => my_host_id,
                            pkt_type => control_pkt,
                            seq => generate_seq_number );
  . . .
end process network_driver;
```

In this model, the process has a variable **next_seq_number**, used by the function **generate_seq_number** to determine the return value each time it is called. The function has the side effect of incrementing this variable, thus changing the value to be returned on the next call. Because of the reference to the variable in the function's parent, the function must be declared to be impure. The advantage of writing the function this way lies in the expressive power of its call. The function call is simply part of an expression, in this case yielding an element in a record aggregate of type **pkt_header**. Writing it this way makes the process body more compact and easily understandable.

6.4.3   The Function **now**

VHDL provides a predefined function, **now**, that returns the current simulation time when it is called. It is defined as

```
impure function now return delay_length;
```

Recall that the type **delay_length** is a predefined subtype of the physical type **time**, constrained to non-negative time values. The function **now** is often used to check that the inputs to a model obey the required timing constraints.

EXAMPLE 6.21   *A process to check hold time*

The process below checks the clock and data inputs of an edge-triggered flipflop for adherence to the minimum hold time constraint, **Thold_d_clk**. When the clock signal changes to '1', the process saves the current simulation time in the variable **last_clk_edge_time**. When the data input changes, the process tests whether the current simulation time has advanced beyond the time of the last clock edge by at least the minimum hold time, and reports an error if it has not.

```
hold_time_checker : process ( clk, d ) is
  variable last_clk_edge_time : time := 0 fs;
begin
  if rising_edge(clk) then
    last_clk_edge_time := now;
  end if;
  if d'event then
    assert now - last_clk_edge_time >= Thold_d_clk
      report "hold time violation";
  end if;
end process hold_time_checker;
```

VHDL-93 and -2002

The function **now** was originally defined to be impure in VHDL-93. As a consequence, it could not be used in an expression that must be globally static. While the need to do this is rare, it did lead to **now** being pure in VHDL-2002. However, that caused more problems than it solved, so the change was reversed in VHDL-2002.

VHDL-87

The function **now** returns a value of type **time** in VHDL-87, since the subtype **delay_length** is not predefined in VHDL-87.

6.5 Overloading

When we are writing subprograms, it is a good idea to choose names for our subprograms that indicate what operations they perform to make it easier for a reader to understand our models. This raises the question of how to name two subprograms that perform the same kind of operation but on parameters of different types. For example, we might wish to write two procedures to increment variables holding numeric values, but in some cases the values are represented as type **integer**, and in other cases they are represented using type **bit_vector**. Ideally, since both procedures perform the same operation, we would like to give them the same name, such as **increment**. But if we did that, would we be able to tell them apart when we wanted to call them? Recall that VHDL strictly enforces the type rules, so we have to refer to the right procedure depending on the type of the variable we wish to increment.

Fortunately, VHDL allows us to define subprograms in this way, using a technique called *overloading* of subprogram names. We can define two distinct subprograms with the same name but with different numbers or types of formal parameters. When we call one of them, the number and types of the actual parameters we supply in the call are used to determine which subprogram to invoke. It is the context of the call that determines how to resolve the apparent ambiguity. We have already seen overloading applied to identifiers used as literals in enumeration types (see Chapter 2). We saw that if two enumeration types included the same identifier, the context of use in a model is used to determine which type is meant.

The precise rules used to disambiguate a subprogram call when the subprogram name is overloaded are quite complex, so we will not enumerate them all here. Fortunately, they are sufficiently complete to sort out most situations that arise in practice. Instead, we look at some examples to show how overloading of procedures and functions works in straightforward cases. First, here are some procedure outlines for the increment operation described above:

```
procedure increment ( a : inout integer;
                      n : in integer := 1 ) is ...

procedure increment ( a : inout bit_vector;
                      n : in bit_vector := B"1" ) is ...

procedure increment ( a : inout bit_vector;
                      n : in integer := 1 ) is ...
```

Suppose we also have some variables declared as follows:

```
variable count_int : integer := 2;
variable count_bv : bit_vector (15 downto 0) := X"0002";
```

If we write a procedure call using **count_int** as the first actual parameter, it is clear that we are referring to the first procedure, since it is the only one whose first formal parameter is an integer. Both of the following calls can be disambiguated in this way:

```
increment ( count_int, 2 );
increment ( count_int );
```

Similarly, both of the next two calls can be sorted out:

```
increment ( count_bv, X"0002");
increment ( count_bv, 1 );
```

The first call refers to the second procedure, since the actual parameters are both bit vectors. Similarly, the second call refers to the third procedure, since the actual parameters are a bit vector and an integer. Problems arise, however, if we try to make a call as follows:

```
increment ( count_bv );
```

This could equally well be a call to either the second or the third procedure, both of which have default values for the second formal parameter. Since it is not possible to determine which procedure is meant, a VHDL analyzer rejects such a call as an error.

6.5.1 Overloading Operator Symbols

When we introduced function subprograms in Section 6.4, we described them as a generalization of operators used in expressions, such as "+", "−", **and**, **or** and so on. Looking at this the other way around, we could say that the predefined operators are specialized functions, with a convenient notation for calling them. In fact, this is exactly what they are. Furthermore, since each of the operators can be applied to values of various types, we see that the functions they represent are overloaded, so the operand types determine the particular version of each operator used in an expression.

Given that we can define our own types in VHDL, it would be convenient if we could extend the predefined operators to work with these types. For example, if we are using bit vectors to model integers using two's-complement notation, we would like to use the addition operator to add two bit vectors in this form. Fortunately, VHDL provides a way for us to define new functions using the operator symbols as names. The extended syntax rules for subprogram declarations are shown in Appendix B. Our bit-vector addition function can be declared as

```
function "+" ( left, right : in bit_vector ) return bit_vector is
begin
   ...
end function "+";
```

We can then call this function using the infix "+" operator with bit-vector operands; for example:

```
variable addr_reg : bit_vector(31 downto 0);
...
addr_reg := addr_reg + X"0000_0004";
```

Operators denoted by reserved words can be overloaded in the same way. For example, we can declare a bit-vector absolute-value function as

```
function "abs" ( right : in bit_vector ) return bit_vector is
begin
  ...
end function "abs";
```

We can use this operator with a bit-vector operand, for example:

```
variable accumulator : bit_vector(31 downto 0);

...

accumulator := abs accumulator;
```

We can overload any of the operator symbols shown in Table 2.2. One important point to note, however, is that overloaded versions of the logical operators **and**, **nand**, **or** and **nor** are not evaluated in the short-circuit manner described in Chapter 2. For any type of operands other than **bit** and **boolean**, both operands are evaluated first, then passed to the function.

EXAMPLE 6.22   *Use of overloaded logical operations in control logic*

The std_logic_1164 package defines functions for logical operators applied to values of type **std_ulogic** and **std_ulogic_vector**. We can use them in functional models to write Boolean equations that represent the behavior of a design. For example, the following model describes a block of logic that controls an input/output register in a microcontroller system. The architecture body describes the behavior in terms of Boolean equations. Its concurrent signal assignment statements use the logical operators **and** and **not**, referring to the overloaded functions defined in the std_logic_1164 package.

```
library ieee;  use ieee.std_logic_1164.all;

entity reg_ctrl is
  port ( reg_addr_decoded,
         rd, wr, io_en, cpu_clk : in std_ulogic;
         reg_rd, reg_wr : out std_ulogic );
end entity reg_ctrl;

-------------------------------------------------

architecture bool_eqn of reg_ctrl is
begin

  rd_ctrl : reg_rd <= reg_addr_decoded and rd and io_en;

  rw_ctrl : reg_wr <= reg_addr_decoded and wr and io_en
                          and not cpu_clk;

end architecture bool_eqn;
```

One particular operator that we can overload is the condition operator, "??", introduced in Section 2.2.5. This operator is predefined for **bit** and **std_ulogic** operands, and we can overload it for operands of other types that we may define. If we overload it in a form that produces a **boolean** result, VHDL can use the overloaded version to implicitly convert a condition value to a **boolean** value. For example, suppose we overload the operator as follows:

```
function "??" ( right : integer ) return boolean is
begin
  return right /= 0;
end function "??";
```

This version treats any non-zero integer as true and 0 as false. We could then write the following:

```
variable m : integer;
...

if m then
  ...
end if;
```

Since there is now an overloaded version of the "??" operator converting the condition type (**integer**) to **boolean**, it is implicitly applied to the condition of the if statement.

VHDL-87, -93, and -2002

These versions of VHDL do not provide the "??" operator and do not perform implicit conversion of conditions.

VHDL-87

Since VHDL-87 does not provide the shift operators **sll**, **srl**, **sla**, **sra**, **rol**, and **ror** and the logical operator **xnor**, they cannot be used as operator symbols.

6.6 Visibility of Declarations

The last topic we need to discuss in relation to subprograms is the use of names declared within a model. We have seen that names of types, constants, variables and other items defined in a subprogram can be used in that subprogram. Also, in the case of procedures and impure functions, names declared in an enclosing process, subprogram or architecture body can also be used. The question we must answer is: What are the limits of use of each name?

To answer this question, we introduce the idea of the *visibility* of a declaration, which is the region of the text of a model in which it is possible to refer to the declared name. We have seen that architecture bodies, processes and subprograms are each divided into two parts: a declarative part and a body of statements. A name declared in a declarative part is visible from the end of the declaration itself down to the end of the corresponding statement part. Within this area we can refer to the declared name. Before the declaration, within it and beyond the end of the statement part, we cannot refer to the name because it is not visible.

EXAMPLE 6.23   *Visibility of declarations within an architecture body*

Figure 6.1 shows an outline of an architecture body of a model. It contains a number of declarations, including some procedure declarations. The visibility of each of the declarations is indicated. The first item to be declared is the type t; its visibility extends

FIGURE 6.1

```
architecture arch of ent is
                                                                              t
    type t is ...;
                                                                          s
    signal s : t;
                                                                     p1
    procedure p1 ( ... ) is
       variable v1 : t;                          v1
    begin
       v1 := s;
    end procedure p1;
begin   -- arch
    proc1 : process is
                                                      v2
       variable v2 : t;
                                                 p2
       procedure p2 ( ... ) is
          variable v3 : t;                   v3
       begin
          p1 ( v2, v3, ... );
       end procedure p2;
    begin   -- proc1
       p2 ( v2, ... );
    end process proc1;

    proc2 : process is
       ...
    begin   -- proc2
       p1 ( ... );
    end process proc2;
end architecture arch;
```

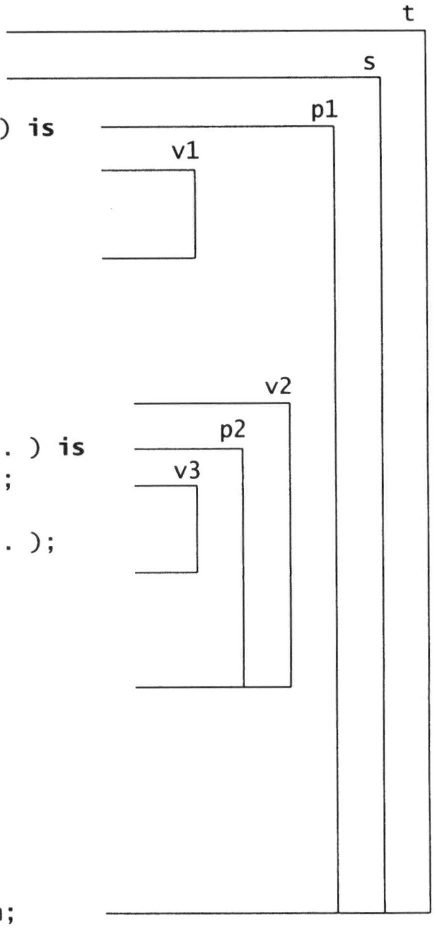

An outline of an architecture body, showing the visibility of declared names within it.

to the end of the architecture body. Thus it can be referred in other declarations, such as the variable declarations. The second declaration is the signal s; its visibility likewise extends to the end of the architecture body. So the assignment within procedure p1 is valid. The third and final declaration in the declarative part of the architecture body is that of the procedure p1, whose visibility extends to the end of the architecture body, allowing it to be called in either of the processes. It includes a local variable, v1, whose visibility extends only to the end of p1. This means it can be referred to in p1, as shown in the signal assignment statement, but neither process can refer to it.

In the statement part of the architecture body, we have two process statements, **proc1** and **proc2**. The first includes a local variable declaration, v2, whose visibility extends to the end of the process body. Hence we can refer to **v2** in the process body and in the procedure **p2** declared within the process. The visibility of **p2** likewise extends to the end of the body of **proc1**, allowing us to call **p2** within **proc1**. The procedure **p2** includes a local variable declaration, v3, whose visibility extends to the end of the statement part of **p2**. Hence we can refer to **v3** in the statement part of **p2**. However, we cannot refer to **v3** in the statement part of **proc1**, since it is not visible in that part of the model.

Finally, we come to the second process, **proc2**. The only items we can refer to here are those declared in the architecture body declarative part, namely, t, s and p1. We cannot call the procedure **p2** within **proc2**, since it is local to **proc1**.

One point we mentioned earlier about subprograms but did not go into in detail was that we can include nested subprogram declarations within the declarative part of a subprogram. This means we can have local procedures and functions within a procedure or a function. In such cases, the simple rule for the visibility of a declaration still applies, so any items declared within an outer procedure before the declaration of a nested procedure can be referred to inside the nested procedure.

EXAMPLE 6.24   *Nested subprograms for memory read operations*

The following is an outline of an architecture of a cache memory for a computer system.

```
architecture behavioral of cache is
begin

  behavior : process is

     ...

  procedure read_block( start_address : natural;
                        entry : out cache_block ) is

     variable memory_address_reg : natural;
     variable memory_data_reg : word;
```

```
        procedure read_memory_word is
        begin
          mem_addr <= memory_address_reg;
          mem_read <= '1';
          wait until mem_ack;
          memory_data_reg := mem_data_in;
          mem_read <= '0';
          wait until not mem_ack;
        end procedure read_memory_word;

    begin  -- read_block
      for offset in 0 to block_size - 1 loop
        memory_address_reg := start_address + offset;
        read_memory_word;
        entry(offset) := memory_data_reg;
      end loop;
    end procedure read_block;

  begin  -- behavior
    ...
    read_block( miss_base_address, data_store(entry_index) );
    ...
  end process behavior;

end architecture behavioral;
```

The entity interface (not shown) includes ports named **mem_addr**, **mem_ready**, **mem_ack** and **mem_data_in**. The process behavior contains a procedure, **read_block**, which reads a block of data from main memory on a cache miss. It has the local variables **memory_address_reg** and **memory_data_reg**. Nested inside of this procedure is another procedure, **read_memory_word**, which reads a single word of data from memory. It uses the value placed in **memory_address_reg** by the outer procedure and leaves the data read from memory in **memory_data_reg**.

Now let us consider a model in which we have one subprogram nested inside another, and each declares an item with the same name as the other, as shown in Figure 6.2. Here, the first variable **v** is visible within all of the procedure **p2** and the statement body of **p1**. However, because **p2** declares its own local variable called **v**, the variable belonging to **p1** is not *directly visible* where **p2**'s **v** is visible. We say the inner variable declaration *hides* the outer declaration, since it declares the same name. Hence the addition within **p2** applies to the local variable **v** of **p2** and does not affect the variable **v** of **p1**. If we need to refer to an item that is visible but hidden, we can use a selected name. For example, within **p2** in Figure 6.2, we can use the name **p1.v** to refer to the variable **v** declared in **p1**. Although the outer declaration is not directly visible, it is *visible by selection*. An important point to note about using a selected name in this way is that it can only be used within the construct containing the declaration. Thus, in Figure 6.2, we can only refer to **p1.v** within **p1**. We cannot use the name **p1.v** to "peek inside" of **p1** from places outside **p1**.

FIGURE 6.2

```
procedure p1 is
    variable v : integer;                    v

    procedure p2 is
        variable v : integer;          v
    begin   -- p2
        . . .
        v := v + 1;
        . . .
    end procedure p2;

begin   -- p1
    . . .
    v := 2 * v;
    . . .
end procedure p1;
```

Nested procedures showing hiding of names. The declaration of v *in* p2 *hides the variable* v *declared in* p1.

The idea of hiding is not restricted to variable declarations within nested procedures. Indeed, it applies in any case where we have one declarative part nested within another, and an item is declared with the same name in each declarative part in such a way that the rules for resolving overloaded names are unable to distinguish between them. The advantage of having inner declarations hide outer declarations, as opposed to the alternative of simply disallowing an inner declaration with the same name, is that it allows us to write local procedures and processes without having to know the names of all items declared at outer levels. This is certainly beneficial when writing large models. In practice, if we are reading a model and need to check the use of a name in a statement against its declaration, we only need to look at successively enclosing declarative parts until we find a declaration of the name, and that is the declaration that applies.

Exercises

1. [❶ 6.2] Write parameter specifications for the following constant-class parameters:

    - an integer, **operand1**,

    - a bit vector, **tag**, indexed from 31 down to 16, and

    - a Boolean, **trace**, with default value **false**.

2. [❶ 6.2] Write parameter specifications for the following variable-class parameters:

    - a real number, **average**, used to pass data back from a procedure, and

    - a string, **identifier**, modified by a procedure.

3. [❶ 6.2] Write parameter specifications for the following signal-class parameters:

   - a bit signal, **clk**, to be assigned to by a procedure, and

   - an unconstrained standard-logic vector signal, **data_in**, whose value is to be read by a procedure.

4. [❶ 6.2] Given the following procedure declaration:

   ```
   procedure stimulate ( signal target : out bit_vector;
                         delay : in delay_length := 1 ns;
                         cycles : in natural := 1 ) is ...
   ```

   write procedure calls using a signal **s** as the actual parameter for **target** and using the following values for the other parameters:

   - **delay** = 5 ns, **cycles** = 3,

   - **delay** = 10 ns, **cycles** = 1 and

   - **delay** = 1 ns, **cycles** = 15.

5. [❶ 6.3] Suppose we have a procedure declared as

   ```
   procedure shuffle_bytes
     ( signal d_in : in std_ulogic_vector(0 to 15);
       signal d_out : out std_ulogic_vector(0 to 15);
       signal shuffle_control : in std_ulogic;
       prop_delay : delay_length ) is ...
   ```

   Write the equivalent process for the following concurrent procedure call:

   ```
   swapper : shuffle_bytes ( ext_data, int_data, swap_control,
                             Tpd_swap );
   ```

6. [❶ 6.4] Suppose we have a function declared as

   ```
   function approx_log_2 ( a : in bit_vector ) return positive is ...
   ```

   that calculates the minimum number of bits needed to represent a binary-encoded number. Write a variable assignment statement that calculates the minimum number of bits needed to represent the product of two numbers in the variables **multiplicand** and **multiplier** and assigns the result to the variable **product_size**.

7. [❶ 6.4] Write an assertion statement that verifies that the current simulation time has not exceeded 20 ms.

8. [❶ 6.5] Given the declarations of the three procedures named **increment** and the variables **count_int** and **count_bv** shown on page 233, which of the three procedures, if any, is referred to by each of the following procedure calls?

   ```
   increment ( count_bv, -1 );
   increment ( count_int );
   ```

```
            increment ( count_int, B"1" );
            increment ( count_bv, 16#10# );
```

9. [❶ 6.6] Show the parts of the following model in which each of the declared items is visible:

    ```
    architecture behavioral of computer_system is

        signal internal_data : bit_vector(31 downto 0);

        interpreter : process is

          variable opcode : bit_vector(5 downto 0);

          procedure do_write is
            variable aligned_address : natural;
          begin
            . . .
          end procedure do_write;

        begin
          . . .
        end process interpreter;

    end architecture behavioral;
    ```

10. [❷ 6.1] Write a procedure that calculates the sum of squares of elements of an array variable **deviations**. The elements are real numbers. Your procedure should store the result in a real variable **sum_of_squares**.

11. [❷ 6.1] Write a procedure that generates a 1 μs pulse every 20 μs on a signal **syn_clk**. When the signal **reset** changes to '1', the procedure should immediately set **syn_clk** to '0' and return.

12. [❷ 6.2] Write a procedure called **align_address** that aligns a binary encoded address in a bit-vector variable parameter. The procedure has a second parameter that indicates the alignment size. If the size is 1, the address is unchanged. If the size is 2, the address is rounded to a multiple of 2 by clearing the least-significant bit. If the size is 4, two bits are cleared, and if the size is 8, three bits are cleared. The default alignment size is 4.

13. [❷ 6.2/6.3] Write a procedure that checks the hold time of a data signal with respect to rising edges of a clock signal. Both signals are of the IEEE standard-logic type. The signals and the hold time are parameters of the procedure. The procedure is invoked by a concurrent procedure call.

14. [❷ 6.2/6.3] Write a procedure, to be invoked by a concurrent procedure call, that assigns successive natural numbers to a signal at regular intervals. The signal and the interval between numbers are parameters of the procedure.

15. [❷ 6.4] Write a function, **weaken**, that maps a standard-logic value to the same value, but with weak drive strength. Thus, '0' and 'L' are mapped to 'L', '1' and 'H' are mapped to 'H', 'X' and 'W' are mapped to 'W' and all other values are unchanged.

16. [❷ 6.4] Write a function, returning a Boolean result, that tests whether a standard-logic signal currently has a valid edge. A valid edge is defined to be a transition from '0' or 'L' to '1' or 'H' or vice versa. Other transitions, such as 'X' to '1', are not valid.

17. [❷ 6.4] Write two functions, one to find the maximum value in an array of integers and the other to find the minimum value.

18. [❷ 6.5] Write overloaded versions of the logical operators to operate on integer operands. The operators should treat the value 0 as logical falsehood and any non-zero value as logical truth.

19. [❸ 6.2] Write a procedure called **scan_results** with an **in**-mode bit-vector signal parameter **results**, and **out**-mode variable parameters **majority_value** of type **bit**, **majority_count** of type **natural** and **tie** of type **boolean**. The procedure counts the occurrences of '0' and '1' values in **results**. It sets **majority_value** to the most frequently occurring value, **majority_count** to the number of occurrences and **tie** to true if there are an equal number of occurrences of '0' and '1'.

20. [❸ 6.2/6.3] Write a procedure that stimulates a bit-vector signal passed as a parameter. The procedure assigns to the signal a sequence of all possible bit-vector values. The first value is assigned to the signal immediately, then subsequent values are assigned at intervals specified by a second parameter. After the last value is assigned, the procedure returns.

21. [❸ 6.2/6.3] Write a passive procedure that checks that setup and hold times for a data signal with respect to rising edges of a clock signal are observed. The signals and the setup and hold times are parameters of the procedure. Include a concurrent procedure call to the procedure in the statement part of a D-flipflop entity.

22. [❸ 6.4] Write a function that calculates the cosine of a real number, using the series

$$\cos\theta = 1 - \frac{\theta^2}{2!} + \frac{\theta^4}{4!} - \frac{\theta^6}{6!} + \cdots$$

Next, write a second function that returns a cosine table of the following type:

type table **is array** (0 **to** 1023) **of** real;

Element i of the table has the value $\cos(i\pi/2048)$. Finally, develop a behavioral model of a cosine lookup ROM. The architecture body should include a constant of type **table**, initialized using a call to the second function.

Chapter 7

Packages and Use Clauses

Packages in VHDL provide an important way of organizing the data and subprograms declared in a model. In this chapter, we describe the basics of packages and show how they may be used. We will return to packages in Chapter 12, where we will see how they can be extended to make them more reusable than the basic form we discuss here.

7.1 Package Declarations

A VHDL package is simply a way of grouping a collection of related declarations that serve a common purpose. They might be a set of subprograms that provide operations on a particular type of data, or they might just be the set of declarations needed to model a particular design. The important thing is that they can be collected together into a separate design unit that can be worked on independently and reused in different parts of a model.

Another important aspect of packages is that they separate the external view of the items they declare from the implementation of those items. The external view is specified in a *package declaration*, whereas the implementation is defined in a separate *package body*. We will look at package declaration first and return to the package body shortly.

The syntax rule for writing a package declaration is

```
package_declaration ⇐
    package identifier is
        { package_declarative_item }
    end [ package ] [ identifier ] ;
```

The identifier provides a name for the package, which we can use elsewhere in a model to refer to the package. Inside the package declaration we write a collection of declarations, including type, subtype, constant, signal and subprogram declarations, as well as several other kinds of declarations that we see in later chapters. These are the declarations that are provided to the users of the package. The advantage of placing them in a package is that they do not clutter up other parts of a model, and they can be shared within and between models without having to rewrite them.

EXAMPLE 7.1   *A package as part of a CPU model*

The following package declares constants and types that we can use in a model of a CPU:

```
package cpu_types is

  constant word_size : positive := 16;
  constant address_size : positive := 24;

  subtype word is bit_vector(word_size - 1 downto 0);
  subtype address is bit_vector(address_size - 1 downto 0);

  type status_value is ( halted, idle, fetch,
                         mem_read, mem_write,
                         io_read, io_write, int_ack );

end package cpu_types;
```

VHDL-87

The keyword **package** may not be included at the end of a package declaration in VHDL-87.

Most of the time, we write a package as another form of design unit, along with entity declarations and architecture bodies. It is separately analyzed and is placed into the working library as a library unit by the analyzer. From there, other library units can refer to an item declared in the package using the *selected name* of the item. The selected name is formed by writing the library name, then the package name and then the name of the item, all separated by dots; for example:

```
work.cpu_types.status_value
```

EXAMPLE 7.2   *An address decoder using the* cpu_types *package*

Suppose the **cpu_types** package, shown in Example 7.1, has been analyzed and placed into the **work** library. We might make use of the declared items when modeling an address decoder to go with a CPU. The entity declaration and architecture body of the decoder are:

```
entity address_decoder is
  port ( addr : in work.cpu_types.address;
         status : in work.cpu_types.status_value;
         mem_sel, int_sel, io_sel : out bit );
end entity address_decoder;
```

```
architecture functional of address_decoder is

  constant mem_low : work.cpu_types.address := X"000000";
  constant mem_high : work.cpu_types.address := X"EFFFFF";
  constant io_low : work.cpu_types.address := X"F00000";
  constant io_high : work.cpu_types.address := X"FFFFFF";

begin

  mem_decoder :
    mem_sel <=
      '1' when ( work.cpu_types."="
                   (status, work.cpu_types.fetch)
                 or work.cpu_types."="(
                     status, work.cpu_types.mem_read)
                 or work.cpu_types."="
                     (status, work.cpu_types.mem_write) )
               and addr >= mem_low
               and addr <= mem_high else
      '0';

  int_decoder :
    int_sel <=
      '1' when work.cpu_types."="
                   (status, work.cpu_types.int_ack) else
      '0';

  io_decoder :
    io_sel <=
      '1' when ( work.cpu_types."="
                   (status, work.cpu_types.io_read)
                 or work.cpu_types."="
                     (status, work.cpu_types.io_write) )
               and addr >= io_low
               and addr <= io_high else
      '0';

end architecture functional;
```

Note that we have to use selected names to refer to the subtype **address**, the type **status_value**, the enumeration literals of **status_value** and the implicitly declared "=" operator, defined in the package **cpu_types**. This is because they are not directly visible within the entity declaration and architecture body. We will see later in this chapter how a use clause can help us avoid long selected names. If we needed to type-qualify the enumeration literals, we would use selected names for both the type name and the literal name; for example:

```
work.cpu_types.status_value'(work.cpu_types.fetch)
```

We have seen that a package, when analyzed, is placed into the working library. Items in the package can be accessed by other library units using selected names starting with **work**. However, if we are writing a package of generally useful declarations, we may wish to place them into a different library, such as a project library, where they can be accessed by other designers. Different VHDL tool suites provide different ways of specifying the library into which a library unit is placed. We must consult the documentation for a particular product to find out what to do. However, once the package has been included in a resource library, we can refer to items declared in it using selected names, starting with the resource library name. As an example, we might consider the IEEE standard-logic package, which must be placed in a resource library called ieee. We can refer to the types declared in that package, for example:

```
variable stored_state : ieee.std_logic_1164.std_ulogic;
```

One kind of declaration we can include in a package declaration is a signal declaration. This gives us a way of defining a signal, such as a master clock or reset signal, that is global to a whole design, instead of being restricted to a single architecture body. Any module that needs to refer to the global signal simply names it using the selected name as described above. This avoids the clutter of having to specify the signal as a port in each entity that uses it, making the model a little less complex. However, it does mean that a module can affect the overall behavior of a system by means other than through its ports, namely, by assigning to global signals. This effectively means that part of the module's interface is implicit, rather than being specified in the port map of the entity. As a matter of style, global signals declared in packages should be used sparingly, and their use should be clearly documented with comments in the model.

EXAMPLE 7.3   *A package containing clock signal declarations*

The following package declares two clock signals for use within an integrated circuit design for an input/output interface controller.

```
library ieee;  use ieee.std_logic_1164.all;

package clock_pkg is

  constant Tpw : delay_length := 4 ns;

  signal clock_phase1, clock_phase2 : std_ulogic;

end package clock_pkg;
```

The top-level architecture of the controller circuit is outlined below. The instance of the **phase_locked_clock_gen** entity uses the **ref_clock** port of the circuit to generate the two-phase clock waveforms on the global clock signals. The architecture also includes an instance of an entity that sequences bus operations using the bus control signals and generates internal register control signals.

```
library ieee;  use ieee.std_logic_1164.all;
```

```
entity io_controller is
  port ( ref_clock : in std_ulogic;  ... );
end entity io_controller;

---------------------------------------------------

architecture top_level of io_controller is

  ...

begin

  internal_clock_gen :
    entity work.phase_locked_clock_gen(std_cell)
      port map ( reference => ref_clock,
                 phi1 => work.clock_pkg.clock_phase1,
                 phi2 => work.clock_pkg.clock_phase2 );

  the_bus_sequencer :
    entity work.bus_sequencer(fsm)
      port map ( rd, wr, sel, width, burst,
                 addr(1 downto 0), ready,
                 control_reg_wr, status_reg_rd,
                 data_fifo_wr, data_fifo_rd,
                 ... );

  ...

end architecture top_level;
```

The architecture body for the sequencer is outlined next. It creates an instance of a register entity and connects the global clock signals to its clock input ports.

```
architecture fsm of bus_sequencer is

  -- This architecture implements the sequencer as
  -- a finite-state machine. NOTE: it uses the clock signals
  -- from clock_pkg to synchronize the fsm.

  signal next_state_vector : ...;

begin

  bus_sequencer_state_register :
    entity work.state_register(std_cell)
      port map ( phi1 => work.clock_pkg.clock_phase1,
                 phi2 => work.clock_pkg.clock_phase2,
                 next_state => next_state_vector,
                 ... );

  ...

end architecture fsm;
```

7.1.1   Subprograms in Package Declarations

Another kind of declaration that may be included in a package declaration is a subprogram declaration—either a procedure or a function declaration. This ability allows us to write subprograms that implement useful operations and to call them from a number of different modules. An important use of this feature is to declare subprograms that operate on values of a type declared by the package. This gives us a way of conceptually extending VHDL with new types and operations, so-called *abstract data types*, a topic we return to in Chapter 12.

An important aspect of declaring a subprogram in a package declaration is that we only write the header of the subprogram, that is, the part that includes the name and the interface list defining the parameters (and result type for functions). We leave out the body of the subprogram. The reason for this is that the package declaration, as we mentioned earlier, provides only the external view of the items it declares, leaving the implementation of the items to the package body. For items such as types and signals, the complete definition is needed in the external view. However, for subprograms, we need only know the information contained in the header to be able to call the subprogram. As users of a subprogram, we need not be concerned with how it achieves its effect or calculates its result. This is an example of a general principle called *information hiding*: making an interface visible but hiding the details of implementation. To illustrate this idea, suppose we have a package declaration that defines a bit-vector subtype:

```
subtype word32 is bit_vector(31 downto 0);
```

We can include in the package a procedure to do addition on **word32** values that represent signed integers. The procedure declaration in the package declaration is

```
procedure add ( a, b : in word32;
                result : out word32;  overflow : out boolean );
```

Note that we do not include the keyword **is** or any of the local declarations or statements needed to perform the addition. These are deferred to the package body. All we include is the description of the formal parameters of the procedure. Similarly, we might include a function to perform an arithmetic comparison of two **word32** values:

```
function "<" ( a, b : in word32 ) return boolean;
```

Again, we omit the local declarations and statements, simply specifying the formal parameters and the result type of the function.

7.1.2   Constants in Package Declarations

Just as we can apply the principle of information hiding to subprograms declared in a package, we can also apply it to constants declared in a package. The external view of a constant is just its name and type. We need to know these in order to use it, but we do not actually need to know its value. This may seem strange at first, but if we recall that the idea of introducing constant declarations in the first place was to avoid scattering literal values throughout a model, it makes more sense. We defer specifying the value of a constant declared in a package by omitting the initialization expression; for example:

```
constant max_buffer_size : positive;
```

This defines the constant to be a positive integer value. However, since we cannot see the actual value, we are not tempted to write the value as an integer literal in a model that uses the package. The specification of the actual value is deferred to the package body, where it is not visible to a model that uses the package. Given the above deferred constant in a package declaration, the corresponding package body must include the full constant declaration, for example:

```
constant max_buffer_size : positive := 4096;
```

Note that we do not have to defer the value in a constant declaration—it is optional.

EXAMPLE 7.4 *The CPU package extended with operations*

We can extend the package specification from Example 7.1, declaring useful types for a CPU model, by including declarations related to opcode processing. The revised package is shown below. It includes a subtype that represents an opcode value, a function to extract an opcode from an instruction word and a number of constants representing the opcodes for different instructions.

```
package cpu_types is

    constant word_size : positive := 16;
    constant address_size : positive := 24;

    subtype word is bit_vector(word_size - 1 downto 0);
    subtype address is bit_vector(address_size - 1 downto 0);

    type status_value is ( halted, idle, fetch,
                           mem_read, mem_write,
                           io_read, io_write, int_ack );

    subtype opcode is bit_vector(5 downto 0);

    function extract_opcode ( instr_word : word ) return opcode;

    constant op_nop : opcode := "000000";
    constant op_breq : opcode := "000001";
    constant op_brne : opcode := "000010";
    constant op_add : opcode := "000011";
    . . .

end package cpu_types;
```

A behavioral model of a CPU that uses these declarations is

```
architecture behavioral of cpu is
begin
```

```
interpreter : process is

  variable instr_reg : work.cpu_types.word;
  variable instr_opcode : work.cpu_types.opcode;
begin
  ...  -- initialize
  loop
    ...  -- fetch instruction
    instr_opcode := work.cpu_types.extract_opcode(instr_reg);
    case instr_opcode is
      when work.cpu_types.op_nop => null;
      when work.cpu_types.op_breq => ...
      ...
    end case;
  end loop;
end process interpreter;

end architecture behavioral;
```

The instruction set interpreter process declares a variable of the **opcode** type and uses the **extract_opcode** function to extract the bits representing the opcode from the fetched instruction word. It then uses the constants from the package as choices in a case statement to decode and execute the instruction specified by the opcode. Note that since the constants are used as choices in the case statement, they must be locally static. If we had deferred the values of the constants to the package body, their value would not be known when the case statement was analyzed. This is why we included the constant values in the package declaration. In general, the value of a deferred constant is not locally static.

7.2   Package Bodies

Now that we have seen how to define the interface to a package, we can turn to the package body. Each package declaration that includes subprogram declarations or deferred constant declarations must have a corresponding package body to fill in the missing details. However, if a package declaration only includes other kinds of declarations, such as types, signals or fully specified constants, no package body is necessary. The syntax rule for a package body is similar to that for the interface, but with the inclusion of the keyword **body**:

package_body ⇐
    package body identifier **is**
        { package_body_declarative_item }
    end [**package body**] [identifier] ;

The items declared in a package body must include the full declarations of all subprograms defined in the corresponding package declaration. These full declarations must include the subprogram headers exactly as they are written in the package declaration, to ensure that the implementation *conforms* with the interface. This means that the names, types, modes and default values of each of the formal parameters must be repeated exactly. There are only two variations allowed. First, a numeric literal may be written differently, for example, in a different base, provided it has the same value. Second, a simple name consisting just of an identifier may be replaced by a selected name, provided it refers to the same item. While this conformance requirement might seem an imposition at first, in practice it is not. Any reasonable text editor used to create a VHDL model allows the header to be copied from the package declaration with little difficulty. Similarly, a deferred constant defined in a package declaration must have its value specified by repeating the declaration in the package body, this time filling in the initialization expression as in a full constant declaration.

In addition to the full declarations of items deferred from the package declaration, a package body may include declarations of additional types, subtypes, constants and subprograms. These items are used to implement the subprograms defined in the package declaration. Note that the items declared in the package declaration cannot be declared again in the body (apart from subprograms and deferred constants, as described above), since they are automatically visible in the body. Furthermore, the package body cannot include declarations of additional signals. Signal declarations may only be included in the interface declaration of a package.

EXAMPLE 7.5   *A package of overloaded arithmetic operators*

The package declaration outlined below declares overloaded versions of arithmetic operators for bit-vector values. The functions treat bit vectors as representing signed integers in binary form. Only the function headers are included in the package declaration.

```
package bit_vector_signed_arithmetic is

  function "+" ( bv1, bv2 : bit_vector )
    return bit_vector;

  function "-" ( bv : bit_vector )
    return bit_vector;

  function "*" ( bv1, bv2 : bit_vector )
    return bit_vector;

  ...

end package bit_vector_signed_arithmetic;
```

The package body contains the full function bodies. It also includes a function, mult_unsigned, not defined in the package declaration. It is used internally in the package body to implement the signed multiplication operator.

```
package body bit_vector_signed_arithmetic is

  function "+" ( bv1, bv2 : bit_vector )
    return bit_vector is ...
```

```vhdl
      function "-" ( bv : bit_vector )
        return bit_vector is ...

      function mult_unsigned ( bv1, bv2 : bit_vector )
        return bit_vector is

        . . .
      begin

        . . .
      end function mult_unsigned;

      function "*" ( bv1, bv2 : bit_vector )
        return bit_vector is
      begin
        if not bv1(bv1'left) and not bv2(bv2'left) then
          return mult_unsigned(bv1, bv2);
        elsif not bv1(bv1'left) and bv2(bv2'left) then
          return -mult_unsigned(bv1, -bv2);
        elsif bv1(bv1'left) and not bv2(bv2'left) then
          return -mult_unsigned(-bv1, bv2);
        else
          return mult_unsigned(-bv1, -bv2);
        end if;
      end function "*";

      . . .

    end package body bit_vector_signed_arithmetic;
```

One further point to mention on the topic of packages relates to the order of analysis. We mentioned before that a package is usually a separate design unit that is analyzed separately from other design units, such as entity declarations and architecture bodies. (We will return to the case of a package not being a design unit shortly.) In most cases, a package declaration and its corresponding package body are each separate design units; hence they may be analyzed separately. A package declaration is a primary design unit, and a package body is a secondary design unit. The package body depends on information defined in the package declaration, so the declaration must be analyzed first. Furthermore, the declaration must be analyzed before any other design unit that refers to an item defined by the package. Once the declaration has been analyzed, it does not matter when the body is analyzed in relation to units that use the package, provided it is analyzed before the model is elaborated. In a large suite of models, the dependency relationships can get quite complex, and a correct order of analysis can be difficult to find. A good VHDL tool suite will provide some degree of automating this process by working out the dependency relationships and analyzing those units needed to build a particular target unit to simulate or synthesize.

VHDL-87

The keywords **package body** may not be included at the end of a package body in VHDL-87.

7.2.1 Local Packages

We have mentioned that package declarations and package bodies are usually separate design units, but that there are cases where that is not so. VHDL allows us to declare a package locally within the declarative part of an entity, architecture body, process, or subprogram. This allows the visibility of the package to be contained to just the enclosing declarative part and the corresponding statement part. If the package declaration requires a package body, we must write the package body in the same declarative part after the package declaration. Since declarations written in a package body are not visible outside the package, we can use a local package to provide controlled access to locally declared items.

EXAMPLE 7.6 *A package to manage unique identification numbers*

Suppose we need to generate test cases in a design, with each test case having a unique identification number. We can declare a package locally within a stimulus-generator process. The package encapsulates a variable that tracks the next identification number to be assigned, and provides an operation to yield the next number. The process outline is:

```
stim_gen : process is

  package ID_manager is
    impure function get_ID return natural;
  end package ID_manager;

  package body ID_manager is
    variable next_ID : natural := 0;
    impure function get_ID return natural is
      variable result : natural;
    begin
      result := next_ID;
      next_ID := next_ID + 1;
      return result;
    end function get_ID;
  end package body ID_manager;

  ...

  type test_case is record
    ...
    ID : natural;
```

```
      end record test_case;
   variable next_test_case : test_case;
begin
   ...
   next_test_case.ID := ID_manager.get_ID;
   ID_manager.next_ID := 0;   -- Illegal
   ...
end process stim_gen;
```

The variable **next_ID** is declared in the package body, and so is not visible outside the package. The only way to access it is using the **get_ID** function provided by the package declaration. This is shown in the first assignment statement within the process body. The package name is used as a prefix in the selected name for the function. The second assignment statement is illegal, since the variable is not visible at that point. The package provides a measure of safety against inadvertent corruption of the data state.

By writing the package locally within the process, it is only available in the process. Thus, we have achieved greater separation of concerns than if we had written the package as a design unit, making it globally visible. Moreover, since the package is local to a process, there can be no concurrent access by multiple processes. Thus, the encapsulated variable can be an ordinary non-shared variable. If the package were declared as a global design unit, there could be concurrent calls to the **get_ID** function. As a consequence, the variable would have to be declared as a shared variable, which would significantly complicate the design. (We describe shared variables in Chapter 23.)

We can also declare a local package within an enclosing package declaration, whether that enclosing package be a design unit or itself a local package. If the nested package is to be accessible outside the enclosing package, the nested package declaration must occur within the enclosing package declaration. The nested package body, if required, must then occur within the body of the enclosing package. As an example, the following outline shows an enclosing package **outer** with a nested package **inner** that declares a function f:

```
package outer is
   ...
   package inner is
      impure function f ( ... ) return natural;
   end package inner;
   ...
end package outer;

package body outer is
   ...
   package body inner is
      ...
      impure function f ( ... ) return natural is
         ...
```

```
      end function f;
    end package body inner;
  ...
end package body outer;
```

Outside the packages, we can refer to the function **f** with the name **outer.inner.f**.

If the enclosing package only requires a local package for private use, we can write both the nested package declaration and nested package body in the body of the enclosing package. The outline is:

```
package outer is
  ...
end package outer;

package body outer is
  ...
  package inner is
    ...
  end package inner;

  package body inner is
    ...
  end package body inner;
  ...
end package body outer;
```

In this case, items in the nested package are only accessible within the enclosing package body. While this scheme would not commonly arise in practice, it does serve to illustrate the consistency and general applicability of the visibility rules we introduced in Section 6.6.

VHDL-87, -93, and -2002

These versions of VHDL do not allow declaration of local packages. A package can only be declared as a design unit.

7.3   Use Clauses

We have seen how we can refer to an item provided by a package by writing its selected name, for example, **work.cpu_types.status_value**. This name refers to the item **status_value** in the package **cpu_types** stored in the library **work**. If we need to refer to this object in many places in a model, having to write the library name and package name becomes tedious and can obscure the intent of the model. We saw in Chapter 5 that we can write a *use clause* to make a library unit directly visible in a model, allowing us to omit the library name when referring to the library unit. Since an analyzed design-unit package is a library unit, use clauses also apply to making such packages directly visible.

So we could precede a model with a use clause referring to the package defined in the example in Example 7.1:

```
use work.cpu_types;
```

This use clause allows us to write declarations in our model more simply; for example:

```
variable data_word : cpu_types.word;
variable next_address : cpu_types.address;
```

In fact, the use clause is more general than this usage indicates and allows us to make any name from a library or package directly visible. Let us look at the full syntax rule for a use clause, then discuss some of the possibilities.

use_clause ⇐ **use** selected_name ⟨ , ₒₒₒ ⟩ ;

selected_name ⇐ name . ⟨ identifier ▯ character_literal ▯ operator_symbol ▯ **all** ⟩

The syntax rule for names, shown in Appendix B, includes the possibility of a name itself being either a selected name or a simple identifier. If we make these substitutions in the above syntax rule, we see that a selected name can be of the form

identifier . identifier . ⟨ identifier ▯ character_literal ▯ operator_symbol ▯ **all** ⟩

One possibility is that the first identifier is a library name, and the second is the name of a package within the library. This form allows us to refer directly to items within a package without having to use the full selected name. For example, we can simplify the above declarations even further by rewriting the use clause as

```
use work.cpu_types.word, work.cpu_types.address;
```

The declarations can then be written as

```
variable data_word : word;
variable next_address : address;
```

We can place a use clause in any declarative part in a model. One way to think of a use clause is that it "imports" the names of the listed items into the part of the model containing the use clause, so that they can be used without writing the library or package name. The names become directly visible after the use clause, according to visibility rules similar to those we discussed in Chapter 6.

The syntax rule for a use clause shows that we can write the keyword **all** instead of the name of a particular item to import from a package. This form is very useful, as it is a shorthand way of importing all of the names defined in the interface of a package. For example, if we are using the IEEE standard-logic package as the basis for the data types in a design, it is often convenient to import everything from the standard-logic package, including all of the overloaded operator definitions. We can do this with a use clause as follows:

```
use ieee.std_logic_1164.all;
```

This use clause means that the model imports all of the names defined in the package std_logic_1164 residing in the library **ieee**. This explains the "magic" that we have used in previous chapters when we needed to model data using the standard-logic types. The keyword **all** can be included for any package where we want to import all of the declarations from the package into a model.

We can also write use clauses for locally declared packages. Thus, in Example 7.6, we could follow the package declaration with the use clause

```
use ID_manager.all;
```

and then rewrite the assignment in the process as

```
next_test_case.ID := get_ID;
```

EXAMPLE 7.7   *A use clause for the CPU types package*

Following is a revised version of the CPU architecture body outlined in Example 7.4. It includes a use clause referring to items declared in the **cpu_types** package. This makes the rest of the model considerably less cluttered and easier to read. The use clause is included within the declarative part of the instruction set interpreter process. Thus the names "imported" from the package are directly visible in the rest of the declarative part and in the body of the process.

```
architecture behavioral of cpu is
begin

  interpreter : process is

    use work.cpu_types.all;

    variable instr_reg : word;
    variable instr_opcode : opcode;

  begin
    ...   -- initialize
    loop
      ...   -- fetch instruction
      instr_opcode := extract_opcode ( instr_reg );
      case instr_opcode is
        when op_nop => null;
        when op_breq => ...
        ...
      end case;
    end loop;
  end process interpreter;

end architecture behavioral;
```

While using **all** to import all declaration from a package is a common case, there are occasions when we would prefer to use just a type declared in a package. This might occur if we need to draw upon several packages, each of which declares numerous other items that could conflict with our own declarations and with items from other packages. However, just importing a type is not particularly useful, since we typically need to operate on values of the type. Fortunately, VHDL helps us by importing more than just the type name when we refer to a type in a use clause. In addition, the following are imported:

- All of the predefined operations on the type, provided they are not hidden by over-loaded versions also declared in the package.

- Overloaded versions of predefined operations on the type declared in the package.

- For an enumeration type or subtype, all of the enumeration literals. This includes any character literals of the type.

- For a physical type or subtype, all of the unit names for the type.

  For example, suppose we declare the following package:

```
package stuff_pkg is

    type color_type is (red, orange, yellow, green, blue, violet);
    subtype warm_color is color_type range red to yellow;

    function "<" ( c1, c2 : color_type ) return boolean;
    function pretty ( c : color_type ) return boolean;

    type resistance is range 0 to 1E9 units
      Ohm;
      kOhm = 1000 Ohm;
      MOhm = 1000 kOhm;
    end units;

    subtype weak_logic is
      IEEE.std_logic_1164.std_ulogic range 'W' to 'H';

end package stuff_pkg;
```

Then the use clause

```
use stuff_pkg.color_type;
```

makes not only the type **color_type** visible, but also the enumeration literals **red** through **violet**, the predefined operations on **color_type** other than "<", and the overloaded "<" operator declared in the package. It does not make the function **pretty** visible, since it is not an overloaded version of a predefined operation. If we write the use clause

```
use stuff_pkg.warm_color;
```

it makes the subtype **warm_color** visible, along with all of the enumeration literals for **color_type** (not just those in the subtype) and the operations for **color_type**.

If we write the use clause

```
use stuff_pkg.resistance;
```

it makes the type **resistance** visible, along with the unit names **Ohm**, **kOhm**, and **MOhm**, and the predefined operations on **resistance**.

Finally, if we write the use clause

```
use stuff_pkg.weak_logic;
```

all we get is the subtype name **weak_logic** made visible, since none of the enumeration literals or operations are declared in the package **stuff_pkg**.

VHDL-87, -93, and -2002

According to a strict reading of the VHDL *Language Reference Manual*, these earlier versions do not import additional operations and other items if a use clause refers to a type name. Only the type name itself is imported. However, since that is not useful in practice, many implementations import at least the predefined operations, enumeration literals, and unit names. Implementations differ on whether they do this and whether they import any additional overloaded operations.

7.3.1 Visibility of Used Declarations

In general, a use clause makes a name directly visible in the enclosing declarative part and in the corresponding statement part. One of the most common places in which we write a use clause is at the beginning of a design unit. We saw in Section 5.4 how we may include library clauses, use clauses, and context references at the head of a design unit, such as an entity interface or architecture body. This area of a design unit is called its *context clause*. The names imported here are made directly visible throughout the design unit. For example, if we want to use the IEEE standard-logic type **std_ulogic** in the declaration of an entity, we might write the design unit as follows:

```
library ieee;  use ieee.std_logic_1164.std_ulogic;

entity logic_block is
  port ( a, b : in std_ulogic;
         y, z : out std_ulogic );
end entity logic_block;
```

The library clause and the use clause together form the context clause for the entity declaration in this example. The library clause makes the contents of the library accessible to the model, and the use clause imports the type name **std_ulogic** declared in the package **std_logic_1164** in the library **ieee**. By including the use clause in the context clause of the entity declaration, the **std_ulogic** type name is available when declaring the ports of the entity.

The names imported by a use clause in this way are made directly visible in the entire design unit after the use clause. In addition, if the design unit is a primary unit (such as an entity declaration or a package declaration), the visibility is extended to any corresponding secondary unit. Thus, if we include a use clause in the primary unit, we do not need to repeat it in the secondary unit, as the names are automatically visible there.

When we write use clauses in a model, there is potential to introduce conflicting names. The rules for dealing with such conflicts are related to those for dealing with over-loading of subprograms and enumeration literals.

One form of conflict arises when we use an item declared in a package, and the de-clarative part containing the use clause declares another item with the same name. For example, our **cpu_types** package in earlier examples declares the subtype **address**. We might use this package in an architecture body as follows:

```
use work.cpu_types.all;
architecture behavior of cpu is
  signal address : bit_vector(15 downto 0);
  ...
end architecture behavior;
```

The subtype name **address** identified by the use clause conflicts with the signal name **address** declared within the architecture body. VHDL resolves this conflict by not making the subtype name directly visible. Any reference to the simple name **address** in the archi-tecture body refers to the signal. Of course, we can still use the selected name **work.cpu_types.address** within the architecture body to refer to the subtype if necessary. In general, this form of conflict arises when an item identified by a use clause has the same name as a locally declared item and either or both are not overloadable (that is, not a sub-program or an enumeration literal). The conflict is resolved by not making the item iden-tified by the use clause directly visible. If both of the items are overloadable, then there is no conflict, since the normal rules for disambiguating subprogram calls apply.

Another form of conflict arises when we use two items of the same name declared in different packages. The rules for dealing with this form of conflict are more involved, and depend on whether the items are overloadable and whether either is a predefined oper-ation. We'll consider the cases in turn.

First, if one item is a predefined operation and the other is explicitly declared, only the explicitly declared item is made directly visible. Explicitly declared items in this case include overloaded versions of predefined operations, as well as other non-overloadable items. For example, suppose we declare two packages as follows:

```
package short_int_types is
  type short_int is range 0 to 255;
end package short_int_types

use work.short_int_types.all;
package short_int_ops is
  function "+" ( L, R : short_int ) return short_int;
  constant maximum : short_int := 255;
  ...
end package short_int_ops;
```

The package **short_int_types** declares the integer type **short_int**. The predefined op-erations "+" and **maximum** are implicitly declared within this package also. The package short_int_ops explicitly declares an overloaded version of the "+" operation to perform modulo addition. It also declares the constant **maximum**, which is a non-overloadable item. Now suppose we include a use clause for both of these packages in an architecture body:

```
use work.short_int_types.all, work.short_int_ops.all;
architecture behavior of alu is
  ...
end architecture behavior;
```

Within the architecture body, the explicitly declared "+" operation and **maximum** constant from the **short_int_ops** package are made directly visible. If we wanted to refer to the original predefined operations, we would have to write **work.short_int_types."+"** and **work.short_int.maximum**.

The second case of conflict from two items of the same name used from different packages arises if both items are explicitly declared and either or both are not overloadable. In this case, neither item is made directly visible. For example, if we declare the following package:

```
package controller_types is
  subtype address is bit_vector(2 downto 0);
  constant int_ack : address := "100";
end package controller_types;
```

and use it together with the **cpu_types** package from earlier examples:

```
use work.cpu_types.all, work.controller_types.all;
architecture rtl of system is
  ...
end architecture rtl;
```

the names address and **int_ack** are not made directly visible in the architecture. Both packages explicitly declare **address** as a subtype, which is a non-overloadable item. The package **cpu_types** declares **int_ack** as an enumeration type, which is overloadable, but **controller_types** declares **int_ack** as a constant which is not overloadable.

There is one case where what we might consider a conflict arises, but in fact a different rule comes into play. Suppose we write two packages, each of which explicitly declares overloaded subprograms, as follows:

```
package int_ops is
  function increment ( a : inout integer; n : in integer := 1 );
end package int_ops;

package counter_ops is
  function increment ( c : inout integer; n : in integer := 1 );
end package counter_ops;
```

If we use these two packages in a given design unit, as follows:

```
use work.int_ops.all, work.counter_ops.all;
```

we might be tempted to say that, since the names and the types of the parameters are the same, the declarations conflict, and so neither version of **increment** would be made directly visible. However, the rules for disambiguating calls to overloaded procedures allow parameter names used in named association to be taken into account. Thus, the following two calls are unambiguous:

```
increment ( a => count_value, n => -1 );
increment ( c => count_value, n => -1 );
```

The first call refers to the **increment** procedure from package **int_ops**, whereas the second call refers to the **increment** procedure from package **counter_ops**. In order for this rule to come into play, the two subprogram names must be directly visible. Rather than there being a conflict, there may be an ambiguity in the call if positional association is used.

VHDL-87, -93, and -2002

According to a strict reading of the VHDL *Language Reference Manual*, these earlier versions did not resolve a conflict between an implicitly declared predefined operation in one package and an explicitly declared overloaded version in another package in the same way as VHDL-2008. Instead, both are made directly visible, and the rules for overload resolution are required to disambiguate calls.

Complicating this is the widely held, but incorrect, view that versions of a subprogram with the same names and parameter types used from different packages do conflict and are not made directly visible. Some tools implement this view, rather than making both subprograms directly visible, as described for the **increment** procedures above.

Exercises

1. [❶ 7.1] Write a package declaration for use in a model of an engine management system. The package contains declarations of a physical type, **engine_speed**, expressed in units of revolutions per minute (RPM); a constant, **peak_rpm**, with a value of 6000 RPM; and an enumeration type, **gear**, with values representing first, second, third, fourth and reverse gears. Assuming the package is analyzed and stored in the current working library, write selected names for each of the items declared in the package.

2. [❶ 7.1] Write a declaration for a procedure that increments an integer, as the procedure declaration would appear in a package declaration.

3. [❶ 7.1] Write a declaration for a function that tests whether an integer is odd, as the function declaration would appear in a package declaration.

4. [❶ 7.1] Write a deferred constant declaration for the real constant $e = 2.71828$.

5. [❶ 7.2] Is a package body required for the package declaration described in Exercise 1?

6. [❶ 7.3] Write a use clause that makes the **engine_speed** type from the package described in Exercise 1 directly visible.

7. [❶ 7.3] Write a context clause that makes a library **DSP_lib** accessible and that makes an entity **systolic_FFT** and all items declared in a package **DSP_types** in the library directly visible.

8. [❸ 7.1/7.2] Develop a package declaration and body that provide operations for dealing with time-of-day values. The package defines a time-of-day value as a record containing hours, minutes and seconds since midnight and provides deferred constants representing midnight and midday. The operations provided by the package are

   - comparison ("<", ">", "<=" and ">="),

   - addition of a time-of-day value and a number of seconds to yield a time-of-day result and

   - subtraction of two time-of-day values to yield a number-of-seconds result.

9. [❸ 7.1/7.2] Develop a package declaration and body to provide operations on character strings representing identifiers. An outline of the package declaration is

   ```
   package identifier_pkg is

      subtype identifier is string(1 to 15);

      constant max_table_size : integer := 50;
      subtype table_index is integer range 1 to max_table_size;
      type table is array (table_index) of identifier;

      . . .

   end package identifier_pkg;
   ```

   The package also declares a procedure to convert alphabetic characters in a string to lowercase and a procedure to search for an occurrence of a given identifier in a table. The search procedure has two **out**-mode parameters: a Boolean value indicating whether the sought string is in the table and a **table_index** value indicating its position, if present.

Chapter 8

Resolved Signals

Throughout the previous chapters we have studiously avoided considering the case of multiple output ports connecting one signal. The problem that arises in such a case is determining the final value of the signal when multiple sources drive it. In this chapter we discuss *resolved signals*, the mechanism provided by VHDL for modeling such cases.

8.1 Basic Resolved Signals

If we consider a real digital system with two outputs driving one signal, we can fairly readily determine the resulting value based on some analog circuit theory. The signal is driven to some intermediate state, depending on the drive capacities of the conflicting drivers. This intermediate state may or may not represent a valid logic state. Usually we only connect outputs in a design if at most one is active at a time, and the rest are in some high-impedance state. In this case, the resulting value should be the driving value of the single active output. In addition, we include some form of "pull-up" that determines the value of the signal when all outputs are inactive.

While this simple approach is satisfactory for some models, there are other cases where we need to go further. One of the reasons for simulating a model of a design is to detect errors such as multiple simultaneously active connected outputs. In this case, we need to extend the simple approach to detect such errors. Another problem arises when we are modeling at a higher level of abstraction and are using more complex types. We need to specify what, if anything, it means to connect multiple outputs of an enumeration type together.

The approach taken by VHDL is a very general one: the language requires the designer to specify precisely what value results from connecting multiple outputs. It does this through *resolved signals*, which are an extension of the basic signals we have used in previous chapters. A resolved signal includes in its definition a function, called the *resolution function*, that is used to calculate the final signal value from the values of all of its sources.

Let us see how this works by developing an example. We can model the values driven by a tristate output using a simple extension to the predefined type **bit**, for example:

```
type tri_state_logic is ('0', '1', 'Z');
```

The extra value, 'Z', is used by an output to indicate that it is in the high-impedance state. Next, we need to write a function that takes a collection of values of this type, representing the values driven by a number of outputs, and return the resulting value to be applied to the connected signal. For this example, we assume that at most one driver is active ('0' or '1') at a time and that the rest are all driving 'Z'. The difficulty with writing the function is that we should not restrict it to a fixed number of input values. We can avoid this by giving it a single parameter that is an unconstrained array of **tri_state_logic** values, defined by the type declaration

```
type tri_state_logic_array is
  array (integer range <>) of tri_state_logic;
```

The declaration of the resolution function is

```
function resolve_tri_state_logic
  ( values : in tri_state_logic_array )
  return tri_state_logic is
  variable result : tri_state_logic := 'Z';
begin
  for index in values'range loop
    if values(index) /= 'Z' then
      result := values(index);
    end if;
  end loop;
  return result;
end function resolve_tri_state_logic;
```

The final step to making a resolved signal is to declare the signal, as follows:

```
signal s1 : resolve_tri_state_logic tri_state_logic;
```

This declaration is almost identical to a normal signal declaration, but with the addition of the resolution function name before the signal type. The signal still takes on values from the type **tri_state_logic**, but inclusion of a function name indicates that the signal is a resolved signal, with the named function acting as the resolution function. The fact that s1 is resolved means that we are allowed to have more than one source for it in the design. (Sources include drivers within processes and output ports of components associated with the signal.) When a transaction is scheduled for the signal, the value is not applied to the signal directly. Instead, the values of all sources connected to the signal, including the new value from the transaction, are formed into an array and passed to the resolution function. The result returned by the function is then applied to the signal as its new value.

Let us look at the syntax rule that describes the VHDL mechanism we have used in the above example. It is an extension of the rules for the subtype indication, which we first introduced in Chapters 2 and 4. The combined rule is

subtype_indication ⇐
    〖 *resolution_function*_name 〗
    type_mark 〖 **range**〖 *range*_attribute_name

〚 simple_expression 〚 **to** 〛 **downto** 〛 simple_expression 〛
〚 (discrete_range { , ... }) 〛

This rule shows that a subtype indication can optionally include the name of a function to be used as a resolution function. Given this new rule, we can include a resolution function name anywhere that we specify a type to be used for a signal. For example, we could write a separate subtype declaration that includes a resolution function name, defining a *resolved subtype*, then use this subtype to declare a number of resolved signals, as follows:

```
subtype resolved_logic is
  resolve_tri_state_logic tri_state_logic;

signal s2, s3 : resolved_logic;
```

The subtype **resolved_logic** is a resolved subtype of **tri_state_logic**, with **resolve_tri_state_logic** acting as the resolution function. The signals **s2** and **s3** are resolved signals of this subtype. Where a design makes extensive use of resolved signals, it is good practice to define resolved subtypes and use them to declare the signals and ports in the design.

The resolution function for a resolved signal is also invoked to initialize the signal. At the start of a simulation, the drivers for the signal are initialized to the expression included in the signal declaration, or to the default initial value for the signal type if no initialization expression is given. The resolution function is then invoked using these driver values to determine the initial value for the signal. In this way, the signal always has a properly resolved value, right from the start of the simulation.

Let us now return to the tristate logic type we introduced earlier. In the previous example, we assumed that at most one driver is '0' or '1' at a time. In a more realistic model, we need to deal with the possibility of driver conflicts, in which one source drives a resolved signal with the value '0' and another drives it with the value '1'. In some logic families, such driver conflicts cause an indeterminate signal value. We can represent this indeterminate state with a fourth value of the logic type, 'X', often called an *unknown* value. This gives us a complete and consistent *multivalued logic* type, which we can use to describe signal values in a design in more detail than we can using just bit values.

EXAMPLE 8.1 *A four-state multivalued logic type and its use*

The following package declaration and body define a four-state multivalued logic type.

```
package MVL4 is

  -- unresolved logic type
  type MVL4_ulogic is ('X', '0', '1', 'Z');

  type MVL4_ulogic_vector is
    array (natural range <>) of MVL4_ulogic;

  function resolve_MVL4
    ( contribution : MVL4_ulogic_vector ) return MVL4_ulogic;
```

```vhdl
    subtype MVL4_logic is resolve_MVL4 MVL4_ulogic;

  function "not" ( r : MVL4_ulogic ) return MVL4_ulogic;
  function "and" ( l, r : MVL4_ulogic ) return MVL4_ulogic;
  function "or"  ( l, r : MVL4_ulogic ) return MVL4_ulogic;
  ...

  function to_X01 ( a : MVL4_ulogic ) return MVL4_ulogic;

  function "??" ( r : MVL4_ulogic ) return boolean;

end package MVL4;
```

```vhdl
package body MVL4 is

  type table is
    array (MVL4_ulogic, MVL4_ulogic) of MVL4_ulogic;

  constant resolution_table : table :=
    --  'X'  '0'  '1'  'Z'
    --  ------------------
    ( ( 'X', 'X', 'X', 'X' ),     -- 'X'
      ( 'X', '0', 'X', '0' ),     -- '0'
      ( 'X', 'X', '1', '1' ),     -- '1'
      ( 'X', '0', '1', 'Z' ) );   -- 'Z'

  function resolve_MVL4 ( contribution : MVL4_ulogic_vector )
                          return MVL4_ulogic is
    variable result : MVL4_ulogic := 'Z';
  begin
    for index in contribution'range loop
      result := resolution_table(result, contribution(index));
    end loop;
    return result;
  end function resolve_MVL4;

  function "not" ( r : MVL4_ulogic ) return MVL4_ulogic is
  begin
    case r is
      when '1' => return '0';
      when '0' => return '1';
      when others => return 'X';
    end case;
  end function "not";

  function "and" ( l, r : MVL4_ulogic ) return MVL4_ulogic is ...

  function "or"  ( l, r : MVL4_ulogic ) return MVL4_ulogic is ...

  ...
```

```
function to_X01 ( a : MVL4_ulogic ) return MVL4_ulogic is
begin
  case a is
    when '0' | '1' => return a;
    when 'X' | 'Z' => return 'X';
  end case;
end function to_X01;

function "??" ( r : MVL4_ulogic ) return boolean is
begin
  return r = '1';
end function "??";
```

end package body MVL4;

The constant **resolution_table** is a lookup table used to determine the value resulting from two source contributions to a signal of the resolved logic type. The resolution function uses this table, indexing it with each element of the array passed to the function. If any source contributes 'X', or if there are two sources with conflicting '0' and '1' contributions, the result is 'X'. If one or more sources are '0' and the remainder 'Z', the result is '0'. Similarly, if one or more sources are '1' and the remainder 'Z', the result is '1'. If all sources are 'Z', the result is 'Z'. The lookup table is a compact way of representing this set of rules. The package also declares overloaded versions of the logical operators and other functions. The **to_X01** function ensures that a value is either a proper logic value ('0' or '1') or an unknown value. The overloaded "**??**" operator allows us to use values of the logic type in conditions with implicit conversion to **boolean**.

We can use this package in a design for a tristate buffer. The entity declaration and a behavioral architecture body are

```
use work.MVL4.all;

entity tri_state_buffer is
  port ( a, enable : in MVL4_ulogic;  y : out MVL4_ulogic );
end entity tri_state_buffer;

--------------------------------------------------

architecture behavioral of tri_state_buffer is
begin

  y <= to_X01(a) when enable else
       'Z'        when not enable else
       'X';

end architecture behavioral;
```

When the buffer is enabled, the buffer copies the input to the output, but with a 'Z' input value changed to 'X' by the **to_X01** function. When the buffer is not enabled, it drives the value 'Z' on its output. If the **enable** port is not a proper logic level, both conditions are false, so the buffer drives the unknown value on its output.

An architecture body for a logic block that uses the tristate buffer is:

```
use work.MVL4.all;

architecture gate_level of misc_logic is

  signal src1, src1_enable : MVL4_ulogic;
  signal src2, src2_enable : MVL4_ulogic;
  signal selected_val : MVL4_logic;
  . . .

begin

  src1_buffer : entity work.tri_state_buffer(behavioral)
    port map ( a => src1, enable => src1_enable,
               y => selected_val );

  src2_buffer : entity work.tri_state_buffer(behavioral)
    port map ( a => src2, enable => src2_enable,
               y => selected_val );

  . . .

end architecture gate_level;
```

The signal **selected_val** is a resolved signal of the multivalued logic type. It is driven by the two buffer output ports. The resolution function for the signal is used to determine the final value of the signal whenever a new transaction is applied to either of the buffer outputs.

8.1.1   Composite Resolved Subtypes

The above examples have all shown resolved subtypes of scalar enumeration types. In fact, VHDL's resolution mechanism is more general. We can use it to define a resolved subtype of any type that we can legally use as the type of a signal. Thus, we can define resolved integer subtypes, resolved composite subtypes and others. In the latter case, the resolution function is passed an array of composite values and must determine the final composite value to be applied to the signal.

EXAMPLE 8.2   *A package for a resolved array subtype*

The package declaration and body below define a resolved array subtype. Each element of an array value of this subtype can be 'X', '0', '1' or 'Z'. The unresolved type **uword** is an unconstrained array of these values. The resolution function has an unconstrained array parameter consisting of elements of type **uword**. The function uses the lookup table to resolve corresponding elements from each of the contributing sources and produces an array result. The subtype **word** is the final resolved array subtype.

package words **is**

```vhdl
    type X01Z is ('X', '0', '1', 'Z');
    type uword is array (natural range <>) of X01Z;

    type uword_vector is array (natural range <>) of uword;

    function resolve_word
      ( contribution : uword_vector ) return uword;

    subtype word is resolve_word uword;

end package words;
```

```vhdl
package body words is

    type table is array (X01Z, X01Z) of X01Z;

    constant resolution_table : table :=
      --  'X'   '0'   '1'   'Z'
      --  ------------------
      ( ( 'X', 'X', 'X', 'X' ),      -- 'X'
        ( 'X', '0', 'X', '0' ),      -- '0'
        ( 'X', 'X', '1', '1' ),      -- '1'
        ( 'X', '0', '1', 'Z' ) );    -- 'Z'

    function resolve_word
      ( contribution : uword_vector ) return uword is
      variable result : uword(contribution'element'range)
                := (others => 'Z');
    begin
      for index in contribution'range loop
        for element in result'range loop
          result(element) :=
            resolution_table( result(element),
                              contribution(index)(element) );
        end loop;
      end loop;
      return result;
    end function resolve_word;

end package body words;
```

We can use these types to declare array ports in entity declarations and resolved array signals with multiple sources. For example, the following CPU entity and memory entity have bidirectional data ports of the unresolved array type.

```vhdl
use work.words.all;

entity cpu is
  port ( address : out uword(23 downto 0);
         data : inout uword(31 downto 0);  ... );
end entity cpu;
```

```
use work.words.all;

entity memory is
  port ( address : in uword(23 downto 0);
            data : inout uword(31 downto 0); ... );
end entity memory;
```

The architecture body for a computer system declares a signal of the resolved subtype and connects it to the data ports of the instances of the CPU and memory:

```
architecture top_level of computer_system is

  use work.words.all;

  signal address : uword(23 downto 0);
  signal data : word(31 downto 0);
  ...

begin

  the_cpu : entity work.cpu(behavioral)
    port map ( address, data, ... );

  the_memory : entity work.memory(behavioral)
    port map ( address, data, ... );

  ...

end architecture top_level;
```

A resolved composite subtype works well provided every source for a resolved signal of the subtype is connected to every element of the signal. For the **data** signal shown in the example, every source must be a 32-element array and must connect to all 32 elements of the data signal. However, in a realistic computer system, sources are not always connected in this way. For example, we may wish to connect an 8-bit-wide device to the low-order eight bits of a 32-bit-wide data bus. We might attempt to express such a connection in a component instantiation statement, as follows:

```
boot_rom : entity work.ROM(behavioral)
  port map ( a => address, d => data(24 to 31), ... );   -- illegal
```

If we add this statement to the architecture body in Example 8.2, we have two sources for elements 0 to 23 of the data signal and three for elements 24 to 31. A problem arises when resolving the signal, since we are unable to construct an array containing the contributions from the sources. For this reason, VHDL does not allow us to write such a description; it is illegal.

The solution to this problem is to describe the data signal as an array of resolved elements, rather than as a resolved array of elements. One way of doing this is to declare an array type whose elements are values of the **MVL4_logic** type, shown in Example 8.2. The array type declaration is

type MVL4_logic_vector **is array** (natural **range** <>) **of** MVL4_logic;

This approach has the advantage that the array type is unconstrained, so we can use it to create signals of different widths, each element of which is resolved. The problem, however, is that the type **MVL4_logic_vector** is distinct from the type **MVL4_ulogic_vector**, since they are defined by separate type declarations. Neither is a subtype of the other. Hence we cannot legally associate a signal of type **MVL4_logic_vector** with a port of type **MVL4_ulogic_vector**, or a signal of type **MVL4_ulogic_vector** with a port of type **MVL4_logic_vector**.

A better way to describe arrays of resolved elements is to declare an array subtype in which we associate the resolution function with the elements. In our example, the array base type is MVL4_ulogic_vector, which has unresolved elements. We declare the subtype as:

subtype MVL4_logic_vector **is** (resolve_MVL4) std_ulogic_vector;

The parentheses around the resolution function name, **resolve_MVL4**, indicates that the resolution function is associated with each element of the array type, rather than with the array type as a whole. Since **MVL4_logic_vector** is now a subtype of **MVL4_ulogic_vector**, not a distinct type, we can freely assign and associate signals and ports of the two types.

This example illustrates a more general form of *resolution indication* to be included in a subtype indication or signal declaration, rather than just naming a resolution function by itself. The syntax rule is:

resolution_indication ⇐
    *resolution_function*_name
    ❙ (resolution_indication
        ❙ ⦅ *record_element*_identifier resolution_indication ⦆ ⦃ , ... ⦄)

If we want to associate a resolution function with an entire subtype, the resolution indication just consists of the resolution function name, as in the declaration of MVL4_logic:

subtype MVL4_logic **is** resolve_MVL4 MVL4_ulogic;

The resolution indication here is just the resolution function name, **resolve_MVL4**. In the case of an array whose elements are to be resolved, we write the resolution function name in parentheses, as in the declaration of **MVL4_logic_vector**. We can also resolve the elements of an array type that is itself an array element type. For example, given the following declaration:

type unresolved_RAM_content_type **is**
    array (natural **range** <>) **of** MVL4_ulogic_vector;

we can declare a subtype with resolved nested elements:

```
subtype RAM_content_type is
  ((resolve_MVL4)) unresolved_RAM_content_type;
```

The degree of nesting of parentheses indicates how deeply nested in the type structure the resolution function is associated. Two levels indicate that the resolution function is associated with the elements of the elements of the type.

If we have a record type, one of whose elements is to be resolved, we include the element name in the resolution indication. For example, given the following record type with no associated resolution information:

```
type unresolved_status_type is record
  valid : MVL4_ulogic;
  dirty : MVL4_ulogic;
  tag : MVL4_ulogic_vector;
end record unresolved_status_type;
```

we can declare a subtype with the **valid** element resolved by the function **wired_and** as follows:

```
subtype status_resolved_valid is
  (valid wired_and) unresolved_status_type;
```

We can include resolution functions with multiple elements of the record type by listing the element names and the resolution function associated with each, for example:

```
subtype status_resolved_flags is
  (valid wired_and, dirty wired_or) unresolved_status_type;
```

For a record element that is itself of a composite type, we can associate a resolution function with subelements of the record element by writing a parenthesized resolution indication for the element. Thus, to resolve the elements of the **tag** element of the above record type, we would declare a subtype as follows:

```
subtype status_resolved_tag is
  (tag(resolve_MVL4)) unresolved_status_type;
```

We could combine all of these examples together, resolving all of the scalar subelements, as follows:

```
subtype resolved_status_type is
  ( tag(resolve_MVL4),
    valid wired_and,
    dirty wired_or ) unresolved_status_type;
```

This declaration illustrates that we do not have to write the resolution indications for the record elements in the same order as the declaration of elements in the record types. The record element names in the resolution indication determine the element with which the resolution function is associated.

EXAMPLE 8.3 *Connection to parts of a bus signal*

Let us assume that the type MVL4_logic_vector described above has been added to the package MVL4. Below are entity declarations for a ROM entity and a single in-line memory module (SIMM), using the MVL4_ulogic_vector type for their data ports. The data port of the SIMM is 32 bits wide, whereas the data port of the ROM is only 8 bits wide.

```
use work.MVL4.all;

entity ROM is
  port ( a : in MVL4_ulogic_vector(15 downto 0);
         d : out MVL4_ulogic_vector(7 downto 0);
         rd : in MVL4_ulogic );
end entity ROM;
```

```
use work.MVL4.all;

entity SIMM is
  port ( a : in MVL4_ulogic_vector(9 downto 0);
         d : inout MVL4_ulogic_vector(31 downto 0);
         ras, cas, we, cs : in MVL4_ulogic );
end entity SIMM;
```

The following architecture body uses these two entities. It declares a signal, internal_data, of the MVL4_logic_vector type, representing 32 individually resolved elements. The SIMM entity is instantiated with its data port connected to all 32 internal data elements. The ROM entity is instantiated with its data port connected to the rightmost eight elements of the internal data signal. When any of these elements is resolved, the resolution function is passed contributions from the corresponding elements of the SIMM and ROM data ports. When any of the remaining elements of the internal data signal are resolved, they have one less contribution, since they are not connected to any element of the ROM data port.

```
architecture detailed of memory_subsystem is

  signal internal_data : MVL4_logic_vector(31 downto 0);
  ...

begin
  boot_ROM : entity work.ROM(behavioral)
    port map ( a => internal_addr(15 downto 0),
               d => internal_data(7 downto 0),
               rd => ROM_select );

  main_mem : entity work.SIMM(behavioral)
    port map ( a => main_mem_addr, d => internal_data, ... );

  ...
```

```
end architecture detailed;
```

8.1.2   Summary of Resolved Subtypes

At this point, let us summarize the important points about resolved signals and their resolution functions. Resolved signals of resolved subtypes are the only means by which we may connect a number of sources together, since we need a resolution function to determine the final value of the signal or port from the contributing values. The resolution function must take a single parameter that is a one-dimensional unconstrained array of values of the signal type, and must return a value of the signal type. The index type of the array does not matter, so long as it contains enough index values for the largest possible collection of sources connected together. For example, an array type declared as follows is inadequate if the resolved signal has five sources:

```
type small_int is range 1 to 4;
type small_array is array (small_int range <>) of ... ;
```

The resolution function must be a pure function; that is, it must not have any side effects. This requirement is a safety measure to ensure that the function always returns a predictable value for a given set of source values. Furthermore, since the source values may be passed in any order within the array, the function should be commutative; that is, its result should be independent of the order of the values. When the design is simulated, the resolution function is called whenever any of the resolved signal's sources is active. The function is passed an array of all of the current source values, and the result it returns is used to update the signal value. When the design is synthesized, the resolution function specifies the way in which the synthesized hardware should combine values from multiple sources for a resolved signal.

8.1.3   IEEE **std_logic_1164** Resolved Subtypes

In previous chapters we have used the IEEE standard multivalued logic package, std_logic_1164. We are now in a position to describe all of the types and subtypes provided by the package. We defer a full description of the operations provided by the package to Chapter 9, in which we also describe other standard packages based on the standard logic types. First, recall that the package provides the basic type **std_ulogic**, defined as

```
type std_ulogic is ('U', 'X', '0', '1', 'Z', 'W', 'L', 'H', '-');
```

and an array type std_ulogic_vector, defined as

```
type std_ulogic_vector is array ( natural range <> ) of std_ulogic;
```

We have not mentioned it before, but the "u" in "ulogic" stands for unresolved. These types serve as the basis for the declaration of the resolved subtype std_logic, defined as follows:

```
function resolved ( s : std_ulogic_vector ) return std_ulogic;
```

```
subtype std_logic is resolved std_ulogic;
```

The standard-logic package also declares an array subtype of standard-logic elements, analogous to the **bit_vector** type, for use in declaring array signals:

```
subtype std_logic_vector (resolved) std_ulogic_vector;
```

The standard defines the resolution function **resolved** as follows:

```
type stdlogic_table is array (std_ulogic, std_ulogic) of std_ulogic;
constant resolution_table : stdlogic_table :=
  -- ---------------------------------------------
  --  'U', 'X', '0', '1', 'Z', 'W', 'L', 'H', '-'
  -- ---------------------------------------------
  ( ( 'U', 'U', 'U', 'U', 'U', 'U', 'U', 'U', 'U' ),   -- 'U'
    ( 'U', 'X', 'X', 'X', 'X', 'X', 'X', 'X', 'X' ),   -- 'X'
    ( 'U', 'X', '0', 'X', '0', '0', '0', '0', 'X' ),   -- '0'
    ( 'U', 'X', 'X', '1', '1', '1', '1', '1', 'X' ),   -- '1'
    ( 'U', 'X', '0', '1', 'Z', 'W', 'L', 'H', 'X' ),   -- 'Z'
    ( 'U', 'X', '0', '1', 'W', 'W', 'W', 'W', 'X' ),   -- 'W'
    ( 'U', 'X', '0', '1', 'L', 'W', 'L', 'W', 'X' ),   -- 'L'
    ( 'U', 'X', '0', '1', 'H', 'W', 'W', 'H', 'X' ),   -- 'H'
    ( 'U', 'X', 'X', 'X', 'X', 'X', 'X', 'X', 'X' )    -- '-'
  );

function resolved ( s : std_ulogic_vector ) return std_ulogic is
  variable result : std_ulogic := 'Z';   -- weakest state default
begin
  if s'length = 1 then
    return s(s'low);
  else
    for i in s'range loop
      result := resolution_table(result, s(i));
    end loop;
  end if;
  return result;
end function resolved;
```

VHDL tools are allowed to provide built-in implementations of this function to improve performance. The function uses the constant **resolution_table** to resolve the driving values. If there is only one driving value, the function returns that value unchanged. If the function is passed an empty array, it returns the value 'Z'. (The circumstances under which a resolution function may be invoked with an empty array will be covered in Section 23.1.) The value of **resolution_table** shows exactly what is meant by "forcing" driving values ('X', '0' and '1') and "weak" driving values ('W', 'L' and 'H'). If one driver of a resolved signal drives a forcing value and another drives a weak value, the forcing value dominates. On the other hand, if both drivers drive different values with the same strength, the result is the unknown value of that strength ('X' or 'W'). The high-impedance value, 'Z', is dominated by forcing and weak values. If a "don't care" value ('–') is to be resolved with any other value, the result is the unknown value 'X'. The interpretation of the "don't care" value is that the model has not made a choice about its output state. Finally, if an "uninitialized" value ('U') is to be resolved with any other value, the result is 'U', indicating that the model has not properly initialized all outputs.

In addition to this multivalued logic subtype, the package **std_logic_1164** declares a number of subtypes for more restricted multivalued logic modeling. The subtype declarations are

```
subtype X01 is
  resolved std_ulogic range 'X' to '1'; -- ('X','0','1')
subtype X01Z is
  resolved std_ulogic range 'X' to 'Z'; -- ('X','0','1','Z')
subtype UX01 is
  resolved std_ulogic range 'U' to '1'; -- ('U','X','0','1')
subtype UX01Z is
  resolved std_ulogic range 'U' to 'Z'; -- ('U','X','0','1','Z')
```

Each of these is a closed subtype; that is, the result of resolving values in each case is a value within the range of the subtype. The subtype **X01Z** corresponds to the type **MVL4** we introduced in Example 8.1.

8.2 Resolved Signals, Ports, and Parameters

In the previous discussion of resolved signals, we have limited ourselves to the simple case where a number of drivers or output ports of component instances drive a signal. Any input port connected to the resolved signal gets the final resolved value as the port value when a transaction is performed. We now look in more detail at the case of ports of mode **inout** being connected to a resolved signal. The question to answer here is, What value is seen by the input side of such a port? Is it the value driven by the component instance or the final value of the resolved signal connected to the port? In fact, it is the latter. An **inout** port models a connection in which the driver contributes to the associated signal's value, and the input side of the component senses the actual signal rather than using the driving value.

EXAMPLE 8.4 *Distributed synchronization using a wired-and signal*

Some asynchronous bus protocols use a distributed synchronization mechanism based on a wired-and control signal. This is a single signal driven by each module using active-low open-collector or open-drain drivers and pulled up by the bus terminator. If a number of modules on the bus need to wait until all are ready to proceed with some operation, they use the control signal as follows. Initially, all modules drive the signal to the '0' state. When each is ready to proceed, it turns off its driver ('Z') and monitors the control signal. So long as any module is not yet ready, the signal remains at '0'. When all modules are ready, the bus terminator pulls the signal up to the '1' state. All modules sense this change and proceed with the operation.

Following is an entity declaration for a bus module that has a port of the unre-solved type **std_ulogic** for connection to such a synchronization control signal.

```
library ieee;  use ieee.std_logic_1164.all;

entity bus_module is
  port ( synch : inout std_ulogic;  ... );
end entity bus_module;
```

The architecture body for a system comprising several such modules is outlined below. The control signal is pulled up by a concurrent signal assignment statement, which acts as a source with a constant driving value of 'H'. This is a value having a weak strength, which is overridden by any other source that drives '0'. It can pull the signal high only when all other sources drive 'Z'.

```
architecture top_level of bus_based_system is

  signal synch_control : std_logic;
  ...

begin

  synch_control_pull_up : synch_control <= 'H';

  bus_module_1 : entity work.bus_module(behavioral)
    port map ( synch => synch_control, ... );

  bus_module_2 : entity work.bus_module(behavioral)
    port map ( synch => synch_control, ... );

  ...

end architecture top_level;
```

An outline of a behavioral architecture body for the bus module is:

```
architecture behavioral of bus_module is
begin

  behavior : process is
    ...
  begin
    synch <= '0'  after Tdelay_synch;
    ...
    -- ready to start operation
    synch <= 'Z' after Tdelay_synch;
    wait until synch = 'H';
    -- proceed with operation
    ...
  end process behavior;

end architecture behavioral;
```

Each instance initially drives its synchronization port with '0'. This value is passed up through the port and used as the contribution to the resolved signal from the entity instance. When an instance is ready to proceed with its operation, it changes its driving value to 'Z', modeling an open-collector or open-drain driver being turned off. The process then suspends until the value seen on the synchronization port changes to 'H'. If other instances are still driving '0', their contributions dominate, and the value of the signal stays '0'. When all other instances eventually change their contributions to 'Z', the value 'H' contributed by the pull-up statement dominates, and the value of the signal changes to 'H'. This value is passed back down through the ports of each instance, and the processes all resume.

8.2.1 Resolved Ports

Just as a signal declared with a signal declaration can be of a resolved subtype, so too can a port declared in an interface list of an entity. This is consistent with all that we have said about ports appearing just like signals to an architecture body. Thus if the architecture body contains a number of processes that must drive a port or a number of component instances that must connect outputs to a port, the port must be resolved. The final value driven by the resolved port is determined by resolving all of the sources within the architecture body. For example, we might declare an entity with a resolved port as follows:

```
library ieee;  use ieee.std_logic_1164.all;

entity IO_section is
  port ( data_ack : inout std_logic;  ... );
end entity IO_section;
```

The architecture body corresponding to this entity might instantiate a number of I/O controller components, each with their data acknowledge ports connected to the **data_ack** port of the entity. Each time any of the controllers updates its data acknowledge port, the standard-logic resolution function is invoked. It determines the driving value for the **data_ack** port by resolving the driving values from all controllers.

EXAMPLE 8.5 *A memory system with tristate buses*

We can write a model for a memory system composed of multiple memory devices with tristate data buses. The entity declaration for the memory devices is

```
library ieee; use ieee.std_logic_1164.all;
entity memory_256Kx8 is
  port ( ce_n, oe_n, we_n : in std_ulogic;
         a : in std_ulogic_vector(17 downto 0);
         d : inout std_ulogic_vector(7 downto 0) );
end entity memory_256Kx8;
```

The **d** port of the entity has unresolved elements because there is only one source for each element within the memory device. The entity declaration for the memory system is

```
library ieee; use ieee.std_logic_1164.all;
entity memory_1Mx8 is
  port ( ce_n, oe_n, we_n : in std_ulogic;
          a : in std_ulogic_vector(19 downto 0);
          d : inout std_logic_vector(7 downto 0) );
end entity memory_1Mx8;
```

In this case, the **d** port is of type **std_logic_vector** with resolved elements, since internally there are multiple sources, one per memory device. The structural architecture is

```
architecture struct of memory_1Mx8 is
  signal ce_decoded_n : std_ulogic_vector(3 downto 0);
begin

  with a(19 downto 18) select
    ce_decoded_n <= "1110" when "00",
                    "1101" when "01",
                    "1011" when "10",
                    "0111" when "11",
                    "XXXX" when others;

  chip0 : component memory_256Kx8
    port map ( ce_n => ce_decoded_n(0),
               oe_n => oe_n, we_n => we_n,
               a => a(17 downto 0), d => d );

  chip1 : component memory_256Kx8
    port map ( ce_n => ce_decoded_n(1),
               oe_n => oe_n, we_n => we_n,
               a => a(17 downto 0), d => d );

  . . .

end architecture struct;
```

We can connect the **d** port of each memory-device instance directly to the **d** port of the memory system. The contributions of all of the instances are resolved to form the driving value of the memory system's **d** port. Had we inadvertently declared the **d** port of the memory system entity to be of type **std_ulogic_vector**, the analyzer would detect the error arising from multiple sources connected to the unresolved elements.

If it happens that the actual signal associated with a resolved port in an enclosing architecture body is itself a resolved signal, then the signal's resolution function will be called separately after the port's resolution function has determined the port's driving value. Note that the signal in the enclosing architecture body may use a different resolution function from the connected port, although in practice most designs use the one function for resolution of all signals of a given subtype.

An extension of the above scenario is a design in which there are several levels of hierarchy, with a process nested at the deepest level generating a value to be passed out through resolved ports to a signal at the top level. At each level, a resolution function is called to determine the driving value of the port at that level. The value finally determined for the signal at the top level is called the *effective value* of the signal. It is passed back down the hierarchy of ports as the effective value of each **in**-mode or **inout**-mode port. This value is used on the input side of each port.

EXAMPLE 8.6   *Hierarchical resolution in a computer system model*

Figure 8.1 shows the hierarchical organization for a single-board computer system, consisting of a frame buffer for a video display, an input/output controller section, a CPU/memory section and a bus expansion block. These are all sources for the re-solved data bus signal. The CPU/memory section in turn comprises a memory block and a CPU/cache block. Both of these act as sources for the data port, so it must be a resolved port. The cache has two sections, both of which act as sources for the data port of the CPU/cache block. Hence, this port must also be resolved.

Let us consider the case of one of the cache sections updating its data port. The new driving value is resolved with the current driving value from the other cache section to determine the driving value of the CPU/cache block data port. This result is then resolved with the current driving value of the memory block to determine the driving value of the CPU/memory section. Next, this driving value is resolved with the current driving values of the other top-level sections to determine the effective value of the data bus signal. The final step involves propagating this signal value back down the hierarchy for use as the effective value of each of the data ports. Thus, a module

FIGURE 8.1

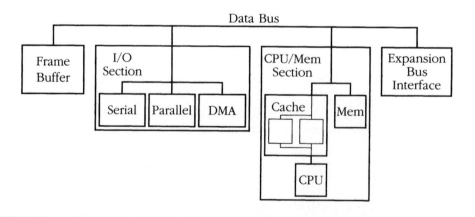

A hierarchical block diagram of a single-board computer system, showing the hierarchical connections of the resolved data bus ports to the data bus signal.

that reads the value of its data port will see the final resolved value of the data bus signal. This value is not necessarily the same as the driving value it contributes.

In Chapter 5, we indicated that we can read the value of a **buffer**-mode or **out**-mode port within an architecture of an entity. There is an important distinction between this and reading the value of an **inout**-mode port. As we mentioned above, the value seen internally for an **inout**-mode port is the effective value of the externally connected signal. Information is passed from the external connection into the architecture, as suggested by the "**in**" part of the **inout** port mode. In contrast, when we read a **buffer**-mode or **out**-mode port, the value seen internally is the driving value of the port. There is no transmission of information into the architecture. We use a **buffer**-mode or **out**-mode port for designs that have buffered internal connections or that read the value of the port for internal verification, for example, using assertions.

EXAMPLE 8.7   *Verification of tristate disconnection timing*

Suppose we wish to verify that the outputs of a device are all 'Z' within a required interval of the device being disabled and remain all 'Z' until the device is enabled. The output values are not required internally to implement any functionality for the device. Hence, we declare the output ports using **out** mode, as follows:

```
entity device is
  port ( en : in std_ulogic;
         d_out : out std_ulogic_vector(7 downto 0); ... );
end entity device;
```

We can read the values driven onto the output ports in verification code in the architecture:

```
architecture verifying of device is
  constant T_z : delay_length := 200 ps;
begin
  d_out <= ... when en else
           ... when not en else
           "XXXXXXXX";
  assert en or (not en'delayed(T_z) and d_out ?= "ZZZZZZZZ");
end architecture verifying;
```

8.2.2   Driving Value Attribute

Since the value seen on a signal or on an **inout**-mode port may be different from the value driven by a process, VHDL provides an attribute, 'driving_value, that allows the process to read the value it contributes to the prefix signal. For example, if a process has a driver for a resolved signal **s**, it may be driving **s** with the value 'Z' from a previously executed signal assignment statement, but the resolution function for **s** may have given it the value

'0'. The process can refer to **s'driving_value** to retrieve the value 'Z'. Note that a process can only use this attribute to determine its own contribution to a signal; it cannot directly find out another process's contribution.

VHDL-87

The **'driving_value** attribute is not provided in VHDL-87.

8.2.3 Resolved Signal Parameters

Let us now return to the topic of subprograms with signal parameters and see how they behave in the presence of resolved signals. Recall that when a procedure with an **out**-mode signal parameter is called, the procedure is passed a reference to the caller's driver for the actual signal. Any signal assignment statements performed within the procedure body are actually performed on the caller's driver. If the actual signal parameter is a resolved signal, the values assigned by the procedure are used to resolve the signal value. No resolution takes place within the procedure. In fact, the procedure need not be aware that the actual signal is resolved.

In the case of reading a signal parameter to a function or procedure, a reference to the actual signal parameter is passed when the subprogram is called, and the subprogram uses the actual value of the signal. If the signal is resolved, the subprogram sees the value determined after resolution. In the case of an **inout** signal parameter, a procedure is passed references to both the signal and its driver, and no resolution is performed internally to the procedure.

EXAMPLE 8.8   *Procedures for distributed synchronization*

We can encapsulate the distributed synchronization protocol described in Example 8.4 in a set of procedures, each with a single signal parameter, as follows:

```
procedure init_synchronize ( signal synch : out std_logic ) is
begin
  synch <= '0';
end procedure init_synchronize;

procedure begin_synchronize ( signal synch : inout std_logic;
                              Tdelay : in delay_length := 0 fs ) is
begin
  synch <= 'Z' after Tdelay;
  wait until synch;
end procedure begin_synchronize;

procedure end_synchronize ( signal synch : inout std_logic;
                            Tdelay : in delay_length := 0 fs ) is
begin
  synch <= '0' after Tdelay;
```

```
  wait until not synch;
end procedure end_synchronize;
```

Suppose a process uses a resolved signal **barrier** of subtype **std_logic** to synchronize with other processes. The process can use the procedures to implement the protocol as follows:

```
synchronized_module : process is
  ...
begin
  init_synchronize(barrier);
  ...
  loop
    ...
    begin_synchronize(barrier);
    ...    -- perform operation, synchronized with other processes
    end_synchronize(barrier);
    ...
  end loop;
end process synchronized_module;
```

The process has a driver for **barrier**, since the procedure calls associate the signal as an actual parameter with formal parameters of mode **out** and **inout**. A reference to this driver is passed to **init_synchronize**, which assigns the value '0' on behalf of the process. This value is used in the resolution of **barrier**. When the process is ready to start its synchronized operation, it calls **begin_synchronize**, passing references to its driver for **barrier** and to the actual signal itself. The procedure uses the driver to assign the value 'Z' on behalf of the process and then waits until the actual signal changes to 'H'. When the transaction on the driver matures, its value is resolved with other contributions from other processes and the result applied to the signal. This final value is used by the wait statement in the procedure to determine whether to resume the calling process. If the value is 'H', the process resumes, the procedure returns to the caller and the operation goes ahead. When the process completes the operation, it calls **end_synchronize** to reset **barrier** back to '0'.

Exercises

1. [❶ 8.1] Suppose there are four drivers connected to a resolved signal that uses the resolution function **resolve_tri_state_logic** shown on page 268. What is the resolved value of the signal if the four drivers contribute these values:

   a. 'Z', '1', 'Z', 'Z'?

   b. '0', 'Z', 'Z, '0'?

   c. 'Z', '1', 'Z', '0'?

2. [❶ 8.1] Rewrite the following resolved signal declaration as a subtype declaration followed by a signal declaration using the subtype.

    signal synch_control : wired_and tri_state_logic := '0';

3. [❶ 8.1] What is the initial value of the following signal of the type **MVL4_logic** defined in Example 8.1? How is that value derived?

    signal int_req : MVL4_logic;

4. [❶ 8.1] Does the result of the resolution function defined in Example 8.1 depend on the order of contributions from drivers in the array passed to the function?

5. [❶ 8.1] Suppose we define a resolved array subtype **byte** that is a subtype of **word**, defined in Example 8.2, with 8 elements. We then declare a signal of type **byte** with three drivers. What is the resolved value of the signal if the three drivers contribute these values:

    a. "ZZZZZZZZ", "ZZZZ0011", "ZZZZZZZZ"?

    b. "XXXXZZZZ", "ZZZZZZZZ", "00000011"?

    c. "00110011", "ZZZZZZZZ", "ZZZZ1111"?

6. [❶ 8.1] Suppose a signal is declared as

    signal data_bus : MVL4_logic_vector(0 **to** 15);

    where **MVL4_logic_vector** is as described on page 275, and the following signal assignments are each executed in different processes:

    data_bus <= "ZZZZZZZZZZZZZZZZ";

    data_bus(0 **to** 7) <= "XXXXZZZZ";

    data_bus(8 **to** 15) <= "00111100";

    What is the resolved signal value after all of the transactions have been performed?

7. [❶ 8.1] Suppose there are four drivers connected to a signal of type **std_logic**. What is the resolved value of the signal if the four drivers contribute these values:

    a. 'Z', '0', 'Z', 'H'?

    b. 'H', 'Z', 'W', '0'?

    c. 'Z', 'W', 'L', 'H'?

    d. 'U', '0', 'Z', '1'?

    e. 'Z', 'Z', 'Z', '–'?

8. [❶ 8.2] Figure 8.2 is a timing diagram for the system with two bus modules using the wired-and synchronization signal described in Example 8.4. The diagram shows the driving values contributed by each of the bus modules to the **synch_control** signal.

Complete the diagram by drawing the resolved waveform for **synch_control**. Indicate the times at which each bus module proceeds with its internal operation.

FIGURE 8.2

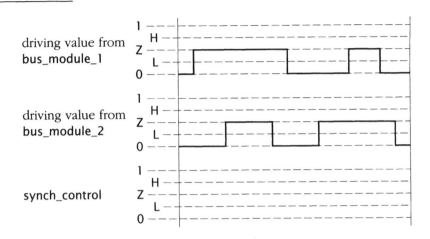

Timing diagram for wired-and synchronization

9. [❶ 8.2] Suppose all of the modules in the hierarchy of Figure 8.1 use resolved ports for their data connections. If the Mem, Cache, Serial and DMA modules all update their data drivers in the same simulation cycle, how many times is the resolution function invoked to determine the final resolved values of the data signals?

10. [❶ 8.2] Suppose a process in a model drives a bidirectional port **synch_T** of type **std_logic**. Write a signal assignment statement that inverts the process's contribution to the port.

11. [❷ 8.1] Develop a model that includes two processes, each of which drives a signal of the type **MVL4_logic** described in Example 8.1. Experiment with your simulator to see if it allows you to trace the invocation and execution of the resolution function.

12. [❷ 8.1] Develop a model of an inverter with an open-collector output of type **std_ulogic**, and a model of a pull-up resistor that drives its single **std_ulogic** port with the value 'H'. Test the models in a test bench that connects the outputs of a number of inverter instances to a signal of type **std_logic**, pulled up with a resistor instance. Verify that the circuit implements the active-low wired-or operation.

13. [❷ 8.1] Develop a behavioral model of an 8-bit-wide bidirectional transceiver, such as the 74245 family of components. The transceiver has two bidirectional data ports, **a** and **b**; an active-low output-enable port, **oe_n**; and a direction port, **dir**. When **oe_n** is low and **dir** is low, data is received from **b** to **a**. When **oe_n** is low and **dir** is high, data is transmitted from **a** to **b**. When **oe_n** is high, both **a** and **b** are high impedance. Assume a propagation delay of 5 ns for all output changes.

14. [❷ 8.1] Many combinatorial logic functions can be implemented in integrated circuits using pass transistors acting as switches. While a pass transistor is, in principle, a bidirectional device, for many circuits it is sufficient to model it as a unidirectional device. Develop a model of a unidirectional pass transistor switch, with an input port, an output port and an enable port, all of type **std_ulogic**. When the enable input is 'H' or '1', the input value is passed to the output, but with weak drive strength. When the enable input is 'L' or '0', the output is high impedance. If the enable input is at an unknown level, the output is unknown, except that its drive strength is weak.

15. [❸ 8.1] Develop a behavioral model of a tristate buffer with data input, data output and enable ports, all of type **std_ulogic**. The propagation time from data input to data output when the buffer is enabled is 4 ns. The turn-on delay from the enable port is 3 ns, and the turn-off delay is 3.5 ns. Use the buffer and any other necessary gate models in a structural model of the 8-bit transceiver described in Exercise 13.

16. [❸ 8.1] Use the unidirectional pass transistor model of Exercise 14 in a structural model of a four-input multiplexer. The multiplexer has select inputs s0 and s1. Pass transistors are used to construct the multiplexer as shown in Figure 8.3.

FIGURE 8.3

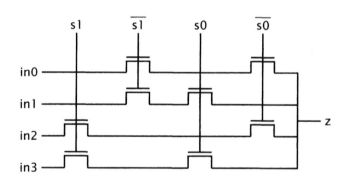

A multiplexer constructed of pass transistors.

17. [❸ 8.1] Develop a model of a distributed priority arbiter for a shared bus in a multiprocessor computer system. Each bus requester has a request priority, R, between 0 and 31, with 0 indicating the most urgent request and 31 indicating no request. Priorities are binary-encoded using 5-bit vectors, with bit 4 being the most-significant bit and bit 0 being the least-significant bit. The standard-logic values 'H' and '1' both represent the binary digit 1, and the standard-logic value '0' represents the binary digit 0. All requesters can drive and sense a 5-bit arbitration bus, A, which is pulled up to 'H' by the bus terminator. The requesters each use A and their own priority to compute the minimum of all priorities by comparing the binary digits of priorities as follows. For each bit position i:

 • if $(R_{4...i+1} = A_{4...i+1})$ and $(R_i = 0)$: drive A_i with '0' after T_{pd}

- if $(R_{4...i+1} \neq A_{4...i+1})$ or $(R_i = 1)$: drive A_i with 'Z' after T_{pd}

T_{pd} is the propagation delay between sensing a value on A and driving a resulting value on A. When the value on A has stabilized, it is the minimum of all request priorities. The requester with $R = A$ wins the arbitration. If you are not convinced that the distributed minimization scheme operates as required, trace its execution for various combinations of priority values.

18. [❹] Develop a behavioral model of a telephone keypad controller. The controller has outputs c1 to c3 and inputs r1 to r4, connected to the 12 switches of a touch-tone telephone as shown in Figure 8.4.

FIGURE 8.4

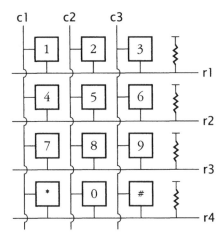

Keypad switch connections for a touch-tone telephone.

Each key in the keypad is a single-pole switch that shorts the row signal to the column signal when the key is pressed. Due to the mechanical construction of the switch, "switch bounce" occurs when the key is pressed. Several intermittent contacts are made between the signals over a period of up to 5 ms before a sustained contact is made. Bounce also occurs when the key is released. Several intermittent contacts may occur over the same period before sustained release is achieved.

The keypad controller scans the keypad by setting each of the column signals to '0' in turn. While a given column signal is '0', the controller examines each of the row inputs. If a row input is 'H', the switch between the column and the row is open. If the row input is '0', the switch is closed. The entire keypad is scanned once every millisecond.

The controller generates a set of column outputs c1_out to c3_out and a set of row outputs r1_out to r4_out. A valid switch closure is indicated by exactly one column output and exactly one row output going to '1' at the same time. The controller filters out spurious switch closures due to switch bounce and ignores multiple concurrent switch closures.

19. [❹] The IEEE standard-logic type models two drive strengths: forcing and weak. This is insufficient to model detailed operation of circuits at the switch level. For example, in circuits that store a charge on the gate terminal of a MOS transistor, we need to distinguish the weaker capacitive drive strength of the stored value from the resistive strength of a value transmitted through a pass transistor. Develop a package that defines a resolved type similar to **std_logic**, with forcing, resistive and capacitive strengths for 0, 1 and unknown values.

20. [❹] Exercise 19 describes a logic type that incorporates three drive strengths. If we need to model switch level circuits in finer detail, we can extend the type to deal with an arbitrary number of drive strengths. Each time a signal is transmitted through a pass transistor, its drive strength is diminished. We can model this by representing a logic value as a record containing the bit value ('0', '1' or unknown) and an integer representing the strength. We use 0 to represent power supply strength and a positive integer n to represent the strength of a signal after being transmitted through n pass transistors from the power supply. A normal driver has strength 1, to reflect the fact that it derives the driving value by turning on a transistor connected to one or the other power supply rail. (This scheme is described by Smith and Acosta in [14].)

   Develop a package that defines a resolved type based on this scheme. Include functions for separating the bit value and strength components of a combined value, for constructing a combined value from separate bit value and strength components and for weakening the strength component of a combined value. Use the package to model a pass transistor component. Then use the pass transistor in a model of an eight-input multiplexer similar to the four-input multiplexer of Exercise 16.

21. [❹] Self-timed asynchronous systems use handshaking to synchronize operation of interacting modules. In such systems, it is sometimes necessary to synchronize a number of modules at a rendezvous. Each module waits until all modules are ready to perform an operation. When all are ready, the operation commences. A scheme for rendezvous synchronization of a number of modules using three wired-and control signals was first proposed by Sutherland et al. for the TRIMOSBUS [15] and was subsequently adopted for use in the arbitration protocol of the IEEE Futurebus [11].

   Develop a high-level model of a system that uses the three-wire synchronization scheme. You should include a package to support your model. The package should include a type definition for a record containing the three synchronization wires and a pair of procedures, one to wait for a rendezvous and another to leave the rendezvous after completion of the operation. The procedures should have a bidirectional signal parameter for the three-wire record and should determine the state of the synchronization protocol from the parameter value.

Chapter 9

Predefined and Standard Packages

In this chapter, we look at several predefined and standard packages, which provide types and operators for use in VHDL models. While we could define all of the data types and operations we need for a given model, we can greatly increase our productivity by reusing the standard packages. Moreover, simulation and synthesis tools often have optimized, built-in implementations of the operations from these packages.

9.1 The Predefined Packages **standard** and **env**

In previous chapters, we have introduced numerous predefined types and operators. We can use them in our VHDL models without having to write type declarations or subprogram definitions for them. These predefined items all come from a special package called **standard**, located in a special design library called **std**. A full listing of the **standard** package is included for reference in Appendix A.

    Because nearly every model we write needs to make use of the contents of this library and package, as well as the library **work**, VHDL includes an implicit context clause of the form

```
library std, work;  use std.standard.all;
```

at the beginning of each design unit. Hence we can refer to the simple names of the predefined items without having to resort to their selected names. In the occasional case where we need to distinguish a reference to a predefined operator from an overloaded version, we can use a selected name, for example:

```
result := std.standard."<" ( a, b );
```

EXAMPLE 9.1 *A comparison operator for signed binary-coded integers*

A package that provides signed arithmetic operations on integers represented as bit vectors might include a relational operator, defined as follows:

```vhdl
function "<" ( a, b : bit_vector ) return boolean is
  variable tmp1 : bit_vector(a'range) := a;
  variable tmp2 : bit_vector(b'range) := b;
begin
  tmp1(tmp1'left) := not tmp1(tmp1'left);
  tmp2(tmp2'left) := not tmp2(tmp2'left);
  return std.standard."<" ( tmp1, tmp2 );
end function "<";
```

The function negates the sign bit of each operand, then compares the resultant bit vectors using the predefined relational operator from the package **standard**. The full selected name for the predefined operator is necessary to distinguish it from the function being defined. If the return expression were written as "tmp1 < tmp2", it would refer to the function in which it occurs, creating a circular definition.

VHDL-87, -93, and -2002

A number of new operations were added to VHDL in the 2008 revision. They are not available in earlier versions of the language. In summary, the changes are

- The types **boolean_vector**, **integer_vector**, **real_vector**, and **time_vector** are predefined (see Section 4.2.1). The predefined operations on **boolean_vector** are the same as those defined for **bit_vector**. The predefined operations on **integer_vector** include the relational operators ("=", "/=", "<", ">", "<=", and ">=") and the concatenation operator ("&"). The predefined operations on **real_vector** and **time_vector** include the equality and inequality operators ("=" and "/=") and the concatenation operator ("&").

- The array/scalar logic operations and logical reduction operations are predefined for **bit_vector** and **boolean_vector**, since they are arrays with **bit** and **boolean** elements, respectively.

- The matching relational operators "?=", "?/=", "?>", "?>=", "?<", and "?<=" are predefined for **bit**. Further, the operators "?=" and "?/=" are predefined for **bit_vector**.

- The condition operator "??" is predefined for **bit**.

- The operators **mod** and **rem** are predefined for **time**, since it is a physical type.

- The **maximum** and **minimum** operations are predefined for all of the predefined types.

- The functions **rising_edge** and **falling_edge** are predefined for **bit** and **boolean**. Prior to VHDL-2008, the **bit** versions of these functions were declared in the package **numeric_bit** (see Section 9.2.3). However, that was mainly to provide consistency with the **std_ulogic** versions defined in the **std_logic_1164** package. They rightly belong with the definition of the type on which they operate; hence, VHDL-2008 includes them in the package **standard**. The VHDL-2008 revision of the **numeric_bit** package redefines the operations there as aliases for the predefined versions. (We discuss aliases in Chapter 11.)

- The **to_string** operations are predefined for all scalar types and for **bit_vector**. Further, the **to_bstring**, **to_ostring**, and **to_hstring** operations and associated aliases are predefined for **bit_vector**.

VHDL also provided a second special package, called **env**, in the **std** library. The **env** package includes operations for accessing the simulation environment provided by a simulator. First, there are procedures for controlling the progress of a simulation:

```
procedure stop (status: integer);
procedure stop;

procedure finish (status: integer);
procedure finish;
```

When the procedure **stop** is called, the simulator stops and accepts further input from the user interface (if interactive) or command file (if running in batch mode). When the procedure **finish** is called, the simulator terminates; simulation cannot continue. The versions of the procedures that have the **status** parameter use the parameter value in an implementation-defined way. They might, for example, provide the value to a control script so that the script can determine what action to take next.

The **env** package also defines a function to access the resolution limit for the simulation:

```
function resolution_limit return delay_length;
```

We described the resolution limit in Section 2.2.4 when we introduced the predefined type **time**. One way in which we might use the **resolution_limit** function is to wait for simulation time to advance by one time step, as follows:

```
wait for env.resolution_limit;
```

Since the resolution limit, and hence the minimum time by which simulation advances, can vary from one simulation run to another, we cannot write a literal time value in such a wait statement. The use of the **resolution_limit** function allows us to write models that adapt to the resolution limit used in each simulation. We need to take care in using this function, however. It might be tempting to compare the return value with a given time unit, for example:

```
if env.resolution_limit > ns then   -- potentially illegal!
    ...  -- do coarse-resolution actions
else
    ...  -- do fine-resolution actions
end if;
```

The problem is that we are not allowed to write a time unit smaller than the resolution limit used in a simulation. If this code were simulated with a resolution limit greater than ns, the use of the unit name ns would cause an error; so the code can only succeed if the resolution limit is less than or equal to ns. We can avoid this problem by rewriting the example as:

```
if env.resolution_limit > 1.0E-9 sec then
   ... -- do coarse-resolution actions
else
   ... -- do fine-resolution actions
end if;
```

For resolution limits less than or equal to ns, the test returns false, so the "else" alternative is taken. For resolution limits greater than ns, the time literal 1.0E-9 sec is truncated to zero, and so the test returns true. Thus, even though the calculation is not quite what appears, it produces the result we want.

VHDL-87, -93, and -2002

These versions do not provide the env package. Some tools might provide equivalent functionality through implementation-defined mechanisms.

9.2   IEEE Standard Packages

When we design models, we can define types and operations using the built-in facilities of VHDL. However, the IEEE has published standards for packages that define commonly used data types and operations. Using these standards can save us development time. Furthermore, many tool vendors provide optimized implementations of the standard packages, so using them makes our simulations run faster. In this section, we outline the types and operations defined in the IEEE standard packages. Complete details of the package declarations are included in Appendix A. Each of these packages is included in a library called ieee. Hence, to use one of the packages in a design, we name the library ieee in a library clause, and name the required package in a use clause. We have seen examples of how to do this for the IEEE standard package std_logic_1164; the same applies for the other IEEE standard packages.

9.2.1   Standard VHDL Mathematical Packages

The IEEE standard packages math_real and math_complex define constants and mathematical functions on real and complex numbers, respectively.

Real Number Mathematical Package

The constants defined in math_real are listed in Table 9.1. The functions, their operand types and meanings are listed in Table 9.2. In the figure, the parameters x and y are of type real, and the parameter n is of type integer.

TABLE 9.1   *Constants defined in the package* math_real

Constant	Value	Constant	Value
math_e	e	math_log_of_2	$\ln 2$
math_1_over_e	$1/e$	math_log_of_10	$\ln 10$
math_pi	π	math_log2_of_e	$\log_2 e$
math_2_pi	2π	math_log10_of_e	$\log_{10} e$
math_1_over_pi	$1/\pi$	math_sqrt_2	$\sqrt{2}$
math_pi_over_2	$\pi/2$	math_1_over_sqrt_2	$1/\sqrt{2}$
math_pi_over_3	$\pi/3$	math_sqrt_pi	$\sqrt{\pi}$
math_pi_over_4	$\pi/4$	math_deg_to_rad	$2\pi/360$
math_3_pi_over_2	$3\pi/2$	math_rad_to_deg	$360/2\pi$

TABLE 9.2   *Functions defined in the package* math_real

Function	Meaning	Function	Meaning
ceil(x)	Ceiling of x (least integer $\geq x$)	sign(x)	Sign of x (−1.0, 0.0 or +1.0)
floor(x)	Floor of x (greatest integer $\leq x$)	"mod"(x, y)	Floating-point modulus of x/y
round(x)	x rounded to nearest integer value (ties rounded away from 0.0)	realmax(x, y)	Greater of x and y
trunc(x)	x truncated toward 0.0	realmin(x, y)	Lesser of x and y
sqrt(x)	$\sqrt{x}$	log(x)	$\ln x$
cbrt(x)	$\sqrt[3]{x}$	log2(x)	$\log_2 x$
"**"(n, y)	n^y	log10(x)	$\log_{10} x$
"**"(x, y)	x^y	log(x, y)	$\log_y x$
exp(x)	e^x		
sin(x)	$\sin x$ (x in radians)	arcsin(x)	$\arcsin x$
cos(x)	$\cos x$ (x in radians)	arccos(x)	$\arccos x$
tan(x)	$\tan x$ (x in radians)	arctan(x)	$\arctan x$
		arctan(y, x)	arctan of point (x, y)

Function	Meaning	Function	Meaning
sinh(x)	sinh x	arcsinh(x)	arcsinh x
cosh(x)	cosh x	arccosh(x)	arccosh x
tanh(x)	tanh x	arctanh(x)	arctanh x

In addition to the functions listed in Table 9.2, the **math_real** package defines the procedure **uniform** as follows:

```
procedure uniform ( variable seed1, seed2 : inout positive;
                    variable x : out real);
```

This procedure generates successive values between 0.0 and 1.0 (exclusive) in a pseudo-random number sequence. The variables **seed1** and **seed2** store the state of the generator and are modified by each call to the procedure. **Seed1** must be initialized to a value between 1 and 2,147,483,562, and **seed2** to a value between 1 and 2,147,483,398, before the first call to **uniform**.

EXAMPLE 9.2   *A random-stimulus test bench for an ALU*

Suppose we need to test a structural implementation of an ALU, whose entity is declared as follows:

```
use ieee.numeric_bit.all;
subtype ALU_func is unsigned(3 downto 0);
subtype data_word is unsigned(15 downto 0);
...

entity ALU is
  port ( a, b : in data_word;  func : in ALU_func;
         result : out data_word;  carry : out bit );
end entity ALU;
```

We can devise a test bench that stimulates an instance of the ALU with randomly generated data and function-code inputs:

```
architecture random_test of test_ALU is

  use ieee.numeric_bit.all;
  use ieee.math_real.uniform;

  signal a, b, result : data_word;
  signal func : ALU_func;
  signal carry : bit;

begin

  dut : entity work.ALU(structural)
    port map ( a, b, func, result, carry );
```

```
stimulus : process is
  variable seed1, seed2 : positive := 1;
  variable a_real, b_real, func_real : real;
begin
  wait for 100 ns;
  uniform ( seed1, seed2, a_real );
  uniform ( seed1, seed2, b_real );
  uniform ( seed1, seed2, func_real );
  a <= to_unsigned(
          natural(a_real
                * real(2**integer'(data_word'length)) - 0.5),
          data_word'length );
  b <= to_unsigned(
          natural(b_real
                * real(2**integer'(data_word'length)) - 0.5),
          data_word'length );
  func <= to_unsigned(
          natural(func_real
                * real(2**integer'(ALU_func'length)) - 0.5),
          ALU_func'length );
  end process stimulus;

  ...    --verification process to check result and carry

end architecture random_test;
```

The **stimulus** process generates new random stimuli for the ALU input signals every 100 ns. The process generates three random numbers in the range (0.0, 1.0) in the variables **a_real**, **b_real** and func_real. It then scales these values to get numbers in the range (–0.5, 65,635.5) for the data values and (–0.5, 15.5) for the function code value. These are rounded and converted to **unsigned** bit vectors for assignment to the ALU input signals.

Complex Number Mathematical Package

The **math_complex** package deals with complex numbers represented in Cartesian and polar form. The package defines types for these representations, as follows:

```
type complex is record
    re : real;      -- Real part
    im : real;      -- Imaginary part
  end record;

subtype positive_real is real range 0.0 to real'high;
subtype principal_value is real range -math_pi to math_pi;

type complex_polar is record
    mag : positive_real;     -- Magnitude
```

```
      arg : principal_value;   -- Angle in radians; -math_pi is illegal
   end record;
```

The constants defined in **math_complex** are

math_cbase_1 $1.0 + j0.0$

math_cbase_j $0.0 + j1.0$

math_czero $0.0 + j0.0$

The package defines a number of overloaded operators, listed in Table 9.3. The curly braces indicate that for each operator to the left of the brace, there are overloaded versions for all combinations of types to the right of the brace. Thus, there are six overloaded versions of each of the "+", "−", "*" and "/" operators.

TABLE 9.3   *Overloaded operators defined in* **math_complex**

Operator	Operation		Left operand	Right operand	Result
=	equality		complex_polar	complex_polar	boolean
/=	inequality		complex_polar	complex_polar	boolean
abs	magnitude			complex	positive_real
				complex_polar	positive_real
−	negation			complex	complex
				complex_polar	complex_polar
+	addition		complex	complex	complex
−	subtraction		real	complex	complex
*	multiplication		complex	real	complex
/	division		complex_polar	complex_polar	complex_polar
			real	complex_polar	complex_polar
			complex_polar	real	complex_polar

Overloaded versions of "=" and "/=" are necessary for numbers in polar form, since two complex numbers are equal if their magnitudes are both 0.0, even if their arguments are different. The predefined equality and inequality operators do not have this behavior. No overloaded versions of these operators are required for Cartesian form, since the predefined operators behave correctly.

In addition to the operators, the **math_complex** package defines a number of mathematical functions, listed in Table 9.4. In the table, the parameters x and y are real, the parameter c is **complex**, the parameter p is **complex_polar**, and the parameter z is either **complex** or **complex_polar**.

TABLE 9.4   *Functions defined in the package* math_complex

Function	Result type	Meaning
cmplx(x, y)	complex	$x + jy$
get_principal_value(x)	principal_value	$x + 2\pi k$ for some k, such that $-\pi < \text{result} \le \pi$
complex_to_polar(c)	complex_polar	c in polar form
polar_to_complex(p)	complex	p in Cartesian form
arg(z)	principal_value	$\arg(z)$ in radians
conj(z)	same as z	complex conjugate of z
sqrt(z)	same as z	$\sqrt{z}$
exp(z)	same as z	e^z
log(z)	same as z	$\ln z$
log2(z)	same as z	$\log_2 z$
log10(z)	same as z	$\log_{10} z$
log(z, y)	same as z	$\log_y z$
sin(z)	same as z	$\sin z$
cos(z)	same as z	$\cos z$
sinh(z)	same as z	$\sinh z$
cosh(z)	same as z	$\cosh z$

9.2.2   The **std_logic_1164** Multivalue Logic System

The IEEE standard package **std_logic_1164** defines types and operations for models that need to deal with strong, weak and high-impedance strengths, and with unknown values. We have already described most of the types and operations in previous chapters and seen their use in examples. For completeness, we draw the information together in this section.

The types declared in **std_logic_1164** are

std_ulogic	The basic multivalued enumeration type (see page 48)
std_ulogic_vector	Array of **std_ulogic** elements (see page 108)
std_logic	Resolved multivalued enumeration subtype (see Section 8.1.3 on page 278)
std_logic_vector	Subtype of **std_ulogic_vector** with resolved elements (see Section 8.1.3 on page 278)

In addition, the package declares the subtypes X01, X01Z, UX01 and UX01Z for cases where we do not need to distinguish between strong and weak driving strengths. Each of these subtypes includes just the values listed in the subtype name.

Since the type **std_ulogic** and the subtype **std_logic** are scalar enumeration types, all of the predefined operations for such types are available. This includes the relational operators, **maximum**, **minimum**, and **to_string**. In addition, the matching relational operators "?=", "?/=", "?>", "?>=", "?<", and "?<=" are predefined for **std_ulogic** and **std_logic**. For the array type **std_ulogic_vector** and the subtype **std_logic_vector**, the predefined operations on one-dimensional arrays of discrete-type elements are available. This includes "&" and the relational operators. In addition, the matching equality ("?=") and inequality ("?/=") operators are defined for these types.

The operations provided by the **std_logic_1164** package include overloaded versions of the logical operators **and**, **nand**, **or**, **nor**, **xor**, **xnor** and **not**, operating on values of each of the scalar and vector types and subtypes listed above. It also includes overloaded versions of the shift operators **sll**, **srl**, **rol** and **ror** operating on the vector types. It does not overload the **sla** and **sra** operators on the premise that they assume a numeric interpretation of vectors. Instead, those operators are overloaded in separate packages, **numeric_std** and **numeric_std_unsigned**, that provide arithmetic operations assuming a numeric interpretation (see Section 9.2.3). The **std_logic_1164** package provides an overloaded version of the "??" operator, allowing **std_ulogic** and **std_logic** value to be used in conditions with implicit conversion.

In Section 4.3.5 we described the **to_ostring** and **to_hstring** operations for converting **bit_vector** operands to strings in octal and hexadecimal form. The **std_logic_1164** package provides overloaded versions of these operations for **std_ulogic_vector** and **std_logic_vector** operands. The operations group elements into threes (octal) or fours (hexadecimal) for conversion into digits. However, if the vector needs to be extended on the left to make a multiple of three or four, the assumed value for the extra elements depends on the leftmost actual element. If the leftmost element is 'Z', then 'Z' elements are assumed; if it is 'X', then 'X' elements are assumed; otherwise, '0' elements are assumed.

Having grouped elements, they are converted to digits. If all of the elements in a group are '0', 'L', '1', or 'H', the group is converted to a normal digit character, with '0' and 'L' elements treated as '0', and '1' and 'H' elements treated as '1'. If all of the elements in a group are 'Z', then 'Z' is used as the digit character for the group. In all other cases, where a group contains one or more non-'0', -'L', -'1', -'H' or -'Z' elements, 'X' is used as the digit character for the group. Some examples are

```
to_ostring(B"01L_1H1") = "27"
to_ostring(B"011_ZZZ") = "3Z"
to_ostring(B"01U_UZZ") = "XX"

to_ostring(B"HH_000") = to_ostring(B"0HH_000") = "30"
to_ostring(B"ZZ_ZZZ") = to_ostring(B"ZZZ_ZZZ") = "ZZ"
to_ostring(B"X1_000") = to_ostring(B"XX1_000") = "X0"
to_ostring(B"1X_000") = to_ostring(B"01X_000") = "X0"
```

As well as providing the overloaded **to_ostring** and **to_hstring** operations, the **std_logic_1164** package provides all of the alternative names: **to_bstring**, **to_binary_string**, **to_octal_string**, and **to_hex_string**. Further, the package provides overloaded versions of the file read and write operations for text-based input/output. We describe files and input/output in Chapter 16.

In addition to the overloaded operations, the package declares a number of functions for conversion between values of different types. In the following lists, the parameter **b** represents a bit value or bit vector, the parameter **s** represents a standard logic value or standard logic vector, and the parameter **x** represents a value of any of these types.

To_Bit(s,xmap)	Convert a standard logic value to a bit value

To_BitVector(s, xmap)	
To_Bit_Vector(s, xmap)	
To_BV(s, xmap)	Convert a standard logic vector to a bit vector

In these functions, the parameter **xmap** is a bit value that is used in the result when a bit to be converted is other than '0', '1', 'L' or 'H'. There are multiple alternative names for the second function, allowing us to choose based on consideration of coding style.

To_StdULogic(b)	Convert a bit value to a standard logic value

To_StdLogicVector(x)	
To_Std_Logic_Vector(x)	
To_SLV(x)	Convert to a **std_logic_vector**

To_StdULogicVector(x)	
To_Std_ULogic_Vector(x)	
To_SULV(x)	Convert to a **std_ulogic_vector**

Note that the **To_StdLogicVector** and **To_StdULogicVector** function perform essentially the same operation. The fact that **std_logic_vector** has resolved elements is not relevant to the conversion. The two forms are provided for backward compatibility with previous versions of VHDL, where the distinction was relevant.

To_01(x, xmap)	Strip strength
To_X01(x)	Strip strength
To_X01Z(x)	Strip strength
To_UX01(x)	Strip strength

These strength-stripping functions remove the driving strength information from the parameter value. **To_01** convert 'L' and 'H' digits in a vector to '0' and '1' digits. The optional second parameter specifies the result value to produce if any digit in the first parameter is other than '0', '1', 'L' or 'H'. In that case, all digits of the result are set to the value specified in the second parameter. The default value of the second parameter is '0'. Some examples are

```
to_01( "LLHH01" ) = "001101"
to_01( "00X11U" ) = "000110"

to_01( "100LLL", 'X' ) = "100000"
to_01( "00W000", 'X' ) = "XXXXXX"
```

To_X01 converts 'U', 'X', 'Z', 'W' and '–' elements to 'X'. To_X01Z is similar, but leaves 'Z' elements intact. To_UX01 is similar to **To_X01**, but leaves 'U' elements intact.

Finally, the **std_logic_1164** package contains the following utility functions:

rising_edge(s)   True when there is a rising edge on s, false otherwise

falling_edge(s)   True when there is a falling edge on s, false otherwise

is_X(s)   True if s contains an unknown value, false otherwise

The edge-detection functions detect changes between low and high values on a scalar signal, irrespective of the driving strengths of the values. The functions are true only during the simulation cycles on which such events occur. They serve the same purpose as the predefined functions of the same name operating on **bit** and **boolean** signals. The unknown-detection function determines whether there is a 'U', 'X', 'Z' or 'W' value in the scalar or vector value s.

VHDL-87, -93, and -2002

_____

In the version of **std_logic_1164** for these versions of VHDL, **std_ulogic_vector** and **std_logic_vector** were declared as distinct array types, rather than one being a subtype of the other. This led to considerable inconvenience when both types were used in a design. The **std_logic_1164** package provided separate overloaded declarations for operations on each of the two types, and additional conversion functions between them were required.

Since VHDL-2008 adds numerous new predefined operations, the VHDL-2008 version of the package provides overloaded versions of them. They are not provided in the version of the package for earlier versions of VHDL.

_____

VHDL-87

_____

The overloaded versions of the **xnor** operator are not included in the VHDL-87 version of the standard-logic package.

_____

9.2.3 Standard Integer Numeric Packages

The IEEE standard packages **numeric_bit** and **numeric_std** define arithmetic operations on integers represented using vectors of **bit** and **std_ulogic** elements respectively. Most synthesis tools accept models that use these types and operations for numeric computations. We discuss the topic of synthesis of VHDL models in more detail in Chapter 21. In this section, we outline the types and operations provided by the IEEE standard integer numeric packages. Full listings of the package declarations are included in Appendix A.

Each of the packages defines two types, **unsigned** and **signed**, to represent unsigned and signed integer values, respectively. In the case of the **numeric_bit** package, the types are unconstrained arrays of **bit** elements:

```
type unsigned is array ( natural range <> ) of bit;
type signed is array ( natural range <> ) of bit;
```

In the case of the **numeric_std** package, the types are defined similarly, but with resolved **std_ulogic** as the element type. This package also defines the types **unresolved_unsigned** and **unresolved_signed** (and the shorter aliases **u_unsigned** and **u_signed**) as arrays of unresolved **std_ulogic** elements. The declarations are

```
type unresolved_unsigned is array (natural range <>) of std_ulogic;
type unresolved_signed   is array (natural range <>) of std_ulogic;

alias u_unsigned is unresolved_unsigned;
alias u_signed   is unresolved_signed;

subtype unsigned is (resolved) unresolved_unsigned;
subtype signed   is (resolved) unresolved_signed;
```

Whichever package and type we use, the leftmost element is the most-significant digit, and the rightmost element is the least-significant digit. Signed numbers are represented using two's-complement encoding.

We declare objects of these types either directly or using a subtype to define the index range. For example:

```
signal head_position : signed ( 0 to 15 );

subtype address is unsigned ( 31 downto 0 );
signal next_PC : address;
constant PC_increment : unsigned := X"4";
```

The operations defined for **unsigned** and **signed** numbers are listed in Table 9.5. The curly braces indicate that for each operator to the left of the brace, there are overloaded versions for all combinations of types to the right of the brace. For example, there are six overloaded versions of each of the "+", "–", "*", "/", **rem** and **mod** operators. The notation "*element type*" refers to the element type (bit or **std_ulogic**) of the vector operand or operands.

The operands of arithmetic and relational operators need not be of the same length. The relational operators determine the result based on the numeric values represented by the operands, rather than using left-to-right lexicographic comparison. For those operators that produce a vector result, the length of the result depends on the length of the operands, as follows.

- **abs** and "–": the length of the operand

- Addition and subtraction of two vectors: the larger of the two operand lengths

- Addition and subtraction of a vector and an integer, natural number or scalar element: the length of the vector operand

- Multiplication of two vectors: the sum of the operand lengths

- Multiplication of a vector and an integer or natural number: twice the length of the vector operand

TABLE 9.5   *Operators defined in the IEEE standard synthesis packages*

Operator	Operation		Left operand	Right operand	Result
abs	absolute value	{		signed	signed
–	negation				
+	addition		unsigned	unsigned	unsigned
–	subtraction		unsigned	natural	unsigned
*	multiplication		natural	unsigned	unsigned
/	division		signed	signed	signed
rem	remainder		signed	integer	signed
mod	modulo		integer	signed	signed
+	addition		unsigned	*element type*	unsigned
–	subtraction		*element type*	unsigned	unsigned
			signed	*element type*	signed
			element type	signed	signed
=	equality		unsigned	unsigned	boolean
/=	inequality		unsigned	natural	boolean
<	less than		natural	unsigned	boolean
<=	less than or equal to		signed	signed	boolean
>	greater than		signed	integer	boolean
>=	greater than or equal to		integer	signed	boolean
?=	matching equality		unsigned	unsigned	*element type*
?/=	matching inequality		unsigned	natural	*element type*
?<	matching less than		natural	unsigned	*element type*
?<=	matching less than or equal to		signed	signed	*element type*
?>	matching greater than		signed	integer	*element type*
?>=	matching greater than or equal to		integer	signed	*element type*
sll	shift-left logical		unsigned	integer	unsigned
srl	shift-right logical		signed	integer	signed
sla	shift-left arithmetic				
sra	shift-right arithmetic				
rol	rotate left				
ror	rotate right				

Operator	Operation		Left operand	Right operand	Result
not	negation	{		unsigned	unsigned
				signed	signed
and	logical and	⌈	unsigned	unsigned	unsigned
or	logical or		unsigned	*element type*	unsigned
nand	negated logical and		*element type*	unsigned	unsigned
nor	negated logical or	⎨	signed	signed	signed
xor	exclusive or		signed	*element type*	signed
xnor	negated exclusive or	⌊	*element type*	signed	signed
and	logical and reduction	⌈		unsigned	*element type*
or	logical or reduction			signed	*element type*
nand	negated logical and reduction				
nor	negated logical or reduction	⎨			
xor	exclusive or reduction				
xnor	negated exclusive or reduction	⌊			

- Division of two vectors: the length of the left operand

- Remainder and modulo of two vectors: the length of the right operand

- Division, remainder and modulo of a vector and an integer or natural number: the length of the vector operand. If the result value is too large to fit in the result vector, the value is truncated and a warning issued during simulation.

- Shift and rotate operators: the length of the vector operand

- **not**: the length of the operand

- Binary logical operators: the length of the operands, whose lengths must be the same

EXAMPLE 9.3 *Addition with carry*

The addition operator that has two vector operands can be used to produce both the sum vector and a carry bit by extending the operands. For example, if we declare unsigned operand signals and a carry signal as

```
signal a, b, sum : unsigned(15 downto 0);
signal c_out : std_ulogic;
```

we can write an assignment that produces both the sum and the carry value:

```
(c_out, sum) <= ('0' & a) + ('0' & b);
```

The concatenations extend the operands by one bit, so the addition produces a 17-bit result. This is assigned to the aggregate of the carry signal (the leftmost bit of the result) and the sum signal (the rightmost 16 bits of the result).

The addition and subtraction operators that have an operand of the scalar element type allow us to describe an addition with carry in or a subtraction with borrow in. For example:

```
signal a, b : unsigned(15 downto 0);
signal sum  : unsigned(16 downto 0);
signal c_in;
...

sum <= ('0' & a) + ('0' & b) + c_in;
```

This can be synthesized as a single 16-bit adder with a carry input and a 17-bit result.

EXAMPLE 9.4 *A conditional incrementer*

We can use the "+" operator to treat a scalar control signal as an operand in a conditional incrementer. If the control signal is '0', an **unsigned** operand value is not incremented; if the control signal is '1", the value is incremented. The declarations and process are:

```
signal inc_en  : std_ulogic;
signal inc_reg : unsigned(7 downto 0);
...

inc_reg_proc : process (clk) is
begin
  if rising_edge(clk) then
    inc_reg <= inc_reg + inc_en;
  end if;
end process inc_reg_proc;
```

If we had written the if statement as follows:

```
if inc_en = '1' then
  inc_reg <= inc_reg + 1;
end if;
```

a synthesis tool might have generated an adder with the vector **"00000001"** as an input, connected to a regsiter with clock enable. By using the control signal as an operand, we more clearly imply an incrementer and a simple register without clock enable. We discuss coding styles for synthesis in more detail in Chapter 21.

For the division, remainder and modulo operators, if the right operand is zero, an assertion violation with severity level error is issued during simulation.

The logical shift operators **sll** and **srl** fill the vacated elements with '0'. Their behavior is the same as that of the predefined operators for bit_vector values. The behavior of the **sla** and **sra** operators, on the other hand, is different from the predefined versions, in that they assume a binary-coded numeric interpretation for a vector. The type of the left operand determines the kind of shift performed. If the operand is **unsigned**, a logical shift is performed, whereas if the parameter is **signed**, an arithmetic shift is performed. An arithmetic shift right (**sra** with a positive right operand or **sla** with a negative right operand) replicates the sign bit, giving the effect of division by a power of 2. An arithmetic shift left (**sra** with a negative right operand or **sla** with a positive right operand) fills the vacated bits on the right with '0', giving the effect of multiplication by a power of 2.

In addition to the overloaded operators, the **numeric_bit** and **numeric_std** packages define a number of functions, listed in Table 9.6. As in Table 9.5, the curly braces indicate that for each function to the left of the brace, there are overloaded versions for all combinations of types to the right of the brace.

TABLE 9.6   *Functions defined in the IEEE standard synthesis packages*

Function	First parameter	Second parameter	Result
minimum maximum	unsigned	unsigned	unsigned
	unsigned	natural	unsigned
	natural	unsigned	unsigned
	signed	signed	signed
	signed	integer	signed
	integer	signed	signed
shift_left shift_right rotate_left rotate_right	unsigned	natural	unsigned
	signed	natural	signed
find_leftmost find_rightmost	unsigned	*element type*	integer
	signed	*element type*	integer
resize	unsigned	natural	unsigned
	unsigned	unsigned	unsigned
	signed	natural	signed
	signed	signed	signed

Function		First parameter	Second parameter	Result
to_integer	{	unsigned		natural
		signed		integer
to_unsigned	{	natural	natural	unsigned
		natural	unsigned	unsigned
to_signed	{	integer	natural	signed
		integer	signed	signed
rising_edge[a] falling_edge[a]	{	**signal** bit		boolean
std_match[b]	{	unsigned	unsigned	boolean
		signed	signed	boolean
		std_ulogic	std_ulogic	boolean
		std_ulogic_vector	std_ulogic_vector	boolean
to_01[b] to_X01[b] to_X01Z[b] to_UX01[b]	{	unsigned	⟦ std_ulogic ⟧	unsigned
		signed	⟦ std_ulogic ⟧	signed
to_string to_bstring to_binary_string to_ostring to_octal_string to_hstring to_hex_string	{	unsigned		string
		signed		string

a. Provided in **numeric_bit** only.
b. Provided in **numeric_std** only.

The **maximum** and **minimum** functions are overloaded to compare their operands based on a numeric interpretation in the same way as the relational operators. For the versions that take two vector operands, the operands do not need to be of the same length.

The shift, rotate and to_01 functions all produce a result that has the same length as the vector parameter. The shift and rotate functions perform similar operations to the overloaded shift and rotate operators. However, their second parameter is constrained to be a non-negative integer. Hence each of these functions can only shift or rotate elements in one direction. These functions were included in the package for earlier versions of VHDL that did not include the shift operators. They are maintained in the package for backward compatibility.

The **find_leftmost** function returns the index of the leftmost occurrence of the right-operand value in the left-operand vector. Similarly, the **find_rightmost** function returns the index of the rightmost occurrence of the right-operand value in the left-operand vector. If there is no such occurrence, the functions return –1. Since the index bound of an unsigned or signed vector are natural numbers, –1 is a clear indication that the value was not found. In both functions, search for the element value is performed using a matching equality test. We can use **find_leftmost** to gauge the magnitude of a number, since the leftmost occurrence of a 1 bit in an unsigned number is approximately $\log_2$ of the number. For a signed number, the leftmost bit that differs from the sign bit is likewise approximately $\log_2$ of the number.

The **resize** functions produce a result whose length is specified by the second parameter. For a second parameter of type natural, the parameter value specifies the number of bits in the result directly. For a second parameter of type unsigned or signed, the result has the same number of bits as the parameter value. Increasing the size of an **unsigned** number zero-extends to the left, whereas increasing the size of a **signed** number replicates the sign bit to the left. Truncating an **unsigned** number to length L keeps the rightmost L bits. Truncating a **signed** number to length L keeps the sign bit and the rightmost $L - 1$ bits.

The **to_integer** functions convert **unsigned** or **signed** values to **natural** or **integer** values, respectively. The **to_unsigned** and **to_signed** functions convert their first parameter to a vector whose length is given by the second parameter, either as a natural number directly, or as a vector whose length is used.

The edge-detection functions **rising_edge** and **falling_edge** are provided as aliases for the predefined edge-detection functions on type bit. They are included here for backward compatibility with earlier versions of VHDL, in which the edge-detection functions were not predefined. The **std_match** functions perform the same operations as the "?=" operator. Again, they are included for backward compatibility with earlier versions of VHDL that did not provide that operator.

The strength-stripping functions **to_01**, **to_X01**, **to_X01Z**, and **to_UX01** are overloaded for unsigned and signed operands, and perform the same operation as the corresponding functions for vectors in the **std_logic_1164** package.

Finally, the packages provide the string conversion operations. The **to_string** operation is actually predefined for **unsigned** and **signed**, since they are one-dimensional arrays of character elements in both packages. The **to_ostring** and **to_hstring** operations work in the same way as the corresponding operations in the **std_logic_1164** package. The only difference is that, when extending a **signed** value to form a complete group of three (for octal) or four (for hexadecimal) bits, the sign bit is replicated rather than '0' bits being added. The same aliases for all of the operations are also defined. Further, the packages provide overloaded versions of the file read and write operations for text-based input/output. We describe files and input/output in Chapter 16.

VHDL-87, -93, and -2002

Since VHDL-2008 adds numerous new predefined operations, the VHDL-2008 version of the package provides overloaded versions of them. They are not provided in the version of the package for earlier versions of VHDL.

VHDL-87

Since VHDL-87 does not include the shift, rotate and **xnor** operators, they should be commented out of the standard synthesis packages when used with VHDL-87 tools. The shift and rotate functions can be used in VHDL-87 models. The **xnor** operator can be expressed as the negation (**not**) of the **xor** operator in VHDL-87.

Since the standard numeric packages are widely used in many models, VHDL defines two standard context declarations (see Section 5.4.2) within the standard library **ieee**:

```
context ieee_bit_context is
  library ieee;
  use ieee.numeric_bit.all;
end context ieee_bit_context;

context ieee_std_context is
  library ieee;
  use ieee.std_logic_1164.all;
  use ieee.numeric_std.all;
end context ieee_std_context;
```

A design based on **bit** values might refer to the first of these context declarations, either in the context clause of a design unit or nested within a project context declaration. Similarly, a design based on **std_ulogic** values might refer to the second of these context declarations.

VHDL-87, -93, and -2002

Since context declarations are not provided in these versions of VHDL, the standard context declaration cannot be used. If a tool does not support VHDL-2008, it would not include the context declarations in its version of the **ieee** library.

As we have mentioned, the **numeric_bit** and **numeric_std** packages define new array types, **unsigned** and **signed**, to represent numbers in binary-coded form. Many designers, however, prefer to use the **bit_vector** and **std_ulogic_vector** types for numeric data. This approach is particularly useful in designs that include components, such as multiplexers, registers, and so on, that do not rely on numeric properties of data; they just store or manipulate arrays of bits. When such designs also include arithmetic elements, having to convert between the types for numeric interpretation and the plain vector type is a distraction. Historically, designers have adopted non-standard packages that provide arithmetic operations on **bit_vector** or **std_ulogic_vector** operands.

In the 2008 revision of VHDL, two new packages were added to library **ieee** for this purpose: **numeric_bit_unsigned** and **numeric_std_unsigned**. They are largely compatible with **numeric_bit** and **numeric_std**, respectively, and provide the corresponding unsigned operations. In summary, the similarities and differences among the packages are

- Numeric_bit_unsigned and numeric_std_unsigned provide the following operations corresponding to operations on unsigned in numeric_bit and numeric_std: arithmetic operators, shift operators, and relational operators; maximum, minimum, shift_left, shift_right, find_leftmost, find_rightmost, resize, and to_integer functions.

- Numeric_bit_unsigned provides a to_bitvector function (plus aliases to_bit_vector and to_bv) instead of to_unsigned.

- Numeric_std_unsigned provides to_stdulogicvector and to_stdlogicvector functions (plus aliases to_std_ulogic_vector, to_sulv, to_std_logic_vector, and to_slv) instead of to_unsigned.

- Numeric_bit_unsigned and numeric_std_unsigned do not provided other operations, since the normal bit_vector and std_ulogic_vector versions can be used. This includes overloaded logical operators; rising_edge, falling_edge, std_match and strength-stripping functions; to_ostring and to_hstring functions and the corresponding aliases; and read and write operations.

VHDL-87, -93, and -2002

These versions of VHDL do not provide the numeric_bit_unsigned and numeric_std_unsigned packages. Many tools for these versions provide a non-standard package, std_logic_unsigned, developed by Syopsys, Inc. It provides many of the same operations as numeric_std_unsigned.

9.2.4 Standard Fixed-Point Packages

Many digital-signal processing applications involve mathematical operations on non-integral data. While we could use floating-point representation and hardware, that would be excessively resource intensive in many cases. Instead, we can use a fixed-point representation, in which the radix point (analogous to the base-10 decimal point) is assumed to have a fixed position. VHDL defines a number of packages for fixed-point math that we introduce in this section. Since the packages provide a large number of overloaded operations and functions, we will not describe them in full detail here. We will simply provide an overview of the data types provided and some basic information on their usage. More details are provided in Appendix A and on the author's companion website for this book. In addition, Section 9.2.6 summarizes the operations provided by these packages and the other standard numeric packages.

For simple cases, fixed-point math amounts to integer math with scaling by a power of 2. More generally, we need to take account of rounding and overflow. The main VHDL fixed-point package, fixed_generic_pkg, is written in such a way that we can choose the rounding and overflow behaviors that are most appropriate for our application. The package uses formal generic constants, a topic that we will cover in detail in Chapter 12. For now, suffice it to say that the package defines four constants as follows:

fixed_round_style : fixed_round_style_type

> This constant determines the rounding behavior for operations in the package. The type fixed_round_style_type is an enumeration type defined in the package fixed_float_types, also in library ieee. The values are fixed_round, if results are to be rounded to the nearest representable value; and fixed_truncate, if results are to be truncated toward zero to the next smallest representable value. The default is fixed_round.

fixed_overflow_style : fixed_overflow_style_type

> This constant determines the behavior on overflow. The type fixed_overlow_style_type is defined in the package fixed_float_types. The values are fixed_saturate, if an overflowing result is to remain at the largest representable value; and fixed_wrap, if modulo-based behavior is required. The default is fixed_saturate.

fixed_guard_bits : natural

> This constant specifies the number of extra bits of precision to use for division operations. The default is 3.

no_warning : boolean

> This constant allows suppression of warning messages on conditions such as non-matching operand lengths and occurrence of metalogical values (values other than '0', 'L', '1' and 'H'). The default is **false**.

If the default values are acceptable for our application, we can use a version of the package named fixed_pkg, located in the ieee library. For other cases, the main package must be instantiated, as described in Chapter 12, to supply alternative values for the constants.

The package fixed_generic_pkg (and any instance of it, such as fixed_pkg) defines types for unsigned and signed fixed-point representation in the form of vectors of std_ulogic elements. The base type for unsigned representation is unresolved_ufixed, declared as

```
type unresolved_ufixed is array (integer range <>) of std_ulogic;
```

The name u_ufixed is defined, for convenience, as an alias to unresolved_ufixed. The package also defines a subtype ufixed with resolved elements. The declarations are

```
alias u_ufixed is unresolved_ufixed;

subtype ufixed is (resolved) unresolved_ufixed;
```

When declaring signals, we should choose between the base type or the subtype with resolved elements, depending on whether the signal has only one source or multiple sources, respectively.

Objects of these types must have descending (**downto**) index ranges. The whole-number part of the value is on the left of the vector, down to index 0, and the fractional part is on the right, starting at index –1. For example, given the following declaration of a fixed-point signal A:

```
signal A : ufixed(3 downto -3) := "0110100";
```

the whole-number part is A(3 **downto** 0), and the fractional part is A(–1 **downto** –3). The range of values represented is 0 to just less than 16 in steps of 0.125 (one-eighth). The value represented by the default initial value is $0110.100_2 = 6.5_{10}$.

This example shows a number with both whole-number and fractional parts. In general, we can declare number with just a whole-number part (the right index being 0) or just a fraction part (the left index being –1). Indeed, we can declare numbers in which the radix point is completely outside the index range of the vector. For example, in the following:

```
variable X : ufixed(9 downto 2);
variable Y : ufixed(-5 downto -14);
```

X is an 8-bit vector representing values in the range 0 to 1020 in steps of 4, and Y is a 10-bit vector representing values in the range 0 to just less than 0.0625 (one-sixteenth) in steps of 2^{-14}.

The base type defined in the package for signed representation is unresolved_sfixed, declared as:

```
type unresolved_sfixed is array (integer range <>) of std_ulogic;
```

As for the unsigned representation, there is an alias, u_sfixed, and a subtype with resolved elements, sfixed:

```
alias u_sfixed is unresolved_sfixed;
```

```
subtype sfixed is (resolved) unresolved_sfixed;
```

Likewise, the index range for a signed value must be descending (**downto**), with the radix point being assumed between index 0 and index –1. The difference is that the signed type and subtypes use 2s-complement binary representation, with the leftmost bit being the sign bit. Thus, for example, the signal:

```
signal S : sfixed(3 downto -3);
```

represents values from –8 to just less than 8 in steps of 0.125.

The fixed-point math packages perform operations with full precision. This is illustrated in the following example:

```
signal A4_2 : ufixed(3 downto -2);
signal B3_3 : ufixed(2 downto -3);
signal Y5_3 : ufixed(4 downto -3);
...

Y5_3 <= A4_2 + B3_3;
```

The whole-number part of the addition result is one bit larger than the larger of the two operand whole-number parts. In this example, the operand whole-number parts are 4 bits and 3 bits, respectively, so the result's whole-number part is 5 bits. The fractional part of the result is the larger fractional part of the operands. In this example, the operands' fractional parts are 2 bits and 3 bits, respectively, so the result has a 3-bit fractional part.

If we want to assign a fixed-point value to an object, one way is to use a string literal, for example:

```
signal A4 : ufixed(3 downto -3);
...

A4 <= "0110100";   -- string literal for 6.5
```

Alternatively, we can apply a conversion function, **to_ufixed** or **to_sfixed**, to an integer or real value. In this case, we need to specify the index range for the conversion result. There are two forms of conversion function. For the first form, we specify the left and right indices for the result, for example:

```
A4 <= to_ufixed(6.5, 3, -3);   -- pass indices
```

For the second form, we provide an object whose index range is used:

```
A4 <= to_ufixed(6.5, A4);   -- sized by A4
```

In this example, the only use of **A4** by the **to_ufixed** function is to read its left and right indices to determine the index range of the result.

The use of a string literal in an arithmetic expression is problematic, since the index range of such a literal is ascending (**to**) and starts with **integer'low**. Fixed-point numbers must have descending index ranges. Instead we can use integer literals, real literals, and qualified string literals, as shown in the following examples:

```
subtype ufixed4_3 is ufixed(3 downto -3);
signal A4, B4 : ufixed4_3;
signal Y5      : ufixed (4 downto -3);
...

Y5 <= A4 + "0110100";                    -- illegal
Y5 <= A4 + ufixed4_3'("0110100");
Y5 <= A4 + 6.5;                          -- overloading with real
Y5 <= A4 + 6;                            -- overloading with integer
```

In the assignment marked "illegal," the index range of the string literal would be **integer'low** to **integer'low** + 6. The type qualification in the next assignment avoids this problem and results in a bit-string value with index bounds taken from the subtype **ufixed4_3**. We can safely apply the addition operator to this value and the operand **A4**, giving a result with index range 4 down to –3.

If we need to change the size of an expression result, we can use a **resize** function. As for the conversion functions, there are two forms, one in which we specify the left and right index values and the other in which we provide an object whose index range is used.

For example, in the following accumulator assignment, since the addition result is one bit larger than the accumulator, we need to resize the result:

```
signal A4_3 : ufixed(3 downto -3);
signal Y7_3 : ufixed(6 downto -3);
...

Y7_3 <= Y7_3 + A4_3;    -- illegal, result too big

Y7_3 <= resize(arg          => Y7_3 + A4_3,
               size_res     => Y7_3,
               overflow_style => fixed_wrap,
               round_style    => fixed_truncate);
```

The **overflow_style** and **round_style** parameters allow us to control the way the value is processed if it cannot be represented exactly. The default values for these parameters are taken from the package constants described on page 314. If those values are satisfactory, we can omit them in the **resize** call. This is shown in the following example, which uses the form of the function specifying left and right index values for the result:

```
Y7_3 <= resize (arg         => Y7_3 + A4_3,
                left_index  => 7,
                right_index => -3);
```

Full-precision arithmetic can lead to some unexpected results in expressions involving multiple operators. Consider, as an example, the following declarations and assignment:

```
signal A4, B4, C4, D4 : ufixed(3 downto 0);
signal Y6             : ufixed(5 downto 0);
signal Y7A, Y7B       : ufixed(6 downto 0);
...

Y6 <= (A4 + B4) + (C4 + D4);
```

The expression in the assignment is built as a balanced tree. Each of the additions A4 + B4 and C4 + D4 yields a 5-bit result, so the final result size is 6 bits. However, if we build the expression in a cascaded fashion, the result size is 7 bits. We can see this most clearly by explicitly parenthesizing the expression:

```
Y7A <= ((A4 + B4) + C4) + D4;
```

The addition A4 + B4 yields a 5-bit result. This added to C4 yields a 6-bit result, and the 6-bit result added to D4 yields a 7-bit result. Since addition is associative, the following unparenthesized expression yields the same 7-bit result:

```
Y7B <= A4 + B4 + C4 + D4;
```

Included in the set of operations provided by the fixed-point packages are overloaded versions of the string conversion operations and file read and write operations for text-based input/output. Since the **ufixed** and **sfixed** types are one-dimensional arrays of character elements, **to_string** is predefined for both types. However, the packages overload the operation with a version that includes a radix point (a '.' character) at the appropriate position. For example, given the following declaration:

```
constant x : ufixed(3 downto -8) := "010000110101";
```

to_string(x) returns the string "0100.00110101".

The packages also define overloaded versions of **to_ostring** and **to_hstring** that behave similarly to the versions for the integer numeric packages (see Section 9.2.3). Elements are grouped into threes (for octal) or fours (for hexadecimal) starting either side of the radix point. Extension on the left to make up a complete group is performed in the same way as for **unsigned** and **signed** numeric values. If extension is required on the right, '0' bits are added. For values in which the radix-point position lies outside the index range, **to_ostring** and **to_hstring** extend the value to include the radix point in the result. For example, a **to_hstring** operation for the value "10100" with index range 7 down to 3 would result in the string "A0.0", corresponding to the binary number 10100000.0. Similarly, a **to_hstring** operation for the value "10100" with index range –3 down to –7 would result in "0.28" (0.0010100 in binary).

9.2.5 Standard Floating-Point Packages

The fixed-point math packages described in the previous section allow us to represent non-integral values with constant absolute precision over a given range. In some applications, however, we would prefer to use a floating-point representation, in which we can represent a greater dynamic range with a given number of bits, and have constant relative precision over the range. VHDL provides abstract floating-point types, including the type **real**, built into the language. However, they are defined to use IEEE 64-bit double-precision representation. That may not be the best choice for all applications. VHDL provides a set of packages for binary-coded floating-point representation and operations in which we can control the range and precision and many aspects of the way arithmetic operations are performed. Floating-point values are represented using the same principles as IEEE-standard floating-point, specified in IEEE Std 754 [9] and IEEE Std 854 [10], with a sign bit, an exponent field, and a fraction field. However, we can choose the field widths that are appropriate for our application.

Since these packages, like the fixed-point packages, provide a large number of overloaded operations and functions, we will not describe them in full detail here. Again, we will simply provide an overview of the data types provided and some basic information on their usage. More details are provided in Appendix A and on the author's companion website; and Section 9.2.6 summarizes the operations provided.

Like the main fixed-point package, the main VHDL fixed-point package, **float_generic_pkg**, is written using formal generic constants so that we can choose the behaviors that are most appropriate for our application. The package defines seven constants as follows:

float_exponent_width : natural

> This constant determines the default width of the exponent field resulting from the **to_float** conversion functions in the package. The default is 8, corresponding to IEEE single-precision representation.

float_fraction_width : natural

> This constant determines the default width of the fraction field resulting from the **to_float** conversion functions in the package. The default is 23, corresponding to IEEE single-precision representation.

float_round_style : round_type

> This constant determines the rounding behavior for operations in the package. The type **round_type** is an enumeration type defined in the package **fixed_float_types**, also in library **ieee**. The values are: **round_nearest**, **round_zero** (truncation), **round_inf** (round up toward infinity) and **round_neginf** (round down toward negative infinity). The default is **round_nearest**.

float_denormalize : boolean

> Denormalized numbers are a form of floating-point numbers that represent very small values near zero. If the constant **float_denormalized** is true, operations in the package deal with denormalized values; otherwise, all numbers are treated as normalized. The default is **true**.

float_check_error : boolean

> This constant controls detection of invalid numbers and overflow. The default is **true**.

float_guard_bits : natural

> This constant specifies the number of extra bits of precision to use within operations prior to rounding the result. The default is 3.

no_warning : boolean

> This constant allows suppression of warning messages. The default is **false**.

The main package also makes use of the fixed-point package, both for conversions between fixed-point and floating-point values and for internal implementation of floating-point operations. Since we can choose the way the fixed-point package behaves by varying the values of its constants, we need to provide the floating-point package with a reference the appropriate version of the fixed-point package for our application. We can do this, if we need to, using the mechanism of formal generic packages, described in Chapter 12. If the default values for the constants are acceptable for our application, we can use a version of the package named **float_pkg**, located in the **ieee** library. For other cases, the main package must be instantiated, as described in Chapter 12, to supply alternative values for the constants.

The package **float_generic_pkg** (and each instance of it, such as **float_pkg**) defines the base type for floating-point numbers, **unresolved_float**, declared as:

```
type unresolved_float is array (integer range <>) of std_ulogic;
```

The alias **u_float** is defined as a convenient shorthand for this type. There is also a subtype, **float**, which has resolved elements. The declarations are:

```
alias u_float is unresolved_float;

subtype float is (resolved) unresolved_float;
```

When declaring signals, we should choose between the base type or the subtype with resolved elements, depending on whether the signal has only one source or multiple sources, respectively.

Objects of these types must have descending (**downto**) index ranges; for example:

```
signal A : float(8 downto -23)
               := "01000000110100000000000000000000";
```

The sign bit is at index A'left (bit 8 in this example), the exponent is indexed from A'left – 1 down to 0 (7 down to 0 in the example), and the fraction is indexed from –1 down to A'right (–1 down to –23 in the example). Unlike fixed-point numbers, floating-point numbers must have the sign, exponent, and fraction all present. The smallest floating-point representation supported by the package has a range of 3 down to –3. In practice, we would expect representations to be 16 bits or more, with at least 6 bits for the exponent and at least 10 bits for the fraction. For the sign bit, 0 is positive, and 1 is negative. The exponent field is an unsigned binary value representing the actual exponent biased by $2^{e-1} - 1$ (where e is the width of the exponent field). Thus, for the signal A declared above, the bias is 127. The actual fraction is normalized to the range of 1.0 to just less than 2.0. Since the bit to the left of the radix point would always be 1, it is not explicitly represented. Instead, the fraction field of a floating-point number just contains the bits to the right of the radix point, with a 1 bit implied to the left of the radix point.

We can use these properties of the representation to analyze the bit string used as the default initial value for the signal A above. The leftmost bit is 0, so the number is positive. The next 8 bits, A(**7 downto 0**), are 10000001. As an unsigned number, this is 129. We subtract the bias, 127, to give an actual exponent of 2. The fraction field is 10100000000000000000000. We include the implied 1 bit to give an actual fraction of 1.101. Thus, the value represented is $+1.101_2 \times 2^2 = 1.625 \times 4 = 6.5$.

The packages declare a number of subtypes and aliases for IEEE standard floating-point representations. The unresolved subtypes are

```
subtype unresolved_float32 is unresolved_float(8 downto -23);

subtype unresolved_float64 is unresolved_float(11 downto -52);

subtype unresolved_float128 is
               unresolved_float (15 downto -112);
```

The **unresolved_float32** subtype correponds to IEEE Std 754 single-precision representation. There are a shorthand alias **u_float32** and a subtype with resolved elements, **float32**. Similarly, **unresolved_float64** corresponds to IEEE Std 754 double-precision representation (like **double float** in C, **float*8** in Fortran, and **real** in VHDL), with an alias **u_float64** and a subtype with resolve elements, **float64**; and **unresolved_float128** corresponds to IEEE Std 854 extended-precision representation (like **long double** in C and **float*16** in Fortran), with an alias **u_float128** and a subtype with resolve elements, **float128**.

The IEEE floating-point number standards reserve a number of representations for special purposes. In particular, numbers with all 0 or all 1 bits in the exponent field have the following meanings:

- Positive zero: 0 00000000 00000000000000000000000

- Negative zero: 1 00000000 00000000000000000000000

- Positive infinity: 0 11111111 00000000000000000000000

- Negative infinity: 1 11111111 00000000000000000000000

Note that there are two representations of 0, one positive and the other negative. Operations on floating-point values generally treat them as equivalent. In addition to these representations, a number with all 1 bits in the exponent field and at least one 1 bit in the fraction field (such as 1 11111111 00000000000000000000001) is called Not-a-Number, or NaN. Such values can result from otherwise illegal operations, such as division of zero by zero, or square root of –1.

Here are some further examples of floating-point numbers. First, the following is a large **float32** value (though not the largest, as that is just less than 2**128).

0 11111110 00000000000000000000000
$$= +1 \times 2^{254-127} \times (1.0 + 0.0)$$
$$= 2^{127} = 1.70141 \times 10^{38}$$

Next, the following is the smallest **float32** value, without using denormals:

0 00000001 00000000000000000000000
$$= +1 \times 2^{1-127} \times (1.0 + 0.0)$$
$$= 2^{-126} = 1.17549 \times 10^{-38}$$

Finally, the following is a small **float32** value using denormals (though not the smallest):

0 00000000 10000000000000000000000
$$= +1 \times 2^{1-127} \times (0.0 + 0.5)$$
$$= +1 \times 2^{-126} \times 0.5$$
$$= 2^{-127} = 5.87747 \times 10^{-39}$$

For floating-point math operations, the result always has the largest of the exponent sizes and fraction sizes of the operands. Most often, the numbers are all of the same size, as in the following example:

```
signal A32, B32, Y32 : float(8 downto -23);
...

Y32 <= A32 + B32;
```

Further details of overloaded operations and result sizes are provided in the tables in Section 9.2.6.

If we want to assign a value to a floating-point object, we can either use a string literal or we can apply a **to_float** conversion function to an integer or real number. This is similar to the way in which we assign values to fixed-point objects (see Section 9.2.4). In the case of conversion functions, we can specify the result size either by specifying the exponent and fraction size, or by providing an object whose index range is used. These approaches are shown in the following example:

```
signal A_fp32 : float32;
...

A_fp32 <= "01000000110100000000000000000000";
A_fp32 <= to_float(6.5, 8, -32);   -- pass sizes
A_fp32 <= to_float(6.5, A_fp32);   -- size using A_fp32
```

As with fixed-point math, use of string literals in an expression is problematic, since their index ranges are ascending (**to**) and start with **integer'low**. The solution is the same, namely, using type-qualified string literals or using overloaded operations that accept integer or real operands. These are shown in the following example:

```
signal A, Y : float32;
...

Y <= A + "01000000110100000000000000000000";    -- illegal
Y <= A + float32'("01000000110100000000000000000000");
Y <= A + 6.5;          -- overloading with real
Y <= A + 6;            -- overloading with integer
```

The floating-point packages also include overloaded versions of the string conversion operations and file read and write operations for text-based input/output. The overload **to_string** operation includes colon characters to separate the sign, exponent, and fraction fields. For example, given the following declaration:

```
constant x : float(6 downto -11) := "011101100010001110";
```

to_string(x) returns the string "0:111011:00010001110".

The floating-point packages also define overloaded versions of **to_ostring** and **to_hstring** that behave similarly to the versions for standard logic vectors (see Section 9.2.2). They do not attempt to include the radix point in the way that the fixed-point versions do, since the radix point is not at a fixed position.

9.2.6 Package Summary

In this section, we summarize the operations defined in the standard packages: std_logic_1164, numeric_std, numeric_bit, numeric_std_unsigned, numeric_bit_unsigned, fixed_generic_pkg, and float_generic_pkg.

Operator Overloading Summary

Table 9.7 summarizes the operand and result types for overloaded operations defined in the standard packages. The table does not include the predefined operations on the various types.

TABLE 9.7   *Operand and result types*

Operators	Left	Right	Result
Binary **and**, **or**, **nand**, **nor**, **xor**, **xnor**	std_ulogic	std_ulogic	std_ulogic
	LogicArrayType	*LogicArrayType*	*LogicArrayType*
	LogicArrayType	std_ulogic	*LogicArrayType*
	std_ulogic	*LogicArrayType*	*LogicArrayType*
not		std_ulogic	std_ulogic
		LogicArrayType	*LogicArrayType*
Unary reduction **and**, **or**, **nand**, **nor**, **xor**, **xnor**		*LogicArrayType*	std_ulogic
=, /=, <, <=, >, >=	*NumericArrayType*	*NumericArrayType*	boolean
	NumericArrayType	integer	boolean
	integer	*NumericArrayType*	boolean
	RealArrayType	real	boolean
	real	*RealArrayType*	boolean
?=, ?/=, ?<, ?<=, ?>, ?>=	*NumericArrayType*	*NumericArrayType*	*ArrayElementType*
	NumericArrayType	integer	*ArrayElementType*
	integer	*NumericArrayType*	*ArrayElementType*
	RealArrayType	real	*ArrayElementType*
	real	*RealArrayType*	*ArrayElementType*
rol, **ror**, **sll**, **srl**	*LogicArrayType*	integer	*LogicArrayType*
sla, **sra**	*NumericArrayType*	integer	*NumericArrayType*

Operators	Left	Right	Result
Binary +, –, *, /, **mod**, **rem**	*NumericArrayType*	*NumericArrayType*	*NumericArrayType*
	NumericArrayType	integer	*NumericArrayType*
	integer	*NumericArrayType*	*NumericArrayType*
	RealArrayType	real	*RealArrayType*
	real	*RealArrayType*	*RealArrayType*
Binary +, –	*NumericArrayType*	std_ulogic	*NumericArrayType*
	std_ulogic	*NumericArrayType*	*NumericArrayType*
Unary –, **abs**		signed, sfixed, float	signed, sfixed, float
maximum, minimum	*NumericArrayType*	*NumericArrayType*	*NumericArrayType*
	NumericArrayType	integer	*NumericArrayType*
	integer	*NumericArrayType*	*NumericArrayType*
	RealArrayType	real	*RealArrayType*
	real	*RealArrayType*	*RealArrayType*

Where overloaded operations are defined, the predefined operations are hidden. In the table, the notation use is as follows:

- *LogicArrayType*: arrays of **std_ulogic** elements
- *NumericArrayType*: **signed**, **unsigned**, **ufixed**, **sfixed**, **float**, **bit_vector** with operations in **numeric_bit_unsigned** visible, or **std_ulogic_vector** with operations in **numeric_std_unsigned** visible
- *RealArrayType*: **ufixed**, **sfixed**, or **float**
- *ArrayElementType*: the element type of the operand array or arrays

Table 9.8 summarizes the result size and/or index range for operations with array results. For arrays representing unsigned or signed integer values, only the size is relevant, as the leftmost bit is the most-significant bit and the rightmost bit is the least-significant bit. For fixed-point and floating-point values, the specific index bounds are relevant, as described in Sections 9.2.4 and 9.2.5. The notation for types is the same as that used in Table 9.7. In addition, *L* represents the left operand, *R* represents the right operand, *A* represents the array operand in the case where the other operand is scalar, and *Result* represents the result of the operation.

TABLE 9.8   *Result sizes and index ranges*

Operator	Result type	Result size and/or range
Array/array **and**, **or**, **nand**, **nor**, **xor**, **xnor**	*ArrayOfBits*	*Result*'length = *L*'length = *R*'length Fixed, Float: *Result*'range = *L*'range
Array/scalar **and**, **or**, **nand**, **nor**, **xor**, **xnor**	*ArrayOfBits*	*Result*'length = *A*'length Fixed, Float: *Result*'range = *A*'range
not	*ArrayOfBits*	*Result*'length = *R*'length Fixed, Float: *Result*'range = *R*'range
rol, **ror**, **sll**, **srl**, **sla**, **sra**	*ArrayOfBits*	*Result*'length = *A*'length Fixed, Float: *Result*'range = *A*'range
+, −, *, /, **rem**, **mod**	float	maximum(*L*'left, *R*'left) down to minimum(*L*'right, *R*'right)
Binary +, −	unsigned, signed	maximum(*L*'length, *R*'length) − 1 down to 0
	ufixed, sfixed	maximum(*L*'left, *R*'left) + 1 down to minimum(*L*'right, *R*'right)
*	unsigned, signed	*L*'length + *R*'length − 1 down to 0
	ufixed, sfixed	*L*'left + *R*'left + 1 down to *L*'right + *R*'right
/	unsigned, signed	*L*'length − 1 down to 0
	ufixed	*L*'left − *R*'right down to *L*'right − *R*'left − 1
	sfixed	*L*'left − *R*'right + 1 down to *L*'right − *R*'left
rem	unsigned, signed	*R*'length − 1 down to 0
	ufixed, sfixed	minimum(*L*'left, *R*'left) down to minimum(*L*'right, *R*'right)
mod	unsigned, signed	*R*'length − 1 down to 0
	ufixed	minimum(*L*'left, *R*'left) down to minimum(*L*'right, *R*'right)
	sfixed	*R*'left down to minimum(*L*'right, *R*'right)
Unary −, **abs**	signed	*R*'length − 1 down to 0
	sfixed	*R*'left + 1 down to *R*'right
minimum, maximum	*DiscreteArrayType*	*Result*'length = *A*'length Fixed, Float: *Result*'range = *A*'range
	unsigned, signed	maximum(*L*'length, *R*'length) − 1 down to 0
	ufixed, sfixed, float	minimum(*L*'left, *R*'left) down to minimum(*L*'right, *R*'right)

Conversion Function Summary

Next, we summarize the conversion functions defined in the standard packages. In order to present the information in more compact form, we have used some abbreviations for types and the packages in which the functions are defined: "bv" for **bit_vector**, "slv" for **std_logic_vector**, "sulv" for **std_ulogic_vector**, "1164" for **std_logic_1164**, "nbu" for **numeric_bit_unsigned**, "nsu" for **numeric_std_unsigned**, "ns/b" for **numeric_std** and **numeric_bit**, "fixed" for **fixed_generic_pkg**, and "float" for **float_generic_pkg**.

Table 9.9 shows the functions that convert between **bit** and **std_ulogic** scalar types, and between vectors of these types. We use the shorthand aliases for the functions here for brevity. The first parameter is the value to be converted. The **to_bit** and **to_bv** functions have a second parameter, **xmap**, as described in Section 9.2.2, to specify how metalogical values should be mapped. Those functions that convert from an abstract numeric value to a vector representation have a second parameter, either a natural value, **size**, specifying the size of the result or a value of the result type, **size_res**, whose size is used for the result.

TABLE 9.9   *Conversions between bit and standard-logic types*

Function	Result type	Parameter 1 type	Parameter 2	Package
to_bit	bit	std_ulogic	xmap	1164
to_std_ulogic	std_ulogic	bit		1164
to_bv	bit_vector	sulv	xmap	1164
		natural	size	nbu
		natural	size_res	nbu
to_sulv	sulv	bv		1164
		slv		1164
		natural	size	nsu
		natural	size_res	nsu
		ufixed		fixed
		sfixed		fixed
		float		float

Function	Result type	Parameter 1 type	Parameter 2	Package
to_slv	slv	bv		1164
		sulv		1164
		natural	natural	nsu
		natural	size_res	nsu
		ufixed		fixed
		sfixed		fixed
		float		float

Table 9.10 shows the functions that convert from the various numeric types to the **unsigned** and **signed** types defined in **numeric_std** and **numeric_bit**. The first parameter is the value to be converted, and the second parameter is either a natural value, **size**, specifying the size of the result or a value of the result type, **size_res**, whose size is used for the result.

The conversions from fixed-point representation have a third parameter, **overflow_style** (abbreviated to "overflow" in the table), of type **fixed_overflow_style_type**. The default value is the value of the generic **fixed_overflow_style**. The fourth parameter, **round_style** (abbreviated to "round"), is of type **fixed_round_style_type** and defaults to the value of the generic **fixed_round_style**.

The conversions from **float** have a third parameter, **round_style** (abbreviated to "round"), of type **round_type**, with the default being the value of the package generic **float_round_style**. The fourth parameter is **check_error** (abbreviated to "chk_err"), of type **boolean**, for controlling error checking during the conversion. The default is the value of the package generic **float_check_error**.

TABLE 9.10   *Conversion functions yielding* unsigned *and* signed *values*

Function	Result type	Param 1 type	Param 2	Param 3	Param 4	Package
to_unsigned	unsigned	natural	size			ns/b
			size_res			
		ufixed	size	overflow	round	fixed
			size_res	overflow	round	
		float	size	round	chk_err	float
			size_res	round	chk_err	

Function	Result type	Param 1 type	Param 2	Param 3	Param 4	Package
to_signed	signed	integer	size			ns/b
			size_res			
		sfixed	size	overflow	round	fixed
			size_res	overflow	round	
		float	size	round	chk_err	float
			size_res	round	chk_err	

Table 9.11 shows the functions that convert from numeric types to the **ufixed** and **sfixed** types defined in the fixed-point packages. In the case of conversion functions defined in the floating-point packages, the definitions of **ufixed** and **sfixed** come from the fixed-point package referenced as a generic package, as outlined in Section 9.2.5. The first parameter of each function is the value to be converted. Following this are either two parameters, **left_index** and **right_index** (abbreviated to "L_index" and "R_index" in the table), to specify the index bounds of the result, or a single parameter of the result type, **size_res**, whose index range is used for the result. For the conversions from **natural** or **unsigned** to **ufixed**, and for the conversions to **integer** or **signed** to **sfixed**, the default for **right_index** is 0. Additional parameters specify overflow and rounding modes (**overflow_style** and **round_style**), the number of guard bits to use (**guard_bits**), whether error checking is required (**check_error**), and whether operands of type float use denormalized representation (**denormalize**). The default values for the **overflow_style**, **round_style**, and **guard_bits** parameters come from the various generic constants of the packages. Note that there are also versions of **to_ufixed** and **to_sfixed** with no parameters beyond the first **unsigned** or **signed** parameter. (This is not an error in the table layout!) These versions simply return the value of the parameter as a fixed-point value with no fractional part (that is, indexed from one less than the length down to 0).

TABLE 9.11   Conversion functions yielding *ufixed* and *sfixed values*

Function	Result type	Param 1 type	Param 2	Param 3	Param 4	Param 5	Param 6	Param 7	Package
to_ufixed	ufixed	sulv	L_index	R_index					fixed
		unsigned	size_res						
		natural	L_index	R_index	overflow	round			
			size_res	overflow	round				
		real	L_index	R_index	overflow	round	guard		
			size_res	overflow	round	guard			
		float	L_index	R_index	overflow	round	chk_err	denorm	float
			size_res	overflow	round	chk_err	denorm		

Function	Result type	Param 1 type	Param 2	Param 3	Param 4	Param 5	Param 6	Param 7	Package
to_sfixed	sfixed	ufixed							fixed
		sulv	L_index	R_index					
			size_res						
		signed							
			L_index	R_index	overflow	round			
			size_res	overflow	round				
		integer	L_index	R_index	overflow	round			
			size_res	overflow	round				
		real	L_index	R_index	overflow	round	guard		
			size_res	overflow	round	guard			
		float	L_index	R_index	overflow	round	chk_err	denorm	float
			size_res	overflow	round	chk_err	denorm		

Table 9.12 shows the functions that convert from numeric types to the **float** type defined in the floating-point packages. Again, the definitions of **ufixed** and **sfixed** come from the fixed-point package referenced as a generic package. The first parameter of each function is the value to be converted. Following this are either two parameters, **exponent_width** and **fraction_width** (abbreviated to "exponent" and "fraction" in the table), to specify the sizes of the corresponding fields in the result, or a single parameter of the result type, **size_res**, whose index range is used for the result. Additional parameters specify the rounding mode (**round_style**) and whether denormalized representation is used (**denormalize**). The default values for the field size, **round_style** and **denormalize** parameters come from the generic constants of the package.

TABLE 9.12   *Conversion functions yielding* float *values*

Function	Result type	Param 1 type	Param 2	Param 3	Param 4	Param 5	Package
to_float	float	sulv	exponent	fraction			float
			size_res				
		unsigned					
			exponent	fraction	round		
			size_res	round			
		signed	exponent	fraction	round		
			size_res	round			
		ufixed	exponent	fraction	round	denorm	
			size_res	round	denorm		
		sfixed	exponent	fraction	round	denorm	
			size_res	round	denorm		
		integer	exponent	fraction	round		
			size_res	round			
		real	exponent	fraction	round	denorm	
			size_res	round	denorm		

The final group of conversion functions is shown in Table 9.13. These functions convert from binary-coded vectors to abstract integer or real types. As in the preceding tables, the first parameter is the value to be converted, and subsequent parameters specify overflow and rounding modes (**overflow_style** and **round_style**), whether error checking is required (**check_error**), and whether operands of type float use denormalized representation (**denormalize**). The default values for these subsequent parameters come from the generic constants of the packages.

TABLE 9.13   *Conversion functions yielding* integer *and* real *values*

Function	Result type	Param 1 type	Param 2	Param 3	Param 4	Package
to_integer	natural	bv				nbu
	natural	sulv				nsu
	natural	unsigned				ns/b
	integer	signed				
	natural	ufixed	overflow	round		fixed
	integer	sfixed	overflow	round		
	integer	float	round	chk_err		float
to_real	real	ufixed				fixed
		sfixed				
		float	round	chk_err	denorm	float

As we have mentioned, the packages provide aliases for the conversion functions for convenience and enhanced readability: for conversion to **bit_vector**, the names we can use are **to_bv**, **to_bitvector** and **to_bit_vector**; for conversion to **std_ulogic_vector**, the names are **to_sulv**, **to_stdulogicvector** and **to_std_ulogic_ vector**; and for conversion to **std_logic_vector**, the names are **to_slv**, **to_stdlogicvector** and **to_std_logic_vector**.

For each binary-coded numeric type, there is a **resize** function, shown in Table 9.14. The versions yielding **bit_vector**, **std_ulogic_vector**, **unsigned**, and **signed** results have a parameter **new_size** to specify the result size, or a parameter **size_res** for an object whose index range is used for that of the result. The versions that yield fixed-point results have either two parameters (**left_index** and **right_index**) to specify the index bounds of the result, or one parameter (**size_res**) for an object whose index range is used for that of the result. They also have parameters to specify overflow and rounding modes (**overflow_style** and **round_style**), with default values coming from the package generics. Similarly, the versions that yield floating-point results have either two parameters to specify the field sizes for the result (**exponent_width** and **fraction_width**), or one parameter (**size_res**) for an object whose index range is used for that of the result. Subsequent parameters specify rounding modes (**round_style**), whether error checking is required (**check_error**), and whether the operand and result use denormalized representation (**denormalize_in** and **denormalize_out**, respectively). The default values for these subsequent parameters come from the generic constants of the package.

TABLE 9.14 Resizing functions

Function	Result type	Param 1 type	Param 2	Param 3	Param 4	Param 5	Param 6	Param 7	Package
resize	bv	bv	new_size						nbu
	bv	bv	size_res						
	sulv	sulv	new_size						nsu
	sulv	sulv	size_res						
	unsigned	unsigned	new_size						ns/b
	unsigned	unsigned	size_res						
	signed	signed	new_size						
	signed	signed	size_res						
	ufixed	ufixed	L_index	R_index	overflow	round			fixed
	ufixed	ufixed	size_res	overflow	round				
	sfixed	sfixed	L_index	R_index	overflow	round			
	sfixed	sfixed	size_res	overflow	round				
	float	float	exponent	fraction	round	chk_err	den_in	den_out	float
	float	float	size_res	round	chk_err	den_in	den_out		

Resizing an unsigned vector of type **bit_vector**, **std_ulogic_vector** or **unsigned** to produce a larger vector involves filling leftmost bits with '0'. Resizing these types to produce a smaller vector involves truncating the leftmost bits. For type **signed**, producing a larger vector involves filling the leftmost bits with copies of the operand's sign bit, and producing a smaller vector involves truncating the leftmost bits while retaining the sign bit.

Resizing a fixed-point value is similar. A **ufixed** vector is extended on the left or right by filling bits with '0'. An **sfixed** vector is extended on the left by replicating the sign bit and extended on the right by filling bits with '0'. Reducing the size of a fixed-point vector is more complicated, and depends on the overflow and rounding modes. If the vector is to be truncated on the right, a rounding mode of **fixed_truncate** causes the truncated bits to be discarded and the rightmost result bit to be unchanged, whereas a rounding mode of **fixed_round** causes the result to be rounded based on the values of the discarded bits and the rightmost result bit. If the vector is to be truncated to the left and the operand value is out of the representable range for the result, the value returned depends on the overflow style. For **fixed_saturate**, the largest representable value (for **ufixed** or for positive **sfixed** values) or the most negative representable value (for negative **sfixed** values) is returned. For **fixed_wrap**, the leftmost bits are simply truncated, which, in the case of **sfixed** values, may result in a change of sign.

Resizing a floating-point value is much more involved than resizing integral and fixed-point values. It involves determining the class of value represented by the operand (normal, denormal, zero, infinity, or NaN), resizing the exponent and fractional parts, rounding according to the **round_style** parameter, renormalizing or representing as a denormal if required, checking for errors, and transforming overflow to infinity.

Strength Reduction Function Summary

The strength reduction functions are defined for the entire family of types based on **std_ulogic**. Functions of the following form are defined:

```
function to_01   (s : Type; xmap : std_ulogic := '0') return Type;

function to_X01  (s : Type) return Type;

function to_X01Z (s : Type) return Type;

function to_UX01 (s : Type) return Type;
```

The type *Type* includes **std_ulogic**, **std_ulogic_vector**, **unresolved_unsigned**, **unresolved_signed**, **unresolved_ufixed**, **unresolved_sfixed**, **unresolved_float**, and the subtypes of the vector types with resolved elements. The value returned by each function for each operand element value is shown in Table 9.15. The functions **to_X01**, **to_X01Z**, and **to_UX01**, when applied to vector operands, convert each operand element according to the table to yield the corresponding result element. The **to_01** function, however, behaves differently. Provided all of the elements are '0', '1', 'L', or 'H', they are converted according to the table. However, if any element is a metalogical value (a value other than '0', '1', 'L', or 'H'), all elements of the result are set to the value of the **xmap** parameter. Thus, we can test any element of the result to determine whether there were any metalogical elements in the operand.

TABLE 9.15   *Strength reduction mappings*

Function	'U'	'X'	'0'	'1'	'Z'	'W'	'L'	'H'	'—'
to_01	xmap	xmap	'0'	'1'	xmap	xmap	'0'	'1'	xmap
to_X01	'X'	'X'	'0'	'1'	'X'	'X'	'0'	'1'	'X'
to_X01Z	'X'	'X'	'0'	'1'	'Z'	'X'	'0'	'1'	'X'
to_UX01	'U'	'X'	'0'	'1'	'X'	'X'	'0'	'1'	'X'

The 'X' detection function is also defined for the entire family of types based on **std_ulogic**. The function definitions are of the form:

function is_X (S : *Type*) **return** boolean;

The version for **std_ulogic** returns true if the operand is a metalogical value, or false otherwise. The versions for vector types return true if any element of the operand is a metalogical value, or false otherwise.

EXAMPLE 9.5   *Unknown detection for a state-machine input*

We can use the **to_X01** function in behavioral models of ASIC and FPGA input cells to promote a resistive strength to a driving level as follows:

ncs_x01 <= to_X01(ncs);

We can use the **is_X** function to detect 'X' values in behavioral models and RTL code, for example, in the input to a state machine:

assert not is_X(ncs) **report** "ncs is X" **severity** error;

Exercises

1. [❶ 9.1] Write an assertion statement that verifies that the resolution limit is at most 1 ns.

2. [❶ 9.2] What string values are produced by **to_hstring** for the following **std_ulogic_vector** values: B"ZZZZ_0100", B"XX_L01H", B"01_00ZZ"?

3. [❶ 9.2] Write declarations for signals **a**, **b**, and **s** representing 24-bit unsigned numbers with **std_ulogic** elements. Write an assignment that adds **a**, **b**, and a **std_ulogic** signal **carry_in**, producing the sum in **s** and a **std_ulogic** signal **carry_out**.

4. [❶ 9.2] Write a process representing an edge-triggered D-flipflop with clock signal **clk**, enable signal **en**, data input **d**, and output **q**, all of type **std_ulogic**. The flipflop should store '0', '1', or 'X' values.

5. [● 9.2] Write declarations for a signed fixed-point signal **a** with 4 pre-binary-point bits and 6 post-binary-point bits. Write an assignment that assigns the square of **a** to a signal **s** that has 8 pre-binary-point bits and 6 post-binary-point bits.

6. [● 9.2] Write a declarations for **float** signals **x** and **y** with 7 exponent bits and 12 fraction bits. Write a concurrent assignment that assigns the value of **x** to **y**, limited to the range −1.0 to +1.0.

7. [❷ 9.1] Integers can be represented in *signed magnitude* form, in which the leftmost bit represents the sign ('0' for non-negative, '1' for negative), and the remaining bits are the absolute value of the number, represented in binary. If we wish to compare bit vectors containing numbers in signed magnitude form, we cannot use the predefined relational operators directly. We must first transform each number as follows: if the number is negative, complement all bits; if the number is non-negative, complement only the sign bit. Write a comparison function, overloading the operator "<", to compare signed-magnitude bit vectors using this method.

8. [❷ 9.2] Write a procedure that uses the uniform random number generator to generate a random value of an enumeration type named **controller_state**.

9. [❷ 9.2] In telephone systems, a signal is compressed before transmission. The formula for μ-law compression of a signal is

$$F(x) = \text{sgn} x \frac{\ln(1 + \mu|x|)}{\ln(1 + \mu)} \qquad -1 \le x \le 1, \mu = 255$$

Develop a functional model of a compressor for values of type **real**.

10. [❷ 9.2] Develop a model of a decade counter with a 4-bit **unsigned** output.

11. [❷ 9.2] Develop a behavioral model of a pipelined multiplier for single-precision (**float32**) operands. On each clock cycle, the multiplier starts a new multiplication, and produces the product at the output of the pipeline five cycles later.

Chapter 10

Case Study: A Pipelined Multiplier Accumulator

Now that we have covered the basic modeling facilities provided by VHDL, we will work through our first case study, the design of a pipelined multiplier accumulator (MAC) for a stream of complex numbers. Many digital signal processing algorithms, such as digital demodulation, filtering and equalization, make use of MACs. We use this design exercise to bring together concepts and techniques introduced in previous chapters.

10.1 **Algorithm Outline**

A complex MAC operates on two sequences of complex numbers, $\{x_i\}$ and $\{y_i\}$. The MAC multiplies corresponding elements of the sequences and accumulates the sum of the products. The result is

$$\sum_{i=1}^{N} x_i y_i$$

where N is the length of the sequences. Each complex number is represented in Cartesian form, consisting of a real and an imaginary part. If we are given two complex numbers x and y, their product is a complex number p, calculated as follows:

$$p_{real} = x_{real} \times y_{real} - x_{imag} \times y_{imag}$$
$$p_{imag} = x_{real} \times y_{imag} + x_{imag} \times y_{real}$$

The sum of x and y is a complex number s calculated as follows:

$$s_{real} = x_{real} + y_{real}$$
$$s_{imag} = x_{imag} + y_{imag}$$

Our MAC calculates its result by taking successive pairs of complex numbers, one each from the two input sequences, forming their complex product and adding it to an accumulator register. The accumulator is initially cleared to zero and is reset after each pair of sequences has been processed.

If we count the operations required for each pair of input numbers, we see that the MAC must perform four multiplications to form partial products, then a subtraction and an addition to form the full product and finally two additions to accumulate the result. This is shown diagrammatically at the top of Figure 10.1. Since the operations must be performed in this order, the time taken to complete processing one pair of inputs is the sum of the delays for the three steps. In a high-performance digital signal processing application, this delay may cause the bandwidth of the system to be reduced below a required minimum.

We can avoid the delay by *pipelining* the MAC, that is, organizing it like an assembly line, as shown at the bottom of Figure 10.1. The first pair of input numbers is stored in the input register on the first clock edge. During the first clock cycle, the multipliers calculate the partial products, while the system prepares the next pair of inputs. On the second clock edge, the partial products are stored in the first pipeline register, and the next pair of inputs is entered into the input register. During the second clock cycle, the subtracter and adder produce the full product for the first input pair, the multipliers produce the partial products for the second input pair and the system prepares the third input pair. On the third clock edge, these are stored, respectively, in the second pipeline register, the first pipeline register and the input register. Then in the third clock cycle, the adders accumulate the product of the first pair with the previous sum, and the preceding stages operate on the second and third pairs, while the system prepares the fourth pair. The sum

FIGURE 10.1

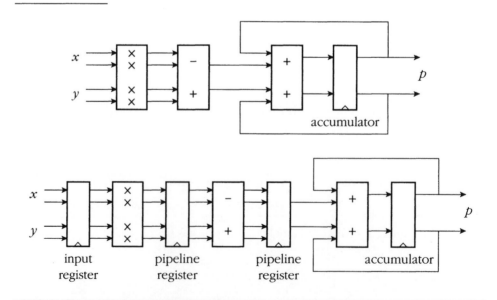

Dataflow diagrams showing order of operations performed by the MAC. Top: combinatorial organization. Bottom: pipelined organization.

in the accumulator is updated on the fourth clock edge. Thus, three clock cycles after the first pair of numbers was entered into the input latch, the sum including this pair is available at the output of the MAC. Thereafter, successive sums are available each clock cycle. The advantage of this approach is that the clock period can be reduced to the slowest of the pipeline stages, rather than the total of their delays.

One detail we have yet to consider is initializing and restarting the pipeline. We need to do this to accumulate sums of products of a number of input sequences, one after another. The simplest approach is to include a "clear" input to the accumulator register that forces its content to zero on the next clock edge. For each pair of sequences to be multiplied and accumulated, we start entering numbers into the input registers on successive clock edges. Then, two clock cycles after we have entered the first pair of numbers, we assert the clear input. This causes the accumulator to reset at the same time as the product of the first pair of numbers reaches the second pipeline register. On the following cycle, this product will be added to the zero value forced into the accumulator. After the last pair in the input sequences has been entered, we must wait three clock cycles until the final sum appears at the output of the MAC. We must separate successive input sequences by at least one idle cycle and reset the accumulator between summations.

The final issue in this outline of the MAC algorithm is the representation of the data. We use a 16-bit, two's-complement, fixed-point binary representation. Each of the real and imaginary parts of the two complex inputs and the complex output of the MAC uses the format shown in Figure 10.2. Bit 0 is the sign bit, and the binary point is assumed to be between bits 0 and -1. Using this format, we can represent numbers in the range -1 (inclusive) to $+1$ (exclusive), with a resolution of 2^{-15}. This raises the possibility of overflow occurring while summing a sequence of numbers, so we include an overflow status signal in our design. Overflow can occur in two cases. First, intermediate partial sums may fall outside of the range -1 to $+1$. We can reduce the likelihood of this happening by expanding the range used to represent intermediate results to -16 to $+16$. However, if an intermediate sum falls outside of the expanded range, the summation for the entire sequence is in error, so the overflow signal must be set. It remains set until the accumulator is cleared, indicating the end of the summation.

The second overflow case occurs if the final sum falls outside the range of values representable by the MAC output. This may be a transient condition, since a subsequent product, when added to the sum, may bring the sum back in range. We assert the overflow signal only during a cycle in which the final sum is out of range, rather than latching the overflow until the end of summation.

Now that we have described the requirements and the algorithm to be performed by the MAC, we can specify its interface. This is defined by the entity declaration:

FIGURE 10.2

bit index	0	-1	-2	-3	-4	-5	-6	-7	-8	-9	-10	-11	-12	-13	-14	-15
bit weight	-2^0	2^{-1}	2^{-2}	2^{-3}	2^{-4}	2^{-5}	2^{-6}	2^{-7}	2^{-8}	2^{-9}	2^{-10}	2^{-11}	2^{-12}	2^{-13}	2^{-14}	2^{-15}

The format of a 16-bit, two's-complement, fixed-point binary number.

```
library ieee;
use ieee.std_logic_1164.all, ieee.fixed_pkg.all;

entity mac is
  port ( clk, reset : in std_ulogic;
          x_real : in u_sfixed(0 downto -15);
          x_imag : in u_sfixed(0 downto -15);
          y_real : in u_sfixed(0 downto -15);
          y_imag : in u_sfixed(0 downto -15);
          s_real : out u_sfixed(0 downto -15);
          s_imag : out u_sfixed(0 downto -15);
          ovf : out std_ulogic );
end entity mac;
```

The **clk** port is used to synchronize operation of the MAC. All data transfers into registers in the pipeline are done on the rising edge (from '0' to '1') of this signal. The **reset** port causes the accumulator registers to be cleared to zero and the overflow condition to be reset. The data ports all use the signed data type from the standard fixed-point package described in Section 9.2.4. The index ranges correspond to the format shown in Figure 10.2. The ports **x_real**, **x_imag**, **y_real** and **y_imag** are the real and imaginary parts of the two input data sequences. These input ports, as well as the **reset** port, are sampled synchronously on the rising edge of the **clk** signal. The ports **s_real** and **s_imag** are the real and imaginary parts of the accumulated sum, and the **ovf** port is the overflow flag, set as described above. These output ports become valid after each rising edge of the **clk** signal.

10.2 A Behavioral Model

Our first implementation of the MAC is a behavioral model. This model allows us to focus on the algorithm without being distracted by other details at this early stage of the design. When we have the behavioral model working, we will be able to use it to generate test data for more detailed implementations. Our behavioral model is expressed as an architecture body, containing a single process that implements the MAC algorithm described in Section 10.1:

```
use ieee.math_complex.all;

architecture behavioral of mac is

  signal x_complex, y_complex, s_complex : complex;

begin

  x_complex <= ( to_real(x_real), to_real(x_imag) );
  y_complex <= ( to_real(y_real), to_real(y_imag) );

  behavior : process (clk) is

    variable input_x, input_y : complex := (0.0, 0.0);
    variable real_part_product_1, real_part_product_2,
             imag_part_product_1, imag_part_product_2 := 0.0;
```

```vhdl
          variable product, sum : complex := (0.0, 0.0);
          variable real_accumulator_ovf,
                   imag_accumulator_ovf : boolean := false;
      begin
        if rising_edge(clk) then
          -- Work from the end of the pipeline back to the start,
          -- so as not to overwrite previous results in pipeline
          -- registers before they are used.

          -- Update accumulator and generate outputs.
          if reset then
            sum := (0.0, 0.0);
            real_accumulator_ovf := false;
            imag_accumulator_ovf := false;
          else
            sum := product + sum;
            real_accumulator_ovf := real_accumulator_ovf
                                    or sum.re < -16.0
                                    or sum.re >= +16.0;
            imag_accumulator_ovf := imag_accumulator_ovf
                                    or sum.im < -16.0
                                    or sum.im >= +16.0;
          end if;
          s_complex <= sum;
          ovf <= '1' when real_accumulator_ovf or imag_accumulator_ovf
                      or sum.re < -1.0 or sum.re >= +1.0
                      or sum.im < -1.0 or sum.im >= +1.0 ) else '0';

          -- Update product registers.
          product.re := real_part_product_1 - real_part_product_2;
          product.im := imag_part_product_1 + imag_part_product_2;

          -- Update partial product registers
          -- (actually with the full product).
          real_part_product_1 := input_x.re * input_y.re;
          real_part_product_2 := input_x.im * input_y.im;
          imag_part_product_1 := input_x.re * input_y.im;
          imag_part_product_2 := input_x.im * input_y.re;

          -- Update input registers using MAC inputs
          input_x := x_complex;
          input_y := y_complex;
        end if;
      end process behavior;

    s_real <= to_sfixed(s_complex.re, s_real);
    s_imag <= to_sfixed(s_complex.im, s_imag);

end architecture behavioral;
```

Note that the algorithm as described involves performing arithmetic on binary vectors representing fixed-point numbers. We can avoid this by converting the input data to the **complex** data type defined by the standard **math_complex** package (see Section 9.2.1), performing the calculations using the overloaded arithmetic operators from that package, then converting the results back to binary vectors. Note that this approach may lead to slightly different results from the ultimate implementation using binary fixed-point representation, due to the difference in precision between the two representations. We should bear this in mind when comparing the output of the behavioral model with the models we develop later, and ignore any small discrepancies.

The signals declared in the architecture are the **complex** representations of the data input and output ports. The inputs are converted using the **to_real** conversion function from the fixed-point package and forming **complex** aggregates. The result calculated by the model on **s_complex** is converted back to fixed point using the **to_sfixed** conversion function.

The process **behavior** implements the MAC algorithm. The variables of type **complex** declared in the process represent the pipeline registers described in Section 10.1. The two Boolean variables represent the overflow conditions arising from the accumulators. The process is sensitive to the **clk** signal and performs a new calculation on each rising edge. It works from the output end of the pipeline back toward the input end to avoid overwriting intermediate results from the previous clock cycle before they have been used in the current cycle.

The process first calculates the new sum and overflow status. If the **reset** input is '1', both the accumulator and overflow variables are reset. Otherwise the process accumulates a new complex sum, based on the previous complex sum and the contents of the product registers, and stores it in the accumulator register variable. It also determines whether the real and imaginary parts are within the range −16.0 to +16.0 and sets the overflow register variables accordingly. The output data signal is assigned the new content of the accumulator, and the overflow signal is set if either of the overflow register variables is set, or if either of the data output parts falls outside the range −1.0 to +1.0. Next, the process updates the product register variables using the previously calculated partial products. It then updates the partial products using the previously stored input values and finally stores the new input data values in the input register variables.

10.2.1 Testing the Behavioral Model

We can test the behavioral model of the MAC by instantiating it in a test bench model that generates stimulus values on signals connected to the MAC inputs. The entity for the test bench has no ports, since it is completely self-contained:

```
entity mac_test is
end entity mac_test;
```

The architecture body is

```
library ieee;
use ieee.std_logic_1164.all, ieee.fixed_pkg.all,
    ieee.math_complex.all;
```

```vhdl
architecture bench_behavioral of mac_test is
  signal clk, reset, ovf : std_ulogic := '0';
  signal x_real, x_imag,
         y_real, y_imag,
         s_real, s_imag : u_sfixed(0 downto -15);

  signal x, y, s : complex := (0.0, 0.0);

  constant Tpw_clk : time := 50 ns;
begin
  x_real <= x.re; x_imag <= x.im;
  y_real <= y.re; y_imag <= y.im;

  dut : entity work.mac(behavioral)
    port map ( clk, reset,
               x_real, x_imag, y_real, y_imag, s_real, s_imag,
               ovf );

  s <= (s_real, s_imag);

  clock_gen : process is
  begin
    clk <= '1' after Tpw_clk, '0' after 2 * Tpw_clk;
    wait for 2 * Tpw_clk;
  end process clock_gen;

  stimulus : process is
  begin
    -- first sequence
                                              reset <= '1';
    wait until not clk;
    x <= (+0.5, +0.5);  y <= (+0.5, +0.5);  reset <= '1';
    wait until not clk;
    x <= (+0.2, +0.2);  y <= (+0.2, +0.2);  reset <= '1';
    wait until not clk;
    x <= (+0.1, -0.1);  y <= (+0.1, +0.1);  reset <= '1';
    wait until not clk;
    x <= (+0.1, -0.1);  y <= (+0.1, +0.1);  reset <= '0';
    wait until not clk;

    -- should be (0.04, 0.58) when it falls out the other end
                                              reset <= '0';
    wait until not clk;
    x <= (+0.5, +0.5);  y <= (+0.5, +0.5);  reset <= '0';
    wait until not clk;
    x <= (+0.5, +0.5);  y <= (+0.1, +0.1);  reset <= '0';
    wait until not clk;
    x <= (+0.5, +0.5);  y <= (+0.5, +0.5);  reset <= '1';
```

```
    wait until not clk;
    x <= (-0.5, +0.5);  y <= (-0.5, +0.5);  reset <= '0';
    wait until not clk;

                                        reset <= '0';
    wait until not clk;

                                        reset <= '0';
    wait until not clk;

                                        reset <= '0';
    wait until not clk;

                                        reset <= '1';
    wait until not clk;

    wait;
  end process stimulus;

end architecture bench_behavioral;
```

The architecture body contains signals that are connected to the input and output ports of the MAC "device under test." In order to simplify the task of writing complex values in the test bench, we use the **complex** type from the **math_complex** package, as we did within the behavioral model. The constant **Tpw_clk** is used to determine the pulse width for the high and low phases of the clock signal, generated by the process **clock_gen**.

Stimulus values for the MAC are provided by the process labeled **stimulus**. During the first four clock cycles, the process keeps the **clr** input active. This clears the accumulator registers while the MAC pipeline starts operation. The stimulus process sets up new complex number data each clock cycle. These enter the pipeline in sequence, and, after a delay due to the pipeline latency, the accumulated sums start appearing on the outputs. Figure 10.3 shows the timing of calculations performed on this first sequence. The final result we should expect is the complex number (0.04, 0.58). The result shown in Figure 10.3 (after the clock-edge at 750 ns) is slightly different due to the reduced precision of the 16-bit fixed-point number representation.

After the result from the first sequence has progressed through to the MAC output, the stimulus process clears the accumulator to prepare it to start accumulating results for the second sequence. This sequence is designed to test the overflow output of the MAC. The sum of the first three pairs produces a number that lies outside of the range −1.0 to +1.0, so we would expect the **ovf** output to be '1' when that sum reaches the output. However, addition of the product of the next pair brings the accumulated sum back into range, so we would expect **ovf** to revert to '0'. This behavior is also shown in Figure 10.3.

The stimulus values, or *test vectors*, used to test the MAC model in this example are synthetically generated by the model designer. While such vectors are useful for small-scale testing, they do not provide high test coverage. There may be errors in the model that are not revealed by the small number of vectors. As the famous computer scientist Nicklaus Wirth commented, "Testing can reveal the presence of bugs, not their absence." We can gain more confidence in the model by providing significantly larger sets of test vectors. One approach is to use a pseudo-random number generator (for example, based on the **uniform** procedure in the **math_real** package) to generate long streams of test vectors. Another approach is to create files of test vectors. A test bench can read such files to stimulate the model under test. We will discuss the use of files in VHDL in Chapter 16.

FIGURE 10.3

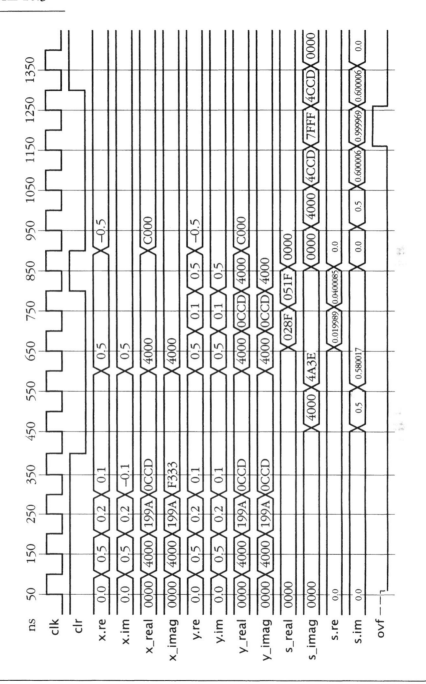

The timing of the MAC operation on the sequences in the test bench.

10.3 A Register-Transfer-Level Model

We now turn to a register-transfer-level implementation of the MAC, based on the pipeline diagram shown in Section 10.1. A more detailed diagram of the MAC at the register-transfer level is shown in Figure 10.4. We will use the overloaded arithmetic operators from **fixed_pkg** to implement the multipliers, adders and the subtractor. Recall that the sizes and index ranges of the results for these operators are shown in Table 9.8 on page 325.

The real and imaginary parts of the two complex inputs are stored in the first set of pipeline registers. The multipliers use the stored values to produce the four partial products. Since the input values are 16-bit fixed-point numbers between −1.0 inclusive and +1.0 exclusive (indexed from 0 down to −15), the partial products are 32-bit fixed-point numbers between −1.0 and +1.0 inclusive (indexed from 1 down to −30), as shown in Figure 10.5. The partial products are stored in the second set of pipeline registers. The subtracter and adder use these values to produce the full 33-bit products in the range −2.0 to +2.0 inclusive (indexed from 2 down to −30). However, only 20 bits of the products are stored in the next set of pipeline registers. The least-significant 13 bits are truncated. This still leaves two extra bits beyond the final precision required for the MAC outputs, in order to reduce the effect of rounding errors during accumulation of the sums. The accumulator adders use an extended range, −16.0 inclusive to +16.0 exclusive, so the pipelined products must be sign-extended by two bits before being added into the previously accumulated sums. The adders also produce overflow status outputs, which are used to set flipflops that record the overflow condition for a sequence of inputs. Finally, the accumulated sums are reduced to 16 bits at the MAC output. The least-significant two bits are truncated and bits 1 to 4 are discarded. The overflow logic must check that these discarded bits are all the same as the sign bit; otherwise the result is outside the range −1.0 to +1.0, and overflow has occurred.

FIGURE 10.4

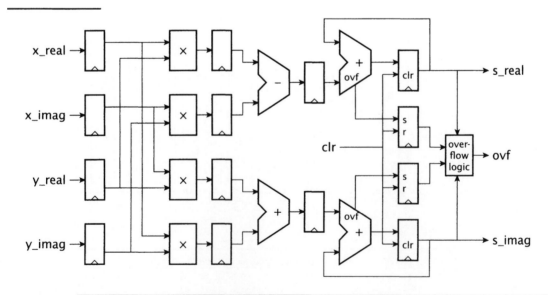

The register-transfer-level organization of the MAC.

FIGURE 10.5

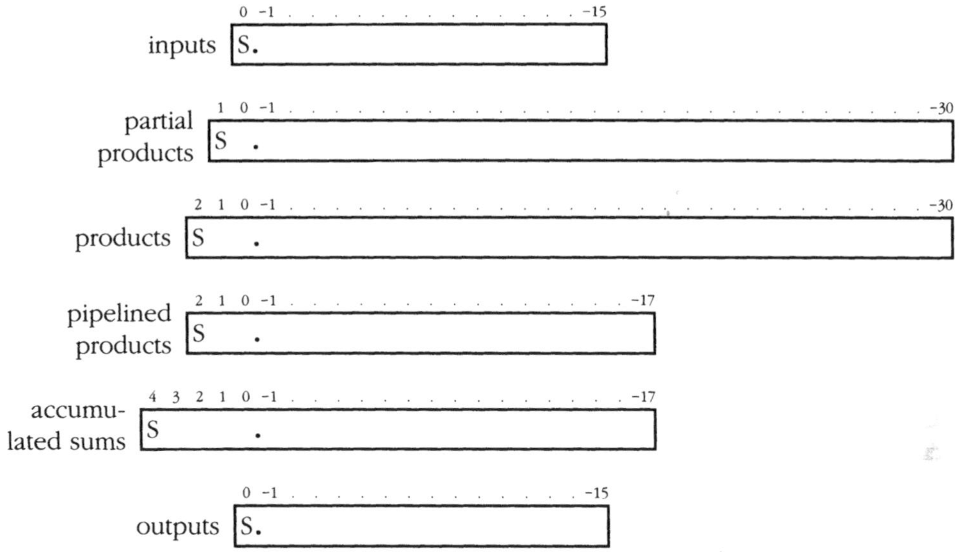

The format of fixed-point intermediate results within the MAC, showing the positions of the sign bits (S) and the binary points.

The register-transfer-level architecture body is

```
architecture rtl of mac is
    signal pipelined_x_real,
           pipelined_x_imag,
           pipelined_y_real,
           pipelined_y_imag : u_sfixed(0 downto -15);
    signal real_part_product_1,
           real_part_product_2,
           imag_part_product_1,
           imag_part_product_2 : u_sfixed(1 downto -30);
    signal pipelined_real_part_product_1,
           pipelined_real_part_product_2,
           pipelined_imag_part_product_1,
           pipelined_imag_part_product_2 : u_sfixed(1 downto -30);
    signal real_product,
           imag_product : u_sfixed(2 downto -30);
    signal pipelined_real_product,
           pipelined_imag_product : u_sfixed(2 downto -17);
    signal extended_real_product,
           extended_imag_product : u_sfixed(4 downto -17);
    signal real_sum,
```

```vhdl
          imag_sum : u_sfixed(4 downto -17);
  signal real_accumulator_ovf,
          imag_accumulator_ovf : std_ulogic;
  signal pipelined_real_sum,
          pipelined_imag_sum : u_sfixed(4 downto -17);
  signal pipelined_real_accumulator_ovf,
          pipelined_imag_accumulator_ovf : std_ulogic;
begin
  input_reg : process (clk) is
  begin
    if rising_edge(clk) then
      pipelined_x_real <= x_real;
      pipelined_x_imag <= x_imag;
      pipelined_y_real <= y_real;
      pipelined_y_imag <= y_imag;
    end if;
  end process input_reg;

  real_part_product_1 <= pipelined_x_real * pipelined_y_real;
  real_part_product_2 <= pipelined_x_imag * pipelined_y_imag;

  imag_part_product_1 <= pipelined_x_real * pipelined_y_imag;
  imag_part_product_2 <= pipelined_x_imag * pipelined_y_real;

  part_product_reg : process (clk) is
  begin
    if rising_edge(clk) then
      pipelined_real_part_product_1 <= real_part_product_1;
      pipelined_real_part_product_2 <= real_part_product_2;
      pipelined_imag_part_product_1 <= imag_part_product_1;
      pipelined_imag_part_product_2 <= imag_part_product_2;
    end if;
  end process part_product_reg;

  real_product <= pipelined_real_part_product_1
                  - pipelined_real_part_product_2;

  imag_product <= pipelined_imag_part_product_1
                  + pipelined_imag_part_product_2;

  product_reg : process (clk) is
  begin
    if rising_edge(clk) then
      pipelined_real_product
        <= resize(real_product, pipelined_real_product);
      pipelined_imag_product
        <= resize(imag_product, pipelined_imag_product);
    end if;
  end process product_reg;
```

```vhdl
    extended_real_product
      <= resize(pipelined_real_product, extended_real_product);

    extended_imag_product
      <= resize(pipelined_imag_product, extended_imag_product);

    real_sum <= extended_real_product + pipelined_real_sum;
    imag_sum <= extended_imag_product + pipelined_imag_sum;

    real_accumulator_ovf
      <= (     not extended_real_product(4)   -- non-negative
          and not pipelined_real_sum(4)       -- non-negative
          and     real_sum(4) )               -- appears negative
        or
          (     extended_real_product(4)      -- negative
          and     pipelined_real_sum(4)       -- negative
          and not real_sum(4) );              -- appears non-negative

    imag_accumulator_ovf
      <= (     not extended_imag_product(4)   -- non-negative
          and not pipelined_imag_sum(4)       -- non-negative
          and     imag_sum(4) )               -- appears negative
        or
          (     extended_imag_product(4)      -- negative
          and     pipelined_imag_sum(4)       -- negative
          and not imag_sum(4) );              -- appears non-negative

    accumulator_reg : process (clk) is
    begin
      if rising_edge(clk) then
        if reset then
          pipelined_real_sum <= (others => '0');
          pipelined_imag_sum <= (others => '0');
          pipelined_real_accumulator_ovf <= '0';
          pipelined_imag_accumulator_ovf <= '0';
        else
          pipelined_real_sum <= real_sum;
          pipelined_imag_sum <= imag_sum;
          pipelined_real_accumulator_ovf
            <= pipelined_real_accumulator_ovf or real_accumulator_ovf;
          pipelined_imag_accumulator_ovf
            <= pipelined_imag_accumulator_ovf or imag_accumulator_ovf;
        end if;
      end if;
    end process accumulator_reg;

    s_real <= resize(pipelined_real_sum, s_real);
    s_imag <= resize(pipelined_imag_sum, s_imag);
```

```
        ovf <= real_accumulator_ovf or imag_accumulator_ovf
            or pipelined_real_sum(4 downto 0) ?= "00000"
            or pipelined_real_sum(4 downto 0) ?= "11111"
            or pipelined_imag_sum(4 downto 0) ?= "00000"
            or pipelined_imag_sum(4 downto 0) ?= "11111";

end architecture rtl;
```

The signals declared in the architecture body represent the values calculated by each pipeline stage and the outputs of the pipeline registers. The **input_reg** process represents the four input pipeline registers connected to the real and imaginary data input ports. This process is followed by four assignments for the partial products according to the formulas given in Section 10.1. The partial products are then stored in the second set of pipeline registers, represented by the **part_product_reg** process. Next, the model includes assignments, one that subtracts partial products to form the real part of the product, and the other that adds partial products to form the imaginary part. The real and imaginary parts of the product are resized for storage in the third set of pipeline registers, represented by the **product_reg** process. Following that are assignments that add the two parts of the product, sign extended, to the previously accumulated sums. There are also assignments that determine whether the additions overflow. Overflow occurs if the two operands are non-negative and the result appears to be negative, or if the two operands are negative and the result appears to be non-negative. The assignments use the sign bits of the operands and the results to check for overflow. The outputs from the adders and the overflow signals are stored in the accumulator registers, represented by the **accumulator_reg** process. These registers differ from the other pipeline registers in that they can be cleared to zero by activating the **reset** input. Also, the stored overflow values, once set, remain set until subsequently reset. The outputs of the accumulator registers, reduced in size to 16 bits each, are used to drive the real and imaginary data outputs of the MAC. The output for the **ovf** port of the MAC is set if either accumulator has overflowed or if reduction of either the real or imaginary output discards significant bits.

10.3.1 Testing the Register-Transfer-Level Model

We could test the register-transfer-level model of the MAC using the same test bench that we used for the behavioral model. This would simply involve replacing the component instance **dut**, as follows:

```
    dut : entity work.mac(rtl)
        port map (clk, reset,
                    x_real, x_imag, y_real, y_imag, s_real, s_imag, ovf );
```

We could then simulate the test bench and manually compare the results with those produced by the behavioral model. However, a better approach is to modify the test bench to include instances of each of the behavioral and register-transfer-level models, as follows:

```vhdl
library ieee;
use ieee.std_logic_1164.all, ieee.fixed_pkg.all,
    ieee.math_complex.all;

architecture bench_verify of mac_test is

  signal clk, reset, behavioral_ovf, rtl_ovf : std_ulogic := '0';
  signal x_real, x_imag,
         y_real, y_imag,
         behavioral_s_real, behavioral_s_imag,
         rtl_s_real, rtl_s_imag : u_sfixed(0 downto -15);

  signal x, y, behavioral_s, rtl_s : complex := (0.0, 0.0);

  constant Tpw_clk : time := 50 ns;

begin

  x_real <= x.re; x_imag <= x.im;
  y_real <= y.re; y_imag <= y.im;

  dut_behavioral : entity work.mac(behavioral)
    port map ( clk, reset,
               x_real, x_imag, y_real, y_imag,
               behavioral_s_real, behavioral_s_imag,
               behavioral_ovf );

  dut_rtl : entity work.mac(rtl)
    port map ( clk, reset,
               x_real, x_imag, y_real, y_imag,
               rtl_s_real, rtl_s_imag, rtl_ovf );

  behavioral_s <= (behavioral_s_real, behavioral_s_imag);
  rtl_s <= (rtl_s_real, rtl_s_imag);

  clock_gen : process is
  begin
    clk <= '1' after Tpw_clk, '0' after 2 * Tpw_clk;
    wait for 2 * Tpw_clk;
  end process clock_gen;

  stimulus : process is
  begin
    -- first sequence
                                                reset <= '1';
    wait until not clk;
    x <= (+0.5, +0.5);  y <= (+0.5, +0.5);  reset <= '1';
    wait until not clk;
    x <= (+0.2, +0.2);  y <= (+0.2, +0.2);  reset <= '1';
    wait until not clk;
    x <= (+0.1, -0.1);  y <= (+0.1, +0.1);  reset <= '1';
    wait until not clk;
```

```vhdl
    x <= (+0.1, -0.1);  y <= (+0.1, +0.1);   reset <= '0';
    wait until not clk;

    -- should be (0.4, 0.58) when it falls out the other end

                                              reset <= '0';
    wait until not clk;
    x <= (+0.5, +0.5);  y <= (+0.5, +0.5);   reset <= '0';
    wait until not clk;
    x <= (+0.5, +0.5);  y <= (+0.1, +0.1);   reset <= '0';
    wait until not clk;
    x <= (+0.5, +0.5);  y <= (+0.5, +0.5);   reset <= '1';
    wait until not clk;
    x <= (-0.5, +0.5);  y <= (-0.5, +0.5);   reset <= '0';
    wait until not clk;

                                              reset <= '0';
    wait until not clk;

                                              reset <= '0';
    wait until not clk;

                                              reset <= '0';
    wait until not clk;

                                              reset <= '1';
    wait until not clk;

    wait;
  end process stimulus;

  verifier : process

    constant epsilon : real := 4.0E-5;   -- 1-bit error
                                         -- in 15-bit mantissa

  begin
    wait until not clk;
    assert behavioral_ovf = rtl_ovf
      report "Overflow flags differ" severity error;
    if not behavioral_ovf and not rtl_ovf then
      assert abs (behavioral_s.re - rtl_s.re) < epsilon
        report "Real sums differ" severity error;
      assert abs (behavioral_s.im - rtl_s.im) < epsilon
        report "Imag sums differ" severity error;
    end if;
  end process verifier;

end architecture bench_verify;
```

The revised test bench stimulates the two instances with the same input data and automatically compares the results they produce. The clock generator and stimulus processes are the same as those in the previous test bench. The additional process, **verifier**, is resumed midway through each clock cycle, after the outputs from each of the MAC devices has had time to stabilize. The process first verifies that the two devices produce the same overflow status outputs. Then, if both device outputs have not overflowed, the process compares the complex outputs of the two devices to verify that they are within a single bit of being equal. Note that since we are comparing real numbers, we must compare the values in this way. Since the behavioral model calculates its result using the **complex** type, which is implemented using the predefined floating-point type **real**, and the MAC calculates its result using fixed-point numbers with less precision than **real**, we should expect the actual results to differ slightly. This difference comes about due to the different round-off errors introduced by the different methods of calculation. If we were to use the equality operator ("=") to compare the results, the test would certainly fail.

Exercises

1. [❶ 10.2] In the behavioral model of the pipelined MAC, results for each stage are computed starting with the last stage and working forward. Show why it would be incorrect to work from the first stage to the last stage.

2. [❶ 10.2] Devise a sequence of input values for the MAC that cause the sum in the accumulator to overflow.

3. [❶ 10.2] Trace the values of the variables in the process **behavior** in the behavioral MAC model during each clock cycle for the first sequence generated by the test bench of Section 10.2.1.

4. [❶ 10.3] For each of the fixed-point formats shown in Figure 10.5, show how the values +0.5 and –0.5 are represented.

5. [❸ 10.1/10.2] Develop a behavioral model of a non-pipelined MAC, based on the dataflow diagram at the top of Figure 10.1. Adapt the test bench of Section 10.2.1 to test your model.

6. [❹] A polynomial function of degree n is

$$p(x) = a_0 + a_1 x + a_2 x^2 + \cdots + a_n x^n$$

where x is the input variable. We can rewrite the polynomial as

$$p(x) = a_0 + x(a_1 + x(a_2 + x(\ldots(a_{n-1} + x a_n)\ldots)))$$

We can evaluate a polynomial in the order implied by this form using n pipeline stages. The first stage evaluates $a_{n-1} + x a_n$ and passes the result and the value of x to the second stage. The second stage multiplies the previous result by x, adds a_{n-2} and passes this result and x to the third stage. The remaining stages continue in like manner.

Develop a behavioral model of a pipelined polynomial evaluator that evaluates polynomials of degree 3. The entity interface is

```
entity polynomial is
  port ( clk, mode, clr : in std_ulogic;
```

```
      coeff_addr : in unsigned(1 downto 0);
      x : in u_sfixed(5 downto -11);
      p : out u_sfixed(5 downto -11);
      ovf : out bit );
end entity polynomial;
```

The **mode** input is used to load coefficient values into internal registers. When **mode** is '1', the value on the **x** input is loaded into the register selected by the **coeff_addr** inputs. When **mode** is '0', the **x** input is used as the next data value for which the polynomial is to be calculated. The **clr** input is used to clear the internal pipeline registers. The **x** input and the **p** output both encode fixed-point values, with 5 pre-binary-point bits and 11 post-binary-point bits. Thus, values in the range −16.0 inclusive to +16.0 exclusive can be represented. The same range is used in the internal pipeline registers. If a calculation in any stage overflows, an overflow flag is propagated through the remainder of the pipeline and presented at the **ovf** output.

Test your model by using it to evaluate the following polynomial functions:

$$e^x \approx 1 + x + \frac{x^2}{2} + \frac{x^3}{6}$$

$$\cos x \approx 1 - \frac{x^2}{2}$$

Refine your model to the register-transfer level, defining entities for the required multipliers, adders, registers and logic blocks.

7. [❹] Develop a behavioral model for a two-stage pipelined floating-point multiplier. The two operands and the result are in IEEE single-precision format. The first stage of the pipeline multiplies the mantissas, subtracts the biases from the exponents and adds the unbiased exponents. The second stage normalizes the result, adds the bias to the result exponent, and determines the sign based on the operand signs. Write a test bench to test your model, comparing its results with those calculated using VHDL's predefined floating-point multiplication operation. Next, refine your pipelined multiplier to the register-transfer level, defining entities for the required multipliers, adders, registers and logic blocks. Do not worry about infinities, NaNs, denormals or rounding according to the IEEE standard. You may wish to add provisions for these aspects of IEEE floating-point arithmetic as an extension to the exercise.

Chapter 11

Aliases

Since the main purpose of a model written in VHDL is to describe a hardware design, it should be made as easy as possible to read and understand. In this chapter, we introduce *aliases* as a means of making a model clearer. As in everyday use, an alias is simply an alternate name for something. We see how we can use aliases in VHDL for both data objects and other kinds of items that do not represent data in a model.

11.1 Aliases for Data Objects

If we have a model that includes a data object, such as a constant, a variable, a signal or, as we see in a later chapter, a file, we can declare an alias for the object with an *alias declaration*. A simplified syntax rule for this is

> alias_declaration ⇐ **alias** identifier **is** name ;

An alias declaration in this form simply defines an alternate identifier to refer to the named data object. We can refer to the object using the new identifier, treating it as being of the type specified in the original object's declaration. Operations we perform using the alias are actually applied to the original object. (The only exceptions are reading the 'simple_name, 'path_name and 'instance_name attributes and the attributes that provide information about the index ranges of an array. In these cases, the attributes refer to the alias name rather than the original object's name.)

EXAMPLE 11.1   *Aliases for package items*

One use of alias declarations is to define simple names for objects imported from packages. Suppose, for example, that we need to use objects from two different packages, **alu_types** and **io_types**, and that each declares a constant named **data_width**, possibly with different values. If we include use clauses for these packages in our model, as follows:

```
use work.alu_types.all, work.io_types.all;
```

neither of the versions of **data_width** becomes directly visible, since they have the same name. Hence we would have to refer to them as **work.alu_types.data_width** and **work.io_types.data_width**. However, we can avoid this long notation simply by introducing two alias declarations into our model, as shown in the following architecture body:

```
library ieee;  use ieee.std_logic_1164.all;
use work.alu_types.all, work.io_types.all;

architecture structural of controller_system is

  alias alu_data_width is work.alu_types.data_width;
  alias io_data_width is work.io_types.data_width;

  signal alu_in1, alu_in2,
         alu_result :
             std_ulogic_vector(0 to alu_data_width - 1);
  signal io_data : std_ulogic_vector(0 to io_data_width - 1);
  ...

begin

  ...

end architecture structural;
```

As well as denoting a whole data object, an alias can denote a single element from a composite data object, such as a record or an array. We write the element name, including a record element selector or an array index, as the name to be aliased. For example, given the following declarations of types and a variable:

```
type register_array is array (0 to 15) of bit_vector(31 downto 0);

type register_set is record
    general_purpose_registers : register_array;
    program_counter : bit_vector(31 downto 0);
    program_status : bit_vector(31 downto 0);
  end record;

variable CPU_registers : register_set;
```

we can declare aliases for the record elements:

```
alias PSW is CPU_registers.program_status;
alias PC is CPU_registers.program_counter;
alias GPR is CPU_registers.general_purpose_registers;
```

We can also declare aliases for individual registers in the register array, for example:

```
alias SP is CPU_registers.general_purpose_registers(15);
```

The name that we are aliasing can itself be an alias. Hence the alias declaration for SP can be written using the alias name GPR:

alias SP **is** GPR(15);

An alias can also be used to denote a slice of a one-dimensional array. For example, given the above declaration for CPU_registers, we can declare an alias for part of the program status register:

alias interrupt_level **is** PSW(30 **downto** 26);

This declares **interrupt_level** to denote a bit vector, with indices from 30 down to 26, being part of the bit vector denoted by PSW. In general, if we declare an alias for an array slice in this way, the alias denotes an array with index range and direction determined by the slice.

In many cases, it would be convenient to use an alias to take a slightly different view of the array being aliased. For example, we would like to view the interrupt_level alias as a bit vector indexed from four down to zero. We can do this by using an extended form of alias declaration, described by the following syntax rule:

alias_declaration ⇐
    alias identifier ⟦ : subtype_indication⟧ **is** name ;

This shows that we can indicate the subtype for the alias. The subtype determines how we view the original object that the alias denotes. We can include a subtype indication in aliases for scalar objects, but the bounds and direction specified must be the same as those of the original object. Hence this only serves as a form of documentation to restate the type information for the object. We can also include an unconstrained or partially constrained composite subtype as the alias subtype when aliasing an array or record object or an array slice. In this case, any index ranges that are unspecified in the subtype come from the corresponding index ranges of the original object. However, we can use the subtype indication to specify different index bounds and direction from the original object. The base type of the subtype indication must be the same as the base type of the original object. (This means that the subtype indication must refer to a composite type with the same element and index types as the original object.) Furthermore, there must be the same number of elements in the alias subtype and the original object. Elements in the alias denote the corresponding elements in the actual object, with array elements corresponding in left-to-right order. For example, if we were to declare the alias interrupt_level as follows:

alias interrupt_level : bit_vector(4 **downto** 0) **is** PSW(30 **downto** 26);

then interrupt_level(4) would denote PSW(30), interrupt_level(3) would denote PSW(29), and so on.

As another example, suppose we declare a register file as follows:

type register_array **is array** (natural **range** <>) **of** bit_vector;
signal register_file : register_array(0 **to** 15)(31 **downto** 0);

We can then declare an alias:

```
alias bigendian_register_file : register_array(open)(0 to 31) is
      register_file;
```

This alias views the register file as an array with the same top-level index range as the original, 0 to 15, since the subtype indication does not specify a top-level index range. Each element, however, is viewed with the index range 0 to 31 specified in the subtype indication.

EXAMPLE 11.2   *Normalization of array index ranges using aliases*

When we write subprograms that take parameters of unconstrained array types, the index bounds and direction of the parameter are not known until actual array objects are passed as arguments during a call. Without this knowledge, the body of the subprogram is difficult to write. For example, suppose we need to implement a function to perform addition on two bit vectors that represent two's-complement, signed integers. The function specification is

```
function "+" ( bv1, bv2 : bit_vector ) return bit_vector;
```

When the function is called it is possible that the first argument is indexed from 0 to 15, while the other argument is indexed from 31 down to 8. We must check that the arguments are of the same size and then index them in a loop running from the rightmost bit to the leftmost. The different ranges, directions and sizes make this difficult.

We can use aliases to make the task easier by viewing the objects as arrays with the same leftmost index and direction. The subprogram body is shown below. The alias declarations create views of the bit-vector arguments, indexed from one up to their length. The function, after checking that the arguments are of the same length, can then use the same index values for corresponding elements of the two arguments and the result.

```
function "+" ( bv1, bv2 : bit_vector ) return bit_vector is

  alias norm1 : bit_vector(1 to bv1'length) is bv1;
  alias norm2 : bit_vector(1 to bv2'length) is bv2;

  variable result : bit_vector(1 to bv1'length);
  variable carry : bit := '0';

begin
  if bv1'length /= bv2'length then
    report "arguments of different length" severity failure;
  else
    for index in norm1'reverse_range loop
      result(index) := norm1(index) xor norm2(index) xor carry;
      carry := ( norm1(index) and norm2(index) )
               or ( carry and ( norm1(index) or norm2(index) ) );
    end loop;
  end if;
```

```
    return result;
  end function "+";
```

EXAMPLE 11.3 *Normalization of element array index ranges*

For subprograms that deal with arrays of arrays, both the top-level and lower-level index ranges may be undefined. Again, we can use aliases to view parameters with consistent index bounds and directions. However, we need to use the 'element attribute to refer to the actual index ranges of the parameter elements. For example, we can write a function that locates the first bit difference between two arrays of bit vectors as follows:

```
type bv_vector is array (natural range <>) of bit_vector;

function find_first_difference ( s1, s2 : in bv_vector)
                                  return natural is
  alias s1_norm : bv_vector(0 to s1'length - 1)
                           (0 to s1'element'length - 1) is s1;
  alias s2_norm : bv_vector(0 to s2'length - 1)
                           (0 to s2'element'length - 1) is s2;
  variable count : natural := 0;
begin
  assert s1'length = s2'length and
         s1'element'length = s2'element'length;
  for i in s1_norm'range loop
    for j in s1_norm'element'range loop
      exit when s1_norm(i)(j) /= s2_norm(i)(j);
      count := count + 1;
    end loop;
  end loop;
  return count;
end function find_first_difference;
```

The two parameters are of an unconstrained type, allowing the function to operate on arrays of various lengths and on arrays with various bit-vector element lengths. The function only requires that, on each call, the two actual parameters have the same shape. In order to deal with the differences, the function declares aliases for the parameters. It views each parameter with an index range starting at 0 and ascending to one less than the length. It views the elements similarly, with an index range starting at 0 and ascending to one less than length of each bit-vector element. The alias declaration uses the 'element attribute to get the constrained subtype for the actual parameter's elements. Within the function body, the inner for loop also uses the 'element attribute to get the index range for the elements of the aliases.

VHDL-87, -93, and -2002

These earlier versions of VHDL did not allow aliases of multidimensional array objects. For example, the following is illegal in earlier versions:

```
type bit_matrix is
  array (natural range <>, natural range <>) of bit;
signal s : bit_matrix(15 downto 0, 15 downto 0);
alias bigendian_s : bit_matrix(0 to 15, 0 to 15) is s;
```

VHDL-87

An alias declaration in VHDL-87 must include a subtype indication.

11.2 Aliases for Non-Data Items

We saw in the previous section that we can declare aliases for data objects such as constants, variables and signals. We can also declare aliases for other named items that do not represent stored data, such as types, subprograms, packages, entities and so on. In fact, the only kinds of items for which we cannot declare aliases are labels, loop parameters and generate parameters (see Chapter 14). The syntax rule for alias declarations for non-data items is

alias_declaration ⇐
    alias 〖 identifier ▯ character_literal ▯ operator_symbol 〗
        is name 〖 signature 〗 ;

We can use character literals as aliases for enumeration literals, and operator symbols as aliases for function subprograms. We will return to the optional signature part shortly.

If we define an alias for a type, we can use the alias in any context where the original type name can be used. Furthermore, all of the predefined operations for values of the original type can be used without being declared. For example, if we define an alias:

```
alias binary_string is bit_vector;
```

we can declare objects to be of type **binary_string** and perform bit-vector operations on them; for example:

```
variable s1, s2 : binary_string(0 to 7);
...

s1 := s1 and not s2;
```

Declaring an alias for a type is different from declaring a new type. In the latter case, new overloaded versions of the operators would have to be declared. The alias, on the other hand, is simply another name for the existing type.

If we define an alias for an enumeration type, all of the enumeration literals of the original type are available for use. We do not need to define aliases for the literals, nor use fully selected names. For example, if a package **system_types** declares an enumeration type as follows:

type system_status **is** (idle, active, overloaded);

and a model defines an alias for this type:

alias status_type **is** work.system_types.system_status;

the model can simply refer to the literals **idle**, **active** and **overloaded**, instead of **work.system_types.overloaded** and so on. Similarly, if we declare an alias for a physical type, all of the unit names are available for use without aliasing or selection.

The optional signature part in an alias declaration is only used in aliases for subprograms and enumeration literals. These items can be overloaded, so it is possible that the name alone is not sufficient to identify which item is being aliased. The signature serves to identify one item uniquely. The syntax rule for a signature is

signature ⇐ [⟦ type_mark ⟨ , ... ⟩ ⟧ ⟦ **return** type_mark ⟧]

Note that the outer square bracket symbols ([...]) are a required part of the signature, whereas the hollow square brackets (⟦ ... ⟧) are part of the EBNF syntax and indicate optional parts of the signature.

When we declare an alias for a subprogram, the signature identifies which overloaded version of the subprogram name is aliased. The signature lists the types of each of the subprogram's parameters, in the same order that they appear in the subprogram's declaration. For example, if a package **arithmetic_ops** declares two procedures as follows:

procedure increment (bv : **inout** bit_vector; by : **in** integer := 1);

procedure increment (int : **inout** integer; by : **in** integer := 1);

we can declare aliases for the procedures as follows:

alias bv_increment **is**
  work.arithmetic_ops.increment [bit_vector, integer];

alias int_increment **is**
  work.arithmetic_ops.increment [integer, integer];

If the subprogram is a function, the signature also includes the type of the return value, after the keyword **return**. For example, we might alias the operator symbols "*", "+" and "−" to the bit operators **and**, **or** and **not**, as follows:

alias "*" **is** "and" [bit, bit **return** bit];

alias "+" **is** "or" [bit, bit **return** bit];

alias "-" **is** "not" [bit **return** bit];

We would then be able to express Boolean equations using these operators. For example, given bit signals s, a, b and c, we could write

```
s <= a * b + (-a) * c;
```

Note that when we alias an operator symbol to a function, the function overloads the operator symbol, so it must have the correct number of parameters for the operator. A binary operator symbol must be aliased to a function with two parameters, and a unary operator symbol must be aliased to a function with one parameter.

If we wish to alias an individual literal of an enumeration type, we must deal with the possibility that the literal may belong to several different enumeration types. We can use a signature to distinguish one particular meaning by noting that an enumeration literal is equivalent to a function with no parameters that returns a value of the enumeration type. For example, when we write the enumeration literal '1', we can think of this as a call to a function with no parameters, returning a value of type **bit**. We can write an alias for this literal as follows:

```
alias high is std.standard.'1' [ return bit ];
```

The signature distinguishes the literal as being of type **bit**, rather than of any other character type. Note that a selected name is required for a character literal, since a character literal by itself is not a syntactically valid name.

EXAMPLE 11.4   *Aliases for composing packages*

One useful application of aliases for non-data items is to compose a package by collecting together a number of items declared in other packages. The following package for use in a DMA controller design defines aliases for two types imported from the **cpu_types** package and for a function imported from a package that provides bit-vector arithmetic operations.

```
package DMA_controller_types_and_utilities is

  alias word is work.cpu_types.word;
  alias status_value is work.cpu_types.status_value;

  alias "+" is work.bit_vector_unsigned_arithmetic."+"
            [ bit_vector, bit_vector return bit_vector ];

  ...

end package DMA_controller_types_and_utilities;
```

The DMA controller architecture body outlined below imports the aliases from the utility package. The reference to the name **word** denotes the type originally defined in the package **cpu_types**, and the operator "+" denotes the bit-vector operator originally defined in the package **bit_vector_unsigned_arithmetic**.

```
architecture behavioral of DMA_controller is

  use work.DMA_controller_types_and_utilities.all;
```

```
    begin

        behavior : process is

            variable address_reg0, address_reg1 : word;
            variable count_reg0, count_reg1 : word;
            ...

        begin
            ...
            address_reg0 := address_reg0 + X"0000_0004";
            ...
        end process behavior;

    end architecture behavioral;
```

VHDL-87

VHDL-87 does not allow aliases for non-data items. Aliases may only be declared for data objects.

Exercises

1. [❶ 11.1] Given the following declarations:

   ```
   subtype byte is bit_vector(0 to 7);
   type data_array is array (0 to 31) of byte;
   type network_packet is record
       source, dest, flags : byte;
       payload : data_array;
       checksum : byte;
     end record network_packet;
   variable received_packet : network_packet;
   ```

   write alias declarations for the individual elements of the variable.

2. [❶ 11.1] The layout of information within the **flags** element of a network packet described in Exercise 1 is shown in Figure 11.1. Write alias declarations for the individual fields of the **flags** element of the **received_packet** variable. The aliases for the ACKNO and SEQNO fields should view the fields as bit vectors indexed from two down to zero.

364   *Chapter 11 — Aliases*

FIGURE 11.1

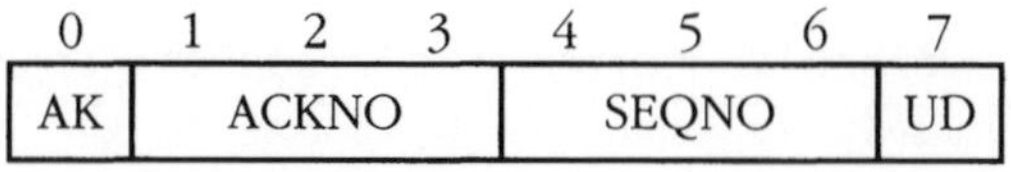

The layout of information within a network packet.

3. [❶ 11.2] Write an alias declaration that defines the name **cons** as an alias for the pre-defined operation "**&**" with a character left argument, a string right argument and a string result. Use the alias in a report statement that reports the string constructed from the value of the variable **grade_char** concatenated to the string "–grade".

4. [❷ 11.1] Develop a behavioral model of a bit-reversing module with the following entity interface:

```
entity reverser is
  port ( d_in : in std_ulogic_vector;
         d_out : out std_ulogic_vector );
end entity reverser;
```

When the entity is instantiated, the actual signals must be of the same length, but may differ in their index bounds and directions. The output is the input delayed by 500 ps using transport delay, and with the bits in reverse order from left to right.

Chapter 12

Generics

The models that we have used as examples in preceding chapters all have fixed behavior and structure. In many respects, this is a limitation, and we would like to be able to write more general, or *generic*, models. VHDL provides us with a mechanism, called generics, for writing parameterized models. We discuss generics in this chapter and show how they may be used to write families of models with varying behavior and structure.

12.1 Generic Constants

The simplest form of generic model is parameterized by one or more *generic constants*. These are constants that cannot be changed within a design unit, but whose values are set when the design unit is initialized. We can write an entity with generic constants by including a *generic interface list* in its declaration, defining the *formal generic constants* that parameterize the entity. The extended syntax rule for entity declarations including generics is

```
entity_declaration ⇐
    entity identifier is
        [ generic ( generic_interface_list ) ; ]
        [ port ( port_interface_list ) ; ]
        { entity_declarative_item }
    [ begin
        { concurrent_assertion_statement
        | passive_concurrent_procedure_call_statement
        | passive_process_statement } ]
    end [ entity ] [ identifier ] ;

interface_list ⇐ interface_declaration { ; … }
```

The difference between this and the simpler rule we saw before is the inclusion of the optional generic interface list before the port interface list. The generic interface list is like any other interface list, but with restrictions on the kinds of interface items that we can include. For now, we will just consider constant-class objects, which must be of mode

in. Since these are the defaults for a generic interface list, we can use a simplified syntax rule for the interface declarations:

interface_declaration ⇐
    identifier ⦃ , ... ⦄ : subtype_indication ⟦ := expression ⟧

A simple example of an entity declaration including a generic interface list is

```
entity and2 is
  generic ( Tpd : time );
  port ( a, b : in bit;  y : out bit );
end entity and2;
```

This entity includes one generic constant, **Tpd**, of the predefined type **time**. The value of this generic constant may be used within the entity statements and any architecture body corresponding to the entity. In this example the intention is that the generic constant specify the propagation delay for the module, so the value should be used in a signal assignment statement as the delay. An architecture body that does this is

```
architecture simple of and2 is
begin

  and2_function :
    y <= a and b after Tpd;

end architecture simple;
```

The visibility of a generic constant extends from the end of the generic interface list to the end of the entity declaration and extends into any architecture body corresponding to the entity declaration.

A generic constant is given an actual value when the entity is used in a component instantiation statement. We do this by including a *generic map*, as shown by the extended syntax rule for component instantiations:

component_instantiation_statement ⇐
    *instantiation*_label :
            entity *entity*_name ⟦ (*architecture*_identifier) ⟧
            ⟦ **generic map** (*generic*_association_list) ⟧
            ⟦ **port map** (*port*_association_list) ⟧ ;

association_list ⇐ assiciation_element ⦃ , ... ⦄

The generic association list is like other forms of association lists, but since generic constants are always of class constant, the actual arguments we supply must be expressions. Thus the simplified syntax rule for generic associations limited to just generic constants is

association_element ⇐
    ⟦ *generic*_name => ⟧ (expression ⎮ **open**)

To illustrate this, let us look at a component instantiation statement that uses the **and2** entity shown above:

```
gate1 : entity work.and2(simple)
  generic map ( Tpd => 2 ns )
  port map ( a => sig1,  b => sig2,  y => sig_out );
```

The generic map specifies that this instance of the **and2** module uses the value 2 ns for the generic constant **Tpd**; that is, the instance has a propagation delay of 2 ns. We might include another component instantiation statement using **and2** in the same design but with a different actual value for **Tpd** in its generic map, for example:

```
gate2 : entity work.and2(simple)
  generic map ( Tpd => 3 ns )
  port map ( a => a1,  b => b1,  y => sig1 );
```

When the design is elaborated we have two processes, one corresponding to the instance **gate1** of **and2**, which uses the value 2 ns for **Tpd**, and another corresponding to the instance **gate2** of **and2**, which uses the value 3 ns.

EXAMPLE 12.1   *Generic constants in an entity for a control logic module*

As the syntax rule for the generic interface list shows, we may define a number of generic constants of different types and include default values for them. An entity declaration for a control logic module illustrates the possibilities:

```
entity control_unit is

  generic ( Tpd_clk_out, Tpw_clk : delay_length;
            debug : boolean := false );

  port ( clk : in bit;
         ready : in bit;
         control1, control2 : out bit );

end entity control_unit;
```

In this example, the generic interface list includes a list of two generic constants that parameterize the propagation delay of the module and a Boolean generic constant, **debug**, with a default value of false. The intention of this last generic constant is to allow a design that instantiates this entity to activate some debugging operation. This operation might take the form of report statements within if statements that test the value of **debug**.

We have the same flexibility in writing a generic map as we have in other association lists. We can use positional association, named association or a combination of both. We can omit actual values for generic constants that have default expressions, or we may explicitly use the default value by writing the keyword **open** in the generic map. To illustrate these possibilities, here are three different ways of writing a generic map for the control_unit entity:

generic map (200 ps, 1500 ps, false)

generic map (Tpd_clk_out => 200 ps, Tpw_clk => 1500 ps)

generic map (200 ps, 1500 ps, debug => **open**)

EXAMPLE 12.2   *Generic constants for specifying timing characteristics*

The entity declaration below for a D-flipflop includes generic constants: Tpd_clk_q to specify the propagation delay from clock rising edge to output, Tsu_d_clk to specify the setup time of data before a clock edge and Th_d_clk to specify the hold time of data after a clock edge.

```
entity D_flipflop is
  generic ( Tpd_clk_q, Tsu_d_clk, Th_d_clk : delay_length );
  port ( clk, d : in bit;  q : out bit );
end entity D_flipflop;
```

The values of these generic constants are used in the architecture body:

```
architecture basic of D_flipflop is
begin

  behavior : q <= d after Tpd_clk_q when rising_edge(clk);

  check_setup : process is
  begin
    wait until clk;
    assert d'last_event >= Tsu_d_clk
      report "setup violation";
  end process check_setup;

  check_hold : process is
  begin
    wait until clk'delayed(Th_d_clk);
    assert d'delayed'last_event >= Th_d_clk
      report "hold violation";
  end process check_hold;

end architecture basic;
```

The entity might be instantiated as follows, with actual values specified in the generic map for the generic constants:

```
request_flipflop : entity work.D_flipflop(basic)
  generic map ( Tpd_clk_q => 4 ns,
                Tsu_d_clk => 3 ns, Th_d_clk => 1 ns )
```

```
    port map ( clk => system_clock,
               d => request, q => request_pending );
```

The second main use of generic constants in entities is to parameterize their structure.
We can use the value of a generic constant to specify the size of an array port. To see why
this is useful, let us look at an entity declaration for a register. A register entity that uses
an unconstrained array type for its input and output ports can be declared as

```
entity reg is
  port ( d : in bit_vector;  q : out bit_vector;  ... );
end entity reg;
```

While this is a perfectly legal entity declaration, it does not include the constraint that the
input and output ports **d** and **q** should be of the same size. Thus, we could write a component instantiation as follows:

```
signal small_data : bit_vector(0 to 7);
signal large_data : bit_vector(0 to 15);
...

problem_reg : entity work.reg
  port map ( d => small_data,  q => large_data, ... );
```

The model is analyzed and elaborated without the error being detected. It is only
when the register tries to assign a small bit vector to a target bit vector of a larger size that
the error is detected. We can avoid this problem by including a generic constant in the
entity declaration to parameterize the size of the ports. We use the generic constant in constraints in the port declarations. To illustrate, here is the register entity declaration rewritten:

```
entity reg is
  generic ( width : positive );
  port ( d : in bit_vector(0 to width - 1);
         q : out bit_vector(0 to width - 1);
         ... );
end entity reg;
```

In this declaration we require that the user of the register specify the desired port
width for each instance. The entity then uses the width value as a constraint on both the
input and output ports, rather than allowing their size to be determined by the signals associated with the ports. A component instantiation using this entity might appear as follows:

```
signal in_data, out_data : bit_vector(0 to bus_size - 1);
...

ok_reg : entity work.reg
  generic map ( width => bus_size )
  port map ( d => in_data,  q => out_data, ... );
```

If the signals used as actual ports in the instantiation were of different sizes, the analyzer would signal the error early in the design process, making it easier to correct. As a matter of style, whenever the sizes of different array ports of an entity are related, generic constants should be considered to enforce the constraint.

EXAMPLE 12.3   *A register model with a generic constant for the port widths*

A complete model for the register, including the entity declaration and an architecture body, is

```
entity reg is
  generic ( width : positive );
  port ( d :   in  bit_vector(0 to width - 1);
         q :   out  bit_vector(0 to width - 1);
         clk, reset : in bit );
end entity reg;

-------------------------------------------------

architecture behavioral of reg is
begin

  behavior : process (clk, reset) is
    constant zero : bit_vector(0 to width - 1) := (others => '0');
  begin
    if reset then
      q <= zero;
    elsif rising_edge(clk) then
      q <= d;
    end if;
  end process behavior;

end architecture behavioral;
```

The generic constant is used to constrain the widths of the data input and output ports in the entity declaration. It is also used in the architecture body to determine the size of the constant bit vector **zero**. This bit vector is the value assigned to the register output when it is reset, so it must be of the same size as the register port. We can create instances of the register entity in a design, each possibly having different-sized ports. For example:

```
word_reg : entity work.reg(behavioral)
  generic map ( width => 32 )
  port map ( ... );
```

creates an instance with 32-bit-wide ports. In the same design, we might include another instance, as follows:

```
subtype state_vector is bit_vector(1 to 5);
...
```

```
state_reg : entity work.reg(behavioral)
  generic map ( width => state_vector'length )
  port map ( ... );
```

This register instance has 5-bit-wide ports, wide enough to store values of the subtype state_vector.

EXAMPLE 12.4   *An adder with generic constants for propagation delays*

Suppose we want to use generic constants to specify the propagation delays for an adder. The entity is declared with input and output ports that are arrays whose sizes are determined by a generic constant. We want to specify individual propagation delays for corresponding input and output port elements. The entity declaration is:

```
entity adder is
  generic ( width    : positive;
            Tpd_ab_s : time_vector(width - 1 downto 0) );
  port    ( a, b  : in  bit_vector(width - 1 downto 0);
            c_in  : in  bit;
            s     : out bit_vector(width - 1 downto 0);
            c_out : out bit );
end entity adder;
```

The generic constant width is used in the declaration of the second generic constant, Tpd_ab_s, to ensure that there is a matching propagation delay for each element of the input and output ports. We can instantiate the entity in a design as follows:

```
subtype byte is bit_vector(7 downto 0);
signal op1, op2, result : byte;
signal c_out : bit;
...

byte_adder : entity work.adder
  generic map ( width    => byte'length,
                Tpd_ab_s => (7 downto 1 => 120 ps,
                             0          => 80 ps) )
  port map    ( a => op1, b => op2, c_in => '0',
                s => result, c_out => c_out );
```

In this instance, an actual value is given for width, and that is used to determine the index range for Tpd_ab_s, as well as for the ports. Note that we don't have to write the actual generics in this order. The values are determined for generics in the order of their occurrence in the generic list, not the generic map. Thus, we could have written the generic map as:

```
generic map ( Tpd_ab_s => (7 downto 1 => 120 ps,
                           0          => 80 ps),
              width    => byte'length )
```

though to do so might look a bit strange.

VHDL-87, -93, and -2002

These versions of VHDL do not allow the value of one generic constant to be used in the declarations of other generic constants in the same generic list. In the preceding example, we would have to declare the **Tpd_ab_s** generic as an unconstrained array and include an assertion in the entity statement part that the length of **Tpd_ab_s** is equal to **width**.

12.2 Generic Types

Generic types allow us to define a type that can be used for ports and internal declarations of an entity, but without specifying the particular type. When we instantiate the entity, we specify the actual type to be used for that instance. As we will see later, generic types can also be specified for packages and subprograms, not just for entities and components.

The syntax rule for declaring a *formal generic type* in a generic list is

interface_declaration ⇐ **type** identifier

The identifier is the name of the formal generic type, and can be used within the rest of the entity in the same way as a normally declared type. When we instantiate the entity, we specify a subtype as the *actual generic type*. The syntax rule for this is

association_element ⇐
    ⟦ *generic*_name => ⟧ subtype_indication

As we have seen before, a subtype indication can take the form of a type or subtype name, a type or subtype name followed by a constraint, or a subtype attribute.

EXAMPLE 12.5   *A generic multiplexer*

A multiplexer selects between two data inputs and copies the value of the selected input to the output. The behavior of the multiplexer is independent of the type of data on the inputs and output. So we can use a formal generic type to represent the type of the data. The entity declaration is

```
entity generic_mux2 is
  generic ( type data_type );
  port    ( sel : in bit; a, b : in data_type;
            z : out data_type );
end entity generic_mux2;
```

The name **data_type** is the formal generic type that stands for some type, as yet unspecified, used for the data inputs **a** and **b** and for the data output **z**. An architecture body for the multiplexer is:

```
architecture rtl of mux2 is
begin
  z <= a when not sel else b;
end architecture rtl;
```

The assignment statement simply copies the value of either **a** or **b** to the output **z**. It is sensitive to all of the inputs. So whenever **a**, **b**, or **sel** change, the assignment will be re-evaluated. In any instance of the multiplexer, changes on **a** and **b** are determined using the predefined equality operator for the actual type in that instance.

We can instantiate the entity to get a multiplexer for **bit** signals as follows:

```
signal sel_bit, a_bit, b_bit, z_bit : bit;
...

bit_mux : entity work.generic_mux2(rtl)
  generic map ( data_type => bit )
  port map    ( sel => sel_bit, a => a_bit, b => b_bit,
                z => z_bit );
```

Similarly, we can instantiate the same entity to get a multiplexer for signals of other types, including user-defined types.

```
type msg_packet is record
  src, dst : unsigned(7 downto 0);
  pkt_type : bit_vector(2 downto 0);
  length   : unsigned(4 downto 0);
  payload  : byte_vector(0 to 31);
  checksum : unsigned(7 downto 0);
end record msg_packet;
signal pkt_sel : bit;
signal pkt_in1, pkt_in2, pkt_out : msg_pkt;
...

pkt_mux : entity work.generic_mux2(rtl)
  generic map ( data_type => msg_packet )
  port map    ( sel => pkt_sel,
                a => pkt_in1, b => pkt_in2, z => pkt_out );
```

VHDL defines a number of rules covering formal generic types and the ways they can be used. The formal generic type name can potentially represent any type, except a file type or a protected type (both of which we describe in later chapters). The entity can only assume that operations available for all such types are applicable. That includes assignment, equality and inequality operations, type qualification, and type conversion. For signals declared to be of a formal generic type, the predefined equality operator of the actual type is used for driver update and event detection.

If we have a formal generic type T, we can use it to declare signals, variables, and constants within the entity and architecture, and we can write signal and variable assignments for objects of the type. For example, the following shows signals declared using T:

```
signal s1, s2 : T;
...

s1 <= s2 after 10 ns;
```

and the following shows variables declared using T:

```
variable v1, v2, temp : T;
...

temp := v1;   v1 := v2;   v2 := temp;
```

Since signal and variable declarations require fully constrained subtypes, the actual type provided in an instance must be a fully constrained type if the formal type is used in this way. If the actual type is not fully constrained, an error occurs in the instantiation. If the formal generic type is not used in any way requiring it to be fully constrained, then the actual type in an instance need not be fully constrained.

For both variables and signals, the default initial value is determined using the actual type in an instance, using the normal rules for the actual type. Thus, if the actual type is a scalar type, the default initial value is the leftmost value of the type, and if the actual type is a composite type, the default initial value is an aggregate of the default initial values for the respective element types.

Declaring constants of a formal generic type might at first seem impossible, since we can't specify an initial value if we don't know the actual type. However, we can use the formal generic type to declare a formal generic constant, and then use that within the entity, for example:

```
entity e is
  generic ( type T; constant init_val : T );
  port    ( ... );
end entity e;

architecture a of e is
begin
  p : process is
    variable v : T := init_val;
  begin
    ...
  end process p;
end architecture a;
```

The actual value for the generic constant is provided when the entity is instantiated, and must be of the type specified as the actual generic type. For example, we might instantiate the entity **e** within a larger design as follows:

```
my_e : entity work.e(a)
  generic map ( T => std_ulogic_vector(3 downto 0),
                init_val => "ZZZZ" );
```

We can also use this technique to provide values for initializing variables and signals declared to be of the formal generic type.

One thing that we cannot do with formal generic types is apply operations that are not defined for all types. For example, we cannot use the "+" operator to add to values of a formal generic type, since the actual type in an instance may not be a numeric type. Similarly, we cannot perform array indexing, or apply most attributes. This may at first seem an onerous restriction, but it does mean that a VHDL analyzer can check the entity and architecture for correctness in isolation, independently of any particular instantiation. It also means we don't get any surprises when we subsequently analyze an instance of the entity. Fortunately, as we will see in Section 12.5, there are ways of providing operations to an instance for use on values of the actual type.

EXAMPLE 12.6 *Illegal use of formal generic types*

Suppose we want to define a generic counter that can be used to count values of types such as integer, unsigned, signed, and so on. We can declare the entity as follows:

```
entity generic_counter is
  generic ( type     count_type;
            constant reset_value : count_type );
  port    ( clk, reset : in  bit;
            data        : out count_type );
end entity generic_counter;
```

We might then try to define an architecture as:

```
architecture rtl of generic_counter is
begin
  count : process (clk) is
  begin
    if rising_edge(clk) then
      if reset then
        data <= reset_value;
      else
        data <= data + 1;   -- Illegal
      end if;
    end if;
  end process count;
end architecture rtl;
```

The problem is that the "+" operator for adding 1 to a value is not defined for all types that might be supplied as actual types. Hence, the analyzer will indicate an error in the expression where the operator is applied. To illustrate why this should be an error, suppose some time after the entity and architecture have been written, we try to instantiate them in a design as follows:

```
type traffic_light_color is (red, yellow, green);
...
```

```
cycle_lights : entity work.generic_counter(rtl)
  generic map ( count_type  => traffic_light_color,
                reset_value => red )
  port map     ( ... );
```

The process in the instance would have to apply the "+" operator to a value of the actual generic type, in this case, **traffic_light_color**. That application would fail, since there is no such operator defined. We will revise this example in Section 12.5 to show how to supply such an operator to the instance.

When we declare a generic constant in a generic list, we can specify a default value that is used if no actual value is provided in an instance. For generic types, there is no means of specifying a default type. That means that we must always specify an actual type in an instance. Since the type of objects in VHDL is considered to be a very important property, the language designers decided to insist on the actual type being explicitly specified.

VHDL-87, -93, and -2002

These versions of VHDL do not allow generic types. Instead, we must declare separate entities for each combination of types we require in a design.

12.3 Generic Lists in Packages

In Chapter 7, we introduced packages and showed how they can be used to declare types and related operations for use throughout a design. In some cases, the type that we wish to declare makes use of some other type in a general way, without relying on the details of that other type. VHDL's composite types exemplify this idea. They are container types, containing elements of other independent types. VHDL provides a mechanism for us to declare our own container type in a package; we can include a generic list in the package and then instantiate the package with different actual generics. The revised syntax rule for a package declaration with a generic list is:

```
package_declaration ⇐
    package identifier is
        〚 generic ( generic_interface_list ) ; 〛
        〔 package_declarative_item 〕
    end 〚 package 〛 〚 identifier 〛 ;
```

The package body corresponding to such a package is unchanged; we don't repeat the generic list there. Within the generic list, we can declare formal generic constants and formal generic types, just as we can in a generic list of an entity or component. We can then use those formal generics in the declarations within the package.

A package with a generic list is called an *uninstantiated package*. (We avoid calling it a generic package, since we use that term for a different purpose; see Section 12.6.) Un-

like a simple package with no generic list, we cannot refer to the declarations in an uninstantiated package with selected names or use clauses. Instead, the uninstantiated package
serves as a form of template that we must instantiate separately. We make an instance with
a *package instantiation*. The syntax rule is:

package_instantiation_declaration ⇐
    package identifier **is new** *uninstantiated_package*_name
        ⟦ **generic map** (*generic*_association_list) ⟧ ;

The identifier is the name for the package instance, and the generic map supplies actual generics for the formal generics defined by the uninstantiated package. If all of the
formal generics have defaults, we can omit the generic map to imply use of the defaults.
(As we mentioned in Section 12.2, if any of the formal generics is a generic type, it cannot
have a default. In that case, we could not omit the generic map in the package instance.)
Once we have instantiated the package, we can then refer to names declared within it
with selected names and use clauses with the instance name as the prefix.

For now, we focus on uninstantiated packages and package instances that are declared as design units and stored in a design library. This suggests that a package instantiation can be written as a further form of design unit, which is indeed the case. All that
we said in Section 5.4 about design units, preceding them with library and use clauses,
and their analysis into design libraries, applies to package instances written as design
units. We will return to locally declared packages and package instances in Section 12.3.1.

EXAMPLE 12.7   *A package for stacks of data*

We can write a package that defines a data type and operations for fixed-sized stacks
of data. A given stack has a specified capacity and stores data of a specified type. The
capacity and type are specified as formal generics of the package, as follows:

```
package generic_stacks is
  generic ( size : positive; type element_type );

  type stack_array is array (0 to size-1) of element_type;
  type stack_type is record
    SP    : integer range 0 to size-1;
    store : stack_array;
  end record stack_type;

  procedure push (s : inout stack_type; e : in  element_type);
  procedure pop  (s : inout stack_type; e : out element_type);

end package generic_stacks;
```

The corresponding package body is

```
package body generic_stacks is

  procedure push (s : inout stack_type; e : in  element_type) is
  begin
    s.store(s.SP) := e;
```

```
    s.SP := (s.SP + 1) mod size;
  end procedure push;

  procedure pop  (s : inout stack_type; e : out element_type) is
  begin
    s.SP := (s.SP - 1) mod size;
    e := s.store(s.SP);
  end procedure pop;

end package body generic_stacks;
```

The uninstantiated package defines types **stack_array** and **stack_type** for representing stacks, and operations to **push** and **pop** elements. The formal generic constant **size** is used to determine the size of the array for storing elements, and the formal generic type **element_type** is the type of elements to be stored, pushed and popped.

We cannot refer to items in this uninstantiated package directly, since there is no specification of the actual size and element type. Thus, for example, we cannot write the following:

```
use work.generic_stacks.all;   -- Illegal
...
variable my_stack : work.generic_stacks.stack_type;   -- Illegal
```

Instead, we must instantiate the package and provide actual generics for that instance. For example, we might declare the following as a design unit for a CPU design:

```
library ieee; use ieee.numeric_std.all;
package address_stacks is new work.generic_stacks
  generic map ( size => 8,
                element_type => unsigned(23 downto 0) );
```

If we analyze this instantiation into our working library, we can refer to it in other design units, for example:

```
architecture behavior of CPU is
  use work.address_stacks.all;
  ...
begin
  interpret_instructions : process is
    variable return_address_stack : stack_type;
    variable PC : unsigned(23 downto 0);
    ...
  begin
    ...
    case opcode is
      when jsb => push(return_address_stack, PC);
                  PC <= jump_target;
      when ret => pop(return_address_stack, PC);
      ...
```

```
        end case;
        ...
    end process interpret_instructions;
  end architecture behavior;
```

This architecture includes a use clause that makes names declared in the package instance **address_stacks** visible. References to **stack_type**, **push** and **pop** in the architecture thus refer to the declarations in the **address_stacks** package instance.

We can declare multiple instances of a given uninstantiated package, each with different actual generics. The packages instances are distinct, even though they declare similarly named items internally. For example, we might declare two instances of the **generic_stacks** package from the preceding example as follows:

```
package address_stacks is new work.generic_stacks
  generic map ( size => 8,
                     element_type => unsigned(23 downto 0) );

package operand_stacks is new work.generic_stacks
  generic map ( size => 16, element_type => real );
```

If we then wrote a use clause in a design unit:

```
use work.address_stacks.all, work.operand_stacks.all;
```

the names from the two package instances would all be ambiguous. This is an application of the rule in VHDL that, if two packages declare the same name and both are "used," we cannot refer to the simple name, since it is ambiguous. Instead, we need to use selected names to distinguish between the versions declared in the two package instances. So, for example, we could write:

```
use work.address_stacks, work.operand_stacks;
```

to make the package names visible without prefixing them with the library name **work**, and then declare variables and use operations as follows:

```
    variable return_address_stack : address_stacks.stack;
    variable PC                    : unsigned(23 downto 0);
    variable FP_operand_stack      : operand_stacks.stack;
    variable TOS_operand           : real;
    ...
    address_stacks.push(return_address_stack, PC);
    operand_stacks.pop(FP_operand_stack, TOS_operand);
```

An important aspect of VHDL's strong-typing philosophy is that two types introduced by two separate type declarations are considered to be distinct, even if they are structurally the same. Thus the two types declared as

```
type T1 is array (1 to 10) of integer;
type T2 is array (1 to 10) of integer;
```

are distinct, and we cannot assign a value of type T1 to an object of type T2. This same principle applies to formal generic types. Within an entity or a package that declares a formal generic type, that type is considered to be distinct from every other type, including other formal generic types. So, for example, we cannot assign a value declared to be of one formal generic type to an object declared to be of another formal generic type.

The fact that two formal generic types are distinct can lead to interesting situations when the actual types provided are the same (or are subtypes of the same base type). Ambiguity can arise between overloaded operations declared using the formal generic types. This kind of situation is not likely to happen in common use cases, but it is worth exploring to demonstrate the way overloading works in the presence of formal generic types.

Suppose we declare a package with two formal generic types, as follows:

```
package generic_pkg is
  generic ( type T1; type T2 );

  procedure proc ( x : T1 );
  procedure proc ( x : T2 );
  procedure proc ( x : bit );

end package generic_pkg;
```

Within the package, T1 and T2 are distinct from each other and from the type **bit**, so the procedure **proc** is overloaded three times. The uninstantiated package can be analyzed without error. If we instantiate the package as follows:

```
package integer_boolean_pkg is new work.generic_pkg
  generic map ( T1 => integer, T2 => boolean );
```

we can successfully resolve the overloading for the following three calls to procedures in the package instance:

```
work.integer_boolean_pkg.proc(3);
work.integer_boolean_pkg.proc(false);
work.integer_boolean_pkg.proc('1');
```

On the other hand, if we instantiate the package as

```
package integer_bit_pkg is new work.generic_pkg
  generic map ( T1 => integer, T2 => bit );
```

the following call is ambiguous:

```
work.integer_bit_pkg.proc('1');
```

It could be a call to the second or third of the three overloaded versions of **proc** in the package instance. Similarly, if we instantiate the package as

```
package integer_integer_pkg is new work.generic_pkg
    generic map ( T1 => integer, T2 => integer );
```

the following call is ambiguous:

```
work.integer_integer_pkg.proc(3);
```

This could be a call to the first or second of the three overloaded versions of **proc**. The point to gain from these examples is that overload resolution depends on the actual types denoted by the formal generic types in the instances. Depending on the actual types, calls to overloaded subprograms may be resolvable for some instances and ambiguous for others.

The final aspect of packages with generic lists is that we can also include a generic map in a package, following the generic list. Such a package is called a *generic-mapped package*, and has the form

package_declaration ⇐
  package identifier **is**
    ⟦ **generic** (*generic*_interface_list) ;
    ⟦ **generic map** (*generic*_association_list) ; ⟧ ⟧
    ⟅ package_declarative_item ⟆
  end ⟦ **package** ⟧ ⟦ identifier ⟧ ;

The generic list defines the generics, and the generic map aspect provides actual values and type for those generics. While VHDL allows us to write a generic-mapped package explicitly, we would not normally do so. Rather, the feature is included in the language as a definitional aid. An instantiation of an uninstantiated package is defined in terms of an equivalent generic-mapped package that is a copy of the uninstantiated package, together with the generic map from the instantiation. Since generic-mapped packages are not a feature intended for regular use, we won't dwell on them further. We simply mention them here to raise awareness, since the occasional error message from an analyzer might hint at them.

12.3.1 Local Packages

In Section 7.2.1 on page 255, we showed how a package can be declared locally within the declarative part of an entity, architecture body, process, or subprogram, allowing the visibility of the package to be contained to just the enclosing declarative part and the corresponding statement part. We presented an example of a package, declared with a stimulus-generator process, that provided a function for generating unique identification numbers for test cases.

A locally declared package need not be just a simple package; it can be an uninstantiated package with a generic list. In that case, we must instantiate the package so that we can refer to items in the instance. The same rules apply to locally declared uninstantiated packages and instances as apply to packages declared as design units.

EXAMPLE 12.8 *A package for wrapping items with unique ID numbers*

We can revise the **ID_manager** package from Section 7.2.1 to make it deal with test cases of generic type, and to wrap each test case in a record together with a unique ID number. The numbers are unique across test cases of all types. We achieve this by keeping the **ID_manager** package as an outer package encapsulating the **next_ID** variable. Within that package, we declare an uninstantiated package for wrapping test cases. The process outline containing the packages is:

```
stim_gen : process is

  package ID_manager is

    package ID_wrappers is
      generic ( type test_case_type );
      type wrapped_test_case is record
        test_case : test_case_type;
        ID        : natural;
      end record wrapped_test_case;
      impure function wrap_test_case
                        ( test_case : test_case_type )
                        return wrapped_test_case;
    end package ID_wrappers;

  end package ID_manager;

  package body ID_manager is

    variable next_ID : natural := 0;

    package body ID_wrappers is
      impure function wrap_test_case
                        ( test_case : test_case_type )
                        return wrapped_test_case is
        variable result : wrapped_test_case;
      begin
        result.test_case := test_case;
        result.ID := next_ID;
        next_ID := next_ID + 1;
        return result;
      end function wrap_test_case;
    end package body ID_wrappers;

  end package body ID_manager;

  use ID_manager.ID_wrappers;

  package word_wrappers is new ID_wrappers
    generic map ( test_case_type => unsigned(32 downto 0) );
  package real_wrappers is new ID_wrappers
    generic map ( test_case_type => real );
```

```
    variable next_word_test : word_wrappers.wrapped_test_case;
    variable next_real_test : real_wrappers.wrapped_test_case;

begin
    ...
    next_word_test := word_wrappers.wrap_test_case(X"0440CF00");
    next_real_test := real_wrappers.wrap_test_case(3.14159);
    ...
end process stim_gen;
```

The process declares two instances of the uninstantiated package ID_wrappers, one for a test-case type of **unsigned**, and another for a test-case type of **real**. The process then refers to the **wrapped_test_case** type and the **wrap_test_case** function declared in each instance.

This example exposes a number of important points about packages. First, a package declared within an enclosing region is just another declared item, and is subject to the normal visibility rules. In the example, the ID_wrappers package is declared within an enclosing package, and so can be referred to with a selected name and made visible by a use clause.

Second, in the case of package instantiations, any name referenced within the uninstantiated package keeps its meaning in each instance. In the example, the name **next_ID** referenced within the uninstantiated package ID_wrappers, refers to the variable declared in the **ID_manager** package. So, within each of the package instances, **word_wrappers** and **real_wrappers**, the same variable is referenced. Importantly, had the process also declared an item called **next_ID** outside the packages but before the instances, that name would not be "captured" by the instances. They still refer to the same variable nested within the **ID_manager** package. The only exception to this rule for interpreting names is that the name of the uninstantiated package itself, when referenced within the package, is interpreted in an instance as a reference to the instance. This allows us to use a selected name within an uninstantiated package to refer to an item declared within the uninstantiated package, and to have it interpreted appropriately in the instance. The rules for name interpretation illustrate quite definitely that package instantiation is different in nature from file inclusion, as is used for C header files. The benefit of the VHDL approach is that names always retain the meaning they are given at the point of declaration, and so we avoid unwanted surprises.

The third point is that local instantiation of an uninstantiated package is a common use case, whether the uninstantiated package be locally declared, as in the example, or globally declared as a design unit. The advantage of local instantiation is that it allows use of a locally declared type as the actual for a formal generic type. Were local instantiation not possible, the actual type would have to be declared in a global package in order to use it in a global package instantiation. Thus, local instantiation improves modularity and information hiding in a design.

EXAMPLE 12.9   *Local instantiation of the stacks package*

In Example 12.7, we declared an uninstantiated package for stacks as a design unit. We can instantiate the package to deal with stacks of a type declared locally within a subprogram that performs a depth-first search of a directed acyclic graph (DAG) consisting of vertices and edges, as follows:

```
subprogram analyze_network ( network : network_type ) is

  type vertex_type is ...;
  type edge_type   is ...;
  constant max_diameter : positive := 30;

  package vertex_stacks is new work.generic_stacks
    generic map ( size => max_diameter,
                  element_type => vertex_type );
  use vertext_stacks.all;

  variable current_vertex   : vertex_type;
  variable pending_vertices : stack_type;

begin
  ...
  push(pending_vertices, current_vertex);
  ...
end subprogram analyze_network;
```

The data types used to represent the DAG for analyzing a network are the local concern of the subprogram. By instantiating the **generic_stacks** package locally, there is no need to expose the data types outside the subprogram.

12.3.2 Abstract Data Types Using Packages

The generic stack package in Example 12.7 is an example of an *abstract data type* (ADT). This is a term we borrow from the discipline of software engineering to refer to a high-level view of a data structure. An ADT is a data type, together with a collection of operations for creating and working with data objects of that type. In a strict implementation of an ADT, the data structure underlying the data type is not visible to users of the ADT. The operations provided are the only way of working with data objects, thus preventing incorrect use of the objects. This is the means of enforcing the abstract view of the data type. Unfortunately, VHDL does not provide a way of hiding the data structure, so we have to rely on conventions and documentation.

The most convenient way to implement an ADT in VHDL is to use a package, as we saw in the example. In the package declaration we write the VHDL type declarations that represent the underlying data structure and declare functions and procedures that perform the ADT operations. We, or other designers, can use these declarations to create data objects and perform operations on them without being concerned about the implementation details of the data structure. Any designer has plenty of other concerns to think about, so the more we can do to ease the task of system modeling, the more productive the designer will be. As implementers of the ADT, we write the details of the operations in the package body. We should make the operations as general as possible, so that we can reuse the ADT in several different designs.

One good application of ADTs is as "container" or "collection" types. As we saw in the stack example, the ADT represents a type that contains values of some element type. We used a formal generic type to represent the element type, allowing us to instantiate the ADT in different places with different element types. We will now consider a more extensive example of a collection ADT.

EXAMPLE 12.10   *A bounded buffer ADT package*

Many high-level designs involve components that operate autonomously of one another, but that need to communicate streams of data. Instead of the sender and receiver synchronizing to transfer each item of data, they can communicate asynchronously using a buffer memory. Items are retrieved from this buffer by the receiver in the same order that they are written by the sender. This is often called a "first in, first out" (FIFO) buffer. Since memory in a real hardware system is not an infinite resource, we specify a bound on the amount of data that can be stored at once.

We can write an ADT to provide types and operations for bounded buffers of items of data. The package declaration is

```
package bounded_buffer_adt is
  generic ( size : positive;
            type element_type );

  type store_array is array (0 to size - 1) of element_type;

  type bounded_buffer is record
      count : natural;
      head, tail : natural;
      store : store_array;
    end record bounded_buffer;

  procedure reset ( b : inout bounded_buffer );
  -- resets a bounded buffer to be empty

  function is_empty ( b : bounded_buffer ) return boolean;
  -- tests whether the bounded buffer is empty
  -- (i.e., no data to read)

  function is_full ( b : bounded_buffer ) return boolean;
  -- tests whether the bounded buffer is full
  --(i.e., no data can be written)

  procedure write ( b : inout bounded_buffer;
                    data : in element_type );
  -- if the bounded buffer is not full, writes the data
  -- if it is full, assertion violation with severity failure
```

```
    procedure read ( b : inout bounded_buffer;
                     data : out element_type );
    -- if the bounded buffer is not empty, read the first item
    -- if it is empty, assertion violation with severity failure

  end package bounded_buffer_adt;
```

The formal generic constant **size** specifies the maximum number of items that can be stored in the buffer at any time, and the formal generic type **element_type** represents the type of the items to be stored. The declaration of the type **store_array** and the structure of the type **bounded_buffer** are private details of the concrete implementation of the bounded buffer ADT. A user of the ADT does not need to know about them. However, we need to include them in the package declaration in order to declare the type **bounded_buffer**, which is the public type. Unfortunately, VHDL does not provide a way of hiding the types that should be private. We will return to the implementation details shortly.

The public information in the package declaration is all that is needed to write a model using bounded buffers. For example, the following process is part of a network receiver model, using the bounded buffer ADT. It instantiates the ADT package locally, specifying a buffer size of 2048 items, each of the subtype **byte**. This process makes no reference to the implementation details of the bounded buffer. It is written using only the operations provided in the public interface of the package. The advantage of separating out the bounded buffer part of the model into an ADT is that the model is more compact, easier to write and easier to understand.

```
receiver : process is

  subtype byte is bit_vector(0 to 7);

  package byte_buffer_adt is new work.bounded_buffer_adt
    generic map ( size => 2048, element_type => byte );

  use work.byte_buffer_adt.all;

  variable receive_buffer : bounded_buffer;
  ...

begin
  reset(receive_buffer);
  ...

  if is_full(receive_buffer) then
    ...   -- buffer overrun
  else
    write(receive_buffer, received_byte);
  end if;
  ...

  if is_empty(receive_buffer) then
    ...   -- buffer underrun
  else
```

```
        read(receive_buffer, check_byte);
    end if;

    ...

end process receiver;
```

We can now turn to the implementation details of the bounded buffer ADT. The converse advantage of the separation is that as the implementer of the ADT, we can concentrate on writing it as a compact, well-defined software module. We are not distracted by the code of the models that use bounded buffers. The private types in the package declaration indicate that the concrete implementation of this bounded buffer ADT is as a *circular buffer*, stored in an array of items, as shown in Figure 12.1. (We can think of the end of the array as being wrapped around to meet the beginning, forming a circle.) Data is stored in successive items in the array, starting from the first element. The record element **tail** contains the index of the next free position in the array, and the element **head** contains the index of the first available item. Each time a new item is written to the buffer, **tail** is incremented, and each time an item is read, **head** is incremented. They are incremented modulo the size of the buffer, so that the space made available when items are read is reused for new items when the end of the array is reached. The record element **count** keeps track of the number of items in the buffer and is used to ensure that the write position does not overtake the read position, and vice versa.

FIGURE 12.1

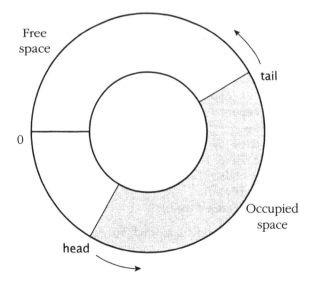

The array used to store data as a circular buffer.

The package body for the bounded buffer ADT implementation is

```
package body bounded_buffer_adt is

  procedure reset ( b : inout bounded_buffer ) is
  begin
    b.count := 0; b.head := 0; b.tail := 0;
  end procedure reset;

  function is_empty ( b : bounded_buffer ) return boolean is
  begin
    return b.count = 0;
  end function is_empty;

  function is_full ( b : bounded_buffer ) return boolean;
  begin
    return b.count = size;
  end function is_full;

  procedure write ( b : inout bounded_buffer;
                    data : in element_type ) is
  begin
    if is_full(b) then
      report "write to full bounded buffer" severity failure;
    else
      b.store(b.tail) := data;
      b.tail := (b.tail + 1) mod size;
      b.count := b.count + 1;
    end if;
  end procedure write;

  procedure read ( b : inout bounded_buffer;
                   data : out element_type ) is
  begin
    if is_empty(b) then
      report "read from empty bounded buffer" severity failure;
    else
      data := b.store(b.head);
      b.head := (b.head + 1) mod size;
      b.count := b.count - 1;
    end if;
  end procedure read;

end package body bounded_buffer_adt;
```

The function **reset** clears the **count**, **head** and **tail** elements to zero. The function **is_empty** simply tests whether the **count** element of the record object is zero, and the function **is_full** tests whether **count** is equal to the size of the array used to store data.

The **write** procedure uses an assertion statement to test whether the buffer is full, using the ADT operation **is_full**. It then writes the data item into the buffer at the tail position and increments the **tail** element of the record. The read procedure similarly uses an assertion statement to test whether the buffer is empty, using the ADT operation **is_empty**. It then reads the data item from the head position of the buffer and increments the **head** element of the record.

The advantages of using ADTs in complex behavioral models are overwhelming, but there is one risk that must be borne. As we mentioned before, VHDL provides no way of hiding the concrete details of the data structure underlying an ADT, as the type declarations must be written in the package declaration. This means that an ADT user can make use of the information to modify the data structures without using the ADT procedures and functions. For example, if an ADT operation simply updates a record element, a user might be tempted to update the record directly and avoid the overhead of a procedure call. However, modern compilers and computers make such "optimizations" unnecessary, and the risk is that the user might inadvertently corrupt the data structure. ADTs in VHDL require that users avoid such temptations and abide by the contract expressed in the ADT interface. A small amount of self-discipline here will yield significant benefits in the modeling process.

12.4 Generic Lists in Subprograms

Just as generic lists in packages allow us to describe container data types that are are reusable, generic lists in subprograms allow us to describe operations that can be reused for operands of various types. The expanded syntax rule for a procedure with a generic list is

subprogram_body ⇐
  procedure identifier
    ⟦ **generic** (*generic*_interface_list) ⟧
    ⟦ ⟦ **parameter** ⟧ (*parameter*_interface_list) ⟧ **is**
    ⦃ subprogram_declarative_part ⦄
  begin
    ⦃ sequential_statement ⦄
  end [**procedure**] [identifier] ;

Similarly, the syntax rule for a function with a generic list is:

subprogram_body ⇐
  ⟦ **pure** ǀ **impure** ⟧
  function identifier
    ⟦ **generic** (*generic*_interface_list) ⟧
    ⟦ ⟦ **parameter** ⟧ (*parameter*_interface_list) ⟧ **return** type_mark **is**
    ⦃ subprogram_declarative_item ⦄
  begin
    ⦃ sequential_statement ⦄
  end ⟦ **function** ⟧ ⟦ identifier ⟧ ;

We use terminology analogous to that for packages to refer to subprograms with generics. Thus, a subprogram with a generic list is called an *uninstantiated subprogram*.

Note that the optional keyword **parameter** allows us to make the demarcation between the generic list and the parameter list clear. We would normally omit it for subprograms without generics and include it or not as a matter of taste for uninstantiated subprograms.

VHDL allows us to declare a subprogram in two parts, one consisting just of the header, and the other consisting of the header together with the body. We can separate a subprogram in this way within a given declarative part, for example, in order to declare mutually recursive subprograms. In the case of subprograms declared in packages, we are required to separate the subprogram header into the package declaration and to repeat the header together with the subprogram body in the package body. In the case of uninstantiated subprograms, the generic list is part of the subprogram header. Thus, if we separate the declaration, we must write the generic list and parameter list in the header, and then repeat both together with the body. Using a text editor to copy and paste the header into the body makes this easy.

We cannot call an uninstantiated subprogram directly. We can think of it as a template that we must instantiate with a *subprogram instantiation* to get a real subprogram that we can call. The syntax rule for instantiating a procedure is:

subprogram_instantiation_declaration ⇐
    procedure identifier **is new** *uninstantiated_subprogram*_name ⟦ signature ⟧
      ⟦ **generic map** (*generic*_association_list) ⟧ ;

and for instantiating a function:

subprogram_instantiation_declaration ⇐
    function identifier **is new** *uninstantiated_subprogram*_name ⟦ signature ⟧
      ⟦ **generic map** (*generic*_association_list) ⟧ ;

In both cases, the identifier is the name for the subprogram instance, and the generic map supplies actual generics for the formal generics defined by the uninstantiated subprogram. If all of the formal generics have defaults, we can omit the generic map to imply use of the defaults. Once we have instantiated the subprogram, we can then use the instance name to call the instance. We will return to the optional signature in an instantiation after a couple of examples.

EXAMPLE 12.11   *A generic swap procedure*

The way in which we swap the values of two variables does not depend on the types of the variables. Hence, we can write a swap procedure with the type as a formal generic, as follows:

```
procedure swap
  generic   ( type T )
  parameter ( a, b : inout T ) is
  variable temp : T;
begin
  temp := a; a := b; b := temp;
end procedure swap;
```

We can now instantiate the procedure to get versions for various types:

```
procedure int_swap is new swap
  generic map ( T => integer );
procedure vec_swap is new swap
  generic map ( T => bit_vector(0 to 7) );
```

and call them to swap values of variables:

```
variable a_int, b_int : integer;
variable a_vec, b_vec : bit_vector(0 to 7);
...

int_swap(a_int, b_int);
vec_swap(a_vec, b_vec);
```

We can't just call the **swap** procedure directly, as follows:

```
swap(a_int, b_int);   -- Illegal
```

since it is an uninstantiated procedure. Note also that we can't instantiate the **swap** procedure with an unconstrained type as the actual generic type, since the procedure uses the type internally to declare a variable. Thus, the following would produce an error:

```
procedure string_swap is new swap generic map ( T => string );
```

since there is no specification of the index bounds for the variable **temp** declared within **swap**.

EXAMPLE 12.12   *A setup timing check procedure*

Suppose we are developing a package of generic operations for timing checks on signals. We include a generic procedure that determines whether a signal meets a setup time constraint. The package declaration is

```
package timing_pkg is

  procedure check_setup
    generic ( type signal_type;
              type clk_type; clk_active_value : clk_type;
              T_su : delay_length )
    ( signal s : signal_type; signal clk : clk_type );

    ...

end package timing_pkg;
```

The package body contains a body for the procedure:

```
package body timing_pkg is
```

```
      procedure check_setup
        generic ( type signal_type;
                  type clk_type; clk_active_value : clk_type;
                  T_su : delay_length )
        ( signal s : signal_type; signal clk : clk_type ) is
      begin
        if clk'event and clk = clk_active_value then
          assert s'last_event >= T_su
            report "Setup time violation" severity error;
        end if;
      end procedure check_setup;

      . . .

  end package body timing_pkg;
```

We can now instantiate the procedure to get versions that check the constraint for signals of different types and for different setup time parameters:

```
  use work.timing_pkg.all;

  procedure check_normal_setup is new check_setup
    generic map ( signal_type => std_ulogic,
                  clk_type => std_ulogic,
                  clk_active_value => '1',
                  T_su => 200ps );

  procedure check_normal_setup is new check_setup
    generic map ( signal_type => std_ulogic_vector,
                  clk_type => std_ulogic,
                  clk_active_value => '1',
                  T_su => 200ps );

  procedure check_long_setup is new check_setup
    generic map ( signal_type => std_ulogic_vector,
                  clk_type => std_ulogic,
                  clk_active_value => '1',
                  T_su => 300ps );
```

Note that the procedure **check_normal_setup** is now overloaded, once for a **std_ulogic** parameter and once for a **std_ulogic_vector** parameter. We can apply these functions to signals of **std_ulogic** and **std_ulogic_vector** types, as follows:

```
  signal status : std_ulogic;
  signal data_in, result : std_ulogic_vector(23 downto 0);
  . . .

  check_normal_setup(status, clk);
  check_normal_setup(result, clk);
  check_long_setup(data_in, clk);
  . . .
```

In each case, the active value for the clock signal and the setup time interval value are bound into the definition of the procedure instance. We do not need to provide the values as separate parameters.

VHDL allows us to overload subprograms, and uses the parameter and result type profiles to distinguish among them based on the types of parameters in a call. Where we need to name a subprogram other than in a call, we can write a signature to indicate which overloaded version we mean. An example of such a place is in an alias declaration. We described the use of signatures in alias declarations in Section 11.2. The signature lists the parameter types and, for functions, the return type, all enclosed in square brackets. This information is sufficient to distinguish one version of an overloaded subprogram from other versions. If an uninstantiated subprogram is overloaded, we can include a signature in an instantiation to indicate which uninstantiated version we mean. In such cases, the uninstantiated subprograms typically have one or more parameters of a formal generic type. We use the formal generic type name in the signature. For example, if we have two uninstantiated subprograms declared as

```
procedure combine
  generic   ( type T )
  parameter ( x : T; value : bit );

procedure combine
  generic   ( type T )
  parameter ( x : T; value : integer );
```

the procedure name **combine** is overloaded. We can use a signature in an instantiation as follows:

```
procedure combine_vec_with_bit is new combine[T, bit]
  generic map ( T => bit_vector );
```

VHDL specifies that a formal generic type name of an uninstantiated subprogram is made visible within a signature in an instantiation of the subprogram. Thus, in this example, the signature distinguishes between the two uninstantiated subprograms, since only one of them has a profile with T for the first parameter and **bit** for the second. The T in the signature refers to the formal generic type for that version of the subprogram.

As with packages, we can also include a generic map in a subprogram, following the generic list. Such a subprogram is called a *generic-mapped subprogram*. The syntax rule for a generic-mapped procedure is

```
subprogram_body ⇐
    procedure identifier
        〖 generic ( generic_interface_list )
        〖 generic map ( generic_association_list ) 〗 〗
        〖 〖 parameter 〗 ( parameter_interface_list ) 〗 is
        { subprogram_declarative_part }
    begin
```

      { sequential_statement }
end [**procedure**] [identifier] ;

and for a generic-mapped function:

subprogram_body ⇐
   ⟦ **pure** ∣ **impure** ⟧
   function identifier
      ⟦ **generic** (*generic*_interface_list)
      ⟦ **generic map** (*generic*_association_list) ⟧ ⟧
      ⟦ ⟦ **parameter** ⟧ (*parameter*_interface_list) ⟧ **return** type_mark **is**
      { subprogram_declarative_item }
   begin
      { sequential_statement }
   end ⟦ **function** ⟧ ⟦ identifier ⟧ ;

In each case, the generic list defines the generics, and the generic map aspect provides actual values and type for those generics. Like generic-mapped packages, we would not normally write a generic-mapped subprogram explicitly, since the feature is included in the language as a definitional aid. Hence, we won't dwell on them further, but simply mention them here to raise awareness in case an analyzer produces a seemingly cryptic error message.

12.5 Generic Subprograms

In the preceding sections of this chapter, we have seen the use of generic constants and generic types for making entities, packages and subprograms reusable. VHDL allows us to declare generic subprograms as a way of providing an operation or action to be performed by a unit. We declare a *formal generic subprogram* in a generic list, representing some subprogram yet to be specified, and include calls to the formal generic subprogram within the unit that has the generic list. When we instantiate the unit, we supply an actual subprogram for that instance. Each call to the formal generic subprogram represents a call to the actual subprogram in the instance. The way we declare a formal generic subprogram is to write a subprogram specification (the header without the body) in the generic list.

A simplified syntax rule for declaring a formal generic procedure in a generic list is

interface_declaration ⇐
   procedure identifier
      ⟦ ⟦ **parameter** ⟧ (*parameter*_interface_list) ⟧

and for declaring a formal generic function:

interface_declaration ⇐
   procedure identifier
      ⟦ ⟦ **parameter** ⟧ (*parameter*_interface_list) ⟧ **return** type_mark

These rules show that the formal generic subprogram must be a simple subprogram; that is, it must not contain a generic list itself.

We will illustrate formal generic subprograms with a number of examples based on typical use cases. One important use case is to supply an operation for use with a formal generic type declared in the same generic list as the subprogram. Recall, from our discussion in Section 12.2, that the only operations we can assume for a formal generic type are those defined for all actual types, such as assignment, equality and inequality. We can use a formal generic subprogram to explicitly provide further operations.

EXAMPLE 12.13 *A generic counter with an increment function*

In Example 12.6 on page 375, we attempted to define a counter that could count with a variety of types. However, our attempt failed because we could not use the "+" operator to increment the count value. We can rectify this by declaring a formal generic function for incrementing the count value:

```
entity generic_counter is
  generic ( type      count_type;
            constant reset_value : count_type;
            function increment ( x : count_type )
                                   return count_type );
    port ( clk, reset : in  bit;
           data        : out count_type );
end entity generic_counter;
```

We can then use the increment function in the architecture:

```
architecture rtl of generic_counter is
begin
  count : process (clk) is
  begin
    if rising_edge(clk) then
      if reset then
        data <= reset_value;
      else
        data <= increment(data);
      end if;
    end if;
  end process count;
end architecture rtl;
```

Having revised the counter in this way, we can instantiate it with various types. For example, to create a counter for **unsigned** values, we define a function, **add1**, to increment using the "+" operator on **unsigned** values and provide it as the actual for the **increment** generic.

```
use ieee.numeric_std.all;
function add1 ( arg : unsigned ) return unsigned is
```

```
begin
  return arg + 1;
end function add1;

signal clk, reset : bit;
signal count_val  : unsigned(15 downto 0);
...

counter : entity work.generic_counter(rtl)
  generic map ( count_type  => unsigned(15 downto 0),
                reset_value => (others => '0'),
                increment   => add1 )   -- add1 is the
                                        -- actual function
    port map ( clk => clk, reset => reset, data => count_val );
```

In the instance, we specify a subtype of **unsigned** as the actual type for the formal generic type **count_type**. That subtype is then used as the subtype of the formal generic constant **reset_value** in the instance, so the actual value is a vector of 16 elements. The subtype is also used for the parameters of the formal generic function **increment** in the instance, so we must provide an actual function with a conforming profile. The **add1** function meets that requirement, since it has **unsigned** as its parameter and result type. Within the instance, whenever the process calls the **increment** function, the actual function **add1** is called.

We can instantiate the same entity to create a counter for the **traffic_light_color** type defined in Example 12.6 on page 375. Again, we define a function, **next_color**, to increment a value of the type, and provide the function as the actual for the **increment** generic.

```
type traffic_light_color is (red, yellow, green);

function next_color ( arg  : traffic_light_color )
                    return traffic_light_color is
begin
  if arg = traffic_light_color'high then
    return traffic_light_color'low;
  else
    return traffic_light_color'succ(arg);
  end if;
end function next_color;

signal east_light : traffic_light_color;
...

east_counter : work.generic_counter(rtl)
  generic map ( count_type  => traffic_light_color,
                reset_value => red,
                increment   => next_color ) -- next_color is the
                                            -- actual function
    port map ( clk => clk, reset => reset, data => east_light );
```

When we declare a formal generic subprogram in a generic list, we can specify a default subprogram that is to be used in an instance if no actual generic subprogram is provided. The expanded syntax rule for declaring a formal generic procedure with a default is

interface_declaration ⇐
    procedure identifier
        ⟦ ⟦ **parameter** ⟧ (*parameter*_interface_list) ⟧
        is *subprogram*_name

and similarly for a formal generic function with a default:

interface_declaration ⇐
    function identifier
        ⟦ ⟦ **parameter** ⟧ (*parameter*_interface_list) ⟧ **return** type_mark
        is *subprogram*_name

The subprogram that we name must be visible at that point. It might be declared before the uninstantiated unit, or it can be another formal generic subprogram declared earlier in the same generic list. In the case of an uninstantiated package, we cannot name a subprogram declared in the package as a default subprogram, since items declared within the package are not visible before they are declared.

EXAMPLE 12.14   *Generic error reporting in a package*

Suppose we are developing a package defining operations to be used in a design and need to report errors that arise while performing operations. We can declare a formal generic procedure in the package to allow separate specification of the error-reporting action. We can also declare a default procedure that simply issues a report message. We need to declare the default action procedure separately from the package so that we can name it in the generic list. We will declare it in a utility package:

```
package error_utility_pkg is
  procedure report_error ( report_string   : string;
                           report_severity : severity_level );
end package error_utility_pkg;

package body error_utility_pkg is
  procedure report_error ( report_string   : string;
                           report_severity : severity_level ) is
  begin
    report report_string severity report_severity;
  end procedure report_error;
end package body error_utility_pkg;
```

We can now declare the operations package:

```
package operations is
  generic ( procedure error_action
                ( report_string   : string;
```

```
                    report_severity : severity_level )
                 is work.error_utility_pkg.report_error );
    procedure step1 ( ... );

    ...

end package operations;

package body operations is
    procedure step1 ( ... ) is
    begin
      ...
      if something_is_wrong then
        error_action("Something is wrong in step1", error);
      end if;
      ...
    end procedure step1;

    ...

end package body operations;
```

If issuing a report message is sufficient for a given design, it can instantiate the operations package without providing an actual generic subprogram:

```
package reporting_operations is new work.operations;
use reporting_operations.all;
...

step1 ( ... );
```

If something goes wrong during execution of **step1** in this instance, the call to **error_action** results in a call to the default generic subprogram **report_error** defined in the utility package.

Another design might need stop simulation using the **stop** procedure from the environment package (see Section 9.2). The design can declare a procedure to deal with error messages as follows:

```
constant stop_status : integer := -1;

procedure stop_on_error ( report_string   : string;
                          report_severity : severity_level ) is
begin
  report report_string severity report_severity;
  std.env.stop(stop_status);
end procedure stop_on_error;
```

The design can then instantiate the operations package with this procedure as the actual generic procedure:

```
package debugging_operations is new work.operations
  generic map ( error_action => stop_on_error );
use debugging_operations.all;

...

step1 ( ... );
```

In this instance, when something goes wrong in **step1**, the call to **error_action** results in a call to the procedure **stop_on_error**, which reports the message and then stops simulation. Since the actual procedure is declared in the context of the instantiating design, it has access to items declared in that context, including the constant **stop_status**. By providing this procedure as the actual generic procedure to the package instance, the instance is able to "import" that context via the actual procedure.

In many use cases where an operation is required for a formal generic type, there may be an overloaded version of the operation defined for the actual generic type at the point of instantiation. VHDL provides a way to indicate that the default for a generic subprogram is a subprogram, directly visible at the point of instantiation, with the same name as the formal generic subprogram and a conforming profile. We use the box symbol ("<>") in place of a default subprogram name in the generic declaration. For example, we might write the following in a generic list of a package:

```
function minimum ( L, R : T ) return T is <>
```

If, when we instantiate the package, we omit an actual generic function, and there is a visible function named **minimum** with the required profile, then that function is used. Normally, the parameter type T used in the declaration of the formal generic subprogram is itself a formal generic type declared earlier in the generic list. We provide an actual type for T in the instance, and that determines the parameter type expected for the visible default subprogram. If we define the formal generic subprogram with the same name and similar profile to a predefined operation, we can often rely on a predefined operation being visible and appropriate for use as the default subprogram. We will illustrate this with an example.

EXAMPLE 12.15 *A generic counter with a default for the increment function*

We can further revise the counter from Example 12.6 on page 375 by using a formal generic "+" operator to perform the increment operation and by specifying a default for the generic:

```
entity generic_counter is
  generic ( type     count_type;
            constant reset_value : count_type;
            function "+" ( L : count_type; R : natural )
                          return count_type is <> );
  port ( clk, reset : in  bit;
```

```
        data          : out count_type );
    end entity generic_counter;
```

We use the "+" operator in the architecture:

```
    architecture rtl of generic_counter is
    begin
      count : process (clk) is
      begin
        if rising_edge(clk) then
          if reset then
            data <= reset_value;
          else
            data <= data + 1;
          end if;
        end if;
      end process count;
    end architecture rtl;
```

When we instantiate the counter, if there is a version of "+" visible for the actual type and we don't provide an actual function, the visible "+" operator is used. For example, in the following:

```
    use ieee.numeric_std.all;

    signal clk, reset : bit;
    signal count_val  : unsigned(15 downto 0);
    ...

    counter : entity work.generic_counter(rtl)
      generic map ( count_type  => unsigned(15 downto 0),
                    reset_value => (others => '0') )
      port map ( clk => clk, reset => reset, data => count_val );
```

the instance uses the "+" operator, defined in **numeric_std**, with an **unsigned** left operand and a **natural** right operand.

EXAMPLE 12.16   *A generic dictionary ADT implemented using arrays*

The following package defines an abstract data type for ordered dictionaries implemented as sorted arrays of elements. (There are more efficient ways to implement dictionaries, but they rely on access types, a language feature that we introduce in Chapter 15. We will illustrate an alternate implementation of ordered dictionaries in that chapter.) A dictionary contains elements that are each identified by a key value. The formal generic constant **size** specifies the maximum number of elements in the dictionary. The formal generic function **key_of** determines the key for a given element. No default function is provided, so we must supply an actual function on instantiation of the package. The formal function "<" is used to compare key values. The default function is specified using the "<>" notation, so if an appropriate function named "<" is directly visible at the point of instantiation, we don't need to specify an actual function.

```
package dictionaries is
  generic ( size : positive;
            type element_type;
            type key_type;
            function key_of ( E : element_type ) return key_type;
            function "<" ( L, R : key_type )
                          return boolean is <> );

  type element_array is array (1 to size) of element_type;

  type dictionary_type is record
      store : element_array;
      count : natural;
    end record dictionary_type;

  procedure initialize ( dictionary : inout dictionary_type );

  procedure lookup ( dictionary : in dictionary_type;
                     lookup_key : in key_type;
                     element : out element_type;
                     found : out boolean );

  procedure search_and_insert
              ( dictionary : in dictionary_type;
                element : in element_type;
                already_present : out boolean );

end package dictionaries;
```

The package body is shown below, with the body of the **search_and_insert** procedure omitted for brevity.

```
package body dictionaries is

  procedure initialize ( dictionary : inout dictionary_type ) is
  begin
    dictionary.count := 0;
  end procedure initialize;

  procedure lookup ( dictionary : in dictionary_type;
                     lookup_key : in key_type;
                     element : out element_type;
                     found : out boolean ) is
    variable left, right, middle : natural;
  begin
    found := false;
    left := 1; right := dictionary.count;
    while left <= right loop
```

```
      middle := (left + right) / 2;
      if lookup_key < key_of( dictionary.store(middle) ) then
        right := middle - 1;
      elsif key_of( dictionary.store(middle) ) < lookup_key then
        left := middle + 1;
      else
        found := true;
        element := dictionary.store(middle);
        return;
      end if;
    end loop;
  end procedure lookup;

  procedure search_and_insert
              ( dictionary : inout dictionary_type;
                element : in element_type;
                already_present : out boolean ) is
    . . .

end package body dictionaries;
```

The function **lookup** uses a binary search algorithm to locate the required element in the array. We use the formal generic function **key_of** to get the key for a candidate element in the dictionary. We compare the key with the value of the **lookup_key** parameter using the formal generic function "<".

Now suppose we require a dictionary of test patterns that use time values as keys. We can instantiate the dictionaries package using our test-pattern type as the actual for **element_type** and **time** as the actual for **key_type**. We need to declare a function to get the time key for a test pattern:

```
type test_pattern_type is ...;

function test_time_of ( test_pattern : in test_pattern_type )
                          return time is
begin
  return ...;
end function test_time_of;
```

We don't need to define a function for use as the actual for the formal generic function "<". Since the predefined function "<" operating on **time** values is directly visible at the point of instantiation, it can be used implicitly as the actual function. As a result, the test patterns will be sorted into ascending order of time in the dictionary. We can write the package instantiation as:

```
package test_pattern_dictionaries is new work.dictionaries
  generic map ( size => 1000,
                element_type => test_pattern_type,
                key_type => time,
                key_of => test_time_of );
```

We can then call the operations defined in the instance:

```
use test_pattern_dictionaries.all;
variable test_set : dictionary_type;
variable generated_test, sought_test : test_pattern_type;
variable was_present : boolean;
...

initialize ( test_set );
...

search_and_insert ( test_set, generated_test, was_present );
assert not was_present
  report "Test at " & to_string(test_time_of(generated_test))
                    & " previously generated";
...
lookup ( test_set, 10 ns, sought_test, was_present );
assert was_present
  report "Test at 10 ns not found in test set";
```

EXAMPLE 12.17   *Dictionary traversal with an action procedure*

We can augment the ordered dictionary abstract data type with an operation for traversing a dictionary to apply an action to each element. We define the traversal procedure as an uninstantiated procedure within the uninstantiated dictionaries package:

```
package dictionaries is
  generic ( ... );

  ...

  procedure traverse
    generic   ( procedure action ( element : in element_type ) )
    parameter ( dictionary : in dictionary_type );

end package dictionaries;

package body dictionaries is

  ...

  procedure traverse
    generic   ( procedure action ( element : in element_type ) )
    parameter ( dictionary : in dictionary_type ) is
  begin
    for i in 1 to dictionary.count loop
      action ( dictionary.store(i) );
    end loop;
  end procedure traverse;

end package body dictionaries;
```

Given this augmented package and the same instance as in the previous example, we can use the **traverse** procedure to find the smallest time interval between successive test patterns in a dictionary. We first declare an action procedure and a number of variables for it to use:

```
variable previous_time : time := time'low;
variable smallest_so_far : time := time'high;
procedure compare_test_pattern
          ( test_pattern : in test_pattern_type ) is
begin
  smallest_so_far
    := minimum(smallest_so_far,
              test_time_of(test_pattern) - previous_time );
  previous_time := test_time_of(test_pattern);
end procedure count_a_test_pattern;
```

We instantiate the **traverse** procedure in the declarative part of the design:

```
procedure find_smallest_interval is new traverse
  generic map ( action => compare_test_pattern );
```

and then call the instance:

```
find_smallest_interval(test_set);
report "The smallest interval between test patterns is "
       & to_string(smallest_so_far);
```

We can use a separate instantiation of the **traverse** procedure to perform a different action. For example, if we need to debug a test bench by displaying a collection of test patterns in order of their time, we would define an action procedure:

```
procedure display_test_pattern
          ( test_pattern : in test_pattern_type ) is
begin
  report "Test at "
         & to_string(test_time_of(test_pattern)) & LF
         & ...;
end procedure display_test_pattern;
```

Again, we instantiate the **traverse** procedure in the declarative part of the design:

```
procedure display_all_test_patterns is new traverse
  generic map ( action => display_test_pattern );
```

and then call the instance:

```
display_all_test_patterns(test_set);
```

In each of the examples we have seen, the subprogram that we provide as an actual, either explicitly or implicitly, for a formal generic subprogram has the same parameter and result type profile as the formal. That means corresponding parameters have the same base types and, in the case of functions, the result types have the same base types. In fact, the rule for generic subprograms is stronger than that. The actual and formal subprograms must have *conforming profiles*, which means both are procedures or both are functions; the parameter and result type profiles of the two subprograms are the same; and corresponding parameters have the same class (**signal**, **variable**, **constant**, or **file**) and mode (**in**, **out**, or **inout**). The purpose of these rules is to ensure that a call to the formal subprogram will be legal for whatever actual subprogram is provided. As a counterexample, suppose the formal subprogram had a signal parameter of a given type, and the actual subprogram had a variable parameter of the same type. A call to the formal subprogram would provide a signal as the actual parameter. However, the actual subprogram would expect a variable, and would perform variable assignments on it. This is clearly an error, even though the parameter and result type profiles of the two subprograms match. The additional requirements for profile conformance avoid this kind of error.

There are two further rules relating to the parameters of generic subprograms. The first is that, if a formal parameter of a formal generic subprogram has a default value, that value is used when an actual parameter is omitted, regardless of whether the corresponding formal parameter of the actual subprogram has a default value. An example will help clarify this. Suppose we declare an entity with a formal generic subprogram, and a corresponding architecture, as follows:

```
entity up_down_counter is
  generic ( type T;
            function add ( x : T; by : integer := 1 ) return T )
  port ( ... );
end entity up_down_counter;

architecture rtl of up_down_counter is
begin
  count : process (clk) is
  begin
    if rising_edge(clk) then
      if mode then
        count_value <= add(count_value);   -- use default value
      else
        count_value <= add(count_value, -1);
      end if;
    end if;
  end process count;
end architecture rtl;
```

The formal generic subprogram **add** has a parameter **by** with the default value 1. In the first call to **add** within the architecture, we omit a value for **by**, so the default value 1 is used. This allows an analyzer to compile the call with the default value independently of any instantiation of the enclosing entity that we might write subsequently. For example, suppose we instantiate the entity with an actual generic subprogram declared as follows:

```
function add_int ( a : integer; incr : integer := 0 )
                    return integer is
begin
  return a + incr;
end function add_int;
...

int_counter : entity work.up_down_counter(rtl)
  generic map ( T => integer; add => add_int )
  port map ( ... );
```

In this instance, the actual generic subprogram associated with **add** has the default value 0 for its second parameter. Despite this, the first call to the subprogram in the architecture still uses the default value 1 for the **by** parameter, since that is what is declared for the formal generic subprogram.

The rule dealing with default values for parameters also applies to the case where the parameter of the formal generic subprogram has no default value. In that case, a call must supply a value, even if the actual generic subprogram in an instance has a default value for the parameter. For example, in the **up_down_counter** entity, had we declared the formal generic function **add** as follows:

```
function add ( x : T; by : integer ) return T
```

the first call within the architecture would have to specify an actual value for the **by** parameter. The fact that the function **add_int** supplied as the actual generic subprogram in the instance has a default value for its second parameter cannot be used within the architecture.

The second rule relating to parameters of generic subprograms is that the parameter-subtype constraints of the actual subprogram apply when the subprogram is called, not the parameter-subtype constraints of the formal subprogram. To illustrate, suppose we instantiate the **up_down_counter** entity with a different function, as follows:

```
function add_nat ( a : natural; incr : natural := 0 )
                    return natural is
begin
  return a + incr;
end function add_nat;
...

nat_counter : entity work.up_down_counter(rtl)
  generic map ( T => natural; add => add_nat )
  port map ( ... );
```

In this instance, the second parameter of the actual generic subprogram is of the base type **integer** with a range constraint requiring the value to be non-negative. The second call within the architecture provides the value –1 for the parameter. While this conforms to the constraint on the **by** parameter of the formal generic subprogram, it does not conform to the constraint on the corresponding parameter of the actual generic subprogram in the instance. Hence, when the function is called with that value in the instance, an error occurs.

12.6 Generic Packages

We have seen that we can include a generic list in a package declaration to make the package reusable for different actual types and operations. This is particularly useful for a package implementing an abstract data type (ADT), as we illustrated in Section 12.3.

Suppose we have an ADT specified in a package with generics, and we want to provide a further package extending the types and operations of the ADT. To make the extension package reusable, we would have to provide a generic type to specify an instance of the ADT's named type, along with generic subprograms for each of the ADT's operations. If the ADT has many operations, specifying them as actual generic subprograms in every instance of the extension package would be extremely onerous. To avoid this, VHDL allows us to specify an instance of the ADT package as a *formal generic package* of the extension package. Once we've instantiated the ADT package, we then provide that instance as the *actual generic package* of the extension package.

There are three forms of formal generic package declaration that we can write in a generic list. The first form has the syntax:

interface_declaration ⇐
    package identifier **is new** *uninstantiated_package_*name
      generic map (⬦)

In this case, the formal package represents an instance of the named uninstantiated package, for use within the enclosing unit containing the generic list. In most use cases, the enclosing unit is itself an uninstantiated package. However, we can also specify formal generic packages in the generic lists of entities and subprograms. When we instantiate the enclosing unit, we provide an actual package corresponding to the formal generic package. The actual package must be an instance of the named uninstantiated packge. The box notation "⬦" written in the generic map of the formal generic package specifies that the actual package is allowed to be any instance of the named uninstantiated package. We use this form when the enclosing unit does not depend on the particular actual generics defined for the actual generic package.

No doubt, all of this discussion of packages within packages and generics at different levels can become confusing. The best way to motivate the need for formal generic packages and to sort out the relationships between the pieces is with an example.

EXAMPLE 12.18 *A package for fixed-point complex numbers*

The standard package **fixed_generic_pkg** (described in Section 9.2.4) defines an ADT for fixed-point numbers represented as vectors of **std_ulogic** elements. The package is an uninstantiated package, with generic constants specifying how to round results, how to handle overflow, the number of guard bits for maintaining precision, and whether to issue warnings. An outline of the package declaration is:

```
library ieee;  use ...;
```

```vhdl
package fixed_generic_pkg is
  generic (
    fixed_round_style     : fixed_round_style_type
                                    := fixed_round;
    fixed_overflow_style  : fixed_overflow_style_type
                                    := fixed_saturate;
    fixed_guard_bits      : natural := 3;
    no_warning            : boolean := false
    );

  ...

end package fixed_generic_pkg;
```

The package defines subtypes **ufixed** and **sfixed** for unsigned and signed fixed-point numbers and numerous arithmetic, conversion and input/output operations. The bodies of the operations use the values of the formal generic constants to govern their behavior. We can instantiate the package with values for the actual generic constants to get a version with the appropriate behavior for our specific design needs.

Now suppose we wish to build upon the fixed-point package to define fixed-point complex numbers, represented in Cartesian form with fixed-point real and imaginary parts. We want the two parts of a complex number to have the same left and right index bounds, implying the same range and precision for the two parts. We can achieve that constraint by defining the complex-number type and operations in a package with formal generic constants for the index bounds. The complex-number type is defined using the **sfixed** type from an instance of the fixed-point package, and the complex-number operations need to use fixed-point operations from that instance. Thus, we include a formal generic package in the generic list of the complex-number package, as follows:

```vhdl
library ieee;
package complex_generic_pkg is
  generic ( left, right : integer;
            package fixed_pkg_for_complex is
              new ieee.fixed_generic_pkg
                generic map (<>) );

  use fixed_pkg_for_complex.all;

  type complex is record
    re, im : sfixed(left downto right);
  end record;

  function "-"  ( z : complex ) return complex;
  function conj ( z : complex ) return complex;
  function "+"  ( l : complex;  r : complex ) return complex;
  function "-"  ( l : complex;  r : complex ) return complex;
  function "*"  ( l : complex;  r : complex ) return complex;
  function "/"  ( l : complex;  r : complex ) return complex;
```

end package complex_generic_pkg;

Within the complex_generic_pkg package, the formal generic package fixed_pkg_for_complex represents an instance of the fixed_generic_pkg package. The box notation in the generic map indicates that any instance of fixed_generic_pkg will be appropriate as an actual package. The use clause makes items defined in the fixed_pkg_for_complex instance visible, so that sfixed can be used in the declaration of type complex. The generic constants left and right are used to specify the index bounds of the two record elements. The operations defined for sfixed in the fixed_pkg_for_complex instance are also used to implement the complex-number operations in the package body for complex_generic_pkg, as follows:

```
package body fixed_complex_pkg is

  function "-" ( z : complex ) return complex is
  begin
    return ( -z.re, -z.im );
  end function "-";

    . . .

end package body fixed_complex_pkg;
```

In the "−" operation for type complex, the "−" operation for type sfixed is applied to each of the real and imaginary parts. The other operations use the sfixed operations similarly.

In a design, we can instantiate both the fixed-point package and the complex-number package according to our design needs, for example:

```
package dsp_fixed_pkg is new ieee.fixed_generic_pkg
  generic map ( fixed_rounding_style => fixed_round,
                fixed_overflow_style => fixed_saturate,
                fixed_guard_bits => 2,
                no_warning => false );

package dsp_complex_pkg is new work.complex_generic_pkg
  generic map ( left => 3, right => -12,
                fixed_pkg_for_complex => dsp_fixed_pkg );
```

The first instantiation defines an instance of the fixed-point package, which provides the type sfixed and operations with the required behavior. The second instantiation defines an instance of the complex-number package with left and right bounds of 3 and −12 for both the real and imaginary parts. The type sfixed and the corresponding operations used within this instance of the complex-number package are provided by the actual generic package dsp_fixed_pkg. We can use the packages to declare variables and apply operations as follows:

```
use dsp_fixed_pkg.all, dsp_complex_pkg.all;
variable a, b, z : complex
```

```
variable c : sfixed;
...

z := a + conj(b);
z := (c * z.re, c * z.im);
```

The second form of formal generic package that we can write in a generic list has the syntax:

interface_declaration ⇐
    package identifier **is new** *uninstantiated_package*_name
      generic map (*generic*_association_list)

Again, the formal package represents an instance of the named uninstantiated package, for use within the enclosing unit containing the generic list. The actual generics provided in the generic map of the formal generic package specify that the actual package must be an instance of the named uninstantiated package with those same actual generics. We generally use this form when the enclosing unit also has another formal generic package defined earlier in its generic list. The latter generic is expected to have a generic package that is the same instance as the actual for the earlier generic package. No doubt that statement is unfathomable due to the packages within packages within packages. We will build on the previous example to help motivate the need for the language feature and to show how it may be used.

EXAMPLE 12.19   *Mathematical operations on fixed-point complex numbers*

In the previous example, we defined a package for complex numbers that provided a complex-number type and basic arithmetic operations. We can build upon this package to define a further package for more advanced mathematical operations on complex values. We will also use a package of advanced mathematical operations defined for fixed-point values:

```
package fixed_math_ops is
  generic ( package fixed_pkg_for_math is
              new ieee.fixed_generic_pkg
                generic map (<>) );

  use fixed_pkg_for_math.all;

  function sqrt ( x : sfixed ) return sfixed;
  function exp ( x : sfixed ) return sfixed;
  ...

end package fixed_math_ops;
```

This package has a formal generic package for an instance of the **fixed_generic_ pkg** package, since the operations it applies to the function parameters of type **sfixed** must be performed using the behavior defined for the **sfixed** type in the package instance proving the type. This is a similar scenario to that described in the previous example.

The advanced complex-number operations must be performed using the same **sfixed** type and basic fixed-point operations used to define the complex-number type and operations. It must also use the advanced fixed-point operations and the complex-number type and operations, with those types and operations being based on the same **sfixed** type and basic fixed-point operations. Thus, the advance complex-number package must have formal generic packages for the fixed-point package, the fixed-point mathematical operations package, and the complex-number package, as follows:

```
package complex_math_ops is
  generic ( left, right : integer;
            package fixed_pkg_for_complex_math is
              new ieee.fixed_generic_pkg
                generic map (<>);
            package fixed_math_ops is
              new work.fixed_math_ops
                generic map ( fixed_pkg_for_math =>
                                fixed_pkg_for_complex_math );
            package complex_pkg is
              new work.complex_generic_pkg
                generic map ( left => left, right => right,
                              fixed_pkg_for_complex =>
                                fixed_pkg_for_complex_math ) );

  use fixed_pkg_for_complex_math.all,
      fixed_math_ops.all, complex_pkg.all;

  function "abs" ( z : complex ) return sfixed;
  function arg   ( z : complex ) return sfixed;
  function sqrt  ( z : complex ) return complex;
  ...

end package complex_math_ops;
```

The package body is

```
package body complex_math_ops is

  function "abs" ( z : complex ) return sfixed is
  begin
    return sqrt(z.re * z.re + z.im * z.im);
  end function "abs";

  ...

end package body complex_math_ops;
```

We can now instantiate the packages for a given design. For example, given the instances **dsp_fixed_pkg** and **dsp_complex_pkg** declared in the previous example, we can also declare instances of the advanced fixed-point operations package and the advanced complex operations package:

```
package dsp_fixed_math_ops is new work.fixed_math_ops
  generic map ( fixed_pkg_for_math => dsp_fixed_pkg );

package dsp_complex_math_ops is new work.complex_math_ops
  generic map ( left => 3, right => -12,
                fixed_pkg_for_complex_math => dsp_fixed_pkg,
                fixed_math_ops => dsp_fixed_math_ops,
                complex_pkg    => dsp_complex_pkg );
```

The third form of formal generic package that we can write in a generic list has the syntax:

interface_declaration ⇐
    package identifier **is new** *uninstantiated_package*_name
    generic map (default)

This form is similar in usage to the second form, but replaces the actual generics with the reserved word **default**. We can use this third form when the named uninstantiated package has defaults for all of its formal generics. The actual package must then be an instance of the named uninstantiated package with all of the actual generics being the same as the defaults. Those actual generics (for the actual generic package) can be either explicitly specified when the actual package is instantiated, or they can be implied by leaving the actual generics unassociated. Thus, this third form is really just a notational convenience, as it saves us writing out the defaults again as actual generics in the generic map of the formal generic package.

While generic packages might seem to be rather complex to put into practice, we envisage that most of the time packages using generic packages will be developed by personnel in support of design teams. They would normally provide source code templates for designers to instantiate the packages, including instantiating any dependent packages as actual generics. Thus, the designers would be largely insulated from the complexity.

For the developers of such packages, however, there are a number of rules relating to formal and actual generic packages. As we have mentioned, the actual package corresponding to a formal generic package must be an instance of the named uninstantiated package. To summarize the rules relating to the generic map in the formal generic package:

- If the generic map of the formal generic package uses the box ("<>") symbol, the actual generic package can be any instance of the named uninstantiated package.

- If the formal generic package declaration includes a generic map with actual generics, then the actual generics in the actual package's instantiation must match the actual generics in the formal generic package declaration.

- If the formal generic package declaration includes a generic map with the reserved word **default**, then the actual generics in the actual package's instantiation must match the default generics in the generic list of the named uninstantiated package.

The meaning of the term "match" applied to actual generics depends on what kind of generics are being matched. For generic constants, the actuals must be the same value. It doesn't matter whether that value is specified as a literal, a named constant, or any other expression. For a generic type, the actuals must denote the same subtype; that is, they must denote the same base type and the same constraints. Constraints on a subtype include range constraints, index ranges and directions, and element subtypes. For generic subprograms, the actuals must refer to the same subprogram, and for generic packages, the actuals must refer to the same instance of a specified uninstantiated package.

In the case of a default generic subprogram implied by a box symbol in the generic list of the named uninstantiated package, the actual subprogram must be the subprogram of the same name and conforming profile directly visible at the point where the formal generic package is declared. For example, if an uninstantiated package is declared as

```
package pkg1 is
  generic ( function "<" ( L, R : integer )
                          return boolean is <> ) );
  ...
end package pkg1;
```

we can declare a second package as follows:

```
package pkg2 is
  generic ( package inst1 is new pkg1 generic map ( default ) );
  ...
end package pkg2;
```

In this case, any package provided as an actual for inst1 must be an instance of pkg1, such as the following:

```
package ascending_pkg1 is new pkg1
  generic map ( T => integer );
```

Since the predefined "<" function for **integer** is visible at the point of declaring ascending_pkg1, that function is used as the actual for the generic function "<" in the instance of pkg1. At the place of declaring the formal generic package inst1 within the generic list of pkg2, the predefined "<" function for integer is also directly visible, so it is this function that must be matched as the actual for "<" in any instance of pkg1 supplied as an actual for inst1. Thus, the following instantiation of pgk2 is legal:

```
package integer_pkg2 is new pkg2
  generic map ( inst1 => ascending_pkg1 );
```

Exercises

1. [● 12.1] Add to the following entity interface a generic clause defining generic constants Tpw_clk_h and Tpw_clk_l that specify the minimum clock pulse width timing. Both generic constants have a default value of 3 ns.

    ```
    entity flipflop is
      port ( clk, d : in bit;  q, q_n : out bit );
    end entity flipflop;
    ```

2. [● 12.1] Write a component instantiation statement that instantiates the following entity from the current working library. The actual value for the generic constant should be 10 ns, and the clk signal should be associated with a signal called master_clk.

    ```
    entity clock_generator is
      generic ( period : delay_length );
      port ( clk : out std_ulogic );
    end entity clock_generator;
    ```

3. [● 12.1] Following is an incomplete entity interface that uses a generic constant to specify the sizes of the standard-logic vector input and output ports. Complete the interface by filling in the types of the ports.

    ```
    entity adder is
      generic ( data_length : positive );
      port ( a, b : in ...;  sum : out ... );
    end entity adder;
    ```

4. [● 12.1] A system has an 8-bit data bus declared as

    ```
    signal data_out : bit_vector(7 downto 0);
    ```

    Write a component instantiation statement that instantiates the reg entity defined in Example 12.3 to implement a 4-bit control register. The register data input connects to the rightmost four bits of data_out, the clk input to io_write, the reset input to io_reset and the data output bits to control signals io_en, io_int_en, io_dir and io_mode.

5. [● 12.2] Write an instantiation of the multiplexer of Example 12.5 connected to a select signal sel of type bit, and to data inputs d_in1 and d_in2 and data output d_out, all of type bit_vector(7 downto 0).

6. [● 12.2] For the entity e and architecture a described on page 374, why is the following instantiation illegal?

    ```
    my_e : entity work.e(a)
      generic map ( T => unsigned,
                    init_val => unsigned'("00000000") )
      port map ( ... );
    ```

7. [❶ 12.3] Write an instantiation of the stacks package of Example 12.7 for stacks of integers up to 100 elements deep. Declare a stack variable and push the number −1 onto it.

8. [❶ 12.3] Given an instance of the package **generic_pkg** described on page 380:

   ```
   package int_std_pkg is new generic_pkg
     generic map ( T1 => integer, T2 => std_ulogic );
   ```

   are the following calls ambiguous? If so, why? If not, which procedure is called?

   ```
   variable b : bit;
   ...

   int_std_pkg(1);
   int_std_pkg('1');
   int_std_pkg(b);
   ```

9. [❶ 12.3] Suppose the variable **test_buffer** is an instance of the bounded buffer ADT described in Section 12.3.2. Write statements that fill the buffer with zero bytes.

10. [❶ 12.4] Write an instantiation of the **check_setup** procedure of Example 12.12 and a call of the instantiation to check that a **bit_vector** signal **s** meets a setup time of 100 ps before a falling edge of a **bit** signal **clk**.

11. [❶ 12.5] Write an instantiation of the **generic_counter** entity of Example 12.13 to connect to a a 10-element **bit_vector** output signal **val_count**. Include any additional declarations you need for the instantiation.

12. [❶ 12.5] Write an instantiation of the **dictionaries** package of Example 12.16 that stores up to 1000 elements, each being a 64-bit **unsigned** value. Use the string representation returned by the **to_hstring** operation as the key type, and sort elements in descending order by providing the ">" operation for comparison of keys.

13. [❶ 12.6] Write a package declaration for a package that provides the mathematical operations **exp** and **log** on values of type **float**. The mathematical operations package should have a formal generic package for the instance of **ieee.float_generic_pkg** providing the definition of **float**. Instantiate the package with **ieee.float_pkg** as the actual generic package.

14. [❷ 12.1] Develop a behavioral model of a D-latch with separate generic constants for specifying the following propagation delays:

    - rising data input to rising data output,
    - falling data input to falling data output,
    - rising enable input to rising data output and
    - rising enable input to falling data output.

15. [❷ 12.1] Develop a behavioral model of a counter with output of type **natural** and clock and reset inputs of type **bit**. The counter has a Boolean generic constant,

trace_reset. When this is true, the counter reports a trace message each time the reset input is activated.

16. [❷ 12.1] Develop a behavioral model of the adder described in Exercise 3.

17. [❷ 12.1] Develop a behavioral model of a multiplexer with n select inputs, 2^n data inputs and one data output.

18. [❷ 12.2] Develop a model of a generic tri-state buffer for use with various types. The entity should have a generic constant for the value used to represent the disabled state. Instantiate your buffer with **std_ulogic** as the data type and 'H' as the disabled-state value, representing a buffer with a resistive pullup on the output.

19. [❷ 12.3] Augment the stacks package of Example 12.7 to include operations to test for an empty stack, a full stack, and to return the number of elements in a stack.

20. [❷ 12.5] Write an instantiation of the operations package of Example 12.14 for which the error-reporting procedure simply counts the number of errors with severity less than **failure**. When an error with severity **failure** occurs, the procedure issues a report message showing the number of accumulated errors, together with the **failure** message, and then finishes simulation.

21. [❷ 12.5] Complete the body of the **search_and_insert** procedure in the dictionaries package body in Example 12.16.

22. [❸ 12.1] Develop a behavioral model of a RAM with generic constants governing the read access time, minimum write time, the address port width and the data port width.

Chapter 13

Components and Configurations

In Chapter 5 we saw how to write entity declarations and architecture bodies that describe the structure of a system. Within an architecture body, we can write component instantiation statements that describe instances of an entity and connect signals to the ports of the instances. This simple approach to building a hierarchical design works well if we know in advance all the details of the entities we want to use. However, that is not always the case, especially in a large design project. In this chapter we introduce an alternative way of describing the hierarchical structure of a design that affords significantly more flexibility at the cost of a little more effort in managing the design.

13.1 Components

The first thing we need to do to describe an interconnection of subsystems in a design is to describe the different kinds of components used. We have seen how to do this by writing entity declarations for each of the subsystems. Each entity declaration is a separate design unit and has corresponding architecture bodies that describe implementations. An alternative approach is to write *component declarations* in the declarative part of an architecture body or package interface. We can then create *instances* of the components within the statement part of the architecture body.

13.1.1 Component Declarations

A component declaration simply specifies the external interface to the component in terms of generic constants and ports. We do not need to describe any corresponding implementation, since all we are interested in is how the component is connected in the current level of the design hierarchy. This makes the architecture completely self-contained, since it does not depend on any other library units except its corresponding entity interface. Let us look at the syntax rule that governs how we write a component declaration.

```
component_declaration ⇐
    component identifier [ is ]
        [ generic ( generic_interface_list ) ; ]
```

```
    [ port ( port_interface_list ) ; ]
  end component [ identifier ] ;
```

A simple example of a component declaration that follows this syntax rule is

```
component flipflop is
  generic ( Tprop, Tsetup, Thold : delay_length );
  port ( clk : in bit;  clr : in bit;  d : in bit;
         q : out bit );
end component flipflop;
```

This declaration defines a component type that represents a flipflop with clock, clear and data inputs, clk, clr and d, and a data output q. It also has generic constants for parameterizing the propagation delay, the data setup time and the data hold time.

Note the similarity between a component declaration and an entity declaration. This similarity is not accidental, since they both serve to define the external interface to a module. Although there is a very close relationship between components and entities, in fact, they embody two different concepts. This may be a source of confusion to newcomers to VHDL. Nevertheless, the flexibility afforded by having the two different constructs is a powerful feature of VHDL, so we will work through it carefully in this section and try to make the distinction clear.

One way of thinking about the difference between an entity declaration and a component declaration is to think of the modules being defined as having different levels of "reality." An entity declaration defines a "real" module: something that ultimately will have a physical manifestation. For example, it may represent a circuit board in a rack, a packaged integrated circuit or a standard cell included in a piece of silicon. An entity declaration is a separate design unit that may be separately analyzed and placed into a design library. A component declaration, on the other hand, defines a "virtual," or "idealized," module that is included within an architecture body. It is as though we are saying, "For this architecture body, we assume there is a module as defined by this component declaration, since such a module meets our needs exactly." We specify the names, types and modes of the ports on the virtual module (the component) and proceed to lay out the structure of the design using this idealized view.

Of course, we do not make these assumptions about modules arbitrarily. One possibility is that we know what real modules are available and customize the virtual reality based on that knowledge. The advantage here is that the idealization cushions us from the irrelevant details of the real module, making the design easier to manage. Another possibility is that we are working "top down" and will later use the idealized module as the specification for a real module. Either way, eventually a link has to be made between an instance of a virtual component and a real entity so that the design can be constructed. In the rest of this section, we look at how to use components in an architecture body, then come back to the question of the binding between component instances and entities.

VHDL-87

The keyword **is** may not be included in the header of a component declaration, and the component name may not be repeated at the end of the declaration.

13.1.2 Component Instantiation

If a component declaration defines a kind of module, then a component instantiation specifies a usage of the module in a design. We have seen how we can instantiate an entity directly using a component instantiation statement within an architecture body. Let us now look at an alternative syntax rule that shows how we can instantiate a declared component:

component_instantiation_statement ⇐
    *instantiation_*label :
        〚 **component** 〛 *component_*name
        〚 **generic map** (*generic_*association_list) 〛
        〚 **port map** (*port_*association_list) 〛 ;

This syntax rule shows us that we may simply name a component declared in the architecture body and, if required, provide actual values for the generics and actual signals to connect to the ports. The label is required to identify the component instance.

EXAMPLE 13.1   *A four-bit register using flipflop components*

We can construct a four-bit register using flipflops and an and gate, similar to the example in Chapter 5. The entity declaration is

```
entity reg4 is
  port ( clk, clr : in bit;  d : in bit_vector(0 to 3);
         q : out bit_vector(0 to 3) );
end entity reg4;
```

The architecture body describing the structure of this register uses the flipflop component shown on page 418.

```
architecture struct of reg4 is

  component flipflop is
    generic ( Tprop, Tsetup, Thold : delay_length );
    port ( clk : in bit;  clr : in bit;  d : in bit;
           q : out bit );
  end component flipflop;

begin
```

```
bit0 : component flipflop
    generic map ( Tprop => 2 ns, Tsetup => 2 ns, Thold => 1 ns )
    port map ( clk => clk, clr => clr, d => d(0), q => q(0) );
bit1 : component flipflop
    generic map ( Tprop => 2 ns, Tsetup => 2 ns, Thold => 1 ns )
    port map ( clk => clk, clr => clr, d => d(1), q => q(1) );
bit2 : component flipflop
    generic map ( Tprop => 2 ns, Tsetup => 2 ns, Thold => 1 ns )
    port map ( clk => clk, clr => clr, d => d(2), q => q(2) );
bit3 : component flipflop
    generic map ( Tprop => 2 ns, Tsetup => 2 ns, Thold => 1 ns )
    port map ( clk => clk, clr => clr, d => d(3), q => q(3) );

end architecture struct;
```

Note that all we have done here is specify the structure of this level of the design hierarchy, without having indicated how the flipflop is implemented. We will see how that may be done in the remainder of this chapter.

VHDL-87

The keyword **component** may not be included in a component instantiation statement in VHDL-87. The keyword is allowed in VHDL-93 and VHDL-2002 to distinguish between instantiation of a component and direct instantiation of an entity. In VHDL-87, the only form of component instantiation statement provided is instantiation of a declared component.

13.1.3 Packaging Components

Let us now turn to the issue of design management for large projects and see how we can make management of large libraries of entities easier using packages and components. Usually, work on a large design is partitioned among several designers, each responsible for implementing one or more entities that are used in the complete system. Each entity may need to have some associated types defined in a utility package, so that entity ports can be declared using those types. When the entity is used, other designers will need component declarations to instantiate components that will eventually be bound to the entity. It makes good sense to include a component declaration in the utility package, along with the types and other related items. This means that users of the entity do not need to rewrite the declarations, thus avoiding a potential source of errors and misunderstanding.

EXAMPLE 13.2 *Packaged component for a serial interface cell*

Suppose we are responsible for designing a serial interface cell for a microcontroller circuit. We can write a package specification that defines the interface to be used in the rest of the design:

```
library ieee;  use ieee.std_logic_1164.all;

package serial_interface_defs is

  subtype reg_address_vector is std_ulogic_vector(1 downto 0);

  constant status_reg_address : reg_address_vector := B"00";
  constant control_reg_address : reg_address_vector := B"01";
  constant rx_data_register : reg_address_vector := B"10";
  constant tx_data_register : reg_address_vector := B"11";

  subtype data_vector is std_ulogic_vector(7 downto 0);

  ...    -- other useful declarations

  component serial_interface is
    port ( clock_phi1, clock_phi2 : in std_ulogic;
           serial_select : in std_ulogic;
           reg_address : in reg_address_vector;
           data : inout data_vector;
           interrupt_request : out std_ulogic;
           rx_serial_data : in std_ulogic;
           tx_serial_data : out std_ulogic );
  end component serial_interface;

end package serial_interface_defs;
```

The component declaration in this package corresponds to our entity declaration for the serial interface:

```
library ieee;  use ieee.std_logic_1164.all;

use work.serial_interface_defs.all;

entity serial_interface is
  port ( clock_phi1, clock_phi2 : in std_ulogic;
         serial_select : in std_ulogic;
         reg_address : in reg_address_vector;
         data : inout data_vector;
         interrupt_request : out std_ulogic;
         rx_serial_data : in std_ulogic;
         tx_serial_data : out std_ulogic );
end entity serial_interface;
```

When other designers working on integrating the entire circuit need to instantiate the serial interface, they only need to import the items in the package, rather than rewriting all of the declarations. An outline of a design that does this is

```
library ieee;  use ieee.std_logic_1164.all;

architecture structure of microcontroller is

  use work.serial_interface_defs.serial_interface;

    ...      -- declarations of other components, signals, etc
begin

  serial_a : component serial_interface
    port map ( clock_phi1 => buffered_phi1,
               clock_phi2 => buffered_phi2,
               serial_select => serial_a_select,
               reg_address => internal_addr(1 downto 0),
               data => internal_data_bus,
               interrupt_request => serial_a_int_req,
               rx_serial_data => rx_data_a,
               tx_serial_data => tx_data_a );

    ...      -- other component instances

end architecture structure;
```

13.2 Configuring Component Instances

Once we have described the structure of one level of a design using components and component instantiations, we still need to flesh out the hierarchical implementation for each component instance. We can do this by writing a *configuration declaration* for the design. In it, we specify which real entity interface and corresponding architecture body should be used for each of the component instances. This is called binding the component instances to design entities. Note that we do not specify any binding information for a component instantiation statement that directly instantiates an entity, since the entity and architecture body are specified explicitly in the component instantiation statement. Thus our discussion in this section only applies to instantiations of declared components.

13.2.1 Basic Configuration Declarations

We start by looking at a simplified set of syntax rules for configuration declarations, as the full set of rules is rather complicated. The simplest case arises when the entities to which component instances are bound are implemented with behavioral architectures. In this case, there is only one level of the hierarchy to flesh out. The simplified syntax rules are

configuration_declaration ⇐
       configuration identifier **of** *entity*_name **is**
          for *architecture*_name

⟦   **for** component_specification
                binding_indication ;
          end for ; ⟧
     end for ;
   end ⟦ **configuration** ⟧ ⟦ identifier ⟧ ;

component_specification ⟸
      (*instantiation*_label ⟨ , … ⟩ ⫾ **others** ⫾ **all**) : *component*_name

binding_indication ⟸ **use entity** *entity*_name ⟦ (*architecture*_identifier) ⟧

The identifier given in the configuration declaration identifies this particular specification for fleshing out the hierarchy of the named entity. There may be other configuration declarations, with different names, for the same entity. Within the configuration declaration we write the name of the particular architecture body to work with (included after the first **for** keyword), since there may be several corresponding to the entity. We then include the binding information for each component instance within the architecture body. The syntax rule shows that we can identify a component instance by its label and its component name, as used in the component instantiation in the architecture body. We bind it by specifying an entity name and a corresponding architecture body name. For example, we might bind instances **bit0** and **bit1** of the component **flipflop** as follows:

```
for bit0, bit1 : flipflop
  use entity work.edge_triggered_Dff(basic);
end for;
```

This indicates that the instances are each to be bound to the design entity **edge_triggered_Dff**, found in the current working library, and that the architecture body basic corresponding to that entity should be used as the implementation of the instances. Note that since we can identify each component instance individually, we have the opportunity to bind different instances of a given component to different entity/architecture pairs. After we have specified bindings for some of the instances in a design, we can use the keyword **others** to bind any remaining instances of a given component type to a given entity/architecture pair. Alternatively, if all instances of a particular component type are to have the same binding, we can use the keyword **all** instead of naming individual instances. The syntax rules also show that the architecture name corresponding to the entity is optional. If it is omitted, a default binding takes place when the design is elaborated for simulation or synthesis. The component instance is bound to whichever architecture body for the named entity has been most recently analyzed at the time of elaboration. A further possibility is that we omit bindings for one or more component instances in the design. In this case, the default binding rule attempts to find an entity with the same name as the component. The entity must be located in the same design library as the design unit in which the instantiated component is declared. If no entity is found, the component instance remains unbound (see Section 13.2.5). Relying on the default binding rules to locate and bind the right entity can make a design difficult to understand and reduces portability. The safest approach is to ensure that we bind all component instances explicitly.

A configuration declaration is a primary design unit, and as such, may be separately analyzed and placed into the working design library as a library unit. If it contains sufficient binding information so that the full design hierarchy is fleshed out down to behavioral architectures, the configuration may be used as the target unit of a simulation. The design is elaborated by substituting instances of the specified architecture bodies for bound component instances in the way described in Section 5.4. The only difference is that when component declarations are instantiated, the configuration must be consulted to find the appropriate architecture body to substitute.

EXAMPLE 13.3   *Configuration of a four-bit register*

Let us look at a sample configuration declaration that binds the component instances in the four-bit register of Example 13.1. Suppose we have a resource library for a project, **star_lib**, that contains the basic design entities that we need to use. Our configuration declaration might be written as follows:

```
library star_lib;
use star_lib.edge_triggered_Dff;

configuration reg4_gate_level of reg4 is

  for struct   -- architecture of reg4

    for bit0 : flipflop
      use entity edge_triggered_Dff(hi_fanout);
    end for;

    for others : flipflop
      use entity edge_triggered_Dff(basic);
    end for;

  end for;   -- end of architecture struct

end configuration reg4_gate_level;
```

The library clause preceding the design unit is required to locate the resource library containing the entities we need. The use clause following it makes the entity names we require directly visible in the configuration declaration. The configuration is called **reg4_gate_level** and selects the architecture **struct** of the **reg4** entity. Within this architecture, we single out the instance **bit0** of the **flipflop** component and bind it to the entity **edge_triggered_Dff** with architecture **hi_fanout**. This shows how we can give special treatment to particular component instances when configuring bindings. We bind all remaining instances of the **flipflop** component to the **edge_triggered_Dff** entity using the **basic** architecture.

VHDL-87

The keyword **configuration** may not be included at the end of a configuration declaration in VHDL-87.

13.2.2 Configuring Multiple Levels of Hierarchy

In the previous section, we saw how to write a configuration declaration for a design in which the instantiated components are bound to behavioral architecture bodies. Most realistic designs, however, have deeper hierarchical structure. The components at the top level have architecture bodies that, in turn, contain component instances that must be configured. The architecture bodies bound to these second-level components may also contain component instances, and so on. In order to deal with configuring these more complex hierarchies, we need to use an alternative form of binding indication in the configuration declaration. The alternative syntax rule is

binding_indication ⇐ **use configuration** *configuration*_name

This form of binding indication for a component instance allows us to bind to a preconfigured entity/architecture pair simply by naming the configuration declaration for the entity. For example, a component instance of **reg4** with the label **flag_reg** might be bound in a configuration declaration as follows:

```
for flag_reg : reg4
  use configuration work.reg4_gate_level;
end for;
```

EXAMPLE 13.4   *Hierarchical configuration of a counter*

In Chapter 5 we looked at a two-digit decimal counter, implemented using four-bit registers. We assume that the type **digit** is defined as follows in a package named counter_types:

```
subtype digit is bit_vector(3 downto 0);
```

The entity declaration for the counter is

```
use work.counter_types.digit;
```

```
entity counter is
  port ( clk, clr : in bit;
         q0, q1 : out digit );
end entity counter;
```

Now that we have seen how to use component declarations, we can rewrite the architecture body using component declarations for the registers, as follows:

```
architecture registered of counter is
```

```
component digit_register is
  port ( clk, clr : in bit;
         d : in digit;
         q : out digit );
end component digit_register;

signal current_val0, current_val1, next_val0, next_val1 : digit;

begin

val0_reg : component digit_register
  port map ( clk => clk, clr => clr, d => next_val0,
             q => current_val0 );

val1_reg : component digit_register
  port map ( clk => clk, clr => clr, d => next_val1,
             q => current_val1 );

-- other component instances
...

end architecture registered;
```

We can configure this implementation of the counter with the following configuration declaration:

```
configuration counter_down_to_gate_level of counter is

  for registered

    for all : digit_register
      use configuration work.reg4_gate_level;
    end for;

    ...       -- bindings for other component instances

  end for;   -- end of architecture registered

end configuration counter_down_to_gate_level;
```

This configuration specifies that each instance of the **digit_register** component is bound using the information in the configuration declaration named **reg4_gate_level** in the current design library, shown in Example 13.3. That configuration in turn specifies the entity to use (**reg4**), a corresponding architecture body (**struct**) and the bindings for each component instance in that architecture body. Thus the two configuration declarations combine to fully configure the design hierarchy down to the process level.

The example above shows how we can use separate configuration declarations for each level of a design hierarchy. As a matter of style this is good practice, since it prevents the configuration declarations themselves from becoming too complex. The alternative approach is to configure an entity and its hierarchy fully within the one configuration declaration. We look at how this may be done, as some models from other designers may take this approach. While this approach is valid VHDL, we recommend the practice of splitting up the configuration information into separate configuration declarations corresponding to the entities used in the design hierarchy.

To see how to configure multiple levels within one declaration, we need to look at a more complex form of syntax rule for configuration declarations. In fact, we need to split the rule into two parts, so that we can write a recursive syntax rule.

configuration_declaration ⟸
    configuration identifier **of** *entity*_name **is**
        block_configuration
    end ⟦ **configuration** ⟧ ⟦ identifier ⟧ ;

block_configuration ⟸
    for *architecture*_name
        ⦃   **for** component_specification
            binding_indication ;
            ⟦ block_configuration ⟧
        end for ; ⦄
    end for ;

The rule for a block configuration indicates how to write the configuration information for an architecture body and its inner component instances. (The reason for the name "block configuration" in the second rule is that it applies to block statements as well as architecture bodies. We discuss block statements in Section 23.1.) Note that we have included an extra part after the binding indication for a component instance. If the architecture that we bind to an instance also contains component instances, we can nest further configuration information for that architecture inside the enclosing block configuration.

EXAMPLE 13.5 *A complete configuration for the counter*

We can write a configuration declaration equivalent to that in Example 13.4 but containing all of the configuration information for the entire hierarchy, as follows:

```
library star_lib;
use star_lib.edge_triggered_Dff;

configuration full of counter is

  for registered   -- architecture of counter

    for all : digit_register
      use entity work.reg4(struct);

      for struct   -- architecture of reg4

        for bit0 : flipflop
          use entity edge_triggered_Dff(hi_fanout);
        end for;
```

```
        for others : flipflop
          use entity edge_triggered_Dff(basic);
        end for;

      end for;   -- end of architecture struct

    end for;

    ...  -- bindings for other component instances

  end for;   -- end of architecture registered

end configuration full;
```

The difference between this configuration declaration and the one in Example 13.4 is that the binding indication for instances of **digit_register** directly refers to the entity **reg4** and the architecture body **struct**, rather than using a separate configuration for the entity. The configuration then includes all of the binding information for component instances within **struct**. This relatively simple example shows how difficult it can be to read nested configuration declarations. Separate configuration declarations are easier to understand and provide more flexibility for managing alternative compositions of a design hierarchy.

13.2.3 Direct Instantiation of Configured Entities

As we have seen, a configuration declaration specifies the design hierarchy for a design entity. We can make direct use of a fully configured design entity within an architecture body by writing a component instantiation statement that directly names the configuration. The alternative syntax rule for component instantiation statements that expresses this possibility is

component_instantiation_statement ⇐
    *instantiation*_label :
        configuration *configuration*_name
        ⟦ **generic map** (*generic*_association_list) ⟧
        ⟦ **port map** (*port*_association_list) ⟧ ;

The configuration named in the statement includes a specification of an entity and a corresponding architecture body to use. We can include generic and port maps in the component instantiation to provide actual values for any generics of the entity and actual signals to connect to the ports of the entity. This is much like instantiating the entity directly, but with all of the configuration information for its implementation included.

EXAMPLE 13.6   *Direct instantiation of the counter configuration*

The following architecture body for an alarm clock directly instantiates the two-digit decimal counter entity. The component instantiation statement labeled **seconds** refers to the configuration **counter_down_to_gate_level**, shown in Example 13.4. That configuration, in turn, specifies the counter entity and architecture to use.

```
architecture top_level of alarm_clock is

  use work.counter_types.digit;

  signal reset_to_midnight, seconds_clk : bit;
  signal seconds_units, seconds_tens : digit;
  ...

begin

  seconds : configuration work.counter_down_to_gate_level
    port map ( clk => seconds_clk, clr => reset_to_midnight,
               q0 => seconds_units, q1 => seconds_tens );

  ...

end architecture top_level;
```

VHDL-87

VHDL-87 does not allow direct instantiation of configured entities. Instead, we must declare a component, instantiate the component and write a separate configuration declaration that binds the instance to the configured entity.

13.2.4   Generic and Port Maps in Configurations

We now turn to a very powerful and important aspect of component configurations: the inclusion of generic maps and port maps in the binding indications. This facility provides a great deal of flexibility when binding component instances to design entities. However, the ideas behind the facility are somewhat difficult to grasp on first encounter, so we will work through them carefully. First, let us look at an extended syntax rule for a binding indication that shows how generic and port maps can be included:

binding_indication ⇐
    use ⟨   **entity** *entity*_name [(*architecture*_identifier)]
           ] **configuration** *configuration*_name ⟩
    [**generic map** (*generic*_association_list)]
    [**port map** (*port*_association_list)]

This rule indicates that after specifying the entity to which to bind (either directly or by naming a configuration), we may include a generic map or a port map or both. We show how this facility may be used by starting with some simple examples illustrating the more common uses. We then proceed to the general case.

    One of the most important uses of this facility is to separate the specification of generic constants used for timing from the structure of a design. We can write component declarations in a structural description without including generic constants for timing. Later, when we bind each component instance to an entity in a configuration declaration, we can specify the timing values by supplying actual values for the generic constants of the bound entities.

EXAMPLE 13.7 *Timing generics in a configuration*

Suppose we are designing an integrated circuit for a controller, and we wish to use the register whose entity declaration includes generic constants for timing and port width, as follows:

```
library ieee;  use ieee.std_logic_1164.all;

entity reg is
  generic ( t_setup, t_hold, t_pd : delay_length;
            width : positive );
  port ( clock : in std_ulogic;
         data_in : in std_ulogic_vector(0 to width - 1);
         data_out : out std_ulogic_vector(0 to width - 1) );
end entity reg;
```

We can write a component declaration for the register without including the generic constants used for timing, as shown in the following architecture body for the controller:

```
architecture structural of controller is

  component reg is
    generic ( width : positive );
    port ( clock : in std_ulogic;
           data_in : in std_ulogic_vector(0 to width - 1);
           data_out : out std_ulogic_vector(0 to width - 1) );
  end component reg;

  . . .

begin

  state_reg : component reg
    generic map ( width => state_type'length )
    port map ( clock => clock_phase1,
               data_in => next_state,
               data_out => current_state );

  . . .

end architecture structural;
```

The component represents a virtual idealized module that has all of the structural characteristics we need, but ignores timing. The component instantiation statement specifies a value for the port width generic constant, but does not specify any timing parameters.

Since we are operating in the real world, we cannot ignore timing forever. Ultimately the values for the timing parameters will be determined from the physical layout of the integrated circuit. Meanwhile, during the design phase, we can use estimates for their values. When we write a configuration declaration for our design, we can configure the component instance as shown below, supplying the estimates in a generic map. Note that we also need to specify a value for the **width** generic of the bound entity. In this example, we supply the value of the **width** generic of the component instance. We discuss this in more detail on page 433.

```
configuration controller_with_timing of controller is

  for structural

    for state_reg : reg
      use entity work.reg(gate_level)
      generic map ( t_setup => 200 ps, t_hold => 150 ps,
                    t_pd => 150 ps, width => width );
    end for;

    . . .

  end for;

end configuration controller_with_timing;
```

When we simulate the design, the estimated values for the generic constants are used by the real design entity to which the component instance is bound. Later, when the integrated circuit has been laid out, we can substitute, or *back annotate*, the actual timing values in the configuration declaration without having to modify the architecture body of the model. We can then resimulate to obtain test vectors for the circuit that take account of the real timing.

Another important use of generic and port maps in a configuration declaration arises when the entity to which we want to bind a component instance has different names for generic constants and ports. The maps in the binding indication can be used to make the link between component generics and ports on the one hand, and entity generics and ports on the other. Furthermore, the entity may have additional generics or ports beyond those of the component instance. In this case, the maps can be used to associate actual values or signals from the architecture body with the additional generics or ports.

EXAMPLE 13.8 *Remapping generics and ports in a configuration*

Suppose we need to use a two-input-to-four-output decoder in a design, as shown in the outline of an architecture body below. The component declaration for the decoder represents a virtual module that meets our needs exactly.

```
architecture structure of computer_system is
```

```vhdl
component decoder_2_to_4 is
  generic ( prop_delay : delay_length );
  port ( in0, in1 : in bit;
           out0, out1, out2, out3 : out bit );
end component decoder_2_to_4;

...

begin

  interface_decoder : component decoder_2_to_4
    generic map ( prop_delay => 4 ns )
    port map ( in0 => addr(4), in1 => addr(5),
                 out0 => interface_a_select,
                 out1 => interface_b_select,
                 out2 => interface_c_select,
                 out3 => interface_d_select );

  ...

end architecture structure;
```

Now suppose we check in our library of entities for a real module to use for this instance and find a three-input-to-eight-output decoder. The entity declaration is:

```vhdl
entity decoder_3_to_8 is
  generic ( Tpd_01, Tpd_10 : delay_length );
  port ( s0, s1, s2 : in bit;
           enable : in bit;
           y0, y1, y2, y3, y4, y5, y6, y7 : out bit );
end entity decoder_3_to_8;
```

We could make use of this entity in our design if we could adapt to the different generic and port names and tie the unused ports to appropriate values. The following configuration declaration shows how this may be done.

```vhdl
configuration computer_structure of computer_system is

  for structure

    for interface_decoder : decoder_2_to_4
      use entity work.decoder_3_to_8(basic)
      generic map ( Tpd_01 => prop_delay, Tpd_10 => prop_delay )
      port map ( s0 => in0, s1 => in1, s2 => '0',
                   enable => '1',
                   y0 => out0, y1 => out1, y2 => out2, y3 => out3,
                   y4 => open, y5 => open, y6 => open, y7 => open );
    end for;

    ...

  end for;
```

end configuration computer_structure;

The generic map in the binding indication specifies the correspondence between entity generics and component generics. In this case, the component generic prop_delay is to be used for both entity generics. The port map in the binding indication similarly specifies which entity ports correspond to which component ports. Where the entity has extra ports, we can specify how those ports are to be connected. In this design, s2 is tied to '0', **enable** is tied to '1', and the remaining ports are left unassociated (specified by the keyword **open**).

The two preceding examples illustrate the most common uses of generic maps and port maps in configuration declarations. We now look at the general mechanism that underlies these examples, so that we can understand its use in more complex cases. We use the terms *local generics* and *local ports* to refer to the generics and ports of a component. Also, in keeping with previous discussions, we use the terms *formal generics* and *formal ports* to refer to the generics and ports of the entity to which the instance is bound.

When we write a component instantiation statement with a generic map and a port map, these maps associate actual values and signals with the *local* generics and ports of the component instance. Recall that the component is just a virtual module used as a template for a real module, so at this stage we have just made connections to the template. Next, we write a configuration declaration that binds the component instance to a real entity. The generic and port maps in the binding indication associate actual values and signals with the *formal* generics and ports of the entity. These actual values and signals may be the locals from the component instance, or they may be values and signals from the architecture body containing the component instance. Figure 13.1 illustrates the mappings. It is this two-stage association mechanism that makes configurations so powerful in mapping a design to real modules.

Figure 13.1 shows that the actual values and signals supplied in the configuration declaration may be local generics or ports from the component instance. This is the case for the formal generics Tpd_01 and Tpd_10 and for the formal ports s0, s1, y0, y1, y2 and y3 in Example 13.8. Every local generic and port of the component instance must be associated with a formal generic or port, respectively; otherwise the design is in error. The figure also shows that the configuration declaration may supply values or signals from the architecture body. Furthermore, they may be any other values or signals visible at the point of the component instantiation statement, such as the literals '0' and '1' shown in the example. Note that while it is legal to associate a signal in the architecture body with a formal port of the entity, it is not good practice to do so. This effectively modifies the structure of the circuit, making the overall design much more difficult to understand and manage. For example, in the configuration in Example 13.8, had we associated the formal port s2 with the signal addr(6) instead of the literal value '0', the operation of the circuit would be substantially altered. Note also that the two-level mapping of generics applies not only to generic constants, but also to generic types, subprograms, and packages.

The preceding examples show how we can use generic and port maps in binding indications to deal with differences between the component and the entity in the number and names of generics and ports. However, if the component and entity have similar interfaces, we can rely on a *default binding rule*. This rule is used automatically if we omit

FIGURE 13.1

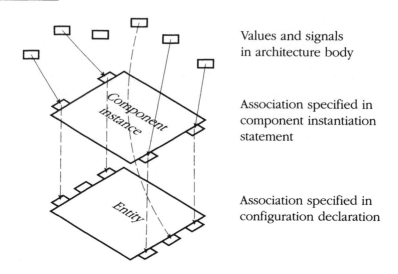

Values and signals
in architecture body

Association specified in
component instantiation
statement

Association specified in
configuration declaration

The generic and port maps in the component instantiation and the configuration declaration define a two-stage association. Values and signals in the architecture body are associated, via the local generics and ports, with the formal generics and ports of the bound entity.

the generic map or the port map in a binding indication, as we did in the earlier examples in this section. The default rule causes each local generic or port of the component to be associated with a formal generic or port of the same name in the entity interface. If the entity interface includes further formal generics or ports, they remain open. If the entity does not include a formal with the same name as one of the locals, the design is in error. So, for example, if we declare a component as

```
component nand3 is
  port ( a, b, c : in bit := '1';  y : out bit );
end component nand3;
```

and instantiate it as

```
gate1 : component nand3
  port map ( a => s1, b => s2, c => open, y => s3 );
```

then attempt to bind to an entity declared as

```
entity nand2 is
  port ( a, b : in bit := '1';  y : out bit );
end entity nand2;
```

with a component configuration

```
    for gate1 : nand3
      use entity work.nand2(basic);
    end for;
```

an error occurs. The reason for the error is that there is no formal port named c to associate with the local port of that name. The default rule requires that such a correspondence be found, even though the local port is unconnected in the architecture body.

13.2.5 Deferred Component Binding

We have seen that we can specify the binding for a component instance either by naming an entity and a corresponding architecture body, or by naming a configuration. A third option is to leave the component instance unbound and to defer binding it until later in the design cycle. The syntax rule for a binding indication that expresses this option is

> binding_indication ⇐ **use open**

If we use this form of binding indication to leave a component instance unbound, we cannot include a generic map or port map. This makes sense: since there is no entity, there are no formal generics or ports with which to associate actual values or signals.

A scenario in which we may wish to defer binding arises in complex designs that can be partially simulated before all subsystems are complete. We can write an architecture body for the system, including component declarations and instances as placeholders for the subsystems. Initially, we write a configuration declaration that defers bindings of the subsystems. Then, as the design of each subsystem is completed, the corresponding component configuration is updated to bind to the new entity. At intermediate stages it may be possible to simulate the system with some of the components unbound. The effect of the deferred bindings is simply to leave the corresponding ports unassociated when the design is elaborated. Thus the inputs to the unbound modules are not used, and the outputs remain undriven.

EXAMPLE 13.9 *Deferred binding in a computer system model*

Following is an outline of a structural architecture for a single-board computer system. The design includes all of the components needed to construct the system, including a CPU, main memory and a serial interface.

```
architecture structural of single_board_computer is

  ...   -- type and signal declarations
  component processor is
    port ( clk : in bit;  a_d : inout word; ... );
  end component processor;

  component memory is
    port ( addr : in bit_vector(25 downto 0); ... );
  end component memory;
```

```
component serial_interface is
  port ( clk : in bit;
         address : in bit_vector(3 downto 0); ... );
end component serial_interface;

begin

cpu : component processor
  port map ( clk => sys_clk, a_d => cpu_a_d, ... );

main_memory : component memory
  port map ( addr => latched_addr(25 downto 0), ... );

serial_interface_a : component serial_interface
  port map ( clk => sys_clk,
             address => latched_addr(3 downto 0), ... );

  ...

end architecture structural;
```

However, if we have not yet designed an entity and architecture body for the serial interface, we cannot bind the component instance for the interface. Instead, we must leave it unbound, as shown in the following configuration declaration:

```
library chips;

configuration intermediate of single_board_computer is

  for structural

    for cpu : processor
      use entity chips.XYZ3000_cpu(full_function)
      port map ( clock => clk, addr_data => a_d, ... );
    end for;

    for main_memory : memory
      use entity work.memory_array(behavioral);
    end for;

    for all : serial_interface
      use open;
    end for;

    ...

  end for;

end configuration intermediate;
```

We can proceed to simulate the design, using the implementations of the CPU and main memory, provided we do not try to exercise the serial interface. If the processor were to try to access registers in the serial interface, it would get no response. Since there is no entity bound to the component instance representing the interface, there is nothing to drive the data or other signals connected to the instance.

13.3 Configuration Specifications

We complete this chapter with a discussion of *configuration specifications*. These provide a way of including binding information for component instances in the same architecture body as the instances themselves, as opposed to separating the information out into a configuration declaration. In some respects, this language feature is a relic of VHDL-87, which did not allow direct instantiation of entities in an architecture body. In VHDL-93 and VHDL-2002, if we know the interface of the entity and want to use it "as is," we can instantiate it directly, without having to write a corresponding component declaration. The main remaining use of configuration specifications is to bind a known entity to component instances in cases where our idealized module is different from the entity. Using a component declaration to describe the idealized module may make the design easier to understand. The syntax rule for a configuration specification is

> configuration_specification ⇐
> **for** component_specification
> binding_indication ;
> ⟦ **end for** ; ⟧

A configuration specification is similar to a component configuration, so we need to take care not to confuse the two. The component specification and binding indication are written in exactly the same way in both cases. However, a configuration specification does not provide an opportunity to configure the internal structure of the architecture to which the component instance is bound. That must be done in a separate configuration declaration. If we write a configuration specification for a component instance, it must be included in the declarative part of the architecture body or block that directly contains the component instance.

The effect of a configuration specification in an architecture body is exactly the same as if the binding indication had been included in a configuration declaration. Thus, we can bind a component instance to a design entity, and we can specify the mapping between the local generics and ports of the component instance and the formal generics and ports of the entity.

EXAMPLE 13.10 *A configuration specification binding a gate component*

Suppose we need to include a two-input nand gate in a model, but our library only provides a three-input nand gate, declared as

```
entity nand3 is
  port ( a, b, c : in bit;  y : out bit );
end entity nand3;
```

We can write our model using a component declaration to show that we really would prefer a two-input gate, and include a configuration specification to handle the difference in interfaces between the component instance and the entity. The architecture is

```
library gate_lib;

architecture ideal of logic_block is

  component nand2 is
    port ( in1, in2 : in bit;  result : out bit );
  end component nand2;

  for all : nand2
    use entity gate_lib.nand3(behavioral)
    port map ( a => in1, b => in2, c => '1', y => result );
  end for;

  ...  -- other declarations

begin

  gate1 : component nand2
    port map ( in1 => s1, in2 => s2, result => s3 );

  ...  -- other concurrent statements

end architecture ideal;
```

VHDL-87, -93, and -2002

In these versions of VHDL, the reserved words **end for** cannot be included in a configuration specification. For example, the configuration specification in Example 13.10 must be written as

```
for all : nand2
  use entity gate_lib.nand3(behavioral)
  port map ( a => in1, b => in2, c => '1', y => result );
```

13.3.1 Incremental Binding

We have now seen that there are two places where we can specify the mapping between the local generics and ports of a component instance and the formal generics and ports of the bound entity. The mappings can be specified either in a configuration specification or in a separate configuration declaration. We must now consider the possibility of having two binding indications for a given component instance, one in each of these places. VHDL does, in fact, allow this. The first binding indication, in the configuration specification in the architecture body, is called the *primary* binding indication. The second binding indication, in the configuration declaration, is called an *incremental* binding indication. The primary binding indication must at least specify the entity or configuration to which the instance is bound and may also include generic and port maps. If there is a primary binding indication, the incremental binding indication can either repeat the entity part exactly as specified in the primary binding indication, or it can omit the entity part. The full syntax rule for a binding indication allows for the entity part to be omitted in this case (see Appendix B). The incremental binding indication can also include generic and port maps, and the associations in them override those made in the primary binding indication, with some restrictions. An incremental binding indication must include at least one of the entity part, a port map or a generic map, since it must not be empty. Further, the only generics that can be incrementally bound are generic constants, not generic types, subprograms, or packages. We look at the various possibilities for incremental binding, with some examples.

The first possibility is that the primary binding indication for a component instance leaves some of the formal generic constants or ports of the entity unassociated. In this case, the incremental binding indication can "fill in the gaps" by associating actual values and with the unassociated generic constants and ports.

EXAMPLE 13.11 *Incremental binding of unassociated generic constants and ports*

Following is an architecture body for the control section of a processor, including a register component to store flag bits.

```
architecture structural of control_section is

  component reg is
    generic ( width : positive );
    port ( clk : in std_ulogic;
           d : in std_ulogic_vector(0 to width - 1);
           q : out std_ulogic_vector(0 to width - 1) );
  end component reg;

  for flag_reg : reg
    use entity work.reg(gate_level)
    port map ( clock => clk, data_in => d, data_out => q );
  end for;

  ...

begin

  flag_reg : component reg
    generic map ( width => 3 )
    port map ( clk => clock_phase1,
               d(0) => zero_result, d(1) => neg_result,
               d(2) => overflow_result,
               q(0) => zero_flag, q(1) => neg_flag,
               q(2) => overflow_flag );

  ...

end architecture structural;
```

The configuration specification binds the register component instance to the following register entity:

```
library ieee;  use ieee.std_logic_1164.all;

entity reg is
  generic ( t_setup, t_hold, t_pd : delay_length;
            width : positive );
  port ( clock : in std_ulogic;
         reset_n : in std_ulogic := 'H';
         data_in : in std_ulogic_vector(0 to width - 1);
         data_out : out std_ulogic_vector(0 to width - 1) );
end entity reg;
```

This entity has additional formal generic constants t_setup, t_hold and t_pd for timing parameters, and an additional port, reset_n. Since the component declaration does not include corresponding local generic constants and ports, and the configuration specification does not specify values or signals for the formal generic constants and ports, they are left open in the architecture body.

The configuration declaration for the design, shown below, contains an incremental binding indication for the register component instance. It does not specify an entity/architecture pair, since that was specified in the primary binding indication. It does, however, include a generic map, filling in values for the formal generic constants that were left open by the primary binding indication. The generic map also associates the value of the local generic constant width with the formal generic constant width. The port map in the incremental binding indication associates the literal value '1' with the formal port reset_n.

```
configuration controller_with_timing of control_section is

  for structural

    for flag_reg : reg
      generic map ( t_setup => 200 ps, t_hold => 150 ps,
                    t_pd => 150 ps, width => width )
      port map ( reset_n => '1' );
    end for;

    . . .

  end for;

end configuration controller_with_timing;
```

The second possibility is that the primary binding indication associates actual values with the formal generic constants of the entity bound to the component instance. In this case, the incremental binding indication can include new associations for these formal generic constants, overriding the associations in the primary binding indication. This may be useful in the back-annotation stage of design processing. Estimates for values of generic constants controlling propagation delay can be included in the primary binding indication and the design simulated before doing physical layout. Later, when actual delay values have been calculated from the physical layout, they can be included in incremental binding indications in a configuration declaration without having to modify the architecture body in any way.

EXAMPLE 13.12 *Incremental rebinding of generic constants*

The following outline of an architecture body for the interlock control logic of a pipelined processor declares a nor-gate component with a generic constant for the input port width, but with no generic constants for timing parameters.

```
architecture detailed_timing of interlock_control is

  component nor_gate is
    generic ( input_width : positive );
    port ( input : in std_ulogic_vector(0 to input_width - 1);
           output : out std_ulogic );
  end component nor_gate;

  for ex_interlock_gate : nor_gate
    use entity cell_lib.nor_gate(primitive)
    generic map ( width => input_width,
                  Tpd01 => 250 ps, Tpd10 => 200 ps );   -- estimates
  end for;

  . . .

begin

  ex_interlock_gate : component nor_gate
    generic map ( input_width => 2 )
    port map ( input(0) => reg_access_hazard,
               input(1) => load_hazard,
               output => stall_ex_n);

  . . .

end architecture detailed_timing;
```

The architecture includes a configuration specification for the instance of the gate component, which binds it to a nor-gate entity that does include generic constants for timing. The generic map in the configuration specification supplies estimates of the timing as actual values for the generic constants.

This model can be simulated with these estimates by configuring it as follows:

```
configuration interlock_control_with_estimates
  of interlock_control is

  for detailed_timing

  end for;
```

. . .

end configuration interlock_control_with_estimates;

Since there is no further configuration information supplied for the nor-gate instance, the estimated timing values are used. After the design has been laid out and the real timing values have been determined, the configuration declaration can be updated as follows:

configuration interlock_control_with_actual
  of interlock_control **is**

  for detailed_timing

    for ex_interlock_gate : nor_gate
      generic map (Tpd01 => 320 ps, Tpd10 => 230 ps);
    end for;

    . . .

  end for;

end configuration interlock_control_with_actual;

An incremental binding indication has been added, supplying the new values for the generic constants. When the design is simulated with this updated configuration, these new values override the estimates specified in the primary binding indication in the architecture body.

The third possibility to consider is that the primary binding indication associates actual signals with the formal ports of the entity. In this case, the incremental binding indication cannot override the associations, since to do so would modify the structure of the design.

The final case that arises is one in which a component instantiation associates actual values and signals with local generic constants and ports, but the primary binding indication does not explicitly associate actual values or signals with formal generic constants or ports of the same name. In this case, the default binding rule normally causes the local generic constants to be associated with formal generic constants of the same name and local ports to be associated with formal ports of the same name. However, we can preempt this default rule by supplying alternative associations for the formal generic constants and ports in the incremental binding indication.

EXAMPLE 13.13   *Incremental binding in place of default association*

Following is an outline of an architecture body for a block of miscellaneous logic. It includes a component declaration for a three-input nand gate and an instance of the component with an actual value supplied for a local timing generic. The primary binding indication binds the instance to a three-input nand gate entity, but does not specify the mappings between the local generic and ports and the formal generic and ports.

architecture gate_level **of** misc_logic **is**

```
      component nand3 is
        generic ( Tpd : delay_length );
        port ( a, b, c : in bit;  y : out bit );
      end component nand3;

      for all : nand3
        use entity project_lib.nand3(basic);
      end for;

      ...

  begin

    gate1 : component nand3
      generic map ( Tpd => 2 ns )
      port map ( a => sig1, b => sig2, c => sig3, y => out_sig );

      ...

  end architecture gate_level;
```

The configuration declaration for this design shown below overrides the default mapping. It supplies an actual value for the formal timing generic **Tpd**, instead of using the value of the local generic of that name. It maps the local port **c** onto the formal port **a**, and the local port **a** onto the formal port **c**. The local ports **b** and **y** map onto the formal ports of the same names.

```
    configuration misc_logic_reconfigured of misc_logic is

      for gate_level

        for gate1 : nand3
          generic map ( Tpd => 1.6 ns )
          port map ( a => c, c => a, b => b, y => y );
        end for;

      end for;

    end configuration misc_logic_reconfigured;
```

VHDL-87

VHDL-87 does not allow incremental binding. It is an error if a design includes both a configuration specification and a component configuration for a given component instance. If we expect to revise the associations in a generic map or port map of a configuration specification, we should omit the configuration specification and write the initial associations in the configuration declaration. Later, when we need to revise the associations, we can simply edit the configuration declaration without changing the architecture containing the component instance.

Exercises

1. [❶ 13.1] List some of the differences between an entity declaration and a component declaration.

2. [❶ 13.1] Write a component declaration for a binary magnitude comparitor, with two standard-logic vector data inputs, **a** and **b**, whose length is specified by a generic constant, and two standard-logic outputs indicating whether **a** = **b** and **a** < **b**. The component also includes a generic constant for the propagation delay.

3. [❶ 13.1] Write a component instantiation statement that instantiates the magnitude comparitor described in Exercise 2. The data inputs are connected to signals **current_position** and **upper_limit**, the output indicating whether **a** < **b** is connected to **position_ok** and the remaining output is open. The propagation delay of the instance is 12 ns.

4. [❶ 13.1] Write a package declaration that defines a subtype of natural numbers representable in eight bits and a component declaration for an adder that adds values of the subtype.

5. [❶ 13.2] Suppose we have an architecture body for a digital filter, outlined as follows:

```
architecture register_transfer of digital_filter is
  ...
  component multiplier is
    port ( ... );
  end component multiplier;
begin
  coeff_1_multiplier : component multiplier
    port map ( ... );
  ...
end architecture register_transfer;
```

   Write a configuration declaration that binds the multiplier component instance to a multiplier entity called **fixed_point_mult** from the library **dsp_lib**, using the architecture **algorithmic**.

6. [❶ 13.2] Suppose the library **dsp_lib** referred to in Exercise 5 includes a configuration of the **fixed_point_mult** entity called **fixed_point_mult_std_cell**. Write an alternative configuration declaration for the filter described in Exercise 5, binding the multiplier instance using the **fixed_point_mult_std_cell** configuration.

7. [❶ 13.2] Modify the outline of the filter architecture body described in Exercise 5 to directly instantiate the **fixed_point_mult_std_cell** configuration described in Exercise 6, rather than using the multiplier component.

8. [❶ 13.2] Suppose we declare and instantiate a multiplexer component in an architecture body as follows:

```
component multiplexer is
  port ( s, d0, d1 : in bit; z : out bit );
```

```
      end component multiplexer;
      ...

      serial_data_mux : component multiplexer
        port map ( s => serial_source_select,
                   d0 => rx_data_0, d1 => rx_data_1,
                   z => internal_rx_data );
```

Write a binding indication that binds the component instance to the following entity in the current working library, using the most recently analyzed architecture and specifying a value of 3.5 ns for the propagation delay.

```
      entity multiplexer is
        generic ( Tpd : delay_length := 3 ns );
        port ( s, d0, d1 : in bit; z : out bit );
      end entity multiplexer;
```

9. [● 13.2] Draw a diagram, based on Figure 13.1, that shows the mapping between entity ports and generics, component ports and generics and other values in the configured computer system model of Example 13.8.

10. [● 13.2] Suppose we have an entity **nand4** with the following interface in a library gate_lib:

```
      entity nand4 is
        generic ( Tpd_01, Tpd_10 : delay_length := 2 ns );
        port ( a, b, c, d : in bit := '1';  y : out bit );
      end entity nand4;
```

We bind the entity to the component instance **gate1** described on page 434 using the following component configuration:

```
      for gate1 : nand3
        use entity get_lib.nand4(basic);
      end for;
```

Write the generic and port maps that comprise the default binding indication used in this configuration.

11. [● 13.3] Rewrite the component configuration information in Example 13.8 as a configuration specification for inclusion in the computer system architecture body.

12. [● 13.3] Assuming that the computer system referred to in Exercise 11 includes the configuration specification, write a configuration declaration that includes an incremental binding indication, specifying values of 4.3 ns and 3.8 ns for the entity generics **Tpd_01** and **Tpd_10**, respectively.

13. [❷ 13.1] Develop a structural model of a 32-bit bidirectional transceiver, implemented using a component based on the 8-bit transceiver described in Exercise 13 in Chapter 8.

14. [❷ 13.1] Develop a structural model for an 8-bit serial-in/parallel-out shift register, assuming you have available a 4-bit serial-in/parallel-out shift register. Include a component declaration for the 4-bit register, and instantiate it as required for the 8-bit register. The 4-bit register has a positive-edge-triggered clock input, an active-low asynchronous reset input, a serial data input and four parallel data outputs.

15. [❷ 13.1] Develop a package of component declarations for two-input gates and an inverter, corresponding to the logical operators in VHDL. Each component has ports of type **bit** and generic constants for rising output and falling output propagation delays.

16. [❷ 13.2] Develop a configuration declaration for the 32-bit transceiver described in Exercise 13 that binds each instance of the 8-bit transceiver component to the 8-bit transceiver entity.

17. [❷ 13.2] Develop a behavioral model of a 4-bit shift register that implements the component interface described in Exercise 14. Write a configuration declaration for the 8-bit shift register, binding the component instances to the 4-bit shift register entity.

18. [❷ 13.2] Suppose we wish to use an XYZ1234A serial interface controller in the microcontroller described in Example 13.2. The entity interface for the XYZ1234A is

```
entity XYZ1234A is
  generic ( T_phi_out, T_d_z : delay_length;
            debug_trace : boolean := false );
  port ( phi1, phi2 : in std_ulogic;    -- 2 phase clock
         cs : in std_ulogic;            -- chip select
         a : in std_ulogic_vector(1 downto 0);      -- address
         d : inout std_ulogic_vector(1 downto 0);  -- data
         int_req : out std_ulogic;      --  interrupt
         rx_d : in std_ulogic;          -- rx serial data
         tx_d : out std_ulogic );       -- tx serial data
end entity XYZ1234A;
```

Write a configuration declaration that binds the **serial_interface** component instance to the **XYZ1234A** entity, using the most recently compiled architecture, setting both timing generics to 6 ns and using the default value for the **debug_trace** generic.

19. [❷ 13.1/13.2] Use the package described in Exercise 15 to develop a structural model of a full adder, described by the Boolean equations

$$S = (A \oplus B) \oplus C_{in}$$
$$C_{out} = A \cdot B + (A \oplus B) \cdot C_{in}$$

Write behavioral models of entities corresponding to each of the gate components and a configuration declaration that binds each component instance in the full adder to the appropriate gate entity.

20. [❸ 13.2] Develop a structural model of a 4-bit adder using instances of a full-adder component. Write a configuration declaration that binds each instance of the full-adder component, using the configuration declaration described in Exercise 19. For

comparison, write an alternative configuration declaration that fully configures the 4-bit adder hierarchy without using the configuration declaration described in Exercise 19.

21. [❸ 13.2] Develop a behavioral model of a RAM with bit-vector address, data-in and data-out ports. The size of the ports should be constrained by generics in the entity interface. Next, develop a test bench that includes a component declaration for the RAM without the generics and with fixed-sized address and data ports. Write a configuration declaration for the test bench that binds the RAM entity to the RAM component instance, using the component local port sizes to determine values for the entity formal generics.

22. [❸ 13.3] The majority function of three inputs can be described by the Boolean equation

$$M(a, b, c) = a \cdot b \cdot c + a \cdot b \cdot \bar{c} + a \cdot \bar{b} \cdot c + \bar{a} \cdot b \cdot c$$

Develop a structural model of a three-input majority circuit, using inverter, and-gate and or-gate components with standard-logic inputs and outputs. Also develop behavioral models for the inverter and gates, including generic constants in the interfaces to specify propagation delays for rising and falling output transitions. Include configuration specifications in the structural model to bind the component instances to the entities. The configuration specifications should include estimated propagation delays of 2 ns for all gates.

Next, develop a configuration declaration for the majority circuit that includes incremental bindings to override the estimated delays with actual propagation delays as shown below.

	rising-output delay	*falling-output delay*
inverter	1.8 ns	1.7 ns
and gate	2.3 ns	1.9 ns
or gate	2.2 ns	2.0 ns

23. [❹] Develop a suite of models of a digital stopwatch circuit. The circuit has three inputs: a 100 kHz clock, a start/stop switch input and a lap/reset switch input. The two switch inputs are normally high and are pulled low when an external push-button switch is pressed. The circuit has outputs to drive an external seven-segment display of minutes, seconds and hundredths of seconds, formatted as shown in the margin. There is a single output to drive the minutes (') and seconds (") indicators. When an output is high, the corresponding segment or indicator is visible. When the output is low, the segment or indicator is blank. The stopwatch circuit contains a time counter that counts minutes, seconds and hundredths of seconds.

The stopwatch counter is initially reset to 00'00"00, with the display showing the counter time and the minute and second indicators on. In this state, pressing the start/stop button starts counting, with the display showing the counter time. Pressing the start/stop button again stops counting. Successive presses of start/stop continue or stop counting, with the display showing the counter time. If the lap/reset button is pressed while the counter is stopped and the display is showing the counter time, the counter is reset to 00'00"00. If the lap/reset button is pressed while the counter is running, the display freezes the time at which the lap/reset button was pressed, the counter continues running and the minutes and seconds indicators flash at a 1 Hz rate to indicate that the counter is still running. If the start/stop button is pressed, the counter stops, the minutes and seconds indicators stop flashing and the displayed time is unchanged. Successive presses of start/stop continue or stop counting, with the displayed time unchanged and the minutes and seconds indicators flashing when the counter is running. Pressing the lap/reset button while the display is frozen causes it to return to displaying the current counter time, whether the counter is running or stopped.

The first model in your suite should be a behavioral model. Test your behavioral model by writing a test bench for it. You should write a configuration declaration for the test bench that binds the unit under test to the behavioral stopwatch model. Next, refine your stopwatch model to a structural design, including a control sequencer, registers, counters, decoders and other components as required. Develop behavioral models corresponding to each of these components, and write a configuration for the stopwatch that binds the behavioral models to the component instances. Revise the test bench configuration to use the structural model, and compare its operation with that of the behavioral model. Continue this process of refinement by implementing the control sequencer as a finite-state machine with next-state logic and a state register and by implementing the other components using successively lower-level components down to the level of flipflops and gates. At each stage, develop configuration declarations to bind entities to component instances, and test the complete model using the test bench.

Chapter 14

Generate Statements

Many digital systems can be implemented as regular iterative compositions of subsystems. Memories are a good example, being composed of a rectangular array of storage cells. Indeed, VLSI designers prefer to find such implementations, as they make it easier to produce a compact, area-efficient layout, thus reducing cost. If a design can be expressed as a repetition of some subsystem, we should be able to describe the subsystem once, then describe how it is to be repeatedly instantiated, rather than describe each instantiation individually. In this chapter, we look at the VHDL facility that allows us to generate such regular structures.

14.1 Generating Iterative Structures

We have seen how we can describe the implementation of a subsystem using concurrent statements such as processes and component instantiations. If we want to replicate a subsystem, we can use a *generate statement*. This is a concurrent statement containing further concurrent statements that are to be replicated during elaboration of a design. Generate statements are particularly useful if the number of times we want to replicate the concurrent statements is not fixed but is determined, for example, from the value of a generic constant. The syntax rule for writing *for-generate statements* is

> for_generate_statement ⇐
>   *generate*_label :
>   **for** identifier **in** discrete_range **generate**
>   generate_statement_body
>   **end generate** [*generate*_label] ;

> generate_statement_body ⇐
>   [{ block_declarative_item }
>   **begin**]
>   { concurrent_statement }
>   [**end** ;]

The generate label is required to identify the generated structure. The header of the for-generate statement looks very similar to that of a for loop and indeed serves a similar purpose. The discrete range specifies a set of values, and for each value, the block declarative items and concurrent statements in the body are replicated once. Within each replication, the value from the range is given by the identifier, called the *generate parameter*. It appears as a constant, with a type that is the base type of the discrete range. We can specify the discrete range using the same notations that we used in for loops. As a reminder, here is the syntax rule for a discrete range:

discrete_range ⇐
    *discrete*_subtype_indication
    ⫾ *range*_attribute_name
    ⫾ simple_expression ⦇ **to** ⫾ **downto** ⦈ simple_expression

The kinds of declarations we can include in the generate statement body are the same kinds that we can declare in the declarative part of the architecture body, including constants, types, subtypes, subprograms and signals. These items are replicated once for each copy of the body and are local to that copy. Note that the syntax rule for a for-generate statement requires us to include the keyword **begin** if we include any declarations. However, if we have no declarations, we may omit the keyword. Also, we can include a closing **end** keyword at the end of the body before the **end generate** keywords. The syntax allows this for consistency with the more elaborate forms of generate statement that we will see in Section 14.2. In practice, we would not normally include the keyword in for-generate statements.

EXAMPLE 14.1   *A register consisting of replicated flipflop cells*

We can implement a register by replicating a flipflop cell. Let us look at how to do this for a register with tristate outputs, conforming to the following entity declaration:

```
library ieee;  use ieee.std_logic_1164.all;

entity register_tristate is
  generic ( width : positive );
  port ( clock : in std_ulogic;
         out_enable : in std_ulogic;
         data_in : in std_ulogic_vector(0 to width - 1);
         data_out : out std_ulogic_vector(0 to width - 1) );
end entity register_tristate;
```

The generic constant **width** specifies the width of the register in bits and is used to determine the size of the data input and output ports. The **clock** port enables data to be stored in the register, and the **out_enable** port controls the tristate data output port.

The architecture body for the register implements this register in terms of a D-flipflop component and a tristate buffer for each bit:

```
architecture cell_level of register_tristate is
```

```
component D_flipflop is
  port ( clk : in std_ulogic;  d : in std_ulogic;
         q : out std_ulogic );
end component D_flipflop;

component tristate_buffer is
  port ( a : in std_ulogic;
         en : in std_ulogic;
         y : out std_ulogic );
end component tristate_buffer;

begin

  cell_array : for bit_index in 0 to width - 1 generate

    signal data_unbuffered : std_ulogic;

  begin

    cell_storage : component D_flipflop
      port map ( clk => clock, d => data_in(bit_index),
                 q => data_unbuffered );

    cell_buffer : component tristate_buffer
      port map ( a => data_unbuffered, en => out_enable,
                 y => data_out(bit_index) );

  end generate cell_array;

end architecture cell_level;
```

The for-generate statement in this structural architecture body replicates the component instantiations labeled **cell_storage** and **cell_buffer**, with the number of copies being determined by **width**. For each copy, the generate parameter bit_index takes on successive values from 0 to **width** − 1. This value is used within each copy to determine which elements of the **data_in** and **data_out** ports are connected to the flipflop input and tristate buffer output. Within each copy there is also a local signal called **data_unbuffered**, which connects the flipflop output to the buffer input.

EXAMPLE 14.2   *Replication in a graphics transformation pipeline*

We can also use for-generate statements to describe behavioral models in which behavioral elements implemented using process statements are replicated. Suppose we are modeling part of a graphics transformation pipeline in which a stream of points representing vertices in a scene is to be transformed by matrix multiplication. The equation describing the transformation is

$$
\begin{bmatrix} p'_1 \\ p'_2 \\ p'_3 \end{bmatrix} = \begin{bmatrix} a_{11} & a_{12} & a_{13} \\ a_{21} & a_{22} & a_{23} \\ a_{31} & a_{32} & a_{33} \end{bmatrix} \begin{bmatrix} p_1 \\ p_2 \\ p_3 \end{bmatrix}
$$

where $[p_1, p_2, p_3]$ is the input point to the pipeline stage, and $[p'_1, p'_2, p'_3]$ is the transformed output three clock cycles later. We can implement the transformation with three identical cells, each producing one result element. The equation is

$$
p'_i = a_{i1} \cdot p_1 + a_{i2} \cdot p_2 + a_{i3} \cdot p_3, \qquad i = 1, 2, 3
$$

An outline of the architecture body implementing the pipeline with this stage is

```
architecture behavioral of graphics_engine is

    type point is array (1 to 3) of real;
    type transformation_matrix is array (1 to 3, 1 to 3) of real;

    signal p, transformed_p : point;
    signal a : transformation_matrix;
    signal clock : bit;
    ...

begin

    transform_stage : for i in 1 to 3 generate
    begin

        cross_product_transform : process is
            variable result1, result2, result3 : real := 0.0;
        begin
            wait until clock = '1';
            transformed_p(i) <= result3;
            result3 := result2;
            result2 := result1;
            result1 :=  a(i, 1) * p(1) + a(i, 2) * p(2) + a(i, 3) * p(3);
        end process cross_product_transform;

    end generate transform_stage;

    ...   -- other stages in the pipeline, etc

end architecture behavioral;
```

The for-generate statement replicates the process statement three times, once for each element of the transformed point signal. Each copy of the process uses its value of the generate parameter i to index the appropriate elements of the point and transformation matrix signals.

If we need to describe a regular two-dimensional structure, we can use nested for-generate statements. Nesting of generate statements is allowed in VHDL, since a generate statement is a kind of concurrent statement, and generate statements contain concurrent statements. Usually we write nested for-generate statements so that the outer statement creates the rows of the structure, and the inner statement creates the elements within each row. Of course, this is purely a convention relating to the way we might draw such a regular structure graphically. However, the convention does help to design and understand such structures.

EXAMPLE 14.3 *A two-dimensional memory structure*

We can use nested generate statements to describe a memory array, constructed from 4-bit-wide static memory (SRAM) circuits. Each SRAM stores 4M words (4×2^{20} words) of four bits each. We can construct a 16M × 32-bit memory array by generating a 4 × 8 array of SRAM circuits. An outline of the architecture body containing the memory array is shown below, and a schematic diagram is shown in Figure 14.1.

```vhdl
architecture generated of memory is

  component SRAM is
    port ( clk : in std_ulogic;
           a :  in std_ulogic_vector(0 to 10);
           d :  inout std_ulogic_vector(0 to 3);
           en, we : in std_ulogic );
  end component SRAM;

  signal buffered_address : std_ulogic_vector(0 to 10);
  signal SRAM_data : std_ulogic_vector(0 to 31);
  signal bank_select : std_ulogic_vector(0 to 3);
  signal buffered_we : std_ulogic;

  ...  -- other declarations

begin

  bank_array : for bank_index in 0 to 3 generate
  begin

    nibble_array : for nibble_index in 0 to 7 generate

      constant data_lo : natural := nibble_index * 4;
      constant data_hi : natural := nibble_index * 4 + 3;

    begin

      an_SRAM : component SRAM
        port map ( clk => clk,
                   a => buffered_address,
                   d => SRAM_data(data_lo to data_hi),
                   en => bank_select(bank_index),
                   we => buffered_we );
```

FIGURE 14.1

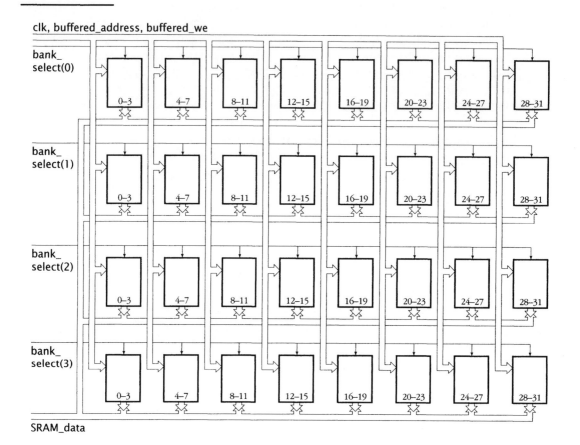

A schematic for a 16M × 32-bit memory array composed of 4M × 4-bit SRAM circuits.

```
        end generate nibble_array;

    end generate bank_array;

    ...   -- other component instances, etc

end architecture generated;
```

VHDL-87, -93, and -2002

These versions do not allow us to include a closing **end** keyword before the **end generate** keywords in a for-generate statement.

VHDL-87

A generate statement may not include a declarative part or the keyword **begin** in VHDL-87. The syntax for a for-generate statement is

for_generate_statement ⟸
   *generate*_label :
    for identifier **in** discrete_range **generate**
      { concurrent_statement }
    end generate ⟦ *generate*_label ⟧ ;

Since it is not possible to declare objects locally within a generate statement, we must declare them in the architecture body containing the generate statement. We can declare an array of objects indexed by the same range as the generate parameter. For example, the signal **data_unbuffered** declared in the generate statement in Example 14.1 can be replaced by an array in the declarative part of the architecture body:

signal data_unbuffered : std_ulogic_vector(0 **to** width - 1);

Each reference to **data_unbuffered** within the generate statement is replaced by a reference to the element **data_unbuffered(bit_index)**.

14.2 Conditionally Generating Structures

In the examples in the previous section, each cell in an iterative structure was connected identically. In some designs, however, there are particular cells that need to be treated differently. This often occurs where cells are connected to their neighbors. The cells at each end do not have neighbors on both sides, but instead are connected to signals or ports in the enclosing architecture body. We can deal with these special cases within an iterative structure using a conditional forms of generate statement. The first of these is an *if-generate statement*, with the syntax rule:

if_generate_statement ⟸
   *generate*_label :
   if condition **generate**
      generate_statement_body
   { **elsif** condition **generate**
      generate_statement_body }
   ⟦ **else generate**
      generate_statement_body ⟧
   end generate ⟦ *generate*_label ⟧ ;

This is like a sequential if statement, except that the conditions control how declarations and concurrent statements are copied in the design, rather than controlling selection of sequential statements for execution. When the model is elaborated, the conditions in the if-generate statement are tested from first to last until one is found that is true. The declarations (if any) and concurrent statements in the corresponding body are then included in the elaborated model. If no condition is true and there is an **else generate** alternative, the declarations and statements from that alternative are included. The **else generate** alternative is optional, allowing for the possibility of no declarations or statements being included if none of the conditions is true. The generate label in an if-generate statement is required to identify the structure that is generated, if any.

When we write an if-generate statement within an enclosing for-generate statement to deal with boundary conditions, we can refer to the values of generic constants or the generate parameter of the enclosing statement in the conditions of the if-generate statement.

EXAMPLE 14.4   *A shift register composed of flipflop cells*

We can construct a serial-to-parallel shift register from master/slave flipflop cells, as follows:

```
library ieee;  use ieee.std_logic_1164.all;

entity shift_reg is
  port ( phi1, phi2 : in std_ulogic;
         serial_data_in : in std_ulogic;
         parallel_data : out std_ulogic_vector );
end entity shift_reg;

-------------------------------------------------

architecture cell_level of shift_reg is

  alias normalized_parallel_data :
          std_ulogic_vector(0 to parallel_data'length - 1)
        is parallel_data;

  component master_slave_flipflop is
    port ( phi1, phi2 : in std_ulogic;
           d : in std_ulogic;
           q : out std_ulogic );
  end component master_slave_flipflop;

begin

  reg_array : for index in normalized_parallel_data'range generate
  begin

    reg : if index = 0 generate
      cell : component master_slave_flipflop
        port map ( phi1, phi2,
                   d => serial_data_in,
                   q => normalized_parallel_data(index) );
    else generate
      cell : component master_slave_flipflop
        port map ( phi1, phi2,
                   d => normalized_parallel_data(index - 1),
```

```
                        q => normalized_parallel_data(index) );
           end generate other_cell;

      end generate reg_array;

   end architecture cell_level;
```

The architecture contains a component declaration for the flipflop, then makes multiple instantiations using a for-generate statement. Within the for-generate statement, an if-generate statement is used to treat the first flipflop cell differently from the other cells. The condition "**index = 0**" identifies this first cell, which takes its input data from the **serial_data_in** port. The remaining cells take their input from the neighboring cell's output.

EXAMPLE 14.5   *A ripple-carry adder composed of full adders and a half adder*

A ripple-carry adder has a half adder at the least-significant end and has different carry in and out connections for the cells at the ends and in the middle. We can use a nested if-generate with three alternatives to deal with the differences:

```
adder: for i in width-1 downto 0 generate

   signal carry_chain : unsigned(width-1 downto 1);

begin
   adder_cell: if i = width-1 generate -- most-significant cell

      add_bit: component full_adder
        port map (a => a(i), b => b(i), s => s(i),
                  c_in => carry_chain(i), c_out => c_out);

   elsif i = 0 generate -- least-significant cell

      add_bit: component half_adder
        port map (a => a(i), b => b(i), s => s(i),
                  c_out => carry_chain(i+1));

   else generate -- middle cell

      add_bit: component full_adder
        port map (a => a(i), b => b(i), s => s(i),
                  c_in => carry_chain(i),
                  c_out => carry_chain(i+1));

   end generate adder_cell;

end generate adder;
```

Another important use of if-generate statements is to conditionally include or omit part of a design, usually depending on the value of a generic constant. A good example is the inclusion or otherwise of *instrumentation*: additional processes or component instances that trace or debug the operation of a design during simulation. When the design is sufficiently tested, a generic constant can be changed to exclude the instrumentation so that it does not slow down a large simulation and is not included when the design is synthesized.

EXAMPLE 14.6 *Conditional inclusion of instrumentation code*

Suppose we wish to measure the relative frequencies of instruction fetches, data reads and data writes made by a CPU accessing memory in a computer system. This information may be important when considering how to optimize a design to improve performance. An entity declaration for the computer system is

```
entity computer_system is
  generic ( instrumented : boolean := false );
  port ( ... );
end entity computer_system;
```

The generic constant **instrumented** is used to determine whether to include the instrumentation to measure relative frequencies of each kind of memory access. An outline of the architecture body is

```
architecture block_level of computer_system is

  ... -- type and component declarations for cpu and memory, etc

  signal clock : bit;      -- the system clock
  signal mem_req : bit;    -- cpu access request to memory
  signal ifetch : bit;     -- indicates access is
                           --   to fetch an instruction
  signal write : bit;      -- indicates access is a write
  ...                      -- other signal declarations

begin

  ... -- component instances for cpu and memory, etc

  instrumentation : if instrumented generate

    signal ifetch_freq, write_freq, read_freq : real := 0.0;

  begin

    access_monitor : process is
      variable access_count, ifetch_count,
               write_count, read_count : natural := 0;
    begin
      wait until mem_req = '1';
      if ifetch then
        ifetch_count := ifetch_count + 1;
      elsif write then
        write_count := write_count + 1;
```

```
      else
        read_count := read_count + 1;
      end if;
      access_count := access_count + 1;
      ifetch_freq <= real(ifetch_count) / real(access_count);
      write_freq <= real(write_count) / real(access_count);
      read_freq <= real(read_count) / real(access_count);
    end process access_monitor;

  end generate instrumentation;

end architecture block_level;
```

The signals **ifetch_freq**, **write_freq** and **read_freq** and the process **access_monitor** are only included in the design if the generic constant **instrumented** is true. The process resumes each time the CPU requests access to the memory and keeps count of the number of each kind of access, as well as the total access count. It uses these values to update the relative frequencies. We can trace these signals using our simulator to see how the relative frequencies converge over the lifetime of a simulation.

We can control whether the instrumentation is included or not when we write a configuration declaration for the design. To include the instrumentation, we configure an instance of the computer system as follows:

```
for system_under_test : computer_system
  use entity work.computer_system(block_level)
  generic map ( instrumented => true )
  ...
end for;
```

To exclude the instrumentation, we change the value of the generic constant in the generic map to false.

VHDL-87, -93, and -2002

These versions do not allow us to include **elsif** or **else** alternatives in an if-generate statement. Instead, we must write each alternative as a separate if-generate statement. For example, the if-generate statement in Example 14.5 would be written as:

```
msb_cell: if i = width-1 generate -- most-significant cell
  ...
end generate msb_cell;

lsb_cell: if i = 0 generate -- least-significant cell
  ...
end generate lsb_cell;

mid_cell: if i > 0 and i < width-1 generate -- middle cell
  ...
end generate lsb_cell;
```

Also, these versions do not permit a closing **end** keyword in each alternative of an if-generate statement.

VHDL-87

A generate statement may not include a declarative part or the keyword **begin** in VHDL-87. The syntax for a conditional generate statement is

generate_statement ⇐
    *generate*_label :
      if condition **generate**
        ⟦ concurrent_statement ⟧
      end generate ⟦ *generate*_label ⟧ ;

Any objects required by the generate statement must be declared in the declarative part of the enclosing architecture body.

The second conditional form of generate statement is a *case-generate statement,* in which we specify alternatives in a similar way to a case statement. We specify a static expression (one whose value can be computed during elaboration), and choice values for each alternative. The syntax rule is:

case_generate_statement ⇐
    *generate*_label :
    case expression **generate**
      ⟦ **when** choices =>
        generate_statement_body ⟧
      ⟦ ⋯ ⟧
    end generate ⟦ *generate*_label ⟧ ;

This is like a sequential case statement, except that the expression and choices control how declarations and concurrent statements are copied in the design. When the model is elaborated, the expression is evaluated (which explains why it must be static) and a matching choice is selected. The declarations (if any) and concurrent statements in the corresponding body are then included in the elaborated model. The rules governing sequential case statement expressions and choices also apply to the expression and choices in a case-generate statement, with the further stipulation that the expression be static.

EXAMPLE 14.7   *Selection among alternative implementations*

Multiplication of complex numbers in Cartesian form involves four scalar multiplications, a subtraction, and an addition. Depending on the constraints that apply to a design, these operations can be implemented in one clock cycle using multiple function units, in multiple clock cycles using fewer function units, or in a pipeline. Suppose we have an enumeration type, defined as follows, for specifying the implementation to use:

```
type implementation_type is
       (single_cycle, multicycle, pipelined);
```

An entity declaration for a complex multiplier has a generic constant of this type controlling the implementation:

```
entity complex_multiplier is
  generic ( implementation : implementation_type; ... );
  port ( ... );
end entity complex_multiplier;
```

Within the architecture, we use the value of the generic constant in a case-generate statement to determine what components to instantiate and how to interconnect them:

```
architecture rtl of complex_multiplier is
  ...
begin

  mult_structure : case implementation generate
    when single_cycle =>
        signal real_pp1, real_pp2 : ...;
        ...
      begin
        real_mult1 : component multiplier
          port map ( ... );
        ...
      end;
    when multicycle =>
        signal real_pp1, real_pp2 : ...;
        ...
      begin
        mult : component multiplier
          port map ( ... );
        ...
      end;
    when pipelined =>
        signal real_pp1, real_pp2 : ...;
        ...
      begin
        mult1 : component multiplier
          port map ( ... );
        ...
      end;
```

```
  end generate mutl_structure;

end architecture rtl;
```

The case-generate statement includes three alternatives, one for each possible implementation style. Each alternative can have local declarations and concurrent statements with the same names and labels as those in other alternatives, as well as differently named declarations and differently labeled statements.

VHDL-87, -93, and -2002

These versions do not provide case-generate statements. Instead, we must write each alternative as a separate if-generate statement in which the condition compares the selector expression with a choice value.

14.2.1 Recursive Structures

A more unusual application of conditional generate statements arises when describing recursive hardware structures, such as tree structures. We can write a description of a recursive structure using a recursive model, that is, one in which an architecture of an entity creates an instance of that same entity. We enclose the recursive instantiation in a conditional generate statement that determines when to terminate the recursion.

EXAMPLE 14.8   *A recursive clock fanout tree structure*

Clock-signal distribution can be a problem in a large integrated circuit. We typically have one clock signal that must be distributed to a very large number of components without overloading the clock drivers and without creating too much skew between different parts of the circuit. One solution is to distribute the clock signal using a fanout tree. A simplified binary fanout tree is shown in Figure 14.2. The clock signal feeds two buffers, each of which in turn feeds two buffers, and so on, until we have generated enough buffered clock signals to drive all elements of the circuit. As the diagram shows, we can think of a tree of height 3 as being constructed from two buffers feeding trees of height 2. Similarly, a tree of height 2 is two buffers feeding trees of height 1. A tree of height 1 is two buffers feeding the outputs of the fanout tree. We can think of these output connections as being degenerate trees of height 0. In general, we can say that a tree of height n consists of two buffers feeding trees of height $n - 1$, where $n > 0$.

We can describe this structure in VHDL by starting with an entity declaration for a fanout tree that includes a generic constant **height** specifying the height of the tree, shown below. The entity has one input and 2^{height} outputs.

```
library ieee;  use ieee.std_logic_1164.all;

entity fanout_tree is
  generic ( height : natural );
```

FIGURE 14.2

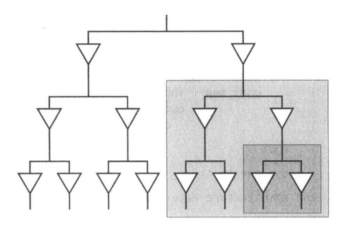

A binary fanout tree for clock distribution. The inner shaded section is a fanout tree of height 1, and the outer shaded section is a tree of height 2. The whole structure is a tree of height 3.

```
port ( input : in std_ulogic;
       output : out std_ulogic_vector (0 to 2**height - 1) );
end entity fanout_tree;
```

The architecture body, shown below, uses an if-generate statement that tests the value of **height** to see if any subtrees are required. If **height** is zero, the output port of the fanout tree is a vector of length one. The body for that alternative creates a connection from the input to the single output element. Otherwise, if **height** is greater than zero, the **else** alternative creates two buffers and two subtrees of reduced height. The local signals **buffered_input_0** and **buffered_input_1** connect the buffers to the inputs of the subtrees. The outputs of the subtrees are of length $2^{height-1}$ and are connected to slices of the output port vector of the enclosing tree.

```
architecture recursive of fanout_tree is
begin
  tree : if height = 0 generate
    output(0) <= input;
  else generate    -- height > 0
    signal buffered_input_0, buffered_input_1 : std_ulogic;
  begin
    buf_0 : entity work.buf(basic)
      port map ( a => input, y => buffered_input_0 );
    subtree_0 : entity work.fanout_tree(recursive)
      generic map ( height => height - 1 )
```

```
            port map ( input => buffered_input_0,
                       output => output(0 to 2**(height - 1) - 1) );

        buf_1 : entity work.buf(basic)
          port map ( a => input, y => buffered_input_1 );

        subtree_1 : entity work.fanout_tree(recursive)
          generic map ( height => height - 1 )
          port map ( input => buffered_input_1,
                     output =>
                        output(2**(height - 1) to 2**height - 1) );

      end;

    end generate tree;

  end architecture recursive;
```

This compact description of a relatively complex structure is fleshed out when the design is elaborated. Suppose we instantiate a fanout tree of height 3 in a design:

```
    clock_buffer_tree : entity work.fanout_tree(recursive)
      generic map ( height => 3 )
      port map ( input => unbuffered_clock,
                 output => buffered_clock_array );
```

In the first stage of elaboration, **height** has the value 3, so the generate statement creates the first two buffers and two instances of the **fanout_tree** entity with **height** having the value 2. In each of these instances, the generate statement creates two more buffers and two instances of the **fanout_tree** entity with **height** having the value 1. Then, in each of these instances, the generate statement creates a further two buffers and two instances of the **fanout_tree** entity with **height** having the value 0. In these last instances, the condition of the generate statement is true, so it creates a connection directly from its input to its output. This alternative is where the recursion terminates, as there are no further instantiations of the **fanout_tree** entity within the alternative.

VHDL-87

Since VHDL-87 does not allow direct instantiation of design entities, descriptions of recursive structures are slightly more complex. In the architecture body, we must declare a component with the same interface as the design entity. Instead of directly instantiating the design entity, we instantiate the declared component and bind it to the design entity using a configuration specification. Note that the configuration specification must be written in the declarative region of the construct immediately enclosing the instantiated component. In our example above, the recursive component instantiation statement is included in an if-generate statement. In VHDL-87, generate statements do not include a declarative part, so we cannot include a configuration specification as part of the generate statement. Instead, we must write the component instantiation statement within a block statement that is in turn nested in the generate statement. (We describe block statements in Section 23.1.) We then write the configuration specification in the declarative part of the block statement. For example, we can rewrite the recursive fanout tree model of Example 14.8 by declaring a component **fanout_tree** with the same interface as the fanout tree entity. The component instantiation statement labeled **subtree_0** is rewritten as shown below.

```
block_0 : block
  for subtree_0 : fanout_tree
    use entity work.fanout_tree(recursive);
begin
  subtree_0 : fanout_tree
    generic map ( height => height - 1 )
    port map (  input => buffered_input_0,
                output => output(0 to 2**(height - 1) - 1) );
end block block_0;
```

14.3 Configuration of Generate Statements

In this section we describe how to write configuration declarations for designs that include generate statements. If a design includes a for-generate statement, we need to be able to identify individual cells from the iteration in order to configure them. If the design includes an if-generate or case-generate statement, we need to be able to include configuration information that is to be used only if a given cell is included in the design. In order to handle these cases, we use an extended form of block configuration. We first introduced block configurations in Section 13.2.2. The syntax rule for the extended form is

```
block_configuration ⇐
    for ⟦ architecture_name
          ▯ block_statement_label
          ▯ generate_statement_label
              ⟦ ( ⟦ static_discrete_range ▯ static_expression ▯ alternative_label ⟧ ) ⟧ ⟧
        ⦃ block_configuration
        ▯ for component_specification
              ⟦ binding_indication ; ⟧
              ⟦ block_configuration ⟧
          end for ; ⦄
    end for ;
```

The new part in this rule is the alternative allowing us to configure a generate statement by writing its label. The optional part after the label allows us to write an expression whose value selects a particular cell from an iterative structure, a range of values that select a collection of cells from an iterative structure, or a label that selects a particular alternative of an if-generate or case-generate statement. We will return to these possibilities shortly. Once we have identified the generate statement using its label, the remaining configuration information within the block configuration specifies how the concurrent statements within the generated cell or cells are to be configured.

Let us first apply this rule to writing configurations for a simple if-generate statement with a single condition and no **elsif** or **else** alternatives. In this case, we simply write the generate statement label in the block configuration and fill in the configuration information for generated component instances. If the generate statement condition is true when the design is elaborated, the configuration information is used to bind entities to the component instances. On the other hand, if the condition is false, no instances are created, and the configuration information is ignored.

EXAMPLE 14.9   *Configuration of conditionally included instrumentation*

Let us return to our model of a computer system that uses a conditional generate statement to include instrumentation. Recall that the entity declaration was

```
entity computer_system is
  generic ( instrumented : boolean := false );
  port ( ... );
end entity computer_system;
```

Suppose we wish to use a general-purpose bus monitor component that collects statistics on bus transactions between the CPU and the memory. An outline of the revised architecture body is

```
architecture block_level of computer_system is

  ... -- type and component declarations for cpu and memory, etc.

  signal clock : bit;     -- the system clock
  signal mem_req : bit;   -- cpu access request to memory
  signal ifetch : bit;    -- indicates access is
                          --    to fetch an instruction
  signal write : bit;     -- indicates access is a write
  ...                     -- other signal declarations

begin

  ... -- component instances for cpu and memory, etc.

  instrumentation : if instrumented generate

    use work.bus_monitor_pkg;
    signal bus_stats : bus_monitor_pkg.stats_type;

  begin

    cpu_bus_monitor : component bus_monitor_pkg.bus_monitor
      port map ( mem_req, ifetch, write, bus_stats );

  end generate instrumentation;

end architecture block_level;
```

We can write a configuration declaration for the computer system as follows:

```
configuration architectural of computer_system is

  for block_level

    ...  -- component configurations for cpu and memory, etc

    for instrumentation

      for cpu_bus_monitor : bus_monitor_pkg.bus_monitor
        use entity work.bus_monitor(general_purpose)
        generic map ( verbose => true, dump_stats => true );
      end for;

    end for;

  end for;

end configuration architectural;
```

This configuration information may be used when the computer system entity is elaborated. If the value of the generic constant **instrumented** is true, the bus monitor is instantiated. In this case, the information in the block configuration starting with "**for** instrumentation" is used to bind an entity to the bus monitor instance. On the other hand, if **instrumented** is false, no instance is created, and the configuration information is ignored.

We now turn to configurations for designs including for-generate statements. The simplest case is a structure in which all cells are to be configured identically. In this case, we just write the generate statement label in the block configuration and include the configuration information to be applied to each cell.

EXAMPLE 14.10 *Configuration of the register composed of flipflops*

In the register model in Example 14.1, each cell consisted of a flipflop and a tristate buffer component. We can write a configuration declaration for this design as shown below. The block configuration starting with "**for** cell_array" identifies the iterative generate statement labeled **cell_array**. Since there is no specification of particular cells within the generated structure, the information in the block configuration is applied to all cells.

```
library cell_lib;

configuration identical_cells of register_tristate is

  for cell_level

    for cell_array

      for cell_storage : D_flipflop
        use entity cell_lib.D_flipflop(synthesized);
      end for;
```

```
      for cell_buffer : tristate_buffer
        use entity cell_lib.tristate_buffer(synthesized);
      end for;

    end for;

  end for;

end configuration identical_cells;
```

Where we have a design that includes nested generate statements to generate a two-dimensional structure, we simply nest block configurations in a configuration declaration.

EXAMPLE 14.11 *Configuration of the two-dimensional memory structure*

The memory array described in Example 14.3 is implemented using two nested iterative generate statements. We can write a configuration declaration for the design as shown below. The block configuration starting with "**for** bank_array" selects the memory array generated by the outer generate statement labeled **bank_array**. Each bank is configured identically, using the inner block configuration starting with "**for** nibble_array". This selects the generate statement that creates a bank of SRAM chips and configures each chip in the bank identically to the rest.

```
library core_lib;  use core_lib.all;

configuration behavioral_SRAM of memory is

  for generated

    for bank_array

      for nibble_array

        for an_SRAM : SRAM
          use entity SRAM_4M_by_4(behavior);
        end for;

      end for;

    end for;

    ...  -- configurations of other component instances

  end for;

end configuration behavioral_SRAM;
```

Before we look at further examples of examples of block configurations for for-generate statements, we need to consider how to write block configurations for the if-generate or case-generate statements that are often nested inside for-generate statements. We saw that the nested generate statement may have different alternatives for cells that are connected differently. We need to be able to write a distinct block configuration for each alternative. In order to do that, we need to label the alternatives so that we can distinguish one from another. Thus, we extend our syntax rules for generate statements to include alternative labels. The extended rule for if-generate statements is:

if_generate_statement ⇐
   *generate*_label :
   if ⟦ *alternative*_label : ⟧ condition **generate**
      generate_statement_body
   ⦃ **elsif** ⟦ *alternative*_label : ⟧ condition **generate**
      generate_statement_body ⦄
   ⟦ **else** ⟦ *alternative*_label : ⟧ **generate**
      generate_statement_body ⟧
   end generate ⟦ *generate*_label ⟧ ;

and for case-generate statements:

case_generate_statement ⇐
   *generate*_label :
   case expression **generate**
      ⦅ **when** ⟦ *alternative*_label : ⟧ choices =>
         generate_statement_body ⦆
      ⦃ ... ⦄
   end generate ⟦ *generate*_label ⟧ ;

We can then use an alternative label in a block configuration for the generate statement to identify a particular alternative of the statement. This is shown in the syntax rule on page 465. If the alternative is included when the design is elaborated, the configuration information is used. If the alternative is not included, the configuration information is ignored. Note that, in the simple case of an if-generate statement with only one alternative, we do not need to label the alternative. It is sufficient to refer to the generate statement label in a corresponding block configuration, as we did in Example 14.9.

EXAMPLE 14.12 *Configuration of the ripple-carry adder*

In Example 14.5 we showed a structure for a ripple-carry adder, in which differences among bit positions were handled by alternatives of an if-generate statement. We can revise the statement to include labels in each alternative:

```
adder: for i in width-1 downto 0 generate

  signal carry_chain : unsigned(width-1 downto 1);

begin

  adder_cell: if most_significant: i = width-1 generate

    add_bit: component full_adder
      port map (...);
```

```
    elsif least_significant: i = 0 generate

      add_bit: component half_adder
        port map (...);

    else middle: generate

      add_bit: component full_adder
        port map (...);

    end generate adder_cell;

  end generate adder;
```

We can now write a configuration declaration for the enclosing entity and architecture:

```
configuration widget_cfg of arith_unit is
  for ripple_adder
    for adder

      for adder_cell(most_significant)
        for add_bit: full_adder
          use entity widget_lib.full_adder(asic_cell);
      end for;

      for adder_cell(middle)
        for add_bit: full_adder
          use entity widget_lib.full_adder(asic_cell);
      end for;

      for adder_cell(least_significant)
        for add_bit: half_adder
          use entity widget_lib.half_adder(asic_cell);
      end for;

    end for; -- adder
  end for; -- ripple_adder
end configuration widget_cfg;
```

The block configuration "**for** adder ... **end for**" applies to all of the repetitions of the for-generate statement. Within it, we have three block configurations, one for each alternative of the if-generate statement. We identify each alternative with a combination of the if-generate statement label (**adder_cell**) and the alternative label (**most_significant**, **least_significant**, and **middle**, respectively). The configuration information for each alternative is only acted upon during elaboration if the corresponding condition is true and the alternative is included in the design hierarchy.

EXAMPLE 14.13 *Configuration of alternative implementations*

We can revise the case-generate statement in Example 14.7 to include alternative labels, allowing the alternatives to be configured:

```
mult_structure : case implementation generate
  when single_cycle_mult: single_cycle =>

    . . .
  when multicycle_mult: multicycle =>

    . . .
  when pipelined_mult: pipelined =>
    . . .;
end generate mutl_structure;
```

We can now write a configuration declaration for the complex multiplier:

```
configuration wallace_tree of complex_multiplier is
  for rtl

    for mult_structure(single_cycle_mult)
      for real_mult1 : multiplier
        use entity work.multiplier(wallace_tree);
      . . .
    end for;

    for mult_structure(multicycle_mult)
      for mult : multiplier
        use entity work.multiplier(wallace_tree);
      . . .
    end for;

    for mult_structure(pipelined_mult)
      for mult1 : multiplier
        use entity work.multiplier(wallace_tree);
      . . .
    end for;

  end for; -- rtl
end for wallage_tree;
```

In some designs using for-generate statements, there may be particular cells or groups of cells that we wish to configure differently from other cells. In these cases we can use an expression or a range of values in parentheses after the generate statement label in the block configuration. The values identify those cells to which the configuration information applies. The rules for specifying the discrete range are the same as those for specifying a discrete range in other contexts.

EXAMPLE 14.14   *Configuration of the shift register composed of flipflops*

The shift register design shown in Example 14.4 is composed of cells indexed from 0 to one less than the length of the **parallel_data** port. Each cell includes an instance of a master/slave flipflop. Suppose we wish to use an ordinary flipflop for all except the last cell of the shift register and a flipflop with high drive capacity for the last cell. In order to be able to refer to the different alternatives in a configuration, we need to add alternative labels, as follows:

```
architecture cell_level of shift_reg is

  ...

begin
  reg_array : for index in normalized_parallel_data'range generate
  begin

    reg : if first_cell : index = 0 generate
      cell : component master_slave_flipflop
        port map ( ... );
    else other_cell : generate
      cell : component master_slave_flipflop
        port map ( ... );
    end generate other_cell;

  end generate reg_array;

end architecture cell_level;
```

A configuration declaration for the shift register that configures the generated flipflop instances as required is:

```
library cell_lib;

configuration last_high_drive of shift_reg is

  for cell_level

    for reg_array ( 0 to parallel_data'length - 2 )

      for reg(first_cell)
        for cell : master_slave_flipflop
          use entity cell_lib.ms_flipflop(normal_drive);
        end for;
      end for;

      for reg(other_cell)
        for cell : master_slave_flipflop
          use entity cell_lib.ms_flipflop(normal_drive);
        end for;
      end for;
```

```
      end for;
    for reg_array ( parallel_data'length - 1 )

      for reg(other_cell)
        for cell : master_slave_flipflop
          use entity cell_lib.ms_flipflop(high_drive);
        end for;
      end for;

    end for;

  end for;

end configuration last_high_drive;
```

The first of the block configurations for **reg_array** identifies those cells generated with index values in the range 0 to **width** − 2. In the first of these cells, the condition labeled **first_cell** of the inner if-generate statement **reg** is true, and the condition labeled **other_cell** is false. In the remaining cells, the condition labeled **first_cell** is false and the condition labeled **other_cell** is true. The two inner block configurations for **reg(first_cell)** and **reg(other_cell)** configure whichever flipflop component instance is created in each of these cells.

The second of the block configurations for **reg_array** singles out the cell generated with index value **parallel_data'length** − 1. This is the cell for which we wish to use a flipflop with high drive capacity. We know that in this cell the condition labeled **first_cell** is false. Hence, we do not need to include a nested block configuration for that alternative. We only include a nested block configuration for the **other_cell** alterntive.

VHDL-87, -93, -2002

These versions of VHDL do not allow alternative labels in if-generate statements. Since an if-generate statement can only include one alternative, the statement label alone is sufficient to identify the alternative in a correspsonding block configuration.

Exercises

1. [**❶** 14.1] Draw a diagram illustrating the circuit described by the following generate statement:

```
synch_delay_line : for stage in 1 to 4 generate
  delay_ff : component d_ff
    port map ( clk => sys_clock,
               d => delayed_data(stage - 1),
               q => delayed_data(stage) );
end generate synch_delay_line;
```

2. [● 14.1] Write a generate statement that instantiates an inverter component for each element of an input bit-vector signal **data_in** to derive an inverted bit-vector output signal **data_out_n**. Use the index range of **data_in** to determine the number of inverters required, and assume that **data_out_n** has the same index range as **data_in**.

3. [● 14.2] Write a conditional generate statement that connects a signal external_clock directly to a signal **internal_clock** if a Boolean generic constant **positive_clock** is true. If the generic is false, the statement should connect **external_clock** to **internal_clock** via an instance of an inverter component.

4. [● 14.3] Write block configurations for the generate statement shown in Exercise 1. The first flipflop (with index 1) should be bound to the entity **d_flipflop** in the library **parts_lib**, using the architecture body **low_input_load**. The remaining flipflops should be bound to the same entity, but use the architecture body **standard_input_load**.

5. [● 14.3] Write a block configuration for the generate statement described in Exercise 3. The inverter component, if generated, should be bound to the entity **inverter** using the most recently analyzed architecture body in the library **parts_lib**.

6. [❷ 14.1] Develop a structural model for an *n*-bit-wide two-input multiplexer composed of single-bit-wide two-input multiplexer components. The width *n* is a generic constant in the entity interface.

7. [❷ 14.1] A first-in/first-out (FIFO) queue can be constructed from the register component shown in Figure 14.3. The bit width of the component is a generic constant in the component interface. The FIFO is constructed by chaining cells together and connecting their **reset** inputs in parallel. The depth of the FIFO is specified by a generic constant in the entity interface. Develop a structural model for a FIFO implemented in this manner.

FIGURE 14.3

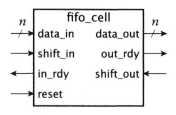

A register component used to construct a FIFO.

8. [❷ 14.1/14.2] Develop a structural model for a binary ripple counter implemented using D-flipflops as shown in Figure 14.4. The width *n* is a generic constant in the entity interface.

FIGURE 14.4

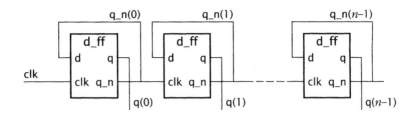

A binary ripple counter.

9. [❷ 14.1/14.2] Develop a structural model for an *n*-bit-wide ripple-carry adder. The least-significant bits are added using a half-adder component, and the remaining bits are added using full-adder components.

10. [❷ 14.3] Develop a behavioral model for a single-bit-wide two-input multiplexer. Write a configuration declaration for the *n*-bit-wide multiplexer described in Exercise 6, binding the behavioral implementation to each component instance.

11. [❷ 14.3] Develop a behavioral model for the D-flipflop described in Exercise 8. Write a configuration declaration for the ripple counter, binding the behavioral implementation to each D-flipflop component instance.

12. [❷ 14.3] Develop a behavioral model for a half adder and a full adder. Write a configuration declaration for the ripple-carry adder described in Exercise 9, binding the behavioral models to the component instances.

13. [❸ 14.1/14.2] Exercises 33 and 37 in Chapter 5 describe the components needed to implement a 16-bit carry-look-ahead adder. The same components can be used to implement a 64-bit carry-look-ahead adder as shown in Figure 14.5. The 64-bit addition is split into four identical 16-bit groups, each implemented with a 16-bit carry-look-ahead adder. The carry-look-ahead generator is augmented to include generate and propagate outputs, calculated in the same way as those calculated by each 4-bit adder. An additional carry-look-ahead generator is used to calculate the carry inputs to each 16-bit group. Develop a structural model of a 64-bit carry-look-ahead adder using nested generate statements to describe the two-level iterative structure of the circuit.

FIGURE 14.5

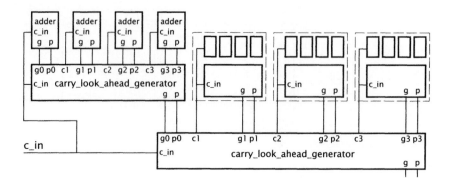

A 64-bit carry-look-ahead adder.

14. [❸ 14.2] A circuit to generate the odd-parity function of an 8-bit word is implemented using a tree of exclusive-or gates as shown in Figure 14.6. This structure can be generalized to an input word size of 2^n, implemented using a tree with n levels of gates. Develop a recursive model that describes such a parity generator circuit. The depth of the tree is a generic constant in the entity interface and is used to constrain the size of the input word.

FIGURE 14.6

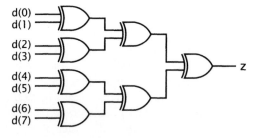

An odd-parity generator implemented using exclusive-or gates.

15. [❸ 14.3] Develop a behavioral model for the FIFO cell described in Exercise 7. The cell contains storage for one n-bit word of data. When reset, the cell sets in_rdy to '1' and out_rdy to '0', indicating that it contains no data. When **shift_in** changes to '1', the cell latches the input data and makes it available at **data_out**, then sets in_rdy to '0' and out_rdy to '1', indicating that the cell contains data. When **shift_out** changes to '1', the cell sets in_rdy to '1' and out_rdy to '0', indicating that the cell no longer contains data. Write a configuration declaration for the FIFO queue described in Exercise 7, binding the behavioral FIFO cell model to each component instance.

16. [❹] Ward and Halstead, in their book *Computation Structures* ([17], pp. 130–134), describe a combinatorial array multiplier that multiplies two unsigned binary numbers. The multiplier consists of an array of cells, each of which contains an and gate to multiply two operand bits and a full adder to form a partial-product bit, as shown in Figure 14.7.

The cells are connected in the multiplier array as shown in Figure 14.8. Develop a structural model of an *n*-bit × *n*-bit array multiplier, in which the word length *n* is a generic constant in the entity interface. Write a behavioral model of the multiplier cell and a configuration declaration that binds the cell model to each cell component instance in the array multiplier. Next, refine the behavioral cell model to a gate-level model, and revise the configuration declaration to use the refined cell model.

FIGURE 14.7

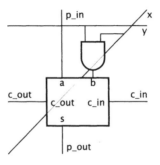

A single-bit multiplier cell.

FIGURE 14.8

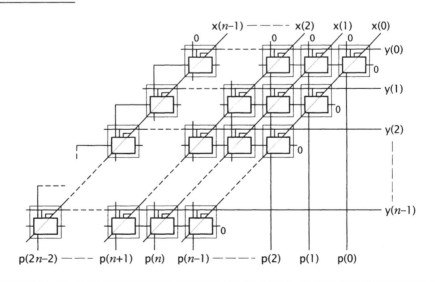

A multiplier array constructed from single-bit multiplier cells.

17. [❹] Weste and Eshraghian, in their book *Principles of CMOS VLSI Design: A Systems Perspective* ([18], pp. 384–407), describe a systolic array processor for dynamic time warping (DTW) pattern-matching operations used in speech recognition. Develop a model of the DTW processing element, and use it to implement the systolic array processor.

18. [❹] A hypercube multicomputer consists of a collection of 2^n processing elements (PEs) arranged at the vertices of an n-dimensional cube. Hypercubes with dimensions 1, 2, 3 and 4 are illustrated in Figure 14.9.

FIGURE 14.9

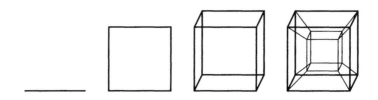

Hypercubes of dimension 1, 2, 3 and 4.

Each PE has a unique address, formed by concatenating the index (0 or 1) in each dimension to derive a binary number. Attached to each PE is a message switch with n bidirectional message channels, one in each dimension. The switches are interconnected along the edges of the hypercube. PEs exchange messages by passing them to the attached switches, which route them through the interconnections from source to destination. A message includes source and destination PE addresses, allowing the switches to determine a route for the message.

The hypercube structure can be described recursively. A hypercube of dimension 1 is simply a line from position 0 to position 1 in the first dimension. A hypercube of dimension n ($n > 1$) is composed of two sub-hypercubes of dimension $n - 1$, one at position 0 in the nth dimension and the other at position 1 in the nth dimension. Each vertex in one sub-hypercube is joined to the vertex with the same address in the other sub-hypercube.

Develop a recursive structural model of an n-dimensional hypercube multicomputer, where the number of dimensions is specified by a generic constant in the entity interface. Your model should include separate component instances for the PEs and the message switches. Also develop behavioral models for the PEs and message switches. The PEs should generate streams of test messages to different destinations to test the switch network. Each switch should implement a simple message-routing algorithm of your devising.

Chapter 15

Access Types

We have seen in previous chapters how we can use variables within processes to create data that is associated with a name. We can write a variable name in a model to read its value in expressions and to update its value in variable assignment statements. In this chapter, we introduce access types as a mechanism in VHDL for creating and managing unnamed data during a simulation.

15.1 Access Types

The scalar and composite data types we are now familiar with can be used to represent either single data items or regular collections of data. However, in some applications, we need to store collections of data whose size is not known in advance. Alternatively, we may need to represent a complex set of relations between individual data objects. In these cases, simple scalar and composite types are not sufficient. Instead, we need to create data objects as they are required during a simulation and to represent the links between these data objects. We do this in VHDL using *access types*. These are similar to pointer types found in many programming languages. In VHDL, access types are used mainly in high-level behavioral models and rarely in low-level models.

We start this section with a description of access types, pointers and mechanisms for creating data objects. Then we look at the way in which these mechanisms are used to create linked data structures during a simulation.

15.1.1 Access Type Declarations and Allocators

We can declare an access type using a new form of type definition, given by the syntax rule

access_type_definition ⇐ **access** subtype_indication

We can include such a type definition in a type declaration, for example:

type natural_ptr **is access** natural;

This defines a new type, named **natural_ptr**, representing values that point to data objects of type **natural**. Values of type **natural_ptr** can only point to natural numbers, not to objects of any other type. In general, we can write access type declarations referring to any VHDL type except file types or protected types.

Once we have declared an access type, we can declare a variable of that type within a process or subprogram. For example, we might declare a variable of the type **natural_ptr** shown above:

variable count : natural_ptr;

This declaration creates a variable, called **count**, that may point to a data object of type **natural** stored in memory. Initially, the variable has the value **null**. This is a special pointer value that does not point to any data object and is the default initial value for any access type. We can represent the null pointer variable pictorially as shown in Figure 15.1(a). The box represents the location in memory where the variable count is stored. Since it is a named variable, we can label the box with the variable name. Note that we cannot declare constants or signals of access types. Variables are the only class of object that may be of an access type.

Next, we can create a new natural number data object and set count to point to it. We do this using an *allocator*, written according to the following syntax rule:

primary ⇐ **new** subtype_indication ⫿ **new** qualified_expression

This rule shows that an allocator, written using the keyword **new**, is a kind of primary. Recall that primaries are the basis of VHDL expressions. The first form of allocator creates a new data object of the specified subtype in memory, initializes it to the default initial value for the subtype and returns a pointer to it. For example, the allocator expression

new natural

creates a natural number data object in memory and initialized to 0 (the leftmost value in the subtype **natural**). The allocator then returns a pointer to the object, as shown in Figure 15.1(b). The box represents the location in memory where the data object is stored, but since it is an unnamed object, there is no label. Instead, the arrow represents the pointer to the object. This is the only way of accessing the object.

FIGURE 15.1

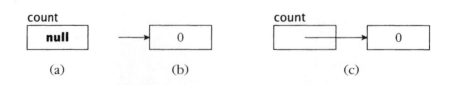

(a)   (b)   (c)

*(a) An access variable initialized to **null**. (b) A data object created by an allocator expression.*
(c) A pointer returned by an allocator assigned to the access variable.

The next step is to assign the pointer to the access variable **count**. Since the allocator is an expression that returns the pointer value, we can write it on the right-hand side of a variable assignment statement, as follows:

```
count := new natural;
```

This statement has the combined effects of creating and initializing the data object and assigning a pointer to it to the variable **count**, as shown in Figure 15.1(c). The pointer overwrites the null pointer previously stored in **count**.

Now that we have an access variable pointing to a data object in memory, we can use and update the value of the object, accessing it via the variable. This use of the variable is the reason for the terms "access type" and "access variable." We access the object using the keyword **all** as a suffix after the access variable name. For example, we can update the object's value as follows:

```
count.all := 10;
```

and we use its value in an expression:

```
if count.all = 0 then
    ...
end if;
```

Note that we need to use the keyword **all** in this way if we wish to use the data object rather than the pointer itself. If we had written the expression "**count = 0**", our VHDL analyzer would report an error, since the value of **count** is a pointer, not a number, so it cannot be compared with the number 0.

The second form of allocator, shown in the syntax rule on page 480, uses a qualified expression to specify both the subtype and the initial value for the created data object. Recall that the syntax rule for a qualified expression is

qualified_expression ⇐ type_mark ' (expression) ▯ type_mark ' aggregate

Thus, instead of writing the two statements

```
count := new natural;
count.all := 10;
```

we could achieve the same effect with this second form of allocator:

```
count := new natural'(10);
```

The qualified expression can also take the form of an array or record aggregate. For example, if we have a record type and access type declared as

```
type stimulus_record is record
    stimulus_time : time;
    stimulus_value : bit_vector(0 to 3);
  end record stimulus_record;

type stimulus_ptr is access stimulus_record;
```

and an access variable declared as

```
variable bus_stimulus : stimulus_ptr;
```

we could create a new stimulus record data object and set **bus_stimulus** to point to it as follows:

```
bus_stimulus := new stimulus_record'( 20 ns, B"0011" );
```

The value in the allocator is a qualified record aggregate that specifies both the type of the data object (**stimulus_record**) and the value for each of the record elements.

15.1.2 Assignment and Equality of Access Values

Let us now look at the effect of assigning one access variable value to another access variable. Suppose we have two access variables declared as follows:

```
variable count1, count2 : natural_ptr;
```

and we create data objects and set the variables to point to them:

```
count1 := new natural'(5);
count2 := new natural'(10);
```

The variables and data objects are illustrated in Figure 15.2(a). Next, we perform the following variable assignment:

```
count2 := count1;
```

The effect of this assignment is to copy the pointer from **count1** into **count2**, making both access variables point to the same object, as shown in Figure 15.2(b). We can see that this is in fact the case by accessing the object via each of the access variables. For example, if we update the object via **count1**,

```
count1.all := 20;
```

FIGURE 15.2

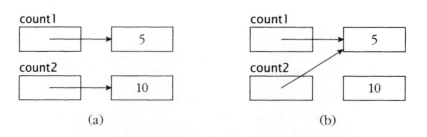

(a)   (b)

The effect of assigning one access variable to another. The two variables point to the same data object.

then the value we get via count2.**all** is 20.

Note that when we copied the pointer from count1 to count2, we overwrote the pointer to the data object 10. The object itself is still stored in memory, but count2 is no longer pointing to it. If we had previously copied the pointer before overwriting it, then we could access the object via that other copy. However, if there is no other pointer to the object, it is inaccessible. This is one of the main differences between named variables and allocated data objects. We can always access a variable by using its name, but an allocated object has no name, so we can only access it via pointers. If there are no pointers to an object, it is lost forever, even though it is still resident in the host computer's memory. We often call such inaccessible objects *garbage*. We return to the topic of dealing with unneeded objects later in this section.

Next, we look at the effect of comparing two access variables using the "=" and "/=" operators. These operators test whether the two pointers point to the same location in memory. For example, after performing the assignment

```
count2 := count1;
```

the expression

```
count1 = count2
```

is true, since, as Figure 15.2(b) shows, the two access variables then point to the same object. However, if we instead set count1 and count2 as follows:

```
count1 := new natural'(30);
count2 := new natural'(30);
```

we create two distinct data objects in memory, each storing the number 30. The variable count1 points to one of them, and count2 points to the other. In this case the result of the equality comparison is false. If we really want to test whether the data objects are equal, as opposed to testing the pointers, we write

```
count1.all = count2.all
```

One very useful pointer comparison is the test for equality with **null**, the special pointer value that does not point to any object. For example, we might write

```
if count1 /= null then
   count1.all := count1.all + 1;
end if;
```

The test in the if statement ensures that we only access the value pointed to by count1 if there is a value to access. If count1 has the value **null**, trying to access count1.**all** results in an error.

15.1.3 Access Types for Records and Arrays

We have introduced access types in this section by concentrating on access types that point to scalars, in order to keep things simple. However, most models that include access

types use them to point to records or arrays. Pointers to records are mainly used for building linked data structures, and pointers to arrays are used if the lengths of the arrays are
not known when the model is written. In both cases, we can use a shorthand notation for
referring to objects via access variables.

Let us start with records and return to the example shown earlier, in which we had
types declared as

```
type stimulus_record is record
    stimulus_time : time;
    stimulus_value : bit_vector(0 to 3);
  end record stimulus_record;

type stimulus_ptr is access stimulus_record;
```

We also declared an access variable as

```
variable bus_stimulus : stimulus_ptr;
```

We have seen that we can access a record object pointed to by bus_stimulus using the
notation "**bus_stimulus.all**". If we want to refer to the **stimulus_time** element, we could
write "**bus_stimulus.all**.stimulus_time". In practice, we usually want to refer either to the
pointer itself or to an element of the record, and rarely to the record as a whole. For this
reason, VHDL allows us to write "**bus_stimulus.stimulus_time**" to refer to the record element. Whenever we select a record element name after an access variable name, we automatically follow the pointer to get to the record.

A similar shorthand notation applies when we use access variables that point to array
data objects. For example, suppose we declare types as follows:

```
type coordinate is array (1 to 3) of real;
type coordinate_ptr is access coordinate;
```

and an access variable:

```
variable origin : coordinate_ptr := new coordinate'(0.0, 0.0, 0.0);
```

This last declaration creates the access variable and initializes it to point to an array object
initialized with the aggregate value. We can refer to the elements of the array using the
notation "**origin(1)**", "**origin(2)**" and "**origin(3)**", instead of having to write "**origin.all**(1)",
and so on. This is similar to accessing elements of records. Whenever we write an array
index after an access variable name, we automatically follow the pointer to the array.

One of the advantages of using access types that point to array objects is that we can
deal with arrays of mixed lengths. This is in contrast to array variables, which have their
length fixed when they are created. For example, if we create an array variable
activation_times as follows:

```
type time_array is array (positive range <>) of time;
variable activation_times : time_array(1 to 100);
```

it is fixed at 100 elements for its entire lifetime. On the other hand, we can create an access type that points to data objects of an unconstrained or partially constrained composite type. For example, if we declare an access type as follows:

```
type time_array_ptr is access time_array;
```

we can declare our variable to be a pointer of this type:

```
variable activation_times : time_array_ptr;
```

Since the variable points to an array object of an unconstrained type, it may point to different array objects of different lengths during the course of a simulation. However, each array object is fully constrained. This means that once an array object is created in memory, its length is fixed. We can create an array object using an allocator that includes a qualified aggregate, for example:

```
activation_times := new time_array'(10 us, 15 us, 40 us);
```

This allocator creates an array object whose length is determined from the length of the aggregate. We can update each of these elements, but we cannot change the size of the array. If we need to add two more elements, we have to create a new array object of length five, with the first three elements being a copy of the elements from the old array. This might be done as follows:

```
activation_times := new time_array'( activation_times.all
                                 & time_array'(70 us, 100 us) );
```

The allocator in this assignment creates an array object whose length is determined by the result of the concatenation operation. If we want to create an array object without initializing the values, we write an allocator that names the array type and includes an index constraint. For example, to create an array object of length 10, we might write

```
activation_times := new time_array(1 to 10);
```

We can also write allocators for more complicated composite types with unconstrained elements. In general, the allocator determines all of the the index ranges of the allocated object. If we write an allocator with just a subtype indication, it must specify a fully constrained subtype, and the index ranges are taken from that subtype. For example, given the following declarations:

```
type RV is record
  v1 : bit_vector;
  v2 : time_vector;
end record RV;
type RV_ptr is access RV;
variable p : RV_ptr;
```

we can write an allocator using a subtype indication:

```
p := new RV_record(v1(0 to 23), v2(0 to 23));
```

The subtype indication specifies index ranges of 0 to 23 for both elements, so they are used for the allocated object. The object is then initialized with the default initial value.

On the other hand, if we write an allocator with a qualified expression, the value in the expression is converted to the named subtype (as described in Section 4.3.7), and that determines the index ranges for the allocated object. Where that subtype specifies index ranges, they are used; and where no index range is specified, an index range is determined from the corresponding index subtype. For example, given the preceding declarations, we can write an allocator:

```
p := new RV_record'(v1 => "010", v2 => (2 ns, 4 ns, 6 ns));
```

Since the subtype **RV_record** does not specify any index ranges, the index subtypes for the record elements are used to determine index ranges for the allocated value. For each element, the index subtype is **natural**, so the index ranges are 0 to 2.

15.2 Linked Data Structures

Suppose we wish to store a list of values to be used to stimulate a signal during a simulation. One possible approach would be to define an array variable of stimulus values. However, a problem arises if we do not know how large to make the array. If we make it too small, we may run out of space. If we make it too large, we may waste space in the host computer's memory and run out of space for other variables. The alternative approach is to use access types and to create values only as they are needed. The values can be linked together with pointers to form an extensible data structure. There are several possible organizations for linked structures, but we look at one of the simplest, a *linked list*, as an example, showing how it is constructed and manipulated.

A linked list of values that might be used as stimuli for a signal is shown in Figure 15.3. To construct this list, we need to compose each cell from a record that has one element for the stimulus value and an extra element for a pointer to the next cell in the list. This pointer must be of an access type used to access record objects. A first attempt to write the type declarations for this structure might be

```
type value_cell is record
    value : bit_vector(0 to 3);
    next_cell : value_ptr;
  end record value_cell;

type value_ptr is access value_cell;
```

FIGURE 15.3

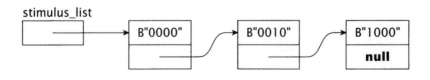

A linked list structure of stimulus records.

The problem here is that the definition of **value_cell** uses the name **value_ptr** as the type of one of the elements, but **value_ptr** is not declared until after the declaration of **value_cell**. If we reverse the two type declarations, the same problem arises in the definition of **value_ptr** when it tries to use the name **value_cell**. To solve this "chicken and egg" problem, VHDL lets us write an *incomplete type declaration* for the record type. The syntax rule is

type_declaration $\Leftarrow$ **type** identifier ;

An incomplete type declaration simply names the type, indicating that it will be fully defined later. Meanwhile, we can use the type name to declare access types. However, we must complete the definition of the incomplete type before the end of the declarative part in which the incomplete declaration appears. Since we can do this after the access type declaration, we can use the name of the access type within the complete type declaration. Thus, we can rewrite our circular type declarations as

```
type value_cell;

type value_ptr is access value_cell;

type value_cell is record
    value : bit_vector(0 to 3);
    next_cell : value_ptr;
  end record value_cell;
```

Next we can declare an access variable to point to the beginning of the list:

```
variable value_list : value_ptr;
```

This declaration creates a variable containing a null pointer, as shown in Figure 15.4(a). We can think of this as representing an empty list. Thus, if we need to determine whether a list is empty, we can test the access variable to see whether it is **null**, for example:

```
if value_list /= null then
    ...  -- do something with the list
end if;
```

We can add a cell to the empty list by allocating a new record and assigning the pointer to the access variable, as follows:

```
value_list := new value_cell'( B"1000", value_list );
```

The second element in the aggregate is a copy of the pointer initially stored in **value_list**. This has the value **null**, so the result of executing the whole statement is as shown in Figure 15.4(b). The reason for using the old value of **value_list** instead of writing in the value **null** is that we can use the same form of statement to add the next cell:

```
value_list := new value_cell'( B"0010", value_list );
```

The allocator creates a new cell in memory, with the **value** element initialized to B"0010" and the **next_cell** element initialized to a copy of the pointer to the old cell. A

FIGURE 15.4

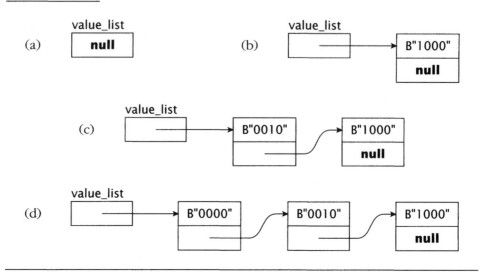

Successive stages in the creation of a list of stimulus values.

pointer to the new cell is then returned and assigned to **value_list**, as shown in Figure 15.4(c). We can create the third cell in the same way:

```
value_list := new value_cell'( B"0000", value_list );
```

This assignment produces the final list as shown in Figure 15.4(d). Note that each cell we create is added on to the front of the list.

Now suppose we have a list of stimulus values of arbitrary length, pointed to by our access variable, and we wish to go through the list applying each value to a signal. We can write a loop to traverse the list as follows. We need to make use of a working variable, **current_cell**, of type **value_ptr**. The statements to perform this traversal are:

```
current_cell := value_list;
while current_cell /= null loop
  s <= current_cell.value;
  wait for 10 ns;
  current_cell := current_cell.next_cell;
end loop;
```

The first assignment sets **current_cell** to point to the first cell in the list, as shown in Figure 15.5(a). The first pass through the loop uses the **value** element of this cell to stimulate the signal, then copies the **next_cell** element of the cell into the working variable. At the end of the first iteration the working variable points to the next element in the list, as shown in Figure 15.5(b). The loop repeats in this way, with **current_cell** being advanced from one cell to the next cell in each iteration. In the last iteration, the variable points to the last cell as shown in Figure 15.5(c). The **next_cell** element of this cell is **null**, and this

FIGURE 15.5

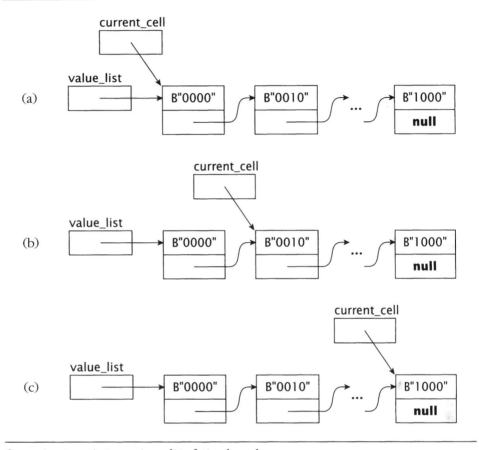

Successive stages in traversing a list of stimulus values.

is copied into **current_cell**. When the loop test is performed again it evaluates to false, and so the loop terminates.

Another operation we may wish to perform on a list is to search for a particular value. Again, we make use of a working access variable to traverse the list, checking each cell to see if its value element matches the value for which we are searching, as follows:

```
current_cell := value_list;
while current_cell /= null
    and current_cell.value /= search_value loop
  current_cell := current_cell.next_cell;
end loop;
assert current_cell /= null
  report "search for value failed";
```

The test for a null pointer in the loop condition is most important. It guards against the possibility that the sought value is not in the list. If the list terminates with the working variable equal to **null**, we know that the value was not found, and we can deal with the condition appropriately. Note that the **and** operator in the loop condition is a "short circuit" operator, so the second part of the test will not proceed if current_cell is **null**, not pointing to any list cell.

The linked list data structure is just one of a number of linked data structures that we can construct using access types. Other examples include queues, trees and network structures. We come across some of these in further examples in this chapter and later in the book. However, the field of data structures is much larger than we can hope to cover in a book that focuses on hardware modeling and simulation. Fortunately, there are numerous good textbooks available that discuss data structures at length. Of these, the books that use the Ada programming language are particularly relevant, as VHDL's access types are based on those of Ada. (See, for example, [6].)

15.2.1 Deallocation and Storage Management

We saw earlier that if we overwrite a pointer to an unnamed data object, we can lose all means of accessing the object, making it "garbage." While this is usually not a problem, if we create too much garbage during a simulation run, the host computer may run out of memory space for allocating new objects. Some computers are able to avoid this problem by periodically scanning memory for inaccessible data and reclaiming the space they occupy, a process called *garbage collection*. However, most computers do not provide this service, so we may have to perform our own storage management.

The mechanism VHDL provides for us to do this is the implicitly defined procedure **deallocate**. Whenever we declare an access type, VHDL automatically provides an overloaded version of **deallocate** to handle pointers of that type. For example, if we declare an access type for pointers to objects of type T as follows:

  type T_ptr **is access** T;

we automatically get a version of **deallocate** declared as

  procedure deallocate (P : **inout** T_ptr);

The purpose of this procedure is to reclaim the memory space used by the data object pointed to by the parameter P. When the procedure returns, it sets P to the null pointer, since the object is no longer stored in memory. Note that if P is **null** to start with, the procedure has no effect. Thus, there is no need to test whether a pointer is **null** before passing it to deallocate.

EXAMPLE 15.1 *Deleting cells from a linked list*

Suppose we wish to delete cells from our list of stimulus values, shown in the previous example. The first cell in the list is pointed to by the access variable **value_list**. We can delete the first cell and reclaim its storage as follows:

```
cell_to_be_deleted := value_list;
value_list := value_list.next_cell;
deallocate(cell_to_be_deleted);
```

The first statement simply copies the pointer to the first cell into the access variable **cell_to_be_deleted**, so that we do not lose access to it. The second statement advances the list head to the second cell. The third statement then reclaims the storage used by the first cell. Note that, if we do not need to reclaim the storage for the first cell, we only need to include the second statement.

If we wish to delete the whole list, we can use a loop to repeat these statements for each cell in the list, as follows:

```
while value_list /= null loop
  cell_to_be_deleted := value_list;
  value_list := value_list.next_cell;
  deallocate(cell_to_be_deleted);
end loop;
```

This loop simply repeats the steps needed to delete the cell at the head of the list until the list is empty, indicated by **value_list** being **null**.

We can use **deallocate** to reclaim memory space, provided we are sure that no other pointer points to the object being deallocated. It is very important that we keep this condition in mind when using **deallocate**. If some other pointer points to an object that we deallocate, that pointer is not set to **null**. Instead, it becomes a "dangling" pointer, possibly pointing to some random piece of data in memory or not pointing to a valid memory location at all. If we try to access data via a dangling pointer, the effects are unpredictable, varying from accessing seemingly random data to crashing the simulation run. Thus, we must take the utmost care to avoid this situation when using **deallocate**. Furthermore, we should document such models very thoroughly, so that other designers using or modifying the models are aware of the potential problems.

15.3 An Ordered-Dictionary ADT Using Access Types

In Chapter 12, we introduced the notion of abstract data types (ATDs) and showed how we could write a reusable ADT in the form of a package with generics. We presented an example of an ordered dictionary ATD implemented by a sorted array of elements. One problem with that implementation is that the maximum size of a dictionary is fixed when the package is instantiated. We may not always be able to determine in advance the maximum number of elements needed in an application. We can avoid this problem by implementing the ADT with a different data structure based on access types and dynamically allocated storage for elements.

The data structure that we will use in this section is called a *binary search tree*, and is described in textbooks on software data structures. The tree consists of a collection of records, each of which stores one elements and has pointers to two subtrees, as shown in Figure 15.6. All of the elements in the left subtree have key values less than that of the element in the root record, and all of the elements in the right subtree have key values greater than that of the element in the root record. This relationship holds recursively in both the left and right subtrees. The elements in the tree of Figure 15.6 have a numeric key, but this need not always be the case. All that is required is that the key type be totally ordered; that is, there must be a "<" relation among key values.

The package declaration for the ADT is shown below. It is much the same as the implementation using a sorted array, apart from the details of the private types and the absence of the **size** generic constant. In particular, the operations are identical, illustrating the separation of interface and implementation afforded by ADTs.

```
package dictionaries is
   generic ( type element_type;
             type key_type;
             function key_of ( E : element_type ) return key_type;
             function "<" ( L, R : key_type )
                          return boolean is <> );

   -- types provided by the package

   type dictionary_object;       -- private

   type dictionary_type is access dictionary_object;

   -- operations on dictionaries

   procedure initialize ( dictionary : inout dictionary_type );
```

FIGURE 15.6

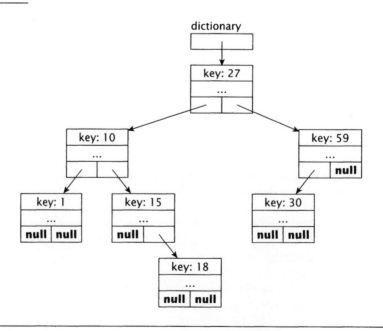

A binary search tree.

```
procedure lookup ( dictionary : in dictionary_type;
                   lookup_key : in key_type;
                   element : out element_type;
                   found : out boolean );

procedure search_and_insert
            ( dictionary : inout dictionary_type;
              element : in element_type;
              already_present : out boolean );

procedure traverse
  generic   ( procedure action ( element : in element_type ) )
  parameter ( dictionary : in dictionary_type );

-- private types: pretend these are not visible

type dictionary_object is
  record
    element : element_type;
    left, right : dictionary_type;
  end record dictionary_object;

end package dictionaries;
```

We can use this implementation of the ADT in exactly the same way as shown in the examples in Chapter 12, apart from the minor difference in the package instantiation. We simply omit the size generic:

```
package test_pattern_dictionaries is new work.dictionaries
  generic map ( element_type => test_pattern_type,
                key_type => time,
                key_of => test_time_of );
```

As implementers of the ADT, we are concerned with the details of the private types. Specifically, the **dictionary_object** record type represents the storage for an element and the left and right subtree pointers. The public type **dictionary_type** is an access type pointing to the root element of a tree. The package body for the ADT is

```
package body dictionaries is

  procedure initialize ( dictionary : inout dictionary_type ) is
  begin
    if dictionary /= null then
      initialize ( dictionary.left );
      initialize ( dictionary.right );
      deallocate ( dictionary );
    end if;
  end function new_dictionary;

  procedure lookup ( dictionary : in dictionary_type;
                     lookup_key : in key_type;
```

```
                             element : out element_type;
                             found : out boolean ) is
        variable current : dictionary_type := dictionary;
begin
    found := false;
    while current /= null loop
        if lookup_key < key_of ( current.element ) then
            current := current.left;
        elsif key_of ( current.element ) < lookup_key then
            current := current.right;
        else
            found := true;
            element := current.element;
            return;
        end if;
    end loop;
end procedure lookup;

procedure search_and_insert
              ( dictionary : inout dictionary_type;
                element : in element_type;
                already_present : out boolean ) is
begin
    if dictionary = null then
        already_present := false;
        dictionary
            := new dictionary_object'( element => element,
                                       left => null, right => null );
    elsif key_of ( element ) < key_of ( dictionary.element ) then
        search_and_insert ( dictionary.left,
                            element, already_present );
    elsif key_of ( dictionary.element ) < key_of ( element ) then
        search_and_insert ( dictionary.right,
                            element, already_present );
    else
        already_present := true;
    end if;
end procedure search_and_insert;

procedure traverse
    generic    ( procedure action ( element : in element_type ) )
    parameter ( dictionary : in dictionary_type ) is
begin
    if dictionary = null then
        return;
    end if;
    traverse ( dictionary.left );
    action    ( dictionary.element );
```

```
        traverse ( dictionary.right );
    end procedure traverse;

end package body dictionaries;
```

The function **initialize** clears the tree representing a dictionary recursively. Provided the tree has at least a root element, the procedure clears the left and right subtrees, then deallocates the root object. This has the effect of setting the **dictionary** pointer to **null**.

The **lookup** procedure searches for an element with the desired key by descending down a path from the root. At a given position in the tree, if the desired key is less than that of the element at the position, the procedure descends into the left subtree. If the desired key is greater than that of the element, the procedure descends into the right subtree. If neither case is true, the keys must be equal, and the element at the given position is the one sought. The descent continues until the sought element is found or until a null pointer is found.

The **search_and_insert** procedure searches the tree for the position in which to insert a given element. While we could write the procedure using a loop, as we did for the lookup procedure, it is more clearly expressed in recursive form. The procedure descends the tree using recursion. If it reaches a null pointer, the tree contains no element with the given key. The procedure allocates a new object for the element, and updates the parent's pointer, sets **already_present** to false, and returns. For example, in Figure 15.6, if the new element has a key value of 12, the procedure would descend via elements with keys 27, 10 and 15, before reaching the null left pointer of the element with key 15. The procedure would then allocate the object for the new element and update the null pointer to refer to the new object. If the procedure reaches an element with the same key as that of the new element, it sets **already_present** to true, does not insert the new element, and returns.

The **traverse** procedure is also recursive. It must apply the **action** procedure to elements in ascending order of their keys. Since all the elements in the left subtree of a given element have lesser keys, and all the elements in the right subtree have greater keys, action must be applied to all of the left subtree elements, then the given element, then the right subtree elements. In the case of an empty subtree (indicated by a null pointer), no application is needed, so the procedure returns immediately.

The recursive **traverse** procedure further illustrates the rules we mentioned in Section 12.3 for interpreting names in uninstantiated units. The reference to the name **traverse** within that procedure is interpreted, in each instance of the procedure, as a reference to the instance. Thus each instance is properly recursive. This is the only situation where we can write a call to an uninstantiated subprogram.

Exercises

1. [❶ 15.1] Write a type declaration for an access type that points to a character data object. Declare a variable of the type, initialized by allocating a character with the value ETX. Write a statement that changes the character value to 'A'.

2. [❶ 15.1] Identify the error in the following VHDL fragment:

    ```
    type real_ptr is access real;
    variable r : real_ptr;
    ```

```
. . .
r := new real;
r := r + 1.0;
```

3. [❶ 15.1] Draw a diagram showing the pointer variables and the data objects to which they refer after execution of the following VHDL fragment:

```
type int_ptr is access integer;
variable a, b, c, d : int_ptr;
. . .
a := new integer'(1);  b := new integer'(2);
c := new integer'(3);  d := new integer'(4);
b := a;  a := b;
c.all := d.all;
```

4. [❶ 15.1] After execution of the fragment shown in Exercise 3, what is the value of each of the following conditions?

```
a = b                c = d

a.all = b.all    c.all = d.all
```

5. [❶ 15.1] Write a type declaration for an access type that points to a string data object. Declare a variable of the type, initialized by allocating a string of four spaces. Write a statement that changes the first character in the string to the character NUL.

6. [❶ 15.1] The following declarations define a type for complex numbers, an access type referring to the complex number type and three pointer variables:

```
type complex is record
    re, im : real;
  end record complex;
type complex_ptr is access complex;
variable x, y, z : complex_ptr;
```

Write statements that assign the complex product of the values pointed to by x and y to the data object pointed to by z. The steps required in complex multiplication are given by the following formulas:

$$z_{real} = x_{real} \times y_{real} - x_{imag} \times y_{imag}$$
$$z_{imag} = x_{real} \times y_{imag} + x_{imag} \times y_{real}$$

7. [❶ 15.2] Write type declarations for use in constructing a linked list of message objects. Each message contains a source and destination number (both of type **natural**) and a 256-bit data field. Declare a variable to point to a list of messages, and write a statement to add to the list a message with source number 1, destination number 5 and a data field of all '0' bits.

8. [❶ 15.2] Why is the following fragment to delete the first object in a linked list incorrect?

```
cell_to_be_deleted := value_list;
deallocate(cell_to_be_deleted);
value_list := value_list.next_cell;
```

9. [❷ 15.2] In Section 15.2 on page 488, we show statements to traverse a linked list of stimulus values and apply them to a signal. Encapsulate these statements in a procedure with the list pointer and the signal as parameters.

10. [❷ 15.2] The algorithm for traversing a list, encapsulated in a procedure as described in Exercise 9, can be expressed recursively. If the list is empty, the procedure has nothing to do, so it returns. Otherwise, the procedure applies the first stimulus value from the list, waits for the delay, then recursively calls itself with the next cell pointer as the list parameter. Thus, recursive invocation of the procedure replaces the iterative traversal of the list. Rewrite the procedure to use this recursive algorithm.

11. [❷ 15.2] Write a recursive procedure to delete all cells from a linked list pointed to by a parameter of type **value_ptr**. Hint: The procedure should call itself to delete the cells after the first cell, then delete the first cell.

12. [❸ 15.2/15.3] Modify the implementation of the dictionary ADT to include a delete procedure to delete an element with a given key and to deallocate the storage for the deleted element.

13. [❸ 15.3] Develop an ADT for last-in/first-out stacks of objects. The ADT should be parameterized by the type of object and should provide operations to create a new stack, to test whether a stack is empty, to push an object onto the stack and to pop the top object from the stack and return the object's value.

14. [❸ 15.3] Develop an ADT for first-in/first-out queues of objects. The ADT should be parameterized by the type of object and maximum number of objects allowed in the queue. The operations are to create a new queue, to test whether a queue is empty or full, to add an object to the tail of a queue and to remove an object from the head of a queue and return the object's value. The queue may be implemented as a linked list, as shown in Figure 15.7. Use the ADT in a behavioral model of an 8-bit-wide FIFO, based on the behavior described in Exercises 7 and 15 in Chapter 14.

FIGURE 15.7

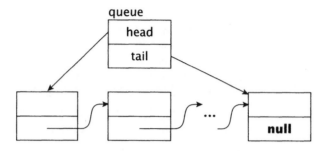

Linked list implementation of a queue.

15. [❹ 15.3] Develop an alternative implementation of the ordered-collection ADT based on a doubly-linked list data structure. A doubly-linked list is a collection of cells, each of which contains pointers to the previous and next cells. Cells are inserted in the list in order of their keys.

16. [❹] Develop an ADT for sparse arrays. A sparse array dynamically allocates storage for array elements as they are accessed. One approach is to allocate chunks of contiguous elements, since array accesses usually exhibit locality of reference. The ADT should have generics for the index bounds, the element type, and the default initial value of each element. It should provide operations to set an element to a given value and to read the value of an element.

Chapter 16

Files and Input/Output

In this chapter we look at the facilities in VHDL for file input and output. Files serve a number of purposes, one of which is to provide long-term data storage. In this context, "long-term" means beyond the lifetime of one simulation run. Files can be used to store data to be loaded into a model when it is run, or to store the results produced by a simulation. VHDL also provides specialized versions of file operations for working with text files. We show how textual input and output can be used to extend the user interface of a simulator with model-specific operations.

16.1 Files

We start our discussion of files by looking at the general-purpose mechanisms provided in VHDL for file input and output. VHDL provides sequential access to files using operations, such as "open", "close", "read" and "write", that are familiar to users of conventional programming languages.

16.1.1 File Declarations

A VHDL file is a class of object used to store data. Hence, as with other classes of objects, we must include file-type definitions in our models. The syntax rule for defining a file type is

    file_type_definition ⇐ **file of** type_mark

A file-type definition simply specifies the type of objects to be stored in files of the given type. For example, the type declaration

    type integer_file **is file of** integer;

defines **integer_file** to be a type of file that can only contain integers. A file can only contain one type of object, but that type can be almost any VHDL type, including scalar types, records and one-dimensional arrays. The only types that cannot be stored in files are multidimensional arrays, access types, protected types and other files.

Once we have defined a file type, we can then declare file objects. We do this with a new form of object declaration, described by the syntax rule

file_declaration ⇐
    file identifier 〖 , ... 〗 : subtype_indication
       〖 〖 **open** *file_open_kind*_expression 〗 **is** *string*_expression 〗 ;

A file declaration creates one or more file objects of a given file type. We can include a file declaration in any declarative part in which we can create objects, such as within architecture bodies, blocks, processes, packages and subprograms.

The optional parts of a file declaration allow us to make an association between the file object and a physical file in the host file system. If we include these parts, the file is automatically opened for access during simulation. The string after the keyword **is** is a file *logical name*, which identifies the host file to access. Since different host operating systems use different formats for naming files, many simulators provide some form of mapping between the logical name strings that we include in our models and the file names used in the host file system. For example, if we declare a file as

```
file lookup_table_file : integer_file is "lookup-values";
```

a simulator running under the UNIX operating system may associate the file object with a physical file named "lookup-values" in the current working directory. A different simulator, running under a Windows operating system, may associate the file object differently, since file names usually include a file-type extension in that operating system. So it might associate the object with a physical file called "lookup-values.dat" in the current working directory.

The optional expression after the keyword **open** allows us to specify how the physical file associated with the file object should be opened. This expression must have a value of the predefined type **file_open_kind**, declared in the package **standard**. The declaration is

```
type file_open_kind is (read_mode, write_mode, append_mode);
```

If we omit the open kind information from a file declaration but include the file logical name, the physical file is opened in read mode. In the rest of this section we discuss each of these modes and see how data is read and written using files opened in each of the modes.

VHDL-87

The syntax rule for file declarations in VHDL-87 is

file_declaration ⇐
    file identifier : subtype_indication **is**
       〖 **in** 〗 **out** 〗 *string*_expression ;

VHDL-87 does not provide the predefined type **file_open_kind**. Instead, the keywords **in** and **out** are used in file declarations to open files in read or write mode, respectively. The default is that a file is opened in read mode. Note that the VHDL-87 syntax for file declarations is not a subset of the VHDL-93 and VHDL-2002 syntax. If a model includes either of the keywords **in** or **out**, it cannot be successfully analyzed with a VHDL-93 or VHDL-2002 analyzer.

16.1.2 Reading from Files

If a file is opened in read mode, successive elements of data are read from the file using the **read** operation. Reading starts from the first element in the file, and each time an element is read the file position advances to the next element. We can use the **endfile** operation to determine when we have read the last element in the file. Given a file type declared as follows:

    type file_type **is file of** element_type;

the **read** and **endfile** operations are implicitly declared as

    procedure read (**file** f : file_type; value : **out** element_type);

    function endfile (**file** f : file_type) **return** boolean;

We explain subprogram file parameters later in this section.

EXAMPLE 16.1   *Initializing the contents of a ROM from a file*

We can use file operations to initialize the contents of a read-only memory (ROM) from a file. Following is an entity declaration for a ROM that includes a generic constant to specify the name of a file from which to load the ROM contents.

```
library ieee;  use ieee.std_logic_1164.all;

entity ROM is
  generic ( load_file_name : string );
  port ( sel : in std_ulogic;
         address : in std_ulogic_vector;
         data : inout std_ulogic_vector );
end entity ROM;
```

The architecture body for the ROM, shown below, uses the file name in a file declaration, creating a file object associated with a physical file of data words. The process that implements the behavior of the ROM loads the ROM storage array by reading successive words of data from the file, using **endfile** to determine when to stop.

```
architecture behavioral of ROM is

begin

  behavior : process is
```

```vhdl
    subtype word is std_ulogic_vector(0 to data'length - 1);
    type storage_array is
      array (natural range 0 to 2**address'length - 1) of word;
    variable storage : storage_array;
    variable index : natural;
    ...  -- other declarations

    type load_file_type is file of word;
    file load_file : load_file_type
            open read_mode is load_file_name;

  begin

    -- load ROM contents from load_file
    index := 0;
    while not endfile(load_file) loop
      read(load_file, storage(index));
      index := index + 1;
    end loop;

    -- respond to ROM accesses
    loop
      ...
    end loop;

  end process behavior;

end architecture behavioral;
```

In the above example, each element of the file is a standard-logic vector of a fixed length, determined by the ROM data port width. However, we are not restricted to fixed-length arrays as file elements. We may declare a file type with an unconstrained or partially constrained array type for the element type, provided the array element type is a scalar type or a fully constrained composite subtype, for example:

```vhdl
type bit_vector_file is file of bit_vector;
```

The data in a file of this type is a sequence of bit vectors, each of which may be of a different length. For such a file, the **read** operation takes a slightly different form, to allow for the fact that we do not know the length of the next element until we read it. The operation is implicitly declared as

```vhdl
procedure read ( file f : file_type;
                 value : out element_type;  length : out natural );
```

When we call this form of **read** operation, we supply an array variable large enough to receive the value we expect to read, and another variable to receive the actual length of the value read. For example, if we make the following declarations:

```
file vectors : bit_vector_file open read_mode is "vectors.dat";
variable next_vector : bit_vector(63 downto 0);
variable actual_len : natural;
```

we can call the read operation as follows:

```
read(vectors, next_vector, actual_len);
```

This allows us to read a bit vector up to 64 bits long. If the next value in the file is less than or equal to 64 bits long, it is placed in the leftmost part of **next_vector**, with the remaining bits being unchanged. If the value in the file is longer than 64 bits, the first 64 bits of the value are placed in **next_vector**, and the remaining bits are discarded. In both cases, **actual_len** is set to the actual length of the value in the file, whether it be shorter or longer than the length of the second argument to **read**. This allows us to test whether information has been lost. If the expression

```
actual_len > next_vector'length
```

is true, the vector variable was not long enough to receive all of the bits.

EXAMPLE 16.2   *Reading stimulus values from a file*

Suppose we have designed a model for a network receiver and we wish to test it. We can generate network packets to stimulate the model by reading variable-length packets from a file. The outline of a process to do this is:

```
stimulate_network : process is

    type packet_file is file of bit_vector;
    file stimulus_file : packet_file
           open read_mode is "test packets";

    variable packet : bit_vector(1 to 2048);
    variable packet_length : natural;

begin

    while not endfile(stimulus_file) loop

        read(stimulus_file, packet, packet_length);
        if packet_length > packet'length then
          report "stimulus packet too long - ignored"
            severity warning;
        else
          for bit_index in 1 to packet_length loop
            wait until stimulus_clock;
            stimulus_network <= not stimulus_network;
            wait until not stimulus_clock;
            stimulus_network <= stimulus_network
                            xor packet(bit_index);
```

```
      end loop;
    end if;

  end loop;

  wait;  -- end of stimulation: wait forever

end process stimulate_network;
```

The process declares a file object, **stimulus_file**, containing variable-length bit vectors. Each file element is read into the bit-vector variable **packet**, with the length of the bit vector read from the file being stored in **packet_length**. If the bit vector in the file is longer than the bit-vector variable, the process reports the fact and ignores that stimulus packet. Otherwise, the value in **packet_length** is used to determine how many bits from **packet** should be used as data bits to stimulate the network.

16.1.3 Writing to Files

If a file is opened in write mode, a new empty file is created in the host computer's file system, and successive data elements are added using the **write** operation. For each file type declared, the **write** operation is implicitly declared as

```
procedure write ( file f : file_type;  value : in element_type );
```

One common use of output files is to save information gathered by instrumentation code. When the simulation is complete, or upon some other trigger condition, the instrumentation code can use write operations to write the data to a file for subsequent analysis.

EXAMPLE 16.3 *Recording CPU instruction frequencies*

When we are designing a new CPU instruction set, it is useful to know how frequently each instruction is used in different programs. We measure this by simulating the CPU running a program and having the CPU keep count of how often it executes each instruction. When it completes the program (for example, by reaching a halt instruction), it writes the accumulated counts to a file.

The architecture body for a CPU shown below illustrates this approach. It contains a file, **instruction_counts**, opened in write mode. There is also a process, **interpreter**, that fetches and interprets instructions. It contains an array of counters, indexed by opcode values. As the instruction interpreter process decodes each instruction, it increments the appropriate counter. When a halt instruction is executed, the interpreter stops execution and writes the counter values as successive elements in the instruction_counts file.

```
architecture instrumented of CPU is

  type count_file is file of natural;
  file instruction_counts : count_file
        open write_mode is "instructions";
```

```vhdl
begin

  interpreter : process is

    variable IR : word;
    alias opcode : byte is IR(0 to 7);
    variable opcode_number : natural;
    type counter_array is
      array (0 to 2**opcode'length - 1) of natural;
    variable counters : counter_array := (others => 0);
    ...

  begin

    ...  -- initialize the instruction set interpreter

    instruction_loop : loop

      ...  -- fetch the next instruction into IR

      -- decode the instruction
      opcode_number := convert_to_natural(opcode);
      counters(opcode_number) := counters(opcode_number) + 1;
      ...

      -- execute the decoded instruction
      case opcode is
        ...
        when halt_opcode => exit instruction_loop;
        ...
      end case;

    end loop instruction_loop;

    for index in counters'range loop
      write(instruction_counts, counters(index));
    end loop;
    wait;  -- program finished, wait forever

  end process interpreter;

end architecture instrumented;
```

If an existing physical file in the host computer's file system is opened in append mode, successive data elements are added to the end of the file using the **write** operation. If there is no host file of the given name in the host file system, opening the file object in append mode creates a new file, so that data elements are written from the beginning. Append mode is used for a file that accumulates log information or simulation results over a number of simulation runs. Each run adds its data to the end of the previously accumulated data in the file.

EXAMPLE 16.4   *Measuring cache performance*

When we are designing a cache memory to attach to a CPU, we need to measure how different cache organizations affect the miss rate, since this influences the average access time seen by the CPU. We measure the miss rate by monitoring the traffic on the buses between the CPU and cache and between the cache and main memory. At the end of a simulation run, the process monitoring the buses appends a record to a data file, storing the parameter values that determine the cache organization and the measured miss rate and average access time. An outline of the process is:

```
cache_monitor : process is

  type measurement_record is
    record
      cache_size, block_size, associativity : positive;
      benchmark_name : string(1 to 10);
      miss_rate : real;
      ave_access_time : delay_length;
    end record;
  type measurement_file is file of measurement_record;
  file measurements : measurement_file
    open append_mode is "cache-measurements";
  . . .

begin
  . . .
  loop
    . . .
    exit when halt;
    . . .
  end loop;

  write ( measurements,
          measurement_record'(
            -- write values of generics for this run
            cache_size, block_size, associativity, benchmark_name,
            -- calculate performance metrics
            miss_rate => real(miss_count) / real(total_accesses),
            ave_access_time => total_delay / total_accesses ) );
  wait;

end process cache_monitor;
```

The process declares a record type that represents the information to be recorded for the simulation run and opens a file of records of this type in append mode. At the end of the simulation run, it creates a record value and appends it to the end of the previously existing data in the file. The record includes the values of generic constants that control the cache organization and identify the benchmark program being run, as well as the calculated values for the miss rate and average access time.

Most file system implementations buffer data written to files in order to improve performance. Data is typically stored in a file in relatively large blocks. Writing a partial block is often more expensive than writing a whole block, since the block must be read from storage, part of it updated, and then the modified block written back to storage. Moreover, when network-based file systems are used, partial-block writes involve more network traffic, further reducing performance. Most file systems thus accumulate complete blocks of data in memory and defer writing to the file storage. While this improves performance, it can cause synchronization problems when the data is read by some other application running concurrently with the VHDL simulation. In particular, if the data consists of textual messages and the "other application" is a human reader, we may be unable to monitor simulation progress properly.

VHDL provide a file **flush** operation to help us avoid these problems. For each file type declared, the operation is implicitly declared as

```
procedure flush ( file f : file_type );
```

The effect of the operation is to request that the host system complete all deferred write operations, writing the buffered data to the file system store. The flush should be completed before any subsequent read operations can proceed. It is important to note that the flush operation is a request to synchronize writes and reads, not a guarantee. A simple computer with local file system storage might be able to satisfy the request reliably. However, simulators in complex design environments may use distributed network-based file systems and interact with many other applications. In such environments, it may not be possible to satisfy all flush requests completely. We must make do with a best effort.

VHDL-87, -93, and -2002

These versions do not provide the **flush** operation.

16.1.4 Files Declared in Subprograms

In all of the previous examples, the file object is declared in an architecture body or a process. In these cases, the file is opened at the start of the simulation and automatically closed again at the end of the simulation. The same applies to files declared in packages. We can also declare files within subprograms, but the behavior in these cases is slightly different. The file is opened when the subprogram is called and is automatically closed again when the subprogram returns. Hence the file object, and its association with a physical file in the host file system, is purely local to the subprogram activation. So, for example, if we declare a file in a subprogram:

```
procedure write_to_file is
  file data_file : data_file_type open write_mode is "datafile";
begin
```

```
  ...
end procedure write_to_file;
```

each time we call the procedure a new physical file is created, replacing the old one.

EXAMPLE 16.5   *Initializing a constant value from a file*

We can initialize the value of a constant array by calling a function that reads element values from a file. Suppose the array is of the following type, containing integer elements:

```
type integer_vector is array (integer range <>) of integer;
```

The function declaration is:

```
impure function read_array ( file_name : string;
                                array_length : natural )
                             return integer_vector is
  type integer_file is file of integer;
  file data_file : integer_file open read_mode is file_name;
  variable result : integer_vector(1 to array_length)
            := (others => 0);
  variable index : integer := 1;
begin
  while not endfile(data_file) and index <= array_length loop
    read(data_file, result(index));
    index := index + 1;
  end loop;
  return result;
end function read_array;
```

The first parameter is the name of the file from which to read data elements, and the second parameter is the size of the array that the function should return. The function creates a file object representing a file of integer values and uses the file name parameter to open the file. It then reads values from the file into an array until it reaches the end of the file or the end of the array. It returns the array as the function result. When the function returns, the file is automatically closed. We can use this function in a constant declaration as follows:

```
constant coeffs : integer_vector := read_array("coeff-data", 16);
```

The length of the constant is determined by the result of the function.

One important point to note about files is that we should be careful not to associate more than one VHDL file object with a single physical file in the host file system. While the language does not expressly prohibit multiple associations, it does not specify what happens when we do several reads or writes to the same physical file through different

VHDL file objects. Hence the results may be unpredictable and may vary from one host to another.

This restriction may seem fairly trivial, but we may violate it inadvertently. For example, we might declare a file object in an architecture body for some entity as follows:

```
file log_info : log_file open write_mode is "logfile";
```

If our design uses multiple instances of the entity, we have multiple instances of the file object, each associated with "**logfile**". Possible consequences include interleaving of writes from different instances and loss of data written from all but one instance. The solution to this problem depends on the desired effect. If we intend to merge log data from all instances into one file, we should declare the file in a package. On the other hand, if we intend each instance to have its own log file, we should compute separate file logical name strings for each instance.

16.1.5 Explicit Open and Close Operations

The syntax rule for a file object declaration, shown on page 500, indicates that the file open mode and logical name are optional. If we include either of them, the physical file is automatically opened when the file object is created. If we omit them, the file object is created but remains unassociated with any physical file. An example of a file declaration in this form is

```
file lookup_table_file, result_file : integer_file;
```

If we declare a file object in this way, we explicitly associate it with a physical file and open the file using the **file_open** operation. Given a file type declared as follows:

```
type file_type is file of element_type;
```

file_open is implicitly declared as

```
procedure file_open ( file f : file_type;
                      external_name : in string;
                      open_kind : in file_open_kind := read_mode );
```

The **external_name** and **open_kind** parameters serve exactly the same purpose as the corresponding information in the optional part of a file object declaration. For example, the declaration

```
file lookup_table_file : integer_file
    open read_mode is "lookup-values";
```

is equivalent to

```
file lookup_table_file : integer_file;
...
```

```
file_open ( lookup_table_file,
            external_name => "lookup-values",
            open_kind => read_mode );
```

The advantage of using an explicit **file_open** operation, as opposed to having the file automatically opened when the file object is created, is that we can first perform some other computation to determine how to open it. For example, we might ask the user to type in a file name.

A problem that arises with both of the previously mentioned ways of opening a file is that the operation may fail, causing the whole simulation to come to an abrupt halt. We can make a model more robust by including some error checking, using a second form of the **file_open** operation, implicitly declared as

```
procedure file_open ( status : out file_open_status;
                      file f : file_type;
                      external_name : in string;
                      open_kind : in file_open_kind := read_mode );
```

The extra parameter, **status**, is used to return information about the success or failure of the operation. Its type is predefined in the package **standard** as

```
type file_open_status is (open_ok, status_error,
                          name_error, mode_error);
```

If the file was successfully opened, the value **open_ok** is returned, and we can proceed with **read**, **write** and **endfile** operations, according to the mode. If there was a problem during the **file_open** operation, one of the remaining values is returned. The value **status_error** indicates that the file object had previously been opened and associated with a physical file. (This error is different from the case in which multiple file objects are associated with the same physical file.) The value **name_error** is returned under different circumstances, depending on the mode in which we attempt to open the file. In read mode, it is returned if the named host file does not exist. In write mode, it is returned if a file of the given name cannot be created. In append mode, it is returned if the named file does not exist and a new file of that name cannot be created. Finally, the value **mode_error** is returned from the **file_open** operation if the file exists but cannot be opened in the specified mode. This error may arise if we attempt to write or append to a file marked read-only in the host file system.

Complementing the **file_open** operation, VHDL also provides a **file_close** operation, which can be used to close a file explicitly. The operation disassociates the file object from the physical file. When a file type is declared, a corresponding version of **file_close** is implicitly declared as

```
procedure file_close ( file f : file_type );
```

We can use **file_open** and **file_close** in combination, either to associate a file object with a number of different physical files in succession, or to access a particular physical file multiple times. While applying the **file_close** operation to a file object that is already closed has no effect, it is good style to make sure that **file_open** and **file_close** operations are always paired. We should open a file in the desired mode, perform the reads and writes required, then close the file. This discipline helps ensure that we do not inadvertently write the wrong data to the wrong file. Finally, a **flush** operation is usually not required before we close a file. The act of closing the file usually includes flushing deferred writes.

EXAMPLE 16.6   *A directory file for stimulus files*

Suppose we wish to apply stimulus vectors from a number of different files to a model during a simulation run. We create a directory file containing a list of file names to be used as stimulus files. Our test bench model then reads the stimulus file names from this directory file and opens the stimulus files one-by-one to read the stimulus data. An outline of a process that reads the stimulus files is:

```
stimulus_generator : process is

  type directory_file is file of string;
  file directory : directory_file
      open read_mode is "stimulus-directory";
  variable file_name : string(1 to 50);
  variable file_name_length : natural;
  variable open_status : file_open_status;

  subtype stimulus_vector is std_ulogic_vector(0 to 9);
  type stimulus_file is file of stimulus_vector;
  file stimuli : stimulus_file;
  variable current_stimulus : stimulus_vector;
  ...

begin
  file_loop : while not endfile(directory) loop
    read( directory, file_name, file_name_length );
    if file_name_length > file_name'length then
      report "file name too long: "
              & file_name & "... - file skipped"
        severity warning;
      next file_loop;
    end if;
    file_open ( open_status, stimuli,
                file_name(1 to file_name_length), read_mode );
    if open_status /= open_ok then
      report file_open_status'image(open_status)
              & " while opening file "
              & file_name(1 to file_name_length)
              & " - file skipped"
        severity warning;
      next file_loop;
    end if;
```

```
            stimulus_loop : while not endfile(stimuli) loop
                read(stimuli, current_stimulus);
                ...  -- apply the stimulus
            end loop stimulus_loop;
            file_close(stimuli);
        end loop file_loop;
        wait;
    end process stimulus_generator;
```

The process has a string variable, **file_name**, into which it reads the name of the next stimulus file to be opened. Note the test to see if the actual file name is longer than this variable. This test guards against the open failing through truncation of a file name. The second form of **file_open** is used to open the stimulus file, using the slice of the **file_name** variable containing the name read from the directory. If the open fails, the stimulus file is skipped. Otherwise, the process reads the stimulus vectors from the file, then closes it. When the end of the directory is reached, all stimulus files have been read, so the process suspends.

VHDL-87

The explicit file open and close operations are not provided in VHDL-87, nor is the predefined type **file_open_status**.

16.1.6 File Parameters in Subprograms

We have seen that the file operations described above take a file object as a parameter. In general, we can include a file parameter in any subprogram we write. Files form a fourth class of parameter, along with constants, variables and signals. The syntax for a file parameter in a subprogram specification is as follows:

interface_file_declaration ⇐ **file** identifier ⟨ , ... ⟩ : subtype_indication

The file parameters in the file operations we have seen conform to this syntax rule. The subtype indication must denote a file type. When the subprogram is called, a file object of that type must be supplied as an actual parameter. This object can be a file object declared by the caller, or, if the caller is itself a subprogram, a formal file parameter of the caller. The file object is passed into the subprogram and any of the file operations can be performed on it (depending on the mode in which the file object is opened).

EXAMPLE 16.7   *Reading two-dimensional array data from a file*

Suppose we need to initialize a number of two-dimensional transformation arrays of real numbers using data stored in a file. We cannot directly declare the file of array objects, as VHDL only allows us to store one-dimensional arrays in a file. Instead, we declare the file to be a file of real numbers and use a procedure to read numbers from the file into an array parameter. First, here are the declarations for the arrays and the file:

```
type transform_array is array (1 to 3, 1 to 3) of real;
variable transform1, transform2 : transform_array;

type transform_file is file of real;
file initial_transforms : transform_file
    open read_mode is "transforms.ini";
```

Next, the declaration of the procedure to read values into an array is:

```
procedure read_transform
  ( file f : transform_file;
    variable transform : out transform_array ) is
begin
  for i in transform'range(1) loop
    for j in transform'range(2) loop
      if endfile(f) then
        report "unexpected end of file in read_transform - "
                & "some array elements not read"
          severity error;
        return;
      end if;
      read ( f, transform(i, j) );
    end loop;
  end loop;
end procedure read_transform;
```

The procedure uses the **endfile** operation to test whether there is an element to read. If not, it reports the fact and returns. Otherwise, it proceeds to use the **read** operation to fetch the next element of the array.

We call this procedure to read values into the two array variables as follows:

```
read_transform ( initial_transforms, transform1 );
read_transform ( initial_transforms, transform2 );
```

The file object **initial_transforms** remains opened between the two calls, so the second call reads values from the file beyond those read by the first call.

VHDL-87

In VHDL-87, files are of the variable class of objects. Hence file parameters in subprograms are specified as variable-class parameters. For example, the procedure **read_transform** in Example 16.7 can be written in VHDL-87 as

```
procedure read_transform
              ( variable f : in transform_file;
                variable transform : out transform_array ) is ...
```

A subprogram that reads a file parameter should declare the parameter to be of mode **in**. A subprogram that writes a file parameter should declare the parameter to be of mode **out**.

16.1.7 Portability of Files

We finish this section on VHDL's file facilities with a few comments about the way in which file data is stored. It is important to note that files of the types we have described store the data in some binary representation. The format is dependent on the host computer system and on the simulator being used. This fact raises the issue of portability of files between different systems. All we can expect is that a file of a given type written by one model can be read as a file of the same type in a different model, provided it is run on the same host computer using the same VHDL simulator. There is no guarantee that it can be read on a different host computer, even using the same simulator retargeted for that host, nor that it can be read on any host using a different simulator.

While this might seem to limit the use of files for storing data, in reality it does not present much of an obstacle. If we do need to transfer files between systems, we can use text files as the interchange medium. As we see in the next section, VHDL provides an extensive set of facilities for dealing with the textual representation of data. Furthermore, tools for transferring text files between different computer systems are commonplace. The other potential problem arises if we wish to use non-VHDL software tools to process files written by VHDL models. For example, we may wish to write a program in some conventional programming language to perform data analysis on a data file produced by an instrumented VHDL model. Again, we can use text files to write data in a form readable by other tools. Alternatively, we can consult the VHDL tool vendor's documentation to learn the details of the binary data representation in a file and write a program to read data in that format.

16.2 The Package Textio

The predefined package **textio** in the library **std** provides a number of useful types and operations for reading and writing text files, that is, files of character strings. In particular, it provides procedures for reading and writing textual representations of the various predefined data types provided in VHDL. These operations make it possible to write files that can be read by other software tools and transferred to other host computer systems. The package specification is:[1]

```
package textio is

  type line is access string;
```

---

1. Derived from IEEE Draft Std 1076-2008/D4.1, Draft Standard VHDL Language Reference Manual.

```
type text is file of string;

type side is (right, left);

subtype width is natural;

file input : text open read_mode is "STD_INPUT";
file output : text open write_mode is "STD_OUTPUT";

procedure readline(file F: text; L: inout line);

procedure read ( L : inout line;   value: out bit;
                                   good : out boolean );

procedure read ( L : inout line;   value: out bit );

procedure read ( L : inout line;   value: out bit_vector;
                                   good : out boolean );

procedure read ( L : inout line;   value: out bit_vector );

procedure read ( L : inout line;   value: out boolean;
                                   good : out boolean );

procedure read ( L : inout line;   value: out boolean );

procedure read ( L : inout line;   value: out character;
                                   good : out boolean );

procedure read ( L : inout line;   value: out character );

procedure read ( L : inout line;   value: out integer;
                                   good : out boolean );

procedure read ( L : inout line;   value: out integer );

procedure read ( L : inout line;   value: out real;
                                   good : out boolean );

procedure read ( L : inout line;   value: out real );

procedure read ( L : inout line;   value: out string;
                                   good : out boolean );

procedure read ( L : inout line;   value: out string );

procedure read ( L : inout line;   value: out time;
                                   good : out boolean );

procedure read ( L : inout line;   value: out time );

procedure sread ( L: inout line;   value : out string;
                                   strlen: out natural);

alias string_read is sread [line, string, natural];

alias bread is read [line, bit_vector, boolean];
alias bread is read [line, bit_vector];
alias binary_read is read [line, bit_vector, boolean];
alias binary_read is read [line, bit_vector];
```

```vhdl
procedure oread ( L: inout line;  value: out bit_vector;
                                  good : out boolean );
procedure oread ( L: inout line;  value: out bit_vector );

alias octal_read is oread [line, bit_vector, boolean];
alias octal_read is oread [line, bit_vector];

procedure hread ( L: inout line;  value: out bit_vector;
                                  good : out boolean );
procedure hread ( L: inout line;  value: out bit_vector );

alias hex_read is hread [line, bit_vector, boolean];
alias hex_read is hread [line, bit_vector];

procedure writeline ( file F : text;  L : inout line );

procedure tee ( file F: text;  L: inout line );

function justify ( value: string;
                   justified: side := right;
                   field: width := 0 ) return string;

procedure write ( L : inout line;  value : in bit;
                  justified: in side := right;
                  field: in width := 0 );

procedure write ( L : inout line;  value : in bit_vector;
                  justified: in side := right;
                  field: in width := 0 );

procedure write ( L : inout line;  value : in boolean;
                  justified: in side := right;
                  field: in width := 0 );

procedure write ( L : inout line;  value : in character;
                  justified: in side := right;
                  field: in width := 0 );

procedure write ( L : inout line;  value : in integer;
                  justified: in side := right;
                  field: in width := 0 );

procedure write ( L : inout line;  value : in real;
                  justified: in side := right;
                  field: in width := 0;
                  digits: in natural := 0 );

procedure write ( L: inout line;  value: in real;
                  format: in string);

procedure write ( L : inout line;  value : in string;
                  justified: in side := right;
                  field: in width := 0 );
```

```
    procedure write ( L : inout line;  value : in time;
                       justified: in side := right;
                       field: in width := 0;
                       unit: in time := ns );

    alias swrite is write [line, string, side, width];
    alias string_write is write [line, string, side, width];

    alias bwrite is write [line, bit_vector, side, width];
    alias binary_write is write [line, bit_vector, side, width];

    procedure owrite ( L: inout line;  value: in bit_vector;
                       justified: in side := right;
                       field: in width := 0 );

    alias octal_write is owrite [line, bit_vector, side, width];

    procedure hwrite ( L: inout line;  value: in bit_vector;
                       justified: in side := right;
                       field: in width := 0 );

    alias hex_write is hwrite [line, bit_vector, side, width];

end package textio;
```

Input and output operations using **textio** are based on dynamic strings, accessed using pointers of the type **line**, declared in the package. We use the **readline** operation to read a complete line of text from an input file. It creates a string object in the host computer's memory and returns a pointer to the string. We then use various versions of the **read** operation to extract values of different types from the string. When we need to write text, we first use various versions of the **write** operation to form a string object in memory, then pass the string to the **writeline** operation via its pointer. The operation writes the complete line of text to the output file and resets the pointer to point to an empty string. If the pointer passed to **writeline** is **null**, the operation writes a blank line to the output file.

The reason that VHDL takes this approach to input and output is to allow multiple processes to read or write to a single file without interfering with each other. Recall that multiple processes that are resumed in the same simulation cycle execute concurrently. If processes were to write directly to the file, partial lines from different processes might be intermixed, making the output unintelligible. By having each process form a line locally, we can write each line as one atomic action. The result is an output file consisting of interleaved lines from the different processes. A similar argument applies to input. If read operations were to read directly from the file, no process would be able to read an entire line without the possibility of interference from another process also reading input. The solution is for a process to read an entire line as one atomic action and then to extract the data from the line locally.

One point to note about the read and write operations provided by **textio** is that they may deallocate storage used by lines of text passed to them as arguments. For example, when a read operation extracts characters from the beginning of a line, the storage for the extracted characters may be deallocated. Alternatively, the whole line may be deallocated and a new line formed from the remaining characters. The trap to be aware of is that if we copy the pointer to a line, using assignment of one value of type **line** to another, we may end up with dangling pointers after doing read or write operations. The best way to avoid problems is to avoid modifying variables of type **line** other than with read and write operations.

The package **textio** declares the file type **text**, representing files of strings. The operations provided by the package act on files of this type. The package also declares the file objects **input** and **output**, respectively associated with physical files using the logical names STD_INPUT and STD_OUTPUT. The intention is that the host simulator associate these file objects with the standard devices used for input and output. For example, the file **input** might be associated with the workstation keyboard and the file **output** with the workstation display. A model then uses the files to interact with the user. Prompts and informational messages are displayed by writing them to **output**, and commands and data typed by the user are read from **input**.

16.2.1 Textio Read Operations

Let us now look at the **read** operations in detail. Each version of **read** has at least two parameters: a pointer to the line of text from which to read and a variable in which to store the value. The operations extract characters from the beginning of the line, looking for characters that form a textual representation of a value of the expected type. The line is modified to contain only the remaining characters, and the value represented by the extracted characters is returned.

The character version of **read** simply extracts the first character in the line and returns it. It does not look for quotation marks around the character. For example, if the line pointed to by L contains

```
a'bcd
```

two successive character **read** operations would return the characters 'a' and '''.

The string version extracts enough characters to fill the actual string argument. This version of **read** does not look for double quotation marks around the string. For example, if **s** is a string variable of length five, and L points to the line

```
fred "cat"
```

a read into **s** returns the string "fred ". A second read into **s** returns the string ""cat"". If the line does not contain enough characters to fill the string variable, the read operation fails. If this possibility could cause problems, we can use the **sread** procedure (or the **string_read** alias) instead. This procedure also heps us divide a line of input into separate tokens, separated by spaced. The procedure starts by skipping over any whitespace characters in the line, and then reads as many non-whitespace characters as are available, up to the length of the **value** parameter. It returns the number of characters actually read in the **strlen** parameter. A whitespace character is a space, a non-breaking space or a horizontal tab character. We might use this procedure as follows:

```
sread(L, s, s_len);
for i in 1 to s_len loop
```

```
    ... s(i) ...
end loop;
```

The versions of **read** for all other types of data skip over any whitespace characters in the line before the textual representation of the data. They then extract as many characters from the line as can be used to form a valid literal of the expected type. Characters are extracted up to the first character that is not valid for a literal of that type or to the end of the line. For example, if L points to the line

        12   -4.27!

an integer read extracts the first two characters and returns the value 12. A subsequent read into a real variable skips the spaces, then extracts the characters up to but not including the '!' and returns the value –4.27.

For time values, the literal should be a number followed by a time unit, with at least one whitespace character between them. For bit-vector values, the literal in the line should be a binary string without quotation marks or a base specifier (that is, just a string of '0' or '1' characters). It may include underline characters ('_'), provided they conform to the rules for bit-string literals. That means they can only occur within the bit string, not at either end, and should only occur singly.

The **textio** package includes **bread** and **binary_read** aliases for the bit-vector **read** operations. These parallel the **oread** and **hread** procedures (and **octal_read** and **hex_read** aliases), which read octal and hexadecimal string representations of bit vectors, respectively. The procedures read octal (or hexadecimal) digits, skipping over properly embedded underline characters, until sufficient digits have been read to fill the **value** vector. Each digit corresponds to a group of three (or four) bits. If the value vector is not a multiple of three (or four) in length, bits from the leftmost group of three (or four) are discarded, provided they are '0'. If any are not '0', the **read** operation fails. This ensures that significant bits are not lost. For example, an **hread** of "3F" into a 6-bit vector succeeds, since the leftmost group is "0011", and the two discarded bits are both '0'. On the other hand, an **hread** of "7F" into the 6-bit vector would fail, since the leftmost group is "0111", and one of the two discarded bits is '1'.

The versions of the **read** operations in **textio** that include the third parameter, **good**, allow for graceful recovery if the next value on the input line is not a valid textual representation of a value of the expected type. In that case, they return with **good** set to false, the line unmodified and the **value** parameter undefined. For example, an integer read from a line containing

        $%@!!&

fails in this way. On the other hand, if the line does contain valid text, **good** is set to true, and the value is extracted as described above. The versions of **read** without the **good** parameter cause an error if the line contains invalid text.

EXAMPLE 16.8   *Reading textual stimulus values*

Suppose we have designed a model for a thermostat system and need to test it. The thermostat has inputs connected to signals **temperature** and setting of type **integer** and **enable** and **heater_fail** of type **bit**. We can use a text editor to write a file that specifies input stimuli to test the thermostat. Each line of the file is formatted as follows

```
time     string     value
```

where *time* is the simulation time at which the stimulus is applied, *string* is a four-character string identifying one of the inputs and *value* is the value to be applied to the input. The allowed *string* values are "temp", "set ", "on " and "fail". We assume that the stimuli are sorted in increasing order of application time. A sample file in this format is

```
0 ms      on      0
2 ms      fail    0
15 ms     temp    56
100 ms    set     70
1.5 sec   on      1
```

We write a process to interpret such a stimulus control file as follows:

```
stimulus_interpreter : process is

  use std.textio.all;

  file control : text open read_mode is "control";

  variable command : line;
  variable read_ok : boolean;
  variable next_time : time;
  variable whitespace : character;
  variable signal_id : string(1 to 4);
  variable temp_value, set_value : integer;
  variable on_value, fail_value : bit;

begin
  command_loop : while not endfile(control) loop

    readline ( control, command );

    -- read next stimulus time, and suspend until then
    read ( command, next_time, read_ok );
    if not read_ok then
      report "error reading time from line: " & command.all
        severity warning;
      next command_loop;
```

```vhdl
  end if;
  wait for next_time - now;

  -- skip whitespace
  while command'length > 0
    and ( command(command'left) = ' '     -- ordinary space
         or command(command'left) = ' '  -- non-breaking space
         or command(command'left) = HT ) loop
    read ( command, whitespace );
  end loop;

  -- read signal identifier string
  read ( command, signal_id, read_ok );
  if not read_ok then
    report "error reading signal id from line: " & command.all
      severity warning;
    next command_loop;
  end if;
  -- dispatch based on signal id
  case signal_id is

    when "temp" =>
      read ( command, temp_value, read_ok );
      if not read_ok then
        report "error reading temperature value from line: "
               & command.all
          severity warning;
        next command_loop;
      end if;
      temperature <= temp_value;

    when "set " =>
      ...  -- similar to "temp"

    when "on  " =>
      read ( command, on_value, read_ok );
      if not read_ok then
        report "error reading on value from line: "
               & command.all
          severity warning;
        next command_loop;
      end if;
      enable <= on_value;

    when "fail" =>
      ...  -- similar to "on  "

    when others =>
      report "invalid signal id in line: " & signal_id
```

```
                    severity warning;
                next command_loop;

            end case;

        end loop command_loop;

        wait;

    end process stimulus_interpreter;
```

The process declares a file object, **control**, associated with the stimulus control file, and an access variable, **command**, to point to a command line read from the file. It also declares a number of variables to store values read from a line. The process body repeatedly reads lines from the file and extracts the fields from it using read operations. We use the forms of **read** with **good** parameters to do error checking. In this way, we make the model less sensitive to formatting errors in the control file and report useful error information when an error is detected. When the end of the command file is reached, the process suspends for the rest of the simulation.

For each command line, the process first extracts the time value and suspends until simulation advances to that time. It then skips over whitespace characters in the line up to the first non-whitespace character, which should represent the signal identifier string. The process skips whitespace characters by repeatedly inspecting the first character in what remains of the command line and, if it is a whitespace character, removing it with a character read operation. Skipping whitespace in this way allows the user some flexibility in formatting the command file. The process next dispatches to different branches of a case statement depending on the signal identifier string. For each string value, the process reads a stimulus value of the appropriate type from the command line and applies it to the corresponding signal.

VHDL-87, -93, and -2002

The **textio** package for these versions does not include the **sread** procedure or **string_read** alias. If a **read** into a string variable could fail for lack of sufficient characters, we can directly access the text line. So, for example, we can test the length of the line and extract fewer characters than the length of the string variable as follows:

```
if L'length < s'length then
    read(L, s(1 to L'length));
else
    read(L, s);
end if;
```

Since L is an access variable to a string, the **'length** attribute applied to L returns the length of the string pointed to by L, provided that L is not **null**.

These versions of VHDL do not provide the **oread** and **hread** procedures and their aliases, nor the **bread** and **binary_read** aliases. Moreover, the **read** procedure for bit vectors does not skip underline characters.

VHDL-87

The VHDL-87 version of the **textio** package declares an additional function:

function endline (L : **in** line) **return** boolean;

This function returns the value **true** if the string pointed to by L is empty and **false** otherwise. The same condition can be tested in VHDL-93 and VHDL-2002 by evaluating the expression L'length = 0.

16.2.2 Textio Write Operations

We now turn to write operations, which form a line of text ready for output. Each version of **write** has two parameters, specifying the pointer to the line being formed and the value whose textual representation is to be added to the line. Subsequent parameters beyond these two are used to control the formatting of the textual representation. The **field** parameter specifies how many characters are used to represent the value. If the field is wider than necessary, space characters are used as padding. The characters representing the value are either left-justified or right-justified within the field, depending on the **justified** parameter. For example, if we write the integer 42 left-justified in a field of five characters, the string "42   " is added to the line. If we write the same value right-justified in a field of five characters, the string "   42" is added. If we specify a field width that is smaller than the minimum required to represent the value, that minimal representation is used with no space padding. Thus, writing the integer 123 with a specified field width of two characters or less results in the three-character string "123" being added to the line. Note that the default values for **justified** and **field** conveniently result in the minimal representation being used.

The write operations for character, string and bit-vector values write representations that do not include quotation marks or a base specifier. Bit-vector values are written in binary. For example, if we perform the following write operations to a line, L, that is initially empty:

```
write ( L, string'( "fred" ) );
write ( L, ' ' );
write ( L, bit_vector'( X"3A" ) );
```

the resulting line is

```
fred 00111010
```

There are two versions of the write operation for real values, corresponding to the two overloaded versions of **to_string** that we described in Section 2.5. One version has an additional parameter, **digits**, that specifies how many digits to the right of the decimal point are to be included in the textual representation of the value. For example, writing the value 3.14159 with **digits** set to 2 results in the string "3.14" being added to the line (without the quotation marks). If **digits** is set to 0 (the default value), the value is represented in exponential notation. For example, writing 123.4567 in this way results in the string "1.234567e+02" (or something similar) being added to the line. The other version of write has a parameter, **format**, for a format specification string of the same form as that used in the C **printf** function. For example, writing 123.4567 with a format string of "%10.3f" results in the string "   123.457" being added to the line.

The write operation for time values has a parameter, **unit**, that specifies the time unit to use to express the value. The output is expressed as a multiple of this unit. For example, writing the value 40 ns with **unit** set to **ps** results in the string "40000 ps" being added to the line. If the value to be written is not an integral multiple of the specified unit, a real literal is used in the textual representation. For example, writing the value 23 μs with **unit** set to **ms** results in the string "0.023 ms" being added to the line.

The package provides **owrite** and **hwrite** operations and the aliases **bwrite**, **binary_write**, **octal_write** and **hex_write**, mirroring the binary, octal and hexadecimal read operations for bit vectors. The values written by the **owrite** and **hwrite** operations are the same as those produced by the **to_ostring** and **to_hstring** operations, respectively.

There are also aliases **swrite** and **string_write** for the string version of the write procedure, mirroring the **sread** operation and **string_read** alias. One of the benefits of using the **swrite** alias is that there are no other overloaded versions of that name. Compare this with **write**, for which there are overloaded version for **string** and **bit_vector**, among others. This means that we need to explicitly qualify a string literal when applying the **write** operation, as we did in the example above. If we write:

```
write ( L, "Trace message " );
write ( L, "0000" );
```

the type rules of the language are not sufficient to distinguish between the **string** and **bit_vector** versions. The rules do not include looking at the characters within a string literal to determine the literal's type. Since writing a literal string value is very common, we can use the **swrite** procedure to avoid type qualification each time:

```
swrite ( L, "Trace message " );
swrite ( L, "0000" );
```

EXAMPLE 16.9   *Writing a textual log file for bus activity*

We can write a bus monitor process for a computer system model that creates a log file of bus activity, similar to that displayed by a bus-state analyzer monitoring real hardware. Suppose the model includes the following signals connecting the CPU with memory and I/O controllers:

```
signal address : bit_vector(15 downto 0);
signal data : resolve_bytes byte;
signal rd, wr, io : bit;      -- read, write, io/mem select
signal ready : resolve_bits bit;
```

Our monitor process is written as follows:

```
bus_monitor : process is
```

```
        constant header : string(1 to 44)
          := FF & "      Time   R/W I/M  Address             Data";

        use std.textio.all;

        file log : text open write_mode is "buslog";
        variable trace_line : line;
        variable line_count : natural := 0;
    begin
        if line_count mod 60 = 0 then
          write ( trace_line, header );
          writeline ( log, trace_line );
          writeline ( log, trace_line );     -- empty line
        end if;
        wait until (rd or wr) and ready;
        write ( trace_line, now,
                justified => right, field => 10, unit => us );
        write ( trace_line, string'("    ") );
        if rd then
          write ( trace_line, 'R' );
        else
          write ( trace_line, 'W' );
        end if;
        write ( trace_line, string'("   ") );
        if io then
          write ( trace_line, 'I' );
        else
          write ( trace_line, 'M' );
        end if;
        write ( trace_line, string'("   ") );
        write ( trace_line, address );
        write ( trace_line, ' ' );
        write ( trace_line, data );
        writeline ( log, trace_line );
        line_count := line_count + 1;

    end process bus_monitor;
```

The process declares an output file **log**, of type **text**, and an access variable
trace_line, of type **line**, for accumulating each line of output. The process is resumed
when the memory or I/O controller responds to a bus read or write request. It generates a formatted line using write operations. It also keeps count of how many lines
are written to the log file and includes a header line after every 60 lines of trace data.
A sample log file showing how the data is formatted is

```
    Time   R/W I/M  Address          Data

    0.4 us    R   M    0000000000000000 10011110
```

```
   0.9 us    R   M    0000000000000001 00010010
     2 us    R   M    0000000000010100 11100111
   2.7 us    W   I    0000000000000111 00000000
```

The **textio** package also declares the **tee** operation, which, like **writeline**, writes a complete line of text to a specified file. It also writes a copy of the line to the **output** file. Thus, using this operation, we can write information to a display device and also record it in a file for subsequent processing or review. For example, we could use the tee procedure in place of writeline in the preceding example as follows:

```
tee ( log, trace_line );
```

This would allow us to see each line of trace data on the display as well as recording it in the log file.

The **write** operations that we have described so far all include the parameters **justified** and **field** to control how the string representation is laid out. Since that form of control is very useful for aligning output, the **textio** package provides a function, **justify**, for controlling justification as a stand-alone operation. We can use it in combination with the predefined and overloaded **to_string** operations on various data types to align text other than for text-file output. For example, if we repeatedly execute a report statement to trace the values of objects during a simulation, we can use the **justify** function to align the content of the report messages:

```
report "%%%TRACE:" &
       justify(to_string(now, ns),  width => 10) &
       justify(to_hstring(out_vec), width =>  6) &
       justify(to_string(count),    width => 10);
```

Successive executions might yield the following messages:

```
%%%TRACE:     20 ns   XXXX          0
%%%TRACE:    120 ns   ZZ00          1
%%%TRACE:    220 ns   FFC0         10
%%%TRACE:    320 ns   0000         31
```

A final aspect of textual output using **textio** operations is that the host system interprets line-feed (**LF**) characters as line breaks in the output, using the appropriate convention for the host operating system. This applies to all writes to files of type **text** (including **output**), whether the writes are done using the **writeline** or **tee** procedures or using the predefined file **write** operation for the **text** file type. For example, we could write a group of output lines using the predefined write operation as follows:

```
write(output, "%%%ERROR data value miscompare." & LF &
              " Actual value = " & to_hstring(data) & LF &
              " Expected value = " & to_hstring(expdata) & LF &
              " at time: " & to_string(now) );
```

One reason for writing lines this way is to ensure that the whole group is written atomically without interleaving output from other processes.

VHDL-87, -93, and -2002

There are a number of features for textual output added to **textio** in VHDL-2008 and not provided in earlier versions. These include: the **swrite** and **string_write** aliases; the **owrite** and **hwrite** procedures, together with the **bwrite**, **binary_write**, **octal_write**, and **hex_write** aliases; the **write** operation for **real** with the C-style **format** parameter; the **tee** procedure; and the **justify** function. Also, line-feed characters written to files of type **text** are not interpreted as line breaks in earlier versions.

16.2.3 Reading and Writing Other Types

We have seen that **textio** provides read and write operations for the predefined types. If we need to read or write values of types we declare, such as new enumeration or physical types, we use the 'image and 'value attributes to convert between the values and their textual representations. For example, if we declare an enumeration type and variable as

```
type speed_category is (stopped, slow, fast, maniacal);
variable speed : speed_category;
```

we can write a value of the type using the 'image attribute to create a string to supply to the string version of **write**:

```
write ( L, speed_category'image(speed) );
```

Alternatively, we can use the predefined **to_string** operations, for example:

```
write ( L, to_string(speed) );
```

Reading a value of a new type we define presents more problems if we want our model to be robust in the face of invalid input. In this case, we must write VHDL code that analyzes the line of text to ensure that it contains a valid representation of a value of the expected type. If we are not so concerned with robustness, we can simply use the 'value attribute to convert the input line to a value of the expected type. For example, the statements

```
readline( input, L );
sread(L, str, str_len
speed := speed_category'value(str(1 to str_len));
```

read a line of input, extract a sequence of characters delimited by whitespace, and convert the sequence to a value of type **speed_category**.

Standard Package Read and Write Operations

In Chapter 9 we described the standard packages in library **ieee** and mentioned that they provide overloaded read and write operations. Now that we have described file types and operations, and textual input/output in particular, we can complete our summary of the operations provided by the standard packages.

Overloaded versions of the operations are defined for the following types, as well as for the corresponding subtypes with resolved elements and other subtypes:

- **std_ulogic_vector** in **std_logic_1164**

- **unresolved_unsigned** and **unresolved_signed** in **numeric_bit** and **numeric_std**

- **unresolved_ufixed** and **unresolved_sfixed** in **fixed_generic_pkg** and instances of it

- **unresolved_float** in **float_generic_pkg** and instances of it

The set of operations is **read**, **oread**, **hread**, **write**, **owrite**, and **hwrite**, each with similar signature to the operations for bit vectors in **textio**. In addition, the same set of aliases if provided in each package as in **textio**.

The **write** operations each produce the same string representation as the **to_string** operation on the value parameter. The effect of the **write** operation is equivalent to:

```
textio.write (L, to_string(value), justified, field);
```

where the **to_string** operation used is that defined for the type of value in the appropriate page. We described the **to_string** operations in Chapter 9. The **owrite** and **hwrite** operations are similarly defined, being equivalent to

```
textio.write (L, to_ostring(value), justified, field);

textio.write (L, to_hstring(value), justified, field);
```

The behavior of the read procedures is somewhat more complicated, as they are designed to provide flexible input formats. The binary **read** procedures for **unsigned** and **signed** in **numeric_bit** behave in the same way as those for **bit_vector** in **textio**. In the remaining packages, where the vector types are based on **std_ulogic**, each binary **read** procedure starts by skipping whitespace. It then reads **std_ulogic** values until it encounters whitespace or a non-**std_ulogic** value, or until it has read **value'length** characters. Underscore characters ("_") embedded within the value are skipped, though it is an error if two underscores appear consecutively. The procedure must read enough characters to fill all of the elements of the **value** array, so it is an error if a space or an invalid character is encountered before **value'length** characters are read. The **read** procedures for **ufixed** and **sfixed** also accept a radix point (".") in the input, though it is an error if the radix point is not at the appropriate position. Specifically, the characters before the radix point must fill elements of the **value** parameter with non-negative indices, and the characters after the radix point must fill elements with negative indices. An error occurs if the radix point is encountered at a position other than between the characters corresponding to indices 0 and −1. Similarly, the **read** procedures for **float** accept ":" or "." delimiters between the sign, exponent, and fraction parts of the input, though it is an error if they are not at the appropriate positions.

The **oread** and **hread** procedures in **std_logic_1164**, **numeric_bit**, and **numeric_std** all have behavior similar to the procedures for **bit_vector** in **textio**. Each operation must read sufficient characters to fill the **value** argument, or an error occurs. Since **value** need not be a multiple of 3 (for **oread**) or 4 (for **hread**) in length, the length is rounded up to the nearest multiple of 3 or 4 to determine how many characters to read. **Oread** (**hread**) starts by skipping whitespace. It then reads octal (hexadecimal) digits until it encounters whitespace or a non-octal (non-hexadecimal) character other than "_", or until it has read sufficient characters to fill the **value** argument. Underscore characters embedded within the octal (hexadecimal) value are skipped. **Oread** converts each octal digit (0–7) to its 3-bit representation, and **hread** converts each hexadecimal digit (0–9, a–f, or A–F) to its 4-bit representation. For array types based on **std_ulogic**, the characters 'X' and 'Z' are also permitted. For octal, these characters are repeated 3 times in the result; hence, a 'Z' input is expanded to "ZZZ". Similarly, for hexadecimal, these characters are repeated 4 times in the result; hence, a 'Z' input is expanded to "ZZZZ". If conversion of characters to groups of 3 or 4 elements result in more elements than the length of the **value** argument, only the rightmost elements are used. Depending on the values of the discarded elements, an error may occur. If the type of the **value** argument is **bit_vector**, **std_ulogic_vector**, or **unsigned**, an error occurs if any of the discarded elements are '1'. For example, an **hread** that reads the characters "82" ("10000010" in binary) into a 6-bit **unsigned** value produces an error, since the two discarded bits are "10". If the type of the value argument is **signed**, an error occurs if the discarded elements are not all the same as the leftmost element used for the **value** argument. For example, an **hread** that read the characters "7F" into a 6-bit **signed** value produces an error, since the two discarded bits are "01", and the leftmost bit used for **value** is '1'.

The **oread** and **hread** operations are also defined for **ufixed** and **sfixed** in the fixed-point packages. **Oread** and **hread** each reads the value prior to the radix point as described above for **unsigned** or **signed** (depending on whether the **value** parameter is **ufixed** or **sfixed**, respectively). For the characters following the radix point, **oread** and **hread** each reads the value as described above for **std_ulogic_vector**; however, instead of discarding elements on the left, the operations discard elements on the right. An error occurs if an element discarded on the right is a '1'. The radix point may be explicitly included in the input, but an error occurs if it is not at the appropriate position (that is, between the characters corresponding to indices 0 and −1 of the **value** parameter). The radix point may also be omitted, in which case it is assumed at the appropriate position.

Finally, the **oread** and **hread** operations are defined for **float** in the floating-point packages. The behavior of the **oread** and **hread** operations depends on whether ":" or "." delimiters are used in the input to separate the sign, exponent, and fraction parts of a floating-point number. When ":" delimiters are used (with the input formatted as "S:EEEE:FFFFFFFF"), the sign bit, the exponent, and the fraction are each read as separate octal or hexadecimal values using the same rules as described above for **std_ulogic_vector** values. When a '.' delimiter is used (with the input formatted as "SEEEE.FFFFFFFF"), the rules described above for reading **ufixed** values are used. The value read before the radix point forms the part of the result comprising the sign and exponent elements, and the value read after the radix point forms the fraction part of the result. When no delimiters are used in the input, the entire **float** value is read as a single hexadecimal value as described above for **std_ulogic_vector** values.

VHDL-87, -93, and -2002

The **std_logic_1164**, **numeric_bit** and **numeric_std** packages for these versions of VHDL do not include the read and write operations.

Exercises

1. [❶ 16.1] Write declarations to define a file of real values associated with the host file "samples.dat" and opened for reading. Write a statement to read a value from the file into a variable **x**.

2. [❶ 16.1] Write declarations to define a file of bit-vector values associated with the host file "/tmp/trace.tmp" and opened for writing. Write a statement to write the concatenation of the values of two signals, **addr** and **d_bus**, to the file.

3. [❶ 16.1] Write statements that attempt to open a file of integers called "waveform" for reading and report any error that results from the attempt.

4. [❶ 16.2] Suppose the next line in a text file contains the characters

    ```
    123 4.5 6789
    ```

    What is the result returned by the following **read** calls?

    ```
    readline(in_file, L);
    read(L, bit_value);    -- read a value of type bit
    read(L, int_value);    -- read a value of type integer
    read(L, real_value);   -- read a value of type real
    read(L, str_value);    -- read a value of type string(1 to 3)
    ```

5. [❶ 16.2] Write declarations and statements for a process to prompt the user to enter a number and to accept the number from the user.

6. [❶ 16.2] What string is written to the output file by the following statements:

    ```
    write(L, 3.5 us, justified => right, field => 10, unit => ns);
    write(L, ' ');
    write(L, bit_vector'(X"3C"));
    write(L, ' ');
    swrite(L, "ok", justified => left, field => 5);
    writeline(output, L);
    ```

7. [❷ 16.1] Develop a behavioral model of a microcomputer system address decoder with the following entity interface:

```
entity address_decoder is
  generic ( log_file_name : string );
  port ( address : in natural;  enable : in bit;
         ROM_sel, RAM_sel, IO_sel, int_sel : out bit );
end entity address_decoder;
```

When **enable** is '1', the decoder uses the value of address to determine which of the output signals to activate. The address ranges are 0 to 16#7FFF# for ROM, 16#8000# to 16#BFFF# for RAM, 16#C000# to 16#EFFF# for I/O and 16#F000# to 16#FFFF# for interrupts. The decoder should write the address to the named log file each time **enable** is activated. The log file should be a binary file rather than a text file.

8. [❷ 16.1] Write a function that may be called to initialize an array of bit-vector words. The words are of the subtype

    ```
    subtype word is std_ulogic_vector(0 to 15);
    ```

    The function should have two parameters, the first being the name of a file from which the words of data are read, and the second being the size of the array to return. If the file contains fewer words than required, the extra words in the array are initialized with all bits set to 'U'.

9. [❷ 16.1] Develop a procedure that writes the contents of a memory array to a file. The memory array is of the type mem_array, declared as

    ```
    subtype byte is bit_vector(7 downto 0);
    type mem_array is array (natural range <>) of byte;
    ```

    The procedure should have two parameters, one being the array whose value is to be written and the other being a file of **byte** elements into which the data is written. The procedure should assume the file has already been opened for writing.

10. [❷ 16.2] Develop a procedure that has a file name and an integer signal as parameters. The file name refers to a text file that contains a delay value and an integer on each line. The procedure should read successive lines, wait for the time specified by the delay value, then assign the integer value to the signal. When the last line has been processed, the procedure should return. Invoke the procedure using a concurrent procedure call in a test bench.

11. [❷ 16.2] Develop a procedure that logs the history of values on a bit-vector signal to a file. The procedure has two parameters, the name of a text file and a bit-vector signal. The procedure logs the initial value and the new values when events occur on the signal. Each log entry in the file should consist of the simulation time and the signal value at that time.

12. [❷ 16.2] Develop a procedure similar to that described in Exercise 11, but which logs values of a signal of type **motor_control**, declared as

    ```
    type motor_state is (idle, forward, reverse);
    type motor_control is record
       state : motor_state;
    ```

```
        speed : natural;
    end record motor_control;
```

The **motor_control** values should be written in the format of a record aggregate using positional association.

13. [❸ 16.1] Experiment with your simulator to determine the format it uses for binary files. Write a model that creates files of various data types, and use operating system utilities (for example, hexadecimal dump utilities) to see how the data is stored. Try to develop programs in a conventional programming language to write files that can be read by VHDL models run by your simulator.

14. [❸ 16.2] A 16L2 programmable logic device (PLD) is organized as shown in Figure 16.1. A programmable fuse connects each of the 32 column wires with each of the 16 row wires. If all fuses for a row are disconnected, the row wire floats high. A row wire is tied low by leaving all fuses in the row intact. The programming of the fuses may be specified in a fuse-map text file. It contains 16 lines, each with 32 '1' or '0' characters. A '1' corresponds to a disconnected fuse, and a '0' corresponds to an intact fuse.

   Develop a behavioral model of a 16L2 PLD, with input and output ports of type **bit** and a generic constant string to specify the fuse-map file for programming an instance. The model should read the fuse-map file during initialization and use the information to perform the programmed logic function during simulation.

FIGURE 16.1

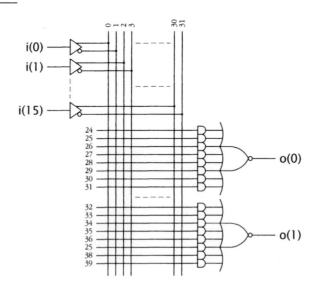

Organization of a 16L2 PLD.

15. [❸ 16.2] Develop a package that provides textual read and write operations for standard-logic scalar and vector values, analogous to the operations provided for types **bit** and **bit_vector** by the **textio** package.

16. [❸ 16.2] Develop a package that provides textual read and write operations for bit-vector values in octal and hexadecimal.

17. [❹ 16.2] Develop a suite of behavioral models of programmable logic devices (PLDs) that read JEDEC format fuse-map files during initialization. Information about the JEDEC format can be found in [5].

18. [❹ 16.2] Develop a behavioral model of a ROM that reads its contents from a file in Intel hex-format.

19. [❹ 16.2] Develop a behavioral model of a ROM that reads its contents from a file in Motorola S-format.

Chapter 17

Case Study:
A Package for Memories

In this case study, we will develop a package of operations on memories, including operations to read and write memories, to implement the behavior of different kinds of memories, to load a simulated memory from a file, and to dump the contents of a simulated memory to a file. Our package is written to be reusable for memories of various sizes and various address and data types. For this case study, we will focus on implementations from which synthesis tools could, in principle, infer block memory resources. Such implementations use array signals for the memory storage. The memory sizes that are fesible range from relatively small (Kbits) to moderately large (100s of Kbits). Test benches, on the other hand, typically use memory models implemented using sparse data structures. Storage for a memory is dynamically allocated as memory locations are accessed. This allows a test bench to deal with much larger memories, which would typically be instantiated as part of a structural model, rather than being inferred by a synthesis tool. Our memories package could be extended to provide such an implementation. However, we omit it here in the interest of brevity.

17.1 The Memories Package

We will make the memory operations package reusable in different designs by specifying the memory width and depth with generic constants, and the types of control, address, and data types with generic types. The package declaration is

```
library ieee;
use ieee.std_logic_1164.std_ulogic_vector;

package memories is
  generic ( width : positive;
            depth : positive;
            type control_type;
            type address_type;
            type data_type;
```

```vhdl
        pure function "??" (c : control_type)
          return boolean is <>;
        function rising_edge(signal c : control_type)
          return boolean is <>;
        pure function to_integer (a : address_type)
          return natural is <>;
        pure function to_address_type (a : natural)
          return address_type is <>;
        pure function to_std_ulogic_vector (d : data_type)
          return std_ulogic_vector is <>;
        pure function to_data_type (d : std_ulogic_vector)
          return data_type is <> );

  type RAM_type is array (0 to 2**depth - 1) of data_type;

  procedure read_RAM (signal   RAM       : in  RAM_type;
                      constant address : in  address_type;
                      signal   data     : out data_type);

  procedure write_RAM (signal   RAM       : out RAM_type;
                       constant address : in  address_type;
                       constant data     : in  data_type);

  procedure asynch_SRAM (signal RAM       : in  RAM_type;
                         signal wr        : in  control_type;
                         signal address   : in  address_type;
                         signal data_in   : out data_type;
                         signal data_out  : out data_type);

  procedure flow_through_SSRAM
            (signal RAM            : in  RAM_type;
             signal clk, en, wr : in  control_type;
             signal address     : in  address_type;
             signal data_in     : out data_type;
             signal data_out    : out data_type);

  procedure pipelined_SSRAM
            (signal RAM            : in  RAM_type;
             signal clk, en, wr : in  control_type;
             signal address     : in  address_type;
             signal data_in     : out data_type;
             signal data_out    : out data_type);

  procedure dump_RAM (signal   RAM            : in  RAM_type;
                      constant file_name      : in  string;
                      constant start_address  : in  address_type
                        := to_address_type(0);
                      constant finish_address : in  address_type
                        := to_address_type(2**depth - 1));
```

```
impure function load_RAM
                      (constant file_name       : in  string;
                       constant start_address   : in  address_type
                          := to_address_type(0);
                       constant finish_address  : in  address_type
                          := to_address_type(2**depth - 1))
                       return RAM_type;

end package memories;
```

The formal generic constants **width** and **depth** specify the bit width of memory data and addresses, respectively. The formal generic types **control_type**, **address_type** and **data_type** are used for control signals, memory addresses and data, respectively. The memory has 2^{depth} locations, indexed from 0 to $2^{\text{depth}-1}$, each storing a **data_type** value. Since we need to test the state of control signals, we need functions to convert a control signal value to **boolean** (the formal generic function "??") and to test for a rising edge (the formal generic function **rising_edge**). Further, since we need to use integer values to index an array storing the memory contents, we need a function to convert an address to an integer; hence, the formal generic function **to_integer**. We also specify formal generic functions for use in the load and dump operations: to convert from an integer to an **address_type** value, to convert from a **std_ulogic_vector** value to a **data_type** value, and to convert from a **data_type** value to a **std_ulogic_vector** value. The reason for the last two is that the load and dump operations will read and write data values using the same formatting as that used for **std_ulogic_vector** values.

The type **RAM_type** provided by the package is an array type used in models for signals representing RAM contents. The procedures **read_RAM** and **write_RAM** each have a signal parameter of this type, as well as parameters for the address and data. A memory model can use these procedures to implement the memory read and write operations, respectively. The package also provides procedures for implementing memories of different kinds: asynchronous SRAM, and flow-through and pipelined SSRAM. Each procedure can be called using a concurrent procedure call, as we will see in Section 17.2.

The **dump_RAM** procedure also has a **RAM_type** signal parameter and writes to a file whose name is specified in the **file_name** parameter. The start and finish addresses are specified as parameters, with default values specified as integers converted to **address_type** values using the formal generic conversion functions.

The **load_RAM** function reads from a file whose name is specified in the **file_name** parameter and returns an array value of type **RAM_type** containing the memory contents. The start and finish addresses for reading are specified as parameters as for the **dump_RAM** procedure. A memory model can call this function to initialize a signal used for memory storage.

The **memories** package body is outlined below. A detailed description of each of the operations follows.

```
package body memories is

   procedure read_RAM (signal   RAM      : in  RAM_type;
                       constant address : in  address_type;
                       signal   data    : out data_type) is
```

```
begin
  assert to_integer(address) <= 2**depth - 1;
  data <= RAM(to_integer(address));
end procedure read_RAM;

procedure write_RAM (signal    RAM       : out RAM_type;
                     constant address : in  address_type;
                     constant data    : in  data_type) is
begin
  assert to_integer(address) <= 2**depth - 1;
  RAM(to_integer(address)) <= data;
end procedure write_RAM;

procedure asynch_SRAM (signal RAM      : in  RAM_type;
                       signal wr       : in  control_type;
                       signal address  : in  address_type;
                       signal data_in  : out data_type;
                       signal data_out : out data_type) is
begin
  loop
    if wr then
      write_RAM(RAM, address, data_in);
      data_out <= data_in;
    else
      read_RAM(RAM, address, data_out);
    end if;
    wait on wr, address, data_in;
  end loop;
end procedure asynch_SRAM;

procedure flow_through_SSRAM
            (signal RAM           : in  RAM_type;
             signal clk, en, wr : in  control_type;
             signal address       : in  address_type;
             signal data_in       : out data_type;
             signal data_out      : out data_type) is
begin
  loop
    if rising_edge(clk) then
      if en then
        if wr then
          write_RAM(RAM, address, data_in);
          data_out <= data_in;
        else
          read_RAM(RAM, address, d_out);
        end if;
      end if;
    end if;
```

```
      wait on clk;
    end loop;
  end procedure flow_through_SSRAM;

procedure pipelined_SSRAM
              (signal RAM           :  in   RAM_type;
               signal clk, en, wr :  in   control_type;
               signal address       :  in   address_type;
               signal data_in       :  out data_type;
               signal data_out      :  out data_type) is
  variable pipelined_en        : control_type;
  variable pipelined_data_out : data_type;
begin
  loop
    if rising_edge(clk) then
      if pipelined_en then
        data_out <= pipelined_data_out;
      end if;
      pipelined_en := en;
      if en then
        if wr then
          write_RAM(RAM, address, data_in);
          pipelined_data_out := data_in;
        else
          assert to_integer(address) <= 2**depth - 1;
          pipelined_data_out := RAM(to_integer(address));
        end if;
      end if;
    end if;
    wait on clk;
  end loop;
end procedure pipelined_SSRAM;

use std.textio.all;
use ieee.numeric_std.all;

procedure dump_RAM (...) is ...

impure function load_RAM (...) return RAM_type is ...

end package body memories;
```

The read_RAM procedure first verifies that the address, converted to integer, is within the valid address range. It then uses the converted address to index the RAM signal parameter and drives the data output signal with the result. The write_RAM procedure is similar. After verifying that the converted address is in range, it uses the converted address as the index into the RAM signal and updates that element with the data input value.

The asynch_SRAM procedure consists of a loop that performs a write or read operation, depending on the state of the wr control signal, and then waits for changes on any of the input signals. When this procedure is called using a concurrent procedure call in an architecture body, the effect is the same as a process containing the procedure's statements. Note that the condition in the if statement involves an implicit conversion of the value of **wr** from **control_type** to **boolean** using the formal generic function "**??**".

The **flow_through_SSRAM** procedure is likewise intended for use in a concurrent procedure call. Its loop tests for a rising edge on the **clk** control signal using the formal generic function **rising_edge**. Then, if the **en** control signal is true, either a write or read operation is performed on the **RAM** signal. The **pipelined_SSRAM** procedure is somewhat more involved, as it must represent the internal pipeline registers of the SSRAM. It does this using the two local variables, **pipelined_en** and **pipelined_data_out**. The loop in the procedure uses the **write_RAM** operation to write data to the **RAM** signal. However, when it reads the memory, it must read into the local variable. Since the **read_RAM** operation has a signal parameter, we cannot use that operation. Instead, the **pipelined_SSRAM** procedure simply performs the same steps as **read_RAM**, but assigns to the local variable using a variable assignment statement.

Before we describe the **dump_RAM** and **load_RAM** procedures, we need to describe the format of data stored in the files. We will use the hexadecimal VMEM format accepted by the **$readmemh** operation in the Verilog hardware description language, since several tools for generating memory contents can write data in that format.

A hexadecimal VMEM format file consists of a sequence of hexadecimal numbers, each representing a location in the memory. The numbers are separate by whitespace characters (space, non-breaking space and tab), form feeds characters, and new lines. Numbers can contain hexadecimal digits, 'Z' and 'X' characters representing groups of 'Z' and 'X' standard-logic values, and '_' characters for readability (as can occur in bit-string literals). In addition, the file may contain addresses, represented by hexadecimal numbers prefixed with the '@' character. Subsequent numbers are located in memory starting from the specified address. For the purpose of this case study, we will assume that the number of digits in a hexadecimal number is exactly that required for an **hread** operation required to read an address or data value. Finally, the file can contain comments: either single-line comments starting with the characters "//", or block comments delimited by "/*" and "*/". Comments can occur anywhere between numbers.

We will describe the **dump_RAM** procedure first, since it is significantly simpler than **load_RAM**. The **dump_RAM** procedure is

```
procedure dump_RAM (signal    RAM                : in  RAM_type;
                    constant file_name       : in  string;
                    constant start_address   : in  address_type
                       := to_address_type(0);
                    constant finish_address  : in  address_type
                       := to_address_type(2**depth - 1)) is

  file dump_file  : text;
  variable status : file_open_status;
  variable L : line;
  constant start_address_int : natural
             := to_integer(start_address);
  constant finish_address_int : natural
```

```
                              := to_integer(finish_address);
              variable address : natural;

          begin
            if start_address_int >= 2**depth - 1 then
              report "dump_RAM: start address "
                      & to_hstring(start_address_int) & " out of range"
                severity error;
              return;
            end if;
            assert finish_address_int <= 2**depth - 1 then
              report "dump_RAM: finish address "
                      & to_hstring(finish_address_int) & " out of range"
                severity error;
              return;
            end if;
            file_open(f => dump_file, external_name => file_name,
                      open_kind => write_mode, status => status);
            if status /= open_ok then
              report "dump_RAM: " & to_string(status)
                      " opening file " & file_name
                severity error;
              return;
            end if;
            -- Write the start address
            write(L, '@');
            hwrite(L, to_unsigned(to_integer(start_address), depth));
            writeline(dump_file, L);
            -- Write the data
            for address in to_integer(start_address)
                          to to_integer(finish_address) loop
              hwrite(L, to_std_ulogic_vector(RAM(address)));
              writeline(dump_file, L);
            end loop;
            file_close(f => dump_file);
          end procedure dump_RAM;
```

The procedure declares two constants, **start_address_int** and **finish_address_int**, that
are converted versions of the two address parameters. The procedure verifies that the converted addresses are within range for the memory size, determined by the **depth** generic
constant. It then attempts to open the named file for writing. If any of these steps fails, the
procedure reports an error message and returns.

Next, **dump_RAM** writes the contents of the file. It writes a line containing the starting
address in hexadecimal, converted first to numeric form using the formal generic conversion function and then to an **unsigned** vector using the **to_unsigned** function from
numeric_std. After writing the address line, the procedure writes successive data lines. It
uses a for loop in which the numeric address iterates from the converted start address to

the converted finish address. For each address, the procedure indexes the RAM array, converts the data to **std_ulogic_vector** form using the formal generic function, and writes the result in hexadecimal form. Once the file has been written, the procedure closes the file.

We now turn to the **load_RAM** function. This function loads data from a VMEM file into successive memory locations starting at the address specified in the **load_RAM** call. When an address is encountered in the file, subsequent numbers are loaded from that address. If an address to be loaded falls outside the range specified in the **load_RAM** call, the operation fails. The numbers in the file are interpreted as representing **std_ulogic_vector** values with length given by the **width** generic constant of the package. The numbers may not contain exactly the right number of digits, so they may need to be resized.

```vhdl
impure function load_RAM
                (constant file_name      : in  string;
                 constant start_address  : in  address_type
                    := to_address_type(0);
                 constant finish_address : in  address_type
                    := to_address_type(2**depth - 1))
                return RAM_type is

  file load_file  : text;
  variable status : file_open_status;
  variable L : line := null;
  variable ok : boolean;
  constant start_address_int : natural
            := to_integer(start_address);
  constant finish_address_int : natural
            := to_integer(finish_address);
  variable next_address : natural := to_integer(start_address);
  variable next_address_unsigned : unsigned;
  variable data_sulv : std_ulogic_vector(0 to width);
  variable RAM : RAM_type;

  procedure skip_whitespace_and_comments is ...

begin
  if start_address_int > 2**depth - 1 then
    report "dump_RAM: start address "
            & to_hstring(start_address_int)
            & " out of range";
    return RAM;
  end if;
  if finish_address_int > 2**depth - 1 then
    report "dump_RAM: finish address "
            & to_hstring(finish_address_int)
            & " out of range";
    return RAM;
  end if;
  file_open(f => load_file, external_name => file_name,
            open_kind => read_mode, status => status);
```

```
        if status /= open_ok then
          report "load_RAM: " & to_string(status)
                 " opening file " & file_name;
          return RAM;
        end if;
        -- code to read and parse memory file contents
        loop
          skip_whitespace_and_comments;
          exit when not ok or endfile(load_file);
          if L(L'left) = '@' then -- address follows
            read(L, at_char);
            hread(L, next_address_unsigned, ok);
            if not ok then
              report "load_RAM: error reading address in file "
                     & file_name;
              exit;
            end if;
            next_address := to_integer(next_address_unsigned);
          else -- data number
            -- read the number
            hread(L, data_sulv, ok);
            if not ok then
              report "load_RAM: error reading data at address "
                     & to_hstring(next_address)
                     & " in file " & file_name;
              exit;
            end if;
            if next_address > finish_address_int
               or next_address < start_address_int then
              report "load_RAM: address " & to_hstring(next_address)
                     & " out of range " & to_hstring(start_address_int)
                     & " to " & to_hstring(finish_address_int) & LF
                     & "  in file " & file_name;
              ok := false;
              exit;
            RAM(next_address) := to_data_type(data_sulv);
            next_address := next_address + 1;
          end if;
        end loop;
        file_close(f => load_file);
        return RAM;
      end procedure load_RAM;
```

The **load_RAM** function declares a number of constants and variables for use in reading the file and loading a memory. Included are **RAM**, into which data is loaded, and **next_address**, for keeping track of the next location to load. **Next_address** is initialized using the **start_address** parameter. **Load_RAM**, like **dump_RAM**, starts by verifying that the address parameters are within range for the memory size. It then attempts to open the named file for reading. If any of these steps fails, the function reports an error message and returns. Otherwise, the function enters a loop to read the file contents. Each iteration of the loop reads either an address or a data value.

The loop starts by calling a local procedure to skip whitespace and comments preceding the address or data value. We will come back to the details of that procedure shortly. If the procedure encounters an error during its operation, it reports the fact and returns with **ok** set to false. In that case, **load_RAM** exits the loop. The procedure might also encounter the end of the file while skipping whitespace and comments, in which case **load_RAM** exits the loop.

If the function does not exit the loop, it is because there is something to read. The function looks ahead at the next character in the string pointed to by **L** to see if there is an address to read, indicated by the character being the '@' prefix. If so, the function reads the prefix character, then attempts to read an **unsigned** address vector using the **hread** procedure from **numeric_std**. If the read fails, the function reports a message and exits the loop. Otherwise, the function converts the **unsigned** address vector using the formal generic conversion function and saves the result as the next memory address to load.

If the look-ahead character is not the '@' prefix, the numer to be read is a data value. The function attempts to read it in **std_ulogic_vector** form. If the read fails, the function reports a message and exits the loop. Otherwise, the function verifies that the next address to be loaded is within the start address to finish address range, and reports a message and exits if not. If these checks all succeed, the function updates the next memory location with the data value converted using the formal generic function and increments the next address variable.

The loop in the **load_RAM** function can be exited for a number of reasons, as we described, either as a result of an error or through the file being completely read. In all cases, the function closes the file and returns the value of the RAM variable.

Part of the operation of the **load_RAM** function involved skipping whitespace characters and comments in the file. The local procedure to do this is

```
procedure skip_whitespace_and_comments is
  variable skipping_block_comment : boolean := false;
  variable whitespace_char, comment_char : character;
begin
  loop
    if L = null or L'length = 0 then
      if endfile(load_file) then
        if skipping_block_comment then
          report "load_RAM: unterminated block comment in file"
            & file_name;
          ok := false;
        else
          ok := true;
        end if;
        return;
      else
        readline(load_file, L);
```

```
      end if;
    elsif L(L'left) = ' ' or L(L'left) = ''
          or L(L'left) = HT or L(L'left) = FF then
      read(L, whitespace_char);
    elsif skipping_block_comment then
      read(L, comment_char);
      if comment_char = '*' and
        L /= null and L'length /= 0 and L(L'left) = '/' then
        read(L, comment_char);
        skipping_block_comment := false;
      end if;
    elsif L(L'left) = '/' then
      read(L, comment_char);
      read(L, comment_char, ok);
      if not ok then
        report "load_RAM: error reading comment in file "
                & file_name;
        return; -- ok is false
      end if;
      if comment_char = '/' then -- single-line comment
        deallocate(L); -- skip rest of comment line
      elsif comment_char = '*' then -- block comment
        skipping_block_comment := true;
      else
        report "load_RAM: malformed comment in file "
                & file_name;
        ok := false;
        return;
      end if;
    else -- not whitespace or comment
      ok := true;
      return;
    end if;
  end loop;
end procedure skip_whitespace_and_comments;
```

The procedure consists of a loop in which it looks ahead at the content of the file and consumes content until it reaches a character that is not whitespace and not within a comment. It leaves that character as the next character to be read from the file. It declares a local state variable, **skipping_block_comment**, to keep track of whether characters being skipped are within a block comment or not. Initially, the variable is false. Recall also that the procedure returns with the variable **ok** set to true or false depending on whether an error is detected.

Each iteration of the loop starts by checking whether the current line of input (if any) has been exhausted. If so, the procedure checks whether the end of the file has been reached. When that condition occurs, the procedure returns. It sets **ok** to false if it was skipping characters of a block comment at the time, since reaching the end of the file would indicate an unterminated comment. Otherwise, it sets **ok** to true, indicating that reading the file is complete. If the end of the file has not been reached, the procedure reads the next line from the file, effectively skipping the whitespace represented by the line end.

The next condition checked in the loop is whether the next character of the line is a whitespace character. If so, the procedure reads the character from the line. If not, the procedure checks whether characters in a block comment are being skipped. If so, the next character is read from the line. If it was a "*" character, it could be the first of the "*/" sequence closing the block comment. The procedure looks ahead to see if the following character is a '/' character, and if is, reads it and clears the state variable to false. If not, block-comment skipping continues.

If the procedure is not skipping block-comment characters, it checks whether the next character in the line is a '/' character, indicating the start of a comment. If it is, it is read from the line and the following character read. If there is no following character (indicated by the read failing, the procedure reports a message and returns with **ok** set to false. If the following character is also '/', the two '/' characters start a single-line comment, so the procedure skips the rest of the line by deallocating the string pointed to by L. On the next iteration of the loop, reading continues with the next line (if any). If the character after '/' is "*", the two characters start a block comment, so the procedure sets the state variable to true. If the character following the first '/' is neither '/' nor "*", the start of the comment is malformed, so the procedure reports a message, sets **ok** to false, and returns.

Finally, if none of the preceding tests lead to characters to be skipped, then the next character in the line is the start of an address or data value. The procedure sets **ok** to true and returns to the **load_RAM** function.

17.2 Using the Memories Package

We can use the **memories** package to model various kinds of memory devices. Suppose we are developing an embedded system that has a 128 K × 16-bit asynchronous ROM for instruction storage and a 32 K × 32-bit flow-through synchronous SRAM for data storage. We will use the package to model these devices. We will load the ROM from a file, and use the **dump_RAM** procedure as part of the test bench for the system.

The entity declaration for the ROM is

```
library ieee;
use ieee.std_logic_1164.all, ieee.numeric_bit.all;

entity instruction_ROM is
  generic ( file_name : string );
  port ( address : in unsigned(16 downto 0);
         d_out : out std_ulogic_vector(15 downto 0) );
end entity instruction_ROM;
```

The entity has a generic constant of type **string** for specifying the file from which to load the ROM content. The address port is a 17-bit **unsigned** vector, and the data output port is a 16-bit **std_ulogic_vector**.

The entity declaration for the RAM is similar:

```
library ieee;
use ieee.std_logic_1164.all, ieee.numeric_bit.all;

entity data_RAM is
  port ( clk, en, wr : in std_ulogic;
         address : in unsigned(14 downto 0);
         d_in : in std_ulogic_vector(31 downto 0);
         d_out : out std_ulogic_vector(31 downto 0) );
end entity data_RAM;
```

For this memory, the address port is an 15-bit **unsigned** vector, and data input and output ports are 32-bit **std_ulogic_vector** values.

For both memories, we need to instantiate the **memories** package to provide operations with the required address and data types. We could include the instantiations within the architecture bodies for the two memories. However, that would hide the data types and operations for the memory signals inside the architecture bodies and make them inaccessible to the test bench. In particular, the test bench would not be able to call the **dump_RAM** operation for the data memory. We can avoid this problem by instantiating the memories package within a separate design-unit package that is accessible both to the memory architectures and to the test bench. The package declaration is

```
library ieee;
use ieee.std_logic_1164.all, ieee.numeric_bit.all;

package embedded_memories is

  constant instruction_width : positive := 16;
  constant instruction_depth : positive := 17;

  constant data_width : positive := 32;
  constant data_depth : positive := 15;

  function to_instruction_address_type ( n : natural )
                                         return unsigned;

  function to_data_address_type ( n : natural )
                                         return unsigned;

  function to_std_ulogic_vector ( d : std_ulogic_vector )
                                         return std_ulogic_vector;

  package instruction_memories is new work.memories
    generic map ( width => instruction_width,
                  depth => instruction_depth,
                  control_type => std_ulogic,
                  address_type => unsigned(19 downto 0),
                  data_type => std_ulogic_vector(15 downto 0),
                  "??" => ieee.std_logic_1164."??",
                  rising_edge => ieee.std_logic_1164.rising_edge,
                  to_integer => ieee.numeric_std.to_integer,
                  to_address_type => to_instruction_address_type,
```

```
                       to_std_ulogic_vector => to_std_ulogic_vector,
                       to_data_type => to_std_ulogic_vector );

      package data_memories is new work.memories
        generic map ( width => data_width, depth => data_depth,
                       control_type => std_ulogic,
                       address_type => unsigned(17 downto 0),
                       data_type => std_ulogic_vector(31 downto 0),
                       "??" => ieee.std_logic_1164."??",
                       rising_edge => ieee.std_logic_1164.rising_edge,
                       to_integer => ieee.numeric_std.to_integer,
                       to_address_type => to_data_address_type,
                       to_std_ulogic_vector => to_std_ulogic_vector,
                       to_data_type => to_std_ulogic_vector );

    end package embedded_memories;
```

The **embedded_memories** package instantiates the **memories** package twice, once each for the instruction ROM and the data RAM. The **control_type** generic is set to **std_ulogic** in both instances, even though the ROM has no control signals, since the package requires a type for this generic. The "**??**" and **rising_edge** operations for **std_ulogic** are provided as actual functions for the corresponding formal generic functions. (Note that we could omit these actual generics, since the package has "**<>**" as the default for "**??**" and **rising_edge**. The actual functions we have specified would be used had we left the formals unassociated.) The **address_type** and **data_type** generics correspond to the types of the address and data ports of the memories. We can use the **to_integer** conversion function from **numeric_std** to convert address values to numeric form. To convert in the other direction, we can use the **to_unsigned** function from **numeric_std**. However, that function has a second parameter to specify the result length. Since we need this to be fixed at 17 for the instruction ROM and 15 for the data RAM, we declare wrapper functions named **to_instruction_address_type** and **to_data_address_type** to convert numbers to **unsigned** values of the required sizes. We use these functions as the actuals for the **to_address_type** generic in the two package instances. Since the data types for the memories are subtypes of **std_ulogic_vector**, we don't actually need to do any conversion. Nonetheless, the **memories** package requires conversion functions between **data_type** and **std_ulogic_vector**. We just declare an identity function, **to_std_ulogic_vector**, and use it as the actual generic for both data-type conversions in each package instance. The package body, showing the implementations of the conversion functions, is

```
    package embedded_memories is

      function to_instruction_address_type ( n : natural )
                                                       return unsigned is
      begin
        return ieee.numeric_std_to_unsigned(n, instruction_depth);
      end function to_instruction_address_type;

      function to_data_address_type ( n : natural )
                                        return unsigned is
```

```
  begin
    return ieee.numeric_std_to_unsigned(n, data_depth);
  end function to_data_address_type;

  function to_std_ulogic_vector ( d : std_ulogic_vector )
                                  return std_ulogic_vector is
  begin
    return d;
  end function to_std_ulogic_vector;

end package body embedded_memories;
```

We now turn to the architecture bodies of the memories. For the ROM, the architecture is

```
architecture file_loaded of instruction_ROM is

  use work.embedded_memories.instruction_memories.all;

  signal ROM : RAM_type := load_RAM(file_name);

begin

  read_RAM(ROM, address, d_out);

end architecture file_loaded;
```

We use declarations from the **instruction_memories** package instance declared in the **embedded_memories** package. We declare a signal, ROM, of the **RAM_type** array type delared by the package instance and initialize the signal by calling the **load_RAM** function. The behavior of the ROM is then implemented by the concurrent procedure call to the **read_RAM** procedure. Whenever the address changes, the procedure is called, reading the addressed element of the ROM and and assigning it to the **d_out** port. Note that the concurrent procedure call is also sensitive to the **ROM** signal, but since it never changes after being initialized, the procedure is only called in response to address changes.

The architecture of the data memory is similar:

```
architecture rtl of data_RAM is

  use work.embedded_memories.data_memories.all;

  signal RAM : RAM_type;

begin

  flow_through_SSRAM(RAM, clk, en, wr, address, d_in, d_out);

end architecture rtl;
```

The concurrent procedure call invokes the **flow_through_SSRAM** procedure from the **data_memories** package instance. Once called, this procedure never returns. The sensivity to clock edges is implemented internally to the procedure.

The instruction ROM and data RAM might be instantiated with the model for the embedded system as follows:

```
entity embedded_system is
  generic ( code_file_name : string );
  port ( ... );
end entity embedded_system;

--------------------------------------------------------------

architecture struct of embedded_system is

  ...

begin

  instr : entity work.instruction_ROM(file_loaded)
    generic map ( file_name => code_file_name )
    port map ( ... );

  data : entity work.data_RAM(rtl)
    port map ( ... );

  ...

end architecture struct;
```

An outline of the test bench architecture for the embedded system is

```
library ieee;
use ieee.std_logic_1164.all, ieee.numeric_bit.all;

architecture dump_data of testbench is

  ...

begin

  dut : entity work.embedded_system(struct)
    generic map ( code_file_name => "testbench.vmem" )
    port map ( ... );

  dumper : process is
    use work.embedded_memories.data_memories.all;
    alias RAM is <<signal dut.data.RAM : RAM_type>>;
  begin
    wait until dump_triggered;
    dump_RAM(RAM, "testbench-dump.vmem");
    wait;
  end process dumper;

end architecture dump_data;
```

The test bench instantiates the embedded system as **dut**. The **dumper** process uses declarations from the **data_memories** package instance in the **embedded_memories** design-unit package. It declares an alias, using an external name, for the **RAM** signal of the data memory. When a trigger condition is true, the process invokes the **dump_RAM** procedure to dump the contents of the data memory signal to the named file.

17.2.1 Common Address and Data Conversions

There are several common cases for address and data types. Specifically, addresses are commonly of type **natural**, **std_ulogic_vector**, or **unsigned**; and data values are commonly of type **std_ulogic_vector**, **unsigned**, or **signed**. As we saw earlier in our example of memories for an embedded system, much of the overhead in using the memories package lies in defining conversion functions for the address and data types. In support of common cases, we can define a package of conversion functions that can be reused, simplifying instantiation of the memories package. Our support package declaration is

```
library ieee; use ieee.all;
use std_logic_1164.std_ulogic_vector;

package memories_support is
  generic ( width : positive;
            depth : positive;
            package fixed_pkg is new fixed_generic_pkg
              generic map ( <> );
            package float_pkg is new float_generic_pkg
              generic map ( <> ) );
  -- Conversions for actual types for address_type:
  -- natural, std_ulogic_vector, unsigned

  pure function to_integer (a: natural) return natural;

  alias to_integer is ieee.numeric_std_unsigned.to_integer;

  alias to_integer is ieee.numeric_std.to_integer
                        [ieee.numeric_std.unsigned return natural];

  pure function to_address_type (a : natural)
                                  return natural;

  pure function to_address_type (a : natural)
                                  return std_ulogic_vector;

  pure function to_address_type (a : natural)
                                  return numeric_std.unsigned;

  -- Conversions for actual types for data_type:
  -- natural, std_ulogic_vector, unsigned, signed,
  -- ufixed, sfixed, float

  pure function to_std_ulogic_vector (d : natural)
                                  return std_ulogic_vector;

  pure function to_std_ulogic_vector (d : std_ulogic_vector)
                                  return std_ulogic_vector;
```

```
    pure function to_std_ulogic_vector (d : numeric_std.unsigned)
                                    return std_ulogic_vector;

    pure function to_std_ulogic_vector (d : numeric_std.signed)
                                    return std_ulogic_vector;

    alias to_std_ulogic_vector is
      fixed_pkg.to_std_ulogic_vector
        [fixed_pkg.ufixed return std_ulogic_vector];

    alias to_std_ulogic_vector is
      fixed_pkg.to_std_ulogic_vector
        [fixed_pkg.sfixed return std_ulogic_vector];

    alias to_std_ulogic_vector is
      float_pkg.to_std_ulogic_vector
        [float_pkg.float return std_ulogic_vector];

    pure function to_data_type (d : std_ulogic_vector)
                               return natural;

    pure function to_data_type (d : std_ulogic_vector)
                               return std_ulogic_vector;

    pure function to_data_type (d : std_ulogic_vector)
                               return numeric_std.unsigned;

    pure function to_data_type (d : std_ulogic_vector)
                               return numeric_std.signed;

    pure function to_data_type_generic_ufixed
      generic ( left_index, right_index : integer )
      parameter (d : std_ulogic_vector) return fixed_pkg.ufixed;

    pure function to_data_type_generic_sfixed
      generic ( left_index, right_index : integer )
      parameter (d : std_ulogic_vector) return fixed_pkg.sfixed;

    pure function to_data_type_generic_float
      generic ( exponent_width : natural
                   := float_pkg.float_exponent_width;
                fraction_width : natural
                   := float_pkg.float_fraction_width)
      parameter (d : std_ulogic_vector) return float_pkg.float;

end package memories_support;
```

The package has **width** and **depth** generic constants, since these are needed to determine the vector length for conversions from integer values to vector values. There are also formal generic packages for fixed-point and floating-point packages, which are needed to define the fixed-point and floating-point data types. The **to_integer** conversions for **std_ulogic_vector** or **unsigned** as address types are just aliases for the functions provided by the **numeric_std_unsigned** and **numeric_std** packages, respectively. Similarly, the to_std_ulogic_vector conversions for **ufixed**, **sfixed**, or **float** as data types are just aliases for the conversion function provided by the packages defining the types. The remaining function bodies are either identities or wrappers around type conversions or conversion functions. The conversion functions from **std_ulogic_vector** to the **ufixed**, **sfixed** and **float** data types are uninstantiated functions, with generic constants for specifying the index range of the result. We declare them in this way to allow particular index ranges to be bound into an instance of each function, giving a conversion function of one parameter, as required by the memories package.

The package body includes the implementations of the conversion functions:

```
package body memories_support is

  pure function to_integer (a: natural) return natural is
  begin
    return a;
  end function to_integer;

  pure function to_address_type (a : natural)
                                     return natural is
  begin
    return a;
  end function to_address_type;

  pure function to_address_type (a : natural)
                                     return std_ulogic_vector is
  begin
    return numeric_std_unsigned.to_std_ulogic_vector(a, depth);
  end function to_address_type;

  pure function to_address_type (a : natural)
                                     return numeric_std.unsigned is
  begin
    return numeric_std.to_unsigned(a, depth);
  end function to_address_type;

  pure function to_std_ulogic_vector (d : natural)
                                     return std_ulogic_vector is
  begin
    return numeric_std_unsigned.to_std_ulogic_vector(d, width);
  end function to_std_ulogic_vector;

  pure function to_std_ulogic_vector (d : std_ulogic_vector)
                                     return std_ulogic_vector is
  begin
    return d;
  end function to_std_ulogic_vector;

  pure function to_std_ulogic_vector (d : numeric_std.unsigned)
                                     return std_ulogic_vector is
  begin
```

```vhdl
    return std_ulogic_vector(d);
  end function to_std_ulogic_vector;

  pure function to_std_ulogic_vector (d : numeric_std.signed)
                                  return std_ulogic_vector is
  begin
    return std_ulogic_vector(d);
  end function to_std_ulogic_vector;

  pure function to_data_type (d : std_ulogic_vector)
                            return natural is
  begin
    return numeric_std_unsigned.to_integer(d);
  end function to_data_type;

  pure function to_data_type (d : std_ulogic_vector)
                            return std_ulogic_vector is
  begin
    return d;
  end function to_data_type;

  pure function to_data_type (d : std_ulogic_vector)
                            return numeric_std.unsigned is
  begin
    return numeric_std.unsigned(d);
  end function to_data_type;

  pure function to_data_type (d : std_ulogic_vector)
                            return numeric_std.signed is
  begin
    return numeric_std.signed(d);
  end function to_data_type;

  pure function to_data_type_generic_ufixed
    generic ( left_index, right_index : integer )
    parameter (d : std_ulogic_vector) return fixed_pkg.sfixed is
  begin
    return fixed_pkg.to_ufixed(d, left_index, right_index);
  end function to_data_type;

  pure function to_data_type_generic_sfixed
    generic ( left_index, right_index : integer )
    parameter (d : std_ulogic_vector) return fixed_pkg.sfixed is
  begin
    return fixed_pkg.to_sfixed(d, left_index, right_index);
  end function to_data_type;

  pure function to_data_type_generic_float
    generic ( exponent_width : natural
                  := float_pkg.float_exponent_width;
              fraction_width : natural
```

```
                            := float_pkg.float_fraction_width)
      parameter (d : std_ulogic_vector) return float_pkg.float is
    begin
      return float_pkg.to_float(d, exponent_width, fraction_width);
    end function to_data_type;

end package body memories_support;
```

With this support package in place, we no longer need to declare our own conversion functions in the **embedded_memories** package of our earlier example. Instead, we can replace them with instantiations of the memories support package. The revised embedded_memories package is

```
library ieee;
use ieee.std_logic_1164.all, ieee.numeric_bit.all;

package embedded_memories is

  constant instruction_width : positive := 16;
  constant instruction_depth : positive := 17;

  constant data_width : positive := 32;
  constant data_depth : positive := 15;

  package instruction_support is new work.memories_support
    generic map ( width => instruction_width,
                  depth => instruction_depth,
                  fixed_pkg => ieee.fixed_pkg,
                  float_pkg => ieee.float_pkg );

  package instruction_memories is new work.memories
    generic map
      ( width => instruction_width, depth => instruction_depth,
        control_type => std_ulogic,
        address_type => unsigned(19 downto 0),
        data_type => std_ulogic_vector(15 downto 0),
        to_integer => instruction_support.to_integer,
        to_address_type => instruction_support.to_address_type,
        to_std_ulogic_vector =>
          instruction_support.to_std_ulogic_vector,
        to_data_type => instruction_support.to_data_type );

  package data_support is new work.memories_support
    generic map ( width => data_width,
                  depth => data_depth,
                  fixed_pkg => ieee.fixed_pkg,
                  float_pkg => ieee.float_pkg );

  package data_memories is new work.memories
    generic map
      ( width => data_width, depth => data_depth,
```

```
                    control_type => std_ulogic,
                    address_type => unsigned(17 downto 0),
                    data_type => std_ulogic_vector(31 downto 0),
                    to_integer => data_support.to_integer,
                    to_address_type => data_support.to_address_type,
                    to_std_ulogic_vector => data_support.to_std_ulogic_vector,
                    to_data_type => data_support.to_data_type );

end package body embedded_memories;
```

In this version, we have also omitted generic associations for the "??" and **rising_edge** operations for **std_ulogic**, relying instead on the generic subprogram default to identify the versions from **std_logic_1164** that are directly visible. Note that we need to supply actual packages for the formal generic packages **fixed_pkg** and **float_pkg**. We just use the standard package instances from the **ieee** library, though they are not used in this example.

As a further illustration, suppose we are designing a digital signal processing application that requires a data memory to store fixed-point data. The memory has $2^{12} = 4$ K locations, each of which stores a 16-bit **sfixed** value indexed from 4 down to –11. Our design instantiates **fixed_generic_pkg** as a design unit for use throughout the design as follows:

```
library ieee;
use ieee.fixed_float_types.all;

package dsp_fixed_pkg is new ieee.fixed_generic_pkg
  generic map ( fixed_round_style => fixed_truncate,
                fixed_overflow_style => fixed_saturate,
                fixed_guard_bits => 2,
                no_warning );
```

We can instantiate the memories support package and memories package as follows:

```
library ieee;
use ieee.std_logic_1164.all, ieee.numeric_bit.all;
use work.dsp_fixed_pkg.all;

package dsp_data_pkg is

  package dsp_data_support is new work.memories_support
    generic map ( width => 16, depth => 12,
                  fixed_pkg => work.dsp_fixed_pkg,
                  float_pkg => ieee.float_pkg );

  use dsp_data_support.all;

  function to_data_type is new to_data_type_generic_sfixed
    generic map ( left_index => 4, right_index => -11 );

  package dsp_data_memories is new work.memories
    generic map
```

```
( width => 16, depth => 12,
  control_type => std_ulogic,
  address_type => unsigned(11 downto 0),
  data_type => sfixed(4 downto -11) );
```

end package dsp_data_pkg;

The instance of the support package has **dsp_fixed_pkg** as the actual fixed-point package for defining the **sfixed** type. The **to_data_type_generic_sfixed** function is instantiated with the index bounds for our application data as the values for the generic constants. This gives a **to_data_type** function that converts a **std_ulogic_vector** value to an **sfixed** value with the required index bounds.

In the instance of the memories package, we provide actual values for the **width** and **depth** and actual types for the control signals, adresses and data. We don't need to specify any of the conversion functions explicitly, as all of the required functions are directly visible and used by default. The "??" and **rising_edge** operations come from **std_logic_1164**. The **to_integer** function on **unsigned** addresses comes from **numeric_std**, and **to_address_type** converting from **natural** to **unsigned** comes from **dsp_data_support**. The **to_std_ulogic_vector** function for the data type **sfixed** comes from **dsp_fixed_pkg**, and **to_data_type** is the instance of **to_data_type_generic_sfixed**.

With this package in place, the data memory design is quite simple:

```
library ieee;
use ieee.std_logic_1164.all, ieee.numeric_bit.all;
use work.dsp_fixed_pkg.all;

entity dsp_data_RAM is
  port ( clk, en, wr : in std_ulogic;
         address : in unsigned(11 downto 0);
         d_in : in sfixed(4 downto -11);
         d_out : out sfixed(4 downto -11) );
end entity dsp_data_RAM;

-------------------------------------------------------------

architecture rtl of dsp_data_RAM is

  use work.dsp_data_pkg.dsp_data_memories.all;

  signal RAM : RAM_type;

begin

  pipelined_SSRAM(RAM, clk, en, wr, address, d_in, d_out);

end architecture rtl;
```

Exercises

1. [● 17.2] Write instantiations of the **memories_support** and **memories** packages for a 1024 × 24-bit memory using **bit** for control signals and **std_ulogic_vector** for addresses and data.

2. [● 17.2] Write instantiations of the **memories_support** and **memories** packages for a 32 × 128-bit memory using **std_ulogic** for control signals, **unsigned** for addresses, and **float128** from **ieee.float_pkg** for data.

3. [❸ 17.1] Develop a function to add to the **memories** package that loads a memory with random values.

4. [❹] For large memories, using an array signal to represent the stored data involves too much overhead during simulation. Instead, we can implement behavioral models of large memories using sparse data structures. A sparse array dynamically allocates chunks of storage as memory locations are accessed during simulation. Develop a package for a sparse array abstract data type, and use it to extend the memories package with procedures for large memories.

Chapter 18

Test Bench and Verification Features

One of the characteristics of VHDL is that it allows a verification test bench to be written in the same language as the design to be verified. We have seen this in numerous examples in earlier chapters. We include the design to be verified as a component instance in a test bench model. We write VHDL statements to apply sequences of test values to the input ports of the design, and verify that the design produces the expected output values. However, some aspects of the language that we have seen so far make it hard to verify designs. The visibility rules are an example. They are intended to help us manage name spaces in complex designs by enforcing abstraction of interfaces and hiding of internal information. While they are good for a design in isolation, they can prevent a test bench from accessing items internal to a design. A test bench may need to monitor the state of internal signals, or force internal signals to particular values.

Another approach to verification relies less on testing and more on formal mathematical proof of correctness of a design. In order to prove correctness, we need to be able to specify what the design is intended to do. VHDL allows us to express design intent in the form of properties written in the Property Specification Language (PSL). While PSL is a separate language, defined by an IEEE standard and applicable to a number of hardware description languages, VHDL allows us to embed PSL specification within a VHDL model. A formal verification tool can then analyze the VHDL behavioral statements to prove that the properties hold.

In this chapter, we will look at features of VHDL that help us write test benches to verify operation of a design. We will start with features that allow a test bench to access objects nested within a design hierarchy, and then look at features for forcing signals. Finally, we will show how PSL specifications can be integrated into VHDL models.

18.1 External Names

VHDL provides a naming feature, *external names*, that allows us to write a test bench that accesses items not normally visible according to the hierarchical scope and visibility rules. An external name specifies a hierarchical path through the design hierarchy to reach a declared object. Thus, a test bench using an external name must have sufficient knowledge of the hierarchical structure of the design for the path to be valid. Validity of the external name is assumed during analysis of the test bench, and is checked during elaboration of the complete design hierarchy.

The syntax rule for an external name is

external_name ⇐
     << **constant** external_pathname : subtype_indication >>
     ⫾ << **signal** external_pathname : subtype_indication >>
     ⫾ << **variable** external_pathname : subtype_indication >>

external_pathname ⇐
     absolute_pathname ⫾ relative_pathname ⫾ package_pathname

As the syntax rule suggests, when we write an external name, we provide more information than just the path to the object. We also specify the class of object (constant, signal, or variable) and the subtype of the object. These specifications are used during analysis to ensure correct usage of the named object and are checked for validity during elaboration. In particular, the subtype indication specifies a view of the object, in a way similar to an alias.

The syntax rule for the pathname shows that there are three forms we can use. We will start with absolute pathnames, and return to the other two forms later in this section. A simplified syntax rule for an absolute pathname is:

absolute_pathname ⇐
     . ⦃ pathname_element . ⦄ *object*_identifier

pathname_element ::=
     *entity*_identifier
     ⫾ *component_instantiation*_label

An absolute pathname starts at the top of a design hierarchy. The first pathname element is the identifier for the top-level entity. In a test bench, where external names typically occur, the top-level entity is usually the test bench entity. Following the entity name, the pathname includes labels of component instantiation statements. The first of these refers to a component instance in the architecture of the top-level entity. The next is a component instance in the bound entity and architecture; the following is a component instance in the bound entity and architecture for the second instance; and so on. Thus, the pathname delves into the design hierarchy to the component instance whose bound entity and architecture contains the named object.

As an example, we might use the following external name in a test bench to monitor the value of a signal within a design under verification:

```
assert <<signal .tb.duv.controller.state :
                std_ulogic_vector(0 to 4)>> /= "00000"
   report "Illegal controller state";
```

Within the test bench, this external name is a reference to a signal nested within the component labeled **controller**, which is nested within the component labeled **duv**, which is within the top-level entity **tb**. The signal is interpreted as a **std_ulogic_vector** indexed from 0 to 4. When the test bench is analyzed, the existence and type of the signal is not checked. However, once the complete design hierarchy is elaborated, the signal must exist and be of an appropriate type to match the subtype specified in the external name.

An external name is just a different form of name for a constant, signal or variable, so we can use an external name at any place where a name is appropriate, subject to some rules that we will return to shortly. That means we can refer to a constant or signal value in an expression, and we can assign to a signal or include it in a port map. The rules for forming a pathname only allow us to refer to items declared in concurrent regions of a design, such as entity and architecture declarative parts, so an external variable name can only refer to a shared variable. Since we have deferred discussion of shared variables to Chapter 19, we will focus on use of external names for constants and signals in this chapter.

If an external name refers to an object of a composite type, we can refer to an element of the object. For example, given an array signal declared within a design, we can index the array with an external name as the prefix:

```
<<signal .tb.duv_rtl.data_bus :
          std_ulogic_vector(0 to 15)>>(8) <= '1';
```

Another way to use an external name is to declare an alias. If we do that, we need only write the full external name in the alias declaration. Thereafter, we can just use the shorter alias name, making the model more succinct. For example, if we need to refer to the **data_bus** signal in several places in a test bench, we could declare an alias for it:

```
alias duv_data_bus is
  <<signal .tb.duv_rtl.data_bus : std_ulogic_vector(0 to 15)>>;
```

and then just use the alias in the assignment and other places:

```
duv_data_bus(8) <= '1';
sign <= duv_data_bus(0);
```

Recall from Chapter 11 that we have the option of specifying a subtype after the alias name in an alias declaration. The syntax rule is:

alias_declaration ⇐
    alias identifier ⟦ : subtype_indication⟧ **is** name ;

An alias declared with a subtype indication gives us a view of the named object as being of that subtype. However, when the name we are aliasing is an external name, the subtype is specified in the external name. We do not repeat the subtype (or specify a conflicting subtype!) after the alias name. So the following two alias declarations are illegal:

```
alias duv_data_bus : std_ulogic_vector(0 to 15) is -- illegal!
  <<signal .tb.duv_rtl.data_bus : std_ulogic_vector(0 to 15)>>;

alias duv_data_bus : std_ulogic_vector(15 downto 0) is -- illegal!
  <<signal .tb.duv_rtl.data_bus : std_ulogic_vector(0 to 15)>>;
```

We can use an external constant name (or an alias of such a name) in an expression, provided the constant has been elaborated and given a value by the time the expression is evaluated. In some cases, expressions are evaluated during elaboration of a design. For example, initial-value expressions and index-bound expressions in declarations are evaluated when the declaration is elaborated, so an external constant name appearing in those places must refer to a constant that has already been elaborated. We can ensure this is the case by writing the part of the design that includes the constant declaration prior to the part of the design that contains the external constant name. VHDL's elaboration rules specify that the design is elaborated in depth-first top-to-bottom order.

EXAMPLE 18.1   *Elaboration order for external names*

To illustrate how we can take account of the elaboration order, suppose we have an entity and architecture for a design that declares a constant, as follows:

```
entity design is
  port ( ... );
end entity design;

architecture rtl is
  constant width : natural := 32;
  ...
begin
  ...
end architecture rtl;
```

Suppose also that we have a test bench entity and architecture:

```
entity test_bench is
end entity test_bench;

architecture directed of test_bench is
  signal test_in :
    bit_vector(0 to <<constant .top.duv.width : natural>> - 1);
  ...
begin
  ...
end architecture directed;
```

We now assemble the design and test bench in a top-level entity and architecture:

```
entity top is
end entity top;

architecture level of top is
begin
  assert false
    report "Width = " &
           to_string(<<constant .top.duv.width : natural>>);
  duv : entity work.design(rtl);
```

```
    tb  : entity work.test_bench(directed);
end architecture level;
```

In this case, the instance of the design under verification is elaborated before the test bench instance. Thus, the declaration of the constant **width** is elaborated and given a value before the external constant name within the **tb** instance is evaluated. Had we written the two instances in the reverse order, the constant would not have been elaborated at the time of elaborating the external constant name, and an error would occur. The external constant name in the concurrent assertion statement, on the other hand, is not evaluated until the model is executed, by which time the model is completely elaborated. Thus, the external constant name is allowed to precede the instance of the design under test in which the constant is declared.

VHDL has a related rule regarding elaboration of a signal referenced by an external signal name. If such a name (or an alias of such a name) is used in a port map, the signal declaration must have been previously elaborated. The reason is that the hierarchy of signal nets and drivers is built during elaboration. If a signal used in a port map is not yet elaborated, the elaborator would have to revisit elaboration of that part of the design hierarchy once the signal declaration was encountered. In general, allowing such use of external signal names would make elaboration of signal nets indefinitely complicated. The rule preventing such use allows elaboration to proceed in a well-defined order, and is not onerous in practice. It usually just requires that the component instance in which the signal is declared be written before the instance referencing the signal in a port map. The typical scenario is that a design under verification be instantiated before the test bench code containing external names.

We will now look at the full syntax rule for an absolute pathname:

absolute_pathname ⇐ . ⟦ pathname_element . ⟧ *object*_identifier

pathname_element ::=
    *entity*_identifier
    ⟦ *component_instantiation*_label
    ⟦ *block*_label
    ⟦ *generate_statement*_label ⟦ (*static*_expression) ⟧
    ⟦ *package*_identifier

As we described earlier, the first element in the pathname names the top-level entity, which has an associated architecture. The entity and architecture can contain declarations of objects. We can identify such an object by writing the object name after the entity name. The entity and architecture can also contain nested declarative parts, such as the component instances we mentioned earlier. Other nested declarative parts can include block statements (which we describe in Chapter 23), generate statements (Chapter 14), and locally declared packages (Chapter 7) and package instances (Section 12.3). These parts can, in turn, contain declarations of objects. We can identify an object in such a nested declarative part by joining the object name onto the name for the declarative part and the name for the top-level entity. In the case of a block, the name for the declarative part is the block label. In the case of a generate statement, the name for the declarative part is the generate label. A for-generate also requires a value to indicate which iteration of the generate to use. In the case of a local package, the name for the declarative part is the package name. Note that we cannot use the name of an uninstantiated package in this way; we can only use the name of an instance of the uninstantiated package. A component instance is somewhat different, in that the declarative part is not written textually inside the enclosing entity or architecture. Instead, the declarative part we refer to is that of the bound entity and architecture. Nonetheless, we use the label of the componenet instantiation statement as the name for the declarative part.

We apply these rules recursively to build up a chain of declarative part names, starting from the entity at the top of the design hierarchy and leading through levels of nesting to identify any object instantiated within the hierarchy. For example, the absolute pathname

```
.tb.duv_rtl.data_bus
```

refers to the object named **data_bus** declared within the entity and architecture bound to the component labeled **duv_rtl** within the top-level entity **tb**. Similarly, the absolute pathname

```
.tb.duv_rtl.memory(3).addr_bus
```

refers to the object named **addr_bus** within the for-generate iteration with index 3 within the component labeled **duv_rtl** within the top-level entity **tb**.

EXAMPLE 18.2   *Monitoring states in an embedded state machine*

Suppose we are verifying a system that includes a finite-state machine control unit embedded as a subcomponent. The control unit is described by the following entity and architecture:

```
library ieee; use ieee.std_logic_1164.all;
entity control is
  port ( clk, reset : in std_ulogic; ... );
end entity control;

architecture fsm of control is
  subtype state_type is std_ulogic_vector(3 downto 0);
  constant idle    : state_type := "0000";
  constant pending1 : state_type := "0001";
  ...
  signal current_state, next_state : state_type;
begin
  state_reg : process (clk) is ...
  fsm_logic : process (all) is ...
end architecture fsm;
```

The entity and architecture for the system being designed are

```
library IEEE; use IEEE.std_logic_1164.all;
entity system is
```

```
      port ( clk, reset : in std_ulogic; ... );
    end entity system;

    architecture rtl of system is
      component control is
        port ( clk, reset : in std_ulogic; ... );
      end component control;
    begin
      control_unit : component control
        port map ( clk => clk, reset => reset, ... );
      ...
    end architecture rtl;
```

We can define a test bench entity and architecture that traces the sequence of states in the control unit, issuing a report message for each:

```
    entity state_monitor is
    end entity state_monitor;

    architecture tracing of state_monitor is
      alias fsm_clk is
        <<signal .tb.system_duv.control_unit.clk : std_ulogic>>;
      alias fsm_state is
        <<signal .tb.system_duv.control_unit.current_state :
                   std_ulogic_vector(3 downto 0)>>;
    begin
      monitor : process (fsm_clk) is
      begin
        if falling_edge(fsm_clk) then
          report to_string(now) & ": " & to_string(fsm_state);
        end if;
      end process monitor;
    end architecture tracing;
```

Note here that the external reference to the **clk** port of the **control_unit** instance treats the port as a signal declared in the region corresponding to the instance. This reflects the rule in VHDL that a port is considered to be a signal declared within an entity. A generic constant of an instance would similarly be referenced using an external constant name with a pathname for the instance.

The external references in this architecture assume that the complete design hierarchy has an entity named **tb** at the root, and that the instance of the system to be monitored is labeled **system_duv** within the top-level architecture. To satisfy those assumptions, we write the top-level entity and architecture as

```
    library IEEE; use IEEE.std_logic_1164.all;
    entity tb is
    end entity tb;
```

```
architecture monitoring of tb is
  signal system_clk, system_reset : std_ulogic;
  ...
begin

  ... -- clock and reset generation

  system_duv : entity work.system(rtl)
    port map ( clk => system_clk, reset => system_reset, ... );

  state_monitor : entity work.state_monitor(tracing);

end architecture monitoring;
```

Within the **tracing** architecture of the **state_monitor** entity, we write an external name for the **current_state** signal with a **std_ulogic_vector** subtype. Normally, we would declare an enumeration type for the states of a finite-state machine. If we declare such a type locally within the control unit architecture, it would not be visible to the external monitor. We would not be able to write an external name with an appropriate subtype for the referenced signal. That is why we used a **std_ulogic_vector** subtype for the state type in this example. If we want to declare an enumeration type for an object that is to be externally monitored, we would have to declare the type in a package that is visible both in the object declaration and in the monitor.

In some test benches, the test bench code is written in the same architecture as an instance of the design under verification. In those cases, there is no need to specify the absolute path starting from the top-level entity. Instead, we can use a *relative pathname*, consisting of the chain of names starting from the immediately enclosing declarative part, without the leading dot symbol. A simplified syntax rule for a relative pathname is:

relative_pathname ⇐ ⦃ pathname_element . ⦄ *object*_identifier

For example, if a test bench architecture includes an instance of the design under verification labeled **duv**, then the architecture could also contain the assertion statement:

```
assert <<signal duv.controller.state :
              std_ulogic_vector(0 to 4)>> /= "00000"
  report "Illegal controller state";
```

Since the starting point for the relative pathname is the enclosing architecture, the first part of the pathname refers to the component instance, and subsequent parts refer to items nested within the bound entity and architecture.

An important point to note when we are talking about the innermost declarative part for a relative pathname is that only *concurrent regions* are considered. By concurrent region, we mean an entity or architecture body, a block statement, a generate statement, or a package declared locally within a concurrent region. If we write an external name with a relative pathname within a process or subprogram, that region does not count, since it is not a concurrent region. Moreover, if the name is within a package that is declared within a process or subprogram, the package does not count either. We need to look outward in the design hierarchy to find an enclosing entity, architecture, block, or generate statement, or a package that is declared in such a region.

EXAMPLE 18.3   *Revised state monitoring for an embedded state machine*

Returning to the test bench of Example 18.2, we can write the state-monitoring code directly in the top-level architecture rather than in an instantiated entity and architecture. In that case, we can use relative pathnames, and so do not have to assume the name of the top-level entity. The revised top-level architecture is

```
architecture monitoring of tb is

  signal system_clk, system_reset : std_ulogic;

  alias fsm_clk is
    <<signal system_duv.control_unit.clk : std_ulogic>>;
  ...

begin

  ... -- clock and reset generation

  system_duv : entity work.system(rtl)
    port map ( clk => system_clk, reset => system_reset, ... );

  monitor : process (fsm_clk) is
    use std.textio.all;
    file state_file : text open write_mode is state_file_name;
    alias fsm_state is
      <<signal system_duv.control_unit.current_state :
                 std_ulogic_vector(3 downto 0)>>;
  begin
    if falling_edge(fsm_clk) then
      report to_string(now) & ": " & to_string(fsm_state);
    end if;
  end process monitor;

end architecture monitoring;
```

In this architecture, the alias declarations refer to external names identified with relative pathnames. The component label **system_duv** is declared in the same enclosing architecture region as the alias declarations, so that label is the one used in the pathnames. Even though the external name aliased to **fsm_state** is written within the process region, the innermost region considered is that of the enclosing architecture.

A further form of relative pathname allows us to identify an outer region as the starting point for the pathname. We write such a pathname using one or more leading "∧" symbols in place of names. The full syntax rule for a relative pathname shows this:

relative_pathname ⇐ ⦃ ∧ . ⦄ ⦃ pathname_element . ⦄ *object*_identifier

As for the relative pathname without the "∧" symbols, we initially start with the inner-most concurrent region enclosing the external name. Then, for each "∧" symbol, we look in the next enclosing region. In the case of instantiated components, the region enclosing an instance of a bound entity and architecture is the region in which the instantiation is written. Thus, if we use this form of pathname in an entity or architecture, we are making a strong assumption about the context in which the entity and architecture are instantiated. Specifically, we are assuming that context also includes the names written in the pathname. The complete design hierarchy must be built in such a way as to ensure the assumption is met, otherwise an error will occur during elaboration.

EXAMPLE 18.4   *Relative pathname in a nested monitor*

Suppose we are verifying a multicore platform, in which each core includes an in-stance of a CPU described by the following entity and architecture:

```
entity CPU is
  . . .
end entity CPU;

architecture BFM of CPU is
  use work.CPU_types.all;
  signal fetched_instruction : instruction_type;
  . . .
begin
  . . .
end architecture BFM;
```

The architecture includes a signal representing a fetched instruction. The multicore platform is described by an entity with a generic constant specifying the number of cores. The architecture of the entity uses a for-generate statement to repli-cate instances of the CPU.

```
entity platform is
  generic ( num_cores : positive );
  port   ( ... );
end entity platform;

architecture BFM_multicore of platform is
  . . .
begin
  cores : for core_num in 1 to num_cores generate
    processor : entity work.CPU(BFM) ...;
    . . .
  end generate cores;
  . . .
end architecture BFM_multicore;
```

We now consider the test bench that instantiates the platform entity and architecture. Again, we use a generic constant to determine the number of cores in the design under verification. We can include a monitor for each instantiated core by writing a for-generate statement in the test bench, mirroring that in the platform architecture.

```
entity test_bench is
  generic ( num_cores : positive );
end entity test_bench;

architecture test_BFM of test_bench is
  . . .
begin

  duv : entity work.platform(BFM_multicore)
    generic map ( num_cored => num_cores )
    port map    ( ... );

  monitors : for core_num in 1 to num_cores generate
    use work.CPU_types.all, work.CPU_trace.all;
    process is
    begin
      . . .
      trace_instruction
        ( <<signal
          ^.duv.cores(core_num).processor.fetched_instruction :
            instruction_type>>,
          ... );
      . . .
    end process;
  end generate monitors;

end architecture test_BFM;
```

The process within the generate statement includes an external name referring to the fetched_instruction signal in the corresponding core instance. The pathname uses the value of the **core_num** generate parameter to identify the corresponding iteration of the generate statement labeled **cores** in the design under verification. Since the external name is in a process nested within a generate statement, the generate statement is the innermost region used as the starting point for the relative pathname. For that reason, the pathname starts with a "∧" symbol to look outside the starting region to the enclosing architecture region. The **duv** component instance is declared in that region, so it can be used as the next part of the pathname.

We now turn to the third form of pathname, a *package pathname*, that we can write in an external name. For objects declared in a package declaration written as a design unit, we can just use the package name as a prefix in a normal selected name to refer to the object. However, objects declared in the package body are not visible to designs. They would normally be referenced indirectly using procedures declared in the package declaration. A test bench, on the other hand, can use an external name to refer to such a hidden object. An object in a design-unit package is not nested within the design hierarchy, but is considered to be nested within the library containing the package. So the chain of region names starts with the library name (the name defined by a library clause) and leads through the top-level package name and any nested package names to the referenced object.

A package pathname takes a similar form to an absolute pathname, but starts with an "@" symbol instead. That is followed by name of the library containing the package, then the package name, then the names of any intervening nested packages, and finally the object name. The syntax rule is:

package_pathname ⇐ @ *library*_identifier . { *package*_identifier . } *object*_identifier

For example, given the following package declaration and body analyzed into the working library:

```
package p1 is
  . . .
end package p1;

package body p1 is
  . . .
  package p2 is
    signal s : bit;
  end package p2;
  . . .
end package body p1;
```

we could write the following external name to refer to the signal:

```
<<signal @work.p1.p2.s : bit>>
```

VHDL-87, -93, and -2002

These versions of VHDL do not allow external names.

18.2 Force and Release Assignments

When verifying a design, we often would like to be able to override the value assigned to a signal in the normal course of design operation and force a different value onto the signal. One reason for doing this is to set up a test scenario by forcing values to a state that would normally be arrived at through a complex initialization sequence. Forcing the values allows us to bypass the sequence and set up the scenario quickly, and so reduce the verification time significantly. Another reason for forcing values is to inject erroneous values into the design to ensure that it detects the error or otherwise responds appropriately.

The forms of signal assignment that we have seen in earlier chapters all contribute to a signal's value. If there are multiple assignment statements in different parts of a model, they constitue separate sources, and their contributed values are resolved to determine the signal's value. For verification purposes, we need a different form of assignment that overrides the normal form. VHDL provides such a form, called a *force assignment*. The syntax rule for a simple force assignment is:

> signal_assignment_statement ⇐
>   ⟦ label : ⟧ name <= **force** ⟦ **in** ⟧ **out** ⟧ expression ;

This is a sequential assignment written within a process forming part of the test bench. The effect is to cause a delta cycle and to force the named signal to take on the value of the expression in that delta cycle, regardless of any value assigned to the signal by any normal signal assignment. The signal is considered to be active during the delta cycle, and if the forcing value is different from the previous value, an event occurs on the signal. Processes sensitive to changes on the signal value would then respond to the value change in the normal way. The usual rules relating the type of the expression to the type of the target signal apply for force assignments. The target signal name can be a normal signal name, or it can be an external signal name or alias.

Once a signal has been forced, we can update the signal with another force assignment to change the overriding value, again causing the signal to become active and possibly to have another event. We can do this as often as needed. Eventually, if we want to stop forcing a signal, we can execute a *release assignment*. The syntax rule is

> signal_assignment_statement ⇐
>   ⟦ label : ⟧ name <= **release** ⟦ **in** ⟧ **out** ⟧ ;

This causes a further delta cycle, with the signal being active. However, since the signal is no longer forced, the current values of its sources are used to determine the signal value in the normal way. We can think of this as the design "taking back control" of the signal.

EXAMPLE 18.5 *Simulated corruption of a state machine's state value*

Clocked sequential systems are usually controlled by a finite-state machine. If the storage for the current state is corrupted, the system may be able to recover by transitioning from the illegal state back to an initial state. A test bench can verify that a design under verification recovers correctly by forcing the signal storing the current state of the state machine to an illegal value. It can then release the signal and monitor recovery. The test bench process is

```
verify_state_recovery : process is
  use work.control_pkg.all;
  alias clk is <<signal duv.clk : std_ulogic>>;
  alias current_state is
        <<signal duv.control.current_state : state_type>>;
begin
  ...
  -- inject corrupt state
  wait until falling_edge(clk);
```

```
        current_state <= force illegal_state_12;
        wait until falling_edge(clk);
        current_state <= release;
        -- monitor recovery activity

        ...
    end process verify_state_recovery;
```

Our discussion of force assignments has so far focused on signals. We can also force and release ports of a design, since they are a form of signal. However, for a port, we distinguish between the driving value and the effective value. The driving value is the value presented externally by an entity, and is determined by the internal sources within the entity. The effective value is the value seen internally by an entity and is determined by whatever is externally connected to the port, whether that be an explicitly declared signal or a port of an enclosing entity. Depending on the port mode and the external connections, the driving and effective values may be different. For example, an **inout**-mode port of type **std_ulogic** might drive a '0' value, but the externally connected signal might have another source driving a '1' value. In that case, the resolved value of the signal is 'X', and that value is seen as the effective value of the **inout**-mode port.

VHDL allows us to force the driving and effective values of a signal or port independently by including a *force mode* in an assignment. For explicitly declared signals, where the driving and effective values are the same, the distinction makes no difference. For ports and signal parameters, we can force the driving value by including the keyword **out** in the force assignment, for example:

```
    duv_bus <= force out "ZZZZZZZZ";
```

Alternatively, we force the effective value by including the keyword **in** in the force assignment, for example:

```
    duv_bus <= force in "XXXXXXXX";
```

Once we've forced a port's or signal parameter's driving value, we can stop forcing it by writing a release assignment with the keyword **out**:

```
    duv_bus <= release out;
```

Similarly, to release a forced effective value, we write a release assignment with the keyword **in**:

```
    duv_bus <= release in;
```

We can force and release driving values of ports and signal parameters of mode **out**, **inout**, and **buffer**, but not those of mode **in**, since they do not have driving values. However, all ports and signal parameters have effective values, so we can force and release the effective value of a port or signal parameter of any mode. (The exception is ports of mode **linkage**, which we describe briefly in Chapter 23.)

If we omit the force mode (**out** or **in**) in a force or release assignment, a default force mode applies. For assignments to ports and signal parameters of mode **in** and to explicitly declared signals, the default force mode is **in**, forcing the effective value. For assignments to ports and signal parameters of mode **out**, **inout**, or **buffer**, the default force mode is **out**, forcing the driving value.

EXAMPLE 18.6 *Forcing disconnection of a port's driving value*

Serial buses such as I^2C, USB and FireWire have bidirectional connections to the bus' physical wires. This allows a device to drive the clock and data wires when transmitting data and to sense the clock and data values when receiving. A test bench can model a broken data driver connection by forcing a 'Z' value on the output part of the bidirectional port, while allowing the input part of the port to operate normally. The code in the test bench is

```
...
-- Test scenario: break in the output connection
<<signal duv.SDA : std_ulogic>> <= force out 'Z';
-- Monitor device operation under this fault condition
...
-- Restore connection for the next scenario
<<signal duv.SDA : std_ulogic>> <= release out;
...
```

In Chapter 5 we introduced conditional and selected forms of sequential signal assignment statements. We can also write conditional and selected force assignments, choosing the value to force onto the target signal from a number of alternatives. The syntax rule for a conditional force assignment is

```
conditional_force_assignment ⇐
    〖 label : 〗
    name <= force 〖 in ▯ out 〗
            expression when condition
            〖 else expression when condition 〗
            〖 else expression 〗 ;
```

and for a selected force assignment:

```
selected_signal_assignment ⇐
    〖 label : 〗
    with expression select 〖 ? 〗
        name <= force 〖 in ▯ out 〗
                { expression when choices , }
                expression when choices ;
```

The force mode is optional, and can be either **in** or **out** to specify forcing of the effective value or the driving value of the target signal, respectively, as described above. The effect of these statements is to allow us to choose the value to force onto the target, depending on a number of conditions or on the value of an expression. They provide a more succinct way of writing the alternatives than embedding a number of simple force assignments in an if statement or case statement.

EXAMPLE 18.7   *Conditional force assignment*

A conditional force assignment can be used to choose between a randomly generated stimulus value or a directed-test stimulus value in a loop that applies successive tests. The stimulus value is used to force the effective value of a bidirectional port of a design under test. The code in the test bench is

```
alias dut_d_bus is
  <<signal dut.d_bus:std_ulogic_vector(15 downto 0)>>;
...

for test_count in 1 to num_tests loop
  dut_d_bus <= force in
    next_random_stim(dut_d_bus'length)
      when test_mode = random else
    directed_stim(test_count);
  wait for test_interval;
end loop;
```

As we have seen earlier, VHDL allows us to assign a composite value to a collection of signals by writing the collection in the form of an aggregate on the left-hand side of the assignment, for example:

```
(carry_out, sum) <= ('0' & a) + ('0' & b);
```

We cannot, however, write an aggregate of signal names as the target of a force or release assignment to force or release each of the signal values. Instead, we must write a separate force or release assignment for each of the signals. For example, if we want to force and release the driving values of the two ports **carry_out** and **sum**, we would have to write:

```
sum <= force out unsigned'("00000000");
carry_out <= force out '1';
...

sum <= release out;
carry_out <= release out;
```

There is a further form of target signal for which we cannot write a force or release assignment. Suppose we define a resolved signal of a composite type, such as an array type. By that, we mean a signal with multiple sources, each of which is a composite value. The resolution function for the signal takes an array of composite values and determines a composite value as the resolved value of the signal. We cannot write a force or release assignment with an element of such a signal as the target. We can only force or release the signal as a whole. This mirrors the requirements that a process driving such a signal have a driver for all elements of the signal, and that sources for such a signal be sources for the entire signal. Note that resolved composite signals are different from signals of resolved elements, for example, signals of type **std_logic_vector**. We can force and release individual elements or slices of those signals, since each element is resolved individually.

Another case to consider is a force or release assignment written in a subprogram. VHDL has a rule, mentioned in Chapter 6, that a signal assignment written in a procedure that is not contained within a process can only assign to a signal parameter of the procedure. The rationale is that assignment to a signal implies a driver for the signal. For signal parameters, the driver used is the driver for the actual signal provided by the process that calls the procedure. For other signals, a driver for the target signal would be implied for every process that calls the procedure. Identifying all of the callers in a large model would be very difficult. Moreover, if the procedure body is written separately from the calling processes, determining what drivers are created for a given process would be difficult. Thus, the restriction makes VHDL designs easier to analyze and understand. Force and release assignments, on the other hand, do not imply drivers. Rather, they would typically occur in test bench code, often referring to the target signals with external names. For these reasons, VHDL allows force and release assignments in procedures outside of processes to signals other than signal parameters.

One final aspect to discuss is the effect of multiple concurrent force and release assignments. Since they are sequential assignments written in processes, it is possible that multiple forces and releases could occur for a given signal during a single simulation cycle. If a force and release both occur, the effect is as though the release is immediately overridden by the force, and so the signal remains forced, but with the new force value. The effect of multiple forces is not defined. We should write our test bench models to avoid that occurring. The effect of multiple releases, however, is the same as a single release, and a release assignment on a signal that is not forced has no effect.

VHDL-87, -93, and -2002

These versions of VHDL do not allow force and release assignments. Simulators for these versions typically provide commands for forcing and releasing signals interactively or as part of a simulation control script.

18.3 Embedded PSL in VHDL

PSL is the IEEE Standard Property Specification Language (IEEE Std 1850). It allows specification of temporal properties of a model that can be verified either statically (using a formal proof tool) or dynamically (using simulation checkers). VHDL allows PSL code to be embedded as part of a VHDL model. This makes design for verification a much more natural activity, and simplifies development and maintenance of models. Since PSL is itself a significant language, we won't describe all of its features in detail in this book. Instead, we will just describe the way in which PSL can be embedded in VHDL. For a full description of PSL and its use in verifying designs, the interested reader is referred to other published books on the subject. (See, for example, *A Practical Introduction to PSL* [4].)

In VHDL we can include PSL property, sequence, and default clock declarations in the declarative part of an entity, architecture, block statement (see Chapter 23), generate statement, or package declaration. We can then use the declared properties and sequences in PSL directives written in the corresponding statement parts.

Any properties that we write in PSL declarations and directives must conform to PSL's simple subset rules. In practice, this means that we can only write properties in which time moves forward from left to right through the property. Two examples from the PSL standard illustrate this. First, the following property is in the simple subset:

always (a -> **next**[3] b)

This property states that if **a** is true, then three cycles later, **b** is true; that is, time moves forward three cycles as we scan the property left to right. In contrast, the following property is not in the simple subset:

always ((a && **next**[3] b) -> c)

This property states that if **a** is true and **b** is true three cycles later, then **c** must have been true at the time **a** was true. The problem with this property is that time goes backward from **b** being true to **c** being true. A tool to check such a property is much more complex than one to check properties in the simple subset.

PSL directives require specification of a clock that determines when temporal expressions are evaluated. We can include a clock expression in a directive. However, since the same clock usually applies to all directives in a design, it is simpler to include a default clock declaration. If we write a default clock declaration in a region of a design, it applies to any PSL directives written in that region. We can include at most one default clock declaration in any given region.

EXAMPLE 18.8   *Pipelined handshake assertion*

In their book *Assertion-Based Design* [7], Foster *et al.* describe a verification pattern for a system in which handshaking is pipelined. In their example, a system can receive up to 16 requests before acknowledging any of them. The system counts the number of requests and acknowledgments and includes an assertion that, for every request with a given request count, there is an acknowledgment with the same count within 100 clock cycles. We can describe the system in VHDL as follows:

```vhdl
library ieee; context ieee.ieee_std_context;
entity slave is
  port ( clk, reset : in  std_ulogic;
         req        : in  std_ulogic;
         ack        : out std_ulogic;
         ... );
end entity slave;

architecture pipelined of slave is

  signal req_cnt, ack_cnt : unsigned(3 downto 0);
```

```
default clock is rising_edge(clk);

property all_requests_acked is
  forall C in {0 to 15}:
    always {req and req_cnt = C} |=>
            {[*0 to 99]; ack and ack_cnt = C};

begin

  req_ack_counter : process (clk) is
  begin
    if rising_edge(clk) then
      if reset = '1' then
        req_cnt <= "0000"; ack_cnt <= "0000";
      else
        if req = '1' then req_cnt <= req_cnt + 1; end if;
        if ack = '1' then ack_cnt <= ack_cnt + 1; end if;
      end if;
    end if;
  end process req_ack_counter;

  . . .

  assert all_requests_acked;

end architecture pipelined;
```

The counters for requests and acknowledgments are implemented using the signals **req_cnt** and **ack_cnt** and the process **req_ack_counter**. We declare a property, **all_requests_acked** that expresses the verification condition for the design. We also include a default clock declaration for the architecture. It applies to the assert directive that we write in the statement part of the architecture, asserting that the verification condition holds.

There is one case where embedding of PSL within VHDL may lead to ambiguity. Both PSL and VHDL include assert statements, but their meanings differ. If we write a statement of the form

assert not (a **and** b) **report** "a and b are both true";

it could be interpreted as a regular VHDL concurrent assertion statement that is to be checked whenever either of **a** or **b** changes value. Alternatively, it could be interpreted as a PSL assert directive that requires the property **not** (a **and** b) to hold at time 0. In the interest of backward compatibility with earlier versions of the language, VHDL interprets such ambiguous statements as regular VHDL concurrent assertion statements. If we really want to write a PSL assert directive of this form, we could modify the property so that it is unambiguously a PSL property, for example:

assert next[0] **not** (a **and** b) **report** "a and b are both true";

In PSL, verification code can be written in verification units (**vunit**, **vprop** and **vmode** units) that are bound to instances of VHDL entities and architectures. VHDL considers such verification units as primary design units. Thus, they can be declared in VHDL design files and analyzed into VHDL design libraries.

A verification unit can include binding information that identifies a component instance to which directives apply. Alternatively, we can bind a verification unit as part of the configuration of a design. One place to do that is in a configuration declaration, introduced in Chapter 13. If we want to bind one or more verification units to the top-level entity in a configuration declaration, we include binding information according to the following synax rule:

configuration_declaration ⇐
    configuration identifier **of** *entity*_name **is**
        ⟦ **use vunit** *verification_unit*_name ⟦ , ₀₀₀ ⟧ ; ⟧
        block_configuration
    end ⟦ **configuration** ⟧ ⟦ identifier ⟧ ;

Whenever the configuration declaration is instantiated, either at the top-level of a design hierarchy or as a component instance within a larger design, the named verification units are bound to the instance of the named entity and architecture. That means the names used in the verification units are interpreted in the context of the entity instance.

EXAMPLE 18.9   *Binding a verification unit in a configuration declaration*

Suppose we have a verification unit that ensures two outputs named Q and Q_n are complementary when sampled on rising edges of a signal named **clk**. The verification unit is

```
vunit complementary_outputs {
  assert always Q = not Q_n;
}
```

We can bind this verification unit to various parts of a design. For example, a gate-level model of a D flipflop might be described as follows:

```
entity D_FF is
  port ( clk, reset, D : in  bit;
         Q, Q_n         : out bit );
end entity D_FF;

architecture gate_level of D_FF is
  component and2 is ...
  ...
begin
  G1 : and2 ...
  ...
end architecture gate_level;
```

A configuration declaration for the D flipflop can bind the verification unit to the top-level entity as follows:

```
configuration fast_sim of D_FF is
  use vunit complementary_outputs;
  for gate_level
    for all : and2
      ...
    end for;
    ...
  end for;
end configuration fast_sim;
```

We could then instantiate the configuration in a design, and for each instance, the verification unit **complementary_outputs** would be bound to the instantiated entity and architecture.

We can also bind verification units to component instances that are configured by a component configuration nested within a configuration declaration. The augmented form of component configuration, assuming the components are bound to an entity and architecture, and the architecture is further configured, is

```
component_configuration ⇐
    for component_specification
        binding_indication ;
        ⟦ use vunit verification_unit_name ⟦ , ... ⟧ ; ⟧
        ⟦ block_configuration ⟧
    end for ;
```

In this case, the named verification units are bound to the instances specified in the component configuration.

EXAMPLE 18.10   *Binding a verification unit in a component configuration*

Suppose we instantiate a parallel-in/serial-out shift register within an RTL design:

```
entity system is
  ...
end entity system;

architecture RTL of system is
  component shift_reg is
    port ( clk, reset, D : in  bit_vector;
           Q, Q_n        : out bit );
  end component shift_reg;
  ...
begin
```

```
      serializer : shift_reg ...;
      ...
  end architecture RTL;
```

We can write a configuration declaration that binds an entity and architecture to the component instance and that also binds the **complementary_outputs** verification unit shown in Example 18.9:

```
  configuration verifying of system is
    for RTL
      for serializer : shift_reg
        use entity work.shift_reg(RTL);
        use vunit complementary_outputs;
      end for;
    end for;
  end configuration verifying;
```

In this case, the assertion in the verification unit applies to the Q and Q_n outputs of the shift register entity bound to the **serializer** component instance.

The third place in which we can bind verification units in a VHDL design is in a configuration specification in the architecture where components are instantiated. The augmented syntax rule for a configuration specification, again assuming components are bound to an entity and architecture, is

```
configuration_specification ⇐
    for component_specification
        binding_indication ;
        { use vunit verification_unit_name { , ... } ; }
    end for ;
```

This is similar to the form in a component configuration, but without the nested configuration for the architecture.

EXAMPLE 18.11   *Binding a verification unit in a configuration specification*

We can revise the architecture of Example 18.10 to include the binding information directly, rather than in a separate configuration. The revised architecture is

```
  architecture RTL of system is
    component shift_reg is
      ...
    end component shift_reg;
    for serializer : shift_reg
      use entity work.shift_reg(RTL);
      use vunit complementary_outputs;
    end for;
```

```
begin
  serializer : shift_reg ...;

  ...
end architecture RTL;
```

Since a verification unit may include binding information as part of its declaration, there is potential for that information to conflict with binding information we write in a configuration. VHDL prevents such conflict by making it illegal to bind a verification unit in a configuration if the declaration of the unit already includes binding information. Hence, we would normally only write verification bindings in configurations for general-purpose verification units, and not for those written with particular instances in mind. In any case, it would be an error if we wrote a verification unit binding for a component instance that had no bound entity and architecture.

In addition to binding verification units directly in their declaration or indirectly in configurations, VHDL allows a tool to bind additional verification units through implementation-defined means. That might include command-line options, script commands, or selection using a graphical user interface.

There are a couple of further points to make about PSL embedded in VHDL. First, PSL has a rich set of reserved words, some of which may conflict with VHDL identifiers. The following PSL keywords are VHDL reserved words, and cannot be used as identifiers:

assert	**assume**	**assume_guarantee**
cover	**default**	**fairness**
property	**restrict**	**restrict_guarantee**
sequence	**strong**	**vmode**
vprop	**vunit**	

Other PSL reserved words are only recognized as such within VHDL code when they occur in PSL declarations and directives. They can be used as VHDL identifiers, but such identifiers are hidden within PSL declarations and directives. For example, we can legally write the following declaration:

function rose (x : boolean) **return** boolean **is** ...;

But if we then declare a sequence:

sequence cover_fifo_empty **is**
  {reset_n && **rose**(cnt = 0)};

The reference to **rose** in the sequence declaration is to the PSL built-in function, not to the declaration written in VHDL.

Second, PSL includes features for declaring and instantiating macros, and allows for preprocessor directives. These features can only be used in PSL verification units, not in other VHDL design units.

VHDL-87, -93, and -2002

These versions of VHDL do not allow PSL to be embedded within VHDL models. PSL code must be written in separate verification units and bound to the VHDL design using tool-defined means.

Exercises

1. [❶ 18.1] The following architecture declares a number of objects:

    ```vhdl
    architecture rtl of datapath is
      constant d_width : positive := 8;
      signal d_bus : std_ulogic_vector(d_width - 1 downto 0);
    begin
      adder : for i in 0 to d_width - 1 generate
        signal carry : std_ulogic;
      begin
        ...
      end generate adder;
      ...
    end architecture datapath;
    ```

    The entity datapath is instantiated in a top-level architecture as follows:

    ```vhdl
    architecture data_test of test_bench is
      ...
    begin
      dp : entity work.datapath(rtl)
        port map ( ... );

      verifier : process is
      begin
        ...
      end process verifier;

    end architecture data_test;
    ```

    Write absolute external names for the objects **d_width** and **d_bus** and for the object **carry** in the instance of the generate statement with i = 3.

2. [❶ 18.1] Write alias declarations for each of the external names described in Exercise 1.

3. [❶ 18.1] Write relative external names for each of the external names described in Exercise 1, assuming the names occur within the process **verifier**.

4. [❶ 18.2] Write statements to force a signal **reset** to the value '1' and then to release it after 200 ns.

5. [❶ 18.1/18.2] Example 8.5 on page 282 describes a memory entity, **memory_1Mx8**, with a bidirectional port **d** of type **std_logic_vector(7 downto** 0). Assuming this entity is instantiated in a test bench architecture using the label mem as the component instantiation label, write statements to force both the driving and effective values of d to all 'Z' values, and subsequently to release both the driving and effective values.

6. [❶ 18.3] Write a configuration declaration that binds the verification unit **verify_protocol** to a top-level entity **bus_interface** with architecture **behavior**.

7. [❶ 18.3] Given an architecture that instantiates a component **ext_interface** using the label **ext**, write a configuration specification that binds the entity **bus_interface** with architecture **behavior** and the verification unit **verify_protocol** to the instance.

8. [❷ 18.1] Write a test bench for the counter of Example 5.22 on page 177. The test bench should verify that the output of each incrementer is one greater than the corresponding register output.

9. [❷ 18.1] Write a test bench for the register of Example 14.1 on page 450. The test bench should verify that, after each clock edge, the internal **data_unbuffered** signals have the same values as the **data_in** values supplied as stimulus.

10. [❷ 18.2] Write a procedure that forces a signal of type **bit_vector** to a random value.

Chapter 19

Shared Variables and Protected Types

When we introduced variables in Chapter 2, we noted that they can only be declared in processes; hence only one process can access each variable. We have also seen variables declared in subprograms, in which case they are local to the invocation of the subprogram. The reason for these restrictions is to prevent indeterminate results arising from a number of processes accessing a variable in an indeterminate order during a simulation cycle. In some circumstances, however, it is desirable to allow a number of processes to share access to a variable. Either the fact of non-determinacy may be irrelevant, or the use of shared variables may allow a more concise and understandable model. In this chapter, we will show how to declare and use shared variables.

19.1 Shared Variables and Mutual Exclusion

VHDL provides a mechanism for sharing variables, shown by the full syntax rule for a variable declaration:

> variable_declaration ⇐
> 〚 **shared** 〛 **variable** identifier 〚 , ... 〛 : subtype_indication 〚 := expression 〛 ;

If we include the keyword **shared** in a variable declaration, the variables defined are called *shared variables* and can be accessed by more than one process. We can only declare shared variables in the places in a model where we cannot declare normal variables, namely, in entity declarations, architecture bodies, block statements (see Chapter 23), generate statements, and packages that are not local to processes or subprograms. Unlike normal variables, there are a number of restrictions on the way we declare and use shared variables. We discuss these restrictions in this section.

Problems with shared variables potentially arise when the processes in the model are executed on a parallel computer, such as a multiprocessor workstation or a parallel supercomputer. They can also occur on a single-processor computer if the simulation kernel preemptively switches between processes. The simplest problem is that two processes trying to update a shared variable might interfere with each other, resulting in an unpredictable final result for the variable. Suppose, for example, that two processes try to increment a counter. Each process executes the statement

```
counter := counter + 1;
```

This statement typically involves reading the variable's value, adding 1 to the value, then storing the result back into the variable's memory location. If one process completes this sequence before the other starts, the variable is incremented by two, as expected. However, if both processes read the initial value before either performs the write, the variable is only incremented by 1. Since the model writer has no control over the interleaving of memory access from multiple processes, the result is non-deterministic. More complex problems arise when the shared variable is not simply a scalar, but is instead a composite or dynamic data structure requiring complex update operations. In these cases, interference between processes can put the data structure in an inconsistent state, resulting in lost or corrupted data.

The key to avoiding interference between processes that concurrently access a shared variable is *mutual exclusion*. A process must acquire exclusive access when it needs to read or update the variable using some sequence of instructions. While the process is performing those instructions, no other process is allowed access to the variable. This rule is enforced by the language implementation.

We achieve mutual exclusion for a shared variable by declaring the variable to be of a *protected type*. There are two parts to the definition of a protected type: a *protected type declaration* and a *protected type body*, each of which is included in a type declaration. The extended syntax rule for type declarations is included in Appendix B. The protected type declaration specifies the interface of the protected type. It contains *methods*, subprograms that will be used by processes to access the shared variable with mutual exclusion. The syntax rule for a protected type declaration is

protected_type_declaration ⇐
   protected
      〔 protected_type_declarative_item 〕
   end protected 〔 identifier 〕

The optional identifier in the syntax rule indicates that the name of the protected type may be repeated at the end of the declaration. The declarations in the declarative part can include subprogram (procedure and function) declarations for the methods of the type, attribute specifications and use clauses. Only the interfaces of subprograms are included; the subprogram bodies are deferred to the protected type body.

EXAMPLE 19.1   *Protected type declaration for a shared counter*

A simple example of a protected type declaration is

```
type shared_counter is protected
  procedure reset;
  procedure increment ( by : integer := 1 );
  impure function value return integer;
end protected shared_counter;
```

This declares a protected type for a shared counter with methods to reset the counter value to zero, to increment the counter by some amount and to read the value of the counter.

We can declare a shared variable to be of this type using a shared variable declaration, for example:

```
shared variable event_counter : shared_counter;
```

A process uses the name of the shared variable as a prefix to a method name to identify the shared variable on which the method is invoked. For example:

```
event_counter.reset;
event_counter.increment (2);
assert event_counter.value > 0;
```

In each case, the process acquires exclusive access to **event_counter** before executing the body of the method. While the process is executing the method, other processes that try to invoke any method on the same shared variable must wait. When the first process finishes executing its method, it releases exclusive access to the shared variable. One of the waiting processes may then resume. The order in which waiting processes are chosen for resumption is not defined.

A protected type body specifies the implementation details of a protected type. The syntax rule for a protected type body is

protected_type_body ⇐
    protected body
        〖 protected_type_body_declarative_item 〗
    end protected body 〖 identifier 〗

Again, the optional identifier in the syntax rule indicates that the name of the protected type may be repeated at the end of the declaration. The declarations in the declarative part can include subprogram declarations and bodies; type, subtype, constant, variable, file and alias declarations; attribute declarations and specifications; group templates and group declarations; and use clauses. We must include subprogram bodies for the methods declared in the protected type declaration. Items declared within a protected type body are not visible outside the protected type, so the only way a process can access the items is by using the methods of the protected type.

Note that we usually include variable declarations in a protected type body. These variables constitute the data stored in a shared variable of the protected type. We might also include file declarations. In this case, the methods of the protected type can be used to ensure that multiple lines of output to a file from a process are performed atomically, rather than being interleaved with output from other processes. We would not normally write a protected type body without variable or file declarations, since there would be no need for mutual exclusion in that case.

EXAMPLE 19.2   *Protected type body for a shared counter*

We can implement the shared counter protected type declared in Example 19.1 as follows:

```
type shared_counter is protected body

  variable count : integer := 0;

  procedure reset is
  begin
    count := 0;
  end procedure reset;

  procedure increment ( by : integer := 1 ) is
  begin
    count := count + by;
  end procedure increment;

  impure function value return integer is
  begin
    return count;
  end function value;

end protected body shared_counter;
```

The variable **count** represents the storage for the counter type. Each shared variable of this type has its own instance of the **count** variable. The methods simply operate on the variable to update or read its variable without interference from other processes.

There are a number of rules governing the use of protected types. In summary, the rules are the following:

- Only method names declared in the protected type declaration are visible outside the protected type definition. Nothing declared in the protected type body is visible outside. However, all names declared in the protected type declaration are visible in the corresponding protected type body. This rule is analogous to the visibility rule for names declared in package declarations and package bodies.

- If a protected type is declared in a package declaration, the protected type body must be declared in the corresponding package body. This is similar to the way in which subprogram specifications and bodies are declared in packages. In other cases, the protected type body must be declared in the same declarative region as the protected type declaration.

- Only variables and variable-class subprogram parameters can be of protected types. Actual values of protected type subprogram parameters are passed by reference. This ensures that when the subprogram invokes a method of the parameter, exclusive access is acquired to the actual shared variable, not to a copy of the variable.

- Shared variables must be of protected types. Other variables may be of protected types, but since they are only accessible to the process in which they are declared, there is no need for mutual exclusion.

- Protected types cannot be used as elements of files, as elements of composite types, or as types designated by access types.

- Variable assignment of one protected-type variable to another is not allowed, since assignment does not provide mutual exclusion for the operands. As a consequence, a protected-type variable must not have an initial value expression in its declaration.

- Similarly, the equality ("=") and inequality ("/=") operators are not predefined for protected types, since they do not provide mutual exclusion.

- A protected type method must not include or execute a wait statement. This ensures that all methods complete and release exclusion within one simulation cycle. Processes that must share information across different simulation cycles should use signals for communication.

- A function method must be declared as impure, since it accesses the variables declared within the protected-type body that are outside the declaration of the function itself.

- A function method that is a unary operator must be declared with no parameters, since the variable on which is is invoked is implicitly the parameter. Similarly, a function method that is a binary operator must be declared with only one parameter. Such function methods can only be invoked using the selected-name notation; they cannot be invoked using infix operator notation.

The preceding examples show that we use a selected name to invoke a method. The prefix is the shared variable name, and the suffix is the method name. In the example, the shared variable is directly visible, so we just used the identifier as the prefix. In a test bench design, we can use an external name to refer to a shared variable declared within the design hierarchy. We can use the external name as a prefix in a selected name to invoke a method of the shared variable.

EXAMPLE 19.3   *Accessing a shared variable from a test bench*

Suppose we declare the shared counter protected type from Examples 19.1 and 19.2 in a package so that it can be used both in a design and in the test bench for the design. The package declaration is

```
package shared_counter_pkg is

  type shared_counter is protected
    procedure reset;
    procedure increment ( by : integer := 1 );
    impure function value return integer;
  end protected shared_counter;

end package shared_counter_pkg;
```

and the package body is

```
package body shared_counter_pkg is

  type shared_counter is protected body
    variable count : integer := 0;
    procedure reset is ...
    procedure increment ( by : integer := 1 ) is ...
    impure function value return integer is
    begin
      return count;
    end function value;
  end protected body shared_counter; ...

end package body shared_counter_pkg;
```

The **event_counter** shared variable is declared within an architecture of a design under verification, and the design is instantiated with the label **duv_behavior** in a test bench. Within the test bench, we can invoke the **reset** method of the event counter variable as follows:

```
<<variable .tb.duv_behavior.event_counter :
            work.shared_counter_pkg.shared_counter>>.reset;
```

We could also declare an alias for the shared variable and use it to invoke the **increment** method:

```
alias duv_event_counter is
  <<variable .tb.duv_behavior.event_counter :
            work.shared_counter_pkg.shared_counter>>;
...

duv_event_counter.increment(4);
```

A typical scenario in which we might use shared variables is to describe passive shared objects that are accessed concurrently by different parts of a design. An example of such an object is a register file in a pipelined CPU, represented simply as an array of bit vectors. The register file must be accessed by the processes representing the operand fetch stage and the result write-back stage. At a lower level of abstraction, we would implement the register file as a component with read and write ports. Read and write operations would take place over signals and would conform to some signaling protocol. However, at the behavioral level of abstraction, we can describe the register file more simply as a shared variable of a protected type with read and write methods. The mutual exclusion afforded by the protected type ensures that reads and writes do not interfere with each other.

EXAMPLE 19.4   *A two-port register file*

Suppose we need to model a two-port register that is read using one port and written
using the other. The entity declaration is

```
entity two_port_reg is
  generic ( width : natural );
  port ( read_clk : in bit;
         read_data : out bit_vector(0 to width-1);
         write_clk : in bit;
         write_data : in bit_vector(0 to width-1) );
end entity two_port_reg;
```

In the architecture body, we represent the reading and writing behavior using
separate processes, as follows:

```
architecture behavioral of two_port_reg is

  subtype word is bit_vector(0 to width-1);

  type protected_reg is protected
    impure function get return word;
    procedure set ( new_value : word );
  end protected protected_reg;

  type protected_reg is protected body

    variable reg : word;

    impure function get return word is
    begin
      return reg;
    end function get;

    procedure set ( new_value : word ) is
    begin
      reg := new_value;
    end procedure set;

  end protected body protected_reg;

  shared variable reg_store : protected_reg;

begin
  reader : process ( read_clk ) is
  begin
    if read_clk then                  -- on rising edge
      read_data <= reg_store.get;     -- read current value
    end if;
  end process reader;
  writer : process ( write_clk ) is
  begin
```

```
        if write_clk then              -- on rising edge
          reg_store.set ( write_data ); -- set to new value
        end if;
     end process writer;
  end architecture behavioral;
```

The protected type **protected_reg** encapsulates the register data and provides **get** and **set** methods to access the storage. The shared variable **reg_store** represents the storage of the register. The two processes **reader** and **writer** implement the behavior for read-port and write-port accesses, respectively.

Another scenario in which we might use shared variables is to instrument a model. We can use a shared variable to collect information about the behavior of processes within a design over the course of a simulation run. Each process updates the variable using update methods as events of interest occur. Mutual exclusion ensures that concurrent updates do not conflict.

EXAMPLE 19.5   *Shared instrumentation in a multiprocessor computer model*

In this example we use shared variables to instrument a behavioral model of a multiprocessor computer system, shown in Figure 19.1. Each processing element (PE) has an attached level 2 (L2) cache. The shared variable cache counters is used to collect sharing statistics for blocks in the address space. Processes representing caches invoke methods to record accesses to memory blocks. The log controller process periodically writes the recorded statistics to a log file.

FIGURE 19.1

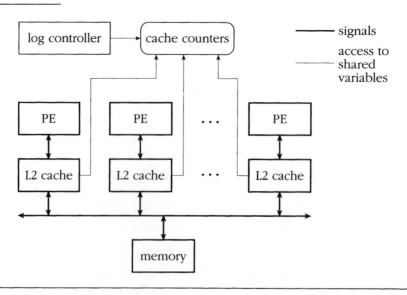

An instrumented multiprocessor computer system with caches.

The package declaration **cache_instrumentation**, shown in below, includes a protected-type declaration for the instrumentation data structure. The protected type includes methods for recording read and write misses and for writing the recorded data to a file. The package also declares a shared variable of the protected type for collecting the data.

```
package cache_instrumentation is

  use work.cache_types.all;

  type shared_counters is protected

    procedure log_read_miss ( block_number : block_range;
                              is_shared : bit );

    procedure log_write_miss ( block_number : block_range );

    procedure dump_log ( file log_file : std.textio.text );

  end protected shared_counters;

  shared variable cache_counters : shared_counters;

end package cache_instrumentation;
```

The implementations of the data structure and the methods are described in the protected-type body, which is declared in the package body:

```
package body cache_instrumentation is

  type shared_counters is protected body

    type counter_record is record   -- counters for a block
      shared_read_misses,
      private_read_misses,
      write_misses : natural;
    end record counter_record;

    type counter_array is array ( block_range ) of counter_record;

    -- instrumentation data structure
    variable counters : counter_array := (others => (0, 0, 0));

    procedure log_read_miss ( block_number : block_range;
                              is_shared : bit ) is
    begin
      if is_shared then
        counters(block_number).shared_read_misses
            := counters(block_number).shared_read_misses + 1;
      else
        counters(block_number).private_read_misses
            := counters(block_number).private_read_misses + 1;
      end if;
    end procedure log_read_miss;
```

```vhdl
    procedure log_write_miss ( block_number : block_range ) is
    begin
      counters(block_number).write_misses
          := counters(block_number).write_misses + 1;
    end procedure log_write_miss;

    procedure dump_log ( file log_file : std.textio.text ) is
      use std.textio.all;
      variable L : line;
    begin
      -- write a line of data for each block in the address space
      for block_number in block_range loop
        swrite ( L, "Block " );
        write  ( L, block_number );
        swrite ( L, ":  shared read misses = " );
        write  ( L, counters(block_number).shared_read_misses );
        . . .
        writeline ( log_file, L );
      end loop;
    end procedure dump_log;

  end protected body shared_counters;

end package body cache_instrumentation;
```

The protected-type body includes an array of records, one for each block of the multiprocessor's memory space. Each record contains counters for the different kinds of misses to be logged.

The individual caches are described by a behavioral architecture body, shown below. The cache controller process invokes the instrumentation methods when cache miss events occur.

```vhdl
architecture behavioral of cache is

  use work.cache_instrumentation.all;

  . . .

begin

  cache_controller : process is
    . . .   -- local variables used by the cache controller
  begin
    . . .
    -- if cache miss, record in the instrumentation shared variable
    if not hit then
      if read then
        cache_counters.log_read_miss ( current_block_number,
                                       block_is_shared );
      else
        cache_counters.log_write_miss ( current_block_number );
```

```vhdl
      end if;
    end if;
    ...
  end process cache_controller;

  ...

end architecture behavioral;
```

The complete multiprocessor system is described by the following architecture body:

```vhdl
architecture system of multiprocessor is

  ... -- signal declarations

begin

  PE_array : for PE_index in 0 to num_PEs - 1 generate

    -- a processing element
    PE : entity work.processor(behavioral)
      port map ( ... );

    -- and its attached level-2 cache
    L2_cache : entity work.cache(behavioral)
      port map ( ... );

    ...

  end generate PE_array;

  -- the instrumentation process that periodically dumps
  -- recorded counts to the output file
  log_controller : process is
    use work.cache_instrumentation.all;
  begin
    wait for 10 ms;
    cache_counters.dump_log ( std.textio.output );
  end process log_controller;

  ...

end architecture system;
```

The generate statement creates multiple instances of a processing element, each with an attached cache. Since each cache instance includes a cache controller process, there are multiple cache controller processes that may concurrently access the instrumentation variable in the instrumentation package. Furthermore, the multiprocessor model includes the **log_controller** process that periodically invokes the **dump_log** method. All of these processes share access to the instrumentation variable, so mutual exclusion is required to prevent interference.

An important point to note in using VHDL protected types is the potential for deadlock. It is possible to write a model in which two processes block waiting for mutual exclusion over shared variables and can never resume. To illustrate the possibility, consider an extension of the shared counter in Examples 19.1 and 19.2. Suppose we augment the protected type to include a method to copy the value from one counter to another. We declare the method in the protected type declaration as

```
procedure copy ( variable from : in shared_counter );
```

We declare the method implementation in the protected-type body as

```
procedure copy ( variable from : in shared_counter ) is
begin
  count := from.value;
end procedure copy;
```

We declare two shared variables as

```
shared variable a, b : shared_counter;
```

Now consider what might happen if process P_1 executes the statement

```
a.copy(b);
```

and process P_2 executes

```
b.copy(a);
```

A possible interleaving of execution involves P_1 acquiring access to **a** and P_2 acquiring access to **b**, before either reaches the body of the method. Note that passing a variable of protected type simply involves passing a reference; it does not involve acquiring access to the variable. When P_1 reaches the invocation of the value method within the copy method, it tries to acquire access to **b**. Since P_2 already has access to **b**, P_1 blocks. Similarly, when P_2 reaches the invocation of the value method, it tries to acquire access to **a**. P_1 already has access to **a**, so P_2 blocks. Neither process can proceed, and execution deadlocks.

While this is a contrived example, it illustrates one situation under which deadlock can arise. VHDL does not prohibit such situations, nor does it require that a simulator detect or resolve deadlock. When writing models using protected types, we must take care not to introduce the potential for deadlock. One way is to ensure that all processes acquire exclusive access to collections of variables in the same order. Of course, if processes only need exclusive access to one variable at a time, the protected-type mechanism does not deadlock.

VHDL-87

Shared variables are not provided in VHDL-87.

VHDL-93

In VHDL-93, shared variables need not be of protected types. Instead, they are like normal unshared variables. A shared variable can be declared of any type except a file type, and the declaration can include an initial value expression. The value of a shared variable can be used in an expression and can be updated using a normal variable assignment statement. The difference is that there is no mutual exclusion for access to a shared variable by different processes in a given simulation cycle. This means that read and assignment operations by different processes might interfere with one another. Hence we must take great care when using shared variables in VHDL-93 to ensure that only one process can access each shared variable in each simulation cycle. The VHDL-93 language specification deems a model to be in error if it depends on the value of a shared variable accessed by more than one process during any simulation cycle.

19.2 Uninstantiated Methods in Protected Types

We now return to a discussion of the relationship between uninstantiated subprograms and protected types. We introduced uninstantiated subprograms, those that include generic lists, in Chapter 12. We will build on our discussion in that chapter to provide motivating examples of the relationships that can occur. There are two cases to consider. The first is declaration of an instance of an uninstantiated subprogram as a method of a protected type, and the second is declaration of an uninstantiated subprogram within a protected type.

Starting with the first case, if we have an uninstantiated subprogram declared outside a protected type, and we declare an instance of the subprogram within the protected type declaration, the instance becomes a method of the protected type. The scheme is

```
procedure uninstantiated_name
   generic ( ... )
   parameter ( ... );

type PT is protected
   ...
   procedure instance_name is new uninstantiated_name
     generic map ( ... );
   ...
end protected PT;
```

We can declare a shared variable of the protected type and call the method:

```
shared variable SV : PT;
...
SV.instance_name ( ... );
```

On the face of it, there seems no purpose to this scheme. The uninstantiated subprogram, being outside the protected type, cannot refer to the items encapsulated within the protected type. So there would appear to be no reason for instantiating the subprogram in the protected type. However, we can provide controlled access to the encapsulated items via a method of the protected type provided as an actual generic subprogram to the instance. The refinement to the scheme is:

```
procedure uninstantiated_name
  generic ( ...; formal_generic_subprogram; ... )
  parameter ( ... );

type PT is protected
  method_declaration;
  procedure instance_name is new uninstantiated_name
    generic map ( ..., method_name, ... );
  ...
end protected PT;
```

In this scheme, the method has access to the encapsulated items within the protected type. When the instance invokes the actual generic subprogram, the method is called.

EXAMPLE 19.6   *Test-vector set with tracing*

Suppose we have an uninstantiated subprogram that gets a test-vector value corresponding to a specified time and that writes the vector value to the standard output file. The procedure has a formal generic subprogram representing the action to perform to get the test vector.

```
procedure trace_test_vector is
  generic ( impure function get_test_vector
                              ( vector_time : time )
                              return test_vector )
  parameter ( vector_time : time ) is

  variable vector : test_vector;
  use std.textio.all;
  variable L : line;

begin
  write(L, now);
  write(L, string'(": "));
  vector := get_test_vector(vector_time);
  ...  -- write test vector
  writeline(output, L);
end procedure trace;
```

We can declare a protected type representing a set of test vectors to be applied at various times. The protected type has a method for getting a test vector for a specific time. We include an instance of the **trace_test_vector** procedure as a method to trace a test vector from the particular set represented by a shared variable of the protected type. The protected type declaration is

```
type test_set is protected
   ...

   impure function get_vector_for_time ( vector_time : time )
                                         return test_vector;

   procedure trace_for_time is new trace_test_vector
      generic map ( get_test_vector => get_vector_for_time );

end protected test_set;
```

We might declare two shared variables of this protected type, representing two distinct sets of test vectors:

```
shared variable main_test_set, extra_test_set : test_set;
```

If we invoke the **trace_for_time** method on one of the shared variables:

```
main_test_set.trace_for_time(100 ns);
```

the instance of the **trace_test_vector** procedure invokes the actual subprogram provided for the instance of the protected type. That is, it invokes the **get_vector_for_time** method associated with the shared variable **main_test_set**. If, on the other hand, we invoke the **trace_for_time** method on the other shared variable:

```
extra_test_set.trace_for_time(100 ns);
```

the instance of the **trace_test_vector** procedure invokes the **get_vector_for_time** method associated with the shared variable **extra_test_set**. What this reveals is that each shared variable of the protected type binds its **get_vector_for_time** method, which has access to the shared variable's state, as the actual generic procedure in its instance of the **trace_test_vector** procedure. That instance, provided as a method of the shared variable, thus has indirect access to the shared variable's state.

The second case to consider is the declaration of an uninstantiated subprogram within a protected type. That uninstantiated procedure is not itself a method, since it cannot be called. However, it can be instantiated within the protected type to provide a method. Moreover, each shared variable of the protected type logically contains a separate declaration of the uninstantiated subprogram. That subprogram can be instantiated, giving a subprogram that has access to the items encapsulated in the shared variable. We will illustrate these mechanisms with an example.

EXAMPLE 19.7 *Stimulus list with visitor traversal*

For a design requiring **signed** stimulus values, we can declare a procedure for displaying a **signed** value to the standard output file, as follows:

```
procedure output_signed ( value : in signed ) is
   use std.textio.all;
```

```
    variable L : line;
begin
  write(L, value);
  writeline(output, L);
end procedure output_signed;
```

We also declare a protected type for a list of **signed** stimulus values:

```
type signed_stimulus_list is protected
  ...

  procedure traverse_with_in_parameter
    generic ( procedure visit ( param : in signed ) );

  procedure output_all is new traverse_with_in_parameter
    generic map ( visit => output_signed );

end protected signed_stimulus_list;
```

The protected type includes an uninstantiated procedure, traverse_with_in_parameter, to apply a visitor procedure to each element in the list of **signed** values. It instantiates the traversal procedure to provide a method that displays each element. We can use this protected type to declare a shared variable and then invoke the method to display its element values:

```
shared variable list1 : signed_stimulus_list;
...

list1.output_all;
```

Suppose now we want to use the traversal procedure to accumulate the sum of elements in a list so that we can calculate the average value. We can provide another action procedure and use it in a further instantiation of the traversal procedure:

```
variable sum, average : signed(31 downto 0);
variable count : natural := 0;

procedure accumulate_signed ( value : in signed ) is
begin
  sum := sum + value;
  count := count + 1;
end procedure accumulate_signed;

procedure accumulate_all_list1 is
  new list1.traverse_with_in_parameter
    generic map ( visit => accumulate_signed );
...

accumulate_all_list1;
average := sum / count;
```

In this case, the instance is a procedure declared externally to the protected type. However, since it is an instance of a subprogram defined within the shared variable list1, the instance has access to the encapsulated items within list1. The instance accumulate_all_list1 thus applies the **accumulate_signed** visitor procedure to each element within list1.

If we want to calculate the average value of any list of elements, we need to wrap these declarations up in a procedure that has a shared variable as a parameter. That includes declaring the instance of the traversal procedure within the outer procedure. The complete procedure would be

```
procedure calculate_average
  ( variable list : inout signed_stimulus_list
    variable average : out signed ) is

  variable sum : signed(average'range);
  variable count : natural := 0;

  procedure accumulate_signed ( value : in signed ) is
  begin
    sum := sum + value;
    count := count + 1;
  end procedure accumulate_signed;

  procedure accumulate_all is
    new list.traverse_with_in_parameter
      generic map ( visit => accumulate_signed );

begin
  accumulate_all;
  average := sum / count;
end procedure calculate_average;
```

In this case, the instance of the traversal procedure is also declared externally to the protected type. However, it is an instance of the subprogram defined within the shared variable list provided as a parameter to the **calculate_average** procedure. Logically, each time the **calculate_average** procedure is called, a new instance of the traversal procedure is defined particular to the actual shared variable provided as the parameter. The instance thus applies the local **accumulate_signed** visitor procedure to each element within the actual shared variable.

Exercises

1. [❶ 19.1] Suppose, using VHDL-93, we have a shared variable declared as follows in a package instrumentation:

```
shared variable multiply_counter : natural := 0;
```

We have two instances, m1 and m2, of a behavioral multiplier model that includes the statement

```
instrumentation.multiply_counter
   := instrumentation.multiply_counter + 1;
```

Show how, in VHDL-93, the variable may be updated incorrectly if we allow the two instances to access the variable in the same simulation cycle.

2. [❶ 19.1] Write a protected type declaration and body that provides mutual exclusion for an integer shared variable. The protected type should include a method called **set** to update the variable value and a method called **get** to retrieve the variable value.

3. [❸ 19.1] The bounded buffer ADT in Section 12.3.2 could be used to send a stream of bytes from one process to another. However, if the sender called the **write** procedure at the same time as the receiver called the **read** procedure, they may interfere and cause corruption of the state of the buffer. Develop a package that defines a protected type for a protected bounded buffer. The protected-type body should encapsulate a variable of the bounded buffer ADT from Section 12.3.2. The protected type should include methods to test whether the buffer is full, whether it is empty, and to read and write bytes. Test your package in an architecture body with producer and consumer processes that communicate using a shared variable of the protected bounded buffer type.

4. [❸ 19.1] We can use a protected type to encapsulate the private data of an abstract data type in a way that hides the data from the user of the type. For example, we can declare a type for random number streams as follows:

```
type random_generator is protected
  impure function next_random return real;
  impure function next_random ( min, max : real )
                               return real;
  impure function next_random ( min, max : integer )
                               return integer;
  ...
end protected random_generator;
```

The protected type body includes variable declarations for storing the random seeds. The methods use the seeds in conjunction with the **math_real.uniform** procedure to generate successive random numbers, with each method scaling and converting the type of the result.

Develop a package that completes the definition of this protected type, including methods that return values of types **bit_vector** and **std_ulogic_vector**. Use your package in a model that generates random stimulus value for a 32-bit adder under test.

Chapter 20

Attributes and Groups

VHDL provides comprehensive facilities for expressing the behavior and structure of a design. VHDL also provides the *attribute* mechanism for annotating a model with additional information. In this chapter, we review the predefined attributes and show how to define new attributes. We also look at the group mechanism, which allows us to describe additional relationships between various items in a model.

20.1 Predefined Attributes

Throughout this book we have seen predefined attributes that are used to retrieve information about types, objects and other items within a model. In this section we summarize the previously introduced attributes and fully describe the remaining predefined attributes.

VHDL-87

VHDL-87 provides two additional attributes beyond those described in the rest of this section. The attributes 'behavior and 'structure can be applied to the names of architecture bodies and that return a Boolean result. The 'behavior attribute indicates whether the architecture body is a behavioral description. It returns true if the architecture body contains no component instantiation statements. The 'structure attribute indicates whether the architecture body is a structural description. It returns true if the architecture body contains only component instantiations and passive processes. If both attributes are false, the architecture body is a mix of behavioral and structural modeling constructs.

20.1.1 Attributes of Scalar Types

The first group of predefined attributes gives information about the values in a scalar type. These were introduced in Chapter 2 and are summarized in Table 20.1.

TABLE 20.1   *The predefined attributes giving information about values in a type*

Attribute	Type of T	Result type	Result
T'left	Any scalar type or subtype	Same as T	Leftmost value in T
T'right	"	"	Rightmost value in T
T'low	"	"	Least value in T
T'high	"	"	Greatest value in T
T'ascending	"	boolean	True if T is an ascending range, false otherwise
T'image(x)	"	string	A textual representation of the value x of type T
T'value(s)	"	base type of T	Value in T represented by the string s
T'pos(s)	Any discrete or physical type or subtype	universal integer	Position number of x in T
T'val(x)	"	Base type of T	Value at position x in T
T'succ(x)	"	"	Value at position one greater than x in T
T'pred(x)	"	"	Value at position one less than x in T
T'leftof(x)	"	"	Value at position one to the left of x in T
T'rightof(x)	"	"	Value at position one to the right of x in T

VHDL-87

The predefined attributes **'ascending**, **'image** and **'value** are not provided in VHDL-87.

20.1.2 Attributes of Array Types and Objects

The second group of predefined attributes gives information about the index values of an array object or type. These were introduced in Chapter 4 and are summarized in Table 20.2. The prefix A in the table refers either to an array type or subtype whose index ranges are defined, to an array object, or to a slice of an array. If A is a variable of an access type pointing to an array object, the attribute refers to the array object, not the pointer value. Each of the attributes optionally takes an argument that selects one of the index dimensions of the array. The default is the first dimension. Note that if the prefix **A** is an alias for an array object, the attributes return information about the index values declared for the alias, not those declared for the original object.

TABLE 20.2 *The predefined attributes giving information about the index range of an array*

Attribute	Result
A'left(n)	Leftmost value in index range of dimension n
A'right(n)	Rightmost value in index range of dimension n
A'low(n)	Least value in index range of dimension n
A'high(n)	Greatest value in index range of dimension n
A'range(n)	Index range of dimension n
A'reverse_range(n)	Index range of dimension n reversed in direction and bounds
A'length(n)	Length of index range of dimension n
A'ascending(n)	True if index range of dimension n is ascending, false otherwise

VHDL-87

The predefined attribute **'ascending** is not provided in VHDL-87.

20.1.3 Attributes Giving Types

The third group of predefined attributes, summarized in Table 20.3, provides type information. The table describes the kind of prefix to which each attribute can be applied.

TABLE 20.3 *The predefined attributes giving type information*

Attribute	Prefix	Result
T'base	Any type or subtype	The base type of T, for use only as prefix of another attribute
O'subtype	Any object or alias of an object	The fully constrained subtype of O, including constraints defining index ranges (if O is an array or has elements that are arrays)
A'element	Any array type, subtype, or object	If A is an array type or subtype, the element subtype of A. If A is an arrray object, the element subtype of A including constraints defining all index ranges

VHDL-87, -93, and -2002

The attributes 'subtype and 'element are not provided in these versions.

20.1.4 Attributes of Signals

The third group of predefined attributes gives information about signals or defines new implicit signals derived from explicitly declared signals. These attributes were introduced in Chapters 5 and 8 and are summarized in Table 20.4. The prefix S in the table refers to any statically named signal. Three of the attributes optionally take a non-negative argument t of type time. The default is 0 fs.

TABLE 20.4   *The predefined attributes giving information about signals and values of signals*

Attribute	Result type	Result
S'delayed(t)	base type of S	Implicit signal, with the same value as S, but delayed by t time units (t $\geq$ 0 ns)
S'stable(t)	boolean	Implicit signal, **true** when no event has occurred on S for t time units, **false** otherwise (t $\geq$ 0 ns)
S'quiet(t)	boolean	Implicit signal, **true** when no transaction has occurred on S for t time units, **false** otherwise (t $\geq$ 0 ns)
S'transaction	bit	Implicit signal, changes value in simulation cycles in which a transaction occurs on S
S'event	boolean	True if an event has occurred on S in the current simulation cycle, **false** otherwise
S'active	boolean	True if a transaction has occurred on S in the current simulation cycle, **false** otherwise
S'last_event	time	Time since last event occurred on S, or **time'high** if no event has yet occurred
S'last_active	time	Time since last transaction occurred on S, or **time'high** if no transaction has yet occurred
S'last_value	base type of S	Value of S before last event occurred on it
S'driving	boolean	True if the containing process is driving S (or every element of a composite signal S), or **false** if the containing process has disconnected its driver for S (or any element of S) with a null transaction
S'driving_value	base type of S	Value contributed by driver for S in the containing process

VHDL-87

The predefined attributes **'driving** and **'driving_value** are not provided in VHDL-87. Note also that the **'last_value** attribute for a composite signal returns the aggregate of last values for each of the scalar elements of the signal. This behavior is different from the VHDL-93 and VHDL-2002 behavior, in which the attribute returns the last value of the entire composite signal. Furthermore, the behavior of the **'last_event** and **'last_active** attributes differs from VHDL-93 and VHDL-2002. In VHDL-87, **'last_event** returns 0 ns if no event has yet occurred, and **'last_active** returns 0 ns if no transaction has yet occurred.

20.1.5 Attributes of Named Items

The remaining predefined attributes are applied to any declared item and return a string representation of the name of the item. These attributes are summarized in Table 20.5. The prefix **X** in the table refers to any declared item. If the item is an alias, the attribute returns the name of the alias itself, not the aliased item.

TABLE 20.5 *The predefined attributes that provide names of declared items*

Attribute	Result
X'simple_name	A string representing the identifier, character or operator symbol defined in the declaration of the item X
X'path_name	A string describing the path through the elaborated design hierarchy, from the top-level entity or package to the item X
X'instance_name	A string similar to that produced by X'path_name, but including the names of the entity and architecture bound to each component instance in the path

The **'simple_name** attribute returns a string representation of the name of an item. For example, if a package **utility_definitions** in a library utilities declares a constant named **word_size**, the attribute

```
utilities.utility_definitions.word_size'simple_name
```

returns the string "**word_size**". We might ask why VHDL provides this attribute, since we need to write the simple name of the item in order to apply the attribute. It would be simpler to write the string literal directly. If nothing else, we can use the attribute to gain consistency of style in reporting item names in messages, since the **'simple_name** attribute always returns a lowercase version of the name.

The **'path_name** and **'instance_name** attributes both return string representations of the path through the design hierarchy to an item. They are especially useful in assertion or report statements to pinpoint exactly which instance of a library unit is the source of a message. VHDL only requires that the message reported to the user by these statements indicate the name of the library unit (entity, architecture body or package) containing the statement. We can use the 'path_name or 'instance_name attribute to determine which particular instance of a process in the design hierarchy is the source of a message.

EXAMPLE 20.1 *Using name attributes in assertion messages*

Suppose we have a design that includes numerous instances of a flipflop component bound to an entity **flipflop** and using an architecture **behavior**. Within this architecture we wish to include timing checks and report an error message if the constraints are violated. An outline of the architecture body incorporating these checks is

```
architecture behavior of flipflop is
begin

  timing_check : process (clk) is
  begin
    if clk = '1' then
      assert d'last_event >= Tsetup
        report "set up violation detected in "
                & timing_check'path_name
        severity error;
    end if;
  end process timing_check;

  ... -- functionality

end architecture behavior;
```

When a flipflop instance in the design detects a timing violation, it will issue an assertion violation message indicating that the problem arose in the architecture **behavior** of **flipflop**. We use the 'path_name attribute in the message string to identify which component instance bound to the **flipflop** entity is the one responsible for issuing the message.

The format of the string produced by the 'path_name and 'instance_name attributes for a library, a design-unit package, or an item declared in a design-unit package is described by the EBNF rule

package_based_path ⟸
    : *library*_logical_name :
        { (*subprogram*_designator signature
            ▯ *variable*_identifier
            ▯ *package*_identifier) : }
        [identifier ▯ character_literal ▯ operator_symbol]

The colon characters serve as punctuation, separating elements within the path string. If the item to which the attribute is applied is a library, the path string includes only the library name. If the item is a design-unit package, the path string includes the library name and the package name. If the item is declared in a package, the path string includes the library name, the package name and the name of the item. If the item is nested within a subprogram, shared variable, or local package in the design-unit package, the string also includes the names of the containing subprogram, shared variable, or local package. The signature of each subprogram is also included to distinguish between possibly overloaded versions of the subprogram name. Recall that the syntax rule for a subprogram signature is

signature $\Leftarrow$ [[type_mark { , ... }] [**return** type_mark]]

If the item is further nested, the names of all the enclosing subprograms, shared variables, or local packages are included in order from outermost to innermost.

Note that, for subprograms that overload operators, the name included in the path string is the operator symbol surrounded by quotation marks. For example, an overloaded **abs** operator declared in a package **pkg** in a library **lib** would have the path string:

```
:lib:pkg:"abs":
```

EXAMPLE 20.2 *Name attributes for items in a package*

Suppose we have a package **mem_pkg** stored in the library **project**. The package declaration is

```
package mem_pkg is

  subtype word is bit_vector(0 to 31);
  type word_array is array (natural range <>) of word;

  procedure load_array ( words : out word_array;
                         file_name : string );

end package mem_pkg;
```

The 'path_name attribute applied to these items gives the following results:

```
mem_pkg'path_name = ":project:mem_pkg:"
word'path_name = ":project:mem_pkg:word"
word_array'path_name = ":project:mem_pkg:word_array"
load_array'path_name = ":project:mem_pkg:load_array:"
```

The 'instance_name attribute returns the same strings for these items. An outline of the package body is

```
package body mem_pkg is

  procedure load_array ( words : out word_array;
                         file_name : string ) is
    -- words'path_name = ":project:mem_pkg:load_array:words"

    use std.textio.all;
    file load_file : text open read_mode is file_name;
```

```
  -- load_file'path_name
  --    = ":project:mem_pkg:load_array:load_file"
procedure read_line is
-- read_line'path_name
--    = ":project:mem_pkg:load_array:read_line:"
  variable current_line : line;
  -- current_line'path_name
  --    = ":project:mem_pkg:load_array:read_line:current_line"
begin
  ...
end procedure read_line;

begin   -- load_array
  ...
end procedure load_array;

end package body mem_pkg;
```

The comments indicate the values of the **'path_name** attribute applied to various names within the package body. Again, the **'instance_name** attribute returns the same strings as the **'path_name** attribute.

In the case of a shared variable, it is the name of the variable that is included in the path string, not the name of the protected type used to declare the variable. This allows us to distinguish between items declared in a protected type but occurring in different shared variables.

EXAMPLE 20.3   *Path name for items in shared variables*

Suppose we have a package in library **project_lib** that declares a protected type, as follows:

```
package counter_pkg is

  type counter is protected
    procedure increment;
    ...
  end protected counter;

end package counter_pkg;

package body counter_pkg is

  type counter is protected body

    constant limit : natural := 100;
    variable count : natural := 0;

    procedure increment is
    begin
```

```
         assert count < limit
           report "Counter overflow in " & increment'path_name;
         count := (count + 1) mod limit;
       end procedure increment;

       ...

     end protected body counter;

 end package body counter_pkg;
```

We now declare a package with two shared variables of the protected type:

```
library project_lib;
package system_counter_pkg is
  use project_lib.counter_pkg.all;
  shared variable test_counter, event_counter : counter;
end package system_counter_pkg;
```

If, as a result of system operation, the assertion fails during execution of the method call **test_counter.increment**, the assertion message includes the path string

```
:work:system_counter_pkg:test_counter:increment:
```

This allows us to identify which instance of the protected type caused the assertion violation.

If an item is declared within an entity or architecture body, the **'path_name** and **'instance_name** attributes return different strings depending on the structure of the elaborated design and the location of the declared item within the design hierarchy. We first look at the string returned by the **'path_name** attribute, as it is the simpler of the two. The format of the string is described by the EBNF rules

instance_based_path ⇐
   : ⟦ path_instance_element : ⟧
      ⟦ simple_name ▯ character_literal ▯ operator_symbol ⟧

path_instance_element ⇐
   *entity*_identifier
   ▯ *component_instantiation*_label
   ▯ *block*_label
   ▯ *generate*_label ⟦ (literal) ⟧
   ▯ ⟦ *process*_label ⟧
   ▯ *subprogram*_designator signature
   ▯ ⟦ *loop*_label ⟧
   ▯ *variable*_identifier
   ▯ *package*_identifier

The string starts with the name of the topmost entity in the design and continues with the labels of any blocks (see Chapter 23), generate statements, processes, subprograms, and so on, between the top and the item. If the design hierarchy includes a component instance bound to an entity and architecture body containing the item, the attribute string includes the label of the component instantiation statement. If the item is contained within a for-generate statement, the string includes the value of the generate parameter for the particular iteration containing the item. The value is included in parentheses after the generate statement label. If the item is included in a process that has no label, the string includes an empty element in place of a process label. If the item is included in a subprogram, the string includes the signature of the subprogram to distinguish between possibly overloaded versions of the subprogram name. If the item is the loop parameter of a for loop, the string includes the label of the loop, or an empty element if the loop is unlabeled. If the item is declared in a protected type used as the type of a variable, the string includes the name of the variable (not the name of the protected type). If the item is included within a locally declared package, the string includes the package name.

The format of the string returned by the **'instance_name** attribute is described by the EBNF rules

full_instance_based_path ⇐
    : ⟦ full_path_instance_element : ⟧
        ⟦ simple_name ⎮ character_literal ⎮ operator_symbol ⟧

full_path_instance_element ⇐
    *entity*_identifier (*architecture*_identifier)
    ⎮ *component_instantiation*_label
        @ *entity*_identifier (*architecture*_identifier)
    ⎮ *block*_label
    ⎮ *generate*_label ⟦ (literal) ⟧
    ⎮ ⟦ *process*_label ⟧
    ⎮ *subprogram*_designator signature
    ⎮ ⟦ *loop*_label ⟧
    ⎮ *variable*_identifier
    ⎮ *package*_identifier

It is the same as that returned by **'path_name**, except that the names of the entity and architecture bound to a component instance are included after the label of the component instantiation statement. Furthermore, the architecture name for the top-level design entity is also included.

EXAMPLE 20.4   *Name attributes in a design hierarchy*

We illustrate the results returned by the **'path_name** and **'instance_name** attributes by looking at a sample design hierarchy. The top level of the hierarchy is formed by the entity **top** and its corresponding architecture **top_arch**, declared as follows:

```
entity top is
end entity top;

--------------------------------------------------

architecture top_arch of top is
```

```
            signal top_sig : ...;                    -- 1

        begin

            stimulus : process is
                variable var : ...;                  -- 2
            begin
                ...
            end process stimulus;

            rep_gen : for index in 0 to 7 generate
            begin

                end_gen : if index = 7 generate
                    signal end_sig : ...;            -- 3
                begin
                    ...
                end generate end_gen;

                other_gen : if index /= 7 generate
                    signal other_sig : ...;          -- 4
                begin
                    other_comp : entity work.bottom(bottom_arch)
                        port map ( ... );
                end generate other_gen;

            end generate rep_gen;

        end architecture top_arch;
```

The numbered comments in this model mark points at which various declared items are visible. The values of the 'path_name and 'instance_name attributes of these items at the marked points are shown in Table 20.6. At point 4, the string returned varies between repetitions created by the generator. Where the table shows *index* in

TABLE 20.6   *The results of applying the path and instance name attributes at the top level*

Point	Item	Item'path_name *and* item'instance_name
1	top	:top: :top(top_arch):
1	top_sig	:top:top_sig :top(top_arch):top_sig
2	stimulus	:top:stimulus: :top(top_arch):stimulus:
2	var	:top:stimulus:var :top(top_arch):stimulus:var
3	end_sig	:top:rep_gen(7):end_gen:end_sig :top(top_arch):rep_gen(7):end_gen:end_sig
4	other_sig	:top:rep_gen(index):other_gen:other_sig :top(top_arch):rep_gen(index):other_gen:other_sig

the attribute value, the value of the generate parameter for that repetition is substituted. For example, in the repetition with the generate parameter set to 4, the result of other_sig'path_name is ":top:rep_gen(4):other_gen:other_sig".

The entity declaration and architecture body for the bottom level of the design hierarchy, instantiated in the preceding architecture body, are

```
entity bottom is
  port ( ... );
end entity bottom;

-----------------------------------------------------

architecture bottom_arch of bottom is

  signal bot_sig : ...;                    -- 5

  procedure proc ( ... ) is
    variable v : ...;                      -- 6
  begin
    ...
  end procedure proc;

begin
  delays : block is
    constant d : integer := 1;             -- 7
  begin
    ...
  end block delays;

  func : block is
  begin

    process is
      variable v : ...;                    -- 8
    begin
      ...
    end process;

  end block func;

end architecture bottom_arch;
```

The architecture includes block statements, which we describe in detail in Chapter 23. Nonetheless, we include them here to illustrate the form of path strings involving them. The values of the 'path_name and 'instance_name attributes of items within this architecture at the marked points are shown in Table 20.7. The values shown are for the instance of the architecture corresponding to the component instantiation statement in the repetition of rep_gen with index set to 4. Point 8 is within a process that has no label, so the strings returned for the item v include an empty element (two consecutive colon characters) where the process label would otherwise be.

TABLE 20.7   *The results of applying the path and instance name attributes at the bottom level*

Point	Item	*Item*'path_name *and item*'instance_name
5	bot_sig	:top:rep_gen(4):other_gen:other_comp:bot_sig :top(top_arch):rep_gen(4):other_gen:other_comp@bottom(bottom_arch):bot_sig
6	v	:top:rep_gen(4):other_gen:other_comp:proc:v :top(top_arch):rep_gen(4):other_gen:other_comp@bottom(bottom_arch):proc:v
7	d	:top:rep_gen(4):other_gen:other_comp:delays:d :top(top_arch):rep_gen(4):other_gen:other_comp@bottom(bottom_arch):delays:d
8	v	:top:rep_gen(4):other_gen:other_comp:func::v :top(top_arch):rep_gen(4):other_gen:other_comp@bottom(bottom_arch):func::v

VHDL-93 and -2002

In these versions, the path element for a subprogram was officially an identifier, so an overloaded operator symbol could not be represented. This was an error of specification in the VHDL standard, so most implementations would have assumed some representation for the operator symbol. Also, in these versions, there was no provision for names of variables (shared or otherwise) of protected types or for loop labels in path strings. Finally, since these versions do not allow locally declared packages, there is no provision for package names other than as design-unit packages in package-based paths.

VHDL-93

The signature for a subprogram is not included in the 'path_name and 'instance_name attributes in VHDL-93. Furthermore, the specification of these attributes in the VHDL-93 *Language Reference Manual* is ambiguous and contains contradictory examples. As a consequence, the results returned by the attributes is somewhat implementation dependent.

VHDL-87

The predefined attributes 'simple_name, 'path_name and 'instance_name are not provided in VHDL-87.

20.2 User-Defined Attributes

The predefined attributes provide information about types, objects or other items in a VHDL model. VHDL also provides us with a way of adding additional information of our own choosing to items in our models, namely, through user-defined attributes. We can use them to add physical design information such as standard cell allocation and placement, layout constraints such as maximum wire delay and inter-wire skew or information for synthesis such as encodings for enumeration types and hints about resource allocation. In general, information of a non-structural and non-behavioral nature can be added using attributes and processed using software tools operating on the design database.

20.2.1 Attribute Declarations

The first step in defining an attribute is to declare the name and type of an attribute, using an *attribute declaration*. The syntax rule describing this is

attribute_declaration ⇐ **attribute** identifier : type_mark ;

An attribute declaration simply defines the identifier as representing a user-defined attribute that can take on values from the specified type. The type can be any VHDL type except an access, file or protected type or a composite type with a subelement that is an access, file or protected type. Some examples of attribute declarations are

```
attribute cell_name : string;
attribute pin_number : positive;
attribute max_wire_delay : delay_length;
attribute encoding : bit_vector;
```

The attribute type need not be a simple scalar. For example, we might define an attribute to represent cell placement as follows:

```
type length is range 0 to integer'high
  units nm;
    um = 1000 nm;
    mm = 1000 um;
    mil = 25400 nm;
  end units length;

type coordinate is record
    x, y : length;
  end record coordinate;

attribute cell_position : coordinate;
```

20.2.2 Attribute Specifications

Once we have defined an attribute name and type, we then use it to decorate items within a design. We write *attribute specifications*, nominating items that take on the attribute with particular values. The syntax rules for an attribute specification are

attribute_specification ⇐
  attribute identifier **of** entity_name_list : entity_class **is** expression ;

entity_name_list ⇐
  ⟦ ⟨ ⟨ identifier ⟦ character_literal ⟦ operator_symbol ⟩ ⟦ signature ⟧ ⟩ ⟨ , ⁙ ⟩
  ⟦ **others**
  ⟦ **all**

entity_class ⇐

entity	⟦ **architecture**	⟦ **configuration**	⟦ **package**
⟦ **procedure**	⟦ **function**	⟦ **type**	⟦ **subtype**
⟦ **constant**	⟦ **signal**	⟦ **variable**	⟦ **file**
⟦ **component**	⟦ **label**	⟦ **literal**	⟦ **units**
⟦ **group**	⟦ **property**	⟦ **sequence**	

The first identifier in an attribute specification is the name of a previously declared attribute. The items to be decorated with this attribute are listed in the entity name list. Note that we use the term "entity" here to refer to any item in the design, not to be confused with an entity interface defined in an entity declaration. We adopt this terminology to remain consistent with the *VHDL Language Reference Manual*, since you may need to refer to it occasionally. However, we use the term as little as possible, preferring instead to refer to "items" in the design, to avoid confusion. The items to be decorated with the attribute are those named items of the particular kind specified by the "entity" class. The list of classes shown covers every kind of item we can name in a VHDL description, so we can decorate any part of a design with an attribute. Finally, the actual value for the attribute of the decorated items is the result of the expression included in the attribute specification. Here are some examples of attribute specifications using the attributes defined earlier:

```
attribute cell_name of std_cell : architecture is "DFF_SR_QQNN";
attribute pin_number of enable : signal is 14;
attribute max_wire_delay of clk : signal is 50 ps;
attribute encoding of idle_state : literal is b"0000";
attribute cell_position of the_fpu : label is ( 540 um, 1200 um );
```

In the case of an attribute declared to be of a composite type, we can write the attribute value in the form of an aggregate or string or bit-string literal. The type can be an unconstrained or partially constrained subtype, in which any index ranges not defined by the subtype are determined from the attribute value. For example, if we declare an attribute of a composite type:

```
type string_vector is array (positive range <>) of string;
attribute key_vector : string_vector;
```

we can decorate an item with the attribute as follows:

```
attribute key_vector of e : entity is
  ("66A6D 7DF3A 88CE1 8DEEB", "012BD 2BEE9 98634 93FE1");
```

Since the subtype for the attribute specifies index ranges in neither the top-level nor the element position, the corresponding index ranges of the aggregate are used to determine the index ranges for the attribute value, giving the ranges 1 to 2 for the top level and 1 to 23 for each element.

We now look at how attribute values may be specified for each of the classes of items shown in the syntax rule. For most classes of items, an attribute specification must appear in the same group of declarations as the declaration for the item being decorated. However, the first three classes shown in the syntax rule are design units that are placed in a design library as library units when analyzed. They are not declared within any enclosing declarative part. Instead, we can consider them as being declared in the context of the design library. The same also applies to packages that are declared as design units, as opposed to being declared locally within a design unit. However, this presents a problem if we wish to decorate an item of one of these classes with an attribute. For entities, architectures, configurations and packages, we solve this problem by placing the attribute specification in the declarative part of the design unit itself. For example, we decorate an architecture **std_cell** with the **cell_name** attribute as follows:

```
architecture std_cell of flipflop is

  attribute cell_name of std_cell : architecture is "DFF_SR_QQNN";

  ...   -- other declarations

begin
  ...
end architecture std_cell;
```

In the case of packages, this rule applies whether a package is declared as a design unit or locally. The attribute specification must be included in the package declaration, not the package body. For example, we can decorate a package **model_utilities** with the **optimize** attribute as follows:

```
package model_utilities is

  attribute optimize : string;
  attribute optimize of model_utilities : package is "level_4";

  ...

end package model_utilities;
```

When we decorate subprograms we may need to distinguish between several overloaded versions. The syntax rule on page 617 shows that we can include a signature to identify one version uniquely by specifying the types of its parameters and return value. Signatures were introduced in Chapter 11.

EXAMPLE 20.5   *Decorating a subprogram*

If we have two overloaded versions of the procedure **add_with_overflow** declared in a process as shown below, we can decorate them using signatures in the attribute specification.

```
process is
    procedure add_with_overflow ( a, b : in integer;
                                  sum : out integer;
                                  overflow : out boolean ) is ...

    procedure add_with_overflow ( a, b : in bit_vector;
                                  sum : out bit_vector;
                                  overflow : out boolean ) is ...

    attribute built_in : string;

    attribute built_in of
        add_with_overflow [ integer, integer,
                              integer, boolean ] : procedure is
        "int_add_overflow";

    attribute built_in of
        add_with_overflow [ bit_vector, bit_vector,
                              bit_vector, boolean ] : procedure is
        "bit_vector_add_overflow";

begin
    ...
end process;
```

The syntax rule also shows that we can identify an overloaded operator by writing the operator symbol as the function name. For example, if we declare a function to concatenate two lists of stimulus vectors:

```
function "&" ( a, b : stimulus_list ) return stimulus_list;
```

we can decorate it with an attribute as follows:

```
attribute debug : string;
attribute debug of "&" [ stimulus_list, stimulus_list
                          return stimulus_list ] : function is
                "source_statement_step";
```

The syntax rules for attribute specifications show the signature to be optional, and indeed, we can omit it when decorating subprograms. In this case, the attribute specification applies to all subprograms with the given name and class declared in the same declarative part as the attribute specification. For example, if we declare the following overloaded subprograms:

```
procedure add ( a, b : in integer; s : out integer );
procedure add ( a, b : in real; s : out real );
function add ( a, b : integer ) return integer;
function add ( a, b : real ) return real;
```

and write and attribute declaration and specifications:

```
attribute built_in : boolean;
atribute built_in of add : procedure is true;
attribute built_in of add : function is false;
```

the two procedures are decorated with the attribute value **true**, and the two functions are decorated with the attribute value **false**.

We can decorate a type, subtype or data object (a constant, variable, signal or file) by including an attribute specification after the declaration of the item. The attribute specification must appear within the same declarative part as the declaration of the item. For example, if we declare a resolved subtype resolved_mvl:

```
type mvl is ('X', '0', '1', 'Z');
type mvl_vector is array ( integer range <>) of mvl;
function resolve_mvl ( drivers : mvl_vector ) return mvl;

subtype resolved_mvl is resolve_mvl mvl;
```

we can decorate it as follows:

```
type builtin_types is (builtin_bit, builtin_mvl, builtin_integer);
attribute builtin : builtin_types;

attribute builtin of resolved_mvl : subtype is builtin_mvl;
```

Generics and ports in the interface of an entity can be decorated with attributes. Generic constants are of **constant** class, generic types are of **type** class, generic subprograms are of **procedure** or **function** class, generic packages are of **package** class, and ports are of **signal** class. The interface list is considered to be in the declarative part of the entity. Hence, we write attribute specifications for generics and ports in the declarative part of the entity.

EXAMPLE 20.6   *Decorating generics and ports of an entity*

Suppose the package physical_attributes declared the following attributes:

```
attribute layout_ignore : boolean;
attribute pin_number : positive;
```

We can declare an entity with decorated generic constants and ports as follows:

```
library ieee;  use ieee.std_logic_1164.all;
use work.physical_attributes.all;
```

```
entity \74x138\ is
  generic ( Tpd : time );
  port ( en1, en2a_n, en2b_n : in std_ulogic;
         s0, s1, s2 : in std_ulogic;
         y0, y1, y2, y3, y4, y5, y6, y7 : out std_ulogic );

  attribute layout_ignore of Tpd : constant is true;

  attribute pin_number of s0 : signal is 1;
  attribute pin_number of s1 : signal is 2;
  attribute pin_number of s2 : signal is 3;
  attribute pin_number of en2a_n : signal is 4;
  ...

end entity \74x138\;
```

Subprogram parameters can also be decorated with attributes. The class is specified or implied in the interface list of the subprogram. We write the attribute specifications for subprogram parameters in the declarative part of the subprogram. Similarly, for uninstantiated subprograms, we can decorate the generics by writing attribute specifications in the declarative part. For uninstantiated packages, we can decorate the generics by writing attribute specification in the package declaration (not the package body). In both cases, the classes of generics are as described above for generics of entities.

EXAMPLE 20.7   *Decorating parameters of a subprogram*

The following procedure has three parameters of different classes. Attribute specifications for the parameters are included in the declarative part of the procedure.

```
procedure mem_read ( address : in natural;
                     result : out byte_vector;
                     signal memory_bus : inout ram_bus ) is

  attribute trace of address : constant is "integer/hex";
  attribute trace of result : variable is "byte/multiple/hex";
  attribute trace of memory_bus : signal is
                      "custom/command=rambus.cmd";
  ...

begin
  ...
end procedure mem_read;
```

We can decorate a component in a model by including an attribute specification along with the component declaration. An important point to realize is that the attribute decorates the template defined by the component declaration. It does not decorate component instances that use that template.

EXAMPLE 20.8   *Decorating a component declaration*

The package below includes a component declaration for an and gate. The package imports two attributes, **graphic_symbol** and **graphic_style**, from a second package **graphics_pkg** in the library **graphics** and decorates the component template with each of these attributes.

```
library ieee;  use ieee.std_logic_1164.all;
library graphics;

package gate_components is

  use graphics.graphics_pkg.graphic_symbol,
      graphics.graphics_pkg.graphic_style;

  component and2 is
    generic ( prop_delay : delay_length );
    port ( a, b : in std_ulogic;  y : out std_ulogic );
  end component and2;

  attribute graphic_symbol of and2 : component is "and2";
  attribute graphic_style of and2 : component is
                                "color:default, weight:bold";

  . . .

end package gate_components;
```

If we wish to decorate a component instance or any other concurrent statement with an attribute, we do so by decorating the label of the statement. The label is implicitly declared in the declarative part of the architecture or block containing the concurrent statement. Hence, we place the attribute specification in that declarative part.

EXAMPLE 20.9   *Decorating a component instance*

We might decorate a component instance in an architecture body with an attribute describing cell placement as follows:

```
architecture cell_based of CPU is

  component fpu is
    port ( ... );
  end component;

  use work.cell_attributes.all;

  attribute cell_position of the_fpu : label is
                                ( 540 um, 1200 um );

  . . .
```

```
begin
  the_fpu : component fpu
    port map ( ... );

  ...

end architecture cell_based;
```

We can decorate sequential statements within a process or a subprogram in a similar way. The syntax rules for sequential statements show that each kind of sequential statement may be labeled. We decorate a sequential statement by specifying an attribute for the label. We place the attribute specification in the declarative part of the process or subprogram containing the sequential statement.

EXAMPLE 20.10   *Decorating a sequential statement*

If we wish to decorate a loop statement in a process with the attribute synthesis_hint, we do so as follows:

```
controller : process is

  attribute synthesis_hint of control_loop : label is
                               "implementation:FSM(clk)";
  ...

begin
  ...  -- initialization
  control_loop : loop
    wait until clk = '1';
    ...
  end loop;
end process controller;
```

When we introduced aliases and signatures in Chapter 11, we mentioned that enumeration literals can be thought of as functions with no parameters that return values of their enumeration types. We can take the same approach when decorating enumeration literals with attributes, in order to distinguish between literals of the same name from different enumeration types.

EXAMPLE 20.11   *Decorating an enumeration literal*

If we have two enumeration types declared as

```
type controller_state is (idle, active, fail_safe);
type load_level is (idle, busy, overloaded);
```

we can decorate the literals of type controller_state as follows:

```
attribute encoding of
  idle [ return controller_state ] : literal is b"00";
attribute encoding of
  active [ return controller_state ] : literal is b"01";
attribute encoding of
  fail_safe [ return controller_state ] : literal is b"10";
```

The signature associated with the literal **idle** indicates that it is of type controller_state, not load_level. As with attribute specifications for subprograms, if a signature is not included for a literal, all literals of the given name declared in the same declarative part as the attribute specification are decorated with the attribute.

When we declare a physical type we introduce a primary unit name and possibly a number of secondary unit names. Each of the unit names is a declared item and so may be decorated with attributes.

EXAMPLE 20.12   *Decorating a physical unit*

The package below defines a physical type **voltage**. It also declares an attribute, **resolution**, and decorates each of the units of **voltage** with this attribute.

```
package voltage_defs is

  type voltage is range -2e9 to +2e9
    units
      nV;
      uV = 1000 nV;
      mV = 1000 uV;
      V = 1000 mV;
    end units voltage;

  attribute resolution : real;

  attribute resolution of nV : units is 1.0;
  attribute resolution of uV : units is 0.01;
  attribute resolution of mV : units is 0.01;
  attribute resolution of V : units is 0.001;

end package voltage_defs;
```

If we embed PSL code in a VHDL model, we can decorate declared properties and sequences. For example:

```
property SingleCycleRequest is
  always req -> next not req;
```

```
sequence ReadCycle is
  { ba; {bb[*]} && {ar[->]; dr[->]}; not bb };

attribute enable_heuristics of
            SingleCycleRequest : propery is true;
attribute enable_heuristics of ReadCycle : sequence is true;
```

The one remaining class of items that can be decorated with attributes is groups. We introduce groups in the next section and show examples of decorated groups.

If we return to the syntax rules for attribute specifications, shown on page 617, we see that we can write the keyword **others** in place of the list of names of items to be decorated. If we do so, the attribute specification applies to all items of the given class in the declarative part that are not otherwise decorated with the attribute. Such an attribute specification must be the last one in the declarative part that refers to the given attribute name and item class.

EXAMPLE 20.13   *Decorating items not previously decorated*

In the following architecture body, signals are decorated with attributes specifying the maximum allowable delays due to the physical layout. The two signals recovered_clk1 and recovered_clk2 are explicitly decorated with the attribute value 100 ps. The remaining signals are decorated with the value 200 ps.

```
library ieee;  use ieee.std_logic_1164.all;
use work.timing_attributes.all;

architecture structural of sequencer is

  signal recovered_clk1, recovered_clk2 : std_ulogic;
  signal test_enable : std_ulogic;
  signal test_data : std_ulogic_vector(0 to 15);

  attribute max_wire_delay of
    recovered_clk1, recovered_clk2 : signal is 100 ps;

  attribute max_wire_delay of others : signal is 200 ps;

  ...

begin
  ...
end architecture structural;
```

The syntax rules also show that we can use the keyword **all** in place of a list of item names. In this case, all items of the given class defined in the declarative part containing the attribute specification are decorated. Such an attribute specification must be the only one in the declarative part to refer to the given attribute name and item class.

Although we can only decorate an item with one value for a given attribute name, we can decorate it with several different attributes. We simply write one attribute specification for each of the attributes decorating the item. For example, a component instance labeled mult might be decorated with several attributes as follows:

```
attribute cell_allocation of
  mult : label is "wallace_tree_multiplier";
attribute cell_position of
  mult : label is ( 1200 um, 4500 um );
attribute cell_orientation of
  mult : label is down;
```

If an item in a design is decorated with a user-defined attribute, we can refer to the attribute value using the same notation that we use for predefined attributes. The syntax rule for an attribute name referring to a user-defined attribute is

attribute_name ⇐ name ⟦ signature ⟧ ' identifier

If the name of the item is unambiguous, we can simply write an apostrophe and the attribute name after the item name. For example:

```
std_cell'cell_name
enable'pin_number
clk'max_wire_delay
v4 := idle_state'encoding
the_fpu'cell_position
```

In the case of attributes decorating subprograms or enumeration literals, it may be necessary to use a signature to distinguish between a number of alternative names. For example, we might refer to attribute values of different versions of an increment function as

```
increment [ bit_vector return bit_vector ] 'built_in
increment [ std_ulogic_vector return std_ulogic_vector ] 'built_in
```

Similarly, we might refer to attribute values of enumeration literals as

```
high [ return speed_range ] 'representation
high [ return coolant_level ] 'representation
```

While it is legal VHDL to refer to attribute values such as these in expressions, it is not good design practice to use attribute values to affect the structure or behavior of the model. It is better to describe structure and behavior using the language facilities intended for that purpose and use attributes to annotate the design with other kinds of information for use by other software tools. For this reason, we do not further discuss the use of attribute values in models. Software tools that use attributes should include documentation describing the required attribute types and their usage.

In Chapter 11, we introduced aliases as a way of defining alternate names for items in a design. In most cases, referring to an item using an alias is exactly the same as referring to it using its original name. The same interpretation holds when decorating items with attributes. When we use an alias of an item in an attribute specification, it is the original object denoted by the alias that is decorated, not the alias. This is the interpretation we saw for the predefined attributes discussed in the previous section. The exceptions are the predefined attributes that return the path name of an item and those that return information about the index ranges of arrays. One restriction on decorating data objects using aliases is that we may only do so using aliases that denote whole objects, not elements or slices of records or arrays. This restriction corresponds to the restriction that an attribute must decorate a whole object. The syntax rule for an attribute specification does not provide for naming parts of objects, since we can only write a simple identifier as an object name.

One final point to mention about user-defined attributes relates to component instantiation statements and to subprogram calls. In a component instantiation statement, actual signals are associated with formal ports of an entity. If the actual signal is decorated with an attribute, the attribute information is only visible in the context of the actual signal, namely, in the architecture body in which the signal is declared. It is not carried through to the instantiated entity. For example, if we have a signal s decorated with an attribute **attr**, we might use it as an actual signal in a component instantiation statement:

```
c1 : entity work.e(arch)
  port map ( p => s );
```

Within the architecture body **arch**, we cannot refer the attribute of the signal using the notation p'attr. This notation instead refers to the attribute **attr** of the port **p**, which can only be defined in the entity declaration.

In a subprogram call an actual parameter (such as a constant, variable, signal or file) is associated with a formal parameter of the subprogram. If the actual parameter is decorated with an attribute, that attribute information is likewise not carried through to the subprogram. The decoration is purely local to the region in which the actual object is declared.

VHDL-87, -93, and -2002

Since these versions of VHDL do not allow PSL code to be embedded within VHDL models, we cannot use the word **property** or **sequence** attribute specifications for the class of an item to be decorated.

VHDL-87

The syntax rules for attribute specifications in VHDL-87 do not allow us to name a character literal as an item to be decorated. Nor may we specify the entity class **literal**, **units**, **group** or **file**. Furthermore, we may not include a signature after an item name. Hence there is no way to distinguish between overloaded subprograms or enumeration literals; all items of the given name are decorated.

20.3 Groups

The user-defined attribute facility discussed in the previous section allows us to annotate individual items in a design with non-structural and non-behavioral information. However, much of the additional information we may need to include can best be expressed as relationships between collections of items, rather than pertaining to individual items. For this reason VHDL provides a grouping mechanism to identify a collection of items over which some relationship holds. The information about the relationship is expressed as an attribute of the group. In this section we see how to define kinds of groups, to identify particular groups of related items and to specify attributes for particular groups.

The first stage in grouping items is to define a template for the classes of items that can be included in the group. We do this with a *group template declaration*, for which the syntax rule is

> group_template_declaration ⇐
>   **group** identifier **is** (〖 entity_class 〖 <> 〗 〗 { , ... }) ;

A group template declaration lists one or more classes of items, in order, that may constitute a group. Note that the syntax rule uses the term "entity" here in the same way as the rules for attribute specifications, namely, to refer to any kind of item in a design. We discuss the meaning of the "<>" notation shortly. An example of a group template declaration is

> **group** signal_pair **is (signal, signal)**;

This defines a template for groups consisting of two signals. We can use this template to define a number of groups using *group declarations*. The syntax rule for a group declaration is

> group_declaration ⇐
>   **group** identifier : *group_template*_name
>   (〖 name 〗 character_literal 〗 { , ... }) ;

A group declaration names a template to use for the group and lists the items that are to be members of the group. Each item in the list must be of the class specified in the corresponding position in the template. For example, if we have two clock signals in a design, **clk_phase1** and **clk_phase2**, we can group them together using the **signal_pair** template defined above by writing

> **group** clock_pair : signal_pair (clk_phase1, clk_phase2);

As we mentioned earlier, the main use of groups is as a mechanism for defining relationships between items by decorating a group of items with an attribute. We decorate a group by naming it in an attribute specification, identifying it as an item of class **group**. For example, if we have an attribute declared as

> **attribute** max_skew : time;

we can decorate the **clock_pair** group with this attribute as follows:

attribute max_skew **of** clock_pair : **group is** 200 ps;

The decoration can be interpreted as an annotation to the design, indicating to a layout tool that the maximum permissible skew between the two signals in the group is 200 ps.

The syntax rule for a group template shows that we may write the box symbol ("<>") after an item class. In fact, we may only include such a class specification once in any template, and it must be in the last position in the list of item classes. It indicates that a group based on that template may have an indefinite number of elements of the given class (including none).

EXAMPLE 20.14 *Groups for physical packaging*

We can define a group template for a group representing component instances to be allocated to the same physical package. The members of such a group are the labels of the component instances. The group template declaration is

```
group component_instances is ( label <> );
```

We can use the template to create groups of instances:

```
group U1 : component_instances ( nand1, nand2, nand3 );
group U2 : component_instances ( inv1, inv2 );
```

We can specify what kind of integrated circuit should be used for each group by defining an attribute and using it to decorate the group:

```
attribute IC_allocation : string;

attribute IC_allocation of U1 : group is "74LS00";
attribute IC_allocation of U2 : group is "74LS04";
```

An individual item in a design can belong to more than one group. We simply include its name in the declaration of each group of which it is a member.

EXAMPLE 20.15 *Groups for port-to-port timing constraints*

We can use groups of signals as the basis for annotating a design entity with port-to-port timing constraints. Suppose we declare a group template **port_pair** and an attribute **max_prop_delay** in a package **constraints**:

```
group port_pair is ( signal, signal );

attribute max_prop_delay : time;
```

We can then use the template to group pairs of ports of an entity and annotate them with constraint attributes, as follows:

```
library ieee;  use ieee.std_logic_1164.all;
use work.constraints.port_pair, work.constraints.max_prop_delay;

entity clock_buffer is
  port ( clock_in : in std_ulogic;
         clock_out1, clock_out2, clock_out3 : out std_ulogic );

    group clock_to_out1 : port_pair ( clock_in, clock_out1 );
    group clock_to_out2 : port_pair ( clock_in, clock_out2 );
    group clock_to_out3 : port_pair ( clock_in, clock_out3 );

    attribute max_prop_delay of clock_to_out1 : group is 2 ns;
    attribute max_prop_delay of clock_to_out2 : group is 2 ns;
    attribute max_prop_delay of clock_to_out3 : group is 2 ns;

end entity clock_buffer;
```

In this entity declaration, the item clock_in is a member of each of the three groups clock_to_out1, clock_to_out2 and clock_to_out3.

VHDL-87

VHDL-87 does not allow declaration of group templates or groups.

Exercises

1. [❶ 20.1] What are the values of the following attributes of items, assuming all design units are analyzed and placed in a library named proj_lib:

   - word'path_name, where word is declared in the cpu_types package in Example 7.1 on page 246.

   - mult_unsigned'path_name, where mult_unsigned is the procedure in the bit_vector_signed_arithmetic package in Example 7.5 on page 253.

   - bv2'path_name, where bv2 is the parameter of mult_unsigned.

   - next_test_case'path_name and next_test_case'instance_name, where next_test_case is the variable in the stim_gen process in Example 7.6 on page 255. Assume the process is an architecture test_rtl of entity test_bench, which is the top-level entity.

   - get_ID'path_name and get_ID'instance_name, where get_ID is the procedure in the ID_manager package in the stim_gen process.

2. [❶ 20.1] Suppose we instantiate the counter entity on page 179 in Example 5.22 in a test bench as follows:

```
dut : entity work.counter(registered)
  port map ( ... );
```

The test bench entity name is **test_bench**, and the architecture body name is **counter_test**. What are the values of the following attributes:

- **val0_reg'path_name** in the architecture **registered** of **counter**,
- **bit0'path_name** in the instance **val1_reg** of the **struct** architecture body of **reg4** and
- **clr'path_name** in the instance **bit2** of the **behavioral** architecture body of **edge_triggered_Dff**, in the instance **val1_reg**?

What are the values of the **'instance_name** attributes of the same items?

3. [❶ 20.2] Given a physical type capacitance, declared as

```
type capacitance is range 0 to integer'high
  units pF;
  end units capacitance;
```

write an attribute declaration that represents a capacitive load and an attribute specification that decorates a signal **d_in** with a load of 3 pF.

4. [❶ 20.2] Write a physical type declaration for areas, with a primary unit of μm^2. Write an appropriate attribute declaration and specification to decorate an architecture body **library_cell** of an entity **and3** with an area of 15 μm^2.

5. [❶ 20.2] Given an attribute declared as

```
attribute optimization : string;
```

decorate the following procedure with the attribute value "inline". Assume that another overloaded version of the procedure, which must not be decorated, is visible.

```
procedure test_empty ( variable list : in list_ptr;
                       is_empty : out boolean ) is ...
```

6. [❶ 20.3] Define a group template that allows two or more statement labels as members. Next, declare a group that includes the labels of the following two statements:

```
step_1 : a := b * c + k;
step_2 : n := a + 4 * j;
```

Then, write an attribute specification that decorates the group with the attribute **resource_allocation** having the value **max_sharing**.

7. [❷ 20.1] Since the definitions in the VHDL-93 and VHDL-2002 versions of the *VHDL Language Reference Manual* of the **'path_name** and **'instance_name** attributes of items declared within packages are ambiguous, different simulators may produce different results. Construct some small examples, such as those shown in Section 20.1, and experiment with your simulator to see how it constructs values for these attributes.

8. [❷ 20.1] Develop an edge-triggered register model that includes generics for setup and hold times in its entity interface and that reports an assertion violation if the timing constraints are not met. The message reported should include the full instance name of the entity instance in which the violation occurs.

9. [❷ 20.2] Write an entity interface that describes a 74x138 three-to-eight decoder. Include an attribute declaration and attribute specifications to decorate the ports with pin-number information for the package shown in Figure 20.1.

FIGURE 20.1

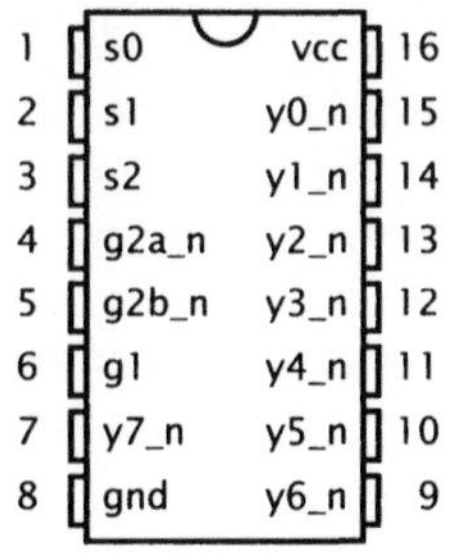

A package for a 74x138 decoder.

10. [❷ 20.3] Write an entity interface for an and-or-invert gate that implements the following function:

$$z = \overline{a_1 \cdot a_2 \cdot a_3 + b_1 \cdot b_2 \cdot b_3}$$

Since the "and" function is commutative and associative, a layout tool should be able to permute the connections within each of the groups a_1, a_2, a_3 and b_1, b_2, b_3 without affecting the function performed by the circuit. Include in the entity interface of the and-or-invert gate a group template declaration and group declarations that encompass ports among which connections may be permuted.

Chapter 21

Design for Synthesis

In this book we have discussed many aspects of VHDL and looked at examples of its use. One very strong motivation for using VHDL is hardware synthesis. The idea behind synthesis is to allow us to think of our design in abstract terms. We need not be so concerned about how best to implement the design in hardware logic—that is the job of the synthesis tool. It converts our abstract description into a structural description at a lower level of abstraction.

This chapter offers a brief introduction to synthesis, based on the IEEE standards that cover synthesis of VHDL models. A full coverage of the topic warrants a complete book in its own right. We refer the interested reader to the large number of books on hardware synthesis (for example, [1]).

21.1 Synthesizable Subsets

There are several synthesis tools available from different design automation tool vendors. While many of them perform the same general process, they differ in their command sets and the way in which we specify synthesis constraints. Hence we discuss synthesis tools only in very general terms. More important, synthesis tools differ in the subsets of VHDL that they accept as input. The majority only accept designs described at the register-transfer level and synthesize to circuits composed of gates, flipflops, registers, and other basic components. A small number of *behavioral synthesis* tools accept designs described at a higher level of abstraction. However, developing behavioral synthesis technology that is usable in practice has proven to be very difficult, so the techniques are not widely adopted. They are most successful in certain specialized application areas, such as digital signal processing.

The disparity between synthesis tools motivated the development of IEEE Standard 1076.6, *Standard for VHDL Register Transfer Level Synthesis*. The first version of this standard, published in 1999, specified a "level-1" lowest common denominator subset of VHDL that was acceptable to most synthesis tools. The intention was to assist designers in writing models that were portable between synthesis tools and to ensure that the behavior of the synthesized designs matched simulation results.

In 2004, a revision of the standard was published, specifying a level-2 synthesis subset of VHDL. The intention of this subset, as described in the Introduction to the standard, is "to include a maximum subset of VHDL that could be used to describe synthesizable RTL logic." It provides considerably more flexibility in the way models can be written. It is much closer to the subsets now implemented by synthesis tools. Nonetheless, there remains variation among tools, so we need to consult the documentation for any particular tool that we might use to find out what forms of input it accepts.

There are two aspects of synthesis subsets of VHDL. The first is the collection of language features that are included in the subset. This comprises the types that can be used to represent data and the declarations, specifications, and statements that can be included in models. The second aspect is the way in which we write code to represent different hardware elements. The task of a synthesis tool is to analyze a VHDL description and infer what hardware elements are represented and how they are connected. A tool cannot infer hardware from any arbitrarily written VHDL model. Instead, we need to write models in a *synthesis style* that is recognized by the tool. Style guidelines include templates for process statements from which various kinds of hardware elements can be inferred, and restrictions on the way in which we write and combine statements. We will look at both of these aspects, as specified by the IEEE 1076.6-2004 standard.

One further point to note about synthesis subsets is that they have historically lagged behind revisions of the VHDL standard. As an illustration, the 1999 version of the IEEE 1076.6 standard specified that models use the VHDL-87 version of the language, despite VHDL-93 having been published six years earlier. In 1999, many synthesis tools still only supported VHDL-87; now, VHDL-93 is widely supported. The changes between VHDL-93 and VHDL-2002 were relatively minor, apart from the addition of protected types. Since these are not supported for synthesis, synthesis tools effectively support VHDL-2002. That is the version of the language referenced in IEEE 1076.6-2004. The changes between VHDL-2002 and VHDL-2008 are much more significant. If past experience is an indicator, it may be some time before synthesis vendors implement the changes in their tools. Again, we should consult the documentation for any particular tool to see whether it supports VHDL-2008 features. The notes throughout this book describing differences between earlier versions of VHDL and VHDL-2008 will also be helpful as we write synthesizable models.

21.2 Use of Data Types

The synthesis standard allows us to use the following data types:

- enumeration types, including the predefined types **boolean**, **bit** and **character**

- integer types, including the predefined type **integer** and the subtypes **natural** and **positive**

- arrays of scalar elements, including the predefined types **bit_vector** and **string**

- **std_ulogic**, **std_ulogic_vector**, **std_logic** and **std_logic_vector**, defined in package **std_logic_1164**

- **unsigned** and **signed**, defined in package **numeric_bit**

• **unsigned** and **signed**, defined in package **numeric_std**

The synthesis standard allows us to use these types for constants, signals, and variables. When we declare a constant, we must include an initial value expression to give the constant a value. We cannot declare a deferred constant in a package. The synthesis standard specifies that any initial value expression in a signal or variable declaration is ignored. This makes sense in some circuits, such as ASICs, where the initial value of a storage location is indeterminate. In FPGAs, however, the storage can be initialized to specified values. Tools for synthesizing to FPGAs may allow an initial value expression in a signal or variable declaration for this purpose.

21.2.1 Scalar Types

Models conforming with the synthesis standard may define and use enumeration types, with some restrictions. The predefined types **boolean** and **bit** and the standard logic types **std_ulogic** and **std_logic** are implemented in hardware as individual bits. Most of the time, we use **std_ulogic** and **std_logic**, since that allows us to represent high-impedance and unknown states, as well as low and high logic levels. User-defined enumeration types may be implemented by tool-dependent encoding. Alternatively, we may specify the encoding by decorating the type with a string attribute, **enum_encoding**, described in Section 21.7.

Models conforming with the synthesis standard may also define and use integer types. Values of these types are implemented in the synthesized design as vectors of bits. If an integer type includes only non-negative values, the synthesized vector uses unsigned binary encoding. If the type includes negative values, two's-complement signed encoding is used. The number of bits in the encoding is determined by the range of values in the type. For example, given the following declarations in a model:

```
type sample is range -64 to 63;
subtype table_index is natural range 0 to 1023;
```

values of type **sample** should be implemented using 7-bit two's-complement encoding, and values of subtype **table_index** should be implemented using 10-bit unsigned encoding. Types that don't include 0 are encoded as though 0 were allowed. For example, the type

```
type index_type is range 4 to 15;
```

would be represented using 4-bit unsigned encoding. Synthesis tools conforming with the standard should support integers within the range -2^{31} to $+2^{31} - 1$, mapping to 32-bit two's-complement encoding.

The synthesis standard also allows use of other predefined enumeration types, including **character**, but they may not be supported by tools. The remaining classes of scalar types, namely, physical and floating-point types, are not supported by the synthesis standard. Definition and use of such types in a model are either ignored or treated as an error.

21.2.2 Composite and Other Types

Models conforming with the synthesis standard may define and use array and record types, but there are some significant restrictions on the use of array types. They must be indexed by an integer range, and the index bounds must be static, so that the synthesis tool can determine how much storage or how many bits of data are required in the hardware. The element type can only be an allowed scalar type, as described above, or a one-dimensional vector of an enumeration type representing individual bits. Thus, for example, the following array types are permissible:

```
type coeffs is array (3 downto 0) of integer;

type channel_states is array (0 to 7) of state;
  -- state is an enumeration type

subtype word is bit_vector(31 downto 0);
type reg_file is array (0 to 15) of word;
```

whereas the following are not:

```
type color is (red, green, blue);
type plane_status is array (color) of boolean;
  -- non-integer index type

type matrix is array (1 to 3, 1 to 3) of real;
  -- 2D, and floating-point elements

type reg_file_set is array (0 to 3) of reg_file;
  -- elements are vectors of non-bits
```

In addition, some tools limit arrays to be one-dimensional. Such tools would not allow the **matrix** type shown above.

The types **unsigned** and **signed** defined in **numeric_bit** and **numeric_std** are array types that meet the requirements for synthesizability, since they are one-dimensional arrays of elements that represent bits. The synthesis standard requires that we use these types if we need to represent unsigned or signed numbers at the bit level. We cannot use array types that we define in our model.

Historically, many designers have used the non-standard packages **std_logic_arith**, **std_logic_signed**, and **std_logic_unsigned**. These packages provide types and operations similar to those now provided by **numeric_std** and **numeric_std_unsigned**. With VHDL-2008, the standard packages incorporate all the operations provided by the non-standard packages. Nonetheless, synthesis tools still support use of the non-standard packages for representing numeric data at the bit level.

The synthesis standard does not support use of access types, file types or incomplete type declarations. Synthesis tools should ignore their declarations and are not required to accept models that use access-type values or file operations. In particular, dynamic allocation of objects using the **new** allocator and deallocation of objects using the **deallocate** procedure are not supported. The synthesis standard does support declaration of subtypes, but ignores user-defined resolution functions within subtype indications.

21.3 Interpretation of Standard Logic Values

If we use the standard logic types **std_ulogic** or **std_logic** in our models, we need to consider how a synthesis tool interprets values of different driving strength and unknown values. The synthesized hardware deals only with logic 0 and 1 values. We use standard logic values other than 0 and 1 to simulate the effects of weak driving strength and indeterminate logic values. We use the term *metalogical* to refer to the values 'U', 'X', 'W', and '–' that do not represent logic levels.

When our model assigns to a signal a value calculated by an expression from other signal values, the synthesis tool generates a hardware circuit that implements the logic of the expression. However, when our model uses a literal standard logic value, the synthesis tool must represent the value as either a logic 0 or a logic 1. The synthesis standard specifies that the standard logic values '0' and 'L', like the bit value '0' and the Boolean value false, are represented as a logic 0. Similarly, the standard logic values '1' and 'H', like the bit value '1' and the Boolean value true, are represented as a logic 1. Thus the synthesis tool does not attempt to interpret the strength information associated with the standard logic value.

When our model assigns the standard logic value 'Z' to a signal, the synthesis tool generates a tristate buffer for the signal. Usually such an assignment is nested within a conditional statement. In that case, hardware generated for the condition is used to enable or disable the tristate buffer. For example, the if statement

```
if request_enable = '1' then
    request <= ready;
else
    request <= 'Z';
end if;
```

would result in synthesis of a tristate buffer driving **request**. The input to the buffer would be connected to **ready**, and the control signal to enable the buffer would be connected to **request_enable**. When our model uses 'Z' in contexts other than a signal assignment (for example, in a comparison expression), the synthesis tool treats it in the same way as a metalogical value.

Use of metalogical values in a model is either ignored or not accepted by the synthesis tool, depending on the context. When values are tested for equality with metalogical values, the result is deemed to be false. Similarly, a test for inequality with metalogical values is deemed to be true. The rationale is that real hardware values are known to be either logic 0 or logic 1, so it does not make sense to synthesize hardware to compare with any other values. Thus any statements controlled by an equality comparison with a metalogical value, such as statements nested in an if statement or a case statement, can be ignored by the synthesis tool. They exist in the model purely for simulation purposes. In the cases of metalogical values appearing as operands of other relational operators and of arithmetic, logical and shift operators, the synthesis tool should not accept the model.

The **std_match** function defined in the **numeric_std** package can be used to compare standard logic values and vectors. It has the advantage of producing the same results in simulation and synthesis, unlike comparison using the "=" operator. Synthesis tools represent the use of **std_match** by an equivalence test. Simulation tools perform the comparison ignoring the driving strength of the parameters. If both values represent the same logic level, the comparison returns true. If either value is a metalogical value other than '–', the comparison returns false. The value '–' is interpreted as "don't care", so comparison with it returns true. Synthesis of a comparison using **std_match** with a literal vector containing "don't care" elements results in comparison hardware that excludes the "don't care" bits from the comparison. VHDL-2008 defines the matching relational operators, including "?=", which has the same behavior for **std_ulogic** and **std_ulogic_vector** values as **std_match**. As synthesis tools evolve to implement VHDL-2008 features, we should expect to see them treat the matching relational operators in a similar way to **std_match**.

21.4 Modeling Combinational Logic

Combinational circuits are those in which the outputs are determined solely by the current values of inputs; the circuit does not maintain any internal state. The simplest way to model combinational logic in synthesizable VHDL is using concurrent signal assignment statements. For example, we can model a Boolean function of inputs as follows:

```
status <= '1' when ready and sample < limit else '0';
```

A synthesis tool would generate hardware composed of a comparator for the input signals **sample** and **limit**, and a gate to combine the comparison output with the signal **ready**. It may optimize the hardware to meet timing and area constraints, but the result would perform the same function. Note that we have written the statement in this way to be compatible with VHDL-2002. As tools include VHDL-2008 features, we can rewrite the statement as

```
status <= ready and sample ?< limit;
```

EXAMPLE 21.1   *Modeling arithmetic circuits*

We can model a combinational arithmetic circuit using a concurrent assignment with an arithmetic expression on the right-hand side. For example, the following assignment in an architecture represents an adder for **unsigned** operands **a** and **b**, producing an unsigned result, **sum**, of the same size:

```
sum <= a + b;
```

If we also need to include a carry input to an addition, we can write the following in VHDL-2008:

```
sum <= a + b + carry_in;
```

In earlier versions of the **numeric_bit** and **numeric_std** packages, there was no overloaded addition operator with a scalar operand. We would have to write this as

```
sum <= a + b + unsigned'(0 => carry_in);
```

to create a one-element vector from the **carry_in** scalar. Alternatively, if we are using a synthesis tool that supports the **std_logic_arith** package, we could use the vector/scalar operator defined there.

If we want the carry result of the addition, we need to extend the operands by one bit, so that the result is one bit longer than the operands. We can then use the extra result bit as the carry:

```
tmp_sum <= ('0' & a) + ('0' & b);
sum <= tmp_sum(7 downto 0);
carry <= tmp_sum(8);
```

In VHDL-2008, we can write this as:

```
(carry, sum) <= ('0' & a) + ('0' & b);
```

EXAMPLE 21.2 *Modeling a multiplexer*

We can use a selected signal assignment statement to describe a multiplexer, for example:

```
with addr(1 downto 0) select
  request <=  request_a when "00",
              request_b when "01",
              request_c when "10",
              request_d when "11";
```

This assumes **bit_vector** signals. If we are using **std_ulogic_vector** signals, the choices do not cover all possible values. We would have to include a further alternative as follows:

```
with addr(1 downto 0) select
  request <=  request_a when "00",
              request_b when "01",
              request_c when "10",
              request_d when "11",
              'X' when others;
```

A synthesis tool would interpret the choices covering valid logic levels as implying hardware, and the **others** choice as representing metalogical values for simulation purposes, to be ignored.

We could also have expressed this behavior using a conditional signal assignment statement, as follows:

```
request <= request_a when addr(1 downto 0) = "00" else
           request_b when addr(1 downto 0) = "01" else
           request_c when addr(1 downto 0) = "10" else
           request_d when addr(1 downto 0) = "11" else
           'X';
```

However, in a conditional signal assignment, the conditions need not be mutually exclusive, so the synthesis tool would infer a priority-encoded chain of multiplexers to conform with the language semantics. This structure would be slower than a simple multiplexer. The tool may be able to optimize the hardware, but it is safer to use a selected signal assignment to imply a multiplexer function if that is our design intent.

One situation in which a conditional signal assignment statement is appropriate is combinational logic with a tristate buffered output, for example:

```
data_bus <= resize(sample_byte, 16)
            when std_match(sample_enable, '1') else
            "ZZZZZZZZZZZZZZZZ";
```

We can also use a process statement to describe combinational logic. The process must be sensitive to all of the inputs, and the combinational outputs must be assigned values in all possible executions of the process. This form of process is most useful when there are multiple outputs.

EXAMPLE 21.3   *A combinational process for multiple outputs*

Suppose we need to model a block of logic that has multiple outputs with tristate drivers all controlled by the same condition. We could use separate conditional signal assignment statements, but that would require us to repeat the condition in each one. Instead, we use a process to represent the logic block, as follows:

```
read_sample : process ( read_enable,
                         sample, limit_exceeded, ready )
begin
  if std_match(read_enable, '1') then
    data <= sample;
    parity <= calc_parity(sample);
    status <= ready and not limit_exceeded;
  else
    data <= "ZZZZZZZZ";
    parity <= 'Z';
    status <= 'Z';
  end if;
end process read_sample;
```

In this process, any change in any of the inputs results in new values being determined for all of the outputs. Thus the design is purely combinational. A synthesis tool would infer combinational network with tristate drivers on the outputs.

When we write a process that declares and uses a variable, a synthesis tool may infer a need for storage in the synthesized hardware. However, if all possible executions of the model in response to input changes involve the variable being assigned a value before being read, no storage is needed.

EXAMPLE 21.4 *Intermediate variables in combinational processes*

Consider the following process containing a variable assignment:

```
adder : process ( sel, a, b, c )
  variable operand : integer;
begin
  if sel = '1' then
    operand := a;
  else
    operand := b;
  end if;
  sum <= operand + c;
end process adder;
```

The process is sensitive to all of the inputs. There are two possible execution paths when an input changes. If **sel** is '1', the value of **a** is assigned to **operand** and subsequently added with **c** to determine the output **sum**. Alternatively, if **sel** is '0', the value of **b** is assigned to **operand** and added with **c**. Thus we can think of **operand** as representing the intermediate node in a combinational network consisting of a multiplexer and an adder. The value assigned to **operand** need not be stored.

Note the importance of including in the sensitivity list all inputs that are read by a combinational process. A synthesis tool will typically issue a warning if an input is read in the process but not mentioned in the sensitivity list. The difficulty in maintaining consistency between the sensitivity list and the set of signals read is the main motivation for allowing the reserved word **all** in sensitivity lists in VHDL-2008. Since this is a relatively minor extension for synthesis vendors to implement, we would hope to see it introduced quickly.

21.5 Modeling Sequential Logic

Sequential circuits are those that maintain an internal state. The outputs they produce in response to given inputs depend on the history of inputs received previously. Most sequential circuits we design are *synchronous*, or *clocked*. They use a rising or falling edge of a clock, or a level of an enable signal, to control advance of state or storage of data. The synthesis standard supports descriptions of these kinds of circuits. Most current design methodologies prefer edge-triggered sequential design, since achieving correct timing is more straightforward. Occasionally we might design an *asynchronous* circuit: a sequential circuit without a clock or enable input. Such circuits store state using combinational feedback loops. The synthesis standard and most synthesis tools do not support synthesis of these kinds of circuits. If we must include them, we must describe them using structural models and instantiate them as components within a synthesizable model.

Unlike signals used for data, which can be of a fairly wide variety of types, clock signals are restricted to be of type **bit**, **std_ulogic** or a subtype such as **std_logic**. A clock signal need not necessarily be a single scalar signal; it may be a scalar element of an array of **bit** or **std_ulogic** values.

21.5.1 Modeling Edge-Triggered Logic

We model edge-triggered sequential logic using processes. The synthesis standard allows considerable flexibility in modeling edge-triggered sequential logic, though not all synthesis tools implement the full generality. The premise is that a signal or variable assignment executed under control of a clock-edge condition implies edge-triggered storage. Clock-edge conditions are expressions of the following forms, for rising clock-edges:

- rising_edge(*clock_signal_*name)

- *clock_signal_*name'event **and** *clock_signal_*name = '1'

- *clock_signal_*name = '1' **and** *clock_signal_*name'event

- **not** *clock_signal_*name'stable **and** *clock_signal_*name = '1'

- *clock_signal_*name = '1' **and not** *clock_signal_*name'stable

and for falling clock-edges:

- falling_edge(*clock_signal_*name)

- *clock_signal_*name'event **and** *clock_signal_*name = '0'

- *clock_signal_*name = '0' **and** *clock_signal_*name'event

- **not** *clock_signal_*name'stable **and** *clock_signal_*name = '0'

- *clock_signal_*name = '0' **and not** *clock_signal_*name'stable

We can write an expression of one of these forms in the condition of an if statement within a process. The process must also have the clock signal name in its sensitivity list. Any signal or variable assignments within the if statement are then said to be *synchronous assignments*, controlled by the clock-edge condition. We can also include assignments in if statements controlled by other conditions involving asynchronous control signals. We must also include the control signals in the sensitivity list of the process. The assignments within such if statements are called *asynchronous assignments*, as they are not controlled by a clock-edge condition. The operators in expressions on the right-hand sides of assignments, whether synchronous or asynchronous, imply combinational logic connected to register inputs. As well as assignments, we can include other sequential statements within the process. These statements govern the flow of control leading to assignments. Thus, they imply combination logic, such as multiplexers, that feed the inputs of registers implied by the assignments. There are some restrictions, however. For example, we cannot arbitrarily include wait statements, nor can we refer to clock-edge expressions in assignments.

EXAMPLE 21.5 *Edge-triggered register*

One of the simplest forms of process represents an edge-triggered register:

```
simple_reg : process ( clk ) is
begin
  if clk'event and clk = '1' then
    reg_out <= data_in;
  end if;
end process simple_reg;
```

In this process, the assignment to **reg_out** is a synchronous assignment, controlled by the rising clock-edge condition. The process represents a register with **clk** as the clock signal, **data_in** as the input, and **reg_out** as the output.

EXAMPLE 21.6 *Edge-triggered register with synchronous control inputs*

We can include more involved statements within the controlling if statement to model registers with synchronous control signals. For example, the following process models a register with synchronous reset and enable controlling storage for two output signals:

```
dual_reg : process ( clk ) is
begin
  if rising_edge(clk) then
    if reset = '1' then
      q1 <= X"00";
      q2 <= X"0000";
    elsif en = '1' then
      q1 <= d1;
      q2 <= d2;
    end if;
  end if;
end process dual_reg;
```

In this case, all of the assignments are synchronous, governed by the rising clock-edge condition. The nested if statement chooses between resetting the outputs, updating them, or leaving them unchanged. A synthesis tool could infer a register, updated on every clock-edge, with a multiplexer at the input selecting between the hardwired 0 values, the data inputs, and the fed-back outputs. Alternatively, if the target technology supports registers with separate reset and enable control signals, the tool may infer use of them. We will see in Section 21.7 how we can use attributes to select the implementation.

EXAMPLE 21.7   *Counters as registers combined with arithmetic*

We can combine computational logic and storage in the one process. The computational logic is represented by the expressions in assignments. Counters are a good illustration of this approach. The following process represents an up/down counter wtih synchronous reset, load and count enable:

```vhdl
signal d_in, count : unsigned(11 downto 0);
...

up_down_counter : process (clk) is
begin
  if rising_edge(clk) then
    if reset = '1' then
      count <= X"000";
    elsif load_en = '1' then
      count <= d_in;
    elsif count_en = '1' then
      if dir = '1' then
        count <= count + 1;
      else
        count <= count - 1;
      end if;
    end if;
  end if;
end process up_down_counter;
```

We could augment this with a concurrent assignment statement to derive a terminal count signal:

```vhdl
tc <= '1' when std_match(count, X"111") else '0';
```

EXAMPLE 21.8   *Register wtih asynchronous control signals*

If our implementation technology provides registers with asynchronous control signals, we can represent them with processes containing asynchronous assignments. For example, the following process represents a register with asynchronous reset:

```vhdl
reg : process (clk, reset) is
begin
  if reset = '1' then
    q <= "0000";
  elsif rising_edge(clk) then
    q <= d;
  end if;
end process reg;
```

In this process, the first assignment is not controlled by a clock-edge condition; hence, it is asynchronous, and the **reset** control signal must be included in the sensitivity list of the process. The second assignment is synchronous and models the edge-triggered behavior of the register.

EXAMPLE 21.9 *Shift register with asynchronous and synchronous control*

We can combine asynchronous and synchronous control in a single process. For example, the we could model a shift register with asyncrhnous reset and synchronous parallel load as follows:

```
shift_reg : process (clk, reset) is
begin
  if reset = '1' then
    q <= "00000000";
  elsif rising_edge(clk) then
    if load_en = '1' then
      q <= d_in
    else
      q <= q(6 downto 0) & d_s;
    end if;
  end if;
end process reg;
```

Note that we do not include the **load_en** signal in the sensitivity list, as it is a synchronous control signal.

The synthesis standard allows more involved structures than those demonstrated by the preceding examples. For example, it lists the following as a legal synthesizable process:

```
RegProc5 : process( clk, reset )
begin
  if (en = '1' and rising_edge(clk)) or reset = '1' then
    if reset = '1' then
      Q <= '0'; -- async assignment
    elsif en = '1' and rising_edge(clk) then -- sync condition
      Q <= D; -- sync assignment
    end if;
  end if ;
end process ;
```

While it is, in principle, possible to analyze the control flow in such a process and determine whether each of the assignments is synchronous or asynchronous, synthesis tools are generally more restrictive in what they will accept. It is often clearer to write processes in the forms illustrated by the preceding examples than in more convoluted forms.

Recall that a process with signals listed in the sensitivity list is equivalent to a process containing a wait statement that is sensitive to the signals. Synthesis tools allow us to express edge-sensitive behavior using explicit wait statements in a process, with some restrictions. The synthesis standard specifies quite complicated rules for the structure of such wait statements, covering the signals that can be listed in the **on** clause and the form of condition that can be written in the **until** clause. (A **for** clause is not allowed, as that would imply specific timing.) Current synthesis tools are more restrictive, since inferring control logic for the general cases allowed by the standard could be arbitrarily complicated.

The simplest case allowed by synthesis tools is a wait statement as the first statement in the process, with a clock-edge condition in the until clause. There must not be any other wait statements or references to clock-edges in the process, nor in any procedures called from the process.

EXAMPLE 21.10   *Edge-triggered register with explicit wait statement*

We can represent and edge-triggered register with synchronous reset as follows:

```
reg : process is
begin
  wait until rising_edge(clk);
  if reset = '1' then
    q <= X"00";
  elsif en = '1' then
    q <= d;
  end if;
end process dual_reg;
```

The assignments to q only occur after a rising edge occurs on the clk signal. Hence, both **reset** and **en** are synchronous control signals.

EXAMPLE 21.11   *Explicit wait and asynchronous control*

The synthesis standard allows us to write the following process to express asynchronous control using an explicit wait statement.

```
reg : process is
begin
  wait until reset = '1' or rising_edge(clk);
  if reset = '1' then
    q <= X"00";
  elsif rising_edge(clk) then
    if en = '1' then
      q <= d;
    end if;
```

```
    end if;
  end process dual_reg;
```

The asynchronous condition is included in the wait statement. The wait statement is followed immediately by an if statement that tests both the asynchronous condition and the clock-edge condition. Thus, in this example, **reset** is an asynchronous control signal and **en** is a synchronous control signal. While this is acceptable according to the synthesis standard, not all tools accept it, instead limiting conditions in wait statements to just clock-edge conditions.

The synthesis standard allows a process to include multiple explicit wait statements, though some tools do not support it. In general, the hardware inferred for such a process includes some form of state machine, since the hardware must keep track of progress through the statements from one cycle to the next. Compare this with a process containing only one wait statement (explicit or implied), which always performs one complete pass through the process statement body for each clock-edge.

The rules for processes with multiple wait statements require that each statement must wait for the same condition, and in particular, must wait for the same edge of a single clock signal. This makes sense, as the wait statements correspond to transitions in the controlling state machine in the inferred hardware. If the wait statements include asynchronous conditions, as in Example 21.11, then each wait statement must be followed by identical tests for those conditions. Again, this corresponds to the control hardware implementation. The asynchronous behavior is that of the state machine.

EXAMPLE 21.12 *Sequential multiplier*

We can describe a multiplier that takes multiple clock cycles to compute its result using a shift-and-add method. The following process, based on an example in the IEEE 1076.6 standard, describes the behavior:

```
MultProc : process is
begin
  wait until rising_edge(clk);
  if start = '1' then
    done <= '0';
    P <= (others => '0');
    for i in A'range loop
      wait until rising_edge(clk);
      if A(i) = '1' then
        P <= (P(6 downto 0) & '0') + B;
      else
        P <= P(6 downto 0) & '0';
      end if;
    end loop;
    done <= '1';
```

```
      end if;
    end process;
```

This process implies an edge-triggered state machine that controls registers for the **done** and P output signals. The state machine tests the **start** signal on each clock-edge. When it is '0', it leaves P unchanged and sets **done**. Otherwise, it resets **done** and P, then sequences through a number of cycles (determined by A'range) to update P with successive partial products. On completion of the sequence, it sets **done** again.

Where wait statements appear in a loop that implements sequential behavior, we can use an exit or next statement after each wait statement to describe reset behavior. For asynchronous reset, this is a case of each wait statement being followed by identical tests for asynchronous conditions.

EXAMPLE 21.13 *Reset in a loop*

The following process from the IEEE 1076.6 synthesis standard models a UART serializer for data transmission:

```
UartTxFunction : process is
begin
  TopLoop : loop
    if nReset = '0' then
      SerialDataOut <= '1';
      TxRdyReg <= '1';
    end if;

    wait until nReset = '0' or
               (rising_edge(UartTxClk) and DataRdy = '1');
    next TopLoop when nReset = '0';
    SerialDataOut <= '0';
    TxRdyReg <= '0';

    -- Send 8 Data Bits
    for i in 0 to 7 loop
      wait until nReset = '0' or rising_edge(UartTxClk);
      next TopLoop when nReset = '0';
      SerialDataOut <= DataReg(i);
      TxRdyReg <= '0';
    end loop;

    -- Send Parity Bit
    wait until nReset = '0' or rising_edge(UartTxClk);
    next TopLoop when nReset = '0';
    SerialDataOut <=
      DataReg(0) xor DataReg(1) xor DataReg(2) xor
      DataReg(3) xor DataReg(4) xor DataReg(5) xor
```

```
          DataReg(6) xor DataReg(7);
        TxRdyReg <= '0';

        -- Send Stop Bit
        wait until nReset = '0' or rising_edge(UartTxClk);
        next TopLoop when nReset = '0';
        SerialDataOut <= '1';
        TxRdyReg <= '1';
    end loop;
end process;
```

Each wait statement in the process includes a test for a rising edge of the UartTxClk signal, as well as for the the asynchronous nReset signal. The identical next statements all restart the outer loop when the asynchronous reset condition is true.

We can rewrite this process to describe synchronous reset, as follows:

```
UartTxFunction : process is
begin
  TopLoop : loop
    wait until rising_edge(UartTxClk);
    if nReset = '0' then
      SerialDataOut <= '1';
      TxRdyReg <= '1';
    elsif DataRdy = '1' then
      SerialDataOut <= '0';
      TxRdyReg <= '0';

      -- Send 8 Data Bits
      for i in 0 to 7 loop
        wait rising_edge(UartTxClk);
        exit TopLoop when nReset = '0';
        SerialDataOut <= DataReg(i);
        TxRdyReg <= '0';
      end loop;

      -- Send Parity Bit
      wait rising_edge(UartTxClk);
      exit TopLoop when nReset = '0';
      SerialDataOut <=
        DataReg(0) xor DataReg(1) xor DataReg(2) xor
        DataReg(3) xor DataReg(4) xor DataReg(5) xor
        DataReg(6) xor DataReg(7);
      TxRdyReg <= '0';

      -- Send Stop Bit
      wait until rising_edge(UartTxClk);
      exit TopLoop when nReset = '0';
      SerialDataOut <= '1';
      TxRdyReg <= '1';
```

```
        end if;
      end loop;
    end process;
```

In this case we have used exit statements instead of next statements, though the latter would work just as well.

The until clause in a wait statement can also be an expression of the following forms, in addition to those listed on page 642:

- *clock_signal_*name = '1'

- *clock_signal_*name = '0'

Since the wait statement waits until the signal changes value, a change to '1' must represent a rising edge, and a change to '0' must represent a falling edge. Thus, we could write the wait statement in Example 21.10 as:

```
wait until clk = '1';
```

While this is allowed, the form using **rising_edge** is a preferred style, since it is more descriptive and deals correctly with weak driving strengths ('L' and 'H') during simulation.

21.5.2 Level-Sensitive Logic and Inferring Storage

Level-sensitive sequential logic maintains state, but does not respond to clock-edges. Instead, state is usually updated under control of an enable signal. While the enable signal is asserted, the state can change according to data inputs. While the enable signal is negated, changes on the data inputs are ignored and the circuit maintains its current state.

EXAMPLE 21.14   *A transparent latch*

Consider the following model for a transparent latch:

```
latch : process ( enable, d )
begin
  if enable = '1' then
    q <= d;
  end if;
end process latch;
```

This process is sensitive to changes on the **enable** and **d** inputs. If **enable** is '1' when either input changes, the data value is used to update the output **q**. If **enable** is '0', the current value is maintained on **q**. This behavior is implied by the semantics of signals and signal assignment in VHDL. When the synthesis tool implements the model as a hardware circuit, it must provide some storage to maintain the value for the output. The tool must infer the need for storage from the model semantics.

In general, a synthesis tool must infer storage if there are possible executions of the process that do not involve assignment to a given signal or variable. If the process does not include clock-edge conditions, then level-sensitive storage is inferred. The process must then include in its sensitivity list all signals that the process reads. The latch example illustrates storage inference due to existence of paths on which a signal is not updated. If the process executes when **enable** is '0', the assignment to **q** is bypassed, so storage is inferred for **q**.

EXAMPLE 21.15   *Transparent latch with reset*

The following process is another example of a latch, in this case involving storage inference for a variable:

```
latch_with_reset : process ( enable, reset, d )
  variable stored_value : bit;
begin
  if reset = '1' then
    stored_value := '0';
  elsif enable = '1' then
    stored_value := d;
  end if;
  q <= stored_value;
end process latch_with_reset;
```

The output signal **q** is assigned on every execution of the process, so no storage is inferred for it. However, the variable **stored_value** is not assigned when **reset** and **enable** are both '0'; hence storage is inferred for the variable.

In principle, storage is also inferred for a process that does not include a reference to a clock-edge expression if there are possible executions in which a signal or variable is read before being assigned. However, such a process represents an asynchronous sequential circuit and is not supported by synthesis tools. Consider the following erroneous process intended to describe a counter with reset:

```
counter : process ( count_en )
  variable count : natural range 0 to 15;
begin
  q <= (q + 1) mod 16;
end process counter;
```

The process is sensitive to changes of **count_en**. Whenever that signal changes, the old value of **q** is read and incremented to determine the new value. The value must be maintained until the next change of **count_en**, implying the need for storage for **q**, even though it is assigned in all possible executions of the process.

The synthesis standard recommends against writing level-sensitive processes in which signals or variables are read before being assigned. Usually such processes are not what we intend, and the inference of storage is inadvertent. On the other hand, a process in which a variable is first assigned and then read in all possible execution paths is legal and useful. Such a process simply models combinational logic, as we discussed in Section 21.4, with the variable denoting an intermediate node in the combinational network.

21.5.3 Modeling State Machines

Many designs expressed at the register-transfer level consist of combinational data paths controlled by finite-state machines. Hence it is important to be able to describe a finite-state machine in such a way that it can be synthesized. The preferred style is to separate the implementation into two processes, one describing the combinational logic that calculates the next state and output values, and the other being a register that stores the state. This style is accepted by all synthesis tools, whereas finite-state machines implied by multiple wait statements in a process, as described in Section 21.5.1, are not uniformly supported.

EXAMPLE 21.16   *Finite-state machine with Mealy and Moore outputs*

The following architecture body represents a finite state machine described using two processes:

```
architecture rtl of state_machine is
  type state is (ready, ack, err);
  signal current_state, next_state : state;
begin

  next_state_and_output : process ( current_state, in1, in2 )
  begin
    case current_state is
      when ready =>
        out1 <= '0';
        if in1 = '1' then
          out2 <= '1';
          next_state <= ack;
        else
          out2 <= '0';
          next_state <= ready;
        end if;
      when ack =>
        out1 <= '0';
        if in2 = '1' then
          out2 <= '0';
          next_state <= ready;
        else
          out2 <= '0';
          next_state <= err;
        end if;
```

```
        when err =>
          out1 <= '1';
          out2 <= '0';
          next_state <= err;
      end case;
    end process next_state_and_output;

    state_reg : process ( clk, reset )
    begin
      if reset = '1' then
        current_state <= ready;
      elsif rising_edge(clk) then
        current_state <= next_state;
      end if;
    end process state_reg;

  end rtl;
```

The architecture body defines an enumeration type for the state values. We can either rely on the synthesis tool to determine the encoding for the state or use the **enum_encoding** attribute to define our own encoding, as described in Section 21.7. The signal **current_state** represents the output of the state register, and the signal **next_state** is the state to be assumed by the state machine on the next clock-edge. The process **next_state_and_output** describes the combinational logic. It uses a case statement to determine the next state and output values, depending on the current state value. Note that the output **out1** is uniquely determined by the current state. It is a Moore machine output. The output **out2**, on the other hand, depends on both the current state and the current inputs to the machine. It is a Mealy machine output. The process **state_reg** describes the state register. It has an asynchronous reset control input that forces the machine into the ready state. When **reset** is inactive, the process updates the current state on each rising clock-edge.

In the process representing the combinational logic, we have included assignments for both output signals in all states. An alternative way of writing the process is:

```
    next_state_and_output : process ( current_state, in1, in2 )
    begin
      out1 <= '0'; out2 <= '0';
      case current_state is
        when ready =>
          if in1 = '1' then
            out2 <= '1';
            next_state <= ack;
          else
            next_state <= ready;
          end if;
        when ack =>
          if in2 = '1' then
            next_state <= ready;
```

```
            else
              next_state <= err;
            end if;
         when err =>
           out1 <= '1';
           next_state <= err;
       end case;
    end process next_state_and_output;
```

In this version, we include "default" assignments to the outputs before the case statement. Then, in the case statement, we only assign to each output when the value differs from the default. This is a much more succinct form, especially when there are many outputs. Moreover, it helps us avoid missing an assignment and inadvertently implying level-sensitive storage for an output. For these reasons, we recommend this style.

21.6 Modeling Memories

Many designs include memories, and many implementation technologies, such as FPGAs, include memory resources that can be used as RAMs and ROMs. Synthesis tools can infer memory hardware for processes written in certain ways. The IEEE 1076.6 synthesis standard specifies that a RAM be modeled in much the same way as a register, but with the storage represented by an array of bits, vectors, or integers. An address vector, converted to integer, is used to index the array for reading or writing.

A RAM model typically includes declaration of an array type and a signal of that type, for example:

```
type mem_array is array (0 to 2**depth - 1) of
                    std_ulogic_vector(width - 1 downto 0);
signal RAM : mem_array;
```

The way we write the process modeling the memory determines what kind of RAM inferred.

EXAMPLE 21.17   *Asynchronous RAM*

An asynchronous RAM is essentially a level-sensitive storage device. Thus, we can model it in a similar way to a latch. Assuming the type and signal declaration given above, the behavior is modeled as follows:

```
asynch_RAM : process (addr, d_in, we) is
begin
  if we = '1' then
    RAM(to_integer(addr)) <= d_in;
  end if;
end process asynch_RAM;
```

```
        d_out <= RAM(to_integer(addr));
```

The process represents the writing part of the behavior; it updates the RAM signal while the **we** input is '1'. The assignment represents the reading part of the behavior. Note that not all implementation technologies provide asynchronous memories, since they do not interface well with clocked synchronous designs.

EXAMPLE 21.18 *Synchronous RAM with asynchronous read*

Most implementation technologies provide RAMs that perform write operations synchronously. They have embedded registers that store the address and data for a write. In some technologies, the read operation is done asynchronously. A model for such a RAM is

```
synch_RAM : process (clk) is
begin
  if rising_edge(clk) then
    if we = '1' then
      RAM(to_integer(addr)) <= d_in;
    end if;
  end if;
end process synch_RAM;

d_out <= RAM(to_integer(addr));
```

EXAMPLE 21.19 *Synchronous RAM with synchronous read*

If a RAM has embedded registers for the read control signals, read operations are also performed synchronously. RAMs differ in the data they read when a write is also performed in the same cycle. One form of RAM reads the old content of the memory location before updating it with the new data. We can model this as follows:

```
synch_RAM : process (clk) is
begin
  if rising_edge(clk) then
    d_out <= RAM(to_integer(addr));
    if we = '1' then
      RAM(to_integer(addr)) <= d_in;
    end if;
  end if;
end process synch_RAM;
```

We could interchange the assignment to **d_out** with the inner if statement without affecting the behavior, since the assignments do not affect the RAM content until after the process suspends. A synthesis tool would infer the same behavior with the statements in either order.

Another form provides the newly written data, modeled as follows:

```vhdl
synch_RAM : process (clk) is
begin
  if rising_edge(clk) then
    if we = '1' then
      RAM(to_integer(addr)) <= d_in;  d_out <= d_in;
    else
      d_out <= RAM(to_integer(addr));
    end if;
  end if;
end process synch_RAM;
```

In both cases, we can add an enable input controlling reading and writing at a new address. For example, the first version above, augmented with an enable input, is

```vhdl
synch_RAM : process (clk) is
begin
  if rising_edge(clk) then
    if en = '1' then
      d_out <= RAM(to_integer(addr));
      if we = '1' then
        RAM(to_integer(addr)) <= d_in;
      end if;
    end if;
  end if;
end process synch_RAM;
```

EXAMPLE 21.20   *Pipelined synchronous RAM*

The RAMs with synchronous read in the preceding examples start a read access on a clock-edge and provide the data after a read-access delay. In some designs, the data may arrive too late in a clock cycle to be used for further computation. We can add a storage register on the RAM output so that the data can be used in the subsequent clock cycle. This amounts to pipelining the RAM access. We can combine the RAM and pipeline registers into a single process representing a pipelined RAM:

```vhdl
pipelined_RAM : process (clk) is
  variable pipelined_en : std_ulogic;
  variable pipelined_d_out :
            std_ulogic_vector(width - 1 downto 0);
begin
  if rising_edge(clk) then
    if pipelined_en = '1' then
      d_out <= pipelined_d_out;
    end if;
    pipelined_en := en;
```

```
        if en = '1' then
          pipelined_d_out := RAM(to_integer(addr));
          if we = '1' then
            RAM(to_integer(addr)) <= d_in;
          end if;
        end if;
      end if;
    end process pipelined_RAM;
```

If we are synthesizing to an implementation technology, such as an FPGA, in which a RAM can be loaded with initial contents on system reset, we may be able to specify the initial content as part of the synthesizable model. Some tools allow us to specify initial contents in an initial value aggregate in the signal declaration, for example:

```
signal RAM : RAM_array := (X"0020", X"FC01", X"101E", X"C000",
                           . . .
                           others => X"0000");
```

Some tools also allow us to write a function that loads data from a file and returns an array of values to assign as the initial value for the signal. We took this approach in the case study in Chapter 17. The IEEE 1076.6 synthesis standard, however, does not specify either of these approaches. We would need to consult our tool vendor's documentation to see how a memory can be initialized, and we should recognize that the approach used may not be portable among different tools.

We can model ROMs using similar techniques to those used for RAMs, but omitting the code representing the write operations. Since ROM content does not change, we can use a constant instead of a signal to model the storage. We specify the ROM content in the form of an array aggregate, for example:

```
constant ROM : mem_array := (X"0020", X"FC01", X"101E", X"C000",
                             . . .
                             others => X"0000");
```

Reading the ROM asynchronously can then be modeled using a concurrent assignment:

```
d_out <= ROM(to_integer(addr));
```

For a small ROM, a synthesis tool could optimize this as combinational logic.

Another way to model a small ROM is using a case statement. We use the address as the selector expression and assign different literal values to the output signal based on the address.

EXAMPLE 21.21 *A small ROM modeled using a case statement*

We can model a lookup ROM giving 7-segment display codes for BCD digits as follows:

```
decoder : process ( bcd ) is
begin
  case bcd is
    when X"0" =>    seg <= "0111111";
    when X"1" =>    seg <= "0000110";
    when X"2" =>    seg <= "1011011";
    when X"3" =>    seg <= "1001111";
    when X"4" =>    seg <= "1100110";
    when X"5" =>    seg <= "1101101";
    when X"6" =>    seg <= "1111101";
    when X"7" =>    seg <= "0000111";
    when X"8" =>    seg <= "1111111";
    when X"9" =>    seg <= "1101111";
    when others => seg <= "1000000";
  end case;
end process decoder;
```

Alternatively, we could use a selected assignment:

```
with bcd select
  seg <= "0111111" when X"0", "0000110" when X"1",
         "1011011" when X"2", "1001111" when X"3",
         "1100110" when X"4", "1101101" when X"5",
         "1111101" when X"6", "0000111" when X"7",
         "1111111" when X"8", "1101111" when X"9",
         "1000000" when others
```

If we want to use a block RAM resource as a ROM, we can use the same form of process as for a RAM, but omit the statements that update the array. For example, we can adapt the first process in Example 21.19 to model a ROM as follows:

```
block_ROM : process (clk) is
begin
  if rising_edge(clk) then
    d_out <= ROM(to_integer(addr));
  end if;
end process block_ROM;
```

21.7 Synthesis Attributes

When we use a synthesis tool to infer a hardware implementation for a design, we can direct it to optimize either the speed or area of the generated circuit. In some applications, that general directive may be sufficient, resulting in an implementation that meets our constraints. Often, however, we need to take finer control of the synthesis process. One way in which we can do so is by including attribute specifications (see Chapter 20) in our models to direct a synthesis tool to infer hardware in particular ways.

Different synthesis tools support different attributes to specify different aspects of hardware inference and different aspects of target technologies. This is possibly an aspect in which tools most widely diverge, since the attributes a given tool supports reflect the particular capabilities and synthesis algorithms implemented by the tool. We need to refer to a tool's documentation to discover what attributes are supported and how to use them.

In an effort to create at least a small amount of harmony, the IEEE 1076.6 synthesis standard defines a minimal set of synthesis attributes. We describe them here, as they are indicative of the kinds of attributes supported by tools. The standard specifies a package of attribute declaration to be analyzed into the **ieee** library. While we could declare the attributes ourselves in each design, using the standard package is more convenient. Synthesis tools usually include similar packages for their implementation-defined attributes. The standard package is

```
package RTL_ATTRIBUTES is
  attribute KEEP : boolean;
  attribute CREATE_HIERARCHY : boolean;
  attribute DISSOLVE_HIERARCHY : boolean;
  attribute SYNC_SET_RESET : boolean;
  attribute ASYNC_SET_RESET : boolean;
  attribute ONE_HOT : boolean;
  attribute ONE_COLD : boolean;
  attribute FSM_STATE : string;
  attribute FSM_COMPLETE : boolean;
  attribute BUFFERED : string;
  attribute INFER_MUX : boolean;
  attribute IMPLEMENTATION : string;
  attribute RETURN_PORT_NAME : string;
  attribute ENUM_ENCODING : string;
  attribute ROM_BLOCK : string;
  attribute RAM_BLOCK : string;
  attribute LOGIC_BLOCK : string;
  attribute GATED_CLOCK : boolean;
  attribute COMBINATIONAL : boolean;
end package RTL_ATTRIBUTES;
```

If we need to use any of these attributes, we can include a use clause in our model to make them directly visible. Several of these attributes are boolean. Decorating an item with the value true for one of these attributes directs a tool to synthesize in a particular way. Decorating with the value false is the same as not decorating the item. Attributes of type string allow us to specify further information for use by the synthesis tool. The meaning of each attribute is described below.

KEEP : boolean

> *Decorates*: entity, component declaration, component instantiation, signal, variable
>
> This attribute directs the tool to preserve the hardware represented by the decorated item in the inferred hardware. The item should not be deleted or

replicated during optimization of the design. We can use this attribute for parts of a design that we have previously synthesized and are re-using.

CREATE_HIERARCHY : boolean

Decorates: entity, block, subprogram, process

This attribute directs the tool to maintain the decorated item as a distinct hierarchical construct in the inferred hardware. It should not be subsumed into an enclosing construct during optimization.

DISSOLVE_HIERARCHY : boolean

Decorates: entity, component declaration, component instantiation

This attribute directs the tool to merge the construct into the hardware in which the item is instantiated. The tool can then globally optimize the construct in the context of its instantiation.

SYNC_SET_RESET : boolean

Decorates: signal, process, block, entity

This attribute is used to identify edge-sensitive storage devices that have separate synchronous set/reset inputs in the target technology. Using those inputs is more efficient than multiplexing the data inputs. We use the attribute to decorate a set/reset signal connected to storage devices or the construct that represents a storage device. For example, in the following:

```
attribute SYNC_SET_RESET of reset : signal is true;
...

reg : process ( clk ) is
begin
  if rising_edge(clk) then
    if reset = '1' then
      q <= (others => '0');
    else
      q <= d;
    end if;
  end if;
end process reg;
```

the tool infers hardware with **reset** connected to the separate synchronous reset input of the register. Without the attribute, the tool could infer a register whose input comes from a multiplexer with **reset** as the select input, a zero vector as one data input, and **d** as the other data input.

ASYNC_SET_RESET : boolean

Decorates: signal, process, block, entity

This attribute is used to identify level-sensitive storage devices that have separate asynchronous set/reset inputs in the target technology. It is used in a similar way to SYNC_SET_RESET.

ONE_HOT : boolean

Decorates: signal

This attribute specifies that, in a collection of signals, each of which is decorated with true for this attribute, at most one scalar value is '1' at any time. The directive allows a synthesis tool to avoid inferring priority logic based on the signals in the collection. For example, if we write

```
ff : process ( clk, reset, set ) is
begin
  if reset = '1' then
    q <= '0';
  elsif set = '1' then
    q <= '1';
  elsif rising_edge(clk) then
    q <= d;
  end if;
end process ff;
```

we are specifying that reset has priority over set. If the target technology includes flipflops with set and reset, but requires that only one be active at a time, the tool would infer a connection to the set input driven with set and not reset. If we add the following attribute specification:

attribute ONE_HOT **of** set, reset : **signal is** true;

we are telling the tool that only one of set and reset is active at a time, so it can make direct connections to the set and reset input of the flipflop. Of course, we must ensure separately that our statement is valid. We might include an assertion to that effect in the model.

ONE_COLD : boolean

Decorates: signal

This attribute specifies that, in a collection of signals, each of which is decorated with true for this attribute, at most one scalar value is '0' at any time. It is used in a similar way to **ONE_HOT**.

FSM_STATE : string

Decorates: type, subtype, signal, variable

This attribute directs the tool to encode the state vector of a finite-state machine in a specified way. The allowed attribute values are:

- "BINARY": unsigned binary encoding with the minimal number of bits

- "GRAY": Gray coding, in which exactly one bit value changes on each tansition

- "ONE_HOT": an encoding in which each code value has exactly one '1' bit

- "ONE_COLD": an encoding in which each code value has exactly one '0' bit

- "AUTO" or empty string: the tool selects an encoding

- A string of the same form as the ENUM_ENCODING attribute, described on page 664

If both the FSM_STATE and ENUM_ENCODING attributes are specified for a given state machine, the FSM_STATE attribute takes precedence.

FSM_COMPLETE : boolean

Decorates: type, subtype, signal, variable

This attribute directs the tool to include default transitions in the finite-state machine that uses the decorated item for its state vector. We typically describe the default transitions with an others clause in a case statement, dealing with state encodings not explicitly described. For example, given the following declarations and attribute specifications:

```
type state_type is (idle, state1, state2);
signal state, next_state : state_type;
attribute FSM_STATE of state, next_state : signal is
            "ONE_HOT";
attribute FSM_COMPLETE of state, next_state : signal is
            true;
...
```

we can write a process for the state machine's combinational logic as follows:

```
fsm_logic : process (state, ...) is
begin
  case state is
    when idle   =>  next_state <= ...;
    when state1 =>  next_state <= ...;
    when state2 =>  next_state <= ...;
    when others =>  next_state <= idle;
  end case;
end process fsm_logic;
```

According to the rules of VHDL, all alternatives for the state value are covered by the first three choices in the case statement. However, there are several synthesized encoding values that are not covered, since one-hot encoding is specified. Given the FSM_COMPLETE attribute, the synthesis tool includes additional transitions based on the **others** clause, as this would be the alternative selected for an illegal state value.

BUFFERED : string

Decorates: signal

This attribute directs the tool to use a particular library cell to buffer the driver for the decorated signal. The attribute value specifies the name of the library cell to use, for example:

attribute BUFFERED **of** clk : **signal is** "BUFG";

Alternatively, the attribute value can be one of the following special strings:

- "HIGH_DRIVE": use a high-drive buffer from the synthesis library
- "CLOCK_BUF": use a clock buffer from the synthesis library
- "RESET_BUF": use a reset buffer from the synthesis library

INFER_MUX : boolean

Decorates: case statement label, selected assignment statement label

This attribute directs the tool to infer a multiplexer implementation for the statement instead of random logic or ROM implementation.

IMPLEMENTATION : string

Decorates: procedure, function, signal or variable assignment label

This attribute directs the tool to use a specified synthesis library cell to implement calls to a subprogram or an assigment to a signal or variable, rather than inferring hardware from the subprogram body or assignment expression. The attribute value specifies the library cell name. We use this attribute when we have a VHDL implementation that we wish to simulate, but want to synthesize using the library cell. For example:

```
procedure multiplier
            ( signal clk, start : in std_ulogic;
              signal a, b : in unsigned(15 downto 0);
              signal done : out std_ulogic;
              signal p : out unsigned(31 downto 0) is
begin
  ...
end procedure multiplier;

attribute IMPLEMENTATION of multiplier : procedure is
            "MULTSEQ_16X16;
...

multiplier(clk, start, data1, data2, done, product);
```

When we simulate this model, the body of the procedure is called. When we synthesize, the procedure body is ignored, and the MULTSEQ_16X16 library cell is included in the inferred hardware.

RETURN_PORT_NAME : string

Decorates: function

This attribute is used in conjunction with the **IMPLEMENTATION** attribute for a function. The **RETURN_PORT_NAME** attribute value specifies the name of the port on the library cell that corresponds to the return value of the function. For example:

```
function and_or_invert (a, b, c, d : in std_ulogic)
                        return std_ulogic is
begin
  return not ( (a and b) or (c and d) );
end function and_or_invert;

attribute IMPLEMENTATION of and_or_invert : function is
        "AOI";
attribute RETURN_PORT_NAME of and_or_invert : function is
        "O";
...

q <= and_or_invert(w, x, y, z);
```

The **IMPLEMENTATION** attribute specifies that the AOI library cell be used for the function. The **RETURN_PORT_NAME** attribute specifies that the O port of that cell is the output port.

ENUM_ENCODING : string

> *Decorates*: type, subtype
>
> This attribute directs the tool to use a specified binary encoding for values of the enumeration type or subtype decorated by the attribute. The value of the attribute is a sequence of bit-vector literals, representing the encoding for the enumeration values in the type. For example, given an enumeration type for states in a finite-state machine:

```
type state is (idle, preamble, data, crc, ok, error);
```

we can define the state encoding as follows:

```
attribute enum_encoding of state : type is
        "000 001 010 011 100 111";
```

> The bit vectors correspond in order to the enumeration values in the type definition. All of the literals must contain the same number of bits, and underscore characters may be included to enhance readability. Note that we should ensure that the enumeration values are listed in ascending order of their encodings. Otherwise synthesized relational operations may not produce the same results as relational operators evaluated during simulation.

ROM_BLOCK : string

> *Decorates*: constant, variable, signal
>
> This attribute directs the tool to implement the decorated item as a ROM. The attribute value specifies the library cell to use for the ROM. For example:

```
constant ROM : mem_array := (X"0020", X"FC01",
                            ...
                            others => X"0000");

attribute ROM_BLOCK of ROM : constant is "ROM_SYNCH";
```

RAM_BLOCK : string

Decorates: variable, signal

This attribute directs the tool to implement the decorated item as a RAM. The attribute value specifies the library cell to use for the RAM. For example:

```
signal RAM : mem_array;

attribute RAM_BLOCK of RAM : signal is "RAM_PIPELINED";
```

LOGIC_BLOCK : string

Decorates: constant, variable, signal

This attribute directs the tool to implement the decorated item as a combinational logic block (instead of a ROM) or discrete sequential logic (for example, using flipflops and registers, instead of a RAM).

GATED_CLOCK : boolean

Decorates: signal, process

This attribute directs the tool to use clock gating rather than separate enable inputs for edge-triggered storage devices. Use of clock gating can reduce power consumption considerably. However, we must ensure that the enable signal used to gate the clock is glitch free and has appropriate timing.

We can use this attribute to decorate a gated clock signal, in which case all storage devices connected to the clock use clock gating. For example:

```
attribute GATED_CLOCK of gclk : signal is true;
...

gclk <= clk and enable;

reg : process (gclk) is
begin
  if rising_edge(gclk) then
    q <= d;
  end if;
end process;
```

Alternatively, we can use the attribute to decorate specific processes that are to use clock gating. For example:

```
attribute GATED_CLOCK of reg : label is true;

reg : process(clk) is
begin
  if rising_edge(clk) then
```

```
      if enable = '1' then
          q <= d;
      end if;
    end if;
  end process;
```

Without gated clocking, the register inferred for this process would be active on each clock-edge. The internal logic transitions would involve dynamic power consumption. By decorating the process with the attribute, we direct the tool to use clock gating instead of using the **enable** signal to select the input to be stored. The register is only active on those clock-edges where **enable** is '1'.

COMBINATIONAL : boolean

Decorates: process, signal assignment label

This attribute directs the tool to infer purely combinational logic for the decorated item. It is an error if the decorated item represents sequential logic. We can use this attribute to force the synthesis tool to issue an error if we inadvertently imply level sensitive storage by omitting an assignment to an object on some path through a process.

21.8 Metacomments

Normally, comments in a model are not interpreted by a tool. We include comments as documentation for the human reader. However, the synthesis standard defines two *metacomments*, that is, comments that are to be interpreted by a synthesis tool. They are

```
-- rtl_synthesis off
```

and

```
-- rtl_synthesis on
```

The metacomments are not case sensitive; they can be in lowercase, uppercase or a combination of the two. Any VHDL code following an **rtl_synthesis off** metacomment and before a subsequent **rtl_synthesis on** metacomment is ignored by the synthesis tool. It is as though the code were also comments. Thus the model, excluding the ignored parts, must still be a valid VHDL model. Other tools, such as simulators, do not ignore the code. We would normally use the metacomments to exclude from synthesis parts of the model that are only intended for simulation. Examples are processes that check timing or that use file input/output for instrumentation.

EXAMPLE 21.22   *Omitting monitoring code from synthesis*

Suppose we include code in an architecture body to monitor operation of the design during simulation. We can exclude the code from analysis by the synthesis tool by surrounding it in metacomments, as follows:

```
architecture rtl of subsystem is
  ...
begin

  ... -- synthesizable processes and assignments

  -- rtl_synthesis off

  monitor : process is
    use std.textio.all;
    file monitor_file : text open write_mode is "monitor.txt";
    variable L : line;
    ...
  begin

    ...

  end process monitor;

  -- rtl_synthesis on

end architecture rtl;
```

Use of files and file operations is not allowed in synthesizable code, so the synthesis tool would produce errors if it tried to interpret the **monitor** process.

We must take care when excluding part of a model from synthesis to ensure that we don't inadvertently omit part of the model that should be synthesized to hardware. Otherwise we can get a mismatch between simulation results and operation of the synthesized hardware.

Exercises

1. [❶ 21.2] Which of the following types does the IEEE 1076.6 standard allow as the type of a variable?

    ```
    type temp is range -60 to 150;
    type temp_vec is array (natural range <>) of temp;
    type location is (inside, outside, buried);
    type local_temp_vec is array (location) of temp;
    type location_vec is array (natural range <>) of location;
    type word_vec is array (natural range <>) of
            std_ulogic_vector(31 downto 0);
    ```

2. [❶ 21.3] What hardware, if any, would be inferred for the following statement?

    ```
    if sel = '0' or sel = 'L' then
      z <= in0;
    elsif sel = '1' or sel = 'H' then
      z <= in1;
    elsif sel = 'U' then
    ```

```
            z <= 'U';
        else
            z <= 'X';
        end if;
```

3. [● 21.4] What hardware would be inferred for the following assignment?

    ```
    with fn select
        z <= a + b         when "00",
             a - b         when "01",
             a and b       when "10",
             a and not b when "11";
    ```

4. [● 21.4] Write a synthesizable process that represents a logic block with tristate
    outputs. If **enable_n** is '0', an 8-bit output **dat_o** is driven with the value of either **reg1**
    (if **adr** is '0') or **reg2** (if **adr** is '1'), and a single-bit output **ack_o** is driven with '1'. If
    enable_n is '1', both outputs are 'Z'.

5. [● 21.4/21.5] Why would a synthesis tool infer storage for the following process?

    ```
    mux_adder : process ( sel, x, y, z, carry_in )
        variable operand : unsigned(15 downto 0);
    begin
        case sel is
            when "00" => operand := x;
            when "01" => operand := y;
            when "10" => operand := z;
            when others =>
                report "Illegal value for sel." severity error;
        end case;
        sum <= operand + carry_in;
    end process mux_adder;
    ```

6. [● 21.5] Write a process representing an edge-triggered register with no reset but with
    synchronous enable.

7. [● 21.6] Write declarations for the storage of an 8K × 16-bit RAM storing values of type
    signed, with all locations initialized to zero.

8. [● 21.6] Rewrite the 7-segment decoder of Example 21.21 using a constant to repre-
    sent the ROM storage.

9. [● 21.7] Write an attribute specification for inclusion in the state machine model of
    Example 21.16 specifying an encoding of "00" for **ready**, "01" for **ack**, and "11" for **err**.

Chapter 22

Case Study: System Design Using the Gumnut Core

In this case study we develop a series of models for the *Gumnut* 8-bit microcontroller core, described in *Digital Design: An Embedded Systems Approach Using VHDL* [2]. We first develop a behavioral model and verify it using a test bench consisting of instrumentation to monitor instruction execution. Next, we refine the model to the synthesizable register-transfer level. We verify this version in a test bench that compares its outputs with those of the behavioral version. Finally, we develop a system model of a digital alarm clock that uses the Gumnut as an embedded soft core, and synthesize the system for implementation in an FPGA target.

22.1 Overview of the Gumnut

The Gumnut is an 8-bit processor core intended for educational purposes. (A gumnut is a small seedpod of an Australian eucalyptus tree. It is something small from which large things grow.) The Gumnut is similar to 8-bit microcontrollers for small embedded applications, but has an instruction set architecture more similar to RISC processors. We use it as the subject of this case study to show how we might develop high-level models of complex devices such as a CPU. We start by describing the view of the processor as seen by the machine language programmer and by the hardware designer interfacing the processor with the rest of a computer system.

22.1.1 Instruction Set Architecture

The Gumnut has separate instruction and data memories. The instruction memory stores up to 4,096 instructions (using 12-bit addresses), and the data memory stores 256 bytes (using 8-bit addresses). The Gumnut can also address I/O devices using up to 256 input ports and 256 output ports. Within the core, there are eight general-purpose registers, named r0 through r7, that can hold data to be operated upon by instructions. Register r0 is special, in that it is hardwired to have the value 0, and any updates to it are ignored. The processor also has two single-bit condition-code registers called Z (zero) and C (carry). They are set to 1 or cleared to 0 depending on the result of certain instructions, and can be tested to decide among alternative courses of action in the program.

Table 22.1 lists the complete Gumnut instruction set in assembly-language format. In the table, *rd* and *rs* are registers, *op2* is a register (*rs2*) or an immediate value (*immed*), *count* is count of number of places to shift or rotate, *disp* is a displacement from the next-instruction address, and *addr* is a jump target address.

TABLE 22.1 *The Gumnut instruction set*

Arithmetic and logical instructions	
add *rd, rs, op2*	Add *rs* and *op2*, result in *rd*
addc *rd, rs, op2*	Add *rs* and *op2* with carry, result in *rd*
sub *rd, rs, op2*	Subtract *op2* from *rs*, result in *rd*
subc *rd, rs, op2*	Subtract *op2* from *rs* with carry, result in *rd*
and *rd, rs, op2*	Logical AND of *rs* and *op2*, result in *rd*
or *rd, rs, op2*	Logical OR of *rs* and *op2*, result in *rd*
xor *rd, rs, op2*	Logical XOR of *rs* and *op2*, result in *rd*
mask *rd, rs, op2*	Logical AND of *rs* and NOT *op2*, result in *rd*
Shift instructions	
shl *rd, rs, count*	Shift *rs* value left *count* places, result in *rd*
shr *rd, rs, count*	Shift *rs* value right *count* places, result in *rd*
rol *rd, rs, count*	Rotate *rs* value left *count* places, result in *rd*
ror *rd, rs, count*	Rotate *rs* value right *count* places, result in *rd*
Memory and I/O instructions	
ldm *rd, (rs)±offset*	Load to *rd* from memory
stm *rd, (rs)±offset*	Store to memory from *rd*
inp *rd, (rs)±offset*	Input to *rd* from input controller register
out *rd, (rs)±offset*	Output to output controller register from *rd*
Branch instructions	
bz ±*disp*	Branch if Z is set
bnz ±*disp*	Branch is Z is not set
bc ±*disp*	Branch if C is set
bnc ±*disp*	Branch if C is not set

Jump instructions	
jmp *addr*	Jump to *addr*
jsb *addr*	Jump to subroutine at *addr*
Miscellaneous instructions	
ret	Return from subroutine
reti	Return from interrupt
enai	Enable interrupts
disi	Disable interrupts
wait	Wait for interrupts
stby	Enter low-power standby mode

The arithmetic and logical instructions operate on 8-bit data values stored in the core's general-purpose registers and store the result in the destination register, *rd*. For each instruction, one value is taken from a source register, *rs*. The other value, *op2*, either comes from a second source register (*rs2*) or is an immediate value (*immed*) specified as part of the instruction.

The addition and subtraction instructions treat the data values as 8-bit unsigned integers. The **addc** instruction includes the value of the C condition code as a carry-in bit, and the **subc** instruction includes the C value as a borrow-in bit. All of the instructions in this group modify the Z and the C bits. They set Z to 1 if the instruction result is 0, and they clear Z to 0 if the result is non-zero. The **add** and **addc** instructions set C to the carry-out bit of the addition, the **sub** and **subc** instruction set C to the borrow out of the subtraction, and the remaining logical instructions clear C to 0.

The shift instructions shift or rotate 8-bit values taken from the general purpose register *rs* and store the result in register *rd*. The number of places to shift or rotate is specified in the instruction as *count*. The shift-left and shift-right instructions discard the bits shifted past the end of the 8-bit byte and fill the vacated bit positions with 0s. The rotate-left and rotate-right instructions copy the bits shifted past the end of the byte around to the other end. All of these instructions set Z to 1 if the instruction result is 0, and they clear Z to 0 if the result is non-zero. They set the C bit to the value of the last bit shifted past the end of the byte.

The Gumnut has separate instructions and separate 8-bit address spaces for accessing data memory and I/O controllers. For all of the Gumnut's memory and I/O instructions, the address to access is computed by adding the current value in *rs* and an *offset* value specified in the instruction. The load from memory instruction reads from the data memory at the computed address and puts the read value in register *rd*. The store to memory instruction writes the value from register *rd* to the data memory at the computed address. The input and output instructions perform similar operations, but read or write to the I/O controller registers at the computed address. None of these instructions affect the values of the Z and C bits.

If we want to specify a particular address to access, we can use r0 as the register for *rs*. Recall that r0 always contains 0, so adding it to the offset value specified in the instruction just gives the offset value. In this case, we usually interpret the offset value as an unsigned 8-bit address. Our assembler tool allows us to imply the specification "(r0)" by omission and just write the address value, for example,

```
inp r3, 156
```

which reads from the I/O controller register at address 156 into r3. Similarly, if a register contains the address we want to access, we can use an offset of 0. Again, our assembler allows us to imply a 0 offset by omission, as in the instruction

```
out r3, (r7)
```

The branch instructions modify the sequential flow of execution by changing value of the program counter (PC) in the Gumnut core. Each form of branch tests a condition, and if the condition is true, adds a signed 8-bit displacement value to the PC. The displacement, specified in the instruction, indicates how many locations forward or backward the next instruction to execute is from the current instruction. (A displacement of 0 refers to the instruction after the branch, since the PC has already been incremented after fetching the branch instruction.) If the condition is false, the PC is unchanged, and execution continues sequentially. The different branch instructions allow us to test each of the Z and C condition code bits for being set to 1 or not set to 1. Since these bits are affected by arithmetic, logical and shift instructions, we often deliberately precede a branch instruction with one of these instructions to compare data values. In other cases, the condition code setting occurs as a serendipitous side effect of data operations that we need to perform anyway. Execution of a branch instruction does not affect the values of the Z and C bits.

The first of the jump instructions, **jmp**, unconditionally breaks the sequential flow of execution by setting the PC to the address specified in the instruction. The second of the jump instructions, **jsb**, allows us to call a subroutine. It is used in tandem with the **ret** instruction, which returns from the subroutine to the place of the call. The **jsb** instruction pushes the incremented PC value (the return address) onto an internal stack and then updates the PC with the subroutine address specified in the instruction. The **ret** instruction pops the saved return address from the stack to the PC. The return-address stack can hold up to eight entries. The **jmp** and **jsb** instructions do not affect the values of the Z and C bits.

The remaining miscellaneous instructions deal with interrupts. The enable-interrupt (**enai**) instruction allows the processor to respond to interrupt events, and the disable-interrupt (**disi**) instruction prevents the processor from responding. When the processor responds to an interrupt event, it saves the incremented PC value and the values of the Z and C condition codes in special registers, disables further interrupts, and then transfers control to the interrupt handler at address 1. The interrupt handler finishes with a return-from-interrupt (**reti**) instruction rather than an **ret** instruction. The **reti** instruction restores the saved PC and condition code values and re-enables interrupts. The **wait** instruction suspends execution until an interrupt occurs, and the **stby** instruction enters a low-power standby mode until an interrupt occurs. The difference is that the CPU would normally be able to respond to an interrupt immediately when suspended using a **wait** instruction, whereas it could take some time to power up from a **stby** instruction. The instructions in this group, apart from the **reti** instruction, do not affect the values of the Z and C bits.

Instructions in the Gumnut are all 18 bits long, and are encoded in several formats, shown in Figure 22.1. The leftmost bits, together with the function code (*fn*), form the opcode. The encoding used for function codes is shown in Table 22.2. Those instructions that specify register numbers have the numbers encoded in 3-bit binary form in the *rd*, *rs*, and *rs2* fields of the instruction word. Similarly, instructions that specify immediate values, offsets, or displacements have those values binary encoded in the rightmost 8 bits of the instruction word. The shaded parts of the instruction word in each format represent bits that are ignored.

FIGURE 22.1

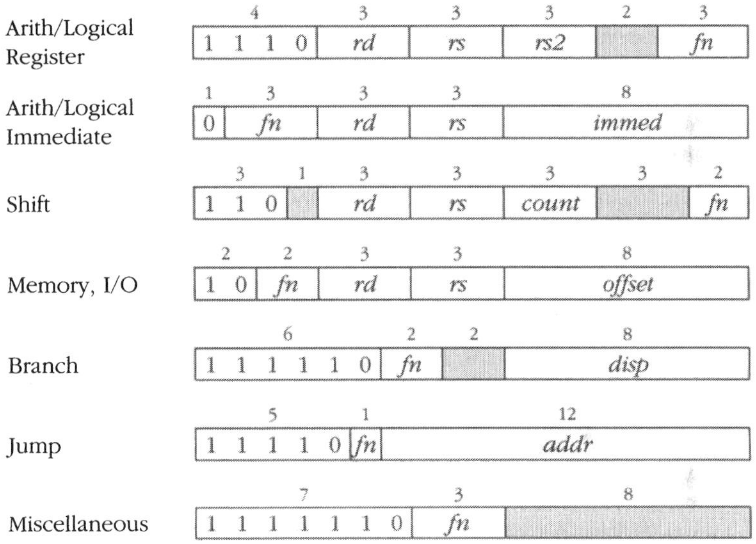

Gumnut instruction formats.

TABLE 22.2 *Function code values*

add	000	addc	001	sub	010	subc	011
and	100	or	101	xor	110	mask	111
shl	00	shr	01	rol	10	ror	11
ldm	00	stm	01	inp	10	out	11
bz	00	bnz	01	bc	10	bnc	11
jmp	0	jsb	1				
ret	000	reti	001	enai	010	disi	011
wait	100	stby	101				

22.1.2 External Interface

The Gumnut interfaces to the rest of the system in which it is embedded via a number of external signals. These are shown in Figure 22.2. Each of the instruction memory, data memory, and I/O ports connect to the core using a simplified version of the Wishbone bus, an open bus specification published by the OpenCores Organization [13]. The **clk_i** signal is the master clock for the Gumnut. All other signals are sampled or set synchronously with the clock. The **rst_i** signal re-initializes the Gumnut to its reset state. When **rst_i** is negated, the Gumnut commences instruction execution, starting from address 0 in the instruction memory.

The **int_req** signal is used to request an interrupt of the Gumnut. When this signal is active and the Gumnut interrupts are enabled, the Gumnut will save state and transfer to the interrupt service code. It asserts the **int_ack** signal for one cycle to indicate start of interrupt service. The I/O port controller must negate **int_req** before the service code returns and re-enables interrupts; otherwise a second spurious interrupt will be received. Usually, an I/O port controller would negate the interrupt request in response to **int_ack** or to the Gumnut reading or writing an I/O port register.

We will describe the bus timing for read and write operations on the data memory bus. The timing for reads and writes on the port bus and for reads on the instruction memory bus is identical. The timing of read operations is shown in Figure 22.3. The Gumnut starts a read operation by driving the **data_adr_o** signals with the address and setting **data_cyc_o** and **data_stb_o** to 1. It also sets **data_we_o** to 0 to indicate that the operation is a read. The memory decodes the address to access the data and drives the data onto the **data_dat_i** signal. If the memory is able to provide the data within the first clock cycle, it sets the **data_ack_i** signal to 1 in that cycle, as shown in Figure 22.3(a). On the next rising clock-edge, the Gumnut sees **data_ack_i** at 1 and completes the operation by setting **data_cyc_o**, **data_stb_o** and **data_we_o** all to 0. If, on the other hand, the memory is slow and is not able to provide the data within the cycle, it leaves **data_ack_i** at 0, as shown in

FIGURE 22.2

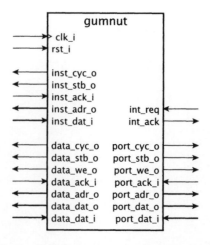

Gumnut core external interface.

FIGURE 22.3

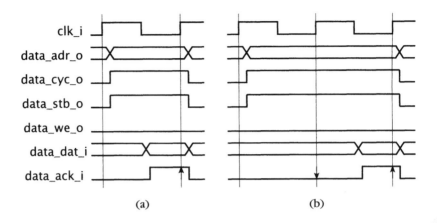

Gumnut data memory read operation: without wait cycles (a), and with one wait cycle (b).

Figure 22.3(b). The Gumnut sees **data_ack_i** at 0 on the rising clock-edge, and extends the operation for a further cycle. The memory can keep **data_ack_i** at 0 for as long as it needs to access the data. Eventually, when it is ready, it drives **data_ack_i** to 1 to complete the operation.

The timing of write operations is similar, shown in Figure 22.4. Again, the Gumnut starts a write operation by driving the **data_adr_o** signal with the address and setting **data_cyc_o** and **data_stb_o** to 1. It sets **data_we_o** to 1 to indicate that the operation is a write and drives the data to be written into the **data_dat_o** signal. The memory decodes the address and updates the selected location with the data. If the memory is able to complete the write within the first clock cycle, it sets the **data_ack_i** signal to 1 in that cycle, as shown in Figure 22.4(a), and the handshake completes as for the read operation. Oth-

FIGURE 22.4

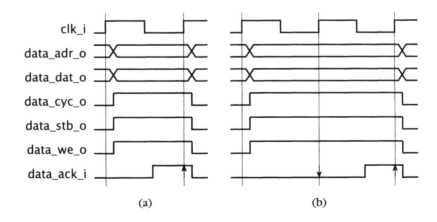

Gumnut data memory write operation: without wait cycles (a), and with one wait cycle (b).

erwise, if the memory is slow, it leaves **data_ack_i** at 0, as shown in Figure 22.4(b), and the operation is extended, as for a read operation.

At first sight, it might appear that the **data_cyc_o** and **data_stb_o** signals are duplicates of each other. However, the Wishbone bus specification defines other more involved operations in which the two control signals serve distinct purposes. While the Gumnut does not use those operations, it includes the signals in order to maintain compatibility with the Wishbone specification. The additional signal is a small cost to pay for compatibility with a large pool of third-party components.

The Gumnut Entity Declaration

We can now write the entity declaration for the Gumnut core, as shown below. The generic constant **debug** controls whether the model writes debugging messages to the standard output stream. The ports of the entity correspond to those shown in Figure 22.2.

```vhdl
library ieee;
use ieee.std_logic_1164.all, ieee.numeric_std.all;

entity gumnut is
  generic ( debug : boolean := false );
  port ( clk_i : in std_ulogic;
         rst_i : in std_ulogic;
         -- Instruction memory bus
         inst_cyc_o : out std_ulogic;
         inst_stb_o : out std_ulogic;
         inst_ack_i : in std_ulogic;
         inst_adr_o : out unsigned(11 downto 0);
         inst_dat_i : in std_ulogic_vector(17 downto 0);
         -- Data memory bus
         data_cyc_o : out std_ulogic;
         data_stb_o : out std_ulogic;
         data_we_o : out std_ulogic;
         data_ack_i : in std_ulogic;
         data_adr_o : out unsigned(7 downto 0);
         data_dat_o : out std_ulogic_vector(7 downto 0);
         data_dat_i : in std_ulogic_vector(7 downto 0);
         -- I/O port bus
         port_cyc_o : out std_ulogic;
         port_stb_o : out std_ulogic;
         port_we_o : out std_ulogic;
         port_ack_i : in std_ulogic;
         port_adr_o : out unsigned(7 downto 0);
         port_dat_o : out std_ulogic_vector(7 downto 0);
         port_dat_i : in std_ulogic_vector(7 downto 0);
         -- Interrupts
         int_req : in std_ulogic;
```

```
              int_ack : out std_ulogic );
  end entity gumnut;
```

Instruction and Data Memories

In systems that use the Gumnut core, we need to provide instruction and data memories. We can provide them as further IP blocks to be instantiated in designs. The entity declaration for the instruction memory is

```
library ieee;
use ieee.std_logic_1164.all, ieee.numeric_std.all;

entity inst_mem is
  generic ( IMem_file_name : string := "gasm_text.dat" );
  port ( clk_i : in std_ulogic;
         cyc_i : in std_ulogic;
         stb_i : in std_ulogic;
         ack_o : out std_ulogic;
         adr_i : in unsigned(11 downto 0);
         dat_o : out std_ulogic_vector(17 downto 0) );
end entity inst_mem;
```

The generic constant **IMem_file_name** specifies the name of a file from which program is loaded. The default file name used by gasm is **gasm_text.dat**, so we use the same default name for the generic constant. The ports of the entity mirror those of the Gumnut entity.

The entity declaration for the data memory is similar:

```
library ieee;
use ieee.std_logic_1164.all, ieee.numeric_std.all;

entity data_mem is
  generic ( DMem_file_name : string := "gasm_data.dat" );
  port ( clk_i : in std_ulogic;
         cyc_i : in std_ulogic;
         stb_i : in std_ulogic;
         we_i : in std_ulogic;
         ack_o : out std_ulogic;
         adr_i : in unsigned(7 downto 0);
         dat_i : in std_ulogic_vector(7 downto 0);
         dat_o : out std_ulogic_vector(7 downto 0) );
end entity data_mem;
```

Again, the entity has a generic for specifying the file name for the initial memory contents, and ports that mirror those of the Gumnut entity. We don't show the architecture bodies for the memories here, in the interest of brevity. They are based on the memory models we described in Chapter 17.

Next, we provide a subsystem model that includes an instance of the core and each of the memories. This subsystem can then be instantiated in a larger design and connected to the required I/O controllers. The subsystem entity declaration is

```vhdl
library ieee;
use ieee.std_logic_1164.all, ieee.numeric_std.all;

entity gumnut_with_mem is
  generic ( IMem_file_name : string := "gasm_text.dat";
            DMem_file_name : string := "gasm_data.dat";
            debug : boolean := false );
  port ( clk_i : in std_ulogic;
         rst_i : in std_ulogic;
         -- I/O port bus
         port_cyc_o : out std_ulogic;
         port_stb_o : out std_ulogic;
         port_we_o : out std_ulogic;
         port_ack_i : in std_ulogic;
         port_adr_o : out unsigned(7 downto 0);
         port_dat_o : out std_ulogic_vector(7 downto 0);
         port_dat_i : in std_ulogic_vector(7 downto 0);
         -- Interrupts
         int_req : in std_ulogic;
         int_ack : out std_ulogic );
end entity gumnut_with_mem;
```

The structural architecture body is shown below. It uses component declarations for the Gumnut core and the memories, allowing alternative architecture bodies to be bound using a separate configuration declaration.

```vhdl
library ieee;
use ieee.std_logic_1164.all, ieee.numeric_std.all;

architecture struct of gumnut_with_mem is

  -- Instruction memory bus
  signal inst_cyc_o : std_ulogic;
  signal inst_stb_o : std_ulogic;
  signal inst_ack_i : std_ulogic;
  signal inst_adr_o : unsigned(11 downto 0);
  signal inst_dat_i : std_ulogic_vector(17 downto 0);
  -- Data memory bus
  signal data_cyc_o : std_ulogic;
  signal data_stb_o : std_ulogic;
  signal data_we_o : std_ulogic;
  signal data_ack_i : std_ulogic;
  signal data_adr_o : unsigned(7 downto 0);
  signal data_dat_o : std_ulogic_vector(7 downto 0);
  signal data_dat_i : std_ulogic_vector(7 downto 0);
```

```vhdl
component gumnut is
  generic ( debug : boolean );
  port ( clk_i : in std_ulogic;
         rst_i : in std_ulogic;
         -- Instruction memory bus
         inst_cyc_o : out std_ulogic;
         inst_stb_o : out std_ulogic;
         inst_ack_i : in std_ulogic;
         inst_adr_o : out unsigned(11 downto 0);
         inst_dat_i : in std_ulogic_vector(17 downto 0);
         -- Data memory bus
         data_cyc_o : out std_ulogic;
         data_stb_o : out std_ulogic;
         data_we_o : out std_ulogic;
         data_ack_i : in std_ulogic;
         data_adr_o : out unsigned(7 downto 0);
         data_dat_o : out std_ulogic_vector(7 downto 0);
         data_dat_i : in std_ulogic_vector(7 downto 0);
         -- I/O port bus
         port_cyc_o : out std_ulogic;
         port_stb_o : out std_ulogic;
         port_we_o : out std_ulogic;
         port_ack_i : in std_ulogic;
         port_adr_o : out unsigned(7 downto 0);
         port_dat_o : out std_ulogic_vector(7 downto 0);
         port_dat_i : in std_ulogic_vector(7 downto 0);
         -- Interrupts
         int_req : in std_ulogic;
         int_ack : out std_ulogic );
end component gumnut;

component inst_mem is
  generic ( IMem_file_name : string );
  port ( clk_i : in std_ulogic;
         cyc_i : in std_ulogic;
         stb_i : in std_ulogic;
         ack_o : out std_ulogic;
         adr_i : in unsigned(11 downto 0);
         dat_o : out std_ulogic_vector(17 downto 0) );
end component inst_mem;

component data_mem is
  generic ( DMem_file_name : string );
  port ( clk_i : in std_ulogic;
         cyc_i : in std_ulogic;
         stb_i : in std_ulogic;
         we_i : in std_ulogic;
         ack_o : out std_ulogic;
```

```vhdl
              adr_i : in unsigned(7 downto 0);
              dat_i : in std_ulogic_vector(7 downto 0);
              dat_o : out std_ulogic_vector(7 downto 0) );
    end component data_mem;

begin

  core : component gumnut
    generic map ( debug => debug )
    port map ( clk_i      => clk_i,
               rst_i      => rst_i,
               inst_cyc_o => inst_cyc_o,
               inst_stb_o => inst_stb_o,
               inst_ack_i => inst_ack_i,
               inst_adr_o => inst_adr_o,
               inst_dat_i => inst_dat_i,
               data_cyc_o => data_cyc_o,
               data_stb_o => data_stb_o,
               data_we_o  => data_we_o,
               data_ack_i => data_ack_i,
               data_adr_o => data_adr_o,
               data_dat_o => data_dat_o,
               data_dat_i => data_dat_i,
               port_cyc_o => port_cyc_o,
               port_stb_o => port_stb_o,
               port_we_o  => port_we_o,
               port_ack_i => port_ack_i,
               port_adr_o => port_adr_o,
               port_dat_o => port_dat_o,
               port_dat_i => port_dat_i,
               int_req    => int_req,
               int_ack    => int_ack );

  core_inst_mem : component inst_mem
    generic map ( IMem_file_name => IMem_file_name )
    port map ( clk_i => clk_i,
               cyc_i => inst_cyc_o,
               stb_i => inst_stb_o,
               ack_o => inst_ack_i,
               adr_i => inst_adr_o,
               dat_o => inst_dat_i );

  core_data_mem : component data_mem
    generic map ( DMem_file_name => DMem_file_name )
    port map ( clk_i => clk_i,
               cyc_i => data_cyc_o,
               stb_i => data_stb_o,
               we_i  => data_we_o,
               ack_o => data_ack_i,
```

```
                    adr_i => data_adr_o,
                    dat_i => data_dat_o,
                    dat_o => data_dat_i );

    end architecture struct;
```

22.2 A Behavioral Model

Our first model for the Gumnut core is a behavioral architecture body. However, before we write this unit, we will look at a package that defines the types and values representing the Gumnut instruction encoding. This package is common to all models of the Gumnut core.

22.2.1 The Gumnut Definitions Package

The following package declaration defines the types for the Gumnut internal signals, based on standard-logic vectors. The package also defines subtypes and constants for encoded instructions and for use within implementations of the processor.

```
library ieee;
use ieee.std_logic_1164.all, ieee.numeric_std.all;

package gumnut_defs is

    constant IMem_addr_width : positive := 12;
    constant IMem_size : positive := 2**IMem_addr_width;
    subtype IMem_addr is unsigned(IMem_addr_width - 1 downto 0);

    subtype instruction is unsigned(17 downto 0);
    type instruction_array is array (natural range <>) of instruction;

    subtype IMem_array is instruction_array(0 to IMem_size - 1);

    constant DMem_size : positive := 256;

    subtype unsigned_byte is unsigned(7 downto 0);
    type unsigned_byte_array is
            array (natural range <>) of unsigned_byte;

    subtype DMem_array is unsigned_byte_array(0 to DMem_size - 1);

    subtype signed_byte is signed(7 downto 0);
    type signed_byte_array is array (natural range <>) of signed_byte;

    subtype reg_addr is unsigned(2 downto 0);
    subtype immed is unsigned(7 downto 0);
    subtype offset is unsigned(7 downto 0);
    subtype disp is unsigned(7 downto 0);
    subtype shift_count is unsigned(2 downto 0);

    subtype alu_fn_code is unsigned(2 downto 0);
    subtype shift_fn_code is unsigned(1 downto 0);
```

```
subtype mem_fn_code is unsigned(1 downto 0);
subtype branch_fn_code is unsigned(1 downto 0);
subtype jump_fn_code is unsigned(0 downto 0);
subtype misc_fn_code is unsigned(2 downto 0);

constant alu_fn_add  : alu_fn_code := "000";
constant alu_fn_addc : alu_fn_code := "001";
constant alu_fn_sub  : alu_fn_code := "010";
constant alu_fn_subc : alu_fn_code := "011";
constant alu_fn_and  : alu_fn_code := "100";
constant alu_fn_or   : alu_fn_code := "101";
constant alu_fn_xor  : alu_fn_code := "110";
constant alu_fn_mask : alu_fn_code := "111";

constant shift_fn_shl : shift_fn_code := "00";
constant shift_fn_shr : shift_fn_code := "01";
constant shift_fn_rol : shift_fn_code := "10";
constant shift_fn_ror : shift_fn_code := "11";

constant mem_fn_ldm : mem_fn_code := "00";
constant mem_fn_stm : mem_fn_code := "01";
constant mem_fn_inp : mem_fn_code := "10";
constant mem_fn_out : mem_fn_code := "11";

constant branch_fn_bz  : branch_fn_code := "00";
constant branch_fn_bnz : branch_fn_code := "01";
constant branch_fn_bc  : branch_fn_code := "10";
constant branch_fn_bnc : branch_fn_code := "11";

constant jump_fn_jmp : jump_fn_code := "0";
constant jump_fn_jsb : jump_fn_code := "1";

constant misc_fn_ret  : misc_fn_code := "000";
constant misc_fn_reti : misc_fn_code := "001";
constant misc_fn_enai : misc_fn_code := "010";
constant misc_fn_disi : misc_fn_code := "011";
constant misc_fn_wait : misc_fn_code := "100";
constant misc_fn_stby : misc_fn_code := "101";
constant misc_fn_undef_6 : misc_fn_code := "110";
constant misc_fn_undef_7 : misc_fn_code := "111";

subtype disassembled_instruction is string(1 to 30);

procedure disassemble ( instr : instruction;
                        result : out disassembled_instruction );

end package gumnut_defs;
```

The constant **IMem_addr_width** specifies the size of instruction memory addresses in a Gumnut implementation, and the related constant **IMem_size** specifies the size of the instruction memory. The subtype **IMem_addr** is used for instruction memory addresses.

The type **instruction** represents an 18-bit encoded instruction word, and **instruction_array** is an array of instruction words. The subtype **IMem_array** is used in an implementation as the type of the instruction memory. A similar set of declarations follows for the data memory, whose size is specified by the constant **DMem_size** and in which each element is of the subtype **unsigned_byte**. There is also a subtype **signed_byte** for an 8-bit signed value, used for arithmetic operations in processor implementations. For each of these subtypes, array types are also defined.

The next group of subtypes, from **reg_addr** through to **misc_fn_code**, represent the fields of encoded instructions. These are followed by constants whose values represent the encodings for function codes.

The string type **disassembled_instruction** is used in conjuction with the **disassemble** procedure, which accepts a word representing a Gumnut instruction in the parameter **instr** and returns a textual representation of the instruction in the parameter **result**.

The package body provides an implementation of the **disassemble** subprogram:

```vhdl
package body gumnut_defs is

  procedure disassemble ( instr : instruction;
                          result : out disassembled_instruction ) is

    subtype name is string(1 to 4);

    type name_table is array (natural range <>) of name;

    constant alu_name_table : name_table(0 to 7)
      := ( 0 => "add ", 1 => "addc", 2 => "sub ", 3 => "subc",
           4 => "and ", 5 => "or  ", 6 => "xor ", 7 => "msk " );

    constant shift_name_table : name_table(0 to 3)
      := ( 0 => "shl ", 1 => "shr ", 2 => "rol ", 3 => "ror " );

    constant mem_name_table : name_table(0 to 3)
      := ( 0 => "ldm ", 1 => "stm ", 2 => "inp ", 3 => "out " );

    constant branch_name_table : name_table(0 to 3)
      := ( 0 => "bz  ", 1 => "bnz ", 2 => "bc  ", 3 => "bnc " );

    constant jump_name_table : name_table(0 to 1)
      := ( 0 => "jmp ", 1 => "jsb " );

    constant misc_name_table : name_table(0 to 7)
      := ( 0 => "ret ", 1 => "reti", 2 => "enai", 3 => "disi",
           4 => "wait", 5 => "stby", 6 => "um_6", 7 => "um_7" );

    variable instr_01 : instruction := to_01(instr);

    alias instr_alu_reg_fn : alu_fn_code is instr_01(2 downto 0);
    alias instr_alu_immed_fn : alu_fn_code is
            instr_01(16 downto 14);
    alias instr_shift_fn : shift_fn_code is instr_01(1 downto 0);
    alias instr_mem_fn : mem_fn_code is instr_01(15 downto 14);
    alias instr_branch_fn : branch_fn_code is
            instr_01(11 downto 10);
```

```vhdl
alias instr_jump_fn : jump_fn_code is instr_01(12 downto 12);
alias instr_misc_fn : misc_fn_code is instr_01(10 downto 8);

alias instr_rd : reg_addr is instr_01(13 downto 11);
alias instr_rs : reg_addr is instr_01(10 downto 8);
alias instr_r2 : reg_addr is instr_01(7 downto 5);
alias instr_immed : immed is instr_01(7 downto 0);
alias instr_count : shift_count is instr_01(7 downto 5);
alias instr_offset : offset is instr_01(7 downto 0);
alias instr_disp : disp is instr_01(7 downto 0);
alias instr_addr : IMem_addr is instr_01(11 downto 0);

procedure disassemble_reg ( reg : reg_addr;
                            index : positive ) is
  constant str : string := to_string(to_integer(reg));
begin
  result(index) := str(str'left);
end procedure disassemble_reg;

procedure disassemble_unsigned ( n : unsigned;
                                 index : positive ) is
  constant str : string := to_string(to_integer(n));
begin
  result(index to index + str'length - 1) := str;
end procedure disassemble_unsigned;

procedure disassemble_signed ( n : signed;
                               index : positive ) is
  constant str : string := to_string(to_integer(n));
begin
  result(index to index + str'length - 1) := str;
end procedure disassemble_signed;

procedure disassemble_effective_addr ( r : reg_addr;
                                       d : offset;
                                       index : positive ) is
  constant signed_str : string
            := to_string(to_integer(signed(d)));
  constant unsigned_str : string
            := to_string(to_integer(d));
begin
  if r = 0 then
    result(index to index + unsigned_str'length - 1)
      := unsigned_str;
  else
    result(index to index + 3) := "(r )";
    disassemble_reg(r, index + 2);
    result(index + 4 to index + 4 + signed_str'length - 1)
      := signed_str;
```

```
      end if;
    end procedure disassemble_effective_addr;

begin
  if is_X(instr) then
    report "disassemble: metalogical value in instruction word"
      severity error;
    result := (others => 'X');
    return;
  end if;
  result := (others => ' ');

  if instr_01(17) = '0' then
    -- Arithmetic/Logical Immediate
    result(1 to name'length)
      := alu_name_table(to_integer(instr_alu_immed_fn));
    result(name'length + 2 to name'length + 8) := "R , R ,";
    disassemble_reg(instr_rd, name'length + 3);
    disassemble_reg(instr_rs, name'length + 7);
    disassemble_unsigned(instr_immed, name'length + 10);
  elsif instr_01(16) = '0' then
    -- Memory I/O
    result(1 to name'length)
      := mem_name_table(to_integer(instr_mem_fn));
    result(name'length + 2 to name'length + 4) := "R ,";
    disassemble_reg(instr_rd, name'length + 3);
    disassemble_effective_addr(instr_rs,
                               instr_offset, name'length + 6);
  elsif instr_01(15) = '0' then
    -- Shift
    result(1 to name'length)
      := shift_name_table(to_integer(instr_shift_fn));
    result(name'length + 2 to name'length + 8) := "R , R ,";
    disassemble_reg(instr_rd, name'length + 3);
    disassemble_reg(instr_rs, name'length + 7);
    disassemble_unsigned(instr_count, name'length + 10);
  elsif instr_01(14) = '0' then
    -- Arithmetic/Logical Register
    result(1 to name'length)
      := alu_name_table(to_integer(instr_alu_reg_fn));
    result(name'length + 2 to name'length + 10) := "R , R , R";
    disassemble_reg(instr_rd, name'length + 3);
    disassemble_reg(instr_rs, name'length + 7);
    disassemble_reg(instr_r2, name'length + 11);
  elsif instr_01(13) = '0' then
    -- Jump
    result(1 to name'length)
      := jump_name_table(to_integer(instr_jump_fn));
```

```
            disassemble_unsigned(instr_addr, name'length + 2);
          elsif instr_01(12) = '0' then
            -- Branch
            result(1 to name'length)
              := branch_name_table(to_integer(instr_branch_fn));
            disassemble_signed(signed(instr_disp), name'length + 2);
          elsif instr_01(11) = '0' then
            -- Miscellaneous
            result(1 to name'length)
              := misc_name_table(to_integer(instr_misc_fn));
          else
            result(1 to 19) := "Illegal Instruction";
          end if;
      end procedure disassemble;

  end package body gumnut_defs;
```

The six constant arrays of fixed-length strings contain textual representations of the instruction names. For example, if we were to convert the binary function code for the **addc** instruction to an integer and use it to index the array **alu_name_table**, we would retrieve the string "addc". Next, the variable **instr_01** is a copy of the instruction word with all bits converted to '0' or '1'. This simplifies disassembly, since we don't need to deal with the strength of bits or with unknown value. The variable is followed by a number of aliases for the fields of the instruction to be disassembled, allowing us to refer to the individual fields easily.

Within the disassemble procedure, there are a number of local procedures. The first of these, **disassemble_reg**, disassembles a register number given by the parameter **reg** into the **result** string at the position given by the parameter **index**. The procedure uses the **to_string** and **to_integer** operations to derive a string representation of the register number. Since register numbers are between 0 and 7, the string is only one character.

The second local procedure, **disassemble_unsigned**, adds the decimal representation of an unsigned binary number to **result**. It also uses the **to_string** and **to_integer** operations to derive the decimal string. The procedure **disassemble_signed** is similar, but deals with signed binary numbers.

The final local procedure, **disassemble_effective_address**, adds an effective address expression to **result**. The parameter **r** is the base register number, and **d** is the displacement. If the base register is r0, its value is 0, so the displacement is interpreted as an unsigned absolute memory address or I/O port number. In this case, the effective address expression is just the displacement value. If the base register is other than r0, the displacement is interpreted as a signed offset to be added to the base register value. In this case, the effective address expression includes the register number and the offset, for example, "(r2)–4". The procedure first uses the **to_string** and **to_integer** operations to derive decimal strings, one for the signed interpretation of the displacement and one for the unsigned interpretation. The procedure then checks the register number. If it is 0, the procedure simply adds the unsigned decimal string to **result**. Otherwise, the procedure adds the template string "(r)" and calls the **disassemble_reg** procedure to fill in the register number in the template. It then adds the signed decimal string after the template.

The body of **disassemble** starts by checking whether the original instruction passed to it has any unknown values in it. If so, the instruction word cannot be properly disassembled, so the procedure reports an error message and returns with the result set to a string of 'X' characters. Otherwise, the procedure clears the result in preparation for disassembly.

The procedure examines successive bits of the instruction word, starting from the left, to determine the kind of instruction. For an arithmetic/logical immediate format instruction, the procedure uses the function code field to index the corresponding instruction name table and inserts the instruction name into the result string. It then inserts a template for the register and immediate operands, then calls the **disassemble_reg** procedure for each register operand and the disassemble_unsigned for the immediate operand.

For a memory-I/O instruction, the procedure similarly uses the function code to include the instruction name in the result string, then includes a template for the source or destination register operand. It then calls **disassemble_effective_address** to process the base address register and the displacement.

The remaining instructions are all processed similarly, using function codes to include the name of the instruction, adding templates for operands, and calling the locally declared procedures to disassemble operands.

After checking for the encoding of all of the valid instructions, the procedure includes an **else** alternative to deal with an illegal instruction, that is, an encoding that does not represent a Gumnut instruction. Such an instruction might be executed in the case of a coding error in a Gumnut program causing a control transfer to an uninitialized part of the instruction memory. The **disassemble** procedure simply returns a result string containing an error message.

22.2.2   The Gumnut Behavioral Architecture Body

We are now in a position to write the behavioral architecture body for the Gumnut core. An outline of the architecture body is shown below. It consists of a single process labeled **interpreter**, which implements the fetch/execute loop common to nearly all basic CPUs. The process also contains variables that represent the internal registers of the Gumnut. We use unsigned-vector types for these internal registers and perform operations using the arithmetic functions from the IEEE standard **numeric_std** package.

```
library ieee;  use ieee.numeric_std.all;

use work.gumnut_defs.all;
use std.textio.all;

architecture behavior of gumnut is
begin

  interpreter : process

    variable PC : IMem_addr;

    variable IR : instruction;

    alias IR_alu_reg_fn : alu_fn_code is IR(2 downto 0);
    alias IR_alu_immed_fn : alu_fn_code is IR(16 downto 14);
```

```vhdl
alias IR_shift_fn : shift_fn_code is IR(1 downto 0);
alias IR_mem_fn : mem_fn_code is IR(15 downto 14);
alias IR_branch_fn : branch_fn_code is IR(11 downto 10);
alias IR_jump_fn : jump_fn_code is IR(12 downto 12);
alias IR_misc_fn : misc_fn_code is IR(10 downto 8);

alias IR_rd : reg_addr is IR(13 downto 11);
alias IR_rs : reg_addr is IR(10 downto 8);
alias IR_r2 : reg_addr is IR(7 downto 5);
alias IR_immed : immed is IR(7 downto 0);
alias IR_count : shift_count is IR(7 downto 5);
alias IR_offset : disp is IR(7 downto 0);
alias IR_disp : disp is IR(7 downto 0);
alias IR_addr : IMem_addr is IR(11 downto 0);

constant stack_depth : positive := 8;
subtype stack_index is natural range 0 to stack_depth - 1;
type stack_array is array (stack_index) of IMem_addr;
variable stack : stack_array;
variable SP : stack_index;

subtype reg_index is natural range 0 to 7;
variable GPR : unsigned_byte_array(reg_index);

variable ALU_result : unsigned_byte;
variable cc_Z : std_ulogic;
variable cc_C : std_ulogic;

variable int_en : std_ulogic;
variable int_PC : IMem_addr;
variable int_Z, int_C : std_ulogic;

variable disassembled_instr : disassembled_instruction;
variable debug_line : line;

-- local procedures for use within the interpreter
...

begin -- interpreter
  perform_reset;
  wait until rising_edge(clk_i) and rst_i = '0';
  -- fetch/decode/execute loop
  fetch_execute_loop : loop
    -- check for interrupts
    if int_en and int_req then
      perform_interrupt;
      exit fetch_execute_loop when rst_i;
      next fetch_execute_loop;
    end if;
```

```
-- fetch next instruction
fetch_instruction;
exit fetch_execute_loop when rst_i;
next fetch_execute_loop when is_X(IR);

-- decode and execute the instruction
if IR(17) = '0' then
  -- Arithmetic/Logical Immediate
  perform_alu_op(fn => IR_alu_immed_fn,
                 a => GPR(to_integer(IR_rs)),
                 b => IR_immed,
                 C_in => cc_C,
                 result => ALU_result,
                 Z_out => cc_Z, C_out => cc_C);
  if IR_rd /= 0 then
    GPR(to_integer(IR_rd)) := ALU_result;
  end if;
elsif IR(16) = '0' then
  -- Memory I/O
  perform_mem;
  exit fetch_execute_loop when rst_i;
elsif IR(15) = '0' then
  -- Shift
  perform_shift_op(fn => IR_shift_fn,
                   a => GPR(to_integer(IR_rs)),
                   count => IR_count,
                   result => ALU_result,
                   Z_out => cc_Z, C_out => cc_C);
  if IR_rd /= 0 then
    GPR(to_integer(IR_rd)) := ALU_result;
  end if;
elsif IR(14) = '0' then
  -- Arithmetic/Logical Register
  perform_alu_op(fn => IR_alu_reg_fn,
                 a => GPR(to_integer(IR_rs)),
                 b => GPR(to_integer(IR_r2)),
                 C_in => cc_C,
                 result => ALU_result,
                 Z_out => cc_Z, C_out => cc_C);
  if IR_rd /= 0 then
    GPR(to_integer(IR_rd)) := ALU_result;
  end if;
elsif IR(13) = '0' then
  -- Jump
  perform_jump;
elsif IR(12) = '0' then
  -- Branch
```

```
        perform_branch;
    elsif IR(11) = '0' then
        -- Miscellaneous
        perform_misc;
        exit fetch_execute_loop when rst_i;
    else
        -- Illegal instruction
        null;
    end if;

  end loop fetch_execute_loop;
 end process interpreter;

end architecture behavior;
```

Overview of the Interpreter

The first group of declarations in the **interpreter** process represents the Gumnut internal registers. The variable **PC** represents the program counter, and **IR** represents the instruction register. The aliases represent the fields of instructions of different formats. These aliases allow us to refer easily to the fields of an instruction in order to interpret it. The declarations **stack_depth**, **stack_index**, **stack_array** and **stack** represent the return address stack used for subroutine linkage, and the variable **SP** is the stack pointer. **Reg_index** and **GPR** represent the general purpose registers described in Section 22.1.1. **ALU_result** is a temporary register used for the result of ALU instructions, and **cc_Z** and **cc_C** represent the Z and C condition codes. **Int_en** is a flag indicating whether interrupts are enabled, and **int_PC**, **int_Z** and **int_C** are used to save the state of the processor during interrupt service. Finally, the variables **disassembled_instr** and **debug_line** are used in conjunction with the **disassemble** procedure described earlier to form debug messages reported by the model. Following the object declarations, we include declarations of local procedures for use within the interpreter. We describe them in detail later in this section.

The statement part of the interpreter process implements the behavior of the Gumnut core. The process first resets the internal state of the Gumnut using the locally declared **perform_reset** procedure, then waits until the next rising edge of **clk_i** at which **rst_i** is '0' before proceeding. The process then enters the fetch/execute loop. While the **rst_i** input remains negated, the loop checks for interrupt requests, fetches the next instruction from the instruction memory, analyzes the fields of the fetched instruction and then performs the appropriate operations to execute the instruction. The analysis is done using an if statement that examines bits of the instruction word, in a similar way to the **disassemble** procedure described earlier. Each class of instruction is performed by calling one of the local procedures. In the case of arithmetic/logical and shift instructions, which produce a result in a destination register, the register is only written if it is not r0. After performing a memory operation, which involves waiting for one or more clock-edges, the process tests the **rst_i** input again and exits the loop if it is '1'. The same is done for miscellaneous group of instructions, which include the **wait** and **stby** instructions. When the loop is terminated, the process starts again from the top of the statement part.

Resetting the Interpreter

The procedure that resets the interpreter before entering the loop is shown below. These statements are executed when the model is initialized, simulating a power-on reset, and when the rst_i input to the processor is activated. The first group of assignments initializes the processor's internal state by clearing all internal registers and flags to 0. The second group resets the external interface of the processor by setting the bus control and interrupt acknowledge signals to '0'.

```
procedure perform_reset is
begin
  -- Reset internal state
  PC := (others => '0');
  SP := 0;
  GPR := (others => X"00");
  cc_Z := '0';
  cc_C := '0';
  int_en := '0';
  -- Reset bus signals
  inst_cyc_o <= '0';
  inst_stb_o <= '0';
  data_cyc_o <= '0';
  data_stb_o <= '0';
  data_we_o <= '0';
  port_cyc_o <= '0';
  port_stb_o <= '0';
  port_we_o <= '0';
  int_ack <= '0';
end perform_reset;
```

Acknowledging an Interrupt

The procedure for acknowledging an interrupt is shown below. It first saves a copy of the current program counter, converted to integer form, for subsequent use in displaying a debug message. It then copies the current values of the program counter and condition codes to the variables used to save them during interrupt service, and disables further interrupts, and forces the PC to instruction address 1. The procedure then sets the int_ack output of the processor core to '1' for a clock cycle, then back to '0', to indicate to the I/O port that the interrupt request is acknowledged. Finally, if debug messages are enabled, the procedure displays a debug message showing the address at which the interrupt occurred.

```
procedure perform_interrupt is
  variable PC_num : natural;
begin
  PC_num := to_integer(PC);
  int_PC := PC;
```

```
        int_Z := cc_Z;
        int_C := cc_C;
        int_en := '0';
        PC := to_unsigned(1, PC'length);
        int_ack <= '1';
        wait until rising_edge(clk_i);
        int_ack <= '0';
        if debug then
          swrite(debug_line, "Interrupt acknowledged at PC = ");
          write(debug_line, PC_num, field => 4, justified => right);
          writeline(output, debug_line);
        end if;
      end procedure perform_interrupt;
```

Fetching an Instruction

The next procedure, shown below, fetches an instruction from the instruction memory. The procedure first saves a copy of the program counter, converted to integer form, for use in a debug message. Next, it starts a read operation on the instruction memory bus, as described in Section 22.1.2. On each subsequent rising clock-edge, the procedure checks whether rst_i is active, and if so, returns so that the main interpreter process can reset the core. When the instruction memory sets inst_ack_i to '1', the procedure exits the loop and completes the read operation, assigns the read data to the instruction register variable, and increments the PC. Finally, if debug messages are enabled, the procedure disassembles the fetched instruction and displays a debug message showing the address of the instruction and its disassembled form.

```
      procedure fetch_instruction is
        variable PC_num : natural;
      begin
        PC_num := to_integer(PC);
        inst_cyc_o <= '1';
        inst_stb_o <= '1';
        inst_adr_o <= PC;
        loop
          wait until rising_edge(clk_i);
          if rst_i then
            return;
          end if;
          exit when inst_ack_i;
        end loop;
        IR := unsigned(inst_dat_i);
        PC := PC + 1;
        inst_cyc_o <= '0';
        inst_stb_o <= '0';
        if debug then
          disassemble(IR, disassembled_instr);
```

```
                write(debug_line, PC_num, field => 4, justified => right);
                swrite(debug_line, ": ");
                write(debug_line, disassembled_instr);
                writeline(output, debug_line);
            end if;
        end procedure fetch_instruction;
```

Performing an Arithmetic/Logical Operation

The procedure **perform_alu_op**, shown below, performs operations needed for arithmetic/logic unit (ALU) instructions and for calculating the effective address in memory-I/O instructions. The parameter **fn** is the function code indicating the operation to perform; a and b are the operands; c_in is the carry-in; **result** is the result of the operation; and Z_out and C_out are the condition code values produced by the operation.

```
        procedure perform_alu_op ( fn : in alu_fn_code;
                                    a, b : in unsigned_byte;
                                    C_in : in std_ulogic;
                                    result : out unsigned_byte;
                                    Z_out, C_out : out std_ulogic ) is
        begin
          case fn is
            when alu_fn_add =>
              (C_out, result) := ('0' & a) + ('0' & b);
            when alu_fn_addc =>
              (C_out, result) := ('0' & a) + ('0' & b) + C_in;
            when alu_fn_sub =>
              (C_out, result) := ('0' & a) - ('0' & b);
            when alu_fn_subc =>
              (C_out, result) := ('0' & a) - ('0' & b) - C_in;
            when alu_fn_and =>
              (C_out, result) := ('0' & a) and ('0' & b);
            when alu_fn_or =>
              (C_out, result) := ('0' & a) or ('0' & b);
            when alu_fn_xor =>
              (C_out, result) := ('0' & a) xor ('0' & b);
            when alu_fn_mask =>
              (C_out, result) := ('0' & a) and not ('0' & b);
            when others =>
              report "Program logic error in interpreter"
                severity failure;
          end case;
          Z_out := result ?= X"00";
        end procedure perform_alu_op;
```

In the body of the procedure, the case statement selects which ALU operation to perform. In each case, the operands are extended with an additional '0' bit so that the carry-out of the operation can be captured in **C_out**. Arithmetic operations are performed using the overloaded operators from the **numeric_std** package. In the case of add and subtract with carry, the carry-in is included in the operation. After computing the result, the procedure compares it with a vector of '0' bits to determine the value of the **Z_out** flag.

Performing a Shift Operation

The procedure for performing shift operations, **perform_shift_op**, is shown below and is similar to **perform_alu_op**. Instead of a second operand parameter, it has the parameter **count** to specify the number of bits by which to shift.

```vhdl
procedure perform_shift_op ( fn : in shift_fn_code;
                             a : in unsigned_byte;
                             count : in shift_count;
                             result : out unsigned_byte;
                             Z_out, C_out : out std_ulogic ) is
begin
  case fn is
    when shift_fn_shl =>
      (C_out, result) := ('0' & a) sll to_integer(count);
    when shift_fn_shr =>
      (result, C_out) := (a & '0') srl to_integer(count);
    when shift_fn_rol =>
      result := a rol to_integer(count);
      C_out := result(unsigned_byte'right);
    when shift_fn_ror =>
      result := a ror to_integer(count);
      C_out := result(unsigned_byte'left);
    when others =>
      report "Program logic error in interpreter"
        severity failure;
  end case;
  Z_out := result ?= X"00";
end procedure perform_shift_op;
```

The statements that perform shift-left and shift-right operations append a '0' bit to the operand, on the left for left shifts and on the right for right shifts, so that the last bit shifted out of the operand can be captured in the **C_out** parameter. In the case of rotate operations, the operand value is simply rotated using VHDL operators. The **C_out** parameter is the last bit value rotated out of the operand vector and back into the opposite end. Thus, it is the rightmost bit of the result for rotate-left operations, and the leftmost bit of the result for rotate-right operations. After computing the result and **C_out** values, the procedure compares the result with a vector of '0' bits to determine the **Z_out** value.

Performing a Memory-I/O Instruction

The procedure **perform_mem**, shown below, performs memory and I/O instructions. The procedure first calculates the effective address by using the **perform_alu_op** procedure to add the values of the base address register and the displacement. The carry-in is set to '0', and the carry-out and zero flag results are unused. The procedure then uses the memory-I/O instruction function code to determine whether to perform a read or write operation using the data memory bus or the I/O port bus. The operations are performed in the same way as described earlier for fetching an instruction. For memory load and I/O input instructions, the procedure assigns the read data to the destination register, provided the register is not r0. For memory store and I/O output instructions, the procedure uses the value from the source register as the data for the bus write operation.

```
procedure perform_mem is
  variable mem_addr : unsigned_byte;
  variable tmp_Z, tmp_C : std_ulogic;
begin
  perform_alu_op(fn => alu_fn_add,
                 a => GPR(to_integer(IR_rs)), b => IR_offset,
                 C_in => '0',
                 result => mem_addr,
                 Z_out => tmp_Z, C_out => tmp_C);
  case IR_mem_fn is
    when mem_fn_ldm =>
      data_cyc_o <= '1';
      data_stb_o <= '1';
      data_we_o <= '0';
      data_adr_o <= mem_addr;
      ldm_loop : loop
        wait until rising_edge(clk_i);
        if rst_i then
          return;
        end if;
        exit ldm_loop when data_ack_i;
      end loop ldm_loop;
      if IR_rd /= 0 then
        GPR(to_integer(IR_rd)) := unsigned(data_dat_i);
      end if;
      data_cyc_o <= '0';
      data_stb_o <= '0';
    when mem_fn_stm =>
      data_cyc_o <= '1';
      data_stb_o <= '1';
      data_we_o <= '1';
      data_adr_o <= mem_addr;
      data_dat_o <= std_ulogic_vector(GPR(to_integer(IR_rd)));
      stm_loop : loop
```

```vhdl
          wait until rising_edge(clk_i);
          if rst_i then
            return;
          end if;
          exit stm_loop when data_ack_i;
        end loop stm_loop;
        data_cyc_o <= '0';
        data_stb_o <= '0';
      when mem_fn_inp =>
        port_cyc_o <= '1';
        port_stb_o <= '1';
        port_we_o <= '0';
        port_adr_o <= mem_addr;
        inp_loop : loop
          wait until rising_edge(clk_i);
          if rst_i then
            return;
          end if;
          exit inp_loop when port_ack_i;
        end loop inp_loop;
        if IR_rd /= 0 then
          GPR(to_integer(IR_rd)) := unsigned(port_dat_i);
        end if;
        port_cyc_o <= '0';
        port_stb_o <= '0';
      when mem_fn_out =>
        port_cyc_o <= '1';
        port_stb_o <= '1';
        port_we_o <= '1';
        port_adr_o <= mem_addr;
        port_dat_o <= std_ulogic_vector(GPR(to_integer(IR_rd)));
        out_loop : loop
          wait until rising_edge(clk_i);
          if rst_i then
            return;
          end if;
          exit out_loop when port_ack_i;
        end loop out_loop;
        port_cyc_o <= '0';
        port_stb_o <= '0';
      when others =>
        report "Program logic error in interpreter"
          severity failure;
    end case;
end procedure perform_mem;
```

For I/O input instructions, the procedure sets the port_addr signal to the effective address and asserts the port_read control signal. It then enters a loop in which it waits for the next clock edge. When that occurs, if the reset input is active, the procedure simply returns, allowing the main interpreter process to reset the processor state. If reset is inactive and the port_ready input is active, the procedure exits from the loop; otherwise, it repeats, waiting for the next clock-edge. On exit from the loop, if the destination register is not R0, the procedure copies the data from the port_data_in signal to the destination register and clears the port_read control signal. I/O write instructions are performed similarly. The difference is that, prior to the loop, data is copied from the source register to the port_data_out signal and the port_write control signal is asserted. After the loop, the port_write signal is cleared.

Performing a Branch Instruction

The procedure for performing branch instructions is shown below. The procedure uses the branch function code to determine which condition code to check to decide whether the branch is taken or not. If the branch is taken, the procedure adds the branch displacement to the program counter.

```
procedure perform_branch is
  variable branch_taken : std_ulogic;
begin
  case IR_branch_fn is
    when branch_fn_bz =>
      branch_taken := cc_Z;
    when branch_fn_bnz =>
      branch_taken := not cc_Z;
    when branch_fn_bc =>
      branch_taken := cc_C;
    when branch_fn_bnc =>
      branch_taken := not cc_C;
    when others =>
      report "Program logic error in interpreter"
        severity failure;
  end case;
  if branch_taken then
    PC := unsigned(signed(PC) + signed(IR_disp));
  end if;
end procedure perform_branch;
```

Performing a Jump Instruction

The procedure for performing jump instructions is shown below. In the case of a **jmp** instruction, the procedure simply copies the target address from the instruction register to the program counter. In the case of a **jsb** instruction, before updating the program counter, the procedure first copies the current program counter value to the stack location indexed by SP, then increments SP. The increment is done using modulo arithmetic. If subroutine calls are nested too deeply, the earlier return address is overwritten with later addresses. The Gumnut does not check for this, as it has no mechanism for dealing with the error. Rather, it relies on the programmer or a compiler to avoid the error condition.

```vhdl
procedure perform_jump is
begin
  case IR_jump_fn is
    when jump_fn_jmp =>
      PC := IR_addr;
    when jump_fn_jsb =>
      stack(SP) := PC;
      SP := (SP + 1) mod stack_depth;
      PC := IR_addr;
    when others =>
      report "Program logic error in interpreter"
        severity failure;
  end case;
end procedure perform_jump;
```

Performing a Miscellaneous Instruction

The procedure for performing the class of miscellaneous instructions is shown below. For **ret** instructions, the procedure decrements SP to index the return address on the stack and copies the address to the program counter. For **reti** instructions, the procedure restores the program counter and condition code flags from the variables in which they were stored during interrupt service and re-enables interrupts. For **enai** and **disi** instructions, the procedure simply sets the interrupt-enable flag to '1' or '0', respectively.

```vhdl
procedure perform_misc is
begin
  case IR_misc_fn is
    when misc_fn_ret =>
      SP := (SP - 1) mod stack_depth;
      PC := stack(SP);
    when misc_fn_reti =>
      PC := int_PC;
      cc_Z := int_Z;
      cc_C := int_C;
      int_en := '1';
    when misc_fn_enai =>
      int_en := '1';
    when misc_fn_disi =>
      int_en := '0';
    when misc_fn_wait | misc_fn_stby =>
      wait_loop : loop
        wait until rising_edge(clk_i);
```

```
                    if rst_i then
                      return;
                    end if;
                    exit wait_loop when int_en and int_req;
                  end loop wait_loop;
                  perform_interrupt;
                when misc_fn_undef_6 | misc_fn_undef_7 =>
                  null;
                when others =>
                  report "Program logic error in interpreter"
                    severity failure;
              end case;
          end procedure perform_misc;
```

For both **wait** and **stby** instructions, the procedure enters a loop in which it checks rst_i and int_req on successive clock cycles. If rst_i is asserted, the procedure returns so that the main interpreter process can reset the core. If int_ack is asserted and interrupts are enabled, the procedure calls the instruction for processing interrupts, thus completing the required operations for the **wait** and **stby** instructions. Note that if interrupts are not enabled, any interrupt request will be ignored. The only way to exit processing for these instructions in that case is to reset the core. Note also that this procedure does not model the low-power standby state of the processor; **wait** and **stby** instructions are modeled identically.

22.2.3 Verifying the Behavioral Model

Now that we have developed our behavioral Gumnut model, we can verify it by writing a test bench model. Since the function performed by the processor is to execute a machine language program stored in memory, we can test the processor by loading the instruction memory with a test program and observing the values in registers and the I/O operations on the external interface. We will use an assembler called *gasm* (provided on the companion website for this book) to generate the machine language program from assembly language source code.

The entity declaration for the test bench model is:

```
entity test is
end test;
```

and the architecture body is:

```
library ieee;
use ieee.std_logic_1164.all, ieee.numeric_std.all;

use work.gumnut_defs.all;
use std.textio.all;

architecture gumnut of test is
```

```vhdl
  signal syscon_clk_o : std_ulogic;
  signal syscon_rst_o : std_ulogic;
  -- I/O port bus
  signal gumnut_port_cyc_o : std_ulogic;
  signal gumnut_port_stb_o : std_ulogic;
  signal gumnut_port_we_o : std_ulogic;
  signal gumnut_port_ack_i : std_ulogic;
  signal gumnut_port_adr_o : unsigned(7 downto 0);
  signal gumnut_port_dat_o : std_ulogic_vector(7 downto 0);
  signal gumnut_port_dat_i : std_ulogic_vector(7 downto 0);
  -- Interrupts
  signal gumnut_int_req : std_ulogic;
  signal gumnut_int_ack : std_ulogic;

  component gumnut_with_mem is
    port ( clk_i : in std_ulogic;
           rst_i : in std_ulogic;
           -- I/O port bus
           port_cyc_o : out std_ulogic;
           port_stb_o : out std_ulogic;
           port_we_o : out std_ulogic;
           port_ack_i : in std_ulogic;
           port_adr_o : out unsigned(7 downto 0);
           port_dat_o : out std_ulogic_vector(7 downto 0);
           port_dat_i : in std_ulogic_vector(7 downto 0);
           -- Interrupts
           int_req : in std_ulogic;
           int_ack : out std_ulogic );
  end component gumnut_with_mem;

begin

  reset_gen : syscon_rst_o <= '0',
                              '1' after  5 ns,
                              '0' after 25 ns;

  clk_gen : process
  begin
    syscon_clk_o <= '0';
    wait for 10 ns;
    loop
      syscon_clk_o <= '1', '0' after 5 ns;
      wait for 10 ns;
    end loop;
  end process clk_gen;

  int_gen : process
  begin
    gumnut_int_req <= '0';
```

```vhdl
        for int_count in 1 to 10 loop
          for cycle_count in 1 to 25 loop
            wait until falling_edge(syscon_clk_o);
          end loop;
          gumnut_int_req <= '1';
          wait until falling_edge(syscon_clk_o)
                     and gumnut_int_ack = '1';
          gumnut_int_req <= '0';
        end loop;
        wait;
      end process int_gen;

      io_control : process
        -- Hard-wired input stream
        constant input_data : unsigned_byte_array
          := ( X"00", X"01", X"02", X"03", X"04", X"05", X"06", X"07",
               X"08", X"09", X"0A", X"0B", X"0C", X"0D", X"0E", X"0F",
               X"10", X"11", X"12", X"13", X"14", X"15", X"16", X"17",
               X"18", X"19", X"1A", X"1B", X"1C", X"1D", X"1E", X"1F" );
        variable next_input : integer := 0;
        variable debug_line : line;
        constant show_actions : boolean := true;
      begin
        gumnut_port_ack_i <= '0';
        loop
          wait until falling_edge(syscon_clk_o);
          if gumnut_port_cyc_o and gumnut_port_stb_o then
            if to_X01(gumnut_port_we_o) = '0' then
              if show_actions then
                swrite(debug_line, "IO: port read; address = ");
                hwrite(debug_line, gumnut_port_adr_o);
                swrite(debug_line, ", data = ");
                hwrite(debug_line, input_data(next_input) );
                writeline(output, debug_line);
              end if;
              gumnut_port_dat_i <=
                std_ulogic_vector(input_data(next_input));
              next_input := (next_input + 1) mod input_data'length;
              gumnut_port_ack_i <= '1';
            else
              if show_actions then
                swrite(debug_line, "IO: port write; address = ");
                hwrite(debug_line, gumnut_port_adr_o );
                swrite(debug_line, ", data = ");
                hwrite(debug_line, gumnut_port_dat_o );
                writeline(output, debug_line);
              end if;
```

```
            gumnut_port_ack_i <= '1';
          end if;
        else
          gumnut_port_ack_i <= '0';
        end if;
      end loop;
    end process io_control;

  dut : component gumnut_with_mem
    port map ( clk_i       => syscon_clk_o,
               rst_i       => syscon_rst_o,
               port_cyc_o => gumnut_port_cyc_o,
               port_stb_o => gumnut_port_stb_o,
               port_we_o  => gumnut_port_we_o,
               port_ack_i => gumnut_port_ack_i,
               port_adr_o => gumnut_port_adr_o,
               port_dat_o => gumnut_port_dat_o,
               port_dat_i => gumnut_port_dat_i,
               int_req     => gumnut_int_req,
               int_ack     => gumnut_int_ack );

end architecture gumnut;
```

The architecture includes a component declaration corresponding to the subsystem containing the Gumnut core and its memories. This component is instantiated as the design under test (**dut**). The architecture also includes an assignment to the reset signal and a clock-generator process, corresponding to a Wishbone system controller ("syscon"). The **int_gen** process generates a sequence of 10 interrupt requests at intervals of 25 clock cycles. The **io_control** process represents an I/O port controller that monitors the external interface of the processor. When the processor performs a read operation, the controller supplies the next in a sequence of input data values and displays a debug message showing the port address and the data value. When the processor performs a write operation, the controller likewise displays a debug message showing the address and data value.

In the test bench, we do not include the debug generic constant in the component declaration for the Gumnut. Instead, we use a separate configuration declaration, shown below, to bind the Gumnut entity to the component instance and to fill in values for the generic constant. We set the **debug** generic constant of the processor to true so that we can trace its operation.

```
    configuration test_gumnut_behavior of test is
      for gumnut

        for dut : gumnut_with_mem
          use entity work.gumnut_with_mem(struct);
          for struct
            for core : gumnut
              use entity work.gumnut(behavior)
                generic map ( debug => true );
            end for;
```

```
      end for;
    end for;

  end for;
end configuration test_gumnut_behavior;
```

In order to execute the test bench, we need to prepare a test program to run on the processor core. The following is a small test program, written in the gasm assembly language:

```
; Counts from 10 downto 1, storing successive values to memory.


         text
         org     0                    ; reset
         jmp     begin

         org     1                    ; interrupt service
         reti


         org     16
begin:   enai
         add     r1, r0, 10           ; initialize count to 10
loop:    stm     r1, 0
         sub     r1, r1, 1            ; decrement count
         bnz     loop
stop:    disi
         stby


         data
         org     0
count:   bss     1
```

We assemble this program to produce the initialization files for the instruction and data memories. (Instructions on using the gasm assembler are provided on the companion website.) We also analyze each of the design units described so far and then invoke our simulator, specifying the configuration declaration as the unit to simulate. We then use the facilities of the simulator to step through the model, to examine the test bench signals to verify that the machine language program in the memory is correctly executed by the CPU. We do not describe the process in detail, as different simulators provide different commands and facilities for executing the model. The companion website provides a number of small test programs, including the one shown above, that can be executed using this test bench. The following is an example of the debugging messages, produced when the program above was run on the author's simulator.

```
#     0: jmp  16
#    16: enai
#    17: add  R1, R0, 10
#    18: stm  R1, 0
```

```
#   19: sub   R1, R1, 1
#   20: bnz   -3
#   18: stm   R1, 0
#   19: sub   R1, R1, 1
#   20: bnz   -3
#   18: stm   R1, 0
#   19: sub   R1, R1, 1
#   20: bnz   -3
#   18: stm   R1, 0
#   19: sub   R1, R1, 1
#   20: bnz   -3
#   18: stm   R1, 0
#   19: sub   R1, R1, 1
#   20: bnz   -3
# Interrupt acknowledged at PC =    18
#    1: reti
#   18: stm   R1, 0
#   19: sub   R1, R1, 1
#   20: bnz   -3
#   18: stm   R1, 0
#   19: sub   R1, R1, 1
#   20: bnz   -3
#   18: stm   R1, 0
#   19: sub   R1, R1, 1
#   20: bnz   -3
#   18: stm   R1, 0
#   19: sub   R1, R1, 1
#   20: bnz   -3
#   18: stm   R1, 0
#   19: sub   R1, R1, 1
#   20: bnz   -3
#   21: disi
#   22: stby
```

22.3 A Register-Transfer-Level Model

We now turn our attention to the next level of refinement of our Gumnut model: a synthesizable register-transfer-level description. At this level, the processor is composed of registers, buses, multiplexers, an ALU and a sequential control section. Figure 22.5 shows the register-transfer-level datapath of the processor upon which we base our VHDL model. It includes a register file (Reg) for the general-purpose registers, storage for the return-address stack (Stack), individual registers for the program counter (PC), instruction register (IR), stack pointer (SP), ALU result (A) and condition codes (CC), the data memory and input/output port data input (D) and the interrupt register (Int Reg). These all correspond to variables defined in the behavioral architecture of the processor. Multiplexers are included to select between inputs to the various components.

FIGURE 22.5

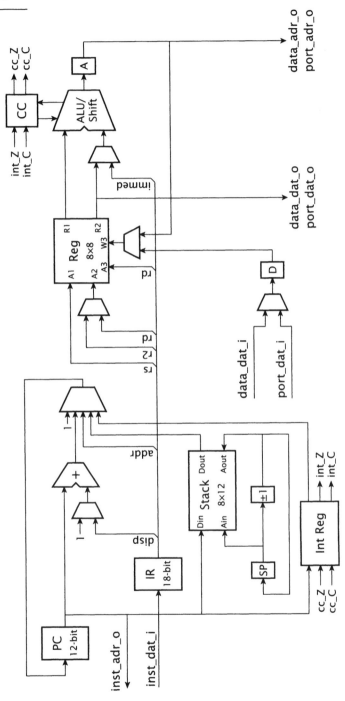

The organization of the Gumnut at the register-transfer level.

Not shown in the figure is the control section, which activates components in the data path in sequence to interpret Gumnut instructions. Each instruction takes multiple clock cycles to execute. Different instructions may take different numbers of cycles, depending on the interpretation steps required.

The first cycle of execution involves checking whether an interrupt request is pending. If so, the current PC and condition code bits are saved in the interrupt register and the PC is set to 1. Execution of the interrupt service code then proceeds in the subsequent cycle. If no interrupt is requested, the first cycle is used to start the instruction memory read operation. When the operation completes (either in the first cycle, or in a subsequent cycle if the memory causes wait cycles) the fetched instruction is stored in the IR, and the PC register is updated with the incremented PC value.

During the second cycle, the register file is accessed to fetch operands, in case they are required. The register file in this implementation has synchronous read ports, and the operands are stored in two output registers. Also in this cycle, control-flow processing is performed. If the instruction in the IR is a conditional branch that is taken, the PC is updated with the sum of its current value and the branch displacement. If the instruction is a **jmp**, the PC is updated with the target address. If the instruction is a **jsb**, the PC is updated with the target address, the current PC value is pushed onto the return-address stack, and the stack pointer is incremented. If the instruction is a **ret**, the PC is updated from the top of the stack, and the stack pointer is decremented. If the instruction is a **reti**, the PC and condition codes are restored from the interrupt register, and interrupts are enabled. If the instruction is an **enai** or **disi**, the interrupt enable bit is set accordingly. In all cases of control flow instructions, processing is complete after the second cycle.

The third cycle (if required) involves computation of a data result or an effective address by the ALU. The result is stored in the A register. Also, for arithmetic, logic and shift instructions, the condition code bits are updated.

For memory and I/O instructions, a further cycle is used to access the data memory or I/O port. The A register is used as the address, and for write operations, the register file r2 output is used for the data. The Gumnut checks the acknowledge input at the end of the cycle. If it is negated, the Gumnut repeats the cycle, allowing the data memory or port controller extra time to read or write the data. When the acknowledge input is active, the operation is complete. For read operations, the data is stored in the D register.

A final cycle is required for instructions that update a destination register in the register file, namely, arithmetic, logic, shift, load and input instructions. The data source is the ALU output register (A) or the data memory and I/O port input register (D), depending on the instruction. The destination register (if not r0) is updated at the end of the cycle.

22.3.1 The Architecture Body

The synthesizable architecture body for the Gumnut core is outlined below. The signals represent the interconnections between the data path components, including both buses and control signals. Within the statement part of the architecture body, the assignments to the signals IR_decode_alu_immed, etc., represent decoding of the instruction word in the instruction regsister. These assignments are followed by a number of processes and further assignments that represent the control section and the data path components. We will describe them in turn in this section.

```vhdl
library ieee;  use ieee.std_logic_1164.all, ieee.numeric_std.all;
use work.gumnut_defs.all;
architecture rtl_unpipelined of gumnut is
  signal PC : IMem_addr;

  signal branch_taken : std_ulogic;

  signal IR : instruction;

  alias IR_alu_reg_fn : alu_fn_code is IR(2 downto 0);
  alias IR_alu_immed_fn : alu_fn_code is IR(16 downto 14);
  alias IR_shift_fn : shift_fn_code is IR(1 downto 0);
  alias IR_mem_fn : mem_fn_code is IR(15 downto 14);
  alias IR_branch_fn : branch_fn_code is IR(11 downto 10);
  alias IR_jump_fn : jump_fn_code is IR(12 downto 12);
  alias IR_misc_fn : misc_fn_code is IR(10 downto 8);

  alias IR_rd : reg_addr is IR(13 downto 11);
  alias IR_rs : reg_addr is IR(10 downto 8);
  alias IR_r2 : reg_addr is IR(7 downto 5);
  alias IR_immed : immed is IR(7 downto 0);
  alias IR_count : shift_count is IR(7 downto 5);
  alias IR_offset : disp is IR(7 downto 0);
  alias IR_disp : disp is IR(7 downto 0);
  alias IR_addr : IMem_addr is IR(11 downto 0);

  signal IR_decode_alu_immed,
         IR_decode_mem,
         IR_decode_shift,
         IR_decode_alu_reg,
         IR_decode_jump,
         IR_decode_branch,
         IR_decode_misc : std_ulogic;

  signal D_state : std_ulogic;

  signal int_PC : IMem_addr;
  signal int_Z : std_ulogic;
  signal int_C : std_ulogic;
  signal int_en : std_ulogic;

  constant SP_length : positive := 3;
  signal SP : unsigned(SP_length - 1 downto 0);
  signal stack_top : IMem_addr;

  signal GPR_rs : unsigned_byte;
  signal GPR_r2 : unsigned_byte;

  signal ALU_result : unsigned_byte;
  signal ALU_Z : std_ulogic;
```

```vhdl
signal ALU_C : std_ulogic;
signal ALU_out : unsigned_byte;

signal cc_Z : std_ulogic;
signal cc_C : std_ulogic;

signal D : unsigned_byte;

type control_state is (fetch_state,
                       decode_state,
                       execute_state,
                       mem_state,
                       write_back_state,
                       int_state);
signal state, next_state : control_state;

begin

IR_decode_alu_immed <= IR(17) ?= '0';
IR_decode_mem       <= IR(17 downto 16) ?= "10";
IR_decode_shift     <= IR(17 downto 15) ?= "110";
IR_decode_alu_reg   <= IR(17 downto 14) ?= "1110";
IR_decode_jump      <= IR(17 downto 13) ?= "11110";
IR_decode_branch    <= IR(17 downto 12) ?= "111110";
IR_decode_misc      <= IR(17 downto 11) ?= "1111110";

...

end architecture rtl_unpipelined;
```

The control section is implemented as a state machine. The state diagram is shown in Figure 22.6. In the transition conditions, the instruction classes shown in italics represent the instruction decode signals, and the instruction names in italic represent values for the function code fields of the instruction register. The term "*int*" represents the condition that interrupts are enabled and the interrupt request input is active. We write the conditions in this way in the diagram for brevity. They are written out in full in the VHDL code for the state machine.

The state machine is represented by two processes, shown below. The **control** process determines the next state value, and the **state_reg** process represents the register for the current state. The state values are defined by the enumeration type **control_state**, each of which corresponds to one of the cycles of instruction interpretation.

```vhdl
control : process (all)
begin
  case state is
    when fetch_state =>
      if not inst_ack_i then
        next_state <= fetch_state;
      else
        next_state <= decode_state;
      end if;
```

FIGURE 22.6

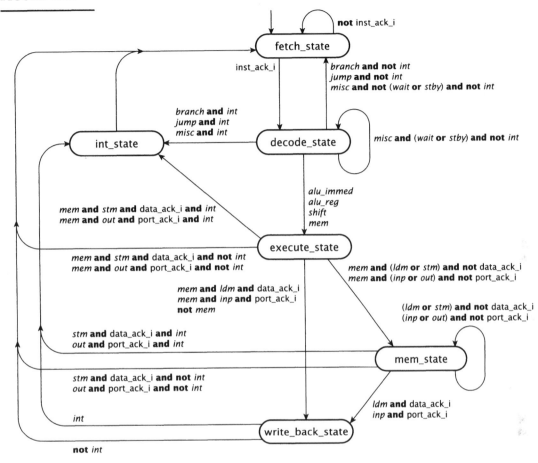

The state diagram for the control section.

```
when decode_state =>
  if IR_decode_branch or IR_decode_jump or IR_decode_misc then
    if IR_decode_misc
      and (IR_misc_fn ?= misc_fn_wait
          or IR_misc_fn ?= misc_fn_stby)
      and not (int_en and int_req) then
    next_state <= decode_state;
  elsif int_en and int_req then
    next_state <= int_state;
  else
    next_state <= fetch_state;
  end if;
  else
    next_state <= execute_state;
```

```
      end if;
    when execute_state =>
      if IR_decode_mem then
        if (IR_mem_fn ?= mem_fn_ldm or IR_mem_fn ?= mem_fn_stm)
          and not data_ack_i then
          next_state <= mem_state;
        elsif (IR_mem_fn ?= mem_fn_inp or IR_mem_fn ?= mem_fn_out)
              and not port_ack_i then
          next_state <= mem_state;
        elsif IR_mem_fn ?= mem_fn_ldm
              or IR_mem_fn ?= mem_fn_inp then
          next_state <= write_back_state;
        else
          if int_en and int_req then
            next_state <= int_state;
          else
            next_state <= fetch_state;
          end if;
        end if;
      else
        next_state <= write_back_state;
      end if;
    when mem_state =>
      if (IR_mem_fn ?= mem_fn_ldm or IR_mem_fn ?= mem_fn_stm)
        and not data_ack_i then
        next_state <= mem_state;
      elsif (IR_mem_fn ?= mem_fn_inp or IR_mem_fn ?= mem_fn_out)
            and not port_ack_i then
        next_state <= mem_state;
      elsif IR_mem_fn ?= mem_fn_ldm
            or IR_mem_fn ?= mem_fn_inp then
        next_state <= write_back_state;
      else
        if int_en and int_req then
          next_state <= int_state;
        else
          next_state <= fetch_state;
        end if;
      end if;
    when write_back_state =>
      if int_en and int_req then
        next_state <= int_state;
      else
        next_state <= fetch_state;
      end if;
    when int_state =>
      next_state <= fetch_state;
```

```
      end case;
    end process control;

    state_reg : process (clk_i)
    begin
      if rising_edge(clk_i) then
        if rst_i then
          state <= fetch_state;
        else
          state <= next_state;
        end if;
      end if;
    end process state_reg;
```

Upon reset, the state machine enters **fetch_state**. In this state, the processor performs a read operation on the instruction memory bus. The state machine remains in **fetch_state** until the bus acknowledge signal is asserted, after which it transitions to **decode_state**.

In **decode_state**, the processor decodes the instruction in the instruction register and fetches register operands. It also computes the next PC value for branch, jump, and return instructions. For **wait** and **stby** instructions, the machine stays in **decode_state** until an interrupt is pending, after which it transitions to **int_state**. For branch, jump, and the remaining miscellaneous statements, the **decode_state** cycle completes their execution. The state machine transitions either to **int_state** if an interrupt is pending, or back to **fetch_state** otherwise. For all other instructions, there is further work to do, so the state machine transitions to **execute_state**.

In **execute_state**, the processor uses the ALU to compute an arithmetic/logical result or an effective address. For non-memory-I/O instructions, the state machine then transitions to **write_back_state** to update the register file. For memory-I/O instructions, a fast memory or I/O controller may be able to complete the operation in the same cycle as that in which the address is calculated. If so, it asserts its acknowledge signal in that cycle, and the state machine transitions to **write_back_state** (for **ldm** and **inp** instructions, requiring a register to be updated), to **int_state** (for **stm** and **out** instructions if an interrupt is pending), or back to **fetch_state** otherwise. A slower memory or I/O controller, on the other hand, requires one or more extra cycles, so it leaves its acknowledge signal negated. In that case, the state machine transitions to **mem_state**.

In **mem_state**, the state machine waits for the relevant acknowledge signal to be asserted. It then makes similar transitions to those described for memory-I/O instructions from **execute_state**.

In **write_back_state**, the processor updates the register file with the result of an arithmetic/logical, shift, **ldm**, or **inp** instruction. The state machine then transitions either to **int_state**, if an interrupt is pending, or to **fetch_state** otherwise.

The state machine transitions to **int_state** in order to acknowledge a pending interrupt. The processor asserts the **int_ack** signal for one cycle in this state. After that, the state machine transitions back to **fetch_state** so that the first instruction of the interrupt handler can be fetched.

The next statement in the RTL architecture body represents the combinational logic for the branch condition decoder:

```
      with IR_branch_fn select
        branch_taken <=        cc_Z when branch_fn_bz,
                           not cc_Z when branch_fn_bnz,
                               cc_C when branch_fn_bc,
                           not cc_C when branch_fn_bnc,
                               'X'        when others;
```

This assignment simply uses the branch function field of the instruction in the IR to determine which condition code (or negation) decides the branch outcome.

The following process represents the program counter and its input combinational
logic:

```
      PC_reg : process (clk_i)
      begin
        if rising_edge(clk_i) then
          if rst_i then
            PC <= (others => '0');
          elsif state = fetch_state and inst_ack_i = '1' then
            PC <= PC + 1;
          elsif state = decode_state then
            if IR_decode_branch and branch_taken then
              PC <= unsigned(signed(PC) + signed(IR_disp));
            elsif IR_decode_jump then
              PC <= IR_addr;
            elsif IR_decode_misc and IR_misc_fn ?= misc_fn_ret then
              PC <= stack_top;
            elsif IR_decode_misc and IR_misc_fn ?= misc_fn_reti then
              PC <= int_PC;
            end if;
          elsif state = int_state then
            PC <= to_unsigned(1, PC'length);
          end if;
        end if;
      end process PC_reg;
```

A synthesis tool will infer storage to drive the PC signal. On reset, the process clears
the PC to 0, the address at which execution commences. When the control section is in
fetch_state and the instruction memory acknowledges completion of the read operation,
the process increments the PC ready for the next instruction. When the control section is
in **decode_state**, the action taken by the process depends on the kind of instruction in the
IR. If it is a branch instruction that is taken, the process adds the branch displacement to
the PC. If the instruction is a jump, the process updates the PC with the target address. For
a return instruction, the process updates the PC with the value on the top of the return
address stack, and for a return from interrupt, the process updates the PC from the interrupt register.

The statements representing the interrupt register and associated logic is

```
      int_ack <= '1' when state = int_state else '0';
```

```
int_reg : process (clk_i)
begin
  if rising_edge(clk_i) then
    if rst_i then
      int_en <= '0';
    elsif state = int_state then
      int_PC <= PC;
      int_Z <= cc_Z;
      int_C <= cc_C;
      int_en <= '0';
    elsif state = decode_state and IR_decode_misc = '1' then
      case IR_misc_fn is
        when misc_fn_reti | misc_fn_enai =>
          int_en <= '1';
        when misc_fn_disi =>
          int_en <= '0';
        when others =>
          null;
      end case;
    end if;
  end if;
end process int_reg;
```

The int_ack signal is asserted when the control section is in int_state, as described earlier. For the int_reg process, a synthesis tool will infer storage for the driven signals, int_en, int_PC, int_Z, and int_C. On reset, the process clears the interrupt enable signal. When the control section is in int_state, the process saves the current PC and condition code values. It also clears the interrupt enable signal to prevent further interrupts. When the control section is in decode_state and the instruction in the IR is a miscellaneous instruction, the process may need to update the interrupt enable signal again. For reti and enai instructions, it sets the signal; for disi instructions, it clears the signal; and for other cases, it leaves the signal unchanged.

The statements representing the instruction memory bus interface and the instruction register are

```
inst_cyc_o <= '1' when state = fetch_state else '0';
inst_stb_o <= '1' when state = fetch_state else '0';
inst_adr_o <= PC;

instr_reg : process (clk_i)
begin
  if rising_edge(clk_i) then
    if state = fetch_state and inst_ack_i = '1' then
      IR <= unsigned(to_X01(inst_dat_i));
    end if;
  end if;
end process instr_reg;
```

The assignments to the inst_cyc_o and inst_stb_o signals cause a bus read to occur when the control section is in fetch_state. The address for the read is the PC value. A synthesis tool will infer a register for the instr_reg process driving the instruction register (IR) signal. The process updates IR with the data returned by the instruction memory when it completes the read operation.

The process representing the return address stack, the stack pointer (SP) and the associated logic is

```vhdl
stack_mem : process (clk_i)
  constant stack_depth : positive := 2**SP_length;
  subtype stack_index is natural range 0 to stack_depth - 1;
  type stack_array is array (stack_index) of IMem_addr;
  variable stack : stack_array;
begin
  if rising_edge(clk_i) then
    if rst_i then
      SP <= (others => '0');
    elsif state = decode_state then
      if IR_decode_jump and IR_jump_fn ?= jump_fn_jsb then
        stack(to_integer(SP)) := PC;
        SP <= SP + 1;
      elsif IR_decode_misc and IR_misc_fn ?= misc_fn_ret then
        SP <= SP - 1;
      end if;
    end if;
    stack_top <= stack(to_integer(SP - 1));
  end if;
end process stack_mem;
```

The storage for saved addresses is represented by the variable **stack**. A synthesis tool will infer a memory for this and registers for the SP and the top-of-stack output. The SP points to the next free location in the stack memory. On reset, the SP is cleared to 0. The stack is updated when the control section is in **decode_state**, depending on the instruction being executed. For **jsb** instructions, the PC value (already incremented after fetching the **jsb** instruction) is written to the next free stack location and the SP is incremented. For **ret** instructions, the SP is decremented. Regardless of the instruction kind, the **stack_top** output is updated with the memory location preceding the free location. Note that the old SP value is used to index the top of stack, not the updated value. The assignments to the SP take effect after the process suspends.

The process representing the general purpose register (GPR) file and its associated logic is

```vhdl
GPR_mem : process (clk_i)
  subtype reg_index is natural range 0 to 7;
  variable r2_addr : reg_addr;
  variable write_data : unsigned_byte;
  variable GPR : unsigned_byte_array(reg_index)
            := (others => X"00");
```

```
begin
  if rising_edge(clk_i) then
    if rst_i then
      GPR := (others => (others => '0'));
    else
      if state = write_back_state and IR_rd /= 0 then
        if IR_decode_alu_reg or IR_decode_alu_immed
            or IR_decode_shift then
          write_data := ALU_out;
        elsif IR_decode_mem
              and (IR_mem_fn ?= mem_fn_ldm
                  or IR_mem_fn ?= mem_fn_inp) then
          write_data := D;
        end if;
        GPR(to_integer(IR_rd)) := write_data;
      end if;
      if state = decode_state then
        if IR_decode_mem
            and (IR_mem_fn ?= mem_fn_stm
                or IR_mem_fn ?= mem_fn_out) then
          r2_addr := IR_rd;
        else
          r2_addr := IR_r2;
        end if;
        GPR_rs <= GPR(to_integer(IR_rs));
        GPR_r2 <= GPR(to_integer(r2_addr));
      end if;
    end if;
  end if;
end process GPR_mem;
```

The storage for the registers is represented by the variable **GPR**. A synthesis tool will infer a memory for the register file with one write port and two read ports. It will infer registers for the outputs **GPR_rs** and **GPR_r2**. However, it will not infer registers for the variables **r2_addr** and **write_data**, as they are both used for intermediate values within any given activation of the process. On reset, the process clears the GPR memory. When the control section is in **write_back_state** and the destination register is other than r0, the process selects the data to be written, depending on the instruction, and updates the GPR memory location indexed by the *rd* field of the IR. For arithmetic/logical and shift instructions, the data is taken from the ALU output register, and for memory load and I/O input instructions, the data is taken from the data input register. When the control section is in **decode_state**, the process selects the register address for the *r2* operand: either the *rd* field of the IR, for memory store or I/O output instructions, or the *r2* field, for other instructions. The process then reads the *rs* and *r2* operands from the GPR memory.

The process representing the arithmetic/logic unit (ALU) and shift logic is

```vhdl
ALU : process (all)
  variable fn : alu_fn_code;
  variable right_operand : unsigned_byte;
  variable shift_result : unsigned_byte;
begin
  if IR_decode_alu_reg or IR_decode_alu_immed
     or IR_decode_mem then
    if IR_decode_alu_reg then
      fn := IR_alu_reg_fn;
      right_operand := GPR_r2;
    elsif IR_decode_alu_immed then
      fn := IR_alu_immed_fn;
      right_operand := IR_immed;
    else
      fn := alu_fn_add;
      right_operand := IR_offset;
    end if;
    case fn is
      when alu_fn_add =>
        (ALU_C, ALU_result) <= ('0' & GPR_rs)
                               + ('0' & right_operand);
      when alu_fn_addc =>
        (ALU_C, ALU_result) <= ('0' & GPR_rs)
                               + ('0' & right_operand)
                               + cc_C;
      when alu_fn_sub =>
        (ALU_C, ALU_result) <= ('0' & GPR_rs)
                               - ('0' & right_operand);
      when alu_fn_subc =>
        (ALU_C, ALU_result) <= ('0' & GPR_rs)
                               - ('0' & right_operand)
                               - cc_C;
      when alu_fn_and =>
        (ALU_C, ALU_result) <= ('0' & GPR_rs)
                               and ('0' & right_operand);
      when alu_fn_or =>
        (ALU_C, ALU_result) <= ('0' & GPR_rs)
                               or ('0' & right_operand);
      when alu_fn_xor =>
        (ALU_C, ALU_result) <= ('0' & GPR_rs)
                               xor ('0' & right_operand);
      when alu_fn_mask =>
        (ALU_C, ALU_result) <= ('0' & GPR_rs)
                               and not ('0' & right_operand);
      when others =>
        (ALU_C, ALU_result) <= 'X'
                               & unsigned_byte'(others => 'X');
```

```
        end case;
      else
        case IR_shift_fn is
          when shift_fn_shl =>
            (ALU_C, ALU_result) <= ('0' & GPR_rs)
                                    sll to_integer(IR_count);
          when shift_fn_shr =>
            (ALU_result, ALU_C) <= (GPR_rs & '0')
                                    srl to_integer(IR_count);
          when shift_fn_rol =>
            shift_result := GPR_rs rol to_integer(IR_count);
            ALU_result <= shift_result;
            ALU_C <= shift_result(unsigned_byte'right);
          when shift_fn_ror =>
            shift_result := GPR_rs ror to_integer(IR_count);
            ALU_result <= shift_result;
            ALU_C <= shift_result(unsigned_byte'left);
          when others =>
            (ALU_C, ALU_result) <= 'X'
                                    & unsigned_byte'(others => 'X');
        end case;
      end if;
    end process ALU;

    ALU_Z <= ALU_result ?= unsigned_byte'(others => '0');
```

A synthesis tool will infer combinational logic for this process, as it is sensitive to all of the signals that are read (specified by **all** in the sensitivity list), and all variables are assigned before being read on all paths through the statements. The process uses the instruction decode signals to determine whether to perform and arithmetic/logic operation or a shift operation. For arithmetic register instructions, the process selects the IR_alu_reg_fn field of the IR as the function code and the *r2* GPR output as the right operand. For arithmetic immediate instructions, the process selects the IR_alu_immed_fn field of the IR as the function code and the IR_immed field of the IR as the right operand. For memory and I/O instructions, the process uses **alu_fn_add** as the function code and the IR_offset field of the IR as the right operand. It then performs the operation indicated by the function code. The statements are largely the same as those in the behavioral model. For shift instructions, the process performs the operation indicated by the IR_shift_fn field of the IR, again using statements that are largely the same as those in the behavioral model. For all instructions, the concurrent assignment to ALU_Z compares the ALU result with 0 to detemine the condition value.

The ALU result and condition values are stored in registers inferred by a synthesis tool from the following processes:

```
    ALU_reg : process (clk_i)
    begin
      if rising_edge(clk_i) then
        if state = execute_state then
```

```
      ALU_out <= ALU_result;
    end if;
  end if;
end process ALU_reg;

cc_reg : process (clk_i)
begin
  if rising_edge(clk_i) then
    if rst_i then
      cc_Z <= '0';
      cc_C <= '0';
    elsif state = execute_state
        and (IR_decode_alu_reg = '1'
          or IR_decode_alu_immed = '1'
          or IR_decode_shift = '1') then
      cc_Z <= ALU_Z;
      cc_C <= ALU_C;
    elsif state = decode_state
        and IR_decode_misc = '1'
        and IR_misc_fn = misc_fn_reti then
      cc_Z <= int_Z;
      cc_C <= int_C;
    end if;
  end if;
end process cc_reg;
```

The ALU_reg process updates the ALU output register whenever the control section is in **execute_state**. The **cc_reg** process, representing the condition-code register, is somewhat more involved. On reset, it clears the condition codes to '0'. When the control section is in **execute_state** and an arithmetic/logical or shift instruction is being executed, the process updates the condition codes with the outputs from the ALU. When the control section is in **decode_state** and an **reti** instruction is being executed, the process restores the saved condition codes from the interrupt register.

The statements representing the memory and I/O data register are

```
D_state <= '1' when (state = execute_state or state = mem_state)
                      and IR_decode_mem = '1' else '0';

D_reg : process (clk_i)
begin
  if rising_edge(clk_i) then
    if D_state and IR_mem_fn ?= mem_fn_ldm and data_ack_i then
      D <= unsigned(data_dat_i);
    elsif D_state and IR_mem_fn ?= mem_fn_inp and port_ack_i then
      D <= unsigned(port_dat_i);
    end if;
  end if;
end process D_reg;
```

The **D_state** signal is asserted when the processor is performing a memory or I/O bus operation, namely, when the control section is in **execute_state** or **mem_state** and the instruction in the IR is a memory-I/O instruction. This signal is used in the enable condition of the data register, represented by the **D_reg** process. A synthesis tool will infer a register driving the D signal. When **D_state** is asserted, the instruction in the IR is **ldm**, and the data memory has completed the data read, the process updates D with the memory data. Similarly, when **D_state** is asserted, the instruction in the IR is **inp**, and the I/O controller has completed the port read, the process updates D with the port data.

The remaining assignments represent the rest of the data memory and I/O bus interfaces:

```
data_cyc_o <= D_state and (IR_mem_fn ?= mem_fn_stm
                        or IR_mem_fn ?= mem_fn_ldm);
data_stb_o <= D_state and (IR_mem_fn ?= mem_fn_stm
                        or IR_mem_fn ?= mem_fn_ldm);
data_we_o  <= D_state and IR_mem_fn ?= mem_fn_stm;

data_adr_o <= ALU_result;
data_dat_o <= std_ulogic_vector(GPR_r2);

port_cyc_o <= D_state and (IR_mem_fn ?= mem_fn_inp
                        or IR_mem_fn ?= mem_fn_out);
port_stb_o <= D_state and (IR_mem_fn ?= mem_fn_inp
                        or IR_mem_fn ?= mem_fn_out);
port_we_o  <= D_state and IR_mem_fn ?= mem_fn_out;

port_adr_o <= ALU_result;
port_dat_o <= std_ulogic_vector(GPR_r2);
```

The **data_cyc_o** and **data_stb_o** signals are asserted when the processor performs a memory bus operation. The **data_we_o** signal is only asserted for an **stm** instruction. The address for a data bus operation is the value previously computed by the ALU, and the data is the value fetched from the GPR register file. The assignments for the I/O bus interface are similar.

The final process, shown below, monitors operation of the processor core and writes debug trace messages.

```
debug_monitor : if debug generate

  debugger : process
    use std.textio.all;
    variable disassembled_instr : disassembled_instruction;
    variable debug_line : line;
    variable PC_num : natural;
  begin
    wait until rising_edge(clk_i);
    loop
      if rst_i then
        swrite(debug_line, "Resetting");
        writeline(output, debug_line);
```

```
                wait until rising_edge(clk_i) and rst_i = '0';
                next;
            elsif state = fetch_state then
                PC_num := to_integer(PC);
            elsif state = decode_state then
                disassemble(IR, disassembled_instr);
                write(debug_line, PC_num,
                        field => 4, justified => right);
                swrite(debug_line, ": ");
                write(debug_line, disassembled_instr);
                writeline(output, debug_line);
                wait until rising_edge(clk_i)
                    and (rst_i ='1' or state /= decode_state);
                next;
            elsif state = int_state then
                PC_num := to_integer(PC);
                swrite(debug_line, "Interrupt acknowledged at PC = ");
                write(debug_line, PC_num,
                        field => 4, justified => right);
                writeline(output, debug_line);
            end if;
            wait until rising_edge(clk_i);
          end loop;
        end process debugger;

    end generate debug_monitor;
```

The enclosing generate statement ensures that the process is only included if the **debug** generic is true. The process writes messages when the Gumnut is reset, when an interrupt is acknowledged, and when an instruction has been fetched into the IR. The output produced is similar to that produced by the behavioral model.

22.3.2 Verifying the RTL Model

We can test our register-transfer-level model using the same test bench that we used to test the behavioral model, as described in Section 22.2.3. We need to modify the configuration declaration for the test bench to bind the register-transfer-level implementation to the processor component in the test bench. The revised configuration declaration is

```
configuration test_gumnut_rtl_unpipelined of test is
  for gumnut

    for dut : gumnut_with_mem
      use entity work.gumnut_with_mem(struct);
      for struct
        for core : gumnut
          use entity work.gumnut(rtl_unpipelined)
            generic map ( debug => true );
```

```
            end for;
          end for;
        end for;

      end for;
    end configuration test_gumnut_rtl_unpipelined;
```

We can then run our simulator, specifying the new configuration as the unit to simulate, to test the register-transfer-level model in the same way as we tested the behavioral model. The sequence of operations should be the same. The only observable difference should be that the CPU takes longer to execute each instruction. This is because our register-transfer-level model accurately describes the cycle-by-cycle operation of the processor.

22.4 A Digital Alarm Clock

In this section we design a digital alarm clock based on the Gumnut as an illustration of a design flow for a complete embedded processor system. We will focus on the hardware design aspects using VHDL, and refer to details of the embedded software design on the companion website. We start with a statement of functional requirements.

Our alarm clock shall have four 7-segment LED digits for displaying the clock time and the alarm time in 24-hour format. The clock shall have four push-button switches:

- Alarm set: When this button is held on, the alarm time can be advanced.

- Time set: When this button is held on, the clock time can be advance.

- Hours advance: When this button is pressed once, the hours of the alarm time or clock time are advanced by one. Whe the button is pressed and held, the hours of the alarm time or clock time are advanced at a a rate of four per second.

- Minutes advance: This button is similar to the hours advance button, but advances the minutes of the alarm time or clock time instead of the hours.

When both the hours advance and minutes advance buttons are pressed while the alarm set or time set button is held on, the alarm time or clock time, respectively, is reset to midnight. When neither the alarm set nor the time set button is held on, the hours advance and minutes advance buttons have no effect.

The clock shall have a two-position switch that enables or disables the alarm, and a LED indicator that shows whether the alarm is enabled. The clock shall also have an output that activates a sounder when the alarm is triggered. (The output could also activate an alternate device, such as a radio or CD player.) The alarm is triggered when the clock time equals the alarm time, provided that the alarm is enabled and neither the clock time nor the alarm time is being set.

These requirements may appear to be quite precise and complete, but as we will see as we proceed through the design, there are some details that are unspecified. We will need to make assumptions based on our common sense and understanding of what a digital alarm clock should do. This is not unusual in system design.

22.4.1 System Design

For the purpose of this case study, we will describe an implementation of the digital alarm clock using an FPGA development board. The author used a Xilinx Spartan-3 starter board with four 7-segment displays, four push buttons, eight 2-position slider switches and eight discrete LEDs. One of the slider switches serves as the alarm enable switch, and one discrete LED serves for the alarm enable indicator. Another discrete LED serves in place of the alarm sounder output.

We start our system design by considering the hardware/software trade-off. We could implement the design entirely in hardware. However, that may not be the most effective, since the clock would need numerous counters, multiplexers and comparators. Since an FPGA includes block memories that can be used for program and data storage, we may be able to fit an implementation based on an embedded processor into a smaller FPGA than would be needed for a hardware implementation. Moreover, developing and maintaining software can be simpler than hardware, so there may be non-technical benefits to an embedded processor implementation. For this case study, we will focus on an implementation that uses an embedded Gumnut core for nearly all of the functionality, and leave comparison with a hardware implementation to an exercise.

Figure 22.7 shows a block diagram of the alarm clock system. It includes

- The Gumnut core, driven by an externally generated 50MHz clock.

- An input port for the push buttons (pb(3 **downto** 0)).

- An input button for the slider switches (sw(7 **downto** 0)).

- A registered output port for the digit anode drivers (digit(3 **downto** 0)).

- A register output port for the digit segment drivers (seg(7 **downto** 0)). Seg(0) is segment *a*, seg(1) is segment *b*, etc., and seg(7) is the decimal point.

- A registered output port for the discrete LEDs (led(7 **downto** 0)).

FIGURE 22.7

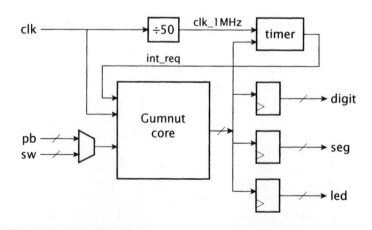

A block diagram of the digital alarm clock using an embedded Gumnut core.

- An output port for a programmable timer for generating periodic interrupts. The timer is driven by a 1 MHz clock divided down from the external clock and has a register to store the divisor written by the processor. The timer repeatedly counts down from the divisor value and generates an interrupt upon reaching 0.

An entity declaration for the alarm clock is shown below. It includes a debug generic constant for use in simulation and ports corresponding to the external connection shown in the block diagram.

```vhdl
library ieee;
use ieee.std_logic_1164.all;

entity alarm_clock is
  generic ( debug : boolean := false );
  port ( clk : in std_ulogic;
         pb : in std_ulogic_vector(3 downto 0);
         sw : in std_ulogic_vector(7 downto 0);
         led : out std_ulogic_vector(7 downto 0);
         digit : out std_ulogic_vector(3 downto 0);
         seg : out std_ulogic_vector(7 downto 0) );
end entity alarm_clock ;
```

The synthesizable architecture body for the alarm clock is based on the block diagram:

```vhdl
library ieee; use ieee.numeric_std.all;

architecture synth of alarm_clock is

  component gumnut_with_mem is
    generic ( debug : boolean := false );
    port ( clk_i : in std_ulogic;
           rst_i : in std_ulogic;
           port_cyc_o : out std_ulogic;
           port_stb_o : out std_ulogic;
           port_we_o : out std_ulogic;
           port_ack_i : in std_ulogic;
           port_adr_o : out unsigned(7 downto 0);
           port_dat_o : out std_ulogic_vector(7 downto 0);
           port_dat_i : in std_ulogic_vector(7 downto 0);
           int_req : in std_ulogic;
           int_ack : out std_ulogic );
  end component gumnut_with_mem;

  signal reset : std_ulogic;

  signal port_cyc_o : std_ulogic;
  signal port_stb_o : std_ulogic;
  signal port_we_o : std_ulogic;
  signal port_ack_i : std_ulogic;
  signal port_adr_o : unsigned(7 downto 0);
```

```vhdl
  signal port_dat_o : std_ulogic_vector(7 downto 0);
  signal port_dat_i : std_ulogic_vector(7 downto 0);
  signal int_req : std_ulogic;
  signal int_ack : std_ulogic;

  signal pb_synch : std_ulogic_vector(pb'range);
  signal sw_synch : std_ulogic_vector(sw'range);
  signal clk_1MHz : std_ulogic;
begin

  -- reset generation

  reset_gen : reset <= '0';

  -- processor core

  core : gumnut_with_mem
    generic map ( debug => debug )
    port map ( clk_i => clk,
               rst_i => reset,
               port_cyc_o => port_cyc_o,
               port_stb_o => port_stb_o,
               port_we_o => port_we_o,
               port_ack_i => port_ack_i,
               port_adr_o => port_adr_o,
               port_dat_o => port_dat_o,
               port_dat_i => port_dat_i,
               int_req => int_req,
               int_ack => int_ack );

  port_ack_i <= '1';

  -- input port:
  --    push buttons 1 and 0 when port_addr = "-------0"
  --    switches when port_addr = "-------1"

  synch : process ( clk )
    variable pb_tmp : std_ulogic_vector(pb'range);
    variable sw_tmp : std_ulogic_vector(sw'range);
  begin
    if rising_edge(clk) then
      pb_synch <= pb_tmp;
      sw_synch <= sw_tmp;
      pb_tmp := pb;
      sw_tmp := sw;
    end if;
  end process synch;

  with port_adr_o(0) select
    port_dat_i <= "0000" & pb_synch when '0',
```

```vhdl
                    sw_synch when '1',
                    "XXXXXXXX" when others;

-- digit output port register (port_addr = "-------1")
digit_reg : process (clk)
begin
  if rising_edge(clk) then
    if reset = '1' then
      digit <= "1111";
    elsif port_cyc_o = '1' and port_stb_o = '1'
          and port_we_o = '1' and port_adr_o(0) = '1' then
      digit <= not port_dat_o(3 downto 0);
    end if;
  end if;
end process digit_reg;

-- segment output port register (port_addr = "------1-")
seg_reg : process (clk)
begin
  if rising_edge(clk) then
    if reset = '1' then
      seg <= "11111111";
    elsif port_cyc_o = '1' and port_stb_o = '1'
          and port_we_o = '1' and port_adr_o(1) = '1' then
      seg <= not port_dat_o;
    end if;
  end if;
end process seg_reg;

-- led output port register (port_addr = "-----1--")
led_reg : process (clk, reset)
begin
  if rising_edge(clk) then
    if reset = '1' then
      led <= "00000000";
    elsif port_cyc_o = '1' and port_stb_o = '1'
          and port_we_o = '1' and port_adr_o(2) = '1' then
      led <= port_dat_o;
    end if;
  end if;
end process led_reg;

-- divide 50MHz clock down to 1MHz

div_50 : process (clk)
  variable count : integer range 0 to 49;
begin
  if rising_edge(clk) then
```

```vhdl
        if reset = '1' then
          clk_1MHz <= '0';
          count := 49;
        elsif count = 0 then
          count := 49;
          clk_1Mhz <= '1';
        else
          count := count - 1;
          clk_1Mhz <= '0';
        end if;
      end if;
    end process div_50;

    -- timer (port_addr = "----1---")

    timer : process (clk)
      variable divisor, count : unsigned(7 downto 0);
    begin
      if rising_edge(clk) then
        if reset = '1' then
          divisor := X"00";
          count := X"00";
          int_req <= '0';
        elsif port_cyc_o = '1' and port_stb_o = '1'
              and port_we_o = '1' and port_adr_o(3) = '1' then
          divisor := unsigned(port_dat_o);
          count := divisor;
        elsif clk_1MHz = '1' then
          if count = 0 then
            int_req <= '1';
            count := divisor;
          else
            count := count - 1;
          end if;
        end if;
        if int_ack = '1' then
          int_req <= '0';
        end if;
      end if;
    end process timer;

  end architecture synth;
```

The architecture includes a component declaration for the Gumnut core with instruction and data memories, described in Section 22.1.2, and an instance of the component. We will rely on the default binding rules to bind the Gumnut entity and architecture to the instance, rather than using a configuration, since not all synthesis tools support use of configurations. The **reset** input to the core is tied to '0', since we don't need an external reset in this design. The **port_ack_i** input is tied high, since all input/output ports transfer data in a single cycle.

The input ports are represented by the process **synch** and the combinational assignment statement. The process synchronizes the push button and switch inputs with the system clock to avoid problems with metastability that might arise from input changes close to a clock-edge. The combinational assignment simply selects between the synchronized push button and switch inputs, based on the port address bus from the Gumnut. The registered output ports for the digit, segment and LED drivers are represented by the processes **digit_reg**, **seg_reg** and **led_reg**, respectively. The digit and segment registers both invert the data, since the circuits on the development board illuminate the display when the driver outputs are low. The circuits for the discrete LEDs, on the other hand, illuminate the LEDs when the driver outputs are high, so the LED register does not not invert the data.

The **div_50** process represents the clock divider. It generates a one-cycle pulse on **clk_1MHz** whenever the **count** variable reaches 0. The **timer** process represents the programmable timer. It has variables for the divisor written by the processor and for the down counter. When the **clk_1MHz** signal is pulsed, the timer decrements the counter. Upon reaching 0, the **int_req** signal is activated and the counter reloaded with the divisor. When the **int_ack** signal is activated, the **int_req** signal is cleared.

While the alarm clock entity and architecture body describe the main hardware required for the system, they do not deal with the specific electrical interface for the FPGA pins. We describe that information in a top-level entity and architecture, as follows:

```vhdl
library ieee;
use ieee.std_logic_1164.all;

entity alarm_clock_top is
  generic ( debug : boolean := false );
  port ( clk50in : in std_ulogic;
         pb_in : in std_ulogic_vector(3 downto 0);
         sw_in : in std_ulogic_vector(7 downto 0);
         led_out : out std_ulogic_vector(7 downto 0);
         digit_out : out std_ulogic_vector(3 downto 0);
         seg_out : out std_ulogic_vector(7 downto 0) );
end entity alarm_clock_top ;

-- ----------------------------------------------------------------

library unisim;
use unisim.vcomponents.all ;

architecture struct of alarm_clock_top is

  signal clkint : std_ulogic;
  signal clk : std_ulogic;

  signal pb : std_ulogic_vector(pb_in'range);
  signal sw : std_ulogic_vector(sw_in'range);
  signal led : std_ulogic_vector(led_out'range);
```

```vhdl
  signal digit : std_ulogic_vector(digit_out'range);
  signal seg : std_ulogic_vector(seg_out'range);

  component alarm_clock is
    generic ( debug : boolean := false );
    port ( clk : in std_ulogic;
           pb : in std_ulogic_vector(3 downto 0);
           sw : in std_ulogic_vector(7 downto 0);
           led : out std_ulogic_vector(7 downto 0);
           digit : out std_ulogic_vector(3 downto 0);
           seg : out std_ulogic_vector(7 downto 0) );
  end component alarm_clock ;

begin

  -- input/output and clock buffers

  clkin_ibuf : component ibufg_lvcmos33
    port map ( i => clk50in, o => clkint );

  clk_bufg : component bufg
    port map ( i => clkint, o => clk ) ;

  pb_buf_gen : for i in 0 to 3 generate
    pb_buf : component ibuf_lvcmos33
      port map ( i => pb_in(i), o => pb(i) );
  end generate pb_buf_gen;

  sw_buf_gen : for i in 0 to 7 generate
    sw_buf : component ibuf_lvcmos33
      port map ( i => sw_in(i), o => sw(i) );
  end generate sw_buf_gen;

  led_buf_gen : for i in 0 to 7 generate
    led_buf : component obuf_lvcmos33
      port map ( i => led(i),   o => led_out(i) );
  end generate led_buf_gen;

  digit_buf_gen : for i in 0 to 3 generate
    digit_buf : component obuf_lvcmos33
      port map ( i => digit(i), o => digit_out(i) );
  end generate digit_buf_gen;

  seg_buf_gen : for i in 0 to 7 generate
    seg_buf : component obuf_lvcmos33
      port map ( i => seg(i),   o => seg_out(i) );
  end generate seg_buf_gen;

  -- the alarm clock core

  alarm_clock_core : alarm_clock
    generic map ( debug => debug )
```

```
    port map ( clk => clk, pb => pb, sw => sw,
              led => led, digit => digit, seg => seg );

end architecture struct;
```

The entity is much the same as that described earlier. The architecture declares a component for the alarm clock internal hardware and instantiates it. The main purpose of the architecture, however, is to instantiate pin buffers connected between the internal hardware signals and the FPGA pins. The component declarations for the buffers are drawn from the package **vcomponents** in library **unisim**, supplied by the FPGA vendor. The pins are programmed to use 3.3V low-voltage CMOS electrical levels, specified using the **ibuf_lvcmos33** input buffer components and the **obuf_lvcmos33** components. The buffered clock signal, **clkint**, is further buffered using a **bufg** component, which drives a global clock tree on the FPGA. These buffer components are all recognized by the FPGA vendor's synthesis and mapping tools. Documentation on use of the buffers and sample models illustrating their use are provided by the vendor.

The hardware we have described so far simply provides an interface to the alarm clock's "user interface." Most of the functionality of the alarm clock is implemented by the embedded software that runs on the Gumnut core. Since we are focusing on the hardware using VHDL in this book, we do not describe the software here. The complete design suite, including the software source code, documentation and tools, is available on the companion website.

22.4.2 Synthesizing and Implementing the Alarm Clock

Now that we have a synthesizable description of the alarm clock, we outline the steps needed to implement it in our FPGA development board. We make use of a number of tools provided by the FPGA vendor, and an assembler for the Gumnut. The basic tool flow is illustrated in Figure 22.8. The inputs to the tool flow are the assembler source code for the emedded software, the VHDL source files, and a constraints file created using a constraints editor program. The constraints include a requirement that the clock frequency be 50 MHz. They also include placement of input and output pins based on the physical connections of FPGA pins to circuits on the development board. These connections are described in the documentation for the development board.

Implementation commences with assembly of the software code. The Gumnut assembler produces two memory data files, one for the program memory image and one for the data memory image. Next, the VHDL files are used by the synthesis tool to generate a low-level representation of the system. The representation also makes use of specialized hardware resources provided by the target FPGA, such as block memories and carry-chains. The mapping tool allocates each of the elements in the low-level representation to specific hardware resources in the target FPGA. The place and route tool then decides which logic blocks, memories and routing connections in the FPGA are used for the hardware resources. This tool makes use of the constraints to ensure that the inputs and outputs are properly connected and that delays do not exceed the maximum specified. The output of the place and route tool is a database (the .ncd file) containing detailed information about how the design is implemented in the FPGA.

FIGURE 22.8

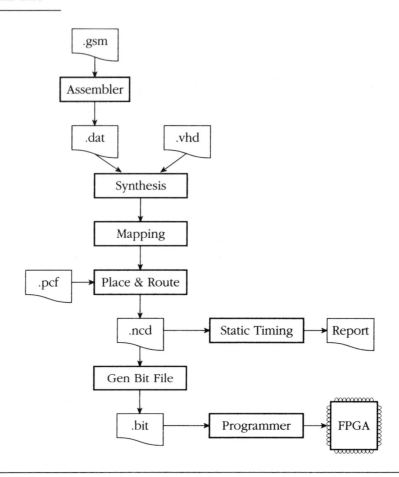

Tool flow for implementing the alarm clock design.

Two tools use the database to generate further information. The static timing analysis tool uses a timing model of the internal paths of the FPGA to determine the delays in the placed and routed design. It predicts that the mimimum clock period for correct operation is 16.989 ns, corresponding to a maximum clock frequency of 58.862 MHz. This is well within the specified clock constraint, so no tuning of the implementation is needed. Were the constraint violated, we could use the tools to identify critical paths and add constraints to guide the tools toward a faster implementation. Alternatively, we could modify the VHDL code to avoid implying hardware with long delays. For example, we could reduce the complexity of combinational logic that processes data between clock-edges.

The second tool that uses the database is the bit-file generator. It creates a file containing information needed to program the target FPGA. In a manufactured design, the bit file would be stored in a flash memory and used to configure the FPGA upon power-up. For our development board, the programmer tool uses a JTAG interface to configure the FPGA from the bit file. Once the FPGA is configured, it commences operation, with the

Gumnut executing the embedded software. The development board becomes a digital alarm clock.

Of course, it is unrealistic to expect to write the VHDL code and embedded software, implement it and have it work correctly first time. For that reason, the flow described above is idealized. In practice, the author developed the alarm clock model and the embedded software incrementally. First, the hardware was developed and simulated using a simple test bench that verified correct operation of the timer. Development of the software started with the interrupt service routine and the periodic task dispatcher. That was followed by code to drive each of the input and output interfaces. The tasks for scanning digits, debouncing buttons and switches, setting the alarm and time values using the buttons, counting the clock time, and managing the alarm function were each added and tested in turn. The behavior of this system is sufficiently simple that debugging could be done by careful code review and by deduction from details of malfunction. An alternative would be to use embedded monitoring hardware that can be automatically inserted into the design using the FPGA vendor's tools.

Exercises

1. [❶ 22.1] Determine the encoding for the following Gumnut instructions:

   a. sub r2, r1, r0

   b. and r4, r4, 0x30

   c. ror r1, r1, 2

   d. ldm r6, (r2)–1

   e. out r4, 0x10

   f. bz +7

   g. jsb 0xD0

2. [❶ 22.1] What Gumnut instructions are encoded by the following 18-bit hexadecimal values?

   a. 0DBC0

   b. 38326

   c. 33D63

   d. 25906

   e. 3EC11

   f. 3DC70

   g. 3F200

3. [❹ 22.2] Develop a "bus functional" architecture body for the Gumnut core. Rather than fetching and interpreting instructions, the model reads a file of commands. A "bus-transaction" command specifies a time at which a bus transaction is to be

initiated by the CPU and includes the address and transaction kind (instruction fetch, data read or write, or I/O port read or write). The time is a delay from completion of the previous command. An "intack" command causes the model to activate the **int_ack** port for one clock cycle. An "include" command causes the model to process a subsidiary command file. Note that the subsidiary file may also contain nested include commands. When the CPU reset port is activated, the model terminates command processing and resets the bus signals. When reset is removed, the model resumes processing the command file from the beginning.

4. [❹ 22.3] Develop a pipelined implementation of the Gumnut core. Implement the instruction fetch, decode, execute, memory, and write back operations as separate pipeline stages.

5. [❹ 22.4] Synthesize and implement the alarm clock design using an FPGA development board.

6. [❹] Develop I/O controllers to interface with the Gumnut core. Possibilities include a serial input/output UART, parallel input/output, USB interface, I^{2}C interface, analog-to-digital and digital-to-analog converters, real-time clock/calendar, and so on. You may wish to contribute your designs to the OpenCores repository at www.opencores.org.

Chapter 23

Miscellaneous Topics

In the preceding chapters we introduced most of the facilities provided by VHDL and showed how they may be used to model a variety of hardware systems at various levels of detail. However, there remain a few VHDL facilities that we have not yet discussed. In this chapter, we tie off these loose ends.

23.1 Guards and Blocks

In this section we look at a number of closely related topics. First, we discuss another kind of resolved signal called a *guarded signal*. We see how we can disconnect drivers from such signals. Next, we introduce the idea of *blocks* in a VHDL design. We show how blocks and guarded signals work together with *guards* and *guard expressions* to cause automatic disconnection of drivers. Finally, we discuss blocks as a mechanism for describing a hierarchical structure within an architecture. While these aspects of VHDL may be useful in some designs, they are not widely used. Hence, we have deferred consideration of the features to this chapter.

23.1.1 Guarded Signals and Disconnection

In Chapter 8 we saw how we can use resolved signals that include values such as 'Z' for modeling high-impedance outputs. However, if we are modeling at a higher level of abstraction, we may wish to use a more abstract type such as an integer type or a simple bit type to represent signals. In such cases, it is not appropriate to include the high-impedance state as a value, so VHDL provides us with an alternative approach, using *guarded signals*. These are resolved signals for which we can *disconnect* the drivers; that is, we can cause the drivers to stop contributing values to the resolved signal. We see why these signals are called "guarded" later in this section. First, let us look at the complete syntax rule for a signal declaration, which includes a means of declaring a signal to be guarded.

signal_declaration ⇐
  signal identifier ⦃ , ... ⦄ : subtype_indication ⟦ **register** ‖ **bus** ⟧
    ⟦ := expression ⟧ ;

The difference between this rule and the simplified rule we introduced earlier is the inclusion of the option to specify the signal kind as either a *register* signal or a *bus* signal. Note that a guarded signal must be a resolved signal. Hence, the subtype indication in the signal declaration must denote a resolved subtype. Some examples of declarations of guarded signals are

```
signal interrupt_request : pulled_up bit bus;

signal stored_state : resolve_state state_type register
        := init_state;
```

The difference between the two kinds of guarded signals lies in their behavior when all of their drivers are disconnected. A bus signal uses the resolution function to determine the signal value by passing it an empty array. The bus kind of guarded signal can be used to model a signal that is "pulled up" to some value dependent on the signal type when all drivers are disconnected. A register signal, on the other hand, keeps the resolved value that it had just before the last disconnection. The register kind of guarded signal can be used to model signals with dynamic storage, for example, signals in CMOS logic that store data as charge on transistor gates when all drivers are disconnected. Note that a signal may be neither a register nor a bus signal, in which case it is a regular (unguarded) signal, from which drivers may not be disconnected.

A process can disconnect a driver for a guarded signal by specifying a *null transaction* in a signal assignment statement. As a reminder, the syntax rule we used to introduce a signal assignment was

signal_assignment_statement ⇐
  ⟦ label : ⟧ name <= ⟦ delay_mechanism ⟧ waveform ;

The waveform is a sequence of transactions, that is, new values to be applied to the signal after given delays. A more complete syntax rule for waveforms includes null transactions:

waveform ⇐
  (*value*_expression ⟦ **after** *time*_expression ⟧
    ‖ **null** ⟦ **after** *time*_expression ⟧) ⦃ , ... ⦄

This rule shows that instead of specifying a value in a transaction, we can use the keyword **null** to indicate that the driver should be disconnected after the given delay. When this null transaction matures, the driver ceases to contribute values to the resolution function used to compute the signal's value. Hence the size of the array of values passed as an argument to the resolution function is reduced by one for each driver that currently has a null transaction determining its contribution. When a driver subsequently performs a non-null transaction, it reconnects and contributes the value in the non-null transaction.

EXAMPLE 23.1 *Disconnection from a bus of type* bit

Following is an outline of an architecture body for a computer system consisting of a CPU, a memory and a DMA controller.

```
architecture top_level of computer_system is

  function resolve_bits ( bits : bit_vector ) return bit is
    variable result : bit := '0';
  begin
    for index in bits'range loop
      result := result or bits(index);
      exit when result;
    end loop;
    return result;
  end function resolve_bits;

  signal write_en : resolve_bits bit bus;
  ...

begin

  CPU : process is
    ...
  begin
    write_en <= '0' after Tpd;
    ...
    loop
      wait until clock;
      if hold_req then
        write_en <= null after Tpd;
        wait on clock until clock and not hold_req;
        write_en <= '0' after Tpd;
      end if;
      ...
    end loop;
  end process CPU;

  ...

end architecture top_level;
```

The architecture body includes a guarded signal of kind bus, **write_en**, representing a control connection to the memory. The resolution function performs the logical "or" operation of all of the contributing drivers and returns '0' if there are no drivers connected. This result ensures that the memory remains inactive when neither the CPU nor the DMA controller is driving the **write_en** control signal.

When the process representing the CPU is initialized, it drives **write_en** with the value '0'. Subsequently, when the DMA controller requests access to the memory by asserting the **hold_req** signal, the CPU schedules a null transaction on **write_en**. This

transaction removes the CPU's driver from the set of drivers contributing to the resolved value of **write_en**. Later, when the DMA controller negates **hold_req**, the CPU
reconnects its driver to **write_en** by scheduling a transaction with the value '0'.

EXAMPLE 23.2   *Disconnection from a bus of type* bit_vector

An outline of a register-transfer-level model of a processor, in which datapath elements are modeled by processes, is

```
architecture rtl of processor is

  subtype word is bit_vector(0 to 31);
  type word_vector is array (natural range <>) of word;

  function resolve_unique ( drivers : word_vector ) return word is
  begin
    return drivers(drivers'left);
  end function resolve_unique;

  signal source1, source2 : resolve_unique word register;
  . . .

begin

  source1_reg : process (phase1, source1_reg_out_en, ...) is
    variable stored_value : word;
  begin
    . . .
    if source1_reg_out_en and phase1 then
      source1 <= stored_value;
    else
      source1 <= null;
    end if;
  end process source1_reg;

  alu : perform_alu_op ( alu_opcode,
                         source1, source2, destination, ... );

  . . .

end architecture rtl;
```

The datapath includes two register signals that represent the source operand connections to the ALU. The source operand buses are register guarded signals driven by
processes during phase 1 of a clock cycle. They retain their values during phase 2. In
this design, only one process should drive each of these signals at a time. The resolution function returns the single contributing value.

The process **source1_reg** represents one of the datapath elements that connects
to the **source1** signal. When its output enable signal and the clock phase 1 signal are
both '1', the process drives the signal with its stored value. The resolution function is

passed an array of one element consisting of this driving value. It is applied to the
source1 signal and is used by the concurrent procedure call representing the ALU. At
the end of the clock phase, the process disconnects from **source1** by scheduling a null
transaction. Since **source1** is a register signal and all drivers are now disconnected,
the resolution function is not called, and **source1** retains its value until some other
driver connects. This models a real system in which the operand value is stored as
electrical charge on the inputs of transistors in the ALU.

When we are dealing with guarded signals of a composite type such as an array type,
it is important to note that within each driver for the signal, all elements must be connected
or all must be disconnected. It is not permissible to disconnect some elements using a null
transaction and leave other elements connected. The reason for this rule is that the complete composite value from each driver is passed as a contribution to the resolution function. For example, it is not possible to pass just half of a bit vector as an element in the
array of values to be resolved. Thus, given a guarded bit-vector signal declared as

```
subtype word is bit_vector(0 to 31);
type word_array is array (integer range <>) of word;

function resolve_words ( words : word_array ) return word;

signal s : resolve_words word bus;
```

we may not write the following signal assignments within one process:

```
s(0 to 15) <= X"003F" after T_delay;
s(16 to 31) <= null after T_delay;
```

If the design requires that only part of a composite driver be connected at some stages
during model execution, then the signal type must be a composite of individually resolved
elements, rather than a resolved composite type. This is similar to the requirement we discussed in Section 8.1.1.

In the above examples, we have assumed that a null transaction is scheduled after all
previously scheduled transactions have been applied. We have yet to consider how null
transactions are scheduled in the general case where there are still transactions pending
in the driver. In Section 5.2.5 we described in detail how the list of transactions previously
scheduled on a driver is edited when a signal assignment is executed. In particular, when
the inertial delay mechanism is used, transactions are deleted if their values differ from
that of the newly scheduled transaction. For the purpose of this editing algorithm, a null
transaction is deemed to have a value that is different from any value of the signal type.
Successive null transactions are deemed to have the same value.

The Driving Attribute

In addition to the 'driving_value attribute for signals that we saw in Chapter 8, VHDL also
provides an attribute, 'driving, that is useful with guarded signals. It returns true if the
driver in the process referring to the attribute currently has its driver connected to the signal. It returns false if the driver is disconnected. Of course, the attribute 'driving_value

should not be used if the driver is disconnected, since there is no driving value in that case. An error will occur if a model tries to do this.

VHDL-87

The 'driving attribute is not provided in VHDL-87.

Guarded Ports

Throughout all the examples in this book, we have seen that the ports of an entity are treated as signals within an architecture body for that entity. Just as we can have guarded signals, so we can have guarded ports as part of an entity's interface. However, there are some important limitations that come about due to the way in which ports are resolved. The main restriction is that a guarded port can only be of the bus kind, not the register kind. A guarded port includes the keyword **bus** in its declaration. For example, given the following declarations to define a resolved subtype resolved_byte:

```
subtype byte is bit_vector(0 to 7);
type byte_array is array (integer range <>) of byte;
function resolve ( bytes : byte_array ) return byte;
subtype resolved_byte is resolve byte;
```

we can declare an entity with a guarded port q as follows:

```
entity tri_state_reg is
  port ( d : in resolved_byte;
         q : out resolved_byte bus;
         clock, out_enable : in bit );
end entity tri_state_reg;
```

Since the port q is declared to be a guarded port, a process in an architecture body for tri_state_reg can disconnect from the port by assigning a null transaction. Here is where the behavior is different from what we might first expect. Since the port is of a resolved subtype, it is resolved independently of any external signal associated with it. This means that even if all processes in the architecture for tri_state_reg are disconnected, the resolution function for the port is still invoked to determine the port's value. The port itself does not become disconnected. It continues to contribute its resolved value to the external signal associated with it. While this may seem counter-intuitive, it follows directly from the way resolved signals and ports behave in VHDL. Hence the entity tri_state_reg declared above does not in fact represent a module that can disconnect its port from an associated signal. There is no mechanism in VHDL for doing that. While some designers argue that this is a limitation of the language, there are often ways to circumvent the problem. The difficulty mainly arises when modeling at a high level of abstraction. At a lower level, we would use some multivalued logic type that includes a representation of the high-impedance state instead of using disconnection, so the problem does not arise.

Guarded Signal Parameters

In Chapter 6 we saw how we can write subprograms that have signal class parameters. We cannot, however, specify that a signal parameter be a bus signal by adding the keyword **bus** in the parameter list, as we can for ports. Instead, the subprogram uses the kind of the actual signal (bus, register or unguarded) associated with a signal parameter. A procedure can include signal assignment statements that assign null transactions to a formal parameter, but if the actual signal is not a guarded signal, the model is in error. Recall that for signal parameters of mode **out** or **inout**, when the procedure is called, it is passed a reference to the driver for the actual signal. Signal assignments within the procedure schedule transactions onto the driver for the actual signal. If the actual signal is a guarded signal, and the procedure assigns a null transaction to it, the driver that is disconnected is the one in the calling process. When the actual signal is resolved, the subprogram, acting on behalf of the process, does not contribute a value. We can take advantage of this behavior when writing high-level models that include processes that disconnect from bus signals. We can use a subprogram as an abstraction for processes, instead of using component instances.

VHDL-87

The VHDL-87 language definition does not disallow the keyword **bus** in the specification of a signal parameter. However, it does not specify whether the kind of signal, guarded or unguarded, is determined by the formal parameter specification or by the actual signal associated with the parameter. Implementations of VHDL-87 make different interpretations. Some require the formal parameter specification to include the keyword **bus** if the procedure includes a null signal assignment to the parameter. The actual signal associated with the parameter in a procedure call must then be a guarded signal. Other implementations follow the approach adopted in VHDL-93 and VHDL-2002, prohibiting the keyword **bus** in the parameter specification and determining the kind of the parameter from the kind of the actual signal.

23.1.2 Blocks and Guarded Signal Assignment

We now introduce the VHDL *block* statement. In their most general form, blocks provide a way of partitioning the concurrent statements within an architecture body. However, we start with a simpler form of block statement that relates to guarded signals and return to the more general form later in this section.

A *block statement* is a concurrent statement that groups together a number of inner concurrent statements. A simplified syntax rule for block statements is

```
block_statement ⇐
    block_label :
    block ⟦ ( guard_expression ) ⟧ ⟦ is ⟧
    begin
        { concurrent_statement }
    end block ⟦ block_label ⟧ ;
```

The block label is required to identify the block statement. The syntax rule shows that we can write a block statement with an optional Boolean *guard expression*. If the guard expression is present, it must be surrounded by parentheses and appear after the keyword **block**. Since a Boolean value is required for the expression, the "??" operator is applied implicitly if necessary to convert the expression value from some other type to **boolean**. It is used to determine the value of an implicitly declared signal called **guard**. This signal is only implicitly declared if the guard expression is present. Its visibility extends over the whole of the block statement. Whenever a transaction occurs on any of the signals mentioned in the guard expression, the expression is reevaluated and the **guard** signal is immediately updated. Since the **guard** signal has its value automatically determined, we may not include a source for it in the block. That means we may not write a signal assignment for it, nor use it as an actual signal for an output port of a component instance.

The main use of guard expressions in a block is to control operation of *guarded signal assignments*. These are special forms of the concurrent signal assignments described in Section 5.2.7. If the target of a concurrent signal assignment is a guarded signal, we must use a guarded signal assignment rather than an ordinary concurrent signal assignment. The extended syntax rules are

> concurrent_simple_signal_assignment ⇐
>   name <= ⟦ **guarded** ⟧ ⟦ delay_mechanism ⟧ waveform ;

> concurrent_conditional_signal_assignment ⇐
>   name <= ⟦ **guarded** ⟧ ⟦ delay_mechanism ⟧
>   waveform **when** condition
>   ⟨ **else** waveform **when** condition ⟩
>   ⟦ **else** waveform ⟧ ;

> concurrent_selected_signal_assignment ⇐
>   **with** expression **select** ⟦ ? ⟧
>   name <= ⟦ **guarded** ⟧ ⟦ delay_mechanism ⟧
>   ⟨ waveform **when** choices , ⟩
>   waveform **when** choices ;

The difference is the inclusion of the keyword **guarded** after the assignment symbol. This denotes that the signal assignment is to be executed when the **guard** signal changes value. The effect depends on whether the target of the assignment is a guarded signal or an ordinary signal. For a guarded target, if **guard** changes from true to false, the driver for the target is disconnected using a null transaction. When **guard** changes back to true, the assignment is executed again to reconnect the driver.

EXAMPLE 23.3   *Distributed multiplexing using guarded assignments*

The architecture body outlined below describes a processor node of a multiprocessor computer.

```
architecture dataflow of processor_node is

  signal address_bus : resolve_unique word bus;
  ...
```

```vhdl
begin
  cache_to_address_buffer :
  block ( cache_miss and dirty ) is
  begin
    address_bus <= guarded
      tag_section0 & set_index & B"0000"
        when replace_section = '0' else
      tag_section1 & set_index & B"0000";
  end block cache_to_address_buffer;

  snoop_to_address_buffer :
  block ( snoop_hit and flag_update ) is
  begin
    address_bus <= guarded snoop_address(31 downto 4) & B"0000";
  end block snoop_to_address_buffer;

  . . .

end architecture dataflow;
```

The signal **address_bus** is a guarded bit-vector signal. The block labeled
cache_to_address_buffer has a guard expression that is true when the cache misses
and a block needs to be replaced. The expression is evaluated whenever either
cache_miss or **dirty** changes value, and the implicit signal **guard** in the block is set to
the result. If it is true, the driver in the concurrent signal assignment statement within
the block is connected. Any changes in the signals mentioned in the statement cause
a new assignment to the target signal **address_bus**. When the **guard** signal changes
to false, the driver in the assignment is disconnected using a null transaction.

The block labeled **snoop_to_address_buffer** also has a guard expression, which
is true when an external bus monitor (the "snoop") needs to update flags in the cache.
The expression is evaluated when either **snoop_hit** or **flag_update** changes. The result
is assigned to a separate **guard** signal for this block, used to control a second concurrent signal assignment statement with **address_bus** as the target. Assuming that the
two guard expressions are mutually exclusive, only one of the drivers is connected to
address_bus at a time.

If the target of a guarded signal assignment is an ordinary unguarded signal, the driver
is not disconnected when **guard** changes to false. Instead, the assignment statement is disabled. No further transactions are scheduled for the target, despite changes that may occur
on signals to which the statement is sensitive. Subsequently, when guard changes to true,
the assignment is executed again and resumes normal operation.

EXAMPLE 23.4 *Latch behavior using guarded assignment*

A simple model for a transparent latch can be written using a guarded signal assignment, as shown below. The architecture body uses a block statement with a guard
expression that tests the state of the **enable** signal. When **enable** is '0', the **guard** signal

is false, and the guarded signal assignment is disabled. Changes in **d** are ignored, so
q maintains its current value. When **enable** changes to '1', the guarded signal assignment is enabled and copies the value of **d** to **q**. So long as **enable** is '1', changes in **d**
are copied to **q**.

```
entity latch is
  generic ( width : positive );
  port ( enable : in bit;
         d : in bit_vector(0 to width - 1);
         q : out bit_vector(0 to width - 1) );
end entity latch;

--------------------------------------------------

architecture behavioral of latch is
begin

  transfer_control : block ( enable ) is
  begin
    q <= guarded d;
  end block transfer_control;

end architecture behavioral;
```

VHDL-87, -93, and -2002

These versions of VHDL do not provide implicit conversion using the "??" operator.
Hence, a guard expression must be of type **boolean** without conversion.

VHDL-87

The keyword **is** may not be included in a block header in VHDL-87.

Explicit Guard Signals

In the preceding examples, the guarded signal assignment statements used the implicitly
declared **guard** signal to determine whether the assignment should be executed. As an
alternative, we can explicitly declare our own Boolean signal called **guard**. Provided it is
visible at the position of a guarded signal assignment, it will be used to control the signal
assignment. The advantage of this approach is that we can use a more complex algorithm
to control the guard signal, rather than relying on a simple Boolean expression. For example, we might use a separate process to drive **guard**. Whenever **guard** is changed to
false, guarded signal assignments are disabled, disconnecting any drivers for guarded signals. When **guard** is changed back to true, the assignments are reenabled.

Disconnection Specifications

One aspect of guarded signal assignments for guarded signals that we have not yet dealt with is timing. In the previous examples illustrating guarded signal assignment, we have only shown zero-delay models. If we need to include delays in signal assignments, we should also include a specification of the delay associated with disconnecting a driver in a guarded signal assignment. The problem is that the null transaction that disconnects a driver in this case is not explicitly written in the model. It occurs as a result of the **guard** signal changing to false. The mechanism in VHDL that we may use if we need to specify a non-zero disconnection delay is a *disconnection specification*. The syntax rule is

disconnection_specification ⇐
 disconnect (*signal*_name { , ... } ▯ **others** ▯ **all**) : type_mark
 after *time*_expression ;

A disconnection specification allows us to identify a particular signal or set of signals by name and type, and to specify the delay associated with any null transactions scheduled for the signals. This delay only applies to the implicit null transactions resulting from guarded signal assignments. It does not apply to null transactions we may write explicitly using the keyword **null** in a signal assignment in a process.

A disconnection specification for a guarded signal must appear in the same list of declarations as the signal declaration for the guarded signal. So, for example, we might include the following in the declarative part of an architecture body:

```
signal memory_data_bus : resolved_word bus;
disconnect memory_data_bus : resolved_word after 3 ns;
```

We might then include the following block in the architecture body:

```
mem_write_buffer : block (mem_sel and mem_write) is
begin
  memory_data_bus <=
    guarded reject 2 ns inertial cache_data_bus after 4 ns;
end block mem_write_buffer;
```

This indicates that so long as the guard expression evaluates to true, the value of cache_data_bus will be copied to **memory_data_bus** with a delay of 4 ns and a pulse rejection interval of 2 ns. When the guard expression changes to false, the driver corresponding to the guarded signal assignment is disconnected with a null transaction. The delay used is 3 ns, as indicated in the disconnection specification, but the pulse rejection limit of 2 ns is still taken from the assignment statement. When the guard expression changes back to true, the assignment is executed again, scheduling a new transaction with 4 ns delay.

If we have a number of guarded signals of the same type in an architecture body, and we wish to use the same disconnection delay for all of them, we can use the **all** keyword in a disconnection specification instead of listing all of the signals. For example, if the following signal declarations are the only ones for guarded signals of type resolved_word:

```
signal source_bus_1, source_bus_2 : resolved_word bus;
signal address_bus : resolved_word bus;
```

we can specify a disconnection delay of 2 ns for all of the signals as follows:

```
disconnect all : resolved_word after 2 ns;
```

The remaining way of identifying which signals a disconnection specification applies to is with the keyword **others**. This identifies all remaining signals of a given type that are not referred to by previous disconnection specifications. For example, suppose that the signal **address_bus** shown above should have a disconnection delay of 3 ns instead of 2 ns. We could write the disconnection specifications for the set of signals as

```
disconnect address_bus : resolved_word after 3 ns;
```

```
disconnect others : resolved_word after 2 ns;
```

If we write a disconnection specification using the keyword **others** in an architecture body, it must appear after any other disconnection specifications referring to signals of the same type and after all declarations of signals of that type. Similarly, if we write a disconnection specification using the keyword **all**, it must be the only disconnection specification referring to signals of the given type and must appear after all declarations of signals of that type.

23.1.3 Using Blocks for Structural Modularity

We now look at the use of blocks to partition the concurrent statements within an architecture body. We can think of a block as a way of drawing a line around a collection of concurrent statements and their associated declarations, so that they can be clearly seen as a distinct aspect of a design. The full syntax rule for a block statement is as follows:

```
block_statement ⇐
    block_label :
    block [ ( guard_expression ) ] [ is ]
        [ generic ( generic_interface_list ) ;
        [ generic map ( generic_association_list ) ; ] ]
        [ port ( port_interface_list ) ;
        [ port map ( port_association_list ) ; ] ]
        { block_declarative_item }
    begin
        { concurrent_statement }
    end block [ block_label ] ;
```

The block label is required to identify the block statement. The guard expression, as we saw earlier, may be used to control guarded signal assignments. If we are only using a block as a means of partitioning a design, we do not need to include a guard expression. The generic and port clauses allow us to define an interface to the block. We return to this shortly.

The declarative part of a block statement allows us to declare items that are local to the block. We can include the same kinds of declarations here as we can in an architecture body, for example, constant, type, subtype, signal and subprogram declarations. Items declared in a block are only visible within that block and cannot be referred to before or after it. However, items declared in the enclosing architecture body remain visible (unless hidden by a local item declared within the block).

EXAMPLE 23.5 *Blocks for partitioning timing and functionality*

To illustrate how blocks can be used for partitioning a design, we develop a model for a counter, including detailed pin-to-pin propagation delays and some error checking. We can specify the propagation delays as combinations of input delays before the function block and output delays after the function block, as shown in Figure 23.1. The function block implements the behavior of the counter with zero delay.

The entity declaration for this counter is

```
entity counter is
    generic ( tipd_reset,          -- input prop delay on reset
              tipd_clk,            -- input prop delay on clk
              topd_q : delay_length;  -- output prop delay on q
              tsetup_reset,        -- setup: reset before clk
              thold_reset :        -- hold time: reset after clk
                  delay_length );
    port ( reset,                  -- synchronous reset input
           clk : in bit;           -- edge-triggered clock input
           q : out bit_vector );   -- counter output
end entity counter;
```

We can separate the delay, function and error-checking aspects of the model into separate blocks within the architecture body, as follows:

```
architecture detailed_timing of counter is
```

FIGURE 23.1

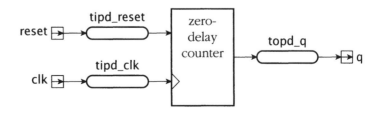

A propagation delay model for a counter.

```vhdl
    signal reset_ipd,                         -- data input port delayed
           clk_ipd : bit;                     -- clock input port delayed
    signal q_zd : bit_vector(q'range);  -- q output with zero delay
begin
  input_port_delay : block is
  begin
    reset_ipd <= reset after tipd_reset;
    clk_ipd <= clk after tipd_clk;
  end block input_port_delay;

  functionality : block is

    function increment ( bv : bit_vector ) return bit_vector is
      variable result : bit_vector(bv'range) := bv;
      variable carry : bit := '1';
    begin
      for index in result'reverse_range loop
        result(index) := bv(index) xor carry;
        carry :=  bv(index) and carry;
        exit when carry = '0';
      end loop;
      return result;
    end function increment;

    signal next_count : bit_vector(q'range);

  begin
    next_count <= increment(q_zd) when reset_ipd = '0' else
                  (others => '0');
    q_zd <= next_count when clk_ipd = '1' and clk_ipd'event;
  end block functionality;

  output_port_delay : block is
  begin
    q <= q_zd after topd_q;
  end block output_port_delay;

  timing_checks : block is
  begin
    -- check setup time: reset before clk
    ...
    -- check hold time: reset after clk
    ...
  end block timing_checks;

end architecture detailed_timing;
```

The first block, **input_port_delay**, derives delayed versions of the input ports. These are used in the second block, **functionality**, the zero-delay behavioral implementation of the counter. This block consists of two concurrent signal assignment statements that together implement a finite-state machine. One statement calculates the next count value using the **increment** function locally declared within the block, and the other implements an edge-triggered register. The signal **next_count**, also locally declared within the block, is used to connect the two statements. The output of the state machine is used in the third block, **output_port_delay**, to apply the delay between the function block and the output port. The final block outlined in the architecture body, **timing_checks**, contains processes that verify correct setup and hold times for the reset signal.

Since a block contains a collection of concurrent statements, and a block statement is itself a concurrent statement, it is perfectly legal to nest blocks one inside another. The same visibility rules that we described for subprograms also apply for items declared in nested blocks. However, in practice, we would rarely write a model with nested blocks. If the design hierarchy is that complex, it is better to use separate entities and component instantiation statements to partition the design. The main reason VHDL allows complex nesting of blocks is that the block structure is used as the underlying mechanism for implementing other VHDL constructs, such as component instantiation (described in Chapter 13) and generate statements (described in Chapter 14). The language definition defines these constructs in terms of the substitution of blocks containing the contents of the architecture body being instantiated or the contents of the generate statement.

External Names and Blocks

As we mentioned in Section 18.1, a block statement forms a concurrent region in a design. When we write external names, we need to include the label of a block in the pathname for any item declared in the block or in an entity instantiated in the block.

EXAMPLE 23.6 *Referring to a block in an external pathname*

Suppose the counter design of Example 23.5 in instantiated with a test bench architecture. We can use an external name in the test bench to refer to the **next_count** signal. The test bench outline is as follows:

```
architecture verifying of test_bench is

  signal clk, reset : bit;
  signal duv_q : bit_vector(7 downto 0);

  alias duv_next_count is
    <<signal duv.functionality.next_count :
             bit_vector(duv_q'range)>>;
  ...

begin
```

```
    duv : entity work.counter(detailed_timing)
      generic map ( ... )
      port map ( clk => clk, reset => reset, q => duv_q );

  ...

end architecture verifying;
```

The counter is instantiated using the label **duv**. In the external name, the relative pathname starts with **duv**, followed by the block label **functionality** and then the signal name **next_count**.

Generics and Ports in Blocks

Another aspect of block statements, also arising from their use as the underlying mechanism for component instantiation, is the possibility of including generic and port interface lists. These allow us to make explicit the interface between the block and its enclosing architecture body or enclosing block. The formal generics and ports can be used within the block in exactly the same way that those of an entity are used within a corresponding architecture body. The actual values for genericts are supplied by a generic map in the block header, and the actual signals associated with the formal ports are supplied by a port map. These are all shown in the syntax rule for block statements on page 744. Since this facility is rarely used in actual model writing, we do not dwell on it.

Configuring Designs with Blocks

In Chapter 13 we showed how to configure a design whose hierarchy was formed by instantiating components. We configure an architecture body containing nested block statements in a similar way. When we write configuration declarations for such architecture bodies, the configuration information must mirror the block structure of the architecture body. We introduce a further level of detail in the syntax rules for configuration declarations, showing how to configure architecture bodies containing blocks.

configuration_declaration ⟸
    configuration identifier **of** *entity*_name **is**
        block_configuration
    end ⟦ **configuration** ⟧ ⟦ identifier ⟧ ;

block_configuration ⟸
    for ⟨ *architecture*_name ⦙ *block_statement*_label ⟩
        ⟨   block_configuration
            ⦙   **for** component_specification
                ⟦ binding_indication ; ⟧
                ⟦ block_configuration ⟧
            end for ; ⟩
    end for ;

The difference here is that we have added a block statement label as an alternative to an architecture name at the point where we specify the region containing concurrent statements. Furthermore, we have allowed a block configuration as an alternative to component configuration information within that region. If we put these together, we can see how to write the configuration information for an architecture body containing block statements. At the top level of the configuration declaration, we write a block configuration naming the architecture body, just as we have done in all of the previous examples. Within it, however, we include block configurations that name and configure each block.

EXAMPLE 23.7   *Configuration of a design partitioned with blocks*

Suppose we need to write a model for an integrated circuit that takes account of propagation delays through input and output pads. The entity declaration and architecture body are shown below. The architecture body is divided into blocks for input delay, function and output delay. The operation of the circuit is described structurally, as an interconnection of cells within the function block.

```
entity circuit is
  generic ( inpad_delay, outpad_delay : delay_length );
  port ( in1, in2, in3 : in bit;  out1, out2 : out bit );
end entity circuit;

-- ------------------------------------------------

architecture with_pad_delays of circuit is

  component subcircuit is
    port ( a, b : in bit;  y1, y2 : out bit );
  end component subcircuit;

  signal delayed_in1, delayed_in2, delayed_in3 : bit;
  signal undelayed_out1, undelayed_out2 : bit;

begin

  input_delays : block is
  begin
    delayed_in1 <= in1 after inpad_delay;
    delayed_in2 <= in2 after inpad_delay;
    delayed_in3 <= in3 after inpad_delay;
  end block input_delays;

  functionality : block is
    signal intermediate : bit;
  begin
    cell1 : component subcircuit
      port map ( delayed_in1, delayed_in2,
                 undelayed_out1, intermediate );
    cell2 : component subcircuit
      port map ( intermediate, delayed_in3,
```

```
                    undelayed_out2, open );
        end block functionality;

    output_delays : block is
    begin
       out1 <= undelayed_out1 after outpad_delay;
       out2 <= undelayed_out2 after outpad_delay;
    end block output_delays;

end architecture with_pad_delays;
```

A configuration declaration for this design is

```
configuration full of circuit is

    for with_pad_delays   -- configure the architecture

      for functionality    -- configure the block

        for all : subcircuit
          use entity work.real_subcircuit(basic);
        end for;

      end for;

    end for;

end configuration full;
```

The configuration binds the instances of the component **subcircuit** within the block **functionality** to an entity **real_subcircuit** with architecture **basic**. The block configuration starting with "**for** with_pad_delays" specifies the architecture of **circuit** that is being configured. Within it, the block configuration starting with "**for** functionality" specifies the configuration of the contents of the block labeled **functionality**. It, in turn, contains a component configuration for the two component instances. Note that there are no block configurations for the other two blocks in the design, since they do not contain any component instances. They only contain concurrent signal assignment statements, which represent leaf nodes of the design hierarchy.

23.2 IP Encryption

As designs become more complex, designers are increasingly using intellectual property (IP) provided by IP vendors. IP providers invest considerable effort in developing their products, and may be loath to release them without protecting their investment. From the IP provider's point of view, there are two potential places where their IP may be compromised. First, the IP provider must transmit the IP to a customer. During that process, a malicious third party could eavesdrop on the transmission and intercept the IP. Second, the customer must receive, store, and use the IP. During that process, an unscrupulous customer could reuse the IP without compensating the provider. Hence, the customer is technically treated as a malicious third party, though it would not be politic to express the relationship in those terms! The real recipient of the IP is the customer's tool, which must use the IP only in ways approved by the IP provider and must avoid disclosing the IP to the customer.

One way of protecting IP is for the provider to encrypt it in a form that can be decrypted and used by a customer's tools, but that cannot be read by the customer. VHDL provides a flexible set of features to support such protection. Before we describe them in detail, we will first review some of the basic principles and protocols for encryption so that we can understand how to use the language features.

Information to be communicated between two parties can be protected by transforming it with a *cipher*. A cipher is a function that takes *plain text* and a string of bits called a *key* as input and produces *cipher text* as output. This process is called *encryption*. The reverse process, *decryption*, takes the cipher text and a key as input and reproduces the original plain text. The quality of a cipher is determined by the difficulty of determining the plain text from the cipher text without the key. A good cipher will yield significantly different cipher text for minor changes to the key used for encryption.

There are two forms of cipher in widespread use. A *symmetric cipher* uses the same key for both encryption and decryption. The key is called a *secret key*, since it must be kept secret between the communicating parties. Should the secret be revealed to a third party, they could decrypt any intercepted encrypted information. Examples of symmetric ciphers are the Data Encryption Standard (DES), and the Advanced Encryption Standard (AES).

An *asymmetric cipher* uses a pair of related keys, one for encryption and the other for decryption. Key pairs are generated in such a way that it is infeasible to determine either key from the other. Information encrypted with one key of a pair can only be decrypted with the other key of the pair. Examples of asymmetric ciphers are RSA and ElGamal. Asymmetric ciphers are used in protocols where each communicating party generates a key pair. They keep one key of the pair, the *private key*, secret. They publish the other key, the *public key*, through some means of dissemination that associates the public key with the communicating party's identity. For example, they might publish it on their website. A sender of information can use an asymmetric cipher to protect information destined for a recipient. The sender encrypts the information using the recipient's public key. Only the recipient can then decrypt the information, since only they have the corresponding private key.

While asymmetric ciphers can yield more secure communication, they involve significantly greater computation than symmetric ciphers. For that reason, most applications involving asymmetric ciphers use a two-stage encryption process called a *digital envelope*. First a *session key* is randomly generated, for use in one communication session only. Next, that session key is used with a symmetric cipher to encrypt the information. In order to communicate the session key to the recipient so that they can decrypt the information, the session key is encrypted using an asymmetric cipher with the recipient's public key, and sent to the recipient. They are the only party able to decrypt the session key, since only they have the right private key. They can then proceed to decrypt the communicated information using the symmetric cipher with the decrypted session key. The advantage of this approach is that only a relatively small amount of information (the session key) need be processed using the computationally intensive asymmetric cipher. The bulk of the information is processed using the lighter-weight symmetric cipher.

One problem that arises in protected communication is the need to verify that received information did in fact originate with the purported sender, and that it was not changed in transit (either by corruption or maliciously) by a third party. This problem is addressed by having the sender transmit a *digital signature* for the information. The sender uses a *hash function* to compute a *digest* of the information. A hash function takes a (potentially large) string of bits as input and produces a small string of bits, the digest, that depends on all of the input bits. A good hash function has the property that two distinct input strings are highly unlikely to yield the same output string. Examples of hash functions include SHA1, MD2, MD5, and RIPEMD. Having computed the digest of the information, the sender encrypts it using an asymmetric cipher with their private key and transmits the result as the digital signature. A recipient decrypts the signature using the purported sender's public key to retrieve the digest. The recipient also independently calculates the digest of the received information using the hash function. If the two digests are the same, the information has been received correctly, since only the real sender's public key could decrypt the digest correctly, and only the real information would yield the same digest. If, on the other hand, the digests differ, then either the transmitted digest was encrypted with someone else's key, or the transmitted message was changed. Either way, the transmission was compromised, and the recipient knows not to trust the received information.

If we are to apply cryptographic techniques to transmission of VHDL models, we need to consider the way in which the encrypted information is encoded. Plain-text VHDL models consist of printable ASCII or Latin-1 characters and are immune to the way ends of lines are encoded. Consequently, we can store and transmit plain-text models through a variety of media without being concerned about encodings. However, the process of encryption produces a string of bits, which cannot be guaranteed to be interpreted as printable characters. We cannot reliably transmit the encrypted model, since some media might transform sequences of bits interpreted as line ends, or might interpret sequences of bits as in-band control codes. To avoid these problems, we can encode the encrypted model using an encoding method that uses printable characters to represent the string of bits. A sender encrypts information and encodes it for transmission, and a recipient decodes the received information and decrypts the result. Examples of encoding methods include uuencode, base64, and quoted-printable, all of which are described by Internet message-transfer standards.

With this overview of cryptography in hand, we can now discuss the features provided in VHDL to support cryptographic protection of IP. The features use a standard set of *tool directives*. A tool directive is an annotation included in a VHDL design file that provides information to a tool processing the VHDL design. It does not logically form part of the design itself. The syntax for a tool directive is

tool_directive $\Leftarrow$ ` identifier ⟦ graphic_character ⟧

A tool directive starts with the "tick" symbol, and ends at the end of the line. The identifier specifies to the tool what form of processing to perform. The remaining text, up to the end of the line, provides any additional information needed.

For IP protection, VHDL defines *protect directives*, in which the identifier is **protect**. (Although the identifier is not a reserved word, we write it and other keywords in protect directives in boldface type to indicate that they have special meanings in the directives.) The protect directives are used by an IP provider's *encryption tool* to govern encryption of sections of a VHDL design and by a customer's *decryption tool* to decrypt those sections.

The decryption tool is typically a simulator, synthesis tool, or some other tool that deals with VHDL code. It uses the decrypted sections of the design, but does not store them in any form that could be revealed to the customer. Protect directives each takes one of three forms:

```
`protect keyword
`protect keyword = value
`protect keyword = ( keyword = value, ... )
```

The keyword or keywords in a protect directive identify the kind of information conveyed by the directive. The values are literal expressions of various types. If we have a number of consecutive protect directives, we can merge them into a single directive. Thus, we can write the sequence of directives

```
`protect keyword1 = value1
`protect keyword2 = value2
`protect keyword3
```

equivalently as

```
`protect keyword1 = value1, keyword2 = value2, keyword3
```

An IP provider starts the process by identifying one or more sections of a VHDL design file that they want to protect. They edit the design file to wrap each section in an *encryption envelope*, consisting of one or more protect directives at the start of the section, and a closing protect directive at the end of the section. The simplest form of encryption envelope is

```
`protect begin
... -- protected source code in plain-text form
`protect end
```

This simply delimits the protected source code, and assumes an encryption tool will use default information about the ciphers, keys and encoding for encryption. More elaborate encryption envelopes precede the **begin** directive with protect directives specifying ciphers, keys, encoding and other optional information.

The IP provider then processes the design file with an encryption tool to produce a version of the design file with each encryption envelope replaced by a corresponding *decryption envelope* of the following form:

```
`protect begin_protected
protect directives and encoded encrypted information
`protect end_protected
```

We will use a series of examples to show how the various directives are used to form encryption and decryption envelopes for various use cases. In each case we will assume that the decryption tool has access to the required keys, and that the encryption tool knows about those keys. We will return to the topic of key exchange in Section 23.2.1.

EXAMPLE 23.8   *Simple encryption envelope with symmetric cipher*

In one of the simplest use cases, an IP provider wants to provide protected IP to a customer for use with a single tool. We can use a symmetric cipher, for which the key is known to both the IP provider and to the customer's decryption tool. The IP provider wraps the protected section in the source code in an encryption envelope, as follows:

```
entity accelerator is
  port ( ... );
end entity accelerator;

architecture RTL of accelerator is
`protect data_keyowner = "ACME IP User"
`protect data_keyname  = "ACME Sim Key"
`protect data_method   = "aes192-cbc"
`protect encoding      = (enctype = "base64")
`protect begin
  signal ...
begin
  process ...
  ...
`protect end
end architecture RTL;
```

The IP provider leaves the information about the entity's interface and the name of the architecture unprotected so that the customer can instantiate the design. The entire inner workings of the architecture, however, are not to be revealed to the customer. The **data_keyowner** and **data_keyname** directives specify identifiers that the encryption and decryption tools can use to retrieve the key. The **data_method** directive specifies the cipher to use for encryption and decryption, and the **encoding** directive specifies the method to use to encode the cipher text produced by the encryption tool.

The IP provider processes the original source code file with an encryption tool, which produces a modified file with the encryption envelope replaced by a decryption envelope:

```
entity accelerator is
  port ( ... );
end entity accelerator;

architecture RTL of accelerator is
`protect begin_protected
`protect encrypt_agent      = "Encryptomatic"
`protect encrypt_agent_info = "2.3.4a"
`protect data_keyowner = "ACME IP User"
`protect data_keyname  = "ACME Sim Key"
`protect data_method   = "aes192-cbc"
```

```
`protect encoding=(enctype="base64", line_length=40, bytes=4006)
`protect data_block
encoded cipher-text

...

`protect end_protected
end architecture RTL;
```

The **encrypt_agent** and **encrypt_agent_info** directives provide information about the encryption tool. This can help in tracking down any problems that might arise. The directives specifying the key, cipher, and encoding method are replicated in the decryption envelope. In the case of the **encoding** directive, further information about the maximum line length for the encoded cipher text and the number of bytes of cipher text is also provided. The encoded cipher text then starts immediately after the **data_block** directive. The **end_protected** directive marks the end of the decryption envelope.

EXAMPLE 23.9   *Digital envelope encrypted for a single customer*

One of the problems with using a symmetric cipher to encrypt IP is that the risk of the secret key being divulged. We can avoid that risk by using a digital envelope to transmit the IP. The IP provider includes directives in the encryption envelope to specify a cipher and key to use to encrypt a session key. The IP provider can also specify the symmetric cipher to use to encrypt the data with the session key. The design file with the revised encryption envelope is

```
entity accelerator is
  port ( ... );
end entity accelerator;

architecture RTL of accelerator is
`protect key_keyowner = "ACME IP User"
`protect key_keyname  = "ACME Sim Key"
`protect key_method   = "rsa"
`protect key_block
`protect data_method  = "aes192-cbc"
`protect encoding     = (enctype = "base64")
`protect begin
  signal ...
begin
  process ...
  ...
`protect end
end architecture RTL;
```

The **key_keyowner** and **key_keyname** directives specify identifiers that the encryption tool can use to retrieve the customer's public key. The **key_method** directive specifies the cipher to use to encrypt the session key. The **key_block** directive marks the end of the key information. Its presence in the encryption envelope specifies use of a digital envelope, since the preceding key directives can be omitted, implying default values. The **data_method** directive specifies the cipher to use for encryption and decryption with the session key. The **encoding** directive specifies the method to use to encode both the encrypted session key and the encrypted section of the model.

The IP provider processes this source code file with an encryption tool, which generates a session key and produces a modified file with the encryption envelope replaced by a decryption envelope specifying a digital envelope:

```
entity accelerator is
  port ( ... );
end entity accelerator;

architecture RTL of accelerator is
`protect begin_protected
`protect encrypt_agent       = "Encryptomatic"
`protect encrypt_agent_info = "2.3.4a"
`protect key_keyowner = "ACME IP User"
`protect key_keyname  = "ACME Sim Key"
`protect key_method   = "rsa"
`protect encoding=(enctype="base64", line_length=40, bytes=256)
`protect key_block
encoded cipher-text for session key
`protect data_method   = "aes192-cbc"
`protect encoding=(enctype="base64", line_length=40, bytes=4006)
`protect data_block
encoded cipher-text for model code
...
`protect end_protected
end architecture RTL;
```

The directives specifying the key and cipher for encrypting the session key are replicated in the decryption envelope. The **encoding** directive is also replicated to specify the encoding for the encrypted session key, augmented with information about the maximum line length for the encoded cipher text and the number of bytes in the encrypted session key. The encoded cipher text for the session key then starts immediately after the **key_block** directive. Next, the **data_method** directive specifying the cipher for the model code is replicated in the decryption envelope. The **encoding** directive is also replicated here, augmented with information about the maximum line length and the number of bytes. The encoded cipher text for the model code starts immediately after the **data_block** directive. The **end_protected** directive marks the end of the decryption envelope.

EXAMPLE 23.10 *Digital envelope encrypted for multiple customers or tools*

In Example 23.8 and Example 23.9, the IP is encrypted in a form that can be decrypted by a single customer or by a single tool. If the IP provider wants to distribute the IP to multiple customers or to a customer for use with multiple tools, he or she would have to generate multiple versions using the encryption tool, once per customer. We can avoid this repetition by using a variation on the digital envelope approach. Again, we specify that a session key be used to encrypt the model code. However, that session key is then encrypted multiple times, once per customer or customer's tool. The revised source file with the encryption envelope is

```
entity accelerator is
  port ( ... );
end entity accelerator;

architecture RTL of accelerator is
`protect key_keyowner = "ACME IP User1"
`protect key_keyname  = "ACME Sim Key"
`protect key_method   = "rsa"
`protect key_block
`protect key_keyowner = "ACME IP User2"
`protect key_keyname  = "ACME Synth Key"
`protect key_method   = "elgamal"
`protect key_block
`protect key_keyowner = "ACME IP User3"
`protect key_keyname  = "ACME P&R Key"
`protect key_method   = "aes192-cbc"
`protect key_block
`protect data_method  = "aes192-cbc"
`protect encoding     = (enctype = "base64")
`protect begin
  signal ...
begin
  process ...
  ...
`protect end
end architecture RTL;
```

Each group of key directives specifies information for encryption of a session key for decryption by a given decryption tool. The first two groups specify encryption using asymmetric ciphers, as is normally done in digital envelopes. However, we can also use a symmetric cipher to encrypt the session key, as specified in the third group of key directives.

As in the earlier examples, the IP provider processes this source code file with an encryption tool, which generates a session key and produces a modified file with the encryption envelope replaced by a decryption envelope specifying a digital envelope:

```
entity accelerator is
  port ( ... );
end entity accelerator;

architecture RTL of accelerator is
`protect begin_protected
`protect encrypt_agent       = "Encryptomatic"
`protect encrypt_agent_info = "2.3.4a"
`protect key_keyowner = "ACME IP User1"
`protect key_keyname  = "ACME Sim Key"
`protect key_method   = "rsa"
`protect encoding=(enctype="base64", line_length=40, bytes=256)
`protect key_block
encoded cipher-text for session key
`protect key_keyowner = "ACME IP User2"
`protect key_keyname  = "ACME Synth Key"
`protect key_method   = "elgamal"
`protect encoding=(enctype="base64", line_length=40, bytes=256)
`protect key_block
encoded cipher-text for session key
`protect key_keyowner = "ACME IP User3"
`protect key_keyname  = "ACME P&R Key"
`protect key_method   = "aes192-cbc"
`protect encoding=(enctype="base64", line_length=40, bytes=256)
`protect key_block
encoded cipher-text for session key
`protect data_method    = "aes192-cbc"
`protect encoding=(enctype="base64", line_length=40, bytes=4006)
`protect data_block
encoded cipher-text for model code
...
`protect end_protected
end architecture RTL;
```

In this case, the decryption envelope includes a group of key directives and a key block corresponding to each group of key directives in the encryption envelope. Each of the targeted decryption tools, when it processes the decryption envelope, checks whether it has access to the key specified by each group of key directives. If it has one of the keys, it can use that key to decrypt the session key, and thus decrypt the model code.

EXAMPLE 23.11 *Digital signature for authentication of the provider*

Suppose our IP provider delivers encrypted IP by making it available for download from a file server. They use our public key to deliver the IP in digital envelope form. An unscrupulous third-party IP provider could seek to besmirch the name of our trusted IP provider by spoofing their server and providing a buggy version of the IP. Since the IP is encrypted using our public key, which is widely known, we would not be aware of the switch.

The solution is for our trusted IP provider to include a digital signature in the delivered model. The encryption envelope, revised from that in Example 23.10, is

```
entity accelerator is
  port ( ... );
end entity accelerator;

architecture RTL of accelerator is
`protect key_keyowner = "ACME IP User"
`protect key_keyname  = "ACME Sim Key"
`protect key_method    = "rsa"
`protect key_block
`protect data_method   = "aes192-cbc"
`protect digest_keyowner    = "GoodGuys IP Author"
`protect digest_keyname     = "GoodGuys Signing Key"
`protect digest_key_method = "rsa"
`protect digest_method      = "sha1"
`protect digest_block
`protect encoding = (enctype = "base64")
`protect begin
  signal ...
begin
  process ...
    ...
`protect end
end architecture RTL;
```

The digest directives in the encryption envelope specify that a digital signature should be generated for the model code contained in the envelope. The **digest_method** directive specifies the hash function for computing the digest, and the **digest_keyowner**, **digest_keyname** and **digest_key_method** directives specify the cipher and key to use to encrypt the digest. The **digest_key_method** directive must specify an asymmetric cipher, since digital signatures are predicated on the use of such ciphers.

The IP provider processes this source code file with an encryption tool, which computes and encrypts the digest to form the digital signature. It uses the private key of the key pair specified by the digest key directives. It includes the digest in the decryption envelope corresponding to the encryption envelope:

```
entity accelerator is
  port ( ... );
end entity accelerator;

architecture RTL of accelerator is
`protect begin_protected
`protect encrypt_agent       = "Encryptomatic"
```

```
`protect encrypt_agent_info = "2.3.4a"
`protect key_keyowner = "ACME IP User"
`protect key_keyname  = "ACME Sim Key"
`protect key_method   = "rsa"
`protect encoding=(enctype="base64", line_length=40, bytes=256)
`protect key_block
```
encoded cipher-text for session key
```
`protect data_method  = "aes192-cbc"
`protect encoding=(enctype="base64", line_length=40, bytes=4006)
`protect data_block
```
encoded cipher-text for model code
```
...
`protect digest_keyowner   = "GoodGuys IP Author"
`protect digest_keyname    = "GoodGuys Signing Key"
`protect digest_key_method = "rsa"
`protect digest_method     = "sha1"
`protect digest_block
`protect encoding=(enctype="base64", line_length=40, bytes=16)
`protect digest_block
```
encoded cipher-text for digest
```
...
`protect end_protected
end architecture RTL;
```

Our trusted IP provider places this model on the file server for us to download. Now suppose the unscrupulous third-party IP provider performs their network hack and substitutes a buggy model. In their first attempt, they substitute the buggy code, encrypted with a session key that they generate, and encrypt the session key with our public key. Our decryption tool successfully decrypts the session key and uses it to decrypt the model. However, since we want to verify that we have the right model, the decryption tool computes the digest of the decrypted model using the hash function specified in the **digest_method** directive. The tool also uses the public key of the key pair identified in the digest key directives to decrypt the transmitted digest. Since the model code is different from the original code provided by the trusted IP provider, the two digests are not the same. Our decryption tool alerts us to this fact, and we contact our trusted IP provider to attempt to remedy the problem.

Now suppose the unscrupulous third-party IP provider realizes their ruse was unsuccessful, and tries a different tack. As well as substituting the buggy model, suitably encrypted, they also generate a digital signature for the buggy model and substitute it for the real digital signature. They use their own private key to encrypt the digest, and include digest key directives that identify their key pair. Again, our decryption tool successfully decrypts the model and calculates the digest. The tool also attempts to decrypt the transmitted digest in order to compare with the computed digest. At this point, there are two possible outcomes. First, if the tool does not have access to the unscrupulous provider's public key, it will be unable to proceed and will warn us that it could not verify the digital signature. Alternatively, if the tool does have access to the unscrupulous provider's public key, it will use it to decrypt the transmitted digest

and compare it with the computed digest. In this case, the digests will match. It will
be up to us to check that signature verification was performed with the key we expected. This illustrates that we need to be vigilant when checking digital signatures,
so that we are not duped by a simple key substitution. We will discuss this more in
Section 23.2.1, where we address the issue of key exchange.

EXAMPLE 23.12   *Viewport for accessing a declaration in a protected model*

An IP provider may wish to allow limited access to some items declared within the
protected source code. In Examples 18.2 and 18.3 in Section 18.1, we showed a test
bench monitoring the internal state of the control section of a design under verification. An IP provider can allow such access by including a **viewport** directive in the
encryption envelope. An example is

```
entity accelerator is
  port ( ... );
end entity accelerator;

architecture RTL of accelerator is
`protect data_keyowner = "ACME IP User"
`protect data_keyname  = "ACME Sim Key"
`protect data_method   = "aes192-cbc"
`protect encoding       = (enctype = "base64")
`protect viewport=(object="accelerator:RTL.state", access="RW");
`protect begin
  signal state : std_ulogic_vector(3 downto 0);
  ...
begin
  process ...
  ...
`protect end
end architecture RTL;
```

While most of the inner workings of the architecture are not revealed to the customer, the **viewport** directive provides the pathname of the object representing the
internal state signal and grants read/write access. The IP provider processes the source
code file with an encryption tool, which includes the **viewport** directive in the decryption envelope:

```
entity accelerator is
  port ( ... );
end entity accelerator;

architecture RTL of accelerator is
`protect begin_protected
`protect encrypt_agent      = "Encryptomatic"
`protect encrypt_agent_info = "2.3.4a"
`protect viewport=(object="accelerator:RTL.state", access="RW");
```

```
`protect data_keyowner = "ACME IP User"
`protect data_keyname  = "ACME Sim Key"
`protect data_method   = "aes192-cbc"
`protect encoding=(enctype="base64", line_length=40, bytes=4006)
`protect data_block
encoded cipher-text
...
`protect end_protected
end architecture RTL;
```

The customer can instantiate the IP in a design and use an external name to refer to the state signal:

```
architecture monitoring of tb is
  ...
begin

  ... -- clock and reset generation

  accelerator_duv : entity work.accelerator(rtl)
    port map ( ... );

  monitor : process (clk) is
    use std.textio.all;
    file state_file : text open write_mode is state_file_name;
    alias accelerator_state is
      <<signal accelerator_duv.state :
              std_ulogic_vector(3 downto 0)>>;
  begin
    if falling_edge(clk) then
      write(L, accelerator_state); writeline(state_file, L);
    end if;
  end process monitor;

end architecture monitoring;
```

While the **viewport** directive provides access to the internal signal, it does not provide complete information. The IP provider would also need to provide documentation describing the type of the signal and other relevant information.

Now that we have seen how protection envelopes are formed in various scenarios, we will describe the details of VHDL's IP protection mechanism. As we have mentioned, it is based on a set of tool directives. The full list of directives is as follows:

`protect begin

Indicates the beginning of the source code to be encrypted in an encryption envelope.

`protect end

>   Indicates the end of an encryption envelope.

`protect begin_protected

>   Indicates the beginning of a decryption envelope.

`protect end_protected

>   Indicates the end of a decryption envelope.

`protect author = "*author name*"

>   Identifies the author of the protected IP. If this directive appears in an encryp-
>   tion envelope, the encryption tool copies it unchanged to the corresponding
>   decryption envelope.

`protect author_info = "*author info*"

>   Provides further information about the author of the protected IP, such as an
>   organization name or contact details. If this directive appears in an encryption
>   envelope, the encryption tool copies it unchanged to the corresponding de-
>   cryption envelope.

`protect encrypt_agent = "*encrypt agent name*"

>   This directive must appear in a decryption envelope, and identifies the en-
>   cryption tool that produced the decryption envelope.

`protect encrypt_agent_info = "*encrypt agent info*"

>   This directive may appear in a decryption envelope, and provides further in-
>   formation about the encryption tool that produced the decryption envelope.

`protect key_keyowner = "*key owner name*"

>   Identifies the owner of a key or key pair used to encrypt a session key.

`protect key_keyname = "*key name*"

>   Used together with the key owner name to identify a particular key or key
>   pair used to encrypt a session key.

`protect key_method = "*cipher name*"

>   Specifies the cipher used to encrypt a session key.

`protect key_block

>   In an encryption envelope, specifies use of a digital envelope. In the corre-
>   sponding decryption envelope, indicates the beginning of the encoded cipher
>   text of the session key.

`protect data_keyowner = "*key owner name*"

>   Identifies the owner of a key or key pair used to encrypt the source code.

`` `protect data_keyname`` = "*key name*"

> Used together with the key owner name to identify a particular key or key pair used to encrypt the source code.

`` `protect data_method`` = "*cipher name*"

> Specifies the cipher used to encrypt the source code.

`` `protect data_block``

> In a decryption envelope, indicates the beginning of the encoded cipher text of the source code.

`` `protect digest_keyowner`` = "*key owner name*"

> Identifies the owner of the key pair used to encrypt the digest in a digital signature.

`` `protect digest_keyname`` = "*key name*"

> Used together with the key owner name to identify a particular key pair used to encrypt the digest in a digital signature.

`` `protect digest_key_method`` = "*cipher name*"

> Specifies the asymmetric cipher used to encrypt the digest in a digital signature.

`` `protect digest_method`` = "*hash function name*"

> Specifies the hash function used to compute the digest in a digital signature.

`` `protect digest_block``

> In an encryption envelope, specifies use of a digital signature. In the corresponding decryption envelope, indicates the beginning of the encoded cipher text of the digest.

`` `protect encoding`` =
(**enctype** = "*encoding name*", **line_length** = *integer*, **bytes** = *integer*)

> In an encryption envelope, this directive specifies the encoding to be used for cipher text in the corresponding decryption envelope. The **line_length** keyword and value are optional and specify the maximum line length for encoded text. Text longer than this amount is split into multiple lines. The **bytes** keyword and value are also optional and are ignored in an encryption envelope in any case.
>
> In a decryption envelope, this directive appears preceding each key, data, and digest block. It specifies the encoding, maximum line length, and number of bytes of cipher text encoded in the block.

`` `protect viewport`` **= (object** = "*object pathname*", **access** = "*access type*")

> Identifies an object declared within the protected source code for which access is granted. If this directive appears in an encryption envelope, the encryption tool copies it unchanged to the corresponding decryption envelope.
>
> The pathname consists of the names of regions enclosing the declaration, starting with the design unit name and continuing with the names of nested regions, separated by "." characters, for example,

```
"my_entity.cycle_monitor.cycle_count"
```

> If the object is declared within an architecture, the design unit name is the combination of the entity name and the architecture name, separated by a colon, for example,

```
"my_entity:RTL.current_state"
```

> If the object is declared within a package body, the design unit name consists of the package name, followed by ":**body**", for example,

```
"IP_pkg:body.trace_file"
```

> The access type string must be one of "R", "W", or "RW" (or the lowercase equivalents), indicating read access, write access, or read/write access, respectively.

`` `protect decrypt_license`` **=**
(library = "*library name*",
 entry = "*acquisition routine name*", **feature** = "*feature name*",
 exit = "*release routine name*", **match** = *integer*)

> This directive specifies information for acquiring a decryption license. If the directive appears in an encryption envelope, the encryption tool copies it unchanged to the corresponding decryption envelope. If the directive appears in a decryption envelope, a decryption tool must attempt to acquire the specified license. If acquisition is successful, it continues decrypting the model. Otherwise, it is expected to stop further decryption.
>
> The library name string identifies the object library in the decryption tool's host file system containing routines for license management. The tool should call the routine identified by the acquisition routine name, passing the feature name string as an argument, to acquire a license. The tool should compare the return value of the routine with the match integer. If they are equal, acquisition succeeded. When the tool has completed decryption, it should relinquish the license by calling the routine identified by the release routine name.

`` `protect runtime_license`` **=**
(library = "*library name*",

entry = "*acquisition routine name*", **feature** = "*feature name*",
exit = "*release routine name*", **match** = *integer*)

> This directive specifies information for acquiring a runtime license. If the directive appears in an encryption envelope, the encryption tool copies it unchanged to the corresponding decryption envelope. If the directive appears in a decryption envelope, a decryption tool must attempt to acquire the specified license. If acquisition is successful, the tool may continue with analysis and execution of the model. Otherwise, it is expected not to execute the model. The information in this directive is the same as that in a decrypt_license directive.

` **protect comment** = "*comment string*"

> This directive allows the IP author to provide comments in the model. If the directive appears in an encryption envelope, either preceding or within the source code, the encryption tool copies it unchanged to the corresponding decryption envelope. If it is within the source code, the encryption tool skips over it when encrypting the source code.

Note, incidentally, that we have split a number of the longer directives over multiple lines, for reasons of presentation here. In practice, each directive must appear entirely on one line in a VHDL model.

Several directives use strings to specify ciphers, encodings, and hash functions. VHDL defines particular string values for these directives. If a tool supports the given cipher, encoding, or hash function, it must use the defined string value to specify it. A tool may also support other methods, in which case it uses an implementation-defined string value. Table 23.1 shows the string values for specifying ciphers. Every tool must support at least the DES cipher. Table 23.2 shows the string values for specifying encodings. Every tool must support at least uuencode and base64. Table 23.3 shows the string values for specifying hash functions. Every tool must support at least SHA1 and MD5.

TABLE 23.1   *Strings for specifying ciphers*

String	Cipher	Cipher type
"des-cbc"	DES in CBC mode.	Symmetric
"3des-cbc"	Triple DES in CBC mode.	Symmetric
"aes128-cbc"	AES in CBC mode with 128-bit key.	Symmetric
"aes192-cbc"	AES in CBC mode with 192-bit key.	Symmetric
"aes256-cbc"	AES in CBC mode with 256-bit key.	Symmetric
"blowfish-cbc"	Blowfish in CBC mode.	Symmetric
"twofish128-cbc"	Twofish in CBC mode with 128-bit key.	Symmetric
"twofish192-cbc"	Twofish in CBC mode with 192-bit key.	Symmetric

String	Cipher	Cipher type
"twofish256-cbc"	Twofish in CBC mode with 256-bit key.	Symmetric
"serpent128-cbc"	Serpent in CBC mode with 128-bit key.	Symmetric
"serpent192-cbc"	Serpent in CBC mode with 192-bit key.	Symmetric
"serpent256-cbc"	Serpent in CBC mode with 256-bit key.	Symmetric
"cast128-cbc"	CAST-128 in CBC mode.	Symmetric
"rsa"	RSA.	Asymmetric
"elgamal"	ElGamal.	Asymmetric
"pgp-rsa"	OpenPGP RSA key.	Asymmetric

TABLE 23.2   *Strings for specifying encodings*

String	Encoding methods
"uuencode"	IEEE Std 1003.1™-2001 (uuencode Historical Algorithm)
"base64"	IETF RFC 2045, also IEEE Std 1003.1 (uuencode -m)
"quoted-printable"	IETF RFC 2045
"raw"	Identity transformation; no encoding is performed, and the data may contain non-printing characters.

TABLE 23.3   *Strings for specifying hash functions*

Digest method string	Required/optional	Hash function
"sha1"	Required	Secure Hash Algorithm 1 (SHA-1).
"md5"	Required	Message Digest Algorithm 5.
"md2"	Optional	Message Digest Algorithm 2.
"ripemd-160"	Optional	RIPEMD-160.

We can now describe the rules for forming an encryption envelope in a model. The rules allow for considerable flexibility, but we must at least include the **begin** and **end** directives to mark out the source code to be encrypted.

We can precede the **begin** directive with a **key_block** directive if we want to specify use of digital envelopes. We can specify the cipher and key to use to encrypt the session key by including a **key_method** and a **key_keyowner** directive (and optionally a **key_keyname** directive). If we don't specify the cipher and key, the encryption tool chooses a default cipher and key. The **key_method**, **key_keyowner** and **key_keyname**

directives can appear in any order, but must immediately precede the **key_block** directive. We can include more than one group of key-related directives, as we described in
Example 23.10.

   We can also precede the **begin** directive with a **data_method** directive if we want
to specify the cipher to use to encrypt the source code. If we are not using digital envelopes and we include a **data_method** directive, we must also include a **data_keyowner**
directive and optionally a **data_keyname** directive to identify the key. If we are using
digital envelopes, the encryption tool generates the session key, so we do not include directives to identify the key. If we omit the **data_method** directive, the encryption tool
chooses a default cipher. All of the directives relating to encryption of the source code
must appear together in an encryption envelope.

   If we want to include a digital signature, we precede the **begin** directive with a
digest_block directive. We can specify the cipher and key to use to encrypt the digest
by including a **digest_key_method** and a **digest_keyowner** directive (and optionally
a **digest_keyname** directive). If we don't specify the cipher and key, the encryption tool
chooses a default cipher and key. Similarly, we can specify the hash function to use by
including a **digest_method** directive. If we don't specify a hash function, the encryption
tool chooses a default hash function. The **digest_key_method**, **digest_keyowner**,
digest_keyname, and **digest_method** directives can appear in any order, but must immediately precede the **digest_block** directive.

   Beyond these specifications, we can include directives to identify the IP author, describe licenses and viewports, and specify the encoding to use. If we don't specify the encoding, the encryption tool chooses a default encoding. We can also include **comment**
directives anywhere within the encryption envelope, including in the source code between the **begin** and **end** directives.

   The rules that an encryption tool must follow to form a decryption envelope are somewhat more prescriptive. Groups of directives must appear in a specified order, even if the
corresponding directives in the encryption envelope appeared in a different order or distributed among other directives, though not all groups are required in every decryption
envelope. The layout of a decryption envelope is

```
`protect begin_protected
author directives
license directives
encrypt agent directives
viewport directives
key block directives
data block directives
digest block directives
`protect end_protected
```

   The author, license, and viewport directives are those that appear in the encryption
envelope, if any. The **encrypt_agent** directive and optionally and **encypt_agent_info**
directive are included by the encryption tool. If a digital envelope is used, there is a group
of key block directives for each encryption of the session key. The directives occur in the
following order, with only the **key_keyname** directive being optional:

```
key_keyowner directive
key_keyname directive
key_keymethod directive
encoding directive
key_block directive
encoded cipher text for session key
```

The data block directives occur in the following order, with the **data_keyowner** and (optional) **data_keyname** directives only appearing if a digital envelope is not being used:

```
data_keyowner directive
data_keyname directive
data_method directive
encoding directive
data_block directive
encoded cipher text for source code
```

If a digital signature is used, the digest block directives occur in the following order, with only the **digest_keyname** directive being optional:

```
digest_keyowner directive
digest_keyname directive
digest_key_method directive
digest_method directive
encoding directive
digest_block directive
encoded cipher text for digest
```

23.2.1 Key Exchange

In our description of IP exchange so far we have assumed that the IP provider's encryption tool and the customer's decryption tool each have the required keys. What we have glossed over is how the tools get the keys. This is a very important topic, since protection of IP from disclosure relies on the security of the encryption and decryption keys. Should a key become known to an unauthorized party, the encrypted IP can be decrypted and disseminated. Normally, when encryption is used to secure communication between two parties, the parties are assumed to have an interest in the security of the encrypted messages and can be trusted to keep the keys secret. However, as we mentioned earlier, when an IP provider delivers a model to a customer, it is the customer's tool that is really the communicating party. The IP provider may not trust the customer not to look at the code or use it in some unauthorized way. A further complication is that the customer may have to provide his or her tool's key to an IP provider, creating an opportunity for the customer to copy the key and subsequently decrypt the code. Given these considerations, we can see that exchange of keys can be quite complicated. VHDL does not specify how keys should be exchanged; that is left to negotiation between IP providers, tool vendors, and customers. The following discussion, drawn from the VHDL standard, explores some of the issues.

Many applications that require secure exchange of keys rely on *public key infrastructure* (PKI). Parties to communication generate, or are given, key pairs for use with asymmetric ciphers. Each party keeps their private key secret, and publishes their public key, for example, in a directory. In order to establish that a public key does, in fact, belong to a given party, the public key is digitally signed by a trusted authority. The signed public key is represented in the form of a digital certificate, containing the key and the signature. The trusted authority is called a certification authority (CA). Many PKI systems have a hierarchy of CAs, allowing a certificate signed by a subordinate CA to be signed by a superior CA, allowing trust to be distributed hierarchically. One or more root CAs are required to be globally trusted.

Key exchange for IP protection may be built upon public key infrastructure. For example, a vendor of a decryption tool may embed a private key of a key pair in the tool and register the public key with a CA. The tool can then generate a key pair for the tool's user, keeping the private key secret and signing the public key with both the vendor's private key and the user's private key. This allows verification that the public key originates with the instance of the vendor's tool owned by the tool user. That public key may then be used by IP authors to provide IP for that use of that tool only. Similar mechanisms might also be employed within tools to allow exchange of private keys among tools without disclosure to the tools' user.

In addition to providing for secure key exchange, a decryption tool must take measures to ensure that stored keys are not disclosed to the tool user. If a tool user could read a tool's stored keys, the user could decrypt IP independently of the tool. One way of ensuring security of a tool's keys is for the tool to encrypt its key store using a secret key embedded in the tool in a disguised manner, and to provide for update and re-encryption of the secret key in case it is compromised.

23.3 VHDL Procedural Interface (VHPI)

VHPI is an application-programming interface (API) to VHDL tools. Using VHPI, a program written in a language such as C or C++ can access information about a VHDL model during analysis, elaboration, and execution of the model. VHPI allows development of add-in tools, such as linters, profilers, code coverage analyzers, timing and power analyzers, and external models, among others. Use of the VHPI to develop such tools is quite complex, and is beyond the scope of this book. Instead, we will describe the way in which we can invoke VHPI programs as part of a VHDL simulation.

VHPI programs are divided into two classes: *foreign models* and *foreign applications*. A foreign model corresponds to an architecture or a subprogram decorated with the **'foreign** attribute, predefined in the package **standard** as follows:

```
attribute foreign : string;
```

An architecture or subprogram decorated with this attribute is not elaborated or executed in the same way as a normal architecture or subprogram. Instead, the value of the attribute is used to identify the VHPI program that implements the behavior of the architecture or subprogram, respectively.

A foreign application does not have a counterpart in the VHDL code. It is executed as part of simulation and performs application-specific processing. Both forms of VHPI program can use API calls to obtain information about the VHDL model, to react to changes in the simulation state, and to cause changes in the simulation state.

VHDL-87, -93, and -2002

These versions of VHDL do not provide a standard API for foreign models or applications. Instead, simulation vendors have provided proprietary APIs for use with their tools. Such APIs are not portable between tools.

VHDL-87

The predefined attribute **'foreign** is not provided in VHDL-87. There is no standard mechanism to define foreign language interfaces.

23.3.1 Direct Binding

If we are to instantiate a foreign model as part of a VHPI design, we need to identify where the VHPI program code is to be found. Typically, the provider of the foreign model would provide documentation listing the names of libraries and functions to which we should refer. The most straightforward method of referring to the VHPI program code is to provide the information in the value of the **'foreign** attribute in a form known as *direct binding*. For a foreign architecture, we write the attribute value in the following form:

> **"VHPIDIRECT** `object_lib_path elab_function exec_function`**"**

The keyword **VHPIDIRECT** specifies standard direct binding, and must be written in uppercase. The *object_lib_path* is a host-dependent path and file name identifying the binary object library in the host file system. It can contain any characters; however, if a space character is required, we must precede it with a backslash character, and if a backslash character is required, we must double the backslash. The *elab_function* is the name of a function within the object library that performs elaboration for the foreign architecture. It is called to elaborate each instance of the foreign architecture during elaboration of the enclosing design. The *exec_function* is similarly the name of a function in the object library that performs simulation for the foreign architecture. It is called once for each instance of the foreign architecture during the initialization phase of simulation.

In the attribute value, we can substitute the keyword **null** for the object library path. In that case, the host system locates the object library in an implementation-dependent way. It might, for example, use an environment variable containing a list of pathnames. We can also substitute the keyword **null** for the elaboration function name if the foreign model does not require any action during elaboration. In both cases, the keyword **null** must be written in lowercase.

EXAMPLE 23.13　*Foreign processor core model*

Suppose a foreign model for a CPU32 processor core is provided in an object library called **cpu32.a** that we have installed in the directory **/usr/local/cpu32**. The elaboration and execution functions for a bus-functional version are named **cpu32_bf_elab_f** and **cpu32_bf_exec_f**, respectively. An entity and architecture that use standard direct binding for the bus-functional version are:

```
entity cpu32 is
  generic ( ... );
  port ( ... );
end entity cpu_32;

architecture bus_functional of cpu32 is
  attribute foreign of bus_functional : architecture is
    "VHPIDIRECT /usr/local/cpu32/cpu32.a " &
    "cpu32_bf_elab_f cpu32_bf_exec_f";
begin
end architecture bus_functional;
```

The attribute value for standard direct binding for a foreign subprogram takes a similar form:

"VHPIDIRECT *object_library_path exec_function*"

In this case, the execution function name identifies a function that performs the action of the foreign subprogram. It is called whenever the foreign subprogram is called during simulation. For foreign subprograms, we can substitute the keyword **null** for the execution function name. In that case, the execution function name is taken to be the same as that of the foreign subprogram declared in the VHDL model, using the case of letters in the VHDL declaration.

EXAMPLE 23.14　*Foreign display subprograms*

Suppose we are given subprograms that show 7-segment display digits graphically on the screen during simulation. The subprograms are in the library **displaylib.a**, and include a function named **create_digit** and a procedure named **update_digit**. We can declare corresponding foreign subprograms in a package as follows:

```
package display_pkg is

  impure function create_digit (title : in string)
                                 return natural;

  attribute foreign of create_digit : function is
    "VHPIDIRECT displaylib.a null";
```

```
        procedure update_digit (id : in natural;
                                val : in bit_vector(0 to 7));

        attribute foreign of update_digit : procedure is
          "VHPIDIRECT displaylib.a null";

      end package display_pkg;
```

23.3.2 Tabular Registration and Indirect Binding

An alternative way of identifying the VHPI program code for a foreign model is to use a *tabular registry*, which is a text file containing the identifying information. A tool can be supplied with any number of tabular registry files, each describing one or more foreign models or applications. The way in which we specify use of a tabular registry file is tool-dependent. It might, for example, involve use of a command-line option or an entry in an options-setting file. Each line of a tabular registry is an entry describing one foreign model, foreign application, or library of VHPI programs. The file can also contain comment lines, starting with characters "--", and blank lines.

A foreign architecture is described by a line of the following form in a tabular registry:

> *object_lib_name model_name* **vhpiArchF** *elab_function exec_function*

The *object_lib_name* is a logical name for the binary object library containing the VHPI program code. The host system maps the logical name to a physical object library in some host-dependent way. The *model_name* is an identifier for the foreign architecture in the object library. Both the library logical name and the model name can be written as a normal identifier or, if non-standard characters are required, as an extended identifier delimited by backslash characters. The keyword **vhpiArchF** indicates that the line in the tabular registry describes a foreign architecture. It must be written using the combination of uppercase and lowercase letters shown here. The *elab_function* and *exec_function* are the names of the elaboration function and execution function, respectively, in the object library. They serve the same purpose as described in Section 23.3.1, and, in a similar way, the elaboration function name can be replaced by the keyword **null**.

Having described a foreign architecture in a tabular registry file, we can specify a 'foreign attribute in the form of an *indirect binding* to use the foreign architecture for a VHDL architecture. This form of attribute value is

> **"VHPI** *object_lib_name model_name***"**

The *object_lib_name* and *model_name* identifiers must correspond to the library logical name and model name identifiers specified in an entry in a tabular registry. The foreign architecture described in that entry is used for each instance of the VHDL architecture decorated with the attribute.

EXAMPLE 23.15   *Foreign processor core model using indirect binding*

Suppose the provider of the CPU32 processor core model described in Example 23.13 also provides a tabular registry file for binding the bus-functional model. The file contains the following entry:

```
cpu32lib \cpu32-bf\ vhpiArchF cpu32_bf_elab_f cpu_bf_exec_f
```

We decorate the architecture with the **'foreign** attribute using indirect binding for the bus-functional model:

```
architecture bus_functional of cpu32 is
  attribute foreign of bus_functional : architecture is
    "VHPI cpu32lib \cpu32-bf\";
begin
end architecture bus_functional;
```

Tabular registration and indirect binding for a foreign subprogram are similar. An entry in a tabular registry file for a foreign procedure takes the form:

```
object_lib_name model_name vhpiProcF null exec_function
```

and for a foreign function:

```
object_lib_name model_name vhpiFuncF null exec_function
```

In both cases, the *object_lib_name* and *model_name* serve the same purpose as for a foreign architecture, and the *exec_function* is the name of the function in the object library that implements the subprogram's actions. The function name can be replaced by the keyword **null**, in which case the execution function is taken to be the same as the model name. The **'foreign** attribute value for indirect binding to a foreign subprogram is the same as that for indirect binding to a foreign architecture, namely,

```
"VHPI object_lib_name model_name"
```

The library name and model name are used in the same way to locate the tabular registry entry for the foreign subprogram.

EXAMPLE 23.16   *Foreign display subprograms using indirect binding*

The provider of the display subprograms described in Example 23.14 might provide a tabular registry file for the subprograms including the following entries:

```
displaylib create_digit vhpiFuncF null null
displaylib update_digit vhpiProcF null null
```

The second **null** in each entry indicates that the execution function names for the subprograms are the same as the foreign model names, namely, **create_digit** and update_digit. We declare the foreign subprograms and use indirect binding in the **'for-eign** attribute values as follows:

```
package display_pkg is

  impure function create_digit (title : in string)
                                    return natural;

  attribute foreign of create_digit : function is
    "VHPI displaylib create_digit";

  procedure update_digit (id : in natural;
                              val : in bit_vector(0 to 7));

  attribute foreign of update_digit : procedure is
    "VHPI displaylib update_digit";

end package display_pkg;
```

23.3.3 Registration of Applications and Libraries

We can use the tabular registration feature described in Section 23.3.2 to describe a VHPI application to be run as part of a simulation. A line in the file for a foreign application takes the form:

```
object_lib_name application_name vhpiAppF reg_function null
```

The *object_lib_name* is a logical name identifying the binary object library containing the program code, and the *application_name* is an identifier for the foreign application in the object library. The rules for these names are the same as those for names identifying foreign models. Thus, they can be written as normal identifiers or extended identifiers. The keyword **vhpiAppF** indicates that the line in the tabular registry describes a foreign application and must be written using the combination of uppercase and lowercase letters shown here. The *reg_function* is the names of a function in the object library that is called at the start of simulation, before elaboration or initialization, to initialize the state of the foreign application. This is all the information we need to supply to the tool to include a foreign application in a simulation. The registration function performs any further application-specific operations required.

EXAMPLE 23.17   *Registration of a power-estimation application*

A third-party tool supplier might provide a tool for estimating dynamic power consumption based on activity during simulation of a model. The tool's program code is installed in a binary object library in the host file system, with a logical name **powerestlib** mapping to the library file. The application is named **powerest**, and the registration function in the library is called **powerest_reg_f**. The supplier provides a tabular registry file with the following contents:

```
-- VHPI tabular registry for the PowerEst foreign application.
-- Map library name powerestlib to the pathname for the
-- powerestlib.a file in your installation.

powerestlib powerestlib vhpiAppF powerest_reg_f null
```

We invoke the simulator with a command-line option identifying this tabular registry file to include the power estimator tool in a simulation.

The final form of entry in a tabular registry file describes a library of VHPI programs, including foreign models or applications. The form of the entry is

```
object_lib_name null vhpiLibF reg_function null
```

As before, the ***object_lib_name*** is a logical name identifying the binary object library containing the program code. The ***reg_function*** is the names of a function in the object library that is called at the start of simulation. It uses the VHPI API to register each foreign model or application. This form of registration is convenient when a large suite of VHPI programs is provided.

23.4 Postponed Processes

VHDL provides a facility, postponed processes, that is useful in delta-delay models. A process is made postponed by including the keyword postponed, as shown by the full syntax rule for a process:

process_statement ⟸
　　⟦ *process*_label : ⟧
　　⟦ **postponed** ⟧ **process** ⟦ ((*signal*_name ⟨ , ... ⟩) ⟦ **all**) ⟧ ⟦ **is** ⟧
　　　　⟨ process_declarative_item ⟩
　　begin
　　　　⟨ sequential_statement ⟩
　　end ⟦ **postponed** ⟧ **process** ⟦ *process*_label ⟧ ;

The difference between a postponed process and a normal process lies in the way in which they are resumed during simulation. In our discussion of the simulation cycle in Chapter 5, we said that a normal process is triggered during a simulation cycle in which one of the signals to which it is sensitive changes value. The process then executes during that same simulation cycle. A postponed process is triggered in the same way, but may not execute in the same cycle. Instead, it waits until the last delta cycle at the current simulation time and executes after all non-postponed processes have suspended. It must wait until the non-postponed processes have suspended in order to ensure that there are no further delta cycles at the current simulation time. In addition, during initialization, a postponed process is started after all normal processes have been started and have suspended.

When we are writing models that use delta delays, we can use postponed processes to describe "steady state" behavior at each simulation time. The normal processes are executed over a series of delta delays, during which signal values are determined incrementally. Then, when all of the signals have settled to their final state at the current simulation time, the postponed processes execute, using these signal values as their input.

EXAMPLE 23.18 *Assertion based on steady-state values of signals*

We can write an entity interface for a set-reset flipflop as follows:

```
entity SR_flipflop is
  port ( s_n, r_n : in bit;  q, q_n : inout bit );
begin

  postponed process (q, q_n) is
  begin
    assert now = 0 fs or q = not q_n
      report "implementation error: q /= not q_n";
  end postponed process;

end entity SR_flipflop;
```

The entity declaration includes a process that verifies the outputs of the flipflop. Every implementation of the interface is required to produce complementary outputs. (The condition "**now = 0 fs**" is included to avoid an assertion violation during initialization.)

A dataflow architecture of the flipflop is

```
architecture dataflow of SR_flipflop is
begin

  gate_1 : q <= s_n nand q_n;
  gate_2 : q_n <= r_n nand q;

end architecture dataflow;
```

The concurrent signal assignment statements **gate_1** and **gate_2** model an implementation composed of cross-coupled gates. Assume that the flipflop is initally in the reset state. When **s_n** changes from '1' to '0', **gate_1** is resumed and schedules a change on **q** from '0' to '1' after a delta delay. In the next simulation cycle, the change on **q** causes **gate_2** to resume. It schedules a change on **q_n** from '1' to '0' after a delta delay. During the first delta cycle, **q** has the new value '1', but **q_n** still has its initial value of '1'. If we had made the verification process in the entity declaration a non-postponed process, it would be resumed in the first delta cycle and report an assertion violation. Since it is a postponed process, it is not resumed until the second delta cycle (the last delta cycle after the change on **s_n**), by which time **q** and **q_n** have stabilized.

It is important to note that the condition that triggers a postponed process may not obtain when the process is finally executed. For example, suppose a signal **s** is updated to the value '1', causing the following postponed process to be triggered:

```
p : postponed process is
  ...
begin
  ...
  wait until s = '1';
  ...   -- s may not be '1'!!
end postponed process p;
```

Because the process is postponed, it is not executed immediately. Instead, some other process may execute, assigning '0' to s with delta delay. This assignment causes a delta cycle during which s is updated to '0'. When **p** is eventually executed, it proceeds with the statements immediately after the wait statement. However, despite the appearance of the condition in the wait statement, **s** does not have the value '1' at that point.

Since each postponed process waits until the last delta cycle at a given simulation time before executing, there may be several postponed processes triggered by different conditions in different delta cycles, all waiting to execute. Since the cycle in which the postponed processes execute must be the last delta cycle at the current simulation time, the postponed processes must not schedule transactions on signals with delta delay. If they did, they would cause another delta cycle at the current simulation time, meaning that the postponed processes should not have executed. The restriction is required to avoid this paradox.

In previous chapters, we described a number of concurrent statements that are equivalent to similar sequential statements encapsulated in processes. We can write postponed versions of each of these by including the keyword **postponed** at the beginning of the statement, as shown by the following syntax rules:

concurrent_procedure_call_statement ⇐
    ⟦ label : ⟧
    ⟦ **postponed** ⟧ *procedure*_name ⟦ (*parameter*_association_list) ⟧ ;

concurrent_assertion_statement ⇐
    ⟦ label : ⟧
    ⟦ **postponed** ⟧ **assert** condition
                 ⟦ **report** expression ⟧ ⟦ **severity** expression ⟧ ;

concurrent_simple_signal_assignment ⇐
    name <= ⟦ **postponed** ⟧ ⟦ **guarded** ⟧ ⟦ delay_mechanism ⟧ waveform ;

concurrent_conditional_signal_assignment ⇐
    name <= ⟦ **postponed** ⟧ ⟦ **guarded** ⟧ ⟦ delay_mechanism ⟧
         waveform **when** condition
         ❴ **else** waveform **when** condition ❵
         ⟦ **else** waveform ⟧ ;

concurrent_selected_signal_assignment ⇐
    with expression **select** ⟦ ? ⟧
         name <= ⟦ **postponed** ⟧ ⟦ **guarded** ⟧ ⟦ delay_mechanism ⟧
             ❴ waveform **when** choices , ❵
             waveform **when** choices ;

Inclusion of the keyword **postponed** simply makes the encapsulating process a postponed process. Thus, we can rewrite the postponed process in Example 23.18 as

```
postponed assert now = 0 fs or q = not q_n
  report "implementation error: q /= not q_n";
```

VHDL-87

Postponed processes are not provided in VHDL-87.

23.5 Conversion Functions in Association Lists

In the preceding chapters, we have seen uses of association lists in generic maps, port maps and subprogram calls. An association list associates actual values and objects with formal objects. Let us now look at the full capabilities provided in association lists, shown by the following full syntax rules:

association_list ⇐ ⟦ ⟦ formal_part => ⟧ actual_part ⟧ ⟨ , ... ⟩

formal_part ⇐
    *generic*_name
    ❙ *port*_name
    ❙ *parameter*_name
    ❙ *function*_name (⟦ *generic*_name ❙ *port*_name ❙ *parameter*_name ⟧)
    ❙ type_mark (⟦ *generic*_name ❙ *port*_name ❙ *parameter*_name ⟧)

actual_part ⇐
    ⟦ **inertial** ⟧ expression
    ❙ *signal*_name
    ❙ *variable*_name
    ❙ *file*_name
    ❙ subtype_indication
    ❙ *subprogram*_name
    ❙ *package*_name
    ❙ **open**
    ❙ *function*_name (⟦ *signal*_name ❙ *variable*_name ⟧)
    ❙ type_mark (⟦ *signal*_name ❙ *variable*_name ⟧)

The simple rules for association lists we used previously allowed us to write associations of the form "**formal => actual**". When we are associating signal and variable object, the new rules allow us to write associations such as

```
f1 ( formal ) => actual

formal => f2 ( actual )

f1 ( formal ) => f2 ( actual )
```

These associations include *conversion functions* or *type conversions*. We discussed type conversions in Chapter 2. They allow us to convert a value from one type to another closely related type. A conversion function, on the other hand, is an explicitly or implicitly declared subprogram or operation. It can be any function with one parameter and can compute its result in any way we choose.

A conversion in the actual part of an association is invoked whenever a value is passed from the actual object to the formal object. For a variable-class subprogram parameter, conversion occurs when the subprogram is called. For a signal associated with a port, conversion occurs whenever an updated signal value is passed to the port. For constant-class subprogram parameters and for generic constants, the actual values are expressions, which may directly take the form of function calls or type conversions. In these cases, the conversion is not considered to be part of the association list; instead, it is part of the expression. Conversions are not allowed in the remaining cases, namely, signal-class and file-class actual subprogram parameters.

EXAMPLE 23.19　　*Conversion in the actual part*

We wish to implement a limit checker, which checks whether a signed integer is out of specified bounds. The integer and bounds are represented as standard-logic vectors of the subtype **word**, declared in the package **project_util** as

```
subtype word is std_ulogic_vector(31 downto 0);
```

We can use a comparison function that compares integers represented as bit vectors. The function is declared in **project_util** as

```
function "<" ( bv1, bv2 : bit_vector ) return boolean;
```

The entity declaration and architecure body for the limit checker are

```
library ieee;  use ieee.std_logic_1164.all;
use work.project_util.all;

entity limit_checker is
  port ( input, lower_bound, upper_bound : in word;
         out_of_bounds : out std_ulogic );
end entity limit_checker;

-------------------------------------------------

architecture behavioral of limit_checker is

  subtype bv_word is bit_vector(31 downto 0);

  function word_to_bitvector ( w : in word ) return bv_word is
  begin
    return To_bitvector ( w, xmap => '0' );
  end function word_to_bitvector;

begin
```

```
algorithm : process (input, lower_bound, upper_bound) is
begin
  if "<" ( bv1 => word_to_bitvector(input),
           bv2 => word_to_bitvector(lower_bound) )
     or "<" ( bv1 => word_to_bitvector(upper_bound),
              bv2 => word_to_bitvector(input) ) then
    out_of_bounds <= '1';
  else
    out_of_bounds <= '0';
  end if;
end process algorithm;

end architecture behavioral;
```

The process performs the comparisons by converting the word values to bit vectors, using the conversion function **word_to_bitvector**. Note that we cannot use the function **To_bitvector** itself in the actual part of the association list, as it has two parameters, not one. Note also that the result type of the conversion function in this example must be a constrained array type in order to specify the array index range for the actual value passed to the comparison function.

Since an actual for a port can take the form of an expression involving one or more signals, and the expression could be application of a function to a signal, there is potential for ambiguity in the association. If it were interpreted as an expression, it would be equivalent to assignment to an intermediate signal and association of that signal as the actual for the port, as described in Section 5.3. That would involve a delta delay between update of the actual signal and update of the port. On the other hand, if the actual were interpreted as a conversion function applied to the actual signal, then the port is updated in the same simulation cycle as the actual signal, with no intervening delta delay. In order to resolve this ambiguity, we can include the reserved word **inertial** in the association to specify that the interpretation involving an implicit signal is the one to use. If we omit the reserved word and the association can be interpreted as application of a conversion function or a type conversion, that interpretation takes precedence.

EXAMPLE 23.20   *Using a single-parameter function in an actual part*

If we were to write a function with one parameter representing some computational logic, for example:

```
function increment ( x : unsigned ) return unsigned;
```

and use it in a port map:

```
op_counter : component reg16
  port map ( d_in => increment(op_count), ... );
```

it would be interpreted as a conversion function, which is not what we want. To make the intention explicit, we include the reserved word **inertial** in the port association to imply an inertial signal assignment of the expression to the anonymous intermediate signal. Thus, we would write the port map as

```
op_counter : component reg16
    port map ( d_in => inertial increment(op_count), ... );
```

A conversion can only be included in the actual part of an association if the interface object is of mode **in** or **inout**. If the conversion takes the form of a type conversion, it must name a subtype that has the same base type as the formal object and is closely related to the type of the actual object. If the conversion takes the form of a conversion function, the function must have only one parameter of the same type as the actual object and must return a result of the same type as the formal object. If the interface object is of an unconstrained or partially constrained type, the type mark of the type conversion or the result type of the conversion function must include constraints that define any index ranges not defined by the interface object's subtype.

A conversion in the formal part of an association is invoked whenever a value is passed from the formal object to the actual object. For a variable-class procedure parameter, conversion occurs when the procedure returns. For a signal associated with a port, conversion occurs whenever the port drives a new value. Conversions are not allowed for signal-class and file-class formal subprogram parameters.

EXAMPLE 23.21   *Conversion in the formal part*

Suppose a library contains the following entity, which generates a random number at regular intervals:

```
entity random_source is
    generic ( min, max : natural;
              seed : natural;
              interval : delay_length );
    port ( number : out natural );
end entity random_source;
```

If we have a test bench including signals of type **bit**, we can use the entity to generate random stimuli. We use a conversion function to convert the numbers to bit-vector values. An outline of the test bench is shown below. The function **natural_to_bv11** has a parameter that is a natural number and returns a bit-vector result. The architecture instantiates the **random_source** component, using the conversion function in the formal part of the association between the port and the signal. Each time the component instance generates a new random number, the function is invoked to convert it to a bit vector for assignment to **stimulus_vector**.

```
architecture random_test of test_bench is

    subtype bv11 is bit_vector(10 downto 0);
```

```
function natural_to_bv11 ( n : natural ) return bv11 is
  variable result : bv11 := (others => '0');
  variable remaining_digits : natural := n;
begin
  for index in result'reverse_range loop
    result(index) := bit'val(remaining_digits mod 2);
    remaining_digits := remaining_digits / 2;
    exit when remaining_digits = 0;
  end loop;
  return result;
end function natural_to_bv11;

signal stimulus_vector : bv11;
...

begin

  stimulus_generator : entity work.random_source
    generic map ( min => 0, max => 2**10 - 1, seed => 0,
                  interval => 100 ns )
    port map ( natural_to_bv11(number) => stimulus_vector );

  ...

end architecture random_test;
```

The type requirements for conversions included in the formal parts of associations mirror those of conversions in actual parts. A conversion can only be included in a formal part if the interface object is of mode **out**, **inout**, or **buffer**. If the conversion takes the form of a type conversion, it must name a subtype that has the same base type as the actual object and is closely related to the type of the formal object. If the conversion takes the form of a conversion function, the function must have only one parameter of the same type as the formal object and must return a result of the same type as the actual object. If the interface object is of an unconstrained or partially constrained type, the type mark of the type conversion or the parameter type of the conversion function must include constraints that define any index ranges not defined by the interface object's subtype.

Note that we can include a conversion in both the formal part and the actual part of an association if the interface object is of mode **inout**. The conversion on the actual side is invoked whenever a value is passed from the actual to the formal, and the conversion on the formal side is invoked whenever a value is passed from the formal to the actual.

VHDL-87, -93, and -2002

In these versions, we cannot use a non-static expression as an actual for a port. Hence, the ambiguity between such an expression and a conversion function does not arise. These versions thus do not allow the reserved word **inertial** to appear in a port map.

Also, in these versions of VHDL, it is not possible to associate a resolution function with the elements of an array type to define a subtype with resolved elements.

Instead, a separate array type must be defined with elements of a resolved subtype. As a consequence, the two array types are distinct, and type conversions are needed to assign a value of one type to and objects of the other.

One important use of type conversions in association lists arises in the earlier versions of VHDL when we mix arrays of unresolved and resolved elements in a model. For example, the standard-logic package declares the two types:

type std_ulogic_vector **is array** (natural **range** <>) **of** std_ulogic;

type std_logic_vector **is array** (natural **range** <>) **of** std_logic;

These are two distinct types, even though the element type of **std_logic_vector** is a subtype of the element type of **std_ulogic_vector**. Thus, we cannot directly associate a **std_ulogic_vector** signal with a **std_logic_vector** port, nor a **std_logic_vector** signal with a **std_ulogic_vector** port. However, we can use type conversions or conversion functions to deal with the type mismatch.

EXAMPLE 23.22   *Array conversions in associations*

Suppose we are developing a register-transfer-level model of a computer system in an earlier version of VHDL. The architecture body for the processor is

```
architecture rtl of processor is
  component latch is
    generic ( width : positive );
    port ( d : in std_ulogic_vector(0 to width - 1);
           q : out std_ulogic_vector(0 to width - 1);
           ... );
  end component latch;

  component ROM is
    port ( d_out : out std_ulogic_vector;  ... );
  end component ROM;

  subtype std_logic_word is std_logic_vector(0 to 31);

  signal source1, source2, destination : std_logic_word;
  ...

begin

  temp_register : component latch
    generic map ( width => 32 )
    port map ( d => std_ulogic_vector(destination),
               std_logic_vector(q) => source1, ... );

  constant_ROM : component ROM
    port map ( std_logic_word(d_out) => source2, ... );

  ...
```

```
end architecture rtl;
```

We declare a latch component and a ROM component, both with unresolved ports. We also declare a constrained array subtype std_logic_word with resolved elements and a number of signals of this subtype representing the internal buses of the processor.

We instantiate the latch component and associate the **destination** bus with the **d** port and the **source1** bus with the **q** port. Since the signals and the ports are of different but closely related types, we use type conversions in the association list. Although the types std_ulogic_vector and std_logic_vector are unconstrained array types, we can name them in the type conversion in this instance, since the component ports are constrained.

We also instantiate the ROM component and associate the **source2** bus with the **d_out** port. Here also we use a type conversion in the association list. However, the port **d_out** is of an unconstrained type. Hence we may not use the name std_logic_vector in the type conversion, since it, too, is unconstrained. Instead, we use the constrained subtype name std_logic_word. The index range of this subtype is used as the index range of the port **d_out** in the component instance.

VHDL-87

VHDL-87 does not allow type conversions in association lists, but does allow conversion functions. If we need to convert between closely related types in an association list, we can write a function that performs the type conversion and use the function as a conversion function in the association list.

23.6 Linkage Ports

When we introduced ports in Chapter 5, we identified four modes, **in**, **out**, **buffer**, and **inout**, that control how data is passed to and from a design entity. VHDL provides a further mode, **linkage**. This mode may only be specified for ports of entities, blocks and components, not for generic constants or subprogram parameters.

Linkage ports were originally included in the language as a means of connecting signals to foreign design entities. If the implementation of an entity is expressed in some language other than VHDL, the way in which values are generated and read within the entity may not conform to the same transaction semantics as those of VHDL. A **linkage**-mode port provides the point of contact between the non-VHDL and the VHDL domains. Unless a simulator provides some additional semantics for generating and reading linkage ports, a model containing linkage ports anywhere in the hierarchy cannot be simulated.

In practice, linkage ports have not been used as originally intended. One alternative that has been developed is an extension to VHDL for analog and mixed-digital/analog modeling, called VHDL-AMS. It defines alternative classes of ports, called quantities and terminals, for dealing with analog connections. The one place where linkage ports have been used is in the Boundary Scan Description Language, which is a subset of VHDL for describing connections to circuits for test equipment. In that language, linkage ports are used for non-functional connections, such as power, ground, and unconnected pins.

Exercises

1. [❶ 23.1] Write signal declarations for

    - a bus-kind signal, **serial_bus**, of the resolved subtype **wired_or_bit**, and

    - a register-kind signal, **d_node**, of the resolved subtype **unique_bit**.

2. [❶ 23.1] A signal **rx_bus** is declared to be a bus-kind signal of type **std_logic**. Trace the value of the signal as transactions from the following two drivers are applied:

    - null, '0' after 10 ns, '1' after 20 ns, '0' after 30 ns, null after 40 ns

    - null, '1' after 35 ns, '0' after 45 ns, null after 55 ns

3. [❶ 23.1] Repeat Exercise 2, this time assuming **rx_bus** is a register-kind signal that is initialized to 'U'.

4. [❶ 23.1] Write a signal assignment statement that schedules the value 3 on an integer signal **vote** after 2 µs, then disconnects from the signal after 5 µs.

5. [❶ 23.1] Suppose a process contains the following signal assignment, executed at time 150 ns:

    ```
    result <= 0 after 10 ns, 42 after 20 ns,
              0 after 100 ns, null after 120 ns;
    ```

    Assuming the driver for **result** is disconnected at time 150 ns, trace the value of **result'driving** resulting from the signal assignment.

6. [❶ 23.1] Write a block with a guard expression that is true when a **std_ulogic** signal **en** is '1' or 'H'. The block should contain a guarded signal assignment that assigns an inverted version of the signal **d_in** to the signal **q_out_n** when the guard expression is true.

7. [❶ 23.1] Write disconnection specifications that specify

    - a disconnection delay of 3.5 ns for a signal **source1** of type **wired_word**,

    - a disconnection delay of 3.2 ns for other signals of type **wired_word** and

    - a disconnection delay of 2.8 ns for all signals of type **wired_bit**.

8. [❶ 23.1] Trace the values on the signal priority resulting from execution of the following statements. The resolution function for the subtype **resolved_integer** selects the leftmost value from the contributing drivers or returns the value 0 if there are no contributions. Assume that no other drivers for **priority** are connected.

    ```
    signal request : integer := 0;
    signal guard : boolean := false;
    ```

```
signal priority : resolved_integer bus := 0;
disconnect priority : resolved_integer after 2 ns;
...

request <= 3 after 40 ns, 5 after 80 ns, 1 after 120 ns;
guard <= true after 50 ns, false after 100 ns;
priority <= guarded request after 1 ns;
```

9. **[❶ 23.1]** Write a block statement that encapsulates component instantiation statements implementing the circuit shown in Figure 23.2. The signal **q_internal**, of type **bit**, should be declared local to the block.

FIGURE 23.2

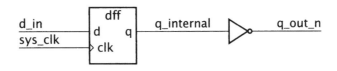

An inverting-register circuit.

10. **[❶ 23.1]** Write a block configuration for the block statement described in Exercise 9, binding the flipflop component instance to an entity **d_flipflop** and architecture **basic**, and the inverter component to the entity **inverter** and architecture **basic**. The entities are in the current working library.

11. **[❶ 23.2]** Add protect directives to form an encryption envelope around the declarations and statements in the following architecture. The encryption envelope should use the triple-DES symmetric cipher with the key owner "IP_werx" and key name "IP_werx_sim".

```
architecture rtl of ethernet_mac is
  signal fifo_enable : std_ulogic;
  ...
begin
  rx_fifo : IP_werx_fifo
    port map ( ... );
  ...
end architecture rtl;
```

12. **[❶ 23.2]** Repeat Exercise 11, this time using a digital envelope with the session key encrypted using the OpenPGP RSA cipher and the data encrypted using the AES cipher with 192-bit key. The recipient key owner is "Aero Industries" and their public key name is "Aero Design". The encrypted information should be encoded using the base64 method.

13. [❶ 23.3] Decorate the following architecture with the 'foreign attribute specifying VHPI direct binding, with the elaboration function **control_elab** and the execution function **control_exec** being stored in the library $VHPIUSERLIB/control.so.

    ```
    architecture vhpi_implementation of control is
    begin
    end architecture vhpi_implementation;
    ```

14. [❶ 23.3] Repeat Exercise 13, this time using VHPI indirect binding. Show the entry in the tabular registry identifying the foreign architecture.

15. [❶ 23.5] Suppose we wish to associate an **out** mode port of type **std_ulogic** with a signal of type **bit**. Why can we not use the function **To_bit** as a conversion function in the association?

16. [❶ 23.5] Suppose we have a gate component declared as

    ```
    component nand2 is
      port ( a, b : in std_ulogic;  y_n : out std_ulogic );
    end component nand2;
    ```

    Write a component instantiation statement that instantiates the gate, with inputs connected to signals **s1** and **s2** and output connected to the signal **s3**. All of the signals are of type **bit**. Use conversion functions where required.

17. [❷ 23.1] Suppose we declare the following subtypes:

    ```
    subtype word is bit_vector(31 downto 0);
    ...
    subtype resolved_word is bitwise_or word;
    ```

    The resolution function performs a bitwise logical "or" operation on the contributing driver values. Write a procedure that encapsulates the behavior of a tristate buffer. The procedure has input signal parameters **oe** of type **bit** and **d** of the subtype **word** and an output signal parameter **z** of type **resolved_word**. When **oe** is '1', the value of **d** is transmitted to **z**. When **oe** is '0', **z** is disconnected. Test the procedure by invoking it with a number of concurrent procedure calls in a test bench.

18. [❷ 23.1] Develop a dataflow model of a latching four-input multiplexer. The multiplexer has four data inputs, two bits of select input, and an enable input. When the enable input is high, the select inputs determine which data input is transmitted to the single data output. When the enable input is low, the value on the data output is latched.

19. [❷ 23.1] A dynamic register can be implemented in NMOS technology as shown in Figure 23.3. Develop a dataflow model for this form of register, using guarded signal assignments to model the pass transistors. The signals should be of a resolved subtype of **bit**, and the signal **latched_d** should be a register-kind signal.

FIGURE 23.3

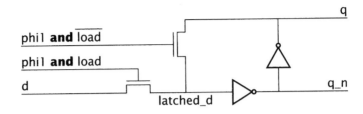

A circuit for a dynamic register.

20. [❷ 23.1] Develop a behavioral model of a three-to-eight decoder with three select inputs, an enable input and eight active-low outputs. The entity interface includes generic constants for

    • input propagation delay for the enable input,

    • input propagation delay for the select inputs and

    • output propagation delay for the outputs.

    Write the architecture body with separate blocks for input delays, function and output delays.

21. [❷ 23.6] Develop a structural model of an SR-flipflop constructed from nor gates as shown in Figure 23.4. Use **buffer** mode ports for q and q_n.

FIGURE 23.4

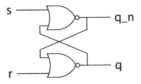

An SR-flipflop constructed from not gates.

22. [❸ 23.4] Exercise 17 in Chapter 8 describes a distributed priority arbiter for a shared-bus multiprocessor system. Each requester computes the minimum of all priorities. Develop a model of the minimization circuit that operates using delta delays. Include a number of instances of the minimizer in a test bench. Also include a process that verifies that the result priority is the minimum of all of the request priorities when the computation is complete.

23. [❸ 23.1] Revise the tristate buffer procedure described in Exercise 17 to make it bidirectional. Include an additional input parameter that determines the direction of data transfer.

24. [❸ 23.1] Develop a behavioral model of a read/write memory with a bidirectional data port of the type **resolved_byte**, defined on page 738. The data port should be a bus-kind signal, and the model should use null signal assignments appropriately to indicate when the memory is not supplying data.

25. [❸ 23.1] A 4-bit carry-look-ahead adder can be implemented in CMOS technology with a Manchester carry chain, shown in Figure 23.5. The signal **c0** is the carry input, **c4** is the carry output, **c̄0** to **c̄4** are active-low intermediate carry signals, **g1** to **g4** are carry generate signals and **p1** to **p4** are carry propagate signals. During the low half of a clock cycle, the intermediate carry signals are precharged to '1'. During the high half of the clock cycle, the pass transistors controlled by the generate and propagate signals conditionally discharge the intermediate carry signals, determining their final value. The Boolean equations for the sum, generate and propagate signals are

$$s_i = a_i \oplus b_i \oplus c_{i-1}$$
$$g_i = a_i b_i$$
$$p_i = a_i + b_i$$

Develop a dataflow model of a 4-bit Manchester carry adder, using register-kind signals for the internal carry signals. All signals should be of a resolved-bit type.

FIGURE 23.5

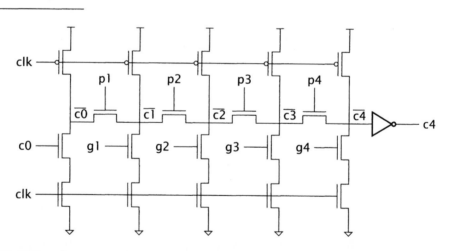

A Manchester carry chain for a carry-look-ahead adder.

26. [❸ 23.1] A 4 × 4 barrel shifter can be constructed from pass transistors as shown in Figure 23.6. The signals **i0** to **i3** are the inputs, and **z0** to **z3** are the outputs. The control signal **s0** causes input bits to be transmitted to the outputs unshifted, **s1** causes them to be shifted by one place, **s2** by two places and **s3** by three places. The outputs must be precharged to '1' on the first half of a clock cycle, then one of the control signals activated on the second half of the clock cycle. Develop a dataflow model of

the barrel shifter, using register-kind signals for the output signals. All signals should be of a resolved-bit type.

FIGURE 23.6

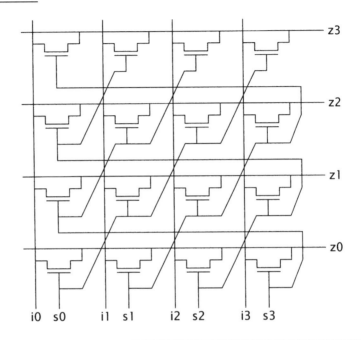

A 4 × 4 barrel shifter.

27. [❸ 23.5] Develop a behavioral model of a counter that counts from 0 to 255 with an output port of type **natural**. In a test bench, define and instantiate an 8-bit counter component. Write a configuration declaration for the test bench, binding the behavioral counter entity to the counter component instance. Use conversion functions in the binding indication as required. You may wish to use the conversion functions from the **numeric_bit** package. Note that a configuration declaration can use items, such as conversion functions, declared in separate packages.

28. [❹ 23.3] If your simulator allows you to call functions written in the C programming language using VHPI, develop a register model that uses the graphical display libraries of your host computer system to create a pop-up window to display the register contents. Instantiate the register model in a test bench, and step through the simulation to verify that the model creates and updates the display.

Appendix A

Standard Packages

The package declarations for the IEEE standard packages are listed in this appendix.[1] The IEEE standard also includes package bodies defining the detailed meaning of each of the operators and functions provided by the packages. However, simulator vendors are allowed to substitute accelerated implementations of the packages rather than compiling the package bodies into simulations. The IEEE standard requires the packages **standard**, **env**, and **textio** to be in a library called **std**, and the remaining packages to be in a library named **ieee**.

A.1 The Predefined Package **standard**

The predefined types, subtypes and functions of VHDL are defined in the package **standard**. Each design unit in a design is automatically preceded by the following context clause:

```
library std, work;  use std.standard.all;
```

so the predefined items are directly visible in the design. The package **standard** is listed here. The comments indicate which operators are implicitly defined for each explicitly defined type. These operators are also automatically made visible in design units. The types *universal_integer* and *universal_real* are anonymous types. They cannot be referred to explicitly.

```
package standard is

  type boolean is (false, true);

    -- implicitly declared for boolean operands:
    -- "and", "or", "nand", "nor", "xor" , "xnor", "not"
    -- "=", "/=", "<", "<=", ">", ">="
    -- minimum, maximum, rising_edge, falling_edge, to_string

  type bit is ('0', '1');
```

---

1. The material in this appendix is derived from IEEE Draft Std 1076-2008/D4.1, Draft Standard VHDL Language Reference Manual.

```
   -- implicitly declared for bit operands:
   -- "and", "or", "nand", "nor", "xor" , "xnor", "not"
   -- "=", "/=", "<", "<=", ">", ">="
   -- "?=", "?/=", "?<", "?<=", "?>", "?>=", "??"
   -- minimum, maximum, rising_edge, falling_edge, to_string

type character is (
  nul,   soh,   stx,   etx,   eot,   enq,   ack,   bel,
  bs,    ht,    lf,    vt,    ff,    cr,    so,    si,
  dle,   dc1,   dc2,   dc3,   dc4,   nak,   syn,   etb,
  can,   em,    sub,   esc,   fsp,   gsp,   rsp,   usp,
  ' ',   '!',   '"',   '#',   '$',   '%',   '&',   ''',
  '(',   ')',   '*',   '+',   ',',   '-',   '.',   '/',
  '0',   '1',   '2',   '3',   '4',   '5',   '6',   '7',
  '8',   '9',   ':',   ';',   '<',   '=',   '>',   '?',
  '@',   'A',   'B',   'C',   'D',   'E',   'F',   'G',
  'H',   'I',   'J',   'K',   'L',   'M',   'N',   'O',
  'P',   'Q',   'R',   'S',   'T',   'U',   'V',   'W',
  'X',   'Y',   'Z',   '[',   '\',   ']',   '^',   '_',
  '`',   'a',   'b',   'c',   'd',   'e',   'f',   'g',
  'h',   'i',   'j',   'k',   'l',   'm',   'n',   'o',
  'p',   'q',   'r',   's',   't',   'u',   'v',   'w',
  'x',   'y',   'z',   '{',   '|',   '}',   '~',   del,
  c128,  c129,  c130,  c131,  c132,  c133,  c134,  c135,
  c136,  c137,  c138,  c139,  c140,  c141,  c142,  c143,
  c144,  c145,  c146,  c147,  c148,  c149,  c150,  c151,
  c152,  c153,  c154,  c155,  c156,  c157,  c158,  c159,
  ' ',   '¡',   '¢',   '£',   '¤',   '¥',   '¦',   '§',
  '¨',   '©',   'ª',   '«',   '¬',   '',   '®',   '¯',
  '°',   '±',   '²',   '³',   '´',   'µ',   '¶',   '·',
  '¸',   '¹',   'º',   '»',   '¼',   '½',   '¾',   '¿',
  'À',   'Á',   'Â',   'Ã',   'Ä',   'Å',   'Æ',   'Ç',
  'È',   'É',   'Ê',   'Ë',   'Ì',   'Í',   'Î',   'Ï',
  'Ð',   'Ñ',   'Ò',   'Ó',   'Ô',   'Õ',   'Ö',   '×',
  'Ø',   'Ù',   'Ú',   'Û',   'Ü',   'Ý',   'Þ',   'ß',
  'à',   'á',   'â',   'ã',   'ä',   'å',   'æ',   'ç',
  'è',   'é',   'ê',   'ë',   'ì',   'í',   'î',   'ï',
  'ð',   'ñ',   'ò',   'ó',   'ô',   'õ',   'ö',   '÷',
  'ø',   'ù',   'ú',   'û',   'ü',   'ý',   'þ',   'ÿ');

   -- implicitly declared for character operands:
   -- "=", "/=", "<", "<=", ">", ">="
   -- minimum, maximum, to_string

type severity_level is (note, warning, error, failure);

   -- implicitly declared for severity_level operands:
   -- "=", "/=", "<", "<=", ">", ">="
   -- minimum, maximum, to_string

-- type universal_integer is range implementation_defined;

   -- implicitly declared for universal_integer operands:
   -- "=", "/=", "<", "<=", ">", ">="
   -- unary "+", "-", "abs"
   -- "+", "-", "*", "/", "mod", "rem"
   -- minimum, maximum, to_string

-- type universal_real is range implementation_defined;
```

```
    -- implicitly declared for universal_real operands:
    -- "=", "/=", "<", "<=", ">", ">="
    -- unary "+", "-", "abs"
    -- "+", "-", "*", "/", "mod", "rem"
    -- minimum, maximum, to_string

type integer is range implementation_defined;

    -- implicitly declared for integer operands:
    -- "=", "/=", "<", "<=", ">", ">="
    -- unary "+", "-", "abs"
    -- "**", "+", "-", "*", "/", "mod", "rem"
    -- minimum, maximum, to_string

subtype natural is integer range 0 to integer'high;
subtype positive is integer range 1 to integer'high;

type real is range implementation_defined;

    -- implicitly declared for real operands:
    -- "=", "/=", "<", "<=", ">", ">="
    -- unary "+", "-", "abs"
    -- "**", "+", "-", "*", "/"
    -- minimum, maximum, to_string

type time is range implementation_defined
  units
      fs;
      ps  = 1000 fs;
      ns  = 1000 ps;
      us  = 1000 ns;
      ms  = 1000 us;
      sec = 1000 ms;
      min = 60 sec;
      hr  = 60 min;
  end units;

    -- implicitly declared for time operands:
    -- "=", "/=", "<", "<=", ">", ">="
    -- unary "+", "-", "abs"
    -- "+", "-", "*", "/", "mod", "rem"
    -- minimum, maximum, to_string

subtype delay_length is time range 0 fs to time'high;

impure impure function now return delay_length;

type string is array (positive range <>) of character;

    -- implicitly declared for string operands:
    -- "=", "/=", "<", "<=", ">", ">=", "&"
    -- minimum, maximum

type boolean_vector is array (natural range <>) of boolean;

    -- implicitly declared for boolean_vector operands:
    -- "and", "or", "nand", "nor", "xor", "xnor", "not"
    -- "sll", "srl", "sla", "sra", "rol", "ror"
    -- "=", "/=", "<", "<=", ">", ">=", "&"
    -- minimum, maximum

type bit_vector is array (natural range <>) of bit;
```

```
    -- implicitly declared for bit_vector operands:
    -- "and", "or", "nand", "nor", "xor" , "xnor", "not"
    -- "sll", "srl", "sla", "sra", "rol", "ror"
    -- "=", "/=", "<", "<=", ">", ">=", "?=", "?/=", "&"
    -- minimum, maximum
    -- to_string, to_bstring, to_binary_string
    -- to_ostring, to_octal_string, to_hstring, to_hex_string

type integer_vector is array (natural range <>) of integer;

    -- implicitly declared for integer_vector operands:
    -- "=", "/=", "<", "<=", ">", ">=", "&"
    -- minimum, maximum

type real_vector is array (natural range <>) of real;

    -- implicitly declared for real_vector operands:
    -- "=", "/=", "&"
    -- minimum, maximum

type time_vector is array (natural range <>) of time;

    -- implicitly declared for time_vector operands:
    -- "=", "/=", "&"
    -- minimum, maximum

type file_open_kind is (read_mode, write_mode, append_mode);

    -- implicitly declared for file_open_kind operands:
    -- "=", "/=", "<", "<=", ">", ">="
    -- minimum, maximum, to_string

type file_open_status is
        (open_ok, status_error, name_error, mode_error);

    -- implicitly declared for file_open_status operands:
    -- "=", "/=", "<", "<=", ">", ">="
    -- minimum, maximum, to_string

attribute foreign: string;

end STANDARD;
```

VHDL-87, -93, and -2002

The following items are not included in standard for these earlier versions: the matching relational operators; the "??" operator; array/scalar and reducing logical operators; the operators **mod** and **rem** on time operands; the functions maximum, minimum, rising_edge, and falling_edge; the functions to_string, to_ostring, to_hstring and associated aliases; the types boolean_vector, integer_vector, real_vector, and time_vector and associated operations.

VHDL-87

The following items are not included in standard in VHDL-87: the types file_open_kind and file_open_status; the subtype delay_length; the attribute foreign;

the operators **xnor**, **sll**, **srl**, **sla**, **sra**, **rol** and **ror**. The result time of the function now is time. The type character includes only the first 128 values, corresponding to the ASCII character set.

A.2   The Predefined Package **env**

```
package env is

  procedure stop (status: integer);
  procedure stop;

  procedure finish (status: integer);
  procedure finish;

  function resolution_limit return delay_length;

end package env;
```

VHDL-87, -93, and -2002

The env package is not provided in these versions.

A.3   The Predefined Package **textio**

```
package textio is

  type line is access string;

    -- implicitly declared for line operands:
    -- "=", "/=", deallocate

  type text is file of string;

    -- implicitly declared for text operands:
    -- file_open, file_close, read, write, flush, endfile

  type side is (right, left);

    -- implicitly declared for side operands:
    -- "=", "/=", "<", "<=", ">", ">="
    -- minimum, maximum, to_string

  subtype width is natural;

  file input : text open read_mode is "STD_INPUT";
  file output : text open write_mode is "STD_OUTPUT";

  procedure readline(file F: text; L: inout line);

  procedure read ( L : inout line;  value: out bit;
                                    good : out boolean );
  procedure read ( L : inout line;  value: out bit );

  procedure read ( L : inout line;  value: out bit_vector;
                                    good : out boolean );
  procedure read ( L : inout line;  value: out bit_vector );
```

```
    procedure read ( L : inout line;   value: out boolean;
                                       good : out boolean );
    procedure read ( L : inout line;   value: out boolean );

    procedure read ( L : inout line;   value: out character;
                                       good : out boolean );
    procedure read ( L : inout line;   value: out character );

    procedure read ( L : inout line;   value: out integer;
                                       good : out boolean );
    procedure read ( L : inout line;   value: out integer );

    procedure read ( L : inout line;   value: out real;
                                       good : out boolean );
    procedure read ( L : inout line;   value: out real );

    procedure read ( L : inout line;   value: out string;
                                       good : out boolean );
    procedure read ( L : inout line;   value: out string );

    procedure read ( L : inout line;   value: out time;
                                       good : out boolean );
    procedure read ( L : inout line;   value: out time );

    procedure sread ( L: inout line;   value : out string;
                                       strlen: out natural);

    alias string_read is sread [line, string, natural];

    alias bread is read [line, bit_vector, boolean];
    alias bread is read [line, bit_vector];
    alias binary_read is read [line, bit_vector, boolean];
    alias binary_read is read [line, bit_vector];

    procedure oread ( L: inout line;   value: out bit_vector;
                                       good : out boolean );
    procedure oread ( L: inout line;   value: out bit_vector );

    alias octal_read is oread [line, bit_vector, boolean];
    alias octal_read is oread [line, bit_vector];

    procedure hread ( L: inout line;   value: out bit_vector;
                                       good : out boolean );
    procedure hread ( L: inout line;   value: out bit_vector );

    alias hex_read is hread [line, bit_vector, boolean];
    alias hex_read is hread [line, bit_vector];

    procedure writeline ( file F : text;   L : inout line );

    procedure tee ( file F: text;  L: inout line );

    function justify ( value: string;
                       justified: side := right;
                       field: width := 0 ) return string;

    procedure write ( L : inout line;  value : in bit;
                      justified: in side := right; field: in width := 0 );

    procedure write ( L : inout line;  value : in bit_vector;
                      justified: in side := right;  field: in width := 0 );
```

```
          procedure write ( L : inout line;  value : in boolean;
                            justified: in side := right;  field: in width := 0 );

          procedure write ( L : inout line;  value : in character;
                            justified: in side := right;  field: in width := 0 );

          procedure write ( L : inout line;  value : in integer;
                            justified: in side := right;  field: in width := 0 );

          procedure write ( L : inout line;  value : in real;
                            justified: in side := right;  field: in width := 0;
                            digits: in natural := 0 );

          procedure write ( L: inout line;  value: in real;
                            format: in string);

          procedure write ( L : inout line;  value : in string;
                            justified: in side := right;  field: in width := 0 );

          procedure write ( L : inout line;  value : in time;
                            justified: in side := right;  field: in width := 0;
                            unit: in time := ns );

      alias swrite is write [line, string, side, width];
      alias string_write is write [line, string, side, width];

      alias bwrite is write [line, bit_vector, side, width];
      alias binary_write is write [line, bit_vector, side, width];

          procedure owrite ( L: inout line;  value: in bit_vector;
                             justified: in side := right;  field: in width := 0 );

      alias octal_write is owrite [line, bit_vector, side, width];

          procedure hwrite ( L: inout line;  value: in bit_vector;
                             justified: in side := right;  field: in width := 0 );

      alias hex_write is hwrite [line, bit_vector, side, width];

    end package textio;
```

VHDL-87, -93, and -2002

The **textio** package in these versions does not provide the following items: the **flush** operation on **text**; the **maximum**, **minimum**, and **to_string** operations on **side**; the **justify** and **tee** operations; the **sread**, **oread**, **hread**, **swrite**, **owrite**, and **hwrite** operations and associated aliases; the **write** operation with C-style formatting.

A.4   Standard VHDL Mathematical Packages

A.4.1   The **math_real** Package

```
package math_real is

  constant copyrightnotice: string
          := "Copyright 1996, 2008 IEEE. All rights reserved.";
```

```
constant  math_e : real := 2.71828_18284_59045_23536;
constant  math_1_over_e : real := 0.36787_94411_71442_32160;
constant  math_pi : real := 3.14159_26535_89793_23846;
constant  math_2_pi : real := 6.28318_53071_79586_47693;
constant  math_1_over_pi : real := 0.31830_98861_83790_67154;
constant  math_pi_over_2 : real := 1.57079_63267_94896_61923;
constant  math_pi_over_3 : real := 1.04719_75511_96597_74615;
constant  math_pi_over_4 : real := 0.78539_81633_97448_30962;
constant  math_3_pi_over_2 : real := 4.71238_89803_84689_85769;
constant  math_log_of_2 : real := 0.69314_71805_59945_30942;
constant  math_log_of_10 : real := 2.30258_50929_94045_68402;
constant  math_log2_of_e : real := 1.44269_50408_88963_4074;
constant  math_log10_of_e : real := 0.43429_44819_03251_82765;
constant  math_sqrt_2: real := 1.41421_35623_73095_04880;
constant  math_1_over_sqrt_2: real := 0.70710_67811_86547_52440;
constant  math_sqrt_pi: real := 1.77245_38509_05516_02730;
constant  math_deg_to_rad: real := 0.01745_32925_19943_29577;
constant  math_rad_to_deg: real := 57.29577_95130_82320_87680;

function sign  (x : in real) return real;
function ceil  (x : in real) return real;
function floor (x : in real) return real;
function round (x : in real) return real;
function trunc (x : in real) return real;

function "mod" (x, y : in real) return real;

function realmax (x, y : in real) return real;
function realmin (x, y : in real) return real;

procedure uniform (variable seed1,seed2 : inout positive;
                   variable x : out real);

function sqrt (x : in real) return real;
function cbrt (x : in real) return real;

function "**" (x : in integer;  y : in real) return real;
function "**" (x : in real;      y : in real) return real;

function exp  (x : in rea) return real;
function log  (x : in rea) return real;
function log2 (x : in rea) return real;
function log10 (x : in rea) return real;
function log  (x : in real; base: in real) return real;

function sin (x : in real) return real;
function cos (x : in real) return real;
function tan (x : in real) return real;

function arcsin (x : in real) return real;
function arccos (x : in real) return real;
function arctan (y : in real) return real;
function arctan (y : in real;  x : in real ) return real;

function sinh (x : in real) return real;
function cosh (x : in real) return real;
function tanh (x : in real) return real;

function arcsinh (x : in real) return real;
function arccosh (x : in real) return real;
function arctanh (x : in real) return real;

end math_real;
```

A.4.2 The **math_complex** Package

```vhdl
use work.math_real.all;
package math_complex is

  constant copyrightnotice: string
            := "Copyright 1996, 2008 IEEE. All rights reserved.";

  type complex is
    record
      re : real;
      im : real;
    end record;

  subtype positive_real is real range 0.0 to real'high;
  subtype principal_value is real range -math_pi to math_pi;

  type complex_polar is
    record
      mag : positive_real;
      arg : principal_value;  -- -math_pi is illegal
    end record;

  constant math_cbase_1 : complex := complex'(1.0, 0.0);
  constant math_cbase_j : complex := complex'(0.0, 1.0);
  constant math_czero   : complex := complex'(0.0, 0.0);

  function "=" (l : in complex_polar;  r : in complex_polar) return boolean;
  function "/=" (l : in complex_polar;  r : in complex_polar) return boolean;

  function cmplx (x : in real;  y: in real:= 0.0) return complex;

  function get_principal_value (x : in real) return principal_value;

  function complex_to_polar (z : in complex)       return complex_polar;
  function polar_to_complex (z : in complex_polar) return complex;

  function "abs" (z: in complex       ) return positive_real;
  function "abs" (z: in complex_polar) return positive_real;

  function arg   (z: in complex       ) return principal_value;
  function arg   (z: in complex_polar) return principal_value;

  function "-"   (z: in complex       ) return complex;
  function "-"   (z: in complex_polar) return complex_polar;

  function conj  (z: in complex       ) return complex;
  function conj  (z: in complex_polar) return complex_polar;

  function sqrt  (z: in complex       ) return complex;
  function sqrt  (z: in complex_polar) return complex_polar;

  function exp   (z: in complex       ) return complex;
  function exp   (z: in complex_polar) return complex_polar;

  function log   (z: in complex       ) return complex;
  function log2  (z: in complex       ) return complex;
  function log10 (z: in complex       ) return complex;
  function log   (z: in complex_polar) return complex_polar;
  function log2  (z: in complex_polar) return complex_polar;
  function log10 (z: in complex_polar) return complex_polar;
  function log   (z: in complex;       base: in real) return complex;
  function log   (z: in complex_polar;  base: in real) return complex_polar;

  function sin   (z : in complex       ) return complex;
  function sin   (z : in complex_polar) return complex_polar;
  function cos   (z : in complex       ) return complex;
  function cos   (z : in complex_polar) return complex_polar;
```

```
function sinh (z : in complex      ) return complex;
function sinh (z : in complex_polar) return complex_polar;
function cosh (z : in complex      ) return complex;
function cosh (z : in complex_polar) return complex_polar;

function "+" (l: in complex;       r: in complex) return complex;
function "+" (l: in real;          r: in complex) return complex;
function "+" (l: in complex;       r: in real   ) return complex;
function "+" (l: in complex_polar; r: in complex_polar) return complex_polar;
function "+" (l: in real;          r: in complex_polar) return complex_polar;
function "+" (l: in complex_polar; r: in real          ) return complex_polar;

function "-" (l: in complex;       r: in complex) return complex;
function "-" (l: in real;          r: in complex) return complex;
function "-" (l: in complex;       r: in real   ) return complex;
function "-" (l: in complex_polar; r: in complex_polar) return complex_polar;
function "-" (l: in real;          r: in complex_polar) return complex_polar;
function "-" (l: in complex_polar; r: in real          ) return complex_polar;

function "*" (l: in complex;       r: in complex) return complex;
function "*" (l: in real;          r: in complex) return complex;
function "*" (l: in complex;       r: in real   ) return complex;
function "*" (l: in complex_polar; r: in complex_polar) return complex_polar;
function "*" (l: in real;          r: in complex_polar) return complex_polar;
function "*" (l: in complex_polar; r: in real          ) return complex_polar;

function "/" (l: in complex;       r: in complex) return complex;
function "/" (l: in real;          r: in complex) return complex;
function "/" (l: in complex;       r: in real   ) return complex;
function "/" (l: in complex_polar; r: in complex_polar) return complex_polar;
function "/" (l: in real;          r: in complex_polar) return complex_polar;
function "/" (l: in complex_polar; r: in real          ) return complex_polar;

end math_complex;
```

A.5 The **std_logic_1164** Multivalue Logic System Package

```
use std.textio.all;

package std_logic_1164 is

    type std_ulogic is ( 'U',    -- Uninitialized
                         'X',    -- Forcing  Unknown
                         '0',    -- Forcing  0
                         '1',    -- Forcing  1
                         'Z',    -- High Impedance
                         'W',    -- Weak     Unknown
                         'L',    -- Weak     0
                         'H',    -- Weak     1
                         '-'     -- Don't care
                       );

    type std_ulogic_vector is array (natural range <>) of std_ulogic;

    function resolved (s : std_ulogic_vector) return std_ulogic;

    subtype std_logic is resolved std_ulogic;
    subtype std_logic_vector is (resolved) std_ulogic_vector;

    subtype X01  is resolved std_ulogic range 'X' to '1';
    subtype X01Z is resolved std_ulogic range 'X' to 'Z';
    subtype UX01 is resolved std_ulogic range 'U' to '1';
    subtype UX01Z is resolved std_ulogic range 'U' to 'Z';
```

```
function "and"  (1 : std_ulogic; r : std_ulogic) return UX01;
function "nand" (1 : std_ulogic; r : std_ulogic) return UX01;
function "or"   (1 : std_ulogic; r : std_ulogic) return UX01;
function "nor"  (1 : std_ulogic; r : std_ulogic) return UX01;
function "xor"  (1 : std_ulogic; r : std_ulogic) return UX01;
function "xnor" (1 : std_ulogic; r : std_ulogic) return ux01;
function "not"  (1 : std_ulogic) return UX01;

function "and"  (1, r : std_ulogic_vector) return std_ulogic_vector;
function "nand" (1, r : std_ulogic_vector) return std_ulogic_vector;
function "or"   (1, r : std_ulogic_vector) return std_ulogic_vector;
function "nor"  (1, r : std_ulogic_vector) return std_ulogic_vector;
function "xor"  (1, r : std_ulogic_vector) return std_ulogic_vector;
function "xnor" (1, r : std_ulogic_vector) return std_ulogic_vector;
function "not"  (1    : std_ulogic_vector) return std_ulogic_vector;

function "and"  (1 : std_ulogic_vector;
                 r : std_ulogic)           return std_ulogic_vector;
function "and"  (1 : std_ulogic;
                 r : std_ulogic_vector) return std_ulogic_vector;
function "nand" (1 : std_ulogic_vector;
                 r : std_ulogic)           return std_ulogic_vector;
function "nand" (1 : std_ulogic;
                 r : std_ulogic_vector) return std_ulogic_vector;
function "or"   (1 : std_ulogic_vector;
                 r : std_ulogic)           return std_ulogic_vector;
function "or"   (1 : std_ulogic;
                 r : std_ulogic_vector) return std_ulogic_vector;
function "nor"  (1 : std_ulogic_vector;
                 r : std_ulogic)           return std_ulogic_vector;
function "nor"  (1 : std_ulogic;
                 r : std_ulogic_vector) return std_ulogic_vector;
function "xor"  (1 : std_ulogic_vector;
                 r : std_ulogic)           return std_ulogic_vector;
function "xor"  (1 : std_ulogic;
                 r : std_ulogic_vector) return std_ulogic_vector;
function "xnor" (1 : std_ulogic_vector;
                 r : std_ulogic)           return std_ulogic_vector;
function "xnor" (1 : std_ulogic;
                 r : std_ulogic_vector) return std_ulogic_vector;

function "and"  (1 : std_ulogic_vector) return std_ulogic;
function "nand" (1 : std_ulogic_vector) return std_ulogic;
function "or"   (1 : std_ulogic_vector) return std_ulogic;
function "nor"  (1 : std_ulogic_vector) return std_ulogic;
function "xor"  (1 : std_ulogic_vector) return std_ulogic;
function "xnor" (1 : std_ulogic_vector) return std_ulogic;

function "sll" (1 : std_ulogic_vector; r : integer) return std_ulogic_vector;
function "srl" (1 : std_ulogic_vector; r : integer) return std_ulogic_vector;
function "rol" (1 : std_ulogic_vector; r : integer) return std_ulogic_vector;
function "ror" (1 : std_ulogic_vector; r : integer) return std_ulogic_vector;

function To_bit       (s : std_ulogic;
                       xmap : bit := '0') return bit;
function To_bitvector (s : std_ulogic_vector;
                       xmap : bit := '0') return bit_vector;

function To_StdULogic       (b : bit)               return std_ulogic;
function To_StdLogicVector   (b : bit_vector)        return std_logic_vector;
function To_StdLogicVector   (s : std_ulogic_vector) return std_logic_vector;
function To_StdULogicVector  (b : bit_vector)        return std_ulogic_vector;
function To_StdULogicVector  (s : std_logic_vector)  return std_ulogic_vector;
```

```
alias To_Bit_Vector is
        To_bitvector[std_ulogic_vector, bit return bit_vector];
alias To_BV is
        To_bitvector[std_ulogic_vector, bit return bit_vector];

alias To_Std_Logic_Vector is
        To_StdLogicVector[bit_vector return std_logic_vector];
alias To_SLV is
        To_StdLogicVector[bit_vector return std_logic_vector];

alias To_Std_Logic_Vector is
        To_StdLogicVector[std_ulogic_vector return std_logic_vector];
alias To_SLV is
        To_StdLogicVector[std_ulogic_vector return std_logic_vector];

alias To_Std_ULogic_Vector is
        To_StdULogicVector[bit_vector return std_ulogic_vector];
alias To_SULV is
        To_StdULogicVector[bit_vector return std_ulogic_vector];

alias To_Std_ULogic_Vector is
        To_StdULogicVector[std_logic_vector return std_ulogic_vector];
alias To_SULV is
        To_StdULogicVector[std_logic_vector return std_ulogic_vector];

function To_01 (s : std_ulogic_vector;
                xmap : std_ulogic := '0') return std_ulogic_vector;
function To_01 (s : std_ulogic;
                xmap : std_ulogic := '0') return std_ulogic;
function To_01 (s : bit_vector;
                xmap : std_ulogic := '0') return std_ulogic_vector;
function To_01 (s : bit;
                xmap : std_ulogic := '0') return std_ulogic;

function To_X01 (s : std_ulogic_vector) return std_ulogic_vector;
function To_X01 (s : std_ulogic)        return X01;

function To_X01 (b : bit_vector) return std_ulogic_vector;
function To_X01 (b : bit)        return X01;

function To_X01Z (s : std_ulogic_vector) return std_ulogic_vector;
function To_X01Z (s : std_ulogic)        return X01Z;

function To_X01Z (b : bit_vector) return std_ulogic_vector;
function To_X01Z (b : bit)        return X01Z;

function To_UX01 (s : std_ulogic_vector) return std_ulogic_vector;
function To_UX01 (s : std_ulogic)        return UX01;

function To_UX01 (b : bit_vector) return std_ulogic_vector;
function To_UX01 (b : bit)        return UX01;

function "??" (l : std_ulogic) return boolean;

function rising_edge  (signal s : std_ulogic) return boolean;
function falling_edge (signal s : std_ulogic) return boolean;

function Is_X (s : std_ulogic_vector) return boolean;
function Is_X (s : std_ulogic)        return boolean;

-- the following operations are predefined

--   function "?=" (l, r : std_ulogic)        return std_ulogic;
--   function "?=" (l, r : std_ulogic_vector) return std_ulogic;

--   function "?/=" (l, r : std_ulogic)        return std_ulogic;
--   function "?/=" (l, r : std_ulogic_vector) return std_ulogic;
```

```
--   function "?<"  (l, r : std_ulogic) return std_ulogic;
--   function "?<=" (l, r : std_ulogic) return std_ulogic;
--   function "?>"  (l, r : std_ulogic) return std_ulogic;
--   function "?>=" (l, r : std_ulogic) return std_ulogic;

-- the following operations are predefined

-- function to_string (value : std_ulogic)        return string;
-- function to_string (value : std_ulogic_vector) return string;

alias to_bstring       is to_string [std_ulogic_vector return string];
alias to_binary_string is to_string [std_ulogic_vector return string];

function to_ostring (value : std_ulogic_vector) return string;
alias to_octal_string is to_ostring [std_ulogic_vector return string];

function to_hstring (value : std_ulogic_vector) return string;
alias to_hex_string is to_hstring [std_ulogic_vector return string];

procedure read (l : inout line; value : out std_ulogic;
                                good : out boolean);
procedure read (l : inout line; value : out std_ulogic);

procedure read (l : inout line; value : out std_ulogic_vector;
                                good : out boolean);
procedure read (l : inout line; value : out std_ulogic_vector);

procedure write (l : inout line; value : in std_ulogic;
                justified : in side := right; field : in width := 0);

procedure write (l : inout line; value : in std_ulogic_vector;
                justified : in side := right; field : in width := 0);

alias bread is read [line, std_ulogic_vector, boolean];
alias bread is read [line, std_ulogic_vector];
alias binary_read is read [line, std_ulogic_vector, boolean];
alias binary_read is read [line, std_ulogic_vector];

procedure oread (l : inout line; value : out std_ulogic_vector;
                                 good : out boolean);
procedure oread (l : inout line; value : out std_ulogic_vector);
alias octal_read is oread [line, std_ulogic_vector, boolean];
alias octal_read is oread [line, std_ulogic_vector];

procedure hread (l : inout line; value : out std_ulogic_vector;
                                 good : out boolean);
procedure hread (l : inout line; value : out std_ulogic_vector);
alias hex_read is hread [line, std_ulogic_vector, boolean];
alias hex_read is hread [line, std_ulogic_vector];

alias bwrite is write [line, std_ulogic_vector, side, width];
alias binary_write is write [line, std_ulogic_vector, side, width];

procedure owrite (l : inout line; value : in std_ulogic_vector;
                justified : in side := right; field : in width := 0);
alias octal_write is owrite [line, std_ulogic_vector, side, width];

procedure hwrite (l : inout line; value : in std_ulogic_vector;
                justified : in side := right; field : in width := 0);
alias hex_write is hwrite [line, std_ulogic_vector, side, width];

end package std_logic_1164;
```

VHDL-87, -93, and -2002

In the std_logic_1164 package for these versions, **std_logic_vector** is declared as a distinct type, not as a subtype of **std_ulogic_vector**. Operations provided by the package for **std_ulogic_vector** are also overloaded for **std_logic_vector**.

   The package for these versions does not provide the following items: array/scalar and reduction logical operators; shift operators; aliases for the conversion functions; the To_01 function; the "??" operator; the matching relational operators; the **to_string**, **to_bstring**, **to_ostring**, and **to_hstring** functions and associated aliases; the **read**, **oread**, **hread**, **write**, **owrite**, and **hwrite** operations and associated aliases.

VHDL-87

The overloaded versions of the xnor operator are not included in the VHDL-87 version of the standard-logic package.

A.6 Standard Integer Numeric Packages

A.6.1 The **numeric_bit** Package

```
use std.textio.all;
package numeric_bit is

  constant CopyRightNotice : string
    := "Copyright © 1997, 2008 IEEE. All rights reserved.";

  type unsigned is array (natural range <>) of bit;
  type signed   is array (natural range <>) of bit;

  function "abs" (arg : signed) return signed;
  function "-"   (arg : signed) return signed;

  function "+" (l, r : unsigned)             return unsigned;
  function "+" (l : unsigned; r : bit)       return unsigned;
  function "+" (l : bit;      r : unsigned)  return unsigned;
  function "+" (l : unsigned; r : natural)   return unsigned;
  function "+" (l : natural;  r : unsigned)  return unsigned;
  function "+" (l, r : signed)               return signed;
  function "+" (l : signed;   r : bit)       return signed;
  function "+" (l : bit;      r : signed)    return signed;
  function "+" (l : integer;  r : signed)    return signed;
  function "+" (l : signed;   r : integer)   return signed;

  function "-" (l, r : unsigned)             return unsigned;
  function "-" (l : unsigned; r : bit)       return unsigned;
  function "-" (l : bit;      r : unsigned)  return unsigned;
  function "-" (l : unsigned; r : natural)   return unsigned;
  function "-" (l : natural;  r : unsigned)  return unsigned;
  function "-" (l, r : signed)               return signed;
  function "-" (l : signed;   r : bit)       return signed;
  function "-" (l : bit;      r : signed)    return signed;
  function "-" (l : signed;   r : integer)   return signed;
  function "-" (l : integer;  r : signed)    return signed;
```

```
function "*" (l, r : unsigned)              return unsigned;
function "*" (l : unsigned; r : natural)    return unsigned;
function "*" (l : natural;  r : unsigned)   return unsigned;
function "*" (l, r : signed)                return signed;
function "*" (l : signed;   r : integer)    return signed;
function "*" (l : integer;  r : signed)     return signed;

function "/" (l, r : unsigned)              return unsigned;
function "/" (l : unsigned; r : natural)    return unsigned;
function "/" (l : natural;  r : unsigned)   return unsigned;
function "/" (l, r : signed)                return signed;
function "/" (l : signed;   r : integer)    return signed;
function "/" (l : integer;  r : signed)     return signed;

function "rem" (l, r : unsigned)             return unsigned;
function "rem" (l : unsigned; r : natural)   return unsigned;
function "rem" (l : natural;  r : unsigned)  return unsigned;
function "rem" (l, r : signed)               return signed;
function "rem" (l : signed;   r : integer)   return signed;
function "rem" (l : integer;  r : signed)    return signed;

function "mod" (l, r : unsigned)             return unsigned;
function "mod" (l : unsigned; r : natural)   return unsigned;
function "mod" (l : natural;  r : unsigned)  return unsigned;
function "mod" (l, r : signed)               return signed;
function "mod" (l : signed;   r : integer)   return signed;
function "mod" (l : integer;  r : signed)    return signed;

function find_leftmost  (arg : unsigned; y : bit) return integer;
function find_leftmost  (arg : signed;   y : bit) return integer;

function find_rightmost (arg : unsigned; y : bit) return integer;
function find_rightmost (arg : signed;   y : bit) return integer;

function "=" (l, r : unsigned)              return boolean;
function "=" (l : unsigned; r : natural)    return boolean;
function "=" (l : natural;  r : unsigned)   return boolean;
function "=" (l, r : signed)                return boolean;
function "=" (l : signed;   r : integer)    return boolean;
function "=" (l : integer;  r : signed)     return boolean;

function "/=" (l, r : unsigned)             return boolean;
function "/=" (l : unsigned; r : natural)   return boolean;
function "/=" (l : natural;  r : unsigned)  return boolean;
function "/=" (l, r : signed)               return boolean;
function "/=" (l : signed;   r : integer)   return boolean;
function "/=" (l : integer;  r : signed)    return boolean;

function "<" (l, r : unsigned)              return boolean;
function "<" (l : unsigned; r : natural)    return boolean;
function "<" (l : natural;  r : unsigned)   return boolean;
function "<" (l, r : signed)                return boolean;
function "<" (l : signed;   r : integer)    return boolean;
function "<" (l : integer;  r : signed)     return boolean;

function "<=" (l, r : unsigned)             return boolean;
function "<=" (l : unsigned; r : natural)   return boolean;
function "<=" (l : natural;  r : unsigned)  return boolean;
function "<=" (l, r : signed)               return boolean;
function "<=" (l : signed;   r : integer)   return boolean;
function "<=" (l : integer;  r : signed)    return boolean;

function ">" (l, r : unsigned)              return boolean;
function ">" (l : unsigned; r : natural)    return boolean;
function ">" (l : natural;  r : unsigned)   return boolean;
function ">" (l, r : signed)                return boolean;
```

```
function ">"  (l : signed;   r : integer) return boolean;
function ">"  (l : integer;  r : signed)  return boolean;

function ">=" (l, r : unsigned)           return boolean;
function ">=" (l : unsigned; r : natural) return boolean;
function ">=" (l : natural;  r : unsigned) return boolean;
function ">=" (l, r : signed)             return boolean;
function ">=" (l : signed;   r : integer) return boolean;
function ">=" (l : integer;  r : signed)  return boolean;

function minimum (l, r : unsigned)            return unsigned;
function minimum (l : unsigned; r : natural)  return unsigned;
function minimum (l : natural;  r : unsigned) return unsigned;
function minimum (l, r : signed)              return signed;
function minimum (l : signed;   r : integer)  return signed;
function minimum (l : integer;  r : signed)   return signed;

function maximum (l, r : unsigned)            return unsigned;
function maximum (l : unsigned; r : natural)  return unsigned;
function maximum (l : natural;  r : unsigned) return unsigned;
function maximum (l, r : signed)              return signed;
function maximum (l : signed;   r : integer)  return signed;
function maximum (l : integer;  r : signed)   return signed;

function "?="  (l, r : unsigned)            return bit;
function "?="  (l : unsigned; r : natural)  return bit;
function "?="  (l : natural;  r : unsigned) return bit;
function "?="  (l, r : signed)              return bit;
function "?="  (l : signed;   r : integer)  return bit;
function "?="  (l : integer;  r : signed)   return bit;

function "?/=" (l, r : unsigned)            return bit;
function "?/=" (l : unsigned; r : natural)  return bit;
function "?/=" (l : natural;  r : unsigned) return bit;
function "?/=" (l, r : signed)              return bit;
function "?/=" (l : signed;   r : integer)  return bit;
function "?/=" (l : integer;  r : signed)   return bit;

function "?<"  (l, r : unsigned)            return bit;
function "?<"  (l : unsigned; r : natural)  return bit;
function "?<"  (l : natural;  r : unsigned) return bit;
function "?<"  (l, r : signed)              return bit;
function "?<"  (l : signed;   r : integer)  return bit;
function "?<"  (l : integer;  r : signed)   return bit;

function "?<=" (l, r : unsigned)            return bit;
function "?<=" (l : unsigned; r : natural)  return bit;
function "?<=" (l : natural;  r : unsigned) return bit;
function "?<=" (l, r : signed)              return bit;
function "?<=" (l : signed;   r : integer)  return bit;
function "?<=" (l : integer;  r : signed)   return bit;

function "?>"  (l, r : unsigned)            return bit;
function "?>"  (l : unsigned; r : natural)  return bit;
function "?>"  (l : natural;  r : unsigned) return bit;
function "?>"  (l, r : signed)              return bit;
function "?>"  (l : signed;   r : integer)  return bit;
function "?>"  (l : integer;  r : signed)   return bit;

function "?>=" (l, r : unsigned)            return bit;
function "?>=" (l : unsigned; r : natural)  return bit;
function "?>=" (l : natural;  r : unsigned) return bit;
function "?>=" (l, r : signed)              return bit;
function "?>=" (l : signed;   r : integer)  return bit;
function "?>=" (l : integer;  r : signed)   return bit;
```

```
function shift_left  (arg : unsigned; count : natural) return unsigned;
function shift_left  (arg : signed;   count : natural) return signed;

function shift_right (arg : unsigned; count : natural) return unsigned;
function shift_right (arg : signed;   count : natural) return signed;

function rotate_left (arg : unsigned; count : natural) return unsigned;
function rotate_left (arg : signed;   count : natural) return signed;

function rotate_right (arg : unsigned; count : natural) return unsigned;
function rotate_right (arg : signed;   count : natural) return signed;

function "sll" (arg : unsigned; count : integer) return unsigned;
function "sll" (arg : signed;   count : integer) return signed;
function "srl" (arg : unsigned; count : integer) return unsigned;
function "srl" (arg : signed;   count : integer) return signed;
function "rol" (arg : unsigned; count : integer) return unsigned;
function "rol" (arg : signed;   count : integer) return signed;
function "ror" (arg : unsigned; count : integer) return unsigned;
function "ror" (arg : signed;   count : integer) return signed;
function "sla" (arg : unsigned; count : integer) return unsigned;
function "sla" (arg : signed;   count : integer) return signed;
function "sra" (arg : unsigned; count : integer) return unsigned;
function "sra" (arg : signed;   count : integer) return signed;

function resize (arg : unsigned; new_size : natural) return unsigned;
function resize (arg : signed;   new_size : natural) return signed;
function resize (arg, size_res : unsigned) return unsigned;
function resize (arg, size_res : signed) return signed;

function to_integer  (arg : unsigned) return natural;
function to_integer  (arg : signed)   return integer;

function to_unsigned (arg : natural; size     : natural) return unsigned;
function to_unsigned (arg : natural; size_res : unsigned) return unsigned;
function to_signed   (arg : integer; size     : natural) return signed;
function to_signed   (arg : integer; size_res : signed) return signed;

function "and"  (l, r : unsigned) return unsigned;
function "and"  (l, r : signed)   return signed;
function "nand" (l, r : unsigned) return unsigned;
function "nand" (l, r : signed)   return signed;
function "or"   (l, r : unsigned) return unsigned;
function "or"   (l, r : signed)   return signed;
function "nor"  (l, r : unsigned) return unsigned;
function "nor"  (l, r : signed)   return signed;
function "xor"  (l, r : unsigned) return unsigned;
function "xnor" (l, r : unsigned) return unsigned;
function "xnor" (l, r : signed)   return signed;
function "xor"  (l, r : signed)   return signed;
function "not"  (l : unsigned)    return unsigned;
function "not"  (l : signed)      return signed;

function "and"  (l : bit;      r : unsigned) return unsigned;
function "and"  (l : unsigned; r : bit)      return unsigned;
function "and"  (l : bit;      r : signed)   return signed;
function "and"  (l : signed;   r : bit)      return signed;
function "nand" (l : bit;      r : unsigned) return unsigned;
function "nand" (l : unsigned; r : bit)      return unsigned;
function "nand" (l : bit;      r : signed)   return signed;
function "nand" (l : signed;   r : bit)      return signed;
function "or"   (l : bit;      r : unsigned) return unsigned;
function "or"   (l : unsigned; r : bit)      return unsigned;
function "or"   (l : bit;      r : signed)   return signed;
function "or"   (l : signed;   r : bit)      return signed;
```

```
function "nor"  (l : bit;       r : unsigned) return unsigned;
function "nor"  (l : unsigned; r : bit)       return unsigned;
function "nor"  (l : bit;       r : signed)   return signed;
function "nor"  (l : signed;   r : bit)       return signed;
function "xor"  (l : bit;       r : unsigned) return unsigned;
function "xor"  (l : unsigned; r : bit)       return unsigned;
function "xor"  (l : bit;       r : signed)   return signed;
function "xor"  (l : signed;   r : bit)       return signed;
function "xnor" (l : bit;       r : unsigned) return unsigned;
function "xnor" (l : unsigned; r : bit)       return unsigned;
function "xnor" (l : bit;       r : signed)   return signed;
function "xnor" (l : signed;   r : bit)       return signed;

function "and"  (l : unsigned) return bit;
function "and"  (l : signed)   return bit;
function "nand" (l : unsigned) return bit;
function "nand" (l : signed)   return bit;
function "or"   (l : unsigned) return bit;
function "or"   (l : signed)   return bit;
function "nor"  (l : unsigned) return bit;
function "nor"  (l : signed)   return bit;
function "xor"  (l : unsigned) return bit;
function "xor"  (l : signed)   return bit;
function "xnor" (l : unsigned) return bit;
function "xnor" (l : signed)   return bit;

alias rising_edge is std.standard.rising_edge
                    [std.standard.bit return std.standard.boolean];
alias falling_edge is std.standard.falling_edge
                    [std.standard.bit return std.standard.boolean];

-- the following operations are predefined

-- function to_string (value : unsigned) return string;
-- function to_string (value : signed)   return string;

alias to_bstring is to_string [unsigned return string];
alias to_bstring is to_string [signed   return string];

alias to_binary_string is to_string [unsigned return string];
alias to_binary_string is to_string [signed   return string];

function to_ostring (value : unsigned) return string;
function to_ostring (value : signed)   return string;

alias to_octal_string is to_ostring [unsigned return string];
alias to_octal_string is to_ostring [signed   return string];

function to_hstring (value : unsigned) return string;
function to_hstring (value : signed)   return string;

alias to_hex_string is to_hstring [unsigned return string];
alias to_hex_string is to_hstring [signed   return string];

procedure read(l : inout line; value : out unsigned; good : out boolean);
procedure read(l : inout line; value : out unsigned);
procedure read(l : inout line; value : out signed;   good : out boolean);
procedure read(l : inout line; value : out signed);

alias bread is read [line, unsigned, boolean];
alias bread is read [line, unsigned];
alias bread is read [line, signed,   boolean];
alias bread is read [line, signed];

alias binary_read is read [line, unsigned, boolean];
alias binary_read is read [line, unsigned];
```

```
alias binary_read is read [line, signed,   boolean];
alias binary_read is read [line, signed];

procedure oread (l : inout line; value : out unsigned; good : out boolean);
procedure oread (l : inout line; value : out unsigned);
procedure oread (l : inout line; value : out signed;   good : out boolean);
procedure oread (l : inout line; value : out signed);

alias octal_read is oread [line, unsigned, boolean];
alias octal_read is oread [line, unsigned];
alias octal_read is oread [line, signed,   boolean];
alias octal_read is oread [line, signed];

procedure hread (l : inout line; value : out unsigned; good : out boolean);
procedure hread (l : inout line; value : out unsigned);
procedure hread (l : inout line; value : out signed;   good : out boolean);
procedure hread (l : inout line; value : out signed);

alias hex_read is hread [line, unsigned, boolean];
alias hex_read is hread [line, unsigned];
alias hex_read is hread [line, signed,   boolean];
alias hex_read is hread [line, signed];

procedure write (l : inout line; value : in unsigned;
                justified : in side := right; field : in width := 0);
procedure write (l : inout line; value : in signed;
                justified : in side := right; field : in width := 0);

alias bwrite is write [line, unsigned, side, width];
alias bwrite is write [line, signed,   side, width];

alias binary_write is write [line, unsigned, side, width];
alias binary_write is write [line, signed,   side, width];

procedure owrite (l : inout line; value : in unsigned;
                justified : in side := right; field : in width := 0);
procedure owrite (l : inout line; value : in signed;
                justified : in side := right; field : in width := 0);

alias octal_write is owrite [line, unsigned, side, width];
alias octal_write is owrite [line, signed,   side, width];

procedure hwrite (l : inout line; value : in unsigned;
                justified : in side := right; field : in width := 0);
procedure hwrite (l : inout line; value : in signed;
                justified : in side := right; field : in width := 0);

alias hex_write is hwrite [line, unsigned, side, width];
alias hex_write is hwrite [line, signed,   side, width];

end package numeric_bit;
```

VHDL-87, -93, and -2002

The numeric_bit package for these versions does not provide the following items: the array/bit addition and subtraction operators; the find_leftmost, find_rightmost, maximum, and minimum operations; the matching relational operators; the **sla** and **sra** operators; the resize, to_unsigned, and to_signed functions with size_res parameter; the array/scalar and reduction logical operators; the to_string, to_bstring, to_ostring, and to_hstring functions and associated aliases; the read, oread, hread, write, owrite, and hwrite operations and associated aliases.

VHDL-87

The overloaded versions of the shift, rotate and **xnor** operators must be commented out of the numeric_bit package if VHDL-87 compatibility is required.

A.6.2 The numeric_std Package

```
use std.textio.all;
library ieee; use ieee.std_logic_1164.all;

package numeric_std is

  constant CopyRightNotice : string
    := "Copyright © 1997, 2008 IEEE. All rights reserved.";

  type unresolved_unsigned is array (natural range <>) of std_ulogic;
  type unresolved_signed   is array (natural range <>) of std_ulogic;

  alias u_unsigned is unresolved_unsigned;
  alias u_signed   is unresolved_signed;

  subtype unsigned is (resolved) unresolved_unsigned;
  subtype signed   is (resolved) unresolved_signed;

  -- The declaration of all operators and functions is exactly the same
  -- as in numeric_bit, but with parameters of types unresolved_unsigned
  -- and unresolved_signed instead of unsigned and signed, respectively.

  -- In addition, the following functions are declared:

  function std_match (l, r : std_ulogic)           return boolean;
  function std_match (l, r : unresolved_unsigned) return boolean;
  function std_match (l, r : unresolved_signed)    return boolean;
  function std_match (l, r : std_ulogic_vector)    return boolean;

  function To_01 (s : unresolved_unsigned;
                  xmap : std_ulogic := '0') return unresolved_unsigned;
  function To_01 (s : unresolved_signed;
                  xmap : std_ulogic := '0') return unresolved_signed;

  function To_X01 (s : unresolved_unsigned) return unresolved_unsigned;
  function To_X01 (s : unresolved_signed)   return unresolved_signed;

  function To_X01Z (s : unresolved_unsigned) return unresolved_unsigned;
  function To_X01Z (s : unresolved_signed)   return unresolved_signed;

  function To_UX01 (s : unresolved_unsigned) return unresolved_unsigned;
  function To_UX01 (s : unresolved_signed)   return unresolved_signed;

  function Is_X   (s : unresolved_unsigned) return boolean;
  function Is_X   (s : unresolved_signed)   return boolean;

end package numeric_std;
```

VHDL-87, -93, and -2002

In the numeric_std package for these versions, unsigned and signed are declared as separate array types with std_logic as the element types. There are no declarations of unresolved_unsigned, unresolved_signed, or the corresponding aliases. The operations provided by the package are declared for the unsigned and signed types.

As for the **numeric_bit** package, the **numeric_std** package for these versions does not provide the following items: the array/bit addition and subtraction operators; the **find_leftmost**, **find_rightmost**, **maximum**, and **minimum** operations; the matching relational operators; the **sla** and **sra** operators; the **resize**, **to_unsigned**, and **to_signed** functions with **size_res** parameter; the array/scalar and reduction logical operators; the **to_string**, **to_bstring**, **to_ostring**, and **to_hstring** functions and associated aliases; the **read**, **oread**, **hread**, **write**, **owrite**, and **hwrite** operations and associated aliases. In addition, the package does not provide the To_X01, To_X01Z, To_UX01, and Is_X functions.

VHDL-87

The overloaded versions of the shift, rotate and **xnor** operators must be commented out of the **numeric_std** package if VHDL-87 compatibility is required.

A.6.3   The **numeric_bit_unsigned** Package

```
package numeric_bit_unsigned is

  constant CopyRightNotice : STRING :=
    "Copyright 2008 IEEE. All rights reserved.";

  function "+" (l, r : bit_vector)          return bit_vector;
  function "+" (l : bit_vector; r : bit)    return bit_vector;
  function "+" (l : bit;        r : bit_vector) return bit_vector;
  function "+" (l : bit_vector; r : natural) return bit_vector;
  function "+" (l : natural;    r : bit_vector) return bit_vector;

  function "-" (l, r : bit_vector)          return bit_vector;
  function "-" (l : bit_vector; r : bit)    return bit_vector;
  function "-" (l : bit;        r : bit_vector) return bit_vector;
  function "-" (l : bit_vector; r : natural) return bit_vector;
  function "-" (l : natural;    r : bit_vector) return bit_vector;

  function "*" (l, r : bit_vector)          return bit_vector;
  function "*" (l : bit_vector; r : natural) return bit_vector;
  function "*" (l : natural;    r : bit_vector) return bit_vector;

  function "/" (l, r : bit_vector)          return bit_vector;
  function "/" (l : bit_vector; r : natural) return bit_vector;
  function "/" (l : natural;    r : bit_vector) return bit_vector;

  function "rem" (l, r : bit_vector)          return bit_vector;
  function "rem" (l : bit_vector; r : natural) return bit_vector;
  function "rem" (l : natural;    r : bit_vector) return bit_vector;

  function "mod" (l, r : bit_vector)          return bit_vector;
  function "mod" (l : bit_vector; r : natural) return bit_vector;
  function "mod" (l : natural;    r : bit_vector) return bit_vector;

  function find_leftmost  (arg : bit_vector; y : bit) return integer;
  function find_rightmost (arg : bit_vector; y : bit) return integer;

  function "=" (l, r : bit_vector)          return boolean;
  function "=" (l : bit_vector; r : natural) return boolean;
  function "=" (l : natural;    r : bit_vector) return boolean;
```

```
function "/=" (l, r : bit_vector)           return boolean;
function "/=" (l : bit_vector; r : natural)     return boolean;
function "/=" (l : natural;     r : bit_vector) return boolean;

function "<"  (l, r : bit_vector)           return boolean;
function "<"  (l : bit_vector; r : natural)     return boolean;
function "<"  (l : natural;     r : bit_vector) return boolean;

function "<=" (l, r : bit_vector)           return boolean;
function "<=" (l : bit_vector; r : natural)     return boolean;
function "<=" (l : natural;     r : bit_vector) return boolean;

function ">"  (l, r : bit_vector)           return boolean;
function ">"  (l : bit_vector; r : natural)     return boolean;
function ">"  (l : natural;     r : bit_vector) return boolean;

function ">=" (l, r : bit_vector)           return boolean;
function ">=" (l : bit_vector; r : natural)     return boolean;
function ">=" (l : natural;     r : bit_vector) return boolean;

function minimum (l, r : bit_vector)            return bit_vector;
function minimum (l : bit_vector; r : natural)      return bit_vector;
function minimum (l : natural;     r : bit_vector)  return bit_vector;

function maximum (l, r : bit_vector)            return bit_vector;
function maximum (l : bit_vector; r : natural)      return bit_vector;
function maximum (l : natural;     r : bit_vector)  return bit_vector;

function "?="  (l, r : bit_vector)          return bit;
function "?="  (l : bit_vector; r : natural)    return bit;
function "?="  (l : natural;     r : bit_vector) return bit;

function "?/=" (l, r : bit_vector)          return bit;
function "?/=" (l : bit_vector; r : natural)    return bit;
function "?/=" (l : natural;     r : bit_vector) return bit;

function "?<"  (l, r : bit_vector)          return bit;
function "?<"  (l : bit_vector; r : natural)    return bit;
function "?<"  (l : natural;     r : bit_vector) return bit;

function "?<=" (l, r : bit_vector)          return bit;
function "?<=" (l : bit_vector; r : natural)    return bit;
function "?<=" (l : natural;     r : bit_vector) return bit;

function "?>"  (l, r : bit_vector)          return bit;
function "?>"  (l : bit_vector; r : natural)    return bit;
function "?>"  (l : natural;     r : bit_vector) return bit;

function "?>=" (l, r : bit_vector)          return bit;
function "?>=" (l : bit_vector; r : natural)    return bit;
function "?>=" (l : natural;     r : bit_vector) return bit;

function shift_left   (arg : bit_vector; count : natural) return bit_vector;
function shift_right  (arg : bit_vector; count : natural) return bit_vector;
function rotate_left  (arg : bit_vector; count : natural) return bit_vector;
function rotate_right (arg : bit_vector; count : natural) return bit_vector;

function "sll" (arg : bit_vector; count : integer) return bit_vector;
function "srl" (arg : bit_vector; count : integer) return bit_vector;
function "rol" (arg : bit_vector; count : integer) return bit_vector;
function "ror" (arg : bit_vector; count : integer) return bit_vector;
function "sla" (arg : bit_vector; count : integer) return bit_vector;
function "sra" (arg : bit_vector; count : integer) return bit_vector;

function resize (arg : bit_vector; new_size : natural)    return bit_vector;
function resize (arg : bit_vector; size_res : bit_vector) return bit_vector;

function to_integer   (arg : bit_vector) return natural;
```

```vhdl
    function to_bitvector (arg : natural;
                           size    : natural)    return bit_vector;
    function to_bitvector (arg : natural;
                           size_res : bit_vector) return bit_vector;

  alias to_bit_vector is
        to_bitvector[natural, natural return bit_vector];
  alias to_bv is
        to_bitvector[natural, natural return bit_vector];

  alias to_bit_vector is
        to_bitvector[natural, bit_vector return bit_vector];
  alias to_bv is
        to_bitvector[natural, bit_vector return bit_vector];

end package numeric_bit_unsigned;
```

VHDL-87, -93, and -2002

The numeric_bit_unsigned package is not provided in these versions.

A.6.4   The **numeric_std_unsigned** Package

```vhdl
library ieee; use ieee.std_logic_1164.all;

package numeric_std_unsigned is

  constant CopyRightNotice : string :=
    "Copyright 2008 IEEE. All rights reserved.";

  -- The declaration of the following operators and functions is exactly
  -- the same as in numeric_bit_unsigned, but with parameters of type
  -- std_ulogic_vector instead of bit_vector:
  --    Operators "+", "-", "*", "/", "rem", "mod"
  --    Functions find_leftmost, find_rightmost, minimum, maximum
  --    Operators "=", "/=", "<", "<=", ">", ">="
  --    Operators "?=", "?/=", "?<", "?<=", "?>", "?>="
  --    Functions shift_left, shift_right, rotate_left, rotate_right

  -- Operators "sll", "srl", "rol", "ror" are not declared in this package,
  -- as the versions in std_logic_1164 perform the required operations.
  -- The following shift operators are declared:

  function "sla" (arg : std_ulogic_vector;
                  count : integer) return std_ulogic_vector;
  function "sra" (arg : std_ulogic_vector;
                  count : integer) return std_ulogic_vector;

  -- In addition, the following functions are declared:

  function to_integer (arg : std_ulogic_vector) return natural;

  function To_StdLogicVector (arg : natural; size : natural)
                             return std_logic_vector;
  function To_StdLogicVector (arg : natural; size_res : std_ulogic_vector)
                             return std_logic_vector;

  alias To_Std_Logic_Vector is
        To_StdLogicVector[natural, natural return std_logic_vector];
  alias To_SLV is
        To_StdLogicVector[natural, natural return std_logic_vector];
  alias To_Std_Logic_Vector is
```

```
                     To_StdLogicVector[natural, std_ulogic_vector return std_logic_vector];
          alias To_SLV is
                     To_StdLogicVector[natural, std_ulogic_vector return std_logic_vector];

          function To_StdULogicVector (arg : natural; size : natural)
                                  return std_ulogic_vector;
          function To_StdULogicVector (arg : natural; size_res : std_ulogic_vector)
                                  return std_ulogic_vector;

          alias To_Std_ULogic_Vector is
                     To_StdULogicVector[natural, natural return std_ulogic_vector];
          alias To_SULV is
                     To_StdULogicVector[natural, natural return std_ulogic_vector];
          alias To_Std_ULogic_Vector is
                     To_StdULogicVector[natural, std_ulogic_vector
                                   return std_ulogic_vector];
          alias To_SULV is
                     To_StdULogicVector[natural, std_ulogic_vector
                                   return std_ulogic_vector];

     end package numeric_std_unsigned;
```

VHDL-87, -93, and -2002

The numeric_std_unsigned package is not provided in these versions.

A.7   Standard Fixed-Point Packages

VHDL-87, -93, and -2002

The standard fixed-point packages are not provided in these versions.

A.7.1   The **fixed_float_types** Package

This package is used to define the types of generic constants for the standard fixed-point packages and the standard floating-point packages (see Section A.8).

```
package fixed_float_types is

  type fixed_round_style_type is (fixed_round, fixed_truncate);

  type fixed_overflow_style_type is (fixed_saturate, fixed_wrap);

  type round_type is (round_nearest, round_inf, round_neginf, round_zero);

end package fixed_float_types;
```

A.7.2   The **fixed_generic_pkg** Package

```
use std.textio.all;
library ieee; use ieee.std_logic_1164.all; use ieee.numeric_std.all;
use ieee.fixed_float_types.all;

package fixed_generic_pkg is
  generic (
```

```
fixed_round_style    : fixed_round_style_type    := fixed_round;
fixed_overflow_style : fixed_overflow_style_type := fixed_saturate;
fixed_guard_bits     : natural                   := 3;
no_warning           : boolean                   := false
);

-- Author David Bishop (dbishop@vhdl.org)
constant CopyRightNotice : string :=
  "Copyright 2008 by IEEE. All rights reserved.";

type unresolved_ufixed is array (integer range <>) of std_ulogic;
type unresolved_sfixed is array (integer range <>) of std_ulogic;

alias u_ufixed is unresolved_ufixed;
alias u_sfixed is unresolved_sfixed;

subtype ufixed is (resolved) unresolved_ufixed;
subtype sfixed is (resolved) unresolved_sfixed;

function "abs" (arg : unresolved_sfixed) return unresolved_sfixed;
function "-"   (arg : unresolved_sfixed) return unresolved_sfixed;

function "+" (l, r : unresolved_ufixed)             return unresolved_ufixed;
function "+" (l : unresolved_ufixed; r : real)      return unresolved_ufixed;
function "+" (l : real; r : unresolved_ufixed)      return unresolved_ufixed;
function "+" (l : unresolved_ufixed; r : natural)   return unresolved_ufixed;
function "+" (l : natural; r : unresolved_ufixed)   return unresolved_ufixed;

function "+" (l, r : unresolved_sfixed)             return unresolved_sfixed;
function "+" (l : unresolved_sfixed; r : real)      return unresolved_sfixed;
function "+" (l : real; r : unresolved_sfixed)      return unresolved_sfixed;
function "+" (l : unresolved_sfixed; r : integer)   return unresolved_sfixed;
function "+" (l : integer; r : unresolved_sfixed)   return unresolved_sfixed;

function "-" (l, r : unresolved_ufixed)             return unresolved_ufixed;
function "-" (l : unresolved_ufixed; r : real)      return unresolved_ufixed;
function "-" (l : real; r : unresolved_ufixed)      return unresolved_ufixed;
function "-" (l : unresolved_ufixed; r : natural)   return unresolved_ufixed;
function "-" (l : natural; r : unresolved_ufixed)   return unresolved_ufixed;

function "-" (l, r : unresolved_sfixed)             return unresolved_sfixed;
function "-" (l : unresolved_sfixed; r : real)      return unresolved_sfixed;
function "-" (l : real; r : unresolved_sfixed)      return unresolved_sfixed;
function "-" (l : unresolved_sfixed; r : integer)   return unresolved_sfixed;
function "-" (l : integer; r : unresolved_sfixed)   return unresolved_sfixed;

function "*" (l, r : unresolved_ufixed)             return unresolved_ufixed;
function "*" (l : unresolved_ufixed; r : real)      return unresolved_ufixed;
function "*" (l : real; r : unresolved_ufixed)      return unresolved_ufixed;
function "*" (l : unresolved_ufixed; r : natural)   return unresolved_ufixed;
function "*" (l : natural; r : unresolved_ufixed)   return unresolved_ufixed;

function "*" (l, r : unresolved_sfixed)             return unresolved_sfixed;
function "*" (l : unresolved_sfixed; r : real)      return unresolved_sfixed;
function "*" (l : real; r : unresolved_sfixed)      return unresolved_sfixed;
function "*" (l : unresolved_sfixed; r : integer)   return unresolved_sfixed;
function "*" (l : integer; r : unresolved_sfixed)   return unresolved_sfixed;

function "/" (l, r : unresolved_ufixed)             return unresolved_ufixed;
function "/" (l : unresolved_ufixed; r : real)      return unresolved_ufixed;
function "/" (l : real; r : unresolved_ufixed)      return unresolved_ufixed;
function "/" (l : unresolved_ufixed; r : natural)   return unresolved_ufixed;
function "/" (l : natural; r : unresolved_ufixed)   return unresolved_ufixed;

function "/" (l, r : unresolved_sfixed)             return unresolved_sfixed;
function "/" (l : unresolved_sfixed; r : real)      return unresolved_sfixed;
function "/" (l : real; r : unresolved_sfixed)      return unresolved_sfixed;
```

```
function "/" (l : unresolved_sfixed; r : integer) return unresolved_sfixed;
function "/" (l : integer; r : unresolved_sfixed) return unresolved_sfixed;

function "rem" (l, r : unresolved_ufixed)            return unresolved_ufixed;
function "rem" (l : unresolved_ufixed; r : real)     return unresolved_ufixed;
function "rem" (l : real; r : unresolved_ufixed)     return unresolved_ufixed;
function "rem" (l : unresolved_ufixed; r : natural)  return unresolved_ufixed;
function "rem" (l : natural; r : unresolved_ufixed)  return unresolved_ufixed;

function "rem" (l, r : unresolved_sfixed)            return unresolved_sfixed;
function "rem" (l : unresolved_sfixed; r : real)     return unresolved_sfixed;
function "rem" (l : real; r : unresolved_sfixed)     return unresolved_sfixed;
function "rem" (l : unresolved_sfixed; r : integer)  return unresolved_sfixed;
function "rem" (l : integer; r : unresolved_sfixed)  return unresolved_sfixed;

function "mod" (l, r : unresolved_ufixed)            return unresolved_ufixed;
function "mod" (l : unresolved_ufixed; r : real)     return unresolved_ufixed;
function "mod" (l : real; r : unresolved_ufixed)     return unresolved_ufixed;
function "mod" (l : unresolved_ufixed; r : natural)  return unresolved_ufixed;
function "mod" (l : natural; r : unresolved_ufixed)  return unresolved_ufixed;

function "mod" (l, r : unresolved_sfixed)            return unresolved_sfixed;
function "mod" (l : unresolved_sfixed; r : real)     return unresolved_sfixed;
function "mod" (l : real; r : unresolved_sfixed)     return unresolved_sfixed;
function "mod" (l : unresolved_sfixed; r : integer)  return unresolved_sfixed;
function "mod" (l : integer; r : unresolved_sfixed)  return unresolved_sfixed;

function divide (l, r       : unresolved_ufixed;
                 round_style : fixed_round_style_type := fixed_round_style;
                 guard_bits  : natural                := fixed_guard_bits)
                return unresolved_ufixed;

function divide (l, r       : unresolved_sfixed;
                 round_style : fixed_round_style_type := fixed_round_style;
                 guard_bits  : natural                := fixed_guard_bits)
                return unresolved_sfixed;

function reciprocal (arg        : unresolved_ufixed;
                     round_style : fixed_round_style_type := fixed_round_style;
                     guard_bits  : natural                := fixed_guard_bits)
                    return unresolved_ufixed;

function reciprocal (arg        : unresolved_sfixed;
                     round_style : fixed_round_style_type := fixed_round_style;
                     guard_bits  : natural                := fixed_guard_bits)
                    return unresolved_sfixed;

function remainder (l, r       : unresolved_ufixed;
                    round_style : fixed_round_style_type := fixed_round_style;
                    guard_bits  : natural                := fixed_guard_bits)
                   return unresolved_ufixed;

function remainder (l, r       : unresolved_sfixed;
                    round_style : fixed_round_style_type := fixed_round_style;
                    guard_bits  : natural                := fixed_guard_bits)
                   return unresolved_sfixed;

function modulo (l, r       : unresolved_ufixed;
                round_style : fixed_round_style_type := fixed_round_style;
                guard_bits  : natural                := fixed_guard_bits)
               return unresolved_ufixed;

function modulo (l, r            : unresolved_sfixed;
                overflow_style : fixed_overflow_style_type := fixed_overflow_style;
                round_style    : fixed_round_style_type    := fixed_round_style;
```

```
                          guard_bits    : natural                    := fixed_guard_bits)
                      return unresolved_sfixed;

procedure add_carry (l, r  : in  unresolved_ufixed; c_in  : in  std_ulogic;
                     result : out unresolved_ufixed; c_out : out std_ulogic);

procedure add_carry (l, r  : in  unresolved_sfixed; c_in  : in  std_ulogic;
                     result : out unresolved_sfixed; c_out : out std_ulogic);

function scalb (y : unresolved_ufixed; N : integer)           return unresolved_ufixed;
function scalb (y : unresolved_ufixed; N : unresolved_signed) return unresolved_ufixed;
function scalb (y : unresolved_sfixed; N : integer)           return unresolved_sfixed;
function scalb (y : unresolved_sfixed; N : unresolved_signed) return unresolved_sfixed;

function Is_Negative (arg : unresolved_sfixed) return boolean;

function "="  (l, r : unresolved_ufixed) return boolean;
function "="  (l : unresolved_ufixed; r : real)    return boolean;
function "="  (l : real; r : unresolved_ufixed)    return boolean;
function "="  (l : unresolved_ufixed; r : natural) return boolean;
function "="  (l : natural; r : unresolved_ufixed) return boolean;

function "="  (l, r : unresolved_sfixed) return boolean;
function "="  (l : unresolved_sfixed; r : real)    return boolean;
function "="  (l : real; r : unresolved_sfixed)    return boolean;
function "="  (l : unresolved_sfixed; r : natural) return boolean;
function "="  (l : natural; r : unresolved_sfixed) return boolean;

function "/=" (l, r : unresolved_ufixed) return boolean;
function "/=" (l : unresolved_ufixed; r : real)    return boolean;
function "/=" (l : real; r : unresolved_ufixed)    return boolean;
function "/=" (l : unresolved_ufixed; r : natural) return boolean;
function "/=" (l : natural; r : unresolved_ufixed) return boolean;

function "/=" (l, r : unresolved_sfixed) return boolean;
function "/=" (l : unresolved_sfixed; r : real)    return boolean;
function "/=" (l : real; r : unresolved_sfixed)    return boolean;
function "/=" (l : unresolved_sfixed; r : natural) return boolean;
function "/=" (l : natural; r : unresolved_sfixed) return boolean;

function "<"  (l, r : unresolved_ufixed) return boolean;
function "<"  (l : unresolved_ufixed; r : real)    return boolean;
function "<"  (l : real; r : unresolved_ufixed)    return boolean;
function "<"  (l : unresolved_ufixed; r : natural) return boolean;
function "<"  (l : natural; r : unresolved_ufixed) return boolean;

function "<"  (l, r : unresolved_sfixed) return boolean;
function "<"  (l : unresolved_sfixed; r : real)    return boolean;
function "<"  (l : real; r : unresolved_sfixed)    return boolean;
function "<"  (l : unresolved_sfixed; r : integer) return boolean;
function "<"  (l : integer; r : unresolved_sfixed) return boolean;

function "<=" (l, r : unresolved_ufixed) return boolean;
function "<=" (l : unresolved_ufixed; r : real)    return boolean;
function "<=" (l : real; r : unresolved_ufixed)    return boolean;
function "<=" (l : unresolved_ufixed; r : natural) return boolean;
function "<=" (l : natural; r : unresolved_ufixed) return boolean;

function "<=" (l, r : unresolved_sfixed) return boolean;
function "<=" (l : unresolved_sfixed; r : real)    return boolean;
function "<=" (l : real; r : unresolved_sfixed)    return boolean;
function "<=" (l : unresolved_sfixed; r : integer) return boolean;
function "<=" (l : integer; r : unresolved_sfixed) return boolean;

function ">"  (l, r : unresolved_ufixed) return boolean;
function ">"  (l : unresolved_ufixed; r : real)    return boolean;
function ">"  (l : real; r : unresolved_ufixed)    return boolean;
```

```
function ">"   (1 : unresolved_ufixed; r : natural) return boolean;
function ">"   (1 : natural; r : unresolved_ufixed) return boolean;

function ">"   (1, r : unresolved_sfixed) return boolean;
function ">"   (1 : unresolved_sfixed; r : real)    return boolean;
function ">"   (1 : real; r : unresolved_sfixed)    return boolean;
function ">"   (1 : unresolved_sfixed; r : integer) return boolean;
function ">"   (1 : integer; r : unresolved_sfixed) return boolean;

function ">="  (1, r : unresolved_ufixed) return boolean;
function ">="  (1 : unresolved_ufixed; r : real)    return boolean;
function ">="  (1 : real; r : unresolved_ufixed)    return boolean;
function ">="  (1 : unresolved_ufixed; r : natural) return boolean;
function ">="  (1 : natural; r : unresolved_ufixed) return boolean;

function ">="  (1, r : unresolved_sfixed) return boolean;
function ">="  (1 : unresolved_sfixed; r : real)    return boolean;
function ">="  (1 : real; r : unresolved_sfixed)    return boolean;
function ">="  (1 : unresolved_sfixed; r : integer) return boolean;
function ">="  (1 : integer; r : unresolved_sfixed) return boolean;

function minimum (1, r : unresolved_ufixed)              return unresolved_ufixed;
function minimum (1 : unresolved_ufixed; r : real)       return unresolved_ufixed;
function minimum (1 : real; r : unresolved_ufixed)       return unresolved_ufixed;
function minimum (1 : unresolved_ufixed; r : natural)    return unresolved_ufixed;
function minimum (1 : natural; r : unresolved_ufixed)    return unresolved_ufixed;

function minimum (1, r : unresolved_sfixed)              return unresolved_sfixed;
function minimum (1 : unresolved_sfixed; r : real)       return unresolved_sfixed;
function minimum (1 : real; r : unresolved_sfixed)       return unresolved_sfixed;
function minimum (1 : unresolved_sfixed; r : integer)    return unresolved_sfixed;
function minimum (1 : integer; r : unresolved_sfixed)    return unresolved_sfixed;

function maximum (1, r : unresolved_ufixed)              return unresolved_ufixed;
function maximum (1 : unresolved_ufixed; r : real)       return unresolved_ufixed;
function maximum (1 : real; r : unresolved_ufixed)       return unresolved_ufixed;
function maximum (1 : unresolved_ufixed; r : natural)    return unresolved_ufixed;
function maximum (1 : natural; r : unresolved_ufixed)    return unresolved_ufixed;

function maximum (1, r : unresolved_sfixed)              return unresolved_sfixed;
function maximum (1 : unresolved_sfixed; r : real)       return unresolved_sfixed;
function maximum (1 : real; r : unresolved_sfixed)       return unresolved_sfixed;
function maximum (1 : unresolved_sfixed; r : integer)    return unresolved_sfixed;
function maximum (1 : integer; r : unresolved_sfixed)    return unresolved_sfixed;

function "?="  (1, r : unresolved_ufixed) return boolean;
function "?="  (1 : unresolved_ufixed; r : real)    return boolean;
function "?="  (1 : real; r : unresolved_ufixed)    return boolean;
function "?="  (1 : unresolved_ufixed; r : natural) return boolean;
function "?="  (1 : natural; r : unresolved_ufixed) return boolean;

function "?="  (1, r : unresolved_sfixed) return boolean;
function "?="  (1 : unresolved_sfixed; r : real)    return boolean;
function "?="  (1 : real; r : unresolved_sfixed)    return boolean;
function "?="  (1 : unresolved_sfixed; r : natural) return boolean;
function "?="  (1 : natural; r : unresolved_sfixed) return boolean;

function "?/=" (1, r : unresolved_ufixed) return boolean;
function "?/=" (1 : unresolved_ufixed; r : real)    return boolean;
function "?/=" (1 : real; r : unresolved_ufixed)    return boolean;
function "?/=" (1 : unresolved_ufixed; r : natural) return boolean;
function "?/=" (1 : natural; r : unresolved_ufixed) return boolean;

function "?/=" (1, r : unresolved_sfixed) return boolean;
function "?/=" (1 : unresolved_sfixed; r : real)    return boolean;
function "?/=" (1 : real; r : unresolved_sfixed)    return boolean;
```

```
function "?/=" (l : unresolved_sfixed; r : natural) return boolean;
function "?/=" (l : natural; r : unresolved_sfixed) return boolean;

function "?<"  (l, r : unresolved_ufixed) return boolean;
function "?<"  (l : unresolved_ufixed; r : real)     return boolean;
function "?<"  (l : real; r : unresolved_ufixed)     return boolean;
function "?<"  (l : unresolved_ufixed; r : natural) return boolean;
function "?<"  (l : natural; r : unresolved_ufixed) return boolean;

function "?<"  (l, r : unresolved_sfixed) return boolean;
function "?<"  (l : unresolved_sfixed; r : real)     return boolean;
function "?<"  (l : real; r : unresolved_sfixed)     return boolean;
function "?<"  (l : unresolved_sfixed; r : integer) return boolean;
function "?<"  (l : integer; r : unresolved_sfixed) return boolean;

function "?<=" (l, r : unresolved_ufixed) return boolean;
function "?<=" (l : unresolved_ufixed; r : real)     return boolean;
function "?<=" (l : real; r : unresolved_ufixed)     return boolean;
function "?<=" (l : unresolved_ufixed; r : natural) return boolean;
function "?<=" (l : natural; r : unresolved_ufixed) return boolean;

function "?<=" (l, r : unresolved_sfixed) return boolean;
function "?<=" (l : unresolved_sfixed; r : real)     return boolean;
function "?<=" (l : real; r : unresolved_sfixed)     return boolean;
function "?<=" (l : unresolved_sfixed; r : integer) return boolean;
function "?<=" (l : integer; r : unresolved_sfixed) return boolean;

function "?>"  (l, r : unresolved_ufixed) return boolean;
function "?>"  (l : unresolved_ufixed; r : real)     return boolean;
function "?>"  (l : real; r : unresolved_ufixed)     return boolean;
function "?>"  (l : unresolved_ufixed; r : natural) return boolean;
function "?>"  (l : natural; r : unresolved_ufixed) return boolean;

function "?>"  (l, r : unresolved_sfixed) return boolean;
function "?>"  (l : unresolved_sfixed; r : real)     return boolean;
function "?>"  (l : real; r : unresolved_sfixed)     return boolean;
function "?>"  (l : unresolved_sfixed; r : integer) return boolean;
function "?>"  (l : integer; r : unresolved_sfixed) return boolean;

function "?>=" (l, r : unresolved_ufixed) return boolean;
function "?>=" (l : unresolved_ufixed; r : real)     return boolean;
function "?>=" (l : real; r : unresolved_ufixed)     return boolean;
function "?>=" (l : unresolved_ufixed; r : natural) return boolean;
function "?>=" (l : natural; r : unresolved_ufixed) return boolean;

function "?>=" (l, r : unresolved_sfixed) return boolean;
function "?>=" (l : unresolved_sfixed; r : real)     return boolean;
function "?>=" (l : real; r : unresolved_sfixed)     return boolean;
function "?>=" (l : unresolved_sfixed; r : integer) return boolean;
function "?>=" (l : integer; r : unresolved_sfixed) return boolean;

function std_match (l, r : unresolved_ufixed) return boolean;
function std_match (l, r : unresolved_sfixed) return boolean;

function "sll" (arg : unresolved_ufixed; count : integer) return unresolved_ufixed;
function "sll" (arg : unresolved_sfixed; count : integer) return unresolved_sfixed;
function "srl" (arg : unresolved_ufixed; count : integer) return unresolved_ufixed;
function "srl" (arg : unresolved_sfixed; count : integer) return unresolved_sfixed;
function "sla" (arg : unresolved_ufixed; count : integer) return unresolved_ufixed;
function "sla" (arg : unresolved_sfixed; count : integer) return unresolved_sfixed;
function "sra" (arg : unresolved_ufixed; count : integer) return unresolved_ufixed;
function "sra" (arg : unresolved_sfixed; count : integer) return unresolved_sfixed;
function "rol" (arg : unresolved_ufixed; count : integer) return unresolved_ufixed;
function "rol" (arg : unresolved_sfixed; count : integer) return unresolved_sfixed;
function "ror" (arg : unresolved_ufixed; count : integer) return unresolved_ufixed;
function "ror" (arg : unresolved_sfixed; count : integer) return unresolved_sfixed;
```

```vhdl
    function shift_left  (arg : unresolved_ufixed; count : natural) return
unresolved_ufixed;
    function shift_right (arg : unresolved_ufixed; count : natural) return
unresolved_ufixed;
    function shift_left  (arg : unresolved_sfixed; count : natural) return
unresolved_sfixed;
    function shift_right (arg : unresolved_sfixed; count : natural) return
unresolved_sfixed;

    function "and"  (l, r : unresolved_ufixed) return unresolved_ufixed;
    function "and"  (l, r : unresolved_sfixed) return unresolved_sfixed;
    function "nand" (l, r : unresolved_ufixed) return unresolved_ufixed;
    function "nand" (l, r : unresolved_sfixed) return unresolved_sfixed;
    function "or"   (l, r : unresolved_ufixed) return unresolved_ufixed;
    function "or"   (l, r : unresolved_sfixed) return unresolved_sfixed;
    function "nor"  (l, r : unresolved_ufixed) return unresolved_ufixed;
    function "nor"  (l, r : unresolved_sfixed) return unresolved_sfixed;
    function "xor"  (l, r : unresolved_ufixed) return unresolved_ufixed;
    function "xor"  (l, r : unresolved_sfixed) return unresolved_sfixed;
    function "xnor" (l, r : unresolved_ufixed) return unresolved_ufixed;
    function "xnor" (l, r : unresolved_sfixed) return unresolved_sfixed;
    function "not"  (l    : unresolved_ufixed) return unresolved_ufixed;
    function "not"  (l    : unresolved_sfixed) return unresolved_sfixed;

    function "and"  (l : std_ulogic; r : unresolved_ufixed) return unresolved_ufixed;
    function "and"  (l : unresolved_ufixed; r : std_ulogic) return unresolved_ufixed;
    function "and"  (l : std_ulogic; r : unresolved_sfixed) return unresolved_sfixed;
    function "and"  (l : unresolved_sfixed; r : std_ulogic) return unresolved_sfixed;
    function "nand" (l : std_ulogic; r : unresolved_ufixed) return unresolved_ufixed;
    function "nand" (l : unresolved_ufixed; r : std_ulogic) return unresolved_ufixed;
    function "nand" (l : std_ulogic; r : unresolved_sfixed) return unresolved_sfixed;
    function "nand" (l : unresolved_sfixed; r : std_ulogic) return unresolved_sfixed;
    function "or"   (l : std_ulogic; r : unresolved_ufixed) return unresolved_ufixed;
    function "or"   (l : unresolved_ufixed; r : std_ulogic) return unresolved_ufixed;
    function "or"   (l : std_ulogic; r : unresolved_sfixed) return unresolved_sfixed;
    function "or"   (l : unresolved_sfixed; r : std_ulogic) return unresolved_sfixed;
    function "nor"  (l : std_ulogic; r : unresolved_ufixed) return unresolved_ufixed;
    function "nor"  (l : unresolved_ufixed; r : std_ulogic) return unresolved_ufixed;
    function "nor"  (l : std_ulogic; r : unresolved_sfixed) return unresolved_sfixed;
    function "nor"  (l : unresolved_sfixed; r : std_ulogic) return unresolved_sfixed;
    function "xor"  (l : std_ulogic; r : unresolved_ufixed) return unresolved_ufixed;
    function "xor"  (l : unresolved_ufixed; r : std_ulogic) return unresolved_ufixed;
    function "xor"  (l : std_ulogic; r : unresolved_sfixed) return unresolved_sfixed;
    function "xor"  (l : unresolved_sfixed; r : std_ulogic) return unresolved_sfixed;
    function "xnor" (l : std_ulogic; r : unresolved_ufixed) return unresolved_ufixed;
    function "xnor" (l : unresolved_ufixed; r : std_ulogic) return unresolved_ufixed;
    function "xnor" (l : std_ulogic; r : unresolved_sfixed) return unresolved_sfixed;
    function "xnor" (l : unresolved_sfixed; r : std_ulogic) return unresolved_sfixed;

    function "and"  (l : unresolved_ufixed) return std_ulogic;
    function "and"  (l : unresolved_sfixed) return std_ulogic;
    function "nand" (l : unresolved_ufixed) return std_ulogic;
    function "nand" (l : unresolved_sfixed) return std_ulogic;
    function "or"   (l : unresolved_ufixed) return std_ulogic;
    function "or"   (l : unresolved_sfixed) return std_ulogic;
    function "nor"  (l : unresolved_ufixed) return std_ulogic;
    function "nor"  (l : unresolved_sfixed) return std_ulogic;
    function "xor"  (l : unresolved_ufixed) return std_ulogic;
    function "xor"  (l : unresolved_sfixed) return std_ulogic;
    function "xnor" (l : unresolved_ufixed) return std_ulogic;
    function "xnor" (l : unresolved_sfixed) return std_ulogic;

    function find_leftmost (arg : unresolved_ufixed; y : std_ulogic) return integer;
    function find_leftmost (arg : unresolved_sfixed; y : std_ulogic) return integer;
```

```
function find_rightmost (arg : unresolved_ufixed; y : std_ulogic) return integer;
function find_rightmost (arg : unresolved_sfixed; y : std_ulogic) return integer;

function resize (arg           : unresolved_ufixed;
                 left_index    : integer;
                 right_index   : integer;
                 overflow_style : fixed_overflow_style_type := fixed_overflow_style;
                 round_style    : fixed_round_style_type    := fixed_round_style)
            return unresolved_ufixed;

function resize (arg           : unresolved_ufixed;
                 size_res      : unresolved_ufixed;
                 overflow_style : fixed_overflow_style_type := fixed_overflow_style;
                 round_style    : fixed_round_style_type    := fixed_round_style)
            return unresolved_ufixed;

function resize (arg           : unresolved_sfixed;
                 left_index    : integer;
                 right_index   : integer;
                 overflow_style : fixed_overflow_style_type := fixed_overflow_style;
                 round_style    : fixed_round_style_type    := fixed_round_style)
            return unresolved_sfixed;

function resize (arg           : unresolved_sfixed;
                 size_res      : unresolved_sfixed;
                 overflow_style : fixed_overflow_style_type := fixed_overflow_style;
                 round_style    : fixed_round_style_type    := fixed_round_style)
            return unresolved_sfixed;

function to_ufixed (arg           : natural;
                    left_index    : integer;
                    right_index   : integer                   := 0;
                    overflow_style : fixed_overflow_style_type := fixed_overflow_style;
                    round_style    : fixed_round_style_type    := fixed_round_style)
               return unresolved_ufixed;

function to_ufixed (arg           : natural;
                    size_res      : unresolved_ufixed;
                    overflow_style : fixed_overflow_style_type := fixed_overflow_style;
                    round_style    : fixed_round_style_type    := fixed_round_style)
               return unresolved_ufixed;

function to_ufixed (arg           : real;
                    left_index    : integer;
                    right_index   : integer;
                    overflow_style : fixed_overflow_style_type := fixed_overflow_style;
                    round_style    : fixed_round_style_type    := fixed_round_style;
                    guard_bits    : natural                    := fixed_guard_bits)
               return unresolved_ufixed;

function to_ufixed (arg           : real;
                    size_res      : unresolved_ufixed;
                    overflow_style : fixed_overflow_style_type := fixed_overflow_style;
                    round_style    : fixed_round_style_type    := fixed_round_style;
                    guard_bits    : natural                    := fixed_guard_bits)
               return unresolved_ufixed;

function to_ufixed (arg           : unresolved_unsigned;
                    left_index    : integer;
                    right_index   : integer                   := 0;
                    overflow_style : fixed_overflow_style_type := fixed_overflow_style;
                    round_style    : fixed_round_style_type    := fixed_round_style)
               return unresolved_ufixed;

function to_ufixed (arg           : unresolved_unsigned;
                    size_res      : unresolved_ufixed;
```

```
                            overflow_style : fixed_overflow_style_type := fixed_overflow_style;
                            round_style    : fixed_round_style_type    := fixed_round_style)
                          return unresolved_ufixed;

    function to_ufixed (arg : unresolved_unsigned) return unresolved_ufixed;

    function to_unsigned (arg            : unresolved_ufixed;
                          size           : natural;
                          overflow_style : fixed_overflow_style_type := fixed_overflow_style;
                          round_style    : fixed_round_style_type    := fixed_round_style)
                        return unresolved_unsigned;

    function to_unsigned (arg            : unresolved_ufixed;
                          size_res       : unresolved_unsigned;
                          overflow_style : fixed_overflow_style_type := fixed_overflow_style;
                          round_style    : fixed_round_style_type    := fixed_round_style)
                        return unresolved_unsigned;

    function to_real (arg : unresolved_ufixed) return real;

    function to_integer (arg            : unresolved_ufixed;
                         overflow_style : fixed_overflow_style_type := fixed_overflow_style;
                         round_style    : fixed_round_style_type    := fixed_round_style)
                        return natural;

    function to_sfixed (arg            : integer;
                        left_index     : integer;
                        right_index    : integer                    := 0;
                        overflow_style : fixed_overflow_style_type := fixed_overflow_style;
                        round_style    : fixed_round_style_type    := fixed_round_style)
                       return unresolved_sfixed;

    function to_sfixed (arg            : integer;
                        size_res       : unresolved_sfixed;
                        overflow_style : fixed_overflow_style_type := fixed_overflow_style;
                        round_style    : fixed_round_style_type    := fixed_round_style)
                       return unresolved_sfixed;

    function to_sfixed (arg            : real;
                        left_index     : integer;
                        right_index    : integer;
                        overflow_style : fixed_overflow_style_type := fixed_overflow_style;
                        round_style    : fixed_round_style_type    := fixed_round_style;
                        guard_bits     : natural                   := fixed_guard_bits)
                       return unresolved_sfixed;

    function to_sfixed (arg            : real;
                        size_res       : unresolved_sfixed;
                        overflow_style : fixed_overflow_style_type := fixed_overflow_style;
                        round_style    : fixed_round_style_type    := fixed_round_style;
                        guard_bits     : natural                   := fixed_guard_bits)
                       return unresolved_sfixed;

    function to_sfixed (arg            : unresolved_signed;
                        left_index     : integer;
                        right_index    : integer                   := 0;
                        overflow_style : fixed_overflow_style_type := fixed_overflow_style;
                        round_style    : fixed_round_style_type    := fixed_round_style)
                       return unresolved_sfixed;

    function to_sfixed (arg            : unresolved_signed;
                        size_res       : unresolved_sfixed;
                        overflow_style : fixed_overflow_style_type := fixed_overflow_style;
                        round_style    : fixed_round_style_type    := fixed_round_style)
                       return unresolved_sfixed;
```

```
function to_sfixed (arg : unresolved_signed) return unresolved_sfixed;

function to_sfixed (arg : unresolved_ufixed) return unresolved_sfixed;

function to_signed (arg           : unresolved_sfixed;
                    size           : natural;
                 overflow_style : fixed_overflow_style_type := fixed_overflow_style;
                    round_style    : fixed_round_style_type    := fixed_round_style)
                 return unresolved_signed;

function to_signed (arg           : unresolved_sfixed;
                    size_res       : unresolved_signed;
                 overflow_style : fixed_overflow_style_type := fixed_overflow_style;
                    round_style    : fixed_round_style_type    := fixed_round_style)
                 return unresolved_signed;

function to_real (arg : unresolved_sfixed) return real;

function to_integer (arg           : unresolved_sfixed;
                 overflow_style : fixed_overflow_style_type := fixed_overflow_style;
                    round_style    : fixed_round_style_type    := fixed_round_style)
                 return integer;

function ufixed_high (left_index, right_index  : integer;
                      operation                 : character := 'X';
                      left_index2, right_index2 : integer   := 0) return integer;

function ufixed_low  (left_index, right_index  : integer;
                      operation                 : character := 'X';
                      left_index2, right_index2 : integer   := 0) return integer;

function sfixed_high (left_index, right_index  : integer;
                      operation                 : character := 'X';
                      left_index2, right_index2 : integer   := 0) return integer;

function sfixed_low  (left_index, right_index  : integer;
                      operation                 : character := 'X';
                      left_index2, right_index2 : integer   := 0) return integer;

function ufixed_high (size_res  : unresolved_ufixed;
                      operation : character := 'X';
                      size_res2 : unresolved_ufixed) return integer;

function ufixed_low  (size_res  : unresolved_ufixed;
                      operation : character := 'X';
                      size_res2 : unresolved_ufixed) return integer;

function sfixed_high (size_res  : unresolved_sfixed;
                      operation : character := 'X';
                      size_res2 : unresolved_sfixed) return integer;

function sfixed_low  (size_res  : unresolved_sfixed;
                      operation : character := 'X';
                      size_res2 : unresolved_sfixed) return integer;

function saturate (left_index, right_index : integer) return unresolved_ufixed;
function saturate (left_index, right_index : integer) return unresolved_sfixed;
function saturate (size_res : unresolved_ufixed) return unresolved_ufixed;
function saturate (size_res : unresolved_sfixed) return unresolved_sfixed;

function To_01 (s    : unresolved_ufixed;
                xmap : std_ulogic := '0') return unresolved_ufixed;
function To_01 (s    : unresolved_sfixed;
                xmap : std_ulogic := '0') return unresolved_sfixed;

function Is_X   (arg : unresolved_ufixed) return boolean;
function Is_X   (arg : unresolved_sfixed) return boolean;
```

```vhdl
function To_X01  (arg : unresolved_ufixed) return unresolved_ufixed;
function To_X01  (arg : unresolved_sfixed) return unresolved_sfixed;
function To_X01Z (arg : unresolved_ufixed) return unresolved_ufixed;
function To_X01Z (arg : unresolved_sfixed) return unresolved_sfixed;
function To_UX01 (arg : unresolved_ufixed) return unresolved_ufixed;
function To_UX01 (arg : unresolved_sfixed) return unresolved_sfixed;

function To_SLV (arg : unresolved_ufixed) return std_logic_vector;

alias To_StdLogicVector    is To_SLV [unresolved_ufixed return std_logic_vector];
alias To_Std_Logic_Vector is To_SLV [unresolved_ufixed return std_logic_vector];

function To_SLV (arg : unresolved_sfixed) return std_logic_vector;

alias To_StdLogicVector    is To_SLV [unresolved_sfixed return std_logic_vector];
alias To_Std_Logic_Vector is To_SLV [unresolved_sfixed return std_logic_vector];

function To_SULV (arg : unresolved_ufixed) return std_ulogic_vector;

alias To_StdULogicVector    is To_SULV [unresolved_ufixed return std_ulogic_vector];
alias To_Std_ULogic_Vector is To_SULV [unresolved_ufixed return std_ulogic_vector];

function To_SULV (arg : unresolved_sfixed) return std_ulogic_vector;

alias To_StdULogicVector    is To_SULV [unresolved_sfixed return std_ulogic_vector];
alias To_Std_ULogic_Vector is To_SULV [unresolved_sfixed return std_ulogic_vector];

function to_ufixed (arg                         : std_ulogic_vector;
                    left_index, right_index : integer) return unresolved_ufixed;

function to_ufixed (arg     : std_ulogic_vector;
                    size_res : unresolved_ufixed) return unresolved_ufixed;

function to_sfixed (arg                         : std_ulogic_vector;
                    left_index, right_index : integer) return unresolved_sfixed;

function to_sfixed (arg     : std_ulogic_vector;
                    size_res : unresolved_sfixed) return unresolved_sfixed;

function to_UFix (arg             : std_ulogic_vector;
                  width, fraction : natural) return unresolved_ufixed;

function to_SFix (arg             : std_ulogic_vector;
                  width, fraction : natural) return unresolved_sfixed;

function UFix_high (width, fraction    : natural;
                    operation          : character := 'X';
                    width2, fraction2 : natural    := 0) return integer;

function UFix_low  (width, fraction    : natural;
                    operation          : character := 'X';
                    width2, fraction2 : natural    := 0) return integer;

function SFix_high (width, fraction    : natural;
                    operation          : character := 'X';
                    width2, fraction2 : natural    := 0) return integer;

function SFix_low  (width, fraction    : natural;
                    operation          : character := 'X';
                    width2, fraction2 : natural    := 0) return integer;

function to_string (value : unresolved_ufixed) return string;
function to_string (value : unresolved_sfixed) return string;

alias to_bstring is to_string [unresolved_ufixed return string];
alias to_bstring is to_string [unresolved_sfixed return string];

alias to_binary_string is to_string [unresolved_ufixed return string];
alias to_binary_string is to_string [unresolved_sfixed return string];
```

```
function to_ostring (value : unresolved_ufixed) return string;
function to_ostring (value : unresolved_sfixed) return string;

alias to_octal_string is to_ostring [unresolved_ufixed return string];
alias to_octal_string is to_ostring [unresolved_sfixed return string];

function to_hstring (value : unresolved_ufixed) return string;
function to_hstring (value : unresolved_sfixed) return string;

alias to_hex_string is to_hstring [unresolved_ufixed return string];
alias to_hex_string is to_hstring [unresolved_sfixed return string];

function from_string (bstring                     : string;
                      left_index, right_index : integer) return unresolved_ufixed;
function from_string (bstring                     : string;
                      left_index, right_index : integer) return unresolved_sfixed;
function from_string (bstring  : string;
                      size_res : unresolved_ufixed)        return unresolved_ufixed;
function from_string (bstring  : string;
                      size_res : unresolved_sfixed)        return unresolved_sfixed;
function from_string (bstring : string)                    return unresolved_ufixed;
function from_string (bstring : string)                    return unresolved_sfixed;

alias from_bstring is from_string [string, integer, integer return unresolved_ufixed];
alias from_bstring is from_string [string, integer, integer return unresolved_sfixed];
alias from_bstring is from_string [string, unresolved_ufixed
                                   return unresolved_ufixed];
alias from_bstring is from_string [string, unresolved_sfixed
                                   return unresolved_sfixed];
alias from_bstring is from_string [string return unresolved_ufixed];
alias from_bstring is from_string [string return unresolved_sfixed];

alias from_binary_string is from_string [string, integer, integer
                                   return unresolved_ufixed];
alias from_binary_string is from_string [string, integer, integer
                                   return unresolved_sfixed];
alias from_binary_string is from_string [string, unresolved_ufixed
                                   return unresolved_ufixed];
alias from_binary_string is from_string [string, unresolved_sfixed
                                   return unresolved_sfixed];
alias from_binary_string is from_string [string return unresolved_ufixed];
alias from_binary_string is from_string [string return unresolved_sfixed];

function from_ostring (ostring                     : string;
                       left_index, right_index : integer) return unresolved_ufixed;
function from_ostring (ostring                     : string;
                       left_index, right_index : integer) return unresolved_sfixed;
function from_ostring (ostring  : string;
                       size_res : unresolved_ufixed)        return unresolved_ufixed;
function from_ostring (ostring  : string;
                       size_res : unresolved_sfixed)        return unresolved_sfixed;
function from_ostring (ostring : string)                    return unresolved_ufixed;
function from_ostring ( ostring : string)                   return unresolved_sfixed;

alias from_octal_string is from_ostring [string, integer, integer
                                   return unresolved_ufixed];
alias from_octal_string is from_ostring [string, integer, integer
                                   return unresolved_sfixed];
alias from_octal_string is from_ostring [string, unresolved_ufixed
                                   return unresolved_ufixed];
alias from_octal_string is from_ostring [string, unresolved_sfixed
                                   return unresolved_sfixed];
alias from_octal_string is from_ostring [string return unresolved_ufixed];
alias from_octal_string is from_ostring [string return unresolved_sfixed];
```

```
function from_hstring (hstring                    : string;
                       left_index, right_index : integer) return unresolved_ufixed;
function from_hstring (hstring                    : string;
                       left_index, right_index : integer) return unresolved_sfixed;
function from_hstring (hstring  : string;
                       size_res : unresolved_ufixed)        return unresolved_ufixed;
function from_hstring (hstring  : string;
                       size_res : unresolved_sfixed)        return unresolved_sfixed;
function from_hstring (hstring : string)                     return unresolved_ufixed;
function from_hstring (hstring : string)                     return unresolved_sfixed;

alias from_hex_string is from_hstring [string, integer, integer
                                        return unresolved_ufixed];
alias from_hex_string is from_hstring [string, integer, integer
                                        return unresolved_sfixed];
alias from_hex_string is from_hstring [string, unresolved_ufixed
                                        return unresolved_ufixed];
alias from_hex_string is from_hstring [string, unresolved_sfixed
                                        return unresolved_sfixed];
alias from_hex_string is from_hstring [string return unresolved_ufixed];
alias from_hex_string is from_hstring [string return unresolved_sfixed];

procedure read(l : inout line; value : out unresolved_ufixed; good : out boolean);
procedure read(l : inout line; value : out unresolved_ufixed);
procedure read(l : inout line; value : out unresolved_sfixed; good : out boolean);
procedure read(l : inout line; value : out unresolved_sfixed);

alias bread is read [line, unresolved_ufixed, boolean];
alias bread is read [line, unresolved_ufixed];
alias bread is read [line, unresolved_sfixed, boolean];
alias bread is read [line, unresolved_sfixed];

alias binary_read is read [line, unresolved_ufixed, boolean];
alias binary_read is read [line, unresolved_ufixed];
alias binary_read is read [line, unresolved_sfixed, boolean];
alias binary_read is read [line, unresolved_sfixed];

procedure oread(l : inout line; value : out unresolved_ufixed; good : out boolean);
procedure oread(l : inout line; value : out unresolved_ufixed);
procedure oread(l : inout line; value : out unresolved_sfixed; good : out boolean);
procedure oread(l : inout line; value : out unresolved_sfixed);

alias octal_read is oread [line, unresolved_ufixed, boolean];
alias octal_read is oread [line, unresolved_ufixed];
alias octal_read is oread [line, unresolved_sfixed, boolean];
alias octal_read is oread [line, unresolved_sfixed];

procedure hread(l : inout line; value : out unresolved_ufixed; good : out boolean);
procedure hread(l : inout line; value : out unresolved_ufixed);
procedure hread(l : inout line; value : out unresolved_sfixed; good : out boolean);
procedure hread(l : inout line; value : out unresolved_sfixed);

alias hex_read is hread [line, unresolved_ufixed, boolean];
alias hex_read is hread [line, unresolved_ufixed];
alias hex_read is hread [line, unresolved_sfixed, boolean];
alias hex_read is hread [line, unresolved_sfixed];

procedure write (l : inout line; value : in unresolved_ufixed;
                 justified : in side := right; field : in width := 0);
procedure write (l : inout line; value : in unresolved_sfixed;
                 justified : in side := right; field : in width := 0);

alias bwrite is write [line, unresolved_ufixed, side, width];
alias bwrite is write [line, unresolved_sfixed, side, width];
```

```
    alias binary_write is write [line, unresolved_ufixed, side, width];
    alias binary_write is write [line, unresolved_sfixed, side, width];

    procedure owrite (l : inout line; value : in unresolved_ufixed;
                      justified : in side := right; field : in width := 0);
    procedure owrite (l : inout line; value : in unresolved_sfixed;
                      justified : in side := right; field : in width := 0);

    alias octal_write is owrite [line, unresolved_ufixed, side, width];
    alias octal_write is owrite [line, unresolved_sfixed, side, width];

    procedure hwrite (l : inout line; value : in unresolved_ufixed;
                      justified : in side := right; field : in width := 0);
    procedure hwrite (l : inout line; value : in unresolved_sfixed;
                      justified : in side := right; field : in width := 0);

    alias hex_write is hwrite [line, unresolved_ufixed, side, width];
    alias hex_write is hwrite [line, unresolved_sfixed, side, width];

  end package fixed_generic_pkg;
```

A.7.3   The **fixed_pkg** Package

```
library ieee;

package fixed_pkg is new ieee.fixed_generic_pkg
  generic map (
    fixed_round_style    => ieee.fixed_float_types.fixed_round,
    fixed_overflow_style => ieee.fixed_float_types.fixed_saturate,
    fixed_guard_bits     => 3,
    no_warning           => false
    );
```

A.8   Standard Floating-Point Packages

The types of generic constants in the standard floating-point packages are defined in the package fixed_float_types (see Section A.7.1).

VHDL-87, -93, and -2002

The standard floating-point packages are not provided in these versions.

A.8.1   The **float_generic_pkg** Package

```
use std.textio.all;
library ieee; use ieee.std_logic_1164.all; use ieee.numeric_std.all;
use ieee.fixed_float_types.all;

package float_generic_pkg is
  generic (
    float_exponent_width : natural    := 8;
    float_fraction_width : natural    := 23;
    float_round_style    : round_type := round_nearest;
    float_denormalize    : boolean    := true;
    float_check_error    : boolean    := true;
    float_guard_bits     : natural    := 3;
```

```
    no_warning            : boolean    := false;
    package fixed_pkg is new ieee.fixed_generic_pkg generic map (<>) );

-- Author David Bishop (dbishop@vhdl.org)
constant CopyRightNotice : string :=
  "Copyright 2008 by IEEE. All rights reserved.";

use fixed_pkg.all;

type unresolved_float is array (integer range <>) of std_ulogic;
alias   u_float is             unresolved_float;
subtype float    is (resolved) unresolved_float;

subtype unresolved_float32 is unresolved_float (8 downto -23);
alias   u_float32 is unresolved_float32;
subtype float32   is float (8 downto -23);

subtype unresolved_float64 is unresolved_float (11 downto -52);
alias   u_float64 is unresolved_float64;
subtype float64   is float (11 downto -52);

subtype unresolved_float128 is unresolved_float (15 downto -112);
alias   u_float128 is unresolved_float128;
subtype float128   is float (15 downto -112);

type valid_fpstate is (nan, quiet_nan,
                       neg_inf, neg_normal, neg_denormal, neg_zero,
                       pos_zero, pos_denormal, pos_normal, pos_inf,
                       isx);

constant fphdlsynth_or_real : boolean;

function Classfp (x           : unresolved_float;
                  check_error : boolean := float_check_error) return valid_fpstate;

function "abs" (arg : unresolved_float)  return unresolved_float;
function "-"   (arg : unresolved_float)  return unresolved_float;

function "+"   (l, r : unresolved_float)                  return unresolved_float;
function "+"   (l : unresolved_float; r : real)           return unresolved_float;
function "+"   (l : real; r : unresolved_float)           return unresolved_float;
function "+"   (l : unresolved_float; r : integer)        return unresolved_float;
function "+"   (l : integer; r : unresolved_float)        return unresolved_float;

function "-"   (l, r : unresolved_float)                  return unresolved_float;
function "-"   (l : unresolved_float; r : real)           return unresolved_float;
function "-"   (l : real; r : unresolved_float)           return unresolved_float;
function "-"   (l : unresolved_float; r : integer)        return unresolved_float;
function "-"   (l : integer; r : unresolved_float)        return unresolved_float;

function "*"   (l, r : unresolved_float)                  return unresolved_float;
function "*"   (l : unresolved_float; r : real)           return unresolved_float;
function "*"   (l : real; r : unresolved_float)           return unresolved_float;
function "*"   (l : unresolved_float; r : integer)        return unresolved_float;
function "*"   (l : integer; r : unresolved_float)        return unresolved_float;

function "/"   (l, r : unresolved_float)                  return unresolved_float;
function "/"   (l : unresolved_float; r : real)           return unresolved_float;
function "/"   (l : real; r : unresolved_float)           return unresolved_float;
function "/"   (l : unresolved_float; r : integer)        return unresolved_float;
function "/"   (l : integer; r : unresolved_float)        return unresolved_float;

function "rem" (l, r : unresolved_float)                  return unresolved_float;
function "rem" (l : unresolved_float; r : real)           return unresolved_float;
function "rem" (l : real; r : unresolved_float)           return unresolved_float;
function "rem" (l : unresolved_float; r : integer)        return unresolved_float;
function "rem" (l : integer; r : unresolved_float)        return unresolved_float;
```

```
function "mod" (l, r : unresolved_float)              return unresolved_float;
function "mod" (l : unresolved_float; r : real)       return unresolved_float;
function "mod" (l : real; r : unresolved_float)       return unresolved_float;
function "mod" (l : unresolved_float; r : integer) return unresolved_float;
function "mod" (l : integer; r : unresolved_float) return unresolved_float;

function add (l, r        : unresolved_float;
              round_style : round_type := float_round_style;
              guard       : natural    := float_guard_bits;
              check_error : boolean     := float_check_error;
              denormalize : boolean     := float_denormalize)
                 return unresolved_float;

function subtract (l, r        : unresolved_float;
                   round_style : round_type := float_round_style;
                   guard       : natural    := float_guard_bits;
                   check_error : boolean     := float_check_error;
                   denormalize : boolean     := float_denormalize)
                      return unresolved_float;

function multiply (l, r        : unresolved_float;
                   round_style : round_type := float_round_style;
                   guard       : natural    := float_guard_bits;
                   check_error : boolean     := float_check_error;
                   denormalize : boolean     := float_denormalize)
                      return unresolved_float;

function divide (l, r        : unresolved_float;
                 round_style : round_type := float_round_style;
                 guard       : natural    := float_guard_bits;
                 check_error : boolean     := float_check_error;
                 denormalize : boolean     := float_denormalize)
                    return unresolved_float;

function remainder (l, r        : unresolved_float;
                    round_style : round_type := float_round_style;
                    guard       : natural    := float_guard_bits;
                    check_error : boolean     := float_check_error;
                    denormalize : boolean     := float_denormalize)
                       return unresolved_float;

function modulo (l, r        : unresolved_float;
                 round_style : round_type := float_round_style;
                 guard       : natural    := float_guard_bits;
                 check_error : boolean     := float_check_error;
                 denormalize : boolean     := float_denormalize)
                    return unresolved_float;

function reciprocal (arg         : unresolved_float;
                     round_style : round_type := float_round_style;
                     guard       : natural    := float_guard_bits;
                     check_error : boolean     := float_check_error;
                     denormalize : boolean     := float_denormalize)
                        return unresolved_float;

function dividebyp2 (l, r        : unresolved_float;
                     round_style : round_type := float_round_style;
                     guard       : natural    := float_guard_bits;
                     check_error : boolean     := float_check_error;
                     denormalize : boolean     := float_denormalize)
                        return unresolved_float;

function mac (l, r, c    : unresolved_float;
              round_style : round_type := float_round_style;
              guard       : natural    := float_guard_bits;
```

```
                    check_error : boolean      := float_check_error;
                    denormalize : boolean      := float_denormalize)
                return unresolved_float;

function sqrt (arg           : unresolved_float;
              round_style : round_type := float_round_style;
              guard       : natural    := float_guard_bits;
              check_error : boolean    := float_check_error;
              denormalize : boolean    := float_denormalize)
          return unresolved_float;

function Is_Negative (arg : unresolved_float) return boolean;

function "="  (l, r : unresolved_float)                  return boolean;
function "="  (l : unresolved_float; r : real)           return boolean;
function "="  (l : real; r : unresolved_float)           return boolean;
function "="  (l : unresolved_float; r : integer)        return boolean;
function "="  (l : integer; r : unresolved_float)        return boolean;

function "/=" (l, r : unresolved_float)                  return boolean;
function "/=" (l : unresolved_float; r : real)           return boolean;
function "/=" (l : real; r : unresolved_float)           return boolean;
function "/=" (l : unresolved_float; r : integer)        return boolean;
function "/=" (l : integer; r : unresolved_float)        return boolean;

function "<"  (l, r : unresolved_float)                  return boolean;
function "<"  (l : unresolved_float; r : real)           return boolean;
function "<"  (l : real; r : unresolved_float)           return boolean;
function "<"  (l : unresolved_float; r : integer)        return boolean;
function "<"  (l : integer; r : unresolved_float)        return boolean;

function "<=" (l, r : unresolved_float)                  return boolean;
function "<=" (l : unresolved_float; r : real)           return boolean;
function "<=" (l : real; r : unresolved_float)           return boolean;
function "<=" (l : unresolved_float; r : integer)        return boolean;
function "<=" (l : integer; r : unresolved_float)        return boolean;

function ">"  (l, r : unresolved_float)                  return boolean;
function ">"  (l : unresolved_float; r : real)           return boolean;
function ">"  (l : real; r : unresolved_float)           return boolean;
function ">"  (l : unresolved_float; r : integer)        return boolean;
function ">"  (l : integer; r : unresolved_float)        return boolean;

function ">=" (l, r : unresolved_float)                  return boolean;
function ">=" (l : unresolved_float; r : real)           return boolean;
function ">=" (l : real; r : unresolved_float)           return boolean;
function ">=" (l : unresolved_float; r : integer)        return boolean;
function ">=" (l : integer; r : unresolved_float)        return boolean;

function eq (l, r          : unresolved_float;
             check_error : boolean := float_check_error;
             denormalize : boolean := float_denormalize)
          return boolean;

function ne (l, r          : unresolved_float;
             check_error : boolean := float_check_error;
             denormalize : boolean := float_denormalize)
          return boolean;

function lt (l, r          : unresolved_float;
             check_error : boolean := float_check_error;
             denormalize : boolean := float_denormalize)
          return boolean;

function gt (l, r          : unresolved_float;
             check_error : boolean := float_check_error;
```

```
                          denormalize : boolean := float_denormalize)
                      return boolean;

      function le (l, r        : unresolved_float;
                   check_error : boolean := float_check_error;
                   denormalize : boolean := float_denormalize)
                   return boolean;

      function ge (l, r        : unresolved_float;
                   check_error : boolean := float_check_error;
                   denormalize : boolean := float_denormalize)
                   return boolean;

      function minimum (l, r : unresolved_float)               return unresolved_float;
      function minimum (l : unresolved_float; r : real)        return unresolved_float;
      function minimum (l : real; r : unresolved_float)        return unresolved_float;
      function minimum (l : unresolved_float; r : integer)     return unresolved_float;
      function minimum (l : integer; r : unresolved_float)     return unresolved_float;

      function maximum (l, r : unresolved_float)               return unresolved_float;
      function maximum (l : unresolved_float; r : real)        return unresolved_float;
      function maximum (l : real; r : unresolved_float)        return unresolved_float;
      function maximum (l : unresolved_float; r : integer)     return unresolved_float;
      function maximum (l : integer; r : unresolved_float)     return unresolved_float;

      function "?="  (l, r : unresolved_float)            return boolean;
      function "?="  (l : unresolved_float; r : real)     return boolean;
      function "?="  (l : real; r : unresolved_float)     return boolean;
      function "?="  (l : unresolved_float; r : integer)  return boolean;
      function "?="  (l : integer; r : unresolved_float)  return boolean;

      function "?/=" (l, r : unresolved_float)            return boolean;
      function "?/=" (l : unresolved_float; r : real)     return boolean;
      function "?/=" (l : real; r : unresolved_float)     return boolean;
      function "?/=" (l : unresolved_float; r : integer)  return boolean;
      function "?/=" (l : integer; r : unresolved_float)  return boolean;

      function "?<"  (l, r : unresolved_float)            return boolean;
      function "?<"  (l : unresolved_float; r : real)     return boolean;
      function "?<"  (l : real; r : unresolved_float)     return boolean;
      function "?<"  (l : unresolved_float; r : integer)  return boolean;
      function "?<"  (l : integer; r : unresolved_float)  return boolean;

      function "?<=" (l, r : unresolved_float)            return boolean;
      function "?<=" (l : unresolved_float; r : real)     return boolean;
      function "?<=" (l : real; r : unresolved_float)     return boolean;
      function "?<=" (l : unresolved_float; r : integer)  return boolean;
      function "?<=" (l : integer; r : unresolved_float)  return boolean;

      function "?>"  (l, r : unresolved_float)            return boolean;
      function "?>"  (l : unresolved_float; r : real)     return boolean;
      function "?>"  (l : real; r : unresolved_float)     return boolean;
      function "?>"  (l : unresolved_float; r : integer)  return boolean;
      function "?>"  (l : integer; r : unresolved_float)  return boolean;

      function "?>=" (l, r : unresolved_float)            return boolean;
      function "?>=" (l : unresolved_float; r : real)     return boolean;
      function "?>=" (l : real; r : unresolved_float)     return boolean;
      function "?>=" (l : unresolved_float; r : integer)  return boolean;
      function "?>=" (l : integer; r : unresolved_float)  return boolean;

      function std_match (l, r : unresolved_float) return boolean;

      function "and"  (l, r : unresolved_float) return unresolved_float;
      function "nand" (l, r : unresolved_float) return unresolved_float;
      function "or"   (l, r : unresolved_float) return unresolved_float;
```

```
function "nor"  (l, r : unresolved_float) return unresolved_float;
function "xor"  (l, r : unresolved_float) return unresolved_float;
function "xnor" (l, r : unresolved_float) return unresolved_float;
function "not"  (l    : unresolved_float) return unresolved_float;

function "and"  (l : std_ulogic; r : unresolved_float) return unresolved_float;
function "and"  (l : unresolved_float; r : std_ulogic) return unresolved_float;
function "nand" (l : std_ulogic; r : unresolved_float) return unresolved_float;
function "nand" (l : unresolved_float; r : std_ulogic) return unresolved_float;
function "or"   (l : std_ulogic; r : unresolved_float) return unresolved_float;
function "or"   (l : unresolved_float; r : std_ulogic) return unresolved_float;
function "nor"  (l : std_ulogic; r : unresolved_float) return unresolved_float;
function "nor"  (l : unresolved_float; r : std_ulogic) return unresolved_float;
function "xor"  (l : std_ulogic; r : unresolved_float) return unresolved_float;
function "xor"  (l : unresolved_float; r : std_ulogic) return unresolved_float;
function "xnor" (l : std_ulogic; r : unresolved_float) return unresolved_float;
function "xnor" (l : unresolved_float; r : std_ulogic) return unresolved_float;

function "and"  (l : unresolved_float) return std_ulogic;
function "nand" (l : unresolved_float) return std_ulogic;
function "or"   (l : unresolved_float) return std_ulogic;
function "nor"  (l : unresolved_float) return std_ulogic;
function "xor"  (l : unresolved_float) return std_ulogic;
function "xnor" (l : unresolved_float) return std_ulogic;

function find_leftmost  (arg : unresolved_float; y : std_ulogic) return integer;
function find_rightmost (arg : unresolved_float; y : std_ulogic) return integer;

function resize (arg              : unresolved_float;
                 exponent_width : natural      := float_exponent_width;
                 fraction_width : natural      := float_fraction_width;
                 round_style    : round_type := float_round_style;
                 check_error    : boolean    := float_check_error;
                 denormalize_in : boolean    := float_denormalize;
                 denormalize    : boolean    := float_denormalize)
            return unresolved_float;

function resize (arg              : unresolved_float;
                 size_res        : unresolved_float;
                 round_style    : round_type := float_round_style;
                 check_error    : boolean    := float_check_error;
                 denormalize_in : boolean    := float_denormalize;
                 denormalize    : boolean    := float_denormalize)
            return unresolved_float;

function to_float32 (arg             : unresolved_float;
                 round_style    : round_type := float_round_style;
                 check_error    : boolean    := float_check_error;
                 denormalize_in : boolean    := float_denormalize;
                 denormalize    : boolean    := float_denormalize)
            return unresolved_float32;

function to_float64 (arg             : unresolved_float;
                 round_style    : round_type := float_round_style;
                 check_error    : boolean    := float_check_error;
                 denormalize_in : boolean    := float_denormalize;
                 denormalize    : boolean    := float_denormalize)
            return unresolved_float64;

function to_float128 (arg            : unresolved_float;
                 round_style    : round_type := float_round_style;
                 check_error    : boolean    := float_check_error;
                 denormalize_in : boolean    := float_denormalize;
                 denormalize    : boolean    := float_denormalize)
            return unresolved_float128;
```

```vhdl
function To_SLV (arg : unresolved_float) return std_logic_vector;

alias To_StdLogicVector   is To_SLV [unresolved_float return std_logic_vector];
alias To_Std_Logic_Vector is To_SLV [unresolved_float return std_logic_vector];

function To_SULV (arg : unresolved_float) return std_ulogic_vector;

alias To_StdULogicVector   is To_SULV [unresolved_float return std_ulogic_vector];
alias To_Std_ULogic_Vector is To_SULV [unresolved_float return std_ulogic_vector];

function to_float (arg           : std_ulogic_vector;
                   exponent_width : natural := float_exponent_width;
                   fraction_width : natural := float_fraction_width)
                  return unresolved_float;

function to_float (arg           : integer;
                   exponent_width : natural    := float_exponent_width;
                   fraction_width : natural    := float_fraction_width;
                   round_style    : round_type := float_round_style)
                  return unresolved_float;

function to_float (arg           : real;
                   exponent_width : natural    := float_exponent_width;
                   fraction_width : natural    := float_fraction_width;
                   round_style    : round_type := float_round_style;
                   denormalize    : boolean    := float_denormalize)
                  return unresolved_float;

function to_float (arg           : unresolved_unsigned;
                   exponent_width : natural    := float_exponent_width;
                   fraction_width : natural    := float_fraction_width;
                   round_style    : round_type := float_round_style)
                  return unresolved_float;

function to_float (arg           : unresolved_signed;
                   exponent_width : natural    := float_exponent_width;
                   fraction_width : natural    := float_fraction_width;
                   round_style    : round_type := float_round_style)
                  return unresolved_float;

function to_float (arg           : unresolved_ufixed;
                   exponent_width : natural    := float_exponent_width;
                   fraction_width : natural    := float_fraction_width;
                   round_style    : round_type := float_round_style;
                   denormalize    : boolean    := float_denormalize)
                  return unresolved_float;

function to_float (arg           : unresolved_sfixed;
                   exponent_width : natural    := float_exponent_width;
                   fraction_width : natural    := float_fraction_width;
                   round_style    : round_type := float_round_style;
                   denormalize    : boolean    := float_denormalize)
                  return unresolved_float;

function to_float (arg        : integer;
                   size_res   : unresolved_float;
                   round_style : round_type := float_round_style)
                  return unresolved_float;

function to_float (arg        : real;
                   size_res   : unresolved_float;
                   round_style : round_type := float_round_style;
                   denormalize : boolean    := float_denormalize)
                  return unresolved_float;

function to_float (arg        : unresolved_unsigned;
                   size_res   : unresolved_float;
```

```vhdl
                    round_style : round_type := float_round_style)
                    return unresolved_float;

  function to_float (arg         : unresolved_signed;
                     size_res    : unresolved_float;
                     round_style : round_type := float_round_style)
                     return unresolved_float;

  function to_float (arg      : std_ulogic_vector;
                     size_res : unresolved_float)
                     return unresolved_float;

  function to_float (arg         : unresolved_ufixed;
                     size_res    : unresolved_float;
                     round_style : round_type := float_round_style;
                     denormalize : boolean    := float_denormalize)
                     return unresolved_float;

  function to_float (arg         : unresolved_sfixed;
                     size_res    : unresolved_float;
                     round_style : round_type := float_round_style;
                     denormalize : boolean    := float_denormalize)
                     return unresolved_float;

  function to_unsigned (arg         : unresolved_float;
                        size        : natural;
                        round_style : round_type := float_round_style;
                        check_error : boolean    := float_check_error)
                        return unresolved_unsigned;

  function to_signed (arg         : unresolved_float;
                      size        : natural;
                      round_style : round_type := float_round_style;
                      check_error : boolean    := float_check_error)
                      return unresolved_signed;

  function to_ufixed (arg            : unresolved_float;
                      left_index     : integer;
                      right_index    : integer;
                      overflow_style : fixed_overflow_style_type := fixed_overflow_style;
                      round_style    : fixed_round_style_type    := fixed_round_style;
                      check_error    : boolean                   := float_check_error;
                      denormalize    : boolean                   := float_denormalize)
                      return unresolved_ufixed;

  function to_sfixed (arg            : unresolved_float;
                      left_index     : integer;
                      right_index    : integer;
                      overflow_style : fixed_overflow_style_type := fixed_overflow_style;
                      round_style    : fixed_round_style_type    := fixed_round_style;
                      check_error    : boolean                   := float_check_error;
                      denormalize    : boolean                   := float_denormalize)
                      return unresolved_sfixed;

  function to_unsigned (arg         : unresolved_float;
                        size_res    : unresolved_unsigned;
                        round_style : round_type := float_round_style;
                        check_error : boolean    := float_check_error)
                        return unresolved_unsigned;

  function to_signed (arg         : unresolved_float;
                      size_res    : unresolved_signed;
                      round_style : round_type := float_round_style;
                      check_error : boolean    := float_check_error)
                      return unresolved_signed;
```

```
function to_ufixed (arg             : unresolved_float;
                    size_res        : unresolved_ufixed;
                    overflow_style : fixed_overflow_style_type := fixed_overflow_style;
                    round_style    : fixed_round_style_type    := fixed_round_style;
                    check_error    : boolean                   := float_check_error;
                    denormalize    : boolean                   := float_denormalize)
              return unresolved_ufixed;

function to_sfixed (arg             : unresolved_float;
                    size_res        : unresolved_sfixed;
                    overflow_style : fixed_overflow_style_type := fixed_overflow_style;
                    round_style    : fixed_round_style_type    := fixed_round_style;
                    check_error    : boolean                   := float_check_error;
                    denormalize    : boolean                   := float_denormalize)
              return unresolved_sfixed;

function to_real (arg         : unresolved_float;
                  check_error : boolean := float_check_error;
                  denormalize : boolean := float_denormalize)
              return real;

function to_integer (arg         : unresolved_float;
                     round_style : round_type := float_round_style;
                     check_error : boolean    := float_check_error)
              return integer;

function realtobits (arg : real)              return std_ulogic_vector;
function bitstoreal (arg : std_ulogic_vector) return real;

function To_01 (arg  : unresolved_float;
                xmap : std_logic := '0') return unresolved_float;

function Is_X    (arg : unresolved_float) return boolean;

function To_X01  (arg : unresolved_float) return unresolved_float;
function To_X01Z (arg : unresolved_float) return unresolved_float;
function To_UX01 (arg : unresolved_float) return unresolved_float;

procedure break_number (arg         : in  unresolved_float;
                        denormalize : in  boolean := float_denormalize;
                        check_error : in  boolean := float_check_error;
                        fract       : out unresolved_unsigned;
                        expon       : out unresolved_signed;
                        sign        : out std_ulogic);

procedure break_number (arg         : in  unresolved_float;
                        denormalize : in  boolean := float_denormalize;
                        check_error : in  boolean := float_check_error;
                        fract       : out unresolved_ufixed;
                        expon       : out unresolved_signed;
                        sign        : out std_ulogic);

function normalize (fract          : unresolved_unsigned;
                    expon          : unresolved_signed;
                    sign           : std_ulogic;
                    sticky         : std_ulogic := '0';
                    exponent_width : natural    := float_exponent_width;
                    fraction_width : natural    := float_fraction_width;
                    round_style    : round_type := float_round_style;
                    denormalize    : boolean    := float_denormalize;
                    nguard         : natural    := float_guard_bits)
              return unresolved_float;

function normalize (fract          : unresolved_ufixed;
                    expon          : unresolved_signed;
                    sign           : std_ulogic;
```

```
                        sticky          : std_ulogic := '0';
                        exponent_width : natural     := float_exponent_width;
                        fraction_width : natural     := float_fraction_width;
                        round_style    : round_type := float_round_style;
                        denormalize    : boolean     := float_denormalize;
                        nguard         : natural     := float_guard_bits)
                     return unresolved_float;

     function normalize (fract       : unresolved_unsigned;
                         expon       : unresolved_signed;
                         sign        : std_ulogic;
                         sticky      : std_ulogic := '0';
                         size_res    : unresolved_float;
                         round_style : round_type := float_round_style;
                         denormalize : boolean     := float_denormalize;
                         nguard      : natural     := float_guard_bits)
                     return unresolved_float;

     function normalize (fract       : unresolved_ufixed;
                         expon       : unresolved_signed;
                         sign        : std_ulogic;
                         sticky      : std_ulogic := '0';
                         size_res    : unresolved_float;
                         round_style : round_type := float_round_style;
                         denormalize : boolean     := float_denormalize;
                         nguard      : natural     := float_guard_bits)
                     return unresolved_float;

  function Copysign (x, y : unresolved_float) return unresolved_float;

  function Scalb (y           : unresolved_float;
                 n           : integer;
                 round_style : round_type := float_round_style;
                 check_error : boolean     := float_check_error;
                 denormalize : boolean     := float_denormalize)
              return unresolved_float;

  function Scalb (y           : unresolved_float;
                 n           : unresolved_signed;
                 round_style : round_type := float_round_style;
                 check_error : boolean     := float_check_error;
                 denormalize : boolean     := float_denormalize)
              return unresolved_float;

  function Logb (x : unresolved_float) return integer;
  function Logb (x : unresolved_float) return unresolved_signed;

  function Nextafter (x, y        : unresolved_float;
                     check_error : boolean := float_check_error;
                     denormalize : boolean := float_denormalize)
                  return unresolved_float;

  function Unordered (x, y : unresolved_float) return boolean;
  function Finite    (x    : unresolved_float) return boolean;
  function Isnan     (x    : unresolved_float) return boolean;

  function zerofp (exponent_width : natural := float_exponent_width;
    fraction_width : natural := float_fraction_width)
    return unresolved_float;

  function nanfp (exponent_width : natural := float_exponent_width;
                 fraction_width : natural := float_fraction_width)
              return unresolved_float;
```

```
function qnanfp (exponent_width : natural := float_exponent_width;
                 fraction_width : natural := float_fraction_width)
            return unresolved_float;

function pos_inffp (exponent_width : natural := float_exponent_width;
                    fraction_width : natural := float_fraction_width)
               return unresolved_float;

function neg_inffp (exponent_width : natural := float_exponent_width;
                    fraction_width : natural := float_fraction_width)
               return unresolved_float;

function neg_zerofp (exponent_width : natural := float_exponent_width;
                     fraction_width : natural := float_fraction_width)
                return unresolved_float;

function zerofp     (size_res : unresolved_float) return unresolved_float;
function nanfp      (size_res : unresolved_float) return unresolved_float;
function qnanfp     (size_res : unresolved_float) return unresolved_float;
function pos_inffp  (size_res : unresolved_float) return unresolved_float;
function neg_inffp  (size_res : unresolved_float) return unresolved_float;
function neg_zerofp (size_res : unresolved_float) return unresolved_float;

function to_string (value : unresolved_float) return string;

alias to_bstring      is to_string [unresolved_float return string];
alias to_binary_string is to_string [unresolved_float return string];

function to_hstring (value : unresolved_float) return string;

alias to_hex_string is to_hstring [unresolved_float return string];

function to_ostring (value : unresolved_float) return string;

alias to_octal_string is to_ostring [unresolved_float return string];

function from_string (bstring        : string;
                      exponent_width : natural := float_exponent_width;
                      fraction_width : natural := float_fraction_width)
                 return unresolved_float;
function from_string (bstring  : string;
                      size_res : unresolved_float) return unresolved_float;

alias from_bstring is from_string [string, natural, natural
                                   return unresolved_float];
alias from_bstring is from_string [string, unresolved_float
                                   return unresolved_float];

alias from_binary_string is from_string [string, natural, natural
                                         return unresolved_float];
alias from_binary_string is from_string [string, unresolved_float
                                         return unresolved_float];

function from_ostring (ostring        : string;
                       exponent_width : natural := float_exponent_width;
                       fraction_width : natural := float_fraction_width)
                  return unresolved_float;
function from_ostring (ostring  : string;
                       size_res : unresolved_float) return unresolved_float;

alias from_octal_string is from_ostring [string, natural, natural
                                         return unresolved_float];
alias from_octal_string is from_ostring [string, unresolved_float
                                         return unresolved_float];

function from_hstring (hstring        : string;
                       exponent_width : natural := float_exponent_width;
                       fraction_width : natural := float_fraction_width)
```

```
                                        return unresolved_float;
          function from_hstring (hstring  : string;
                                 size_res : unresolved_float) return unresolved_float;

          alias from_hex_string is from_hstring [string, natural, natural
                                              return unresolved_float];
          alias from_hex_string is from_hstring [string, unresolved_float
                                              return unresolved_float];

          procedure read (l : inout line; value : out unresolved_float; good : out boolean);
          procedure read (l : inout line; value : out unresolved_float);

          alias bread is read [line, unresolved_float, boolean];
          alias bread is read [line, unresolved_float];

          alias binary_read is read [line, unresolved_float, boolean];
          alias binary_read is read [line, unresolved_float];

          procedure oread (l : inout line; value : out unresolved_float; good : out boolean);
          procedure oread (l : inout line; value : out unresolved_float);

          alias octal_read is oread [line, unresolved_float, boolean];
          alias octal_read is oread [line, unresolved_float];

          procedure hread (l : inout line; value : out unresolved_float; good : out boolean);
          procedure hread (l : inout line; value : out unresolved_float);

          alias hex_read is hread [line, unresolved_float, boolean];
          alias hex_read is hread [line, unresolved_float];

          procedure write (l : inout line; value : in unresolved_float;
                          justified : in side := right; field : in width := 0);

          alias bwrite       is write [line, unresolved_float, side, width];
          alias binary_write is write [line, unresolved_float, side, width];

          procedure owrite (l : inout line; value : in unresolved_float;
                           justified : in side := right; field : in width := 0);

          alias octal_write is owrite [line, unresolved_float, side, width];

          procedure hwrite (l : inout line; value : in unresolved_float;
                           justified : in side := right; field : in width := 0);

          alias hex_write is hwrite [line, unresolved_float, side, width];

       end package float_generic_pkg;
```

A.8.2   The **float_pkg** Package

```
       library ieee;

       package float_pkg is new ieee.float_generic_pkg
         generic map (
           float_exponent_width => 8,
           float_fraction_width => 23,
           float_round_style    => ieee.fixed_float_types.round_nearest,
           float_denormalize    => true,
           float_check_error    => true,
           float_guard_bits     => 3,
           no_warning           => false,
           fixed_pkg            => ieee.fixed_pkg
           );
```

Appendix B

VHDL Syntax

In this appendix we present the full set of syntax rules for VHDL using the EBNF notation introduced in Chapter 1. The form of EBNF used in this book differs from that of the VHDL Language Reference Manual (LRM) in order to make the syntax rules more intelligible to the VHDL user. The LRM includes a separate syntax rule for each minor syntactic category. In this book, we condense the grammar into a smaller number of rules, each of which defines a larger part of the grammar. We introduce the EBNF symbols "⟮", "⟯" and "◦◦◦" as part of this simplification. Our aim is to avoid the large amount of searching required when using the LRM rules to resolve a question of grammar.

Those parts of the syntax rules that were introduced in VHDL-2008 are underlined in this appendix. A model written using earlier versions of the language may not use these features. In addition, there are some entirely new rules introduced in VHDL-2008 that have no predecessors in earlier versions. We identify these rules by underlining the rule name on the left-hand side of the "⇐"symbol.

Some of the rules refer to the syntax rules for the PSL. Such references are identified by the italicized prefix "*PSL_*". The PSL syntax rules are not included here. The interested reader should refer to the PSL standard or to books on PSL for details.

Index to Syntax Rules

B.1 Design File

design_file ⇐ design_unit {{ ⚬⚬⚬ }}

design_unit ⇐
    { library_clause I use_clause I context_reference }
    library_unit

library_unit ⇐
    entity_declaration   I architecture_body
    I package_declaration   I package_body
    I package_instantiation_declaration
    I configuration_declaration   I context_declaration
    I *PSL*_Verification_Unit

library_clause ⇐ **library** identifier {{ , ⚬⚬⚬ }} ;

context_reference ⇐ **context** selected_name {{ , ⚬⚬⚬ }} ;

B.2 Library Unit Declarations

entity_declaration ⇐
    entity identifier **is**
        [**generic** (*generic*_interface_list) ;]
        [**port** (*port*_interface_list) ;]
        { entity_declarative_item }
    [**begin**
        { concurrent_assertion_statement
        I *passive*_concurrent_procedure_call_statement
        I *passive*_process_statement }]
        I *PSL*_PSL_Directive
    end [**entity**] [identifier] ;

entity_declarative_item ⇐
    subprogram_declaration   ⎮ subprogram_body
    ⎮ subprogram_instantiation_declaration
    ⎮ package_declaration   ⎮ package_body
    ⎮ package_instantiation_declaration
    ⎮ type_declaration   ⎮ subtype_declaration
    ⎮ constant_declaration   ⎮ signal_declaration
    ⎮ *shared*_variable_declaration   ⎮ file_declaration
    ⎮ alias_declaration
    ⎮ attribute_declaration   ⎮ attribute_specification
    ⎮ disconnection_specification   ⎮ use_clause
    ⎮ group_template_declaration   ⎮ group_declaration
    ⎮ *PSL*_Property_Declaration   ⎮ *PSL*_Sequence_Declaration
    ⎮ *PSL*_Clock_Declaration

architecture_body ⇐
    architecture identifier **of** *entity*_name **is**
        ❴ block_declarative_item ❵
    begin
        ❴ concurrent_statement ❵
    end ⟦ **architecture** ⟧ ⟦ identifier ⟧ ;

configuration_declaration ⇐
    configuration identifier **of** *entity*_name **is**
        ❴ use_clause ⎮ attribute_specification ⎮ group_declaration ❵
        ❴ **use vunit** *verification_unit*_name ❴ , ... ❵ ; ❵
        block_configuration
    end ⟦ **configuration** ⟧ ⟦ identifier ⟧ ;

block_configuration ⇐
    for ❨   *architecture*_name
          ⎮ *block_statement*_label
          ⎮ *generate_statement*_label
             ⟦ ❨ ❨ *static*_discrete_range ⎮ *static*_expression ⎮ *alternative*_label ❩ ❩ ⟧ ❩
        ❴ use_clause ❵
        ❴ block_configuration ⎮ component_configuration ❵
    end for ;

component_configuration ⇐
    for component_specification
        ⟦ binding_indication ; ⟧
        ❴ **use vunit** *verification_unit*_name ❴ , ... ❵ ; ❵
        ⟦ block_configuration ⟧
    end for ;

context_declaration ⇐
    context identifier **is**
        ❴ library_clause ⎮ use_clause ⎮ context_reference ❵
    end ⟦ **context** ⟧ ⟦ identifier ⟧ ;

B.3 Declarations and Specifications

package_declaration ⇐
    package identifier is
        〚 **generic** (*generic*_interface_list) ;
        〚 **generic map** (*generic*_association_list) ; 〛〛
        { package_declarative_item }
    end 〚 **package** 〛 〚 identifier 〛 ;

package_declarative_item ⇐
    subprogram_declaration
    〚 subprogram_instantiation_declaration
    〚 package_declaration
    〚 package_instantiation_declaration
    〚 type_declaration   〚 subtype_declaration
    〚 constant_declaration   〚 signal_declaration
    〚 variable_declaration   〚 file_declaration
    〚 alias_declaration   〚 component_declaration
    〚 attribute_declaration   〚 attribute_specification
    〚 disconnection_specification   〚 use_clause
    〚 group_template_declaration   〚 group_declaration
    〚 *PSL*_Property_Declaration   〚 *PSL*_Sequence_Declaration

package_body ⇐
    package body identifier **is**
        { package_body_declarative_item }
    end 〚 **package body** 〛 〚 identifier 〛 ;

package_body_declarative_item ⇐
    subprogram_declaration   〚 subprogram_body
    〚 subprogram_instantiation_declaration
    〚 package_declaration   〚 package_body
    〚 package_instantiation_declaration
    〚 type_declaration   〚 subtype_declaration
    〚 constant_declaration   〚 variable_declaration
    〚 file_declaration   〚 alias_declaration
    〚 attribute_declaration   〚 attribute_specification
    〚 use_clause
    〚 group_template_declaration   〚 group_declaration

package_instantiation_declaration ⇐
    package identifier **is new** *uninstantiated_package*_name
        〚 **generic map** (*generic*_association_list) 〛 ;

subprogram_specification ⇐
    procedure_specification 〚 function_specification

procedure_specification ⇐
    procedure identifier
        ⟦ **generic** (*generic*_interface_list)
        ⟦ **generic map** (*generic*_association_list) ⟧ ⟧
        ⟦ ⟦ **parameter** ⟧ (*parameter*_interface_list) ⟧

function_specification ⇐
    ⟦ **pure** ▯ **impure** ⟧ **function** (identifier ▯ operator_symbol)
        ⟦ **generic** (*generic*_interface_list)
        ⟦ **generic map** (*generic*_association_list) ⟧ ⟧
        ⟦ ⟦ **parameter** ⟧ (*parameter*_interface_list) ⟧ **return** type_mark

subprogram_declaration ⇐ subprogram_specification ;

subprogram_body ⇐
    subprogram_specification **is**
        ❴ subprogram_declarative_item ❵
    begin
        ❴ sequential_statement ❵
    end ⟦ **procedure** ▯ **function** ⟧ ⟦ identifier ▯ operator_symbol ⟧ ;

subprogram_declarative_item ⇐
    subprogram_declaration   ▯ subprogram_body
    ▯ subprogram_instantiation_declaration
    ▯ package_declaration   ▯ package_body
    ▯ package_instantiation_declaration
    ▯ type_declaration   ▯ subtype_declaration
    ▯ constant_declaration   ▯ variable_declaration
    ▯ file_declaration   ▯ alias_declaration
    ▯ attribute_declaration   ▯ attribute_specification
    ▯ use_clause
    ▯ group_template_declaration   ▯ group_declaration

subprogram_instantiation_declaration ⇐
    ⦇ **procedure** ▯ **function** ⦈ identifier **is**
        new *uninstantiated_subprogram*_name ⟦ signature ⟧
            ⟦ **generic map** (*generic*_association_list) ⟧ ;

type_declaration ⇐
    type identifier **is** type_definition ;
    ▯ **type** identifier ;

type_definition ⇐
    enumeration_type_definition   ▯ integer_type_definition
    ▯ floating_type_definition   ▯ physical_type_definition
    ▯ array_type_definition   ▯ record_type_definition
    ▯ access_type_definition   ▯ file_type_definition
    ▯ protected_type_declaration   ▯ protected_type_body

constant_declaration ⇐
    constant identifier ❴ , ... ❵ : subtype_indication ⟦ := expression ⟧ ;

signal_declaration ⟸
    signal identifier 〚 , ... 〛 : subtype_indication 〚 **register** Ⅰ **bus** 〛
                     〚 := expression 〛 ;

variable_declaration ⟸
    〚 **shared** 〛 **variable** identifier 〚 , ... 〛 : subtype_indication 〚 := expression 〛 ;

file_declaration ⟸
    file identifier 〚 , ... 〛 : subtype_indication
        〚 〚 **open** *file_open_kind*_expression 〛 **is** *string*_expression 〛 ;

alias_declaration ⟸
    alias 〚 identifier Ⅰ character_literal Ⅰ operator_symbol 〛
        〚 : subtype_indication〛 **is** name 〚 signature 〛 ;

component_declaration ⟸
    component identifier 〚 **is** 〛
        〚 **generic** (*generic*_interface_list) ; 〛
        〚 **port** (*port*_interface_list) ; 〛
    end component 〚 identifier 〛 ;

attribute_declaration ⟸ **attribute** identifier : type_mark ;

attribute_specification ⟸
    attribute identifier **of** entity_name_list : entity_class **is** expression ;

entity_name_list ⟸
    〚 〚 identifier Ⅰ character_literal Ⅰ operator_symbol 〛 〚 signature 〛 〛 〚 , ... 〛
    Ⅰ **others**
    Ⅰ **all**

entity_class ⟸

entity	Ⅰ **architecture**	Ⅰ **configuration**	Ⅰ **package**
Ⅰ **procedure**	Ⅰ **function**	Ⅰ **type**	Ⅰ **subtype**
Ⅰ **constant**	Ⅰ **signal**	Ⅰ **variable**	Ⅰ **file**
Ⅰ **component**	Ⅰ **label**	Ⅰ **literal**	Ⅰ **units**
Ⅰ **group**	Ⅰ **property**	Ⅰ **sequence**	

configuration_specification ⟸
    for component_specification
        binding_indication ;
        〚 **use vunit** *verification_unit*_name 〚 , ... 〛 ; 〛
    〚 **end for** ; 〛

component_specification ⟸
    〚 *instantiation*_label 〚 , ... 〛 Ⅰ **others** Ⅰ **all** 〛 : *component*_name

binding_indication ⇐
    use ⟨ **entity** *entity*_name 〚 (*architecture*_identifier) 〛
               〗 **configuration** *configuration*_name
               〗 **open** ⟩
    〚 **generic map** (*generic*_association_list) 〛
    〚 **port map** (*port*_association_list) 〛

disconnection_specification ⇐
    disconnect ⟨ *signal*_name ⦃ , ... ⦄ 〗 **others** 〗 **all** ⟩ : type_mark
        after *time*_expression ;

group_template_declaration ⇐
    group identifier **is** (⟨ entity_class 〚 <> 〛 ⟩ ⦃ , ... ⦄) ;

group_declaration ⇐
    group identifier : *group_template*_name (⟨ name 〗 character_literal ⟩ ⦃ , ... ⦄) ;

use_clause ⇐ **use** selected_name ⦃ , ... ⦄ ;

B.4 Type Definitions

enumeration_type_definition ⇐ (⟨ identifier 〗 character_literal ⟩ ⦃ , ... ⦄)

integer_type_definition ⇐
    range ⟨ *range*_attribute_name
          〗 simple_expression ⟨ **to** 〗 **downto** ⟩ simple_expression ⟩

floating_type_definition ⇐
    range ⟨ *range*_attribute_name
          〗 simple_expression ⟨ **to** 〗 **downto** ⟩ simple_expression ⟩

physical_type_definition ⇐
    range ⟨ *range*_attribute_name
          〗 simple_expression ⟨ **to** 〗 **downto** ⟩ simple_expression ⟩
        units
           identifier ;
           ⦃ identifier = physical_literal ; ⦄
        end units 〚 identifier 〛

array_type_definition ⇐
    array (⟨ type_mark **range** <> ⟩ ⦃ , ... ⦄) **of** *element*_subtype_indication
    〗 **array** (discrete_range ⦃ , ... ⦄) **of** *element*_subtype_indication

record_type_definition ⇐
    record
        ⟨ identifier ⦃ , ... ⦄ : subtype_indication ; ⟩
        ⦃ ... ⦄
    end record 〚 identifier 〛

access_type_definition ⇐ **access** subtype_indication

file_type_definition ⇐ **file of** type_mark

protected_type_declaration ⇐
   protected
       ❴ protected_type_declarative_item ❵
   end protected ⟦ identifier ⟧

protected_type_declarative_item ⇐
   subprogram_declaration   ❘ <u>subprogram_instantiation_declaration</u>
   ❘ attribute_specification   ❘ use_clause

protected_type_body ⇐
   protected body
       ❴ protected_type_body_declarative_item ❵
   end protected body ⟦ identifier ⟧

protected_type_body_declarative_item ⇐
   subprogram_declaration   ❘ subprogram_body
   ❘ <u>subprogram_instantiation_declaration</u>
   ❘ <u>package_declaration</u>   ❘ <u>package_body</u>
   ❘ <u>package_instantiation_declaration</u>
   ❘ type_declaration   ❘ subtype_declaration
   ❘ constant_declaration   ❘ variable_declaration
   ❘ file_declaration   ❘ alias_declaration
   ❘ attribute_declaration   ❘ attribute_specification
   ❘ use_clause
   ❘ group_template_declaration   ❘ group_declaration

subtype_declaration ⇐ **subtype** identifier **is** subtype_indication ;

subtype_indication ⇐
   ⟦ resolution_indication ⟧ type_mark ⟦ constraint ⟧

resolution_indication ⇐
   *resolution_function*_name
   ❘ (<u>resolution_indication</u>
       ❘ ⟮ *record_element*_identifier resolution_indication ⟯ ❴ , ∘∘∘ ❵)

constraint ⇐
   range ⟮ *range*_attribute_name
          ❘ simple_expression ⟮ **to** ❘ **downto** ⟯ simple_expression ⟯
   ❘ array_constraint
   ❘ <u>record_constraint</u>

array_constraint ⇐
   (discrete_range ❴ , ... ❵) ⟦ array_constraint ❘ record_constraint ⟧
   ❘ (**open**) ⟦ array_constraint ❘ record_constraint ⟧

<u>record_constraint</u> ⇐
   (⟮ *record_element*_identifier ⟮ array_constraint ❘ record_constraint ⟯ ⟯ ❴ , ∘∘∘ ❵)

discrete_range ⇐
  *discrete*_subtype_indication
  ⫾ *range*_attribute_name
  ⫾ simple_expression ⟦ **to** ⫾ **downto** ⟧ simple_expression

type_mark ⇐ *type*_name ⫾ *subtype*_name

B.5 Concurrent Statements

concurrent_statement ⇐
  block_statement
  ⫾ process_statement
  ⫾ concurrent_procedure_call_statement
  ⫾ concurrent_assertion_statement
  ⫾ concurrent_signal_assignment_statement
  ⫾ component_instantiation_statement
  ⫾ generate_statement
  ⫾ *PSL*_PSL_Directive

block_statement ⇐
  *block*_label :
  block ⟦ (*guard*_expression) ⟧ ⟦ **is** ⟧
    ⟦ **generic** (*generic*_interface_list) ;
    ⟦ **generic map** (*generic*_association_list) ; ⟧ ⟧
    ⟦ **port** (*port*_interface_list) ;
    ⟦ **port map** (*port*_association_list) ; ⟧ ⟧
    { block_declarative_item }
  begin
    { concurrent_statement }
  end block ⟦ *block*_label ⟧ ;

block_declarative_item ⇐
  subprogram_declaration   I subprogram_body
  ⫾ subprogram_instantiation_declaration
  ⫾ package_declaration   ⫾ package_body
  ⫾ package_instantiation_declaration
  I type_declaration   I subtype_declaration
  I constant_declaration   I signal_declaration
  I *shared*_variable_declaration   I file_declaration
  I alias_declaration   I component_declaration
  I attribute_declaration   I attribute_specification
  I configuration_specification   I disconnection_specification
  I use_clause
  I group_template_declaration   I group_declaration
  ⫾ *PSL*_Property_Declaration   ⫾ *PSL*_Sequence_Declaration
  ⫾ *PSL*_Clock_Declaration

process_statement ⇐
    ⟦ *process*_label : ⟧
    ⟦ **postponed** ⟧ **process** ⟦ ((*signal*_name ⟨ , ⁓ ⟩) ∣ **all**) ⟧ ⟦ **is** ⟧
        ⟨ process_declarative_item ⟩
    begin
        ⟨ sequential_statement ⟩
    end ⟦ **postponed** ⟧ **process** ⟦ *process*_label ⟧ ;

process_declarative_item ⇐
    subprogram_declaration   I subprogram_body
    <u>∣ subprogram_instantiation_declaration</u>
    <u>∣ package_declaration</u>   <u>∣ package_body</u>
    <u>∣ package_instantiation_declaration</u>
    I type_declaration   I subtype_declaration
    I constant_declaration   I variable_declaration
    I file_declaration   I alias_declaration
    I attribute_declaration   I attribute_specification
    I use_clause
    I group_template_declaration   I group_declaration

concurrent_procedure_call_statement ⇐
    ⟦ label : ⟧ ⟦ **postponed** ⟧ *procedure*_name ⟦ (*parameter*_association_list) ⟧ ;

concurrent_assertion_statement ⇐
    ⟦ label : ⟧ ⟦ **postponed** ⟧ **assert** condition
                            ⟦ **report** expression ⟧ ⟦ **severity** expression ⟧ ;

concurrent_signal_assignment_statement ⇐
    ⟦ label : ⟧ ⟦ **postponed** ⟧ concurrent_simple_signal_assignment
    ∣ ⟦ label : ⟧ ⟦ **postponed** ⟧ concurrent_conditional_signal_assignment
    ∣ ⟦ label : ⟧ ⟦ **postponed** ⟧ concurrent_selected_signal_assignment

concurrent_simple_signal_assignment ⇐
    target <= ⟦ **guarded** ⟧ ⟦ delay_mechanism ⟧ waveform ;

concurrent_conditional_signal_assignment ⇐
    target <= ⟦ **guarded** ⟧ ⟦ delay_mechanism ⟧
        waveform **when** condition
        ⟨ **else** waveform **when** condition ⟩
        ⟦ **else** waveform ⟧ ;

concurrent_selected_signal_assignment ⇐
    with expression **select** ⟦ ? ⟧
        target <= ⟦ **guarded** ⟧ ⟦ delay_mechanism ⟧
            ⟨ waveform **when** choices , ⟩
            waveform **when** choices ;

component_instantiation_statement ⇐
   *instantiation*_label :
      (⟦ **component** ⟧ *component*_name
        ⫲ **entity** *entity*_name ⟦ (*architecture*_identifier) ⟧
        ⫲ **configuration** *configuration*_name)
    ⟦ **generic map** (*generic*_association_list) ⟧
    ⟦ **port map** (*port*_association_list) ⟧ ;

generate_statement ⇐
   for_generate_statement ⫲ if_generate_statement ⫲ case_generate_statement

for_generate_statement ⇐
   *generate*_label :
   for identifier **in** discrete_range **generate**
      generate_statement_body
   end generate ⟦ *generate*_label ⟧ ;

if_generate_statement ⇐
   *generate*_label :
   if ⟦ *alternative*_label : ⟧ condition **generate**
      generate_statement_body
   ⦃ **elsif** ⟦ *alternative*_label : ⟧ condition **generate**
      generate_statement_body ⦄
   ⟦ **else** ⟦ *alternative*_label : ⟧ **generate**
      generate_statement_body ⟧
   end generate ⟦ *generate*_label ⟧ ;

case_generate_statement ⇐
   *generate*_label :
   case expression **generate**
      (**when** ⟦ *alternative*_label : ⟧ choices =>
         generate_statement_body)
      ⦃ ... ⦄
   end generate ⟦ *generate*_label ⟧ ;

generate_statement_body ⇐
     ⟦ ⦃ block_declarative_item ⦄
   begin ⟧
     ⦃ concurrent_statement ⦄
   ⟦ **end** ⟦ *alternative*_label ⟧ ; ⟧

B.6 Sequential Statements

sequential_statement ⇐
   wait_statement   ⫲ assertion_statement
   ⫲ report_statement   ⫲ signal_assignment_statement
   ⫲ variable_assignment_statement   ⫲ procedure_call_statement
   ⫲ if_statement   ⫲ case_statement
   ⫲ loop_statement   ⫲ next_statement

     ⟦ exit_statement   ⟦ return_statement
     ⟦ null_statement

wait_statement ⟸
    ⟦ label : ⟧ **wait** ⟦ **on** *signal*_name ⟨ , ₀₀₀ ⟩ ⟧
                        ⟦ **until** condition ⟧
                        ⟦ **for** *time*_expression ⟧ ;

assertion_statement ⟸
    ⟦ label : ⟧ **assert** condition
            ⟦ **report** expression ⟧ ⟦ **severity** expression ⟧ ;

report_statement ⟸
    ⟦ label : ⟧ **report** expression ⟦ **severity** expression ⟧ ;

signal_assignment_statement ⟸
    ⟦ label : ⟧ simple_signal_assignment
    ⟦ ⟦ label : ⟧ conditional_signal_assignment
    ⟦ ⟦ label : ⟧ selected_signal_assignment

simple_signal_assignment ⟸
    (name ⟦ aggregate) <= ⟦ delay_mechanism ⟧ waveform ;
    ⟦ name <= **force** ⟦ **in** ⟦ **out** ⟧ expression ;
    ⟦ name <= **release** ⟦ **in** ⟦ **out** ⟧ ;

conditional_signal_assignment ⟸
    conditional_waveform_assignment ⟦ conditional_force_assignment

conditional_waveform_assignment ⟸
    ⟦ label : ⟧
    (name ⟦ aggregate) <= ⟦ delay_mechanism ⟧
        waveform **when** condition
        ⟨ **else** waveform **when** condition ⟩
        ⟦ **else** waveform ⟧ ;

conditional_force_assignment ⟸
    ⟦ label : ⟧
    name <= **force** ⟦ **in** ⟦ **out** ⟧
        expression **when** condition
        ⟨ **else** expression **when** condition ⟩
        ⟦ **else** expression ⟧ ;

selected_signal_assignment ⟸
    selected_waveform_assignment ⟦ selected_force_assignment

selected_waveform_assignment ⟸
    ⟦ label : ⟧
    with expression **select** ⟦ ? ⟧
        (name ⟦ aggregate) <= ⟦ delay_mechanism ⟧
            ⟨ waveform **when** choices , ⟩
            waveform **when** choices ;

selected_signal_assignment ⇐
    〚 label : 〛
    with expression **select** 〚 ? 〛
        name <= **force** 〚 **in** ▯ **out** 〛
            { expression **when** choices , }
            expression **when** choices ;

delay_mechanism ⇐ **transport** ▯ 〚 **reject** *time*_expression 〛 **inertial**

waveform ⇐
    (*value*_expression 〚 **after** *time*_expression 〛
        ▯ **null** 〚 **after** *time*_expression 〛) { , ₀₀₀ }
    ▯ **unaffected**

variable_assignment_statement ⇐
    〚 label : 〛 simple_variable_assignment
    ▯ 〚 label : 〛 conditional_variable_assignment
    ▯ 〚 label : 〛 selected_variable_assignment

simple_variable_assignment ⇐
    (name ▯ aggregate) := expression ;

conditional_variable_assignment ⇐
    (name ▯ aggregate) :=
        expression **when** condition
        { **else** expression **when** condition }
        〚 **else** expression 〛 ;

selected_variable_assignment ⇐
    with expression **select** 〚 ? 〛
        (name ▯ aggregate) :=
        { expression **when** choices , }
        expression **when** choices ;

procedure_call_statement ⇐
    〚 label : 〛 *procedure*_name 〚 (*parameter*_association_list) 〛 ;

if_statement ⇐
    〚 *if*_label : 〛
    if condition **then**
        { sequential_statement }
    { **elsif** condition **then**
        { sequential_statement } }
    〚 **else**
        { sequential_statement } 〛
    end if 〚 *if*_label 〛 ;

case_statement ⟸
   ⟦ *case*_label : ⟧
   case ⟦ ? ⟧ expression **is**
      (**when** choices => ⟨ sequential_statement ⟩)
      ⟨ ... ⟩
   end case ⟦ ? ⟧ ⟦ *case*_label ⟧ ;

loop_statement ⟸
   ⟦ *loop*_label : ⟧
   ⟦ **while** condition ⎮ **for** identifier **in** discrete_range ⟧ **loop**
      ⟨ sequential_statement ⟩
   end loop ⟦ *loop*_label ⟧ ;

next_statement ⟸ ⟦ label : ⟧ **next** ⟦ *loop*_label ⟧ ⟦ **when** condition ⟧ ;

exit_statement ⟸ ⟦ label : ⟧ **exit** ⟦ *loop*_label ⟧ ⟦ **when** condition ⟧ ;

return_statement ⟸ ⟦ label : ⟧ **return** ⟦ expression ⟧ ;

null_statement ⟸ ⟦ label : ⟧ **null** ;

B.7   Interfaces and Associations

interface_list ⟸
   (interface_constant_declaration
   ⎮ interface_signal_declaration
   ⎮ interface_variable_declaration
   ⎮ interface_file_declaration
   ⎮ interface_type_declaration
   ⎮ interface_subprogram_declaration
   ⎮ interface_package_declaration) ⟨ ; ... ⟩

interface_constant_declaration ⟸
   ⟦ **constant** ⟧ identifier ⟨ , ... ⟩ : ⟦ **in** ⟧ subtype_indication
                                  ⟦ := *static*_expression ⟧

interface_signal_declaration ⟸
   ⟦ **signal** ⟧ identifier ⟨ , ... ⟩ : ⟦ mode ⟧ subtype_indication ⟦ **bus** ⟧
                                  ⟦ := *static*_expression ⟧

interface_variable_declaration ⟸
   ⟦ **variable** ⟧ identifier ⟨ , ... ⟩ : ⟦ mode ⟧ subtype_indication
                                  ⟦ := *static*_expression ⟧

mode ⟸ **in** ⎮ **out** ⎮ **inout** ⎮ **buffer** ⎮ **linkage**

interface_file_declaration ⟸
   file identifier ⟨ , ... ⟩ : subtype_indication

interface_type_declaration ⟸ **type** identifier

interface_subprogram_declaration ⇐
      ⦅   **procedure** identifier
          ⟦ ⟦ **parameter** ⟧ (*parameter*_interface_list) ⟧
    ❙ ⟦ **pure** ❙ **impure** ⟧ **function** ⦅ identifier ❙ operator_symbol ⦆
          ⟦ ⟦ **parameter** ⟧ (*parameter*_interface_list) ⟧ **return** type_mark
    ⦆ ⟦ **is** ⦅ *subprogram*_name ❙ <> ⦆ ⟧

interface_package_declaration ⇐
    package identifier **is new** *uninstantiated_package*_name
        generic map (⦅ *generic*_association_list ❙ <> ❙ **default** ⦆)

association_list ⇐ ⦅ ⟦ formal_part => ⟧ actual_part ⦆ ⦃ , ... ⦄

formal_part ⇐
    *generic*_name
    ❙ *port*_name
    ❙ *parameter*_name
    ❙ *function*_name (⦅ *generic*_name ❙ *port*_name ❙ *parameter*_name ⦆)
    ❙ type_mark (⦅ *generic*_name ❙ *port*_name ❙ *parameter*_name ⦆)

actual_part ⇐
    ⟦ **inertial** ⟧ expression
    ❙ *signal*_name
    ❙ *variable*_name
    ❙ *file*_name
    ❙ subtype_indication
    ❙ *subprogram*_name
    ❙ *package*_name
    ❙ **open**
    ❙ *function*_name (⦅ *signal*_name ❙ *variable*_name ⦆)
    ❙ type_mark (⦅ *signal*_name ❙ *variable*_name ⦆)

B.8 Expressions and Names

condition ⇐ expression

expression ⇐ ⦅ **??** primary ⦆ ❙ logical_expression

logical_expression ⇐
    relation ⦃ **and** relation ⦄   ❙ relation ⟦ **nand** relation ⟧
    ❙ relation ⦃ **or** relation ⦄   ❙ relation ⟦ **nor** relation ⟧
    ❙ relation ⦃ **xor** relation ⦄   ❙ relation ⦃ **xnor** relation ⦄

relation ⇐
    shift_expression
        ⟦ ⦅ = ❙ /= ❙ < ❙ <= ❙ > ❙ >= ❙ ?= ❙ ?/= ❙ ?< ❙ ?<= ❙ ?> ❙ ?>= ⦆ shift_expression ⟧

shift_expression ⇐
    simple_expression ⟦ ⦅ **sll** ❙ **srl** ❙ **sla** ❙ **sra** ❙ **rol** ❙ **ror** ⦆ simple_expression ⟧

simple_expression ⇐ ⟦ + ⫴ – ⟧ term ⦃ (+ ⫴ – ⫴ &) term ⦄

term ⇐ factor ⦃ (* ⫴ / ⫴ **mod** ⫴ **rem**) factor ⦄

factor ⇐
    primary ⟦ ** primary ⟧
    ⫴ **abs** primary
    ⫴ **not** primary
    ⫴ **and** primary
    ⫴ **nand** primary
    ⫴ **or** primary
    ⫴ **nor** primary
    ⫴ **xor** primary
    ⫴ **xnor** primary

primary ⇐
    name   ⫴ literal
    ⫴ aggregate   ⫴ function_call
    ⫴ qualified_expression   ⫴ type_mark (expression)
    ⫴ **new** subtype_indication   ⫴ **new** qualified_expression
    ⫴ (expression)

function_call ⇐ *function*_name ⟦ (*parameter*_association_list) ⟧

qualified_expression ⇐ type_mark ' (expression) ⫴ type_mark ' aggregate

name ⇐
    identifier
    ⫴ operator_symbol
    ⫴ character_literal
    ⫴ selected_name
    ⫴ ⦅ name ⫴ function_call ⦆ (expression ⦃ , ₒₒₒ ⦄)
    ⫴ ⦅ name ⫴ function_call ⦆ (discrete_range)
    ⫴ attribute_name
    ⫴ external_name

selected_name ⇐
    ⦅ name ⫴ function_call ⦆ . ⦅ identifier ⫴ character_literal ⫴ operator_symbol ⫴ **all** ⦆

operator_symbol ⇐ " ⦃ graphic_character ⦄ "

attribute_name ⇐
    ⦅ name ⫴ function_call ⦆ ⟦ signature ⟧ ' identifier ⟦ (expression) ⟧

signature ⇐ [⟦ type_mark ⦃ , ₒₒₒ ⦄ ⟧ ⟦ **return** type_mark ⟧]

external_name ⇐
    << **constant** external_pathname : subtype_indication >>
    ⫴ << **signal** external_pathname : subtype_indication >>
    ⫴ << **variable** external_pathname : subtype_indication >>

external_pathname ⇐
    absolute_pathname ⫴ relative_pathname ⫴ package_pathname

<u>absolute_pathname</u> ⇐ . {{ pathname_element . }} *object*_identifier

<u>relative_pathname</u> ⇐ {{ ^ . }} {{ pathname_element . }} *object*_identifier

<u>pathname_element</u> ::=
 *entity*_identifier
 ⫚ *component_instantiation*_label
 ⫚ *block*_label
 ⫚ *generate_statement*_label ⟦ (*static*_expression) ⟧
 ⫚ *package*_identifier

<u>package_pathname</u> ⇐ @ *library*_identifier . { *package*_identifier . } *object*_identifier

literal ⇐
 decimal_literal ⫚ based_literal
 ⫚ physical_literal ⫚ identifier
 ⫚ character_literal ⫚ string_literal
 ⫚ bit_string_literal ⫚ **null**

physical_literal ⇐ ⟦ decimal_literal ⫚ based_literal ⟧ *unit*_name

decimal_literal ⇐ integer ⟦ . integer ⟧ ⟦ E ⟦ + ⟧ integer ⫚ E – integer ⟧

based_literal ⇐
 integer # based_integer ⟦ . based_integer ⟧ # ⟦ E ⟦ + ⟧ integer ⫚ E – integer ⟧

integer ⇐ digit {{ ⟦ _ ⟧ ₀₀₀ }}

based_integer ⇐ (digit ⫚ letter) {{ ⟦ _ ⟧ ₀₀₀ }}

character_literal ⇐ ' graphic_character '

string_literal ⇐ " {{ graphic_character }} "

bit_string_literal ⇐
 ⟦ integer ⟧ (B ⫚ O ⫚ X ⫚ <u>UB</u> ⫚ <u>UO</u> ⫚ <u>UX</u> ⫚ <u>SB</u> ⫚ <u>SO</u> ⫚ <u>SX</u> ⫚ <u>D</u>)
 " ⟦ graphic_character {{ ⟦ _ ⟧ ₀₀₀ }} ⟧ "

aggregate ⇐ ((⟦ choices => ⟧ expression) {{ , ₀₀₀ }})

choices ⇐ (simple_expression ⫚ discrete_range ⫚ identifier ⫚ **others**) {{ | ₀₀₀ }}

label ⇐ identifier

identifier ⇐ letter {{ ⟦ _ ⟧ (letter ⫚ digit) }} ⫚ \ graphic_character {{ ₀₀₀ }} \

<u>tool_directive</u> ⇐ ` identifier {{ graphic_character }}

Appendix C

Answers to Exercises

In this appendix, we provide sample answers to the quiz-style exercises marked with the symbol "❶". Readers are encouraged to test their answers to the other, more involved, exercises by running the models on a VHDL simulator.

Chapter 1

1. Entity declaration: defines the interface to a module, in terms of its ports, their data transfer direction and their types. Behavioral architecture body: defines the function of a module in terms of an algorithm. Structural architecture body: defines an implementation of a module in terms of an interconnected composition of sub-modules. Process statement: encapsulates an algorithm in a behavioral description, contains sequential actions to be performed. Signal assignment statement: specifies values to be applied to signals at some later time. Port map: specifies the interconnection between signals and component instance ports in a structural architecture.

2.
```
apply_transform : process is
begin
    d_out <= transform(d_in) after 200 ps;
    -- debug_test <= transform(d_in);
    wait on enable, d_in;
end process apply_transform;
```

3. Basic identifiers: last_item. Reserved words: **buffer**. Invalid: prev item, value–1 and element#5 include characters that may not occur within identifiers; _control starts with an underscore; 93_999 starts with a digit; entry_ ends with an underscore.

4. 16#1#   16#22#   16#100.0#   16#0.8#

5. 12   132   44   250000   32768   0.625

6. The literal 16#23DF# is an integer expressed in base 16, whereas the literal X"23DF" is a string of 16 bits.

7. O"747" = B"111_100_111"
   O"377" = B"011_111_111"
   O"1_345" = B"001_011_100_101"

   X"F2" = B"1111_0010"
   X"0014" = B"0000_0000_0001_0100"
   X"0000_0001" = B"0000_0000_0000_0000_0000_0000_0000_0001"

8. 10UO"747" = B"0_111_100_111"
   10UO"377" = B"0_011_111_111"
   10UO"1_345" = B"1_011_100_101"

   10SO"747" = B"1_111_100_111"
   10SO"377" = B"0_011_111_111"
   10SO"1_345" = B"1_011_100_101"

   12UX"F2" = B"0000_1111_0010"
   12SX"F2" = B"1111_1111_0010"
   10UX"F2" *is illegal due to truncation of leading 1 bits*
   10SX"F2" = B"11_1111_0010"

9. D"24" = B"11000"
   12D"24" = B"0000_0001_1000"
   4D"24" *is illegal due to truncation of a leading 1 bit*

Chapter 2

1. **constant** bits_per_word : integer := 32;
   constant pi : real := 3.14159;

2. **variable** counter : integer := 0;
   variable busy_status : boolean;
   variable temp_result : std_ulogic;

3. counter := counter + 1;
   busy_status := true;
   temp_result := 'W';

4. **package** misc_types **is**
       type small_int **is range** 0 **to** 255;
       type fraction **is range** -1.0 **to** +1.0;
       type current **is range** integer'low **to** integer'high
         units nA;
           uA = 1000 nA;
           mA = 1000 uA;
            A = 1000 mA;
         end units;
       type colors **is** (red, yellow, green);
   end package misc_types;

5. a. Legal

   b. Illegal, should be: a = '1' **and** b = '0' **and** state = idle

   c. Illegal, should be: a = '0' **and** b = '1' **and** state = idle

   d. Legal

6. ```
   pulse_range'left =  pulse_range'low = 1 ms
   pulse_range'right = pulse_range'high = 100 ms
   pulse_range'ascending = true

   word_index'left = 31word_index'right = 0
   word_index'low = 0word_index'high  = 31
   word_index'ascending = false
   ```

7. ```
   state'pos(standby) = 1          state'val(2) = active1
   state'succ(active2) is undefined    state'pred(active1) = standby
   state'leftof(off) is undefined      state'rightof(off) = standby
   ```

8. ```
   2 * 3 + 6 / 4 = 7
   3 + -4 is syntactically incorrect
   "cat" & character'('0') = "cat0"
   true and x and not y or z is syntactically incorrect
   B"101110" sll 3 = B"110000"
   (B"100010" sra 2) & X"2C"= B"11100000101100"
   ```

Chapter 3

1. ```
   if n mod 2 = 1 then
      odd := '1';
   else
      odd := '0';
   end if;

   odd := '1' when n mod 2 = 1 else
          '0';
   ```

2. ```
   if year mod 400 = 0 then
      days_in_February := 29;
   elsif year mod 100 = 0 then
      days_in_February := 28;
   elsif year mod 4 = 0 then
      days_in_February := 29;
   else
      days_in_February := 28;
   end if;

   days_in_February := 29 when year mod 400 = 0 else
                       28 when year mod 100 = 0 else
   ```
   ```

```
                               29 when year mod 4 = 0 else
                               28;
```

3.  ```
    case x is
        when '0' | 'L' => x := '0';
        when '1' | 'H' => x := '1';
        when others => x := 'X';
    end case;

    with x select
        x := '0' when '0' | 'L',
             '1' when '1' | 'H',
             'X' when others;
    ```

4.  ```
    case ch is
        when 'A' to 'Z' | 'a' to 'z' |
             'À' to 'Ö' | 'Ø' to 'ß' | 'à' to 'ö' | 'ø' to 'ÿ' =>
            character_class := 1;
        when '0' to '9' => character_class := 2;
        when nul to usp | del | c128 to c159 => character_class := 4;
        when others => character_class := 3;
    end case;

    with ch select
        character_class :=
            1 when 'A' to 'Z' | 'a' to 'z' |
                   'À' to 'Ö' | 'Ø' to 'ß' | 'à' to 'ö' | 'ø' to 'ÿ',
            2 when '0' to '9',
            4 when nul to usp | del | c128 to c159,
            3 when others;
    ```

5.  ```
    loop
        wait until clk;
        exit when d;
    end loop;
    ```

6.  ```
    sum := 1.0;
    term := 1.0;
    n := 0;
    while abs term > abs (sum / 1.0E5) loop
      n := n + 1;
      term := term * x / real(n);
      sum := sum + term;
    end loop;
    ```

7.  ```
    sum := 1.0;
    term := 1.0;
    for n in 1 to 7 loop
      term := term * x / real(n);
    ```
    ```

```
      sum := sum + term;
   end loop;
```

8. **assert** to_X01(q) = **not** to_X01(q_n)
      report "flipflop outputs are not complementary";

9. Insert the statement after the comment "*-- at this point, reset = '1'*":

```
      report "counter is reset";
```

Chapter 4

1. **type** num_vector **is array** (1 **to** 30) **of** integer;
   variable numbers : num_vector;
   . . .
   sum := 0;
   for i **in** numbers'range **loop**
      sum := sum + numbers(i);
   end loop;
   average := sum / numbers'length;

2. **type** std_ulogic_to_bit_array **is array** (std_ulogic) **of** bit;
   constant std_ulogic_to_bit : std_ulogic_to_bit_array
      := ('U' => '0', 'X' => '0', '0' => '0', '1' => '1', 'Z' => '0',
           'W' => '0', 'L' => '0', 'H' => '1', '-' => '0');
   . . .
   for index **in** 0 **to** 15 **loop**
      v2(index) := std_ulogic_to_bit(v1(index));
   end loop;

3. **type** free_map_array **is array** (0 **to** 1, 0 **to** 79, 0 **to** 17) **of** bit;
   variable free_map : free_map_array;
   . . .
   found := false;
   search_loop : **for** side **in** 0 **to** 1 **loop**
      for track **in** 0 **to** 79 **loop**
         for sector **in** 0 **to** 17 **loop**
            if free_map(side, track, sector) **then**
               found := true; free_side := side;
               free_track := track; free_sector := sector;
               exit search_loop;
            end if;
         end loop;
      end loop;
   end loop;
```
```

4. **subtype** std_ulogic_byte **is** std_ulogic_vector(7 **downto** 0);
    constant Z_byte : std_ulogic_byte := "ZZZZZZZZ";

5. **type** times_array **is array** (positive **range** <>) **of** time_vector;
    subtype times4_array **is** times_array(1 **to** 4);
    variable times4_10 : times4_array(**open**)(0 **to** 9);

6. count := 0;
    for index **in** v'range **loop**
      if v(index) **then**
        count := count + 1;
      end if;
    end loop;

7. Assuming the declarations

        variable v1 : bit_vector(7 **downto** 0);
        variable v2 : bit_vector(31 **downto** 0);
        ...

        v2(31 **downto** 24) := v1;
        v2 := v2 sra 24;

8. **type** test_record **is record**
        stimulus : bit_vector(0 **to** 2);
        delay : delay_length;
        expected_response : bit_vector(0 **to** 7);
      end record test_record;

Chapter 5

1. **entity** lookup_ROM **is**
        port (address : **in** lookup_index;   data : **out** real);

        type lookup_table **is array** (lookup_index) **of** real;
        constant lookup_data : lookup_table
                  := (real'high, 1.0, 1.0/2.0, 1.0/3.0, 1.0/4.0, ...);

    end entity lookup_ROM;

2. **architecture** functional **of** lookup_ROM **is**
    begin
      data <= lookup_data(address) **after** 200 ps;
    end architecture functional;

3. Transactions are 'Z' at 0 ns, '0' at 10 ns, '1' at 30 ns, '1' at 55 ns, 'H' at 65 ns and 'Z' at 100 ns. The signal is active at all of these times. Events occur at each time except 55 ns, since the signal already has the value '1' at that time.

4. s'delayed(5 ns): 'Z' at 5 ns, '0' at 15 ns, '1' at 35 ns, 'H' at 70 ns, 'Z' at 105 ns. s'stable(5 ns): false at 0 ns, true at 5 ns, false at 10 ns, true at 15 ns, false at 30 ns, true at 35 ns, false at 65 ns, true at 70 ns, false at 100 ns, true at 105 ns. s'quiet(5 ns): false at 0 ns, true at 5 ns, false at 10 ns, true at 15 ns, false at 30 ns, true at 35 ns, false at 55 ns, true at 60 ns, false at 65 ns, true at 70 ns, false at 100 ns, true at 105 ns. s'transaction (assuming an initial value of '0'): '1' at 0 ns, '0' at 10 ns, '1' at 30 ns, '0' at 55 ns, '1' at 65 ns, '0' at 100 ns. At time 60 ns, s'last_event is 30 ns, s'last_active is 5 ns, and s'last_value is '0'.

5. **wait on** s **until not** s **and** en;

6. **wait until** ready **for** 5 ms;

7. The variable v1 is assigned **false**, since s is not updated until the next simulation cycle. The variable v2 is assigned **true**, since the wait statement causes the process to resume after s is updated with the value '1'.

8. At 0 ns: schedule '1' for 6 ns. At 3 ns: schedule '0' for 7 ns. At 8 ns: schedule '1' for 14 ns. At 9 ns: delete transaction scheduled for 14 ns, schedule '0' for 13 ns. The signal z takes on the values '1' at 6 ns and '0' at 7 ns. The transaction scheduled for 13 ns does not result in an event on z.

9. At 0 ns: schedule 1 for 7 ns, 23 for 9 ns, 5 for 10 ns, 23 for 12 ns and –5 for 15 ns. At 6 ns: schedule 23 for 13 ns, delete transactions scheduled for 15 ns, 10 ns and 9 ns. The signal x takes on the values 1 at 7 ns and 23 at 12 ns.

10. The process is sensitive to current_state and in1, as these are the signals read by the process.

11.
```
mux_logic : process is
begin
    if enable and sel then
      z <= a and not b after 5 ns;
    elsif enable and sel then
      z <= x or y after 6 ns;
    else
      z <= '0' after 4 ns;
    end if;
    wait on a, b, enable, sel, x, y;
end process mux_logic;
```

12.
```
process is
begin
  case bit_vector'(s, r) is
    when "00"         => q <= unaffected;
    when "01"         => q <= '0';
    when "10" | "11" => q <= '1';
  end case;
  wait on s, r;
end process;
```

13. **assert** (**not** clk'event) **or** clk'delayed'last_event >= T_pw_clk
        report "interval between changes on clk is too small";

14. bit_0 : **entity** work.ttl_74x74(basic)
        port map (pr_n => '1', d => q0_n, clk => clk, clr_n => reset,
                q => q0, q_n => q0_n);

    bit_1 : **entity** work.ttl_74x74(basic)
        port map (pr_n => '1', d => q1_n, clk => q0_n, clr_n => reset,
                q => q1, q_n => q1_n);

15.

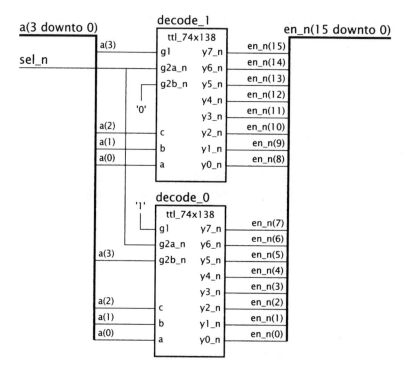

16. One possible order is suggested: analyzing all entity declarations first, followed by all
    architecture bodies:

        entity **edge_triggered_Dff**
        entity **reg4**
        entity **add_1**
        entity **buf4**
        entity **counter**
        architecture **behav** of **edge_triggered_Dff**
        architecture **struct** of **reg4**
        architecture **boolean_eqn** of **add_1**
        architecture **basic** of **buf**
        architecture **registered** of **counter**

    An alternative is

```
entity counter
entity buf4
entity add_1
entity reg4
architecture registered of counter
architecture basic of buf
architecture boolean_eqn of add_1
entity edge_triggered_Dff
architecture struct of reg4
architecture behav of edge_triggered_Dff
```

17. **library** company_lib, project_lib;
    use company_lib.in_pad, company_lib.out_pad, project_lib.**all**;

18. **context** phantom_context **is**
       library ieee, IP_worx, phantom_lib;
      use ieee.std_logic_1164.**all**;
      use IP_worx.**all**, phantom_lib.**all**;
    end context phantom_context;

Chapter 6

1. **constant** operand1 : **in** integer
   operand1 : integer
   constant tag : **in** bit_vector(31 **downto** 16)
   tag : bit_vector(31 **downto** 16)
   constant trace : **in** boolean := false
   trace : boolean := false

2. **variable** average : **out** real
   average : **out** real
   variable identifier : **inout** string
   identifier : **inout** string

3. **signal** clk : **out** bit
   signal data_in : **in** std_ulogic_vector
   signal data_in : std_ulogic_vector

4. Some alternatives are

   ```
   stimulate ( s, 5 ns, 3 );
   stimulate ( target => s, delay => 5 ns, cycles => 3 );

   stimulate ( s, 10 ns, 1 );
   stimulate ( s, 10 ns );
   stimulate ( target => s, delay => 10 ns, cycles => open );
   stimulate ( target => s, cycles => open, delay => 10 ns );
   stimulate ( target => s, delay => 10 ns );
   ```

```
            stimulate ( s, 1 ns, 15 );
            stimulate ( target => s, delay => open, cycles => 15 );
            stimulate ( target => s, cycles => 15 );
            stimulate ( s, cycles => 15 );
```

5.  ```
    swapper : process is
    begin
        shuffle_bytes ( ext_data, int_data, swap_control, Tpd_swap );
        wait on ext_data, swap_control;
    end process swapper;
    ```

6.  ```
    product_size := approx_log_2(multiplicand)
                    + approx_log_2(multiplier);
    ```

7.  ```
    assert now <= 20 ms
        report "simulation time has exceeded 20 ms";
    ```

8. The third, first, none and third, respectively.

9.  ```
    architecture behavioral of computer system is
        signal internal_data : bit_vector(31 downto 0);
        interpreter : process is
            variable opcode : bit_vector(5 downto 0);
            procedure do_write is
                variable aligned_address : natural;
            begin
                ...
            end procedure do_write;
        begin
            ...
        end process interpreter;
    end architecture behavioral;
    ```

Chapter 7

1.  ```
    package EMS_types is
        type engine_speed is range 0 to integer'high
            units rpm;
            end units engine_speed;
        constant peak_rpm : engine_speed := 6000 rpm;
        type gear is (first, second, third, fourth, reverse);
    end package EMS_types;
    ```

    ```
    work.EMS_types.engine_speed
    work.EMS_types.rpm              work.EMS_types.peak_rpm
    work.EMS_types.gear            work.EMS_types.first
    ```
    ```

```
    work.EMS_types.second          work.EMS_types.third
    work.EMS_types.fourth          work.EMS_types.reverse
```

2. **procedure** increment (num : **inout** integer);

3. **function** odd (num : integer) **return** boolean;

4. **constant** e : real;

5. No. The package does not contain any subprogram declarations or deferred constant declarations.

6. **use** work.EMS_types.engine_speed;

7. **library** DSP_lib;
   use DSP_lib.systolic_FFT, DSP_lib.DSP_types.**all**;

Chapter 8

1. a. '1'.

   b. '0'.

   c. Either '1' or '0'. The order of contributions within the array passed to the resolution function is not defined. This particular resolution function returns the leftmost non-'Z' value in the array, so the result depends on the order in which the simulator assembles the contributions.

2. **subtype** wired_and_logic **is** wired_and tri_state_logic;
   signal synch_control : wired_and_logic := '0';

3. The initial value is 'X'. The default initial value of type **MVL4**, 'X', is used as the initial value of each driver of **int_req**. These contributions are passed to the resolution function, which returns the value 'X'.

4. No, since the operation represented by the table in the resolution function is commutative and associative, with 'Z' as its identity.

5. a. "ZZZZ0011"

   b. "XXXX0011"

   c. "0011XX11"

6. "XXXXZZZZ00111100"

7. a. '0'

   b. '0'

   c. 'W'

   d. 'U'

    e. 'X'

8.

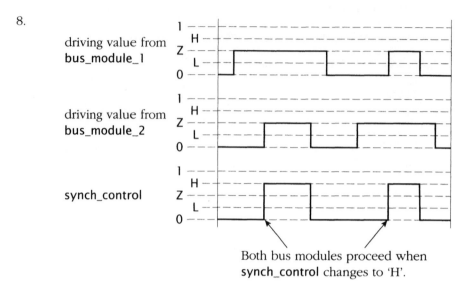

driving value from
bus_module_1

driving value from
bus_module_2

synch_control

Both bus modules proceed when
synch_control changes to 'H'.

9. The resolution function is invoked seven times: for the Mem port, the Cache port, the CPU/Mem Section port, the Serial port, the DMA port, the I/O Section port and the Data Bus signal.

10. We cannot simply invert the value read from the port, since the value may differ from that driven by the process. Instead, we use the **'driving_value** attribute:

```
synch_T <= not synch_T'driving_value;
```

Chapter 9

1. **assert** std.env.resolution_limit < 1.0E-9 sec;

2. to_hstring(B"ZZZZ_0100") = "Z4"
   to_hstring(B"XX_L01H") = to_hstring(B"XXXX_L01H") = "X3"
   to_hstring(B"01_00ZZ") = to_hstring(B"0001_00ZZ") = "1X"

3. **use** ieee.numeric_std.**all**;
   ...
   signal a, b, s : unsigned(23 **downto** 0);
   signal carry_in, carry_out : std_ulogic;
   ...

   (carry_out, s) <= a + b + carry_in;

4. One approach, using implicit conversion of **en** to **boolean**, is

```
D_ff : process (clk) is
begin
  if rising_edge(clk) then
```

```
        if en then
            q <= To_X01(d);
        end if;
      end if;
    end process D_ff;
```

Alternatively:

```
    D_ff : process (clk) is
    begin
      if rising_edge(clk) and To_X01(en) = '1' then
          q <= To_X01(d);
      end if;
    end process D_ff;
```

5. `signal a : sfixed(3 downto -6);`

   `...`

   ```
   s <= resize(a*a, 7, -6);   -- resize using left and right bounds
   s <= resize(a*a, s);       -- resize using bounds of s
   ```

6. `signal x, y : float(7 downto -12)`

   `...`

   ```
   y <= to_float(-1.0, y) when x < -1.0 else
        to_float(+1.0, y) when x > +1.0 else
        x;
   ```

Chapter 10

1. The variable **partial_product** is used to hold the results of the first pipeline stage computation. It is also used as the source operand for the second pipeline stage computations. If computation is performed for the first stage first, the variable is overwritten before being used for the second stage computation. The same argument applies for the variables used to hold results for subsequent stages in the pipeline.

2. Since the real part and the imaginary part of the accumulator are each restricted to the range −16.0 to +16.0, any sequence that causes either accumulator part to fall out of this range results in an overflow. An example is the sequence

   $(-1.0, 0.0) \times (-1.0, 0.0) + (-1.0, 0.0) \times (-1.0, 0.0) + \ldots$

   Each product is the complex value (1.0, 0.0), so after 16 terms, the real part of the accumulator reaches the value 16.0 and overflows.

3. The values in successive clock cycles after the first rising clock-edge are shown in the following table:

Variable	Value in successive clock cycles						
input_x.re	+0.50	+0.20	+0.10	+0.10			
input_x.im	+0.50	+0.20	−0.10	−0.10			
input_y.re	+0.50	+0.20	+0.10	+0.10			
input_y.im	+0.50	+0.20	+0.10	+0.10			
real_part_product_1	?	+0.25	+0.04	+0.01	+0.01		
real_part_product_2	?	+0.25	+0.04	−0.01	−0.01		
imag_part_product_1	?	+0.25	+0.04	+0.01	+0.01		
imag_part_product_2	?	+0.25	+0.04	−0.01	−0.01		
product.re	?	?	0.00	0.00	+0.02	+0.02	
product.im	?	?	+0.50	+0.08	0.00	0.00	
sum.re	0.00	0.00	0.00	0.00	0.00	+0.02	+0.04
sum.im	0.00	0.00	0.00	+0.50	+0.58	+0.58	+0.58
real_accumulator_ovf	false	false	false	false	false	false	false
imag_accumulator_ovf	false	false	false	false	false	false	false

4.   The values +0.5 and −0.5 are represented as shown in the following table:

Format	+0.5	−0.5
inputs	0100 ... 0	1100 ... 0
partial products	00100 ... 0	11100 ... 0
products	000100 ... 0	111100 ... 0
pipelined products	000100 ... 0	111100 ... 0
accumulated sums	00000100 ... 0	11111100 ... 0
outputs	0100 ... 0	1100 ... 0

Chapter 11

1.   **alias** received_source **is** received_packet.source;
     alias received_dest **is** received_packet.dest;
     alias received_flags **is** received_packet.flags;

```
    alias received_payload is received_packet.payload;
    alias received_checksum is received_packet.checksum;
```

2.  ```
    alias received_AK is received_packet.flags(0);
    alias received_ACKNO : bit_vector(2 downto 0)
           is received_packet.flags(1 to 3);
    alias received_SEQNO : bit_vector(2 downto 0)
           is received_packet.flags(4 to 6);
    alias received_UD is received_packet.flags(7);
    ```

3.  ```
    alias cons is "&" [ character, string return string ];

    report cons ( grade_char, "-grade" );
    ```

Chapter 12

1.  ```
    entity flipflop is
       generic ( Tpw_clk_h, T_pw_clk_l : delay_length := 3 ns );
       port ( clk, d : in bit;  q, q_n : out bit );
    end entity flipflop;
    ```

2.  ```
    clk_gen : entity work.clock_generator
       generic map ( period => 10 ns )
       port map ( clk => master_clk );
    ```

3.  ```
    entity adder is
       generic ( data_length : positive );
       port ( a, b : in std_ulogic_vector(data_length - 1 downto 0);
              sum : out std_ulogic_vector(data_length - 1 downto 0) );
    end entity adder;
    ```

4.  ```
    io_control_reg : entity work.reg
       generic map ( width => 4 )
       port map ( d => data_out(3 downto 0),
                  q(0) => io_en, q(1) => io_int_en,
                  q(2) => io_dir, q(3) => io_mode,
                  clk => io_write, reset => io_reset );
    ```

5.  ```
    bv_mux : entity work.generic_mux2(rtl)
       generic map ( data_type => bit_vector(7 downto 0) )
       port map ( sel => sel, a => d_in1, b => d_in2,
                            z => d_out );
    ```

6. The formal generic type is used as the type of the variable v, so the actual generic type must be a fully constrained type. The type **unsigned** is unconstrained, and so cannot be used as the type of a variable.

7.  ```
    package int_stacks is new work.generic_stacks
       generic map ( size => 100, element_type => integer );
    ```

```
use int_stacks.all;
variable int_stack : stack_type;
...

push(int_stack, -1);
```

8. The call with the actual parameter 1 is unambiguous, and calls the first overloaded
    version (with formal parameter of type T1). The call with the actual parameter '1' is
    ambiguous, since the literal '1' could be interpreted as a **std_ulogic** or a **bit** value. The
    second overloaded version has a formal parameter of type T2, which represents
    std_ulogic, and the third overloaded version has a formal parameter of type **bit**. So
    the call could refer to either of these versions. The call with the actual parameter **b** is
    unambiguous, and calls the third overloaded version.

9.  ```
    while not is_full(test_buffer) loop
       write(test_buffer, "00000000");
    end loop;
    ```

10. ```
    procedure check_bv_setup is new check_setup
       generic map ( signal_type => bit_vector,
                     clk_type => bit, clk_active_value => '0',
                     T_su => 100ps );
       ...

    check_bv_setup(s, clk);
    ```

11. ```
    function bv_increment(bv : bit_vector) return bit_vector is
       use ieee.numeric_bit_unsigned.all;
    begin
       return bv + 1;
    end function bv_increment;
    ...

    val_counter : work.generic_counter(rtl)
       generic map ( count_type => bit_vector(9 downto 0),
                     reset_value => (others => '0'),
                     increment => bv_increment )
       port map ( clk => clk, reset => reset, data => val_count );
    ```

12. ```
    use ieee.numeric_std.all;
    package unsigned_dictionaries is new work.dictionaries
       generic map ( size => 1000,
                     element_type => unsigned(63 downto 0),
                     key_type => string,
                     key_of => to_hstring,
                     "<" => ">" );
    ```

13. ```
    package float_generic_math_ops is
       generic ( package float_pkg_for_math is
                    new ieee.float_generic_pkg generic map (<>) );
    ```
    ```

```
    use fixed_pkg_for_math.all;

    function exp ( x : float ) return float;
    function log ( x : float ) return float;
    ...

  end package float_generic_math_ops;

  package float_math_ops is new float_generic_math_ops
    generic map ( float_pkg_for_math => ieee.float_pkg );
```

Chapter 13

1. An entity declaration uses the keyword **entity** where a component declaration uses the keyword **component**. An entity declaration is a design unit that is analyzed and placed into a design library, whereas a component declaration is simply a declaration in an architecture body or a package. An entity declaration has a declarative part and a statement part, providing part of the implementation of the interface, whereas a component declaration simply declares an interface with no implementation information. An entity declaration represents the interface of a "real" electronic circuit, whereas a component declaration represents a "virtual" or "template" interface.

2.  ```
    component magnitude_comparator is
      generic ( width : positive;  Tpd : delay_length );
      port ( a, b : in std_ulogic_vector(width - 1 downto 0);
             a_equals_b, a_less_than_b : out std_ulogic );
    end component magnitude_comparator;
    ```

3.  ```
    position_comparator : component magnitude_comparator
      generic map ( width => current_position'length, Tpd => 12 ns )
      port map ( a => current_position, b => upper_limit,
                 a_less_than_b => position_ok, a_equals_b => open );
    ```

4.  ```
    package small_number_pkg is
      subtype small_number is natural range 0 to 255;
      component adder is
        port ( a, b : in small_number;  s : out small_number );
      end component adder;
    end package small_number_pkg;
    ```

5.  ```
    library dsp_lib;
    configuration digital_filter_rtl of digital_filter is
      for register_transfer
        for coeff_1_multiplier : multiplier
          use entity dsp_lib.fixed_point_mult(algorithmic);
        end for;
      end for;
    end configuration digital_filter_rtl;
    ```

6. **library** dsp_lib;
    configuration digital_filter_std_cell **of** digital_filter **is**
       for register_transfer
          for coeff_1_multiplier : multiplier
             use configuration dsp_lib.fixed_point_mult_std_cell;
          end for;
       end for;
    end configuration digital_filter_std_cell;

7. **library** dsp_lib;
    architecture register_transfer **of** digital_filter **is**
       ...
    begin
       coeff_1_multiplier :
          configuration dsp_lib.fixed_point_mult_std_cell
             port map (...);
       ...
    end architecture register_transfer;

8. **use entity** work.multiplexer
       generic map (Tpd => 3.5 ns);

9.

10. **generic map** (Tpd_01 => **open**, Tpd_10 => **open**)
    port map (a => a, b => b, c => c, d => **open**, y => y)

11. **for** interface_decoder : decoder_2_to_4
        use entity work.decoder_3_to_8(basic)
          generic map (Tpd_01 => prop_delay, Tpd_10 => prop_delay)
          port map (s0 => in0, s1 => in1, s2 => '0',
                      enable => '1',
                      y0 => out0, y1 => out1, y2 => out2, y3 => out3,
                      y4 => **open**, y5 => **open**, y6 => **open**, y7 => **open**);
    end for;

12. **configuration** rebound **of** computer_system **is**
        for structure
          for interface_decoder : decoder_2_to_4
            generic map (Tpd_01 => 4.3 ns, Tpd_10 => 3.8 ns);
          end for;
        end for;
    end configuration rebound;

Chapter 14

1.

2. inverter_array : **for** index **in** data_in'range **generate**
        inv : **component** inverter
          port map (i => data_in(index), y_n => data_out_n(index));
    end generate inverter_array;

3. direct_clock : **if** positive_clock **generate**
        internal_clock <= external_clock;
    else generate
        clock_inverter : **component** inverter
          port map (i => external_clock, y => internal_clock);
    end generate inverting_clock;

4. **for** synch_delay_line(1)
        for delay_ff : d_ff
          use entity parts_lib.d_flipflop(low_input_load);

```
    end for;
  end for;

for synch_delay_line(2 to 4)
  for delay_ff : d_ff
    use entity parts_lib.d_flipflop(standard_input_load);
  end for;
end for;
```

5. A block configuration is not required for the alternative that directly connects the signals, since the statement does not include any component instances. In order to write a block configuration for the other alternative, we need to revise the generate statement to include an alternative label:

```
direct_clock : if noninverting : positive_clock generate
  internal_clock <= external_clock;
else inverting : generate
  clock_inverter : component inverter
    port map ( i => external_clock, y => internal_clock );
end generate inverting_clock;
```

The required block configuration is:

```
for inverting_clock(inverting)
  for clock_inverter : inverter
    use entity parts_lib.inverter;
  end for;
end for;
```

This block configuration is only used if the generic **positive_clock** is false when the design is elaborated.

Chapter 15

1.
```
type character_ptr is access character;
variable char : character_ptr := new character'(ETX);
...

char.all := 'A';
```

2. The statement "r := r + 1.0;" should be "**r.all** := r**.all** + 1.0;". The name r in the statement denotes the pointer, rather than the value pointed to. It is an error to perform an arithmetic operation on a pointer value.

3.

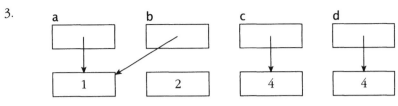

4. a = b is true, a.**all** = b.**all** is true, c = d is false, c.**all** = d.**all** is true.

5.
```
type string_ptr is access string;
variable str : string_ptr := new string'("    ");
...

str(1) := NUL;
```

6.
```
z.re := x.re * y.re - x.im * y.im;
z.im := x.re * y.im + x.im * y.re;
```

7.
```
type message_cell;
type message_ptr is access message_cell;
type message_cell is record
    source, destination : natural;
    data : bit_vector(0 to 255);
    next_cell : message_ptr;
  end record message_cell;
variable message_list : message_ptr;
...

message_list := new message_cell'( source => 1, destination => 5,
                                   data => (others => '0'),
                                   next_cell => message_list );
```

8. The first statement copies the pointer to the first cell to the access variable cell_to_be_deleted and leaves value_list also pointing to that cell. The call to deallocate reclaims the storage and sets cell_to_be_deleted to the null pointer, but leaves value_list unchanged. The host computer system is free to reuse or remove the reclaimed storage, so the access using value_list in the third statement may not be valid.

Chapter 16

1.
```
type real_file is file of real;
file sample_file : real_file open read_mode is "samples.dat";
...

read ( sample_file, x );
```

2.
```
type bv_file is file of bit_vector;
file trace_file : bv_file open write_mode is "/tmp/trace.tmp";
...
```

```
       write ( trace_file, addr & d_bus );
```

3.   `file_open ( status => waveform_status, f => waveform_file,`
              `external_name => "waveform", open_kind => read_mode);`
     `assert` `waveform_status = open_ok`
       `report` `file_open_status'image(waveform_status)`
                `& " occurred opening waveform file"` **`severity`** `error;`

4.   The first call returns the bit value '1'. The second call returns the integer value 23. The
     third call returns the real value 4.5. The fourth call returns the three-character string
     " 67".

5.   **`use`** `std.textio.`**`all`**`;`
     `variable` `prompt_line, input_line : line;`
     `variable` `number : integer;`

     `...`

     `write(prompt_line, string'("Enter a number:"));`
     `writeline(output, prompt_line);`
     `readline(input, input_line);`
     `read(input_line, number);`

6.   "   3500 ns 00111100 ok   "

Chapter 17

1.   **`package`** `memories_support_1Kx24` **`is new`** `work.memories_suppor`
       `generic map` `( width => 24, depth => 10,`
                     `fixed_pkg => ieee.fixed_pkg,`
                     `float_pkg => ieee.float_pkg );`

     `use` `memories_support_1Kx24.`**`all`**`;`

     `package` `memories_1Kx24` **`is new`** `work.memories`
       `generic map`
         `( width => 24, depth => 10,`
           `control_type => bit,`
           `address_type => std_ulogic_vector(9 downto 0),`
           `data_type => std_ulogic_vector(23 downto 0) );`

2.   **`package`** `memories_support_32x128` **`is new`** `work.memories_suppor`
       `generic map` `( width => 128, depth => 5,`
                     `fixed_pkg => ieee.fixed_pkg,`
                     `float_pkg => ieee.float_pkg );`

     `use` `memories_support_32x128.`**`all`**`;`

     `package` `memories_32x128` **`is new`** `work.memories`
       `generic map`
         `( width => 128, depth => 5,`

```
                control_type => std_ulogic,
                address_type => unsigned(4 downto 0),
                data_type => float128 );
```

Chapter 18

1. `<<constant .test_bench.dp.d_width : positive>>`
   `<<signal .test_bench.dp.d_bus : std_ulogic_vector(7 downto 0)>>`
   `<<signal .test_bench.dp.adder(3).carry : std_ulogic>>`

2. ```
   alias d_width is
       <<constant .test_bench.dp.d_width : positive>>;
   alias d_bus is
       <<signal .test_bench.dp.d_bus : std_ulogic_vector(7 downto 0)>>;
   alias carry is
       <<signal .test_bench.dp.adder(3).carry : std_ulogic>>;
   ```

3. `<<constant dp.d_width : positive>>`
   `<<signal dp.d_bus : std_ulogic_vector(7 downto 0)>>`
   `<<signal dp.adder(3).carry : std_ulogic>>`

4. ```
   reset <= force '1';
   wait for 200 ns;
   reset <= release;
   ```

5. ```
   alias mem_d is <<signal mem.d : std_logic_vector(7 downto 0)>>;
   ...

   mem_d <= force out "ZZZZZZZZ";
   mem_d <= force in "ZZZZZZZZ";
   ...

   mem_d <= release out;
   mem_d <= release in;
   ```

6. ```
   configuration verifying of bus_interface is
     use vunit verify_protocol;
     for behavior
     end for;
   end configuration verifying;
   ```

7. ```
   for ext : ext_interface
     use entity bus_interface(behavior);
     use vunit verify_protocol;
   end for;
   ```
   ```

Chapter 19

1. If the host computer system has multiple processors, m1 and m2 may be resumed concurrently on different processors. Suppose the variable starts with the value 0. A possible sequence of events is the following: m1 reads the variable and gets the value 0, m2 reads the variable and gets the value 0, m1 updates the variable with the value 1, m1 updates the variable with the value 1. Thus, the final value of the variable is 1, even though there were two increments performed.

2.  ```
    type shared_integer is protected
      procedure set ( i : integer );
      impure function get return integer;
    end protected shared_integer;

    type shared_integer is protected body

      variable value : integer;

      procedure set ( i : integer ) is
      begin
        value := i;
      end procedure set;

      impure function get return integer is
      begin
        return value;
      end function get;

    end protected body shared_integer;
    ```

Chapter 20

1.  ```
    word'path_name = ":proj_lib:cpu_types:word"

    mult_unsigned'path_name =
        ":proj_lib:bit_vector_signed_arithmetic:"
        & "mult_unsigned[bit_vector, bit_vector return bit_vector]:"

    bv2'path_name =
        ":proj_lib:bit_vector_signed_arithmetic:"
        & "mult_unsigned[bit_vector, bit_vector return bit_vector]:bv2"

    next_test_case'path_name =
        ":tes_tbench:stim_gen:next_test_case"
    next_test_case'instance_name =
        ":test_bench(test_rtl):stim_gen:next_test_case"

    get_ID'path_name =
        ":test_bench:stim_gen:ID_manager:"
        & "get_ID[return natural:"
    get_ID'instance_name =
    ```

```
        ":test_bench(test_rtl):stim_gen:ID_manager:"
      & "get_ID[return natural]:"
```

2. val0_reg'path_name = ":test_bench:dut:val0_reg:"
   val0_reg'instance_name =
       ":test_bench(counter_test)"
     & ":dut@counter(registered):val0_reg@reg4(struct):"
   bit0'path_name = ":test_bench:dut:val1_reg:bit0"
   bit0'instance_name =
       ":test_bench(counter_test)"
     & ":dut@counter(registered):val1_reg@reg4(struct)"
     & ":bit0@edge_triggered_dff(behavioral):"
   clr'path_name = ":test_bench:dut:val1_reg:bit0:clr"
   clr'instance_name =
       ":test_bench(counter_test)"
     & ":dut@counter(registered):val1_reg@reg4(struct)"
     & ":bit0@edge_triggered_dff(behavioral):clr"

3. **attribute** load : capacitance;
   attribute load **of** d_in : **signal is** 3 pF;

4. **type** area **is range** 0 **to** integer'high
     units um_2;
     end units area;
   attribute cell_area : area;
   attribute cell_area **of** library_cell : **architecture is** 15 um_2;

5. **attribute** optimization **of**
       test_empty [list_ptr, boolean] : **procedure is** "inline";

6. **group** statement_set **is** (**label**, **label** <>);
   group steps_1_and_2 : statement_set (step_1, step_2);
   attribute resource_allocation **of** steps_1_and_2 : **group is**
                 max_sharing;

Chapter 21

1. Temp is allowed, as it is an integer type. Temp_vec is allowed, as it is an array type
   indexed by an integer type and has an integer element type. Location is allowed, as
   it is an enumeration type. Local_temp_vec is not allowed, as its index type is not an
   integer type. Location_vec is allowed, as it is an array type indexed by an integer type
   and has an enumeration element type. Word_vec is allowed, as it is an array type indexed by an integer type and has as its element type a one-dimensional array of an
   enumeration type representing bits.

2. A 2-to-1 multiplexer with sel as its select input, in0 and in1 as the data inputs, and z
   as the data output. The tests for 'U' and other metalogical values would be ignored by
   the synthesis tool.

3.   A synthesis tool might infer the hardware shown at the left below. This might subsequently be optimized as shown at the right.

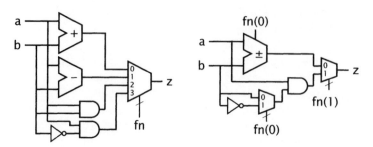

4.  ```
logic_block : process (enable_n, adr, reg1, reg2) is
begin
    if std_match(enable_n, '0') then
        if std_match(adr, '0') then
            dat_o <= reg1;
        else
            dat_o <= reg2;
        end if;
        ack_o <= '1';
    else
        dat_o <= "ZZZZZZZZ";
        ack_o <= 'Z';
    end if;
end process logic_block;
```

5.   The process makes no assignment to **operand** on the path in which **sel** is "11". Hence, the previous value of operand must be stored when **sel** changes to "11". The storage takes the form of a transparent latch enabled by **sel** being other than "11". We can eliminate this storage by adding an assignment of a default value to **operand** in the **others** alternative of the case statement.

6.  ```
en_reg : process ( clk ) is
begin
    if rising_edge(clk) then
        if en = '1' then
            reg_out <= data_in;
        end if;
    end if;
end process en_reg;
```

7.  ```
type RAM_type is array (0 to 8191) of signed(15 downto 0);
signal RAM : RAM_type := (others => X"0000");
```

8.  ```
type decoder_array is array (0 to 15) of std_ulogic_vector(1 to 7);
constant decoder_ROM : decoder_array :=
    ( 0 => "0111111", 1 => "0000110",
```

```
      2 => "1011011",  3 => "1001111",
      4 => "1100110",  5 => "1101101",
      6 => "1111101",  7 => "0000111",
      8 => "1111111",  9 => "1101111",
      others => "1000000" );
...

decoder : seg <= decoder_ROM(to_integer(bcd));
```

9. We can decorate the type **state** with the **enum_encoding** attribute as follows:

   attribute enum_encoding **of** state : **type is** "00 01 11";

   Alternatively, we could decorate **state** with the **fsm_state** attribute:

   attribute fsm_state **of** state : **type is** "00 01 11";

   or we could decorate the signals current_state and next_state:

   attribute fsm_state **of**
       current_state, next_state : **signal is** "00 01 11";

Chapter 22

1. a. **sub r2, r1, r0**: 11 1001 0001 0000 0010 = 0x39102

    b. **and r4, r4, 0x30**: 01 0010 0100 0011 0000 = 0x12430

    c. **ror r1, r1, 2**: 11 0000 1001 0100 0011 = 0x30943

    d. **ldm r6, (r2)–1**: 10 0011 0010 1111 1111 = 0x232FF

    e. **out r4, 0x10**: 10 1110 0000 0000 1010 = 0x2E00A

    f. **bz +7**: 11 1110 0000 0000 0111 = 3E007

    g. **jsb 0xD0**: 11 1101 0000 1101 0000 = 3D0D0

2. a. 0DBC0 = 00 1101 1011 1100 0000: **subc r3, r3, 0xC0**

    b. 38326 = 11 1000 0011 0010 0110: **xor r0, r3, r1**

    c. 33D63 = 11 0011 1101 0110 0011: **ror r7, r5, 3**

    d. 25906 = 10 0101 1001 0000 0110: **stm r3, (r1)+6**

    e. 3EC11 = 11 1110 1100 0001 0001: **bnc +17**

    f. 3DC70 = 11 1101 1100 0111 0000: **jsb 0xC70**

    g. 3F200 = 11 1111 0010 0000 0000: **enai**

Chapter 23

1. **signal** serial_bus : wired_or_bit **bus**;
    signal d_node : unique_bit **register**;

2. When the resolution function for a standard-logic signal is passed an empty vector, it returns the value 'Z'. Thus, the values on **rx_bus** are 'Z', '0' after 10 ns, '1' after 20 ns, '0' after 30 ns, 'X' after 35 ns, '1' after 40 ns, '0' after 45 ns and 'Z' after 55 ns.

3. 'U', '0' after 10 ns, '1' after 20 ns, '0' after 30 ns, 'X' after 35 ns, '1' after 40 ns, '0' after 45 ns.

4. vote <= 3 **after** 2 us, **null after** 5 us;

5. Initially false, true at 160 ns, false at 270 ns.

6. inverting_latch : **block** (en) **is**
    begin
        q_out_n <= **guarded not** d_in;
    end block inverting_latch;

7. **disconnect** source1 : wired_word **after** 3.5 ns;
    disconnect others : wired_word **after** 3.2 ns;
    disconnect all : wired_bit **after** 2.8 ns;

8. Initially 0, 3 at 51 ns, 5 at 81 ns, 0 at 102 ns.

9. inverting_ff : **block is**
        signal q_internal : bit;
    begin
        the_dff : **component** dff
            port map (clk => sys_clk, d => d_in, q => q_internal);
        the_inverter : **component** inverter
            port map (i => q_internal, y => q_out_n);
    end block inverting_ff;

10. **for** inverting_ff
        for the_dff : dff
            use entity work.d_flipflop(basic);
        end for;
        for the_inverter : inverter
            use entity work.inverter(basic);
        end for;
    end for;

11. **architecture** rtl **of** ethernet_mac **is**
    `**protect data_keyowner** = "IP_werx"
    `**protect data_keyname** = "IP_werx_sim"
    `**protect data_method** = "3des-cbc"
    `**protect begin**

```
         signal fifo_enable : std_ulogic;
         ...
      begin
        rx_fifo : IP_werx_fifo
          port map ( ... );
          ...
       `protect end
      end architecture rtl;
```

12.
```
    architecture rtl of ethernet_mac is
      `protect key_keyowner = "Aero Industries"
      `protect key_keyname = "Aero Design"
      `protect key_method = "pgp-rsa"
      `protect key_block
      `protect data_method = "aes192-cbc"
      `protect encoding = (enctype="base64")
      `protect begin
        signal fifo_enable : std_ulogic;
        ...
      begin
        rx_fifo : IP_werx_fifo
          port map ( ... );
          ...
       `protect end
      end architecture rtl;
```

13.
```
    architecture vhpi_implementation of control is
        attribute foreign of vhpi_implementation : architecture is
          "VHPIDIRECT $VHPIUSERLIB/control.so control_elab control_exec";
    begin
    end architecture vhpi_implementation;
```

14.
```
    architecture vhpi_implementation of control is
        attribute foreign of vhpi_implementation : architecture is
          "VHPI control_lib control_model";
    begin
    end architecture vhpi_implementation;
```

   The line in the tabular registry is

   `control_lib control_model vhpiArchF control_elab control_exe`

   This requires that the logical name control_lib be mapped to the object library.

15. The function **To_bit** has two parameters: the value to be converted and the parameter **xmap** that indicates how an unknown logic level should be converted. A conversion function in an association list must have only one parameter.

16. We need to define a conversion function from **std_ulogic** to bit:

```
function cvt_to_bit ( s : std_ulogic ) return bit is
begin
  return To_bit(s);
end function cvt_to_bit;
```

We can use this function and the standard-logic conversion function To_stdulogic in the association list:

```
gate1 : component nand2
  port map ( a => To_stdulogic(s1), b => To_stdulogic(s2),
             cvt_to_bit(y_n) => s3 );
```

References

[1] R. Airiau, J.-M. Bergé and V. Olive, *Circuit Synthesis with VHDL*, Kluwer, Dordrecht, The Netherlands, 1994.

[2] P. J. Ashenden, *Digital Design: An Embedded Systems Approach Using VHDL*, Morgan Kaufmann Publishers, Boston, MA, 2008.

[3] C. G. Bell and A. Newell, *Computer Structures: Readings and Examples*, McGraw-Hill, New York, 1971.

[4] C. Eisner and D. Fisman, *A Practical Introduction to PSL*, Springer, 2006.

[5] Electronic Industries Association, *Standard Data Transfer Format between Data Preparation System and Programmable Logic Device Programmer*, JEDEC Standard JESD3-C, EIA, Washington, DC, 1994.

[6] M. B. Feldman, *Data Structures with Ada*, Prentice-Hall, Englewood Cliffs, NJ, 1985.

[7] Harry D. Foster, Adam C. Krolnik, and David J. Lacey, *Assertion-Based Design*, Kluwer Academic Publishers, 2003.

[8] D. D. Gajski and R. H. Kuhn, "New VLSI Tools," IEEE Computer, Vol. 16, no. 12 (December 1983), pp. 11–14.

[9] Institute for Electrical and Electronic Engineers, *IEEE Standard for Binary Floating-Point Arithmetic*, ANSI/IEEE Std 754–1985, IEEE, New York, 1985.

[10] Institute for Electrical and Electronic Engineers, *IEEE Standard for Radix-Independent Floating-Point Arithmetic*, ANSI/IEEE Std 854–1987, IEEE, New York, 1987.

[11] Institute for Electrical and Electronic Engineers, *Information Technology—Microprocessor Systems—Futurebus+—Logical Protocol Specification*, ISO/IEC 10857, ANSI/IEEE Std. 896.1, IEEE, New York, 1994.

[12] R. Jain, *The Art of Computer System Performance Analysis: Techniques for Experimental Design, Measurement, Simulation and Modeling*, Wiley, New York, 1991.

[13] OpenCores Organization, *WISHBONE System-on-Chip (SoC) Interconnection Architecture for Portable IP Cores, Revision B.3*, 2002, www.opencores.org/projects.cgi/web/wishbone/wbspec_b3.pdf.

[14] S. P. Smith and R. D. Acosta, "Value System for Switch-Level Modeling," IEEE Design & Test of Computers, Vol. 7, No. 3 (June 1990), pp. 33–41.

[15] I. E. Sutherland, C. E. Molnar, R. F. Sproull and J. C. Mudge, "The TRIMOSBUS," Proceedings of the Caltech Conference on VLSI, January 1979.

[16] A. S. Tanenbaum, *Structured Computer Organization, 3rd edition*, Prentice-Hall, Englewood Cliffs, NJ, 1990.

[17] S. A. Ward and R. H. Halstead Jr., *Computation Structures*, MIT Press, Cambridge, MA, and McGraw-Hill, New York, 1990.

[18] N. Weste and K. Eshraghian, *Principles of CMOS VLSI Design: A Systems Perspective*, Addison-Wesley, Reading, MA, 1985.

Index

Printed and bound by CPI Group (UK) Ltd, Croydon, CR0 4YY

22/10/2024

01777361-0001